The Bahía Blanca Estuary

Sandra M. Fiori • Paula D. Pratolongo
Editors

The Bahía Blanca Estuary

Ecology and Biodiversity

Springer

Editors
Sandra M. Fiori
Instituto Argentino de Oceanografía IADO
(Universidad Nacional del Sur-CONICET)
Departamento de Biología, Bioquímica y
Farmacia, Universidad Nacional del Sur
Bahía Blanca, Argentina

Paula D. Pratolongo
Centro de Recursos Naturales Renovables
de la Zona Semiárida, CERZOS
(Universidad Nacional del Sur-CONICET)
Departamento de Biología, Bioquímica y
Farmacia, Universidad Nacional del Sur
Bahía Blanca, Argentina

ISBN 978-3-030-66485-5 ISBN 978-3-030-66486-2 (eBook)
https://doi.org/10.1007/978-3-030-66486-2

This Springer imprint is published by the registered company Springer Nature Switzerland AG
The registered company address is: Gewerbestrasse 11, 6330 Cham, Switzerland

Preface

The Bahía Blanca Estuary is among the most ecologically rich and complex areas along the Argentine coast. It is composed of 2,300 km^2 of intertidal flats, salt marshes, shallow channels, and emerged islands, arranged to create intricate patterns of land-sea interfaces where terrestrial and marine realms converge. Ecosystems in the estuary are shaped by complex exchanges of energy, water, nutrients, sediments, and biota. These natural environments sustain a high concentration of marine and terrestrial species, including endemic, threatened, and endangered fish, shorebirds, and marine mammals, whose major resting and reproductive habitats have been included within the several nature reserves established in the area. Puerto Cuatreros, in the inner zone of the estuary hosts a permanent station of marine research, whose records span more than 30 years of biophysical variables and represent one of the largest time series of ecological data in South America.

Large scale deterioration of coastal ecosystem is a global issue, and the Bahía Blanca Estuary is no exception. Despite its large ecological relevance, the estuary is under increasing anthropogenic pressure by large urban settlements, industrial installations, and port expansion projects, raising the question of how we can balance both conservation and human development. Overfishing, invasive species that establish and thrive, and runoff of nutrients and pollutants from cities and land-based industries are major environmental issues that arise from conflicts of interest between the various users of the coastal zone. This book deals primarily with the biological aspects of ecosystem health, but we recognize the many dimensions of the problem and offer an insight into some relevant social challenges for the environmental management in the region.

This book presents an updated revision of the biology and ecology of the major groups of living organisms inhabiting the coastal zone of Bahía Blanca, including new material on current research revealing the presence of previously unrecognized species and changes in ecosystem functions. The different chapters throughout the book include additions on how humans adversely impact the natural communities, but specific chapters have been also incorporated to deal with major anthropogenic pressures and conservation efforts in the area.

The book is organized into 20 chapters written by the local experts on each topic. The first four chapters introduce readers to the regional context, geographical settings, and physical and chemical processes that affect the distribution and abundance of living organisms. Chapters 5, 6, 7, 8, 9, 10, 11, 12, 13, 14, 15 and 16 comprise the core of the book, on the composition, organization, management issues, and conservation status of the relevant biological communities of major habitat types within the estuary and the area of influence. The final four chapters in the book deal with socio-environmental aspects relevant for the ecosystem health of the estuary.

At the regional level, we hope this book will help young students and citizens to appreciate the many valuable ecosystems around us and promote a new generation to continue efforts towards sensible management and conservation of our natural environments. From a broader perspective, we expect to offer access to the international readership seeking for updated information on the biology and ecology of the Bahía Blanca Estuary.

Bahía Blanca, Argentina

Sandra M. Fiori
Paula D. Pratolongo

Acknowledgement

This book would not be complete without the acknowledgement of the exceptional job carried out by the authors who contributed their expertise to each chapter. They are all renowned local specialists in the different topics covered within the biological, social, conservation, and education sciences, and their distinctive views of the Bahía Blanca Estuary are all represented in this book. As editors, we want to express our gratitude to the authors who produced high-quality material, mostly based on their own research in the region, and contributed original photographs, maps, and illustrations. Through their contributions they have made this book a high-quality education resource for a broad community.

Sandra M. Fiori and Paula D. Pratolongos
Bahía Blanca, July 2020

Contents

About the Editors

Sandra M. Fiori, PhD is a Professor of Ecology of Communities and Ecosystems at Universidad Nacional del Sur, Argentina, and also holds a position as Independent Researcher at CONICET (National Research Council, Argentina). She obtained a Master in Wild Life Management from Universidad de Córdoba in 1997 and PhD in Biological Sciences from Universidad de Nacional del Sur in 2003, and completed postdoctoral training at Instituto Politécnico Nacional – Unidad Mérida (Yucatán, México).

Paula D. Pratolongo, PhD is a Professor of Marine Ecology at Universidad Nacional del sur, Argentina, and also holds a position as Independent Researcher at CONICET (National Research Council, Argentina). She obtained a PhD in Biological Sciences from Universidad de Buenos Aires in 2005, and completed postdoctoral training at East Carolina University, North Carolina, under a Fulbright Award. Dra. Pratolongo received the MaB Young Scientists Award on Ecosystems and Water, granted by the MaB-UNESCO Program.

Chapter 1
The Bahía Blanca Estuary in a Regional Context

Paula D. Pratolongo and Sandra M. Fiori

1.1 Introduction

The Bahía Blanca Estuary is one of the largest coastal systems in Atlantic South America, shaped by a unique combination of large interannual climatic variations and a transgressive sea level during the Holocene. This mesotidal estuary is placed in a sharp transition between humid subtropical and semiarid climates and encompasses about 2300 km^2 of wide intertidal flats, salt marshes, and emerged islands, arranged to create intricate landscape patterns. If we are to study the ecosystems and living organisms of the Bahía Blanca Estuary, we need to first describe the regional context and recognize the physical and sociocultural attributes that influence species distributions, life histories, and ecosystem's health. Thus, the best way to open this book is to provide the readers with an introduction to the regional context along the Atlantic coast of temperate South America.

Many estuaries and coastal wetland systems occur along the Atlantic coast of South America, from the temperate southern coasts of Brazil to Tierra del Fuego Island in Argentina (Fig. 1.1). All these systems share a similar configuration of the coast, located on a passive margin far back from the seafloor spreading at the mid-Atlantic ridge. The east coast of both South and North America is geologically

P. D. Pratolongo (✉)
Centro de Recursos Naturales Renovables de la Zona Semiárida, CERZOS (Universidad Nacional del Sur-CONICET), Bahía Blanca, Argentina

Departamento de Biología, Bioquímica y Farmacia, Universidad Nacional del Sur, Bahía Blanca, Argentina

S. M. Fiori
Departamento de Biología, Bioquímica y Farmacia, Universidad Nacional del Sur, Bahía Blanca, Argentina

Instituto Argentino de Oceanografía, IADO (Universidad Nacional del Sur-CONICET), Bahía Blanca, Argentina

S. M. Fiori, P. D. Pratolongo (eds.), *The Bahía Blanca Estuary*,
https://doi.org/10.1007/978-3-030-66486-2_1

Fig. 1.1 The Atlantic coast of South America from southern Brazil to Tierra del Fuego Island in Argentina. (Map by Walter D. Melo)

ancient and has been stable for millions of years. During this time, the coastline eroded, and continental landforms wore down, leaving wide continental margins of accumulated sediments. Although South and North American Atlantic shorelines are similar in their geographical configuration, they went through different relative sea-level histories during the Holocene, which modeled a distinct evolution of coastal landscapes.

Worldwide, estuaries adapted to the changing sea levels during the Holocene. At any time, the balance between the regional rates of relative sea-level change and the local sediment budget determined the evolution of an estuary. Under a rising sea level, the evolution of estuaries was mainly controlled by vertical accretion at rates that kept pace with the sea-level rise. For instance, in the Chesapeake Bay (eastern North America), shallow estuaries fringed by wetlands formed at locations with abundant sediment inputs and slow rates of relative sea-level rise over the late Holocene. Sediment-starved estuaries in the same region, like those on the mainland edge of the Delmarva Peninsula, moved upland along with their coastal wetlands, creating a varied succession of wetland habitats. Further north, where the relative sea level fell and new land emerged constantly (e.g., along the Hudson Bay, Canada), estuaries migrated seaward and the old coastal features became part of the terrestrial landscape. So let us first introduce the concept of relative sea level and describe the different regional evolutions of the relative sea level during the Holocene.

1.2 The Relative Sea Level During the Holocene

The globally averaged (also called eustatic) sea level has been rising from the Last Glacial Maximum (LGM) to the present. However, the relative sea level (relative height of the sea with respect to land) can vary from place to place because of eustatic, isostatic, ocean circulation, tectonic, and local factors (Khan et al. 2015). After the LGM, the extensive melting of massive ice sheets produced a transfer of mass from the continents to the ocean. Ice melting increased the ocean volume (thus augmenting the eustatic sea level), but also uplifted the Earth's crust, in a process called glacio-isostatic adjustment (Peltier 2002).

Near-field locations are those regions that were close to the margins of large ice sheets during the LGM. In these regions, the rate of glacio-isostatic uplift was higher than the eustatic sea-level rise, and the relative sea level monotonically felt from the LGM to the present. In these areas, typically high-latitude coastal regions (e.g., Fennoscandia, Finland, Labrador), along with the falling sea level, new land emerged, and coastal settings expanded seaward.

Intermediate-field regions extend for several thousand kilometers away from ice sheet edges (Clark et al. 1978). In these areas, the mantle material flowed off the forebulge, which is the lump of the lithosphere bordering the ice sheet. This effect typically produced the glacio-isostatic subsidence that characterizes intermediate-field coastal regions (Fleming et al. 1998). In coastal settings along the temperate Atlantic coasts of North America, for instance, the relative sea level was continually rising from the LGM to the present due to the combination of isostatic and eustatic effects. In these and other intermediate-field regions, Holocene estuaries drowned, and new estuaries formed landward.

Finally, far-field locations are away from large ice sheets. In these regions, eustatic contributions first exceeded glacio-isostatic effects, and the relative sea level raised during the early Holocene. Far-field locations commonly present a mid-Holocene highstand (e.g., relative sea-level maximum) above the elevation of present shorelines. After the Holocene highstand, the relative sea level in far-field regions begun to fall to the current position. In low-latitude regions, the late

Holocene relative sea-level fall was mainly due to the migration of water away from the tropics into subsiding forebulges (Mitrovica and Milne 2002). In far-field locations at higher latitudes, a crustal tilting, upward toward the continent, produced the observed relative sea-level fall during the late Holocene (Kopp et al. 2015). Examples of these far-field locations are the east coast of South America, including the Bahía Blanca Estuary, West Australia, East China, and South Africa. In these regions, the coastline initially retreated along with a rising sea level until it reached a transgressive maximum. After the highstand, the coast prograded as the relative sea level felt to its present position.

1.3 Major Coastal Systems Along Temperate Atlantic South America

The southern coast of Brazil comprises an extensive sequence of barrier islands and coastal lagoons, which extend to the northern coast of Uruguay. The Merin lagoon, in northern Uruguay, is a 6000 km^2 freshwater body draining into the Dos Patos lagoon, a 10,360 km^2 estuary in southern Brazil, which is connected to the sea by a single and narrow inlet (Fig. 1.1). The relative sea level in this region reached 2 m above present at the end of the postglacial marine transgression (about 5000 years BP) and then started a progressively falling trend to the present position (Angulo et al. 1999). Barriers along these coastlines comprise a series of four barrier-lagoon depositional systems, each barrier representing the landward limit of a maximum transgression event, and with the youngest Holocene barrier located seaward. Along with the falling sea-level trend after the highstand, vast areas of tidal flats and marshes developed, which are outstanding elements of the coastal landscape. In the Dos Patos Estuary, irregularly flooded salt marshes extend over 70 km^2. *Spartina alterniflora* is the only species in the low marsh, forming dense stands. In the middle marsh, the southern cordgrass *Spartina densiflora* is one of the dominant species, commonly associated with the Cyperaceae *Bolboschoenus maritimus*. Taller species like *Myrsine parvifolia* and *Acrostichum danaeifolium* gradually displace *S. densiflora* at higher elevations (Azevedo 2000). In hypersaline areas, *Sarcocornia ambigua* is a common species (Costa and Neves 2006).

The Río de la Plata Estuary is a significant coastal wetland system in northern coasts of Argentina. The area comprises the Paraná River Delta, located at the head of the estuary, and extensive coastal plains that formed after the postglacial transgression (Figs. 1.1 and 1.2). Several authors locate a highstand for this region about 6000 years BP, when the relative sea level reached around 6 m above present (Isla 1989; Violante and Parker 2000; Cavallotto et al. 2004). During the regressive stage, minor deltas developed at the mouths of small rivers, along with beach ridges, barriers, and spits that began to prograde. Progradation gave place to the formation of coastal lagoons and tidal flats in the protected areas behind the new formed barriers and spits. About 3000 years BP, the relative sea level was around 3 m above

Fig. 1.2 Major coastal features of northern Argentina and geographical location of the Bahía Blanca Estuary (Map by Walter D. Melo)

present, and a period of drier climate interrupted the progradation of small deltas. Instead, a sequence of successive estuarine beaches began to form, as sea level continued to fall. During the final stage of the regression, from about 1600 BP to present, the increase in the Paraná River discharge induced the formation of the modern delta and displaced the maximum salinity gradient seaward (Cavallotto et al. 2005).

A complex mosaic of tidal and nontidal freshwater wetlands occupies the modern delta, along with the extensive area of former innermost estuarine environments, including shelly and sandy beach ridges, coastal lagoons, and tidal flats. This freshwater wetland system covers an area of about 17,500 km^2 (Malvárez 1997). Close to the mouth of the Paraná River, the Lower Delta is a freshwater tidal system that covers 2700 km^2, which is aggrading into Rio de la Plata Estuary at an estimated rate of 2.64 km^2 year^{-1} (Medina and Codignotto 2013). The Cyperaceae *Schoenoplectus californicus* colonizes the new bars, stabilizes sediments, and establishes monospecific marshes. As these new bars mature, a mixture of forbs like *Hymenachne grumosa*, *Ludwigia* spp., and *Senecio bonariensis* replace *S. californicus*. In a more advanced stage, a surrounding levee begins to develop.

Mature islands in the frontal zone finally assume the form of a central depression of marsh, where the Cyperaceae *Scirpus giganteus* grows in nearly monospecific stands, and a surrounding levee covered with forest (Kandus and Malvárez 2004). Tidal influences attenuate upstream, and the pulsing hydrology of the Paraná River becomes dominant (Neiff 1999). Upstream, nontidal freshwater wetlands progressively replace tidal freshwater marshes. These wetlands form a large corridor that extends along 3400 km, from the Paraná River Delta to the Gran Pantanal, through the Paraná River alluvial valley, and the Paraguay River lowlands (Petean and Cappato 2005).

The coastal plain southward Río de la Plata developed in the same depositional environment after the Holocene transgression (Violante and Parker 2000). Samborombón Bay is an extensive coastal system composed of low-energy prograding landforms, hosting more than 800 km^2 of salt marshes (Isacch et al. 2006). In the low marsh, *S. alterniflora* grows in monospecific stands and *S. densiflora* marshes establish at higher elevations, along with typical accompanying species like *Apium sellowianum*, *Limonium brasiliense*, and *Cortaderia selloana*. In some places, monospecific stands of *S. ambigua* occur between *S. alterniflora* and *S. densiflora* marshes. Mixed marshes with *S. densiflora*, *S. ambigua*, *Juncus acutus*, and *Distichlis spicata* are also common in elevated sites (Cagnoni 1999). Although there is a general arrangement of communities from the seashore to the uplands, the plant zonation is not conspicuous. Instead, there is a complex mosaic of different salt marsh species, intermixed with patches and narrow strips of *Celtis tala* forest, associated with shelly ridges parallel to the coast (Vilanova et al. 2006).

Southward, in a gradient of increasing aridity, salt marsh vegetation becomes sparse. In the Bahía Blanca Estuary, at the northern limit of the Patagonian Region, most of the intertidal fringe is covered by extensive barren mudflats. Pure stands of *S. alterniflora* are commonly restricted to low marshes in the middle reach of the estuary, but do not appear in the inner zone (Pratolongo et al. 2016). Through the shallow inner section of the estuary, *S. ambigua* marshes constrain to elevations close to the mean high tide level (Pratolongo et al. 2016). Above the elevation of the mean high tide, irregularly inundated by seawater, land cover is a mosaic of salt flats, halophytic steppes, and shrubs. Salt flats occupy large areas beyond the limits of the tidal influence. *Sabkha* is an Arabic term that describes the extensive, barren, salt-encrusted, and periodically flooded coastal flats as well as inland salt flats that form by capillary rise of saline groundwater. Inland and coastal sabkhas (i.e., salt flats) of a considerable extension also occur further south in San Blas Bay (Pratolongo et al. 2016). However, despite the vast area that they cover, salt flats in the region are virtually absent from the literature.

South of the Colorado River extends the Patagonian Region, the largest dry region of South America. Rivers debouching in the coast of Patagonia typically have broad alluvial valleys, disproportionate for their little discharge and sediment loads (Kokot 2004). Cliffed coasts are common along Patagonia and Holocene marine deposits are widespread, raised at different elevations (Rostami et al. 2000). San Sebastián Bay, in Tierra del Fuego Island, is a macrotidal embayment characterized by a 17 km long gravel spit called *Península del Páramo*. The spit protects extensive tidal flats that formed by sedimentation during the regressive phase after the

Holocene highstand. Chenier ridges and muddy tidal flats developed by rapid progradation, leaving a sequence that spans about 8 km (Vilas et al. 1999). Vegetation is virtually absent in the intertidal zone except for the circular mounds of *Sarcocornia magellanica* (ex *Salicornia ambigua*) that colonize the upper portion and scattered plants of *Puccinellia* spp. occurring at some locations. At higher elevations, pastures dominated by *Puccinellia magellanica*, *Puccinellia biflora*, and *S. magellanica* usually cover the abandoned tidal flats. *Lepidophyllum cupressiforme*, an endemic species, appears on the crests of cheniers, forming open scrublands (Collantes and Faggi 1999).

1.4 The Regional Climate and Vegetation Patterns of Central East Argentina

The Bahía Blanca Estuary is located in central east Argentina, southern South America. According to the Köppen-Geiger climate classification (Peel et al. 2007) (Fig. 1.3), central east Argentina is a transitional zone between temperate oceanic and cold desert climates and includes zones of humid subtropical and cold semiarid

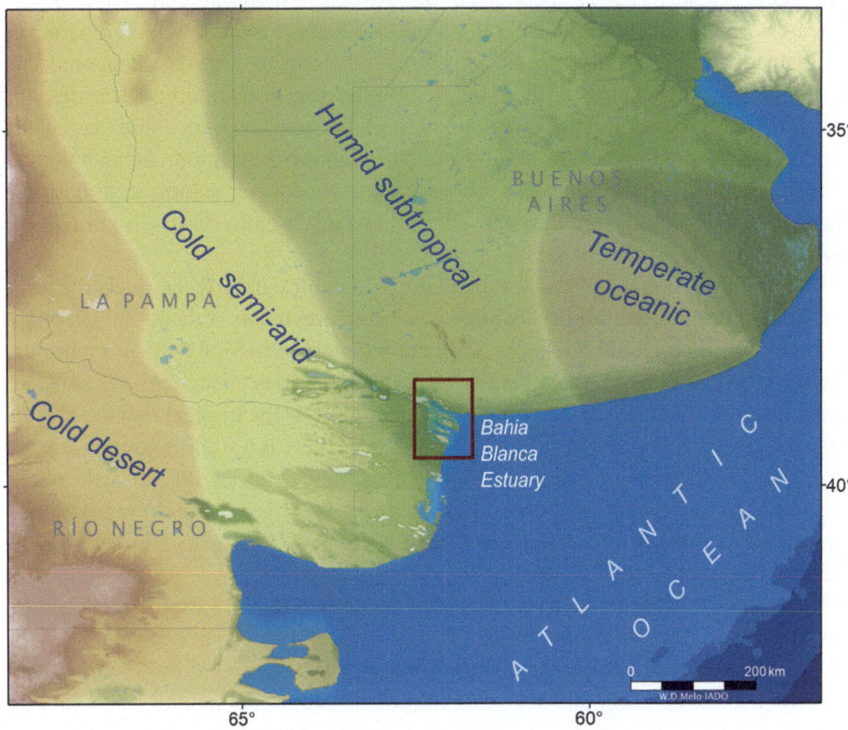

Fig. 1.3 Climates of central east Argentina according to the Köppen-Geiger climate classification. (Map by Walter D. Melo)

climate. This gradient results from the interaction of humid Atlantic air masses from the northeast with drier and colder masses from the southwest (Burgos and Vidal 1951). Climate in the Southern Hemisphere is controlled by semipermanent pressure fields of the subtropical South Atlantic and South Pacific anticyclones and temperate low-pressure cells (Clapperton 1993). Because of the interaction between the general atmospheric circulation and the Andes Mountain, there are two principal climatic dominions in southern South America: the Atlantic and the Pacific. A fringe called the Arid Diagonal crosses southern South America from coastal Peru to Patagonia and connects zones of arid and semiarid climate. The Arid Diagonal encompasses several dry lands like Atacama, Monte, and the Patagonian Region, which isolate the temperate and subtropical forests of Chile and southern Argentina from other humid regions of South America. In central east Argentina, this climatic transition is characterized by a strong gradient of decreasing rainfall from the northeast to the southwest (Bruniard 1982).

Iglesias (1981) proposed a climatic characterization of Argentina. In this classification, climate in the *Pampas* (plains of central east Argentina) was typified as Humid Temperate of the Plains in the east with mean annual rainfall ranging 1200–800 mm, and Subhumid Temperate of the Plains in the west with 800–500 mm. The mean annual temperature ranges 14–20 °C, there is no dry season, but a decrease in precipitation occurs at the end of summer and during winter. In northern Patagonia, Iglesias (1981) recognized the climates Semiarid Temperate of the Plateau, with mean annual rainfall ranging 200–300 mm, and the Arid Temperate of the Plateau, with 100–200 mm. Both climates belong to the Arid Diagonal and define the transition between the Atlantic and the Pacific Dominion. Besides this well-established pattern in the total annual rainfall, there is a noticeable decreasing seasonality from northeast to southwest. Throughout the region, mean monthly precipitation is higher between October and March, and lower values occur between June and August, but differences between wet and dry months diminish through the southwest (Celleri et al. 2018).

One of the most distinctive climatic features of central east Argentina is the large interannual rainfall variability, which has been previously related to El Niño-Southern Oscillation (ENSO) (e.g., Rivera and Penalba 2015). ENSO is a naturally occurring phenomenon that results from interactions between sea surface temperature anomalies and atmospheric circulation over the tropical Pacific. During the warm phase of the ENSO cycle (El Niño years), there is a warming in sea surface temperatures across the central and east-central equatorial Pacific. In central east Argentina, El Niño years have been associated with positive (above average) rainfall anomalies. Conversely, La Niña years have been associated with negative (below average) rainfall anomalies

Celleri et al. (2018) found a significant periodic interannual variability for central east Argentina, characterized by an 8-year cycle, with a strong signal in the southwestern Pampas. Even though the ENSO cycle explains a large part of this variability, extremely wet and dry years may also occur under neutral ENSO conditions. These discrepancies may respond to exceptional atmospheric circulations and unusual water vapor fluxes over the area (Scian et al. 2006). According to more recent studies (Celleri et al. 2018), summer rainfall (November to January) has a lagged response to the Southern Oscillation Index (SOI), and total values show a significant negative correlation with the average SOI estimated from August to October (3-month lag). According to Kayano and Andreoli (2007), ENSO teleconnections for the regional rainfall are considerably strong when ENSO and the Pacific Decadal Oscillation (PDO) are in the same phases, but weak correlations characterize those years with oscillations out of phase. While summer rainfall is significantly associated to El Niño events, the annual totals may not reflect the teleconnection, and the observed discrepancies may be due to the interaction of ENSO, PDO, and anomalous circulation patterns.

Vegetation types along central east Argentina are strongly associated with the regional precipitation gradient, ranging from grasslands in the northeast to xerophytic shrublands and bushy steppes in the southwest. According to Cabrera (1978), three different phytogeographic regions can be distinguished within central east Argentina: Pampas, Espinal, and Monte. The Pampas Region (Fig. 1.4), which represents the eastern half of the area, is mostly covered by grasslands. Based on its

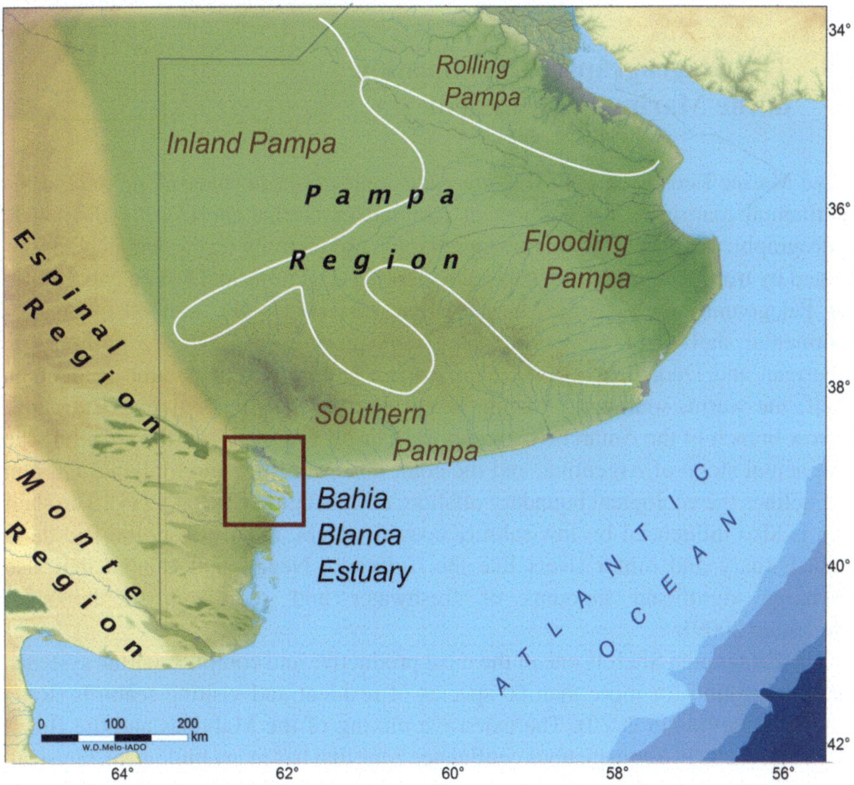

Fig. 1.4 Phytogeographic regions in central east Argentina and suregions within the Pampas Region. (Map by Walter D. Melo)

microtopography and soil characteristics, this ecoregion can be further divided into four subregions: Flooding Pampa, Inland Pampa, Southern Pampa, and Rolling Pampa (Soriano et al. 1992). Surrounding the Pampas, the Espinal represents a phytogeographic transition to the Monte Region, and vegetation is composed of a mixture of grasslands, xerophytic shrublands, and bushy steppes. Finally, the Monte Region extends through the southwest and is covered by xerophytic shrublands. Psammophytic and halophytic communities also occur under specific soil conditions (Cabrera 1978).

Most of the area is dedicated to agriculture and cattle raising. In the Rolling Pampa, a combination of humid climate, deep and well drained soils allow extensive agriculture and continuous cropping of soybean, wheat, and maize under rain-fed regime (INTA 1990). Shallow saline soils, poor drainage, and water erosion impose limitations to crop production in the Flooding Pampa, and the area is primarily dedicated to livestock production, mostly sustained by natural pastures (Bilenca et al. 2012; Frank and Viglizzo 2012). Lands in the rest of central east Argentina are suitable to cattle and cattle-crop mixed production schemes, including pastures in rotation with wheat, sunflower, and maize (Hall et al. 1992).

1.5 The Biogeographic Classification of the Marine Environment

Large Marine Ecosystems (LMEs) are wide areas of ocean space along the Earth's continental margins. LMEs represent regional units characterized by distinctive hydrographic regimens, submarine topography, productivity levels, and populations related by trophic interactions. Coastal waters off Argentina and Uruguay belong to the Patagonian Shelf Large Marine Ecosystem (Fig. 1.1), one of the widest continental shelves in the world, that covers an area of about 1.2 million km^2 (Sherman and Duda 1999). Two major surface currents influence the Patagonian Shelf: the warm, southward-flowing Brazil Current and the Malvinas Current, a narrow branch of the Antarctic Circumpolar Current that flows northward along the continental slope of Argentina, and the cold, northward-flowing Malvinas Current that defines the ecological boundary offshore of this region (Bisbal 1995). The shelf area is also influenced by low-salinity coastal waters, mainly from the Río de la Plata Estuary and minor rivers like the Colorado, Negro, and Chubut that also discharge significant amounts of freshwater and sediments to the shelf (Heileman 2009).

The Patagonian Shelf is one of the most productive and complex marine systems, foraging ground for more than 60 species of resident and visiting seabirds (Raya Rey and Huettmann 2020). The extensive mixing of the Malvinas and the Brazil Currents results in a conspicuous confluence zone that has many biological, physical, and meteorological consequences over the entire region. The high productivity of this marine area is modulated by the presence of several estuarine, tidal, and shelf-break frontal systems that support high concentrations of phytoplankton and feed trophic webs. Major frontal systems include the Shelf-Break Front, the Northern Patagonian Front, and the Southern Patagonian Front.

The Shelf-Break Front is a permanent thermohaline frontal system that extends along the shelf break from the convergence between the Brazil and Malvinas Currents, south to the Malvinas Islands (Acha et al. 2004). The Southern Patagonian Front extends southward from the San Jorge Gulf. Through the south, the front is characterized by a strong salinity gradient, due to the influence of low-salinity waters from the Magellan Strait and the Cape Horn Current (Acha et al. 2004). The Northern Patagonian Frontal System includes the Peninsula Valdes Tidal Front, which forms during spring and summer and represents the boundary between stratified shelf waters during the warm season and tidally mixed coastal waters (Acha et al. 2004). Frontal areas commonly have high values of surface chlorophyll concentration, which can be linked to their primary productivity (Rivas et al. 2006; Romero et al. 2006). This high productivity supports a diverse marine biota, with species from warm, temperate, and cold waters. In addition, coastal areas within the Patagonian Shelf provide reproductive habitats for many species, some of them of outstanding global ecological, economic, and social importance, including seabird and marine mammals, as well as fish and invertebrates (Miloslavich et al. 2011).

From a zoogeographic perspective, the Patagonian Shelf can be divided into the Argentine and the Magellanic Provinces (Miloslavich et al. 2011). Provinces are large areas defined by the presence of particular biotas, with some cohesion from an evolutionary perspective and endemism, mainly at the level of species (Spalding et al. 2007). The Argentine Biogeographic Province extends from a fluctuating northern limit between 30° and 32° S (off Rio Grande do Sul State, Brazil) to northern Patagonia at 43° S. The Magellanic Province comprises the complete continental shelf between 55° and 43° S and ascends along the shelf border up to between 36° and 35° S (Balech and Ehrlich 2008). The transition between both provinces corresponds to the southernmost extension of the convergence zone between the Malvinas and Brazil Currents. The extreme temperature gradients, intense horizontal and vertical mixing, and changes in habitat features have a large influence on species distributions and set the transition between provinces (Boltovskoy 1979; Piola et al. 2000; Wieters et al. 2012).

Biogeographic studies considering benthic macroinvertebrate assemblages of mollusks, echinoderms, and bryozoans over the Argentine Shelf (Bastida et al. 1992) confirmed that two major faunal groups can be distinguished, in agreement with the traditional biogeographic division into Argentine and Magellanic provinces. Furthermore, these authors suggested that the Magellanic province could be further divided into the Patagonian and Malvinean districts, which would cover the warmer inner shelf, and the deeper and colder outer shelf, respectively.

1.6 El Rincón Coastal System

Waters of the Bahía Blanca Estuary belong to the Argentine Province, more specifically to a coastal oceanographic area called El Rincón that extends from 39° S to 41° S, up to a depth of 40 m (Acha et al. 2004). El Rincón is characterized by vertically homogeneous waters due to a shallow bathymetry and strong tidal forcing. Coastal waters have lower salinities because of the influence of both the Negro and Colorado Rivers, and a coastal front separates El Rincón from high-salinity shelf waters (Fig. 1.2). The salinity gradient in the frontal zone is increased by the influence of high saline waters flowing north from the San Matías Gulf. Bathymetry and the mean shelf circulation contribute to maintain the front, oriented north-south, which encloses an area of about 10,000 km^2 (Guerrero 1998; Lucas et al. 2005).

El Rincón is regarded as a highly productive area due to the presence of elevated densities of fish and zooplankton species (Macchi and Acha 1998; Viñas et al. 2013). It has been ascribed as a very complex oceanographic system, influenced by variable inputs of continental runoff from the Negro and Colorado Rivers, local generation of high-salinity cells, and winds that dominate the inner shelf dynamics (Lucas et al. 2005). The circulation in the area is not yet fully understood, but a gyre-like pattern has been detected in numerical models, which would favor fish larval retention and the presence of their predatory fish (Piola and Rivas 1997). The availability of adequate food, suitable thermal and salinity ranges, and the oceanic circulation that favors a retention mechanism would facilitate fish larval recruitment and growth of valuable commercial species (Marrari et al. 2004; Delgado et al. 2015). Also, this area is essential for spawning, calving, and breeding of different emblematic species of the coastal fish complex (Molina and Lopez Cazorla 2011; Militelli et al. 2013).

Rich oceanic conditions bring nutrients, plankton, and fish into Argentine waters, but unsustainable practices have led to the collapse of several commercial fisheries. However, there is an increasing effort directed to improve the conservation status of the Argentine Sea by creating protected areas and developing strategies for the ecosystem-based management of fisheries. The National System of Marine Protected Areas (MPAs) was created in 2014, to protect and conserve marine spaces that are representative of habitats and ecosystems under environmental policy objectives. Within this frame, El Rincón Marine National Park and National Marine Reserve has been proposed as a protected area, along with another five marine regions, because of their great ecological significance. In the specific case of El Rincón, the area has been proposed as MPA based on the presence of many commercially important fish species, as well as feeding resources for top predatory birds and mammals. The area also entails a large biogeographic relevance, since it connects two coastal protected areas (Reserva Natural de Uso Múltiple Bahía San Blas and Reserva Natural de Uso Múltiple Bahía Blanca, Falsa y Verde). The high mortality of keystone predators by incidental catch and the poor condition of fishing stocks have been pointed as major threats for this area (Falabella 2014; DNC/APN 2017).

References

Acha EM, Mianzan HW, Guerrero R et al (2004) Marine fronts at the continental shelves of austral South America. J Mar Syst 44:83–105

Angulo RJ, Giannini PC, Suguio K et al (1999) Relative sea-level changes in the last 5500 years in southern Brazil (Laguna–Imbituba region, Santa Catarina State) based on vermetid ^{14}C ages. Mar Geol 159:323–339

Azevedo AMG (2000) Hábitats, associações vegetais e fenologia das plantas das marismas da Ilha da Pólvora, Estuário da Lagoa dos Patos (RS, Brasil). Doctoral dissertation, Fundação Universidade Federal do Rio Grande

Balech E, Ehrlich MD (2008) Esquema biogeográfico del mar Argentino. Rev Investig Desarro Pesq 19:45–75

Bastida R, Roux A, Martínez DE (1992) Benthic communities of the argentine continental-shelf. Oceanol Acta 15:687–698

Bilenca D, Codesido M, González Fischer C et al (2012) Impactos de la transformación agropecuaria sobre la biodiversidad en la provincia de Buenos Aires. Rev Mus Argent Cienc Nat 14:189–198

Bisbal GA (1995) The Southeast South American shelf large marine ecosystem: evolution and components. Mar Policy 19:21–38

Boltovskoy D (1979) Zooplankton of the South-western Atlantic. S Afr J Sci 75:541–544

Bruniard E (1982) La Diagonal Árida Argentina: un límite climático real. Rev Geogr 95:5–20

Burgos JJ, Vidal AL (1951) The climates of the Argentine Republic according to the new Thornthwaite classification. Ann Assoc Am Geogr 41:237–263

Cabrera AL (1978) Flora de la provincia de Jujuy, República Argentina: Compositae. INTA, Buenos Aires

Cagnoni M (1999) Espartillares de la costa bonaerense de la República Argentina. Un caso de humedales costeros. In Malvárez AI (ed) Tópicos sobre Humedeles Subtropicales y templados de Sudamérica. MAB-UNESCO, Montevideo

Cavallotto JL, Violante RA, Parker G (2004) Sea-level fluctuations during the last 8600 years in the de la Plata river (Argentina). Quat Int 114:155–165

Cavallotto JL, Violante RA, Colombo F (2005) Evolución y cambios ambientales de la llanura costera de la cabecera del río de la Plata. Rev Asoc Geol Argent 60:353–367

Celleri C, Zapperi G, González Trilla G et al (2018) Spatial and temporal patterns of rainfall variability and its relationship with land surface phenology in central east Argentina. Int J Climatol 38:3963–3975

Clapperton CM (1993) Quaternary geology and geomorphology of South America. Elsevier, Amsterdam

Clark JA, Farrell WE, Peltier WR (1978) Global changes in postglacial sea level: a numerical calculation. Q Rev 9:265–287

Collantes MB, Faggi AM (1999) Los humedales del sur de Sudamérica. In: Malvárez AI (ed) Tópicos sobre humedales subtropicales y templados de Sudamérica. NESCO, Montevideo, pp 15–25

Costa CSB, Neves LSD (2006) Respostas cromáticas de Salicornia gaudichaudiana Mog. (Chenopodiaceae) a diferentes níveis de radiação UV-B e salinidade. Atlantica 28:25–31

Delgado AL, Loisel H, Jamet C et al (2015) Seasonal and inter-annual analysis of chlorophyll-a and inherent optical properties from satellite observations in the inner and mid-shelves of the south of Buenos Aires Province (Argentina). Remote Sens 7:11821–11847

DNC/APN (Dirección Nacional de Conservación/Administración de Parques Nacionales) (2017) Sistema Nacional de Áreas Marinas Protegidas (Ley N° 27.037): Justificación técnica para la creación de seis áreas Marinas Protegidas. Dirección Nacional de Conservación, Ministerio de Ambiente y Desarrollo Sustentable, p 52

Falabella V (2014) Identificación de áreas de alto valor de conservación como potenciales áreas marinas protegidas. Informe elaborado durante la fase preparatoria del Proyecto GEF 5112-FAO-Secretaria de Ambiente y Desarrollo Sustentable, p 97
Fleming K, Johnston P, Zwartz D et al (1998) Refining the eustatic sea-level curve since the last glacial maximum using far and intermediate-field sites. Earth Planet Sci Lett 163:327–342
Frank FC, Viglizzo EF (2012) Water use in rain-fed farming at different scales in the Pampas of Argentina. Agric Syst 109:35–42
Guerrero RA (1998) Oceanografía física del estuario del Río de la Plata y el sistema costero de El Rincón. Noviembre, 1994. In: Lasta CA (ed) Resultados de una campaña de evaluación de recursos demersales costeros de la provincia de Buenos Aires y del litoral uruguayo, vol 21. INIDEP Informe Técnico, pp 29–54
Hall AJ, Rebella CM, Ghersa C et al (1992) Field-crop systems of the Pampas. In: Pearson CJ (ed) Field crop ecosystems, Serie: Ecosystems of the World. Elsevier, Amsterdam, pp 413–450
Heileman S (2009) XVI-55 Patagonian Shelf LME in the UNEP large marine ecosystem report: a perspective on changing conditions in LMES of the world's regional seas. In: Sherman K, Hempel G (eds) UNEP regional seas report and studies N°182. United Nations Environment Programme, Nairobi, pp 735–746
Iglesias de Cuello A (1981) Tipos de clima. In: Chiozza E, Figueira E (eds) Atlas Total la República Argentina. Centro Editor de América Latina, Buenos Aires, pp 193–200
INTA (1990) Atlas de suelos de la República Argentina. Ediciones INTA, Buenos Aires
Isla FI (1989) Holocene sea-level fluctuation in the southern hemisphere. Quat Sci Rev 8:359–368
Isacch JP, Costa CSB, Rodriguez-Gallego L, Conde D, et al (2006) Distribution of saltmarsh plant communities associated with environmental factors along a latitudinal gradient on the south-west Atlantic coast. J Biogeogr 33:888–900
Kandus P, Malvárez AI (2004) Vegetation patterns and change analysis in the lower delta islands of the Paraná River (Argentina). Wetlands, 24:620–632
Kayano MT, Andreoli RV (2007) Relations of South American summer rainfall interannual variations with the Pacific decadal oscillation. Int J Climatol 27:531–540
Khan NS, Vane CH, Horton BP (2015) Stable carbon isotope and C/N geochemistry of coastal wetland sediments as a sea-level indicator. In Shennan I, Long AJ, Horton BP (eds) Handb Sea-Level Res. John Wiley & Sons, Ltd, Oxford, UK
Kokot RR (2004) Erosión en la costa patagónica por cambio climático. Rev Asoc Geol Argent 59:715–726
Kopp RE, Horton BP, Kemp AC et al (2015) Past and future sea-level rise along the coast of North Carolina, USA. Clim Chang 132:693–707
Lucas A, Guerrero RA, Mianzán H et al (2005) Coastal oceanographic regimes of the Northern Argentine Continental Shelf (34–43°S). Estuar Coast Shelf Sci 65:405–420
Macchi GJ, Acha ME (1998) Aspectos reproductivos de las principales especies de peces en la Zona Común de Pesca Argentino-Uruguaya y en El Rincón. In: Lasta CA (ed) Resultados de una campaña de evaluación de recursos demersales costeros de la provincia de Buenos Aires y del litoral uruguayo. Noviembre 1994, vol 21. INIDEP Inf. Téc, pp 67–89
Malvárez AI (1997) Las comunidades vegetales del Delta del río Paraná. Su relación con factores ambientales y patrones del paisaje. PhD thesis. Facultad de Ciencias Exactas y Naturales, Universidad de Buenos Aires
Marrari M, Viñas MD, Martos P et al (2004) Spatial patterns of mesozooplankton distribution in the Southwestern Atlantic Ocean (34°–41°S) during austral spring: relationship with the hydrographic conditions. J Mar Sci 61:667–679
Medina R, Codignotto J (2013) Evolución del delta del río Paraná y su posible vinculación con el calentamiento global. Rev Mus Argent Cienc Nat 15:191–200
Militelli MI, Macchi GJ, Rodrigues KA (2013) Comparative reproductive biology of Sciaenidae family species in the Río de la Plata and Buenos Aires Coastal Zone, Argentina. J Mar Biol Assoc UK 93:413–423

Miloslavich P, Klein E, Díaz JM et al (2011) Marine biodiversity in the Atlantic and Pacific coasts of South America: knowledge and gaps. PLoS One 6(1):e14631. https://doi.org/10.1371/journal.pone.0014631
Mitrovica JX, Milne GA (2002) On the origin of late Holocene sea-level highstands within equatorial ocean basins. Quat Sci Rev 21:2179–2190
Molina JM, Lopez Cazorla A (2011) Trophic ecology of *Mustelus schmitti* in a nursery area of northern Patagonia. J Sea Res 65:381–389
Neiff JJ (1999) El régimen de pulsos en ríos y grandes humedales de Sudamérica. In Malvarez AI, Kandus P (eds) Tópicos sobre Humedales Sudamericanos ORCYT-MAB (UNESCO), Buenos Aires, Argentina
Peel MC, Finlayson BL, McMahon TA (2007) Updated world map of the Köppen-Geiger climate classification. Hydrol Earth Syst Sci 11:1633–1644
Peltier WR, (2002) On eustatic sea level history: Last Glacial Maximum to Holocene. Quat Sci Rev 21:377–396
Peteán J, Cappato J (2005) Humedales fluviales de América del Sur. Hacia un manejo Sustentable. Imprenta Lux SA, Santa Fé, Argentina
Piola AR, Rivas AL (1997) Corrientes en la plataforma continental. In: Boschi EE (ed) El Mar Argentino y Sus Recursos Pesqueros 1: Antecedentes históricos de las exploraciones en el mar y las características ambientales. Instituto Nacional de Investigación y Desarrollo Pesquero, Mar del Plata, pp 119–132
Piola AR, Campos EJD, Möller OO Jr et al (2000) Subtropical shelf front off eastern South America. J Geophys Res Oceans 105(C3):6565–6578
Pratolongo PD, Piovan MJ, Cuadrado DG (2016) Coastal environments in the Bahía Blanca estuary, Argentina. In: Khan M, Boër B, Őzturk M et al (eds) Sabkha ecosystems, Tasks for Vegetation Science, vol 48. Springer, Cham, pp 205–224
Raya Rey A, Huettmann F (2020) Telecoupling analysis of the Patagonian Shelf: a new approach to study global seabird-fisheries interactions to achieve sustainability. J Nat Conserv 53:125748
Rivas AL, Dogliotti AI, Gagliardini DA (2006) Seasonal variability in satellite-measured surface chlorophyll in the Patagonian Shelf. Cont Shelf Res 26:703–720
Rivera J, Penalba O (2015) El Niño/La Niña events as a tool for regional drought monitoring in Southern South America. In: Andreu J, Solera A, Paredes-Arquiola J et al (eds) Drought: research and science-policy interfacing. CRC Press, Valencia, pp 293–299
Romero SI, Piola AR, Charo M et al (2006) Chlorophyll-a variability off Patagonia based on SeaWiFS data. J Geophys Res 111:C05021
Rostami K, Peltier WR, Mangini A (2000) Quaternary marine terraces, sea-level changes and uplift history of Patagonia, Argentina: comparisons with predictions of the ICE-4G (VM2) model of the global process of glacial isostatic adjustment. Quat Sci Rev 19:1495–1525
Scian B, Labraga JC, Reimers W et al (2006) Characteristics of large-scale atmospheric circulation related to extreme monthly rainfall anomalies in the Pampa region, Argentina, under non-ENSO conditions. Theor Appl Climatol 85:89–106
Sherman K, Duda AM (1999) Large marine ecosystems: an emerging paradigm for fishery sustainability. Fisheries 24:15–26
Soriano A, León RJC, Sala OE et al (1992) Río de la Plata grasslands. In: Coupland RT (ed) Natural grasslands: introduction and western hemisphere, Serie: Ecosystems of the world. Elsevier, New York
Spalding MD, Fox HE, Allen GR et al (2007) Marine ecoregions of the world: a bioregionalization of coastal and shelf areas. Bioscience 57:573–583
Viñas MD, Marrari M, Di Mauro RP et al (2013) El zooplancton del hábitat reproductivo de la población bonaerense de anchoíta (*Engraulis anchoita*), con especial énfasis en crustáceos. Rev Investig Desarr Pesq 23:125–144
Violante RA, Parker G (2000) El Holoceno en las regiones marinas y costeras del nordeste de Buenos Aires. Rev Asoc Geol Argent 55:337–351

Vilanova I, Prieto AR, Stutz S (2006) Historia de la vegetación en relación con la evolución geomorfológica de las llanuras costeras del este de la provincia de Buenos Aires durante el Holoceno. Ameghiniana 43:147–159

Vilas F, Arche A, Ferrero M, Isla F (1999) Subantarctic macrotidal flats, cheniers and beaches in San Sebastián Bay, Tierra del Fuego, Argentina. Mar Geol 160:301–326

Wieters EA, McQuaid C, Palomo G et al (2012) Biogeographical boundaries, functional group structure and diversity of Rocky Shore communities along the Argentinean coast. PLoS One 7(11):e49725. https://doi.org/10.1371/journal.pone.0049725

Chapter 2
Geography of the Bahía Blanca Estuary

Walter Daniel Melo

2.1 Introduction

From the physiographic point of view, the Bahía Blanca Estuary and associated wetlands represent one of the most remarkable features within the environmental diversity that characterizes the Argentine coast. The system encompasses a wide variety of landscape elements like sandy beaches, coastal dunes, tidal plains, marshes, islands, tidal channels, and extensive salt flats, inherited from the transgressive past.

Bahía Blanca is located at an inflection point in the coast of the Buenos Aires Province, central east Argentina, where the coastline turns from the predominant north-south direction to the east-west (Fig. 1.1; Chap. 1). The coastal system is a 2,500 km^2 embayment that extends from the cliffs at "Baliza Monte Hermoso" (an old beacon northwestward) to Laberinto Point on the southern shore of Brightman Creek (Fig. 2.1). The estuarine system comprises three major first-order channels, almost parallelly oriented northwest to southeast: Principal Channel, Bahía Falsa, and Bahía Verde. The Principal Channel, the main navigation channel, extends along 120 km, naturally reaching around 10 m depth before dredging. Besides main channels, the estuarine system is composed of more than 128,000 ha of tidal mudflats and salt marshes, dissected by countless tidal channels of smaller order and varying sizes (Perillo and Piccolo 1999a, b). The southern limit of the system is defined by the Colorado River, the only permanent watercourse with a substantial discharge (Melo and Limbozzi 2008). Within the Bahía Blanca Estuary, continental runoff restricts to the northern shore of Principal Channel (Píccolo and Perillo

W. D. Melo (✉)
Instituto Argentino de Oceanografía, IADO (Universidad Nacional del Sur-CONICET), Bahía Blanca, Argentina

Departamento de Geografía y Turismo, Universidad Nacional del Sur, Bahía Blanca, Argentina
e-mail: wdmelo@iado-conicet.gob.ar

S. M. Fiori, P. D. Pratolongo (eds.), *The Bahía Blanca Estuary*,
https://doi.org/10.1007/978-3-030-66486-2_2

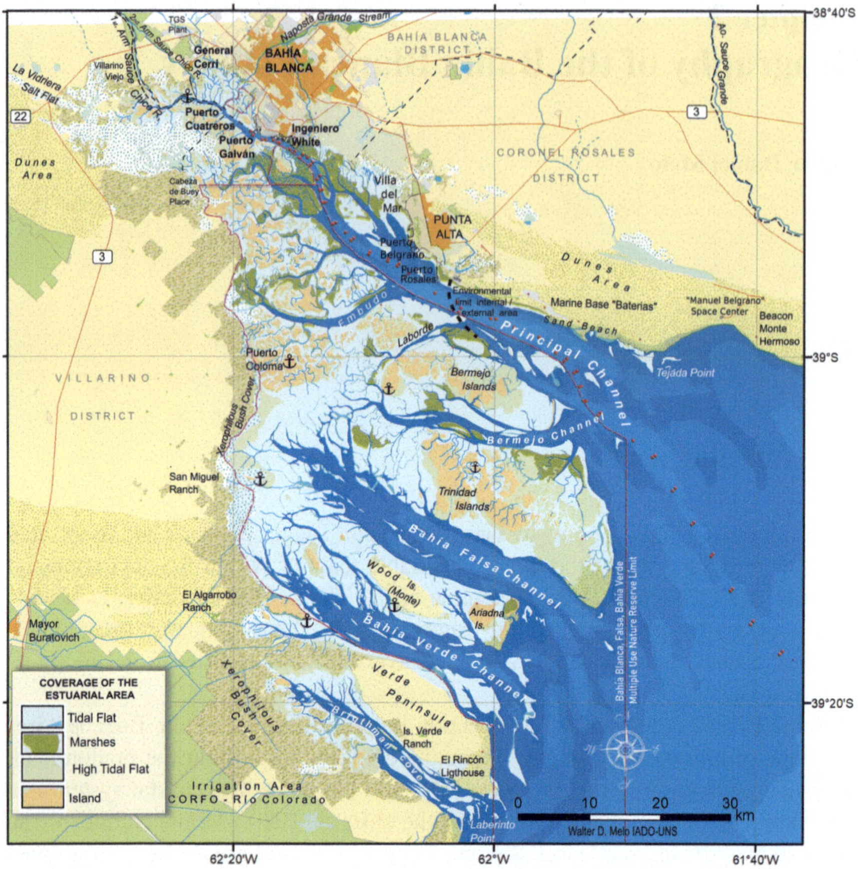

Fig. 2.1 Bahía Blanca Estuary. District, cities, and main coastal environments. (Map by Walter D. Melo)

1990). Two permanent small rivers represent the main freshwater input to the estuary (Fig. 2.1). The Sauce Chico River has a hydrographic basin that covers 1450 km^2 and an average flow of 1,807 m^3/s. The Napostá Grande Stream has a hydrographic basin of 1100 km^2 and an average flow rate of 0.8 m^3/s (Piccolo et al. 2008). Besides these two permanent sources, there are intermittent streams that drain rainfall excess during wet periods. Main intermittent streams are those in the group named Saladillo, which encompasses three different streams (Saladillo or Dulce, Saladillo de Lazaga, and Saladillo de García) and the Maldonado Channel, built in 1950 to manage excess rainfall on the Bahía Blanca City and overflows of the Napostá Grande River (Fig. 2.2).

Large salt flats inherited from the transgressive past extend through the northwest of Principal Channel, becoming dominant landscape features beyond tidal influences. These gently sloped plains, currently unconnected to tides, performed as intertidal mudflats under a higher relative sea level during the Holocene. Nowadays,

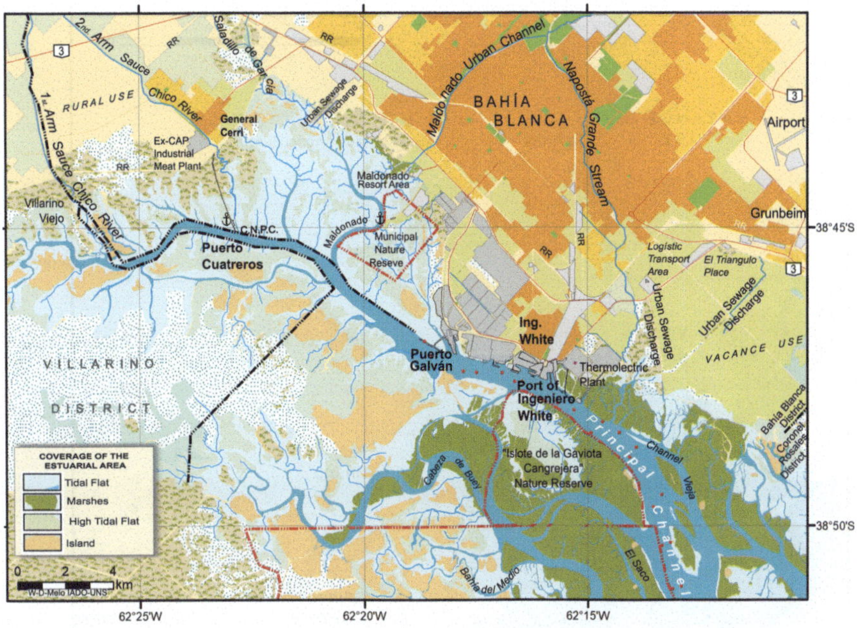

Fig. 2.2 Coastal uses and environments in Bahía Blanca District. (Map by Walter D. Melo)

these extended salt flats function as transitional areas between the marine and continental domains. "La Vidriera" is the largest salt flat in the area (Fig. 2.1), extending along 8 km through the northwest, following the axis Chasicó Lagoon-La Vidriera-Principal Channel. This alignment defines a morphologic transition that separates North and South Continental Regions. The northern shore of Principal Channel, eastward from La Vidriera Salt Flat, hosts a large environmental variability, including minor estuarine areas at the mouth of the Sauce Chico River, Napostá Grande Stream, and small stream Saladillo de García. Through the mouth and after a salt flat area, the small village Villa Del Mar, Puerto Belgrano, and the Puerto Rosales are located in the transition between tide- and wave-dominated systems (Melo and Carbone 2012) (Fig. 2.3). East from Puerto Rosales, fine-grained mudflats give place to sandy beaches that develop at the edge of coastal dunes, like those in Tejada Point. Sandy beaches extend eastward up to the cliffs at Baliza Monte Hermoso.

The total area floodable by extraordinary tides in storm situations totals approximately 128,000 ha, of which 28,000 ha would correspond to high tidal flats, the regular tidal flats cover about 90,000 ha, and the marsh areas about 18,000 ha (Fig. 2.1).

Regarding island morphology, two different zones can be distinguished. North of the Bahía Falsa Channel, tidal flats of considerable extension cover most of the estuarine area, and plentiful isles of variable sizes form disaggregated clusters (Melo et al. 2003). Despite its name, Trinidad Island is actually a group of 46 islands covering 4,300 ha of emerged land (Fig. 2.1). The surface area of the largest island

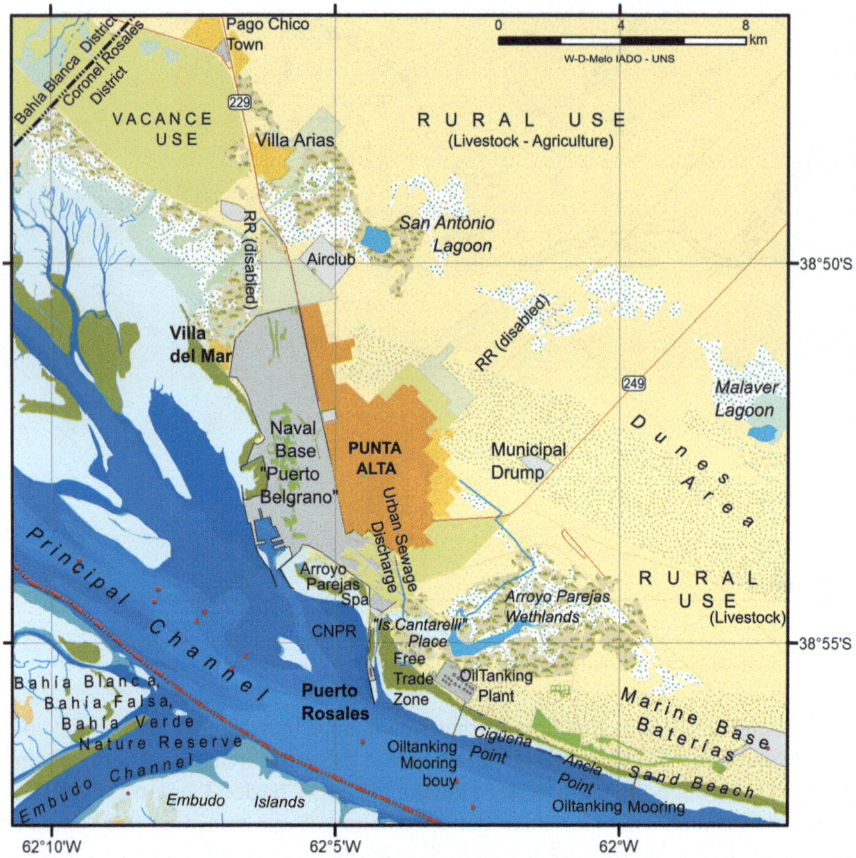

Fig. 2.3 Main coastal uses and environments in Coronel Rosales District. (Map by Walter D. Melo)

is 1,340 ha, but smaller isles cover less than half a hectare. Intertidal environments cover most of Trinidad Islands, including about 15,000 ha of tidal flats, 3,300 ha of high tidal flats, and 1000 ha of marshes. Through the southeast, under a wave-dominated regime, sand flats develop instead of tidal flats (Carbone and Melo 2016). Similarly, the group Bermejo Island (2,200 ha of emerged land) comprises 52 separated isles, the largest island covering about 470 ha, surrounded by more than 8,900 ha of tidal flats and 700 ha of marshes. Finally, the group Embudo Islands is composed of about 40 islands covering 675 ha of emerged land. About 250 additional islands and islets form smaller groups, covering 6,800 ha along with small channels and tidal flats. The emerged land of these islands presents highly saline soils, and vegetation restricts to halophytic steppes and shrubs.

South of the Bahía Falsa Channel, single larger isles conform geomorphic units instead of groups. The higher elevations and sandy soils allow the formation of sand dunes, and sand flats commonly replace mudflats in the intertidal zone. Major isles are Wood (also called Monte) and Ariadna, both aligned between the Bahía Falsa and Bahía Verde Channels, and Verde Peninsula, formerly an island that got attached to the continent once the relative sea level dropped after the Holocene highstand. Verde Peninsula covers about 9,000 ha, and extensive dune fields develop near the eastern coast.

This complex geomorphic setting was shaped by the recent geological history of the Bahía Blanca Estuary, during the late Pleistocene and Holocene (Melo 2004). There is a paleomorphologic relationship between the estuary and the adjacent continental environment. Bahía Blanca Estuary is the reception basin of a drainage basin having a surface of the order of 3000 km^2. However, in the current drainages of its hydrographic basin, no large watercourses are observed that are directly related to the geomorphological succession that determined the origin of the estuary. The existence of geomorphologic units such as the Colorado River, dune fields, sand strandflats, and a series of the valleys and axes of depressions areas in the SW of La Pampa Province (Fig. 1.2; Chap. 1) they are connected since the Pleistocene to the genesis and evolution of the estuary. The latter are aligned and related to previous faulting which may be related directly to the origin of the estuary. Part of these alignments are below mean sea level (−20 m) and formed by canyons and valleys containing dunes of varied shapes. These valleys have been arranged in groups. The first group starts on the NW of Lihué Calel Hills and formed two alignments having a NW-SE and E-W trends. Both follow the Chasicó-Salinas Chicas depression, continuing into the La Vidriera Salt Flat and the Principal Channel. The second group is 120 km in length formed by the alignment passing through the Blanca Grande and Callaquéo lagoons, ending in the Chasicó-Salinas Chicas depression (Fig. 1.2; Chap. 1). The alignments in the third group are mostly trending in the E-W direction starting at the confluence between the Colorado and Salado Rivers ending in the Chasicó-Salinas Chicas depression. This group is independent of the other two, and its alignments are related to the Colorado River directed toward the southern portion of the estuary (González Uriarte 1984). The most significant axis has a length of 140 km and relates the Anzoátegui and La Gotera Salt Lagoons (Fig. 1.2; Chap. 1) with Bahía Falsa and Bahía Verde. The fluvial discharge along these groups provided the hydric connection between the continent and the littoral environments developing the initial stages of the formation of the Bahía Blanca Estuary.

The area presently occupied by the estuary was subaerial. Later on, the climate changed to warmer and more humid conditions, and the mean sea level increased to approximately the present conditions (Aguirre 1995). The drain introduced a large sediment load coming from the three first groups, developing a delta. Major channels formed from old continental drains that followed the three different drainage axes, ending in the present area of the Bahía Blanca Estuary. Evidence of these former drainages persists in the major estuarine channels aligned with continental depressions. According to Aguirre (1995), a sharp change in climate occurred during the early Holocene, establishing a period of warmer and more humid conditions, before the transgressive maximum. Additional evidences of humid climate during the transgressive phase were found in the coastal zone that extends up to 100 km west from Bahía Blanca. In this area, continental lagoons showed a rise in water levels and higher abundances of freshwater diatoms, fishes, and vascular plants (Gutiérrez-Téllez and Schillizzi 2002). Under humid conditions during the early Holocene, rivers may have introduced large amounts of fine sediments (mostly loessic pampean silts) to the estuarine area, using these previous topographic depressions.

According to Spalletti and Isla (2003), during the mid-Holocene transgressive stage, the deltaic deposits were partially covered with medium sand and biogenic remains, as well as compacted silty clay sediments in shallower zones. The advance of marine sediments as well as the rework and redistribution of fluvial-estuarine sediments filled paleochannels, until the establishment of a large estuarine-marine environment. Because of the higher sea level, the delta was fully covered by the sea reaching the continent through the La Vidriera Salt Flat (Melo 2004). Humid conditions may have continued until about 5000 years BP, when the establishment of psammophytic and halophytic communities indicates a climatic change toward sub-humid to dry conditions (Prieto 1996). The increase of the mean sea level reached values of 5–12 m above present conditions (González Uriarte 1984). These values are related to similar studies made by Cavallotto et al. (1999) for the Río de la Plata and Aliotta et al. (2000) and Spagnuolo et al. (2000) for Arroyo Parejas. Toward 3000 years BP, climate changed to temperate-dry, similar to the present conditions (Aguirre 1995), which determined the disappearance of the rivers being replaced by Aeolic environments. Consequently, the Colorado River progressively migrated southward. In this migration, the river occupied and later released the axis Anzoátegui Salt Lake-La Gotera-Bahía Falsa.

Under a falling sea level and diminished river discharge, the present islands may have emerged, and large continental drainages disappeared, or transformed in the tidal channels Bahía Verde, Bahía Falsa, and Principal Channel. About 2000 years BP, the connection between the continental and transitional environments (La Vidriera) was transformed in a coastal lagoon.

As sea level continued to fall, low elevation isles connected and reshaped to form extensive mudflats. Coastal lagoons got progressively isolated from the marine influence and transformed in salt flats, and the increased aridity enhanced wind transport of fine sediments, allowing the formation of large dune fields that presently cover southern isles. Wind action transported eastward the original sand deposits from southeast of La Pampa Province, forming extensive dune fields. This material also covered the Verde, Ariadna, and Monte Island, located to the south of the estuary. González and Weiler (1983) and Spalletti and Isla (2003) found at the mouth of the Colorado Nuevo River, to the south of the indicated islands, the presence of a lobular prograding shelf delta originated by marine reworking.

2.2 Occupation Processes and Land Use in Bahía Blanca

In 1520 the european explorers first entered at the Principal Channel, and Brightman Cove and the "Cabeza de Buey" Channel were used as a refuge by the navigators. However, studies on the survey of canals and islands would begin almost 300 years later (Melo and Piccolo 2006). The complex environmental situations articulated with historical processes led to a late survey of the coast, of the islands, and of the plains being in 1804 when the first geographical recognitions are made. The contribution of fresh water and the Principal Channel presence were the main reasons that allowed Bahía Blanca City to be founded in 1828, which, as a small military garrison and a smaller civilian population, survived almost alone until 1880 with the effective occupation of Patagonia. Until those years, its port lacked relevant infrastructure, and in 1884 with the presence of the railroad at "Ingeniero White" town, the port works began to be significant (Fig. 2.2). Toward the end of that century, in 1898, the creation of the Naval Base "Puerto Belgrano" resulted in the establishment of a small population and which were the origins of the "Punta Alta" city (Fig. 2.3). At the beginning of the twentieth century, it already established the conditions for the consolidation of the railway activity and the establishment of local industries, creating Puerto Galván (1902), Puerto Cuatreros (1903), and Puerto Rosales (1912) (Coleman 1948). Land uses such as way urban, industrial and railway occupation was carried out only along the north coast of the Principal Channel, between Puerto Cuatreros and Puerto Rosales. In this strip of the channel, the permanent watercourses formed former alluvial cones where its elevated terrain did not give rise to the formation of extensive tidal plains and thus facilitates the continental proximity to the greater depths of the Principal Channel, which allowed port filing.

The territorial organization of the Bahía Blanca Estuary area and its wetlands on the coast, from the political and administrative aspect, three different districts integrate the area: (i) Municipality of Villarino; (ii) Municipality of Bahía Blanca, including the localities of General Daniel Cerri and Ingeniero White (hereinafter referred to as Cerri and White, respectively); and (iii) the Municipality of Coronel de Marina Leonardo Rosales which includes Villa del Mar and Pehuen Co.

2.2.1 Villarino District

This district has a coastline extension of approximately 250 km (Fig. 2.1), of which about 200 km is of jurisdiction on the estuarial coastal front, which develops from north to south between the 1st Arm of the Sauce Chico River and Laberinto Point, at south of Brightman Cove (Figs. 2.1 and 2.2). Although this district has the greatest extension over the wetlands of the Bahía Blanca, it does not have any localities or minor settlements on the coast and the activity is rural only. Only one mention, "Villarino Viejo" Place, which is about 2 km from the head of the Principal Channel is a small sector of farms next to the aforementioned arm and that initially functioned as the district's header (Fig. 2.2). However, its activity of coastal use is by only sports anglres that to arrive at the coast are used neighborhood roads. In pioneer times, there were smaller ports or berths such as "Villarino Viejo," "Puerto Coloma," "San Miguel" Farm, and "El Algarrobo" farm were used by rural establishments on the continent. Today, they are disjointed, and they are used as sport fishing activity points for private use. The Villarino District dominance reaches

Verde Peninsula. In its origins of productive occupation, the “Isla Verde” Farm of 9000 ha worked. Since 1947, the occupation belongs to the Argentine Navy, where a limited sector is for military use, but its use remains predominantly rural (Melo et al. 2005). In turn, the isthmus of the peninsula belongs to a private farm. In “Bahía Verde” Channel and Brightman Cove, the drainage channels of the “Corfo Rio Colorado” corporation area have their mouth at the time of irrigation (Fig. 2.1).

2.2.2 *Bahía Blanca District*

The Bahía Blanca jurisdiction covers the entire coastal front of the district, from the 1st Arm of the Sauce Chico River to the cadastral boundary with the Coronel Rosales District, occupying about 35 km from the north veril of the Principal Channel (Fig. 2.1). Basically, three sectors are identified on the Bahía Blanca District coastal front. To the west of said front, the space of 5000 m that covers from the 1st Arm of the “Sauce Chico” River to the “General Cerri” Town the are lands in which spaces for rural use are interspersed with spaces without uses. In this sector, the presence of Puerto Cuatreros stands out, next to the 2nd Arm of the Sauce Chico River. It is a great breakwater, which is reached by a consolidated road on the salt flats the tidal plains. This functioned as an export point for the former “Sansinena” Industrial Meat Plant (then CAP Corporation, today abandoned) (CGPBB 2020) and which determined the growth of the town of “General Cerri.” This port today functions as a recreation, environmental observation, and fishing club (Fig. 2.2). In this west sector, the smaller sector rural use, 400 ha, is interrupted by the installation of the urban effluent discharge treatment plant and one waste treatment plant of the meat industry. Between the “Maldonado” Spa and the mouth of the Napostá Grande Stream, the greatest urban and industrial pressure develops over the entire coastal area (Melo and Limbozzi 2008) (Fig. 2.4). The mentioned spa is the only coastal summer recreation center in the district and operates on the east veril of the “Maldonado” Urban Channel. To the southeast of the same, at the estuary in the east border of Maldonado Channel, there is a coastal front project that contains the Municipal Natural Reserve and that the “Belisario Roldán” Municipal Dump was formerly operating. Neighboring this reserve is the most important port complex on the Argentine coast that covers 9 km between Puerto Galván and Ingeniero White Town (Fig. 2.4).

2.2.2.1 Puerto Galván

It was founded as a cereal terminal in 1905 by the company “Buenos Aires al Pacífico” (BAP), a former British railroad company (CGPBB 2020). At present, it has diversified its operational activity toward oil and gas activity. Among its original facilities is the specialized terminal for the management of cereals and by-products by the companies “Oleaginosa Moreno Hnos. S.A.” and the incorporation of the

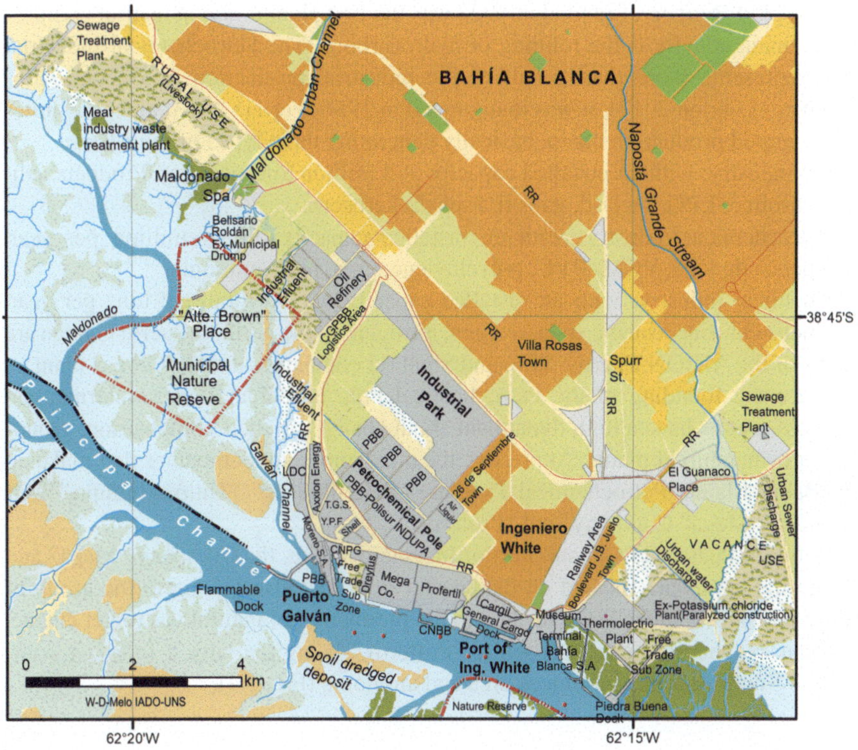

Fig. 2.4 Coastal uses and environments in main pressure area of Bahía Blanca District. (Map by Walter D. Melo)

companies "Los Grobo Inversora S.A." and "Louis Dreyfus Commodities (LDC)." At the west end of Puerto Galván is the Flammables dock (CGPBB 2020) (Fig. 2.4). It is composed of two berthing sites, site No. 1, destined to the operation of liquid fuels by oil companies and caustic soda produced by the firm "Indupa S.A.," and site No. 2, assigned to the operation with gaseous and petrochemical products by the companies of the Bahía Blanca Petrochemical Pole and "Transportadora de Gas del Sur" (TGS). Also, a floating petrochemical plant ("Polisur S.A.") dedicated to the production of high- and low-density polyethylene is located. To the west of the main pier is a sector of 6 hectares dedicated to general cargoes. In that same sector, the "Puerto Galván" Yacht Club (CNPG) has its facilities (Fig. 2.4).

2.2.2.2 Petrochemical Pole

Between Puerto Galván and the Port of Ingeniero White, the largest petrochemical complex in South America is located, which together with the Industrial Park occupies some 600 hectares (Figs. 2.2 and 2.4). To the north of the sector is the "Ricardo

Eliçabe" Oil Refinery, founded in 1922 was the first plant of the industrial complex whose main activity is the refining of hydrocarbons. The Petrochemical Industrial Pole represents more than a half of the industrial production of Buenos Aires Province (London 2017), it was inaugurated in 1986 (Ferrera 2003) and constitutes an integrated production chain, basically generating three types of industries:

Oil industry: with an installed capacity of 4 million tons per year. Products: ethane, gasoline, LPG, fuel oil, gas oil, asphalt, kerosene.

Petrochemical industry: with an installed capacity of 3.4 million per year. Products: ethylene, VCM, PVC, polyethylene, urea, pure ammonia.

Chemical industry: with an installed capacity of 350 thousand tons per year. Products: chlorine, caustic soda, oxygen, nitrogen. In the north of Puerto Galván are plants of the AXION Energy, YPF, and Shell oil companies and a TGS gas plant.

Since the founding of the Petrochemical Pole of Bahía Blanca, international business groups have a high dynamism about changes in their composition, grouping, or ownership of plants. At present, the most important companies that make up the Bahía Blanca Petrochemical Pole are "PBB-Polisur," "Unipar-Indupa," "Air Liquide," "Mega" Company, and "Profertil."

Between Puerto Galván and Port of Ingeniero White, the Company "Mega" and "Profertil" plants are located on the coastal front and also belong to the group of port companies. "Mega" is dedicated to the production of natural gas to get ethane, propane, butane, and gasoline, while "Profertil," being the largest fertilizer company in South America, produces urea and ammonia (Fig. 2.4).

2.2.2.3 Port of Ingeniero White

The Port of Ingeniero White, originally called "Puerto Commercial Bahía Blanca," was founded in 1884 as the railroad head of the "Ferrocarril del Sud," a company with capitals of British origin. The port consisted of a series of iron docks, of a structure called transparent, that crossed the tidal plains to the edge of the Principal Channel (CGPBB 2020). In its origins, the so-called fruits of the country were exported, consisting of cereals, leather, and wool. Subsequently, fruits and fishing products were added. Since the end of the nineteenth century and throughout the twentieth century, artisanal fishing was carried out in the estuary canals. The large number of dock workers, railways workers, and fishermen also determined the origin of the Ingeniero White Town (Fig. 2.4). Between 1948 and 1993, the National State administered the port through the "Administración General de Puertos" (AGP). In 1993, the "Consorcio General de Puertos de Bahía Blanca" (CGPBB), an administrative entity composed of representatives of private port companies, the provincial and local state, replaced this administration.

Its access navigation channel, the Principal Channel, has 40 feet of the draft that makes it the most important deepwater port in Argentina. Cereals are exported through the "Cargill" and "Terminal Bahía Blanca" companies. Containers enter the general cargo dock and tie up boats of various uses, such as tourism, defense, and deep-sea fishing. The boat dock dedicated to artisanal fishing is still maintained in a small sector. In the western sector are the facilities of the "Bahía Blanca" Yacht Club (Fig. 2.4).

To the east and bordering on Coronel Rosales District, a salt area, abandoned tidal plains and bush land coverage that total some 2,500 hectares of vacant land are linked to the future port, industrial, and services expansion. This sector is also crossed by an area of urban water discharges from Grünbeim Town.

The area occupied by the "Great Bahía Blanca," which involves the city of Bahía Blanca, General Cerri, and Ingeniero White Towns, is 150 km^2. The total population of the districts is 305,000 inhabitants, discriminated for Bahía Blanca 285,000 habitants, Ingeniero White 12,489, and General Cerri 6,745 habitants (INDEC 2010).

2.2.3 Coronel Rosales District

This district totals a coastal front of about 78 km, of which about 60 km of extension are on coastal of the Bahía Blanca Estuary. On the border with the Bahía Blanca District, the vacant of salt areas, abandoned tidal plains, and shrub coverage continues, where in Coronel Rosales District it covers about 2000 hectares. "Villa del Mar" town stands out as the only residential urban coastal center of the Villarino, Bahía Blanca, and Coronel Rosales District (Fig. 2.1). Born from lots as a seaside resort in 1944 (Cinti 2017), today and with 327 inhabitants, it behaves like a neighborhood center, the result of the urban expansion of Punta Alta city (Fig. 2.3).

Neighboring to "Villa del Mar" is the Naval Base "Puerto Belgrano," which with about 8 km of development and about 1,200 hectares is imposed with a strong presence on the coastal front (Fig. 2.3). Being the main military settlement of the Argentine Navy, it determined the foundation and growth of the town of Punta Alta (Chalier 2019), which has about 58,000 inhabitants (INDEC 2010). However, this presence also prevented the direct geographical linking of this town with the coastal area, which was only achieved with its urban expansion toward "Villa del Mar" (Fig. 2.3).

The Puerto Rosales area is presented as a complex sector (Fig. 2.3). It is a large dock and breakwater of 700 m built as a poorly planned railroad project in 1912 (Melo et al. 1997). In the western sector of the breakwater, the "Arroyo Parejas" Spa, the main summer recreation site immediately to the city of Punta Alta, operates. While in the east sector this operates a minor port linked to fishing activities and the maintenance and service of a system of boats that serve the maintenance of oil monobuoys of "Oiltanking" company. The access channel of this port is the "Arroyo Parejas" Stream, which also functions as a discharge channel for the urban effluents of Punta Alta. To the east of the access channel to dock of Puerto Rosales, in the spikes and coastal dunes sector called "Cantarelli Island," there is a place dedicated to the future Free Trade Zone of Coronel Rosales District and next to it is the oil pumping and shipping plant of "Oiltanking" company (Fig. 2.3).

In the last 35 km of coastal development, up to the cliffs of "Beacon Monte Hermoso," the sandy beaches are under the total control of the Marine Base

"Baterias" and its access is restricted. In the east sector of said base, in the area near the mentioned cliffs, the "Manuel Belgrano" Space Center is built whose function will be to be a launching platform for national satellites.

2.2.4 The Islands

In the Buenos Aires Province, the islands are provincial property, where the districts have their municipal jurisdiction. Since the end of the nineteenth century, the "Bermejo," "Trinidad," and "Wood" Islands were given in concession for productive purposes, where they worked from factories to process sharks and sea lions as well as sheep, swine, and bovine livestock exploitation (Cinti 2017). Since 1992, all the islands were declared Natural Reserves and their commercial exploitation ceased. Only "Wood" Island is maintained with livestock activity, while "Ariadne" Island, like the northern islands, has no commercial use (Fig. 2.1).

References

Aguirre ML (1995) Cambios ambientales en la región costera bonaerense durante el Cuaternario tardío. Evidencias Malacológicas. IV Jornadas Geológicas y Geofísicas Bonaerenses. Actas I: 35–40

Aliotta S, Schnack EJ, IslaF et al (2000) Desarrollo secuencial de formas de fondo en un régimen macromareal. Latin American Journal of Sedimentology and Basin Analysis 7(12):95–107

Carbone ME, Melo WD (2016) Aplicación de geotecnologias en el análisis de procesos morfodinámicos del sector costero externo del estuario de Bahía Blanca. 3ras Jornadas de Nuevas Tecnologías de la Información Geográfica del Sur Argentino. Departamento de Geografía y Turismo UNS. Bahía Blanca

Cavallotto JL, Violante R, Parker G (1999) Historia Evolutiva del Río de la Plata durante el Holoceno. Actas XIV Congreso Geológico Argentino, I: 508–511

CGPBB (2020) La Bahía… Razones para una Identidad. Consorcio General de Puerto de Bahía Blanca. Disponible en: https://puertobahiablanca.com

Chalier G (2019) Algunas consideraciones acerca de Puerto Belgrano como área patrimonial: concientización y revalorización para el fortalecimiento del sentido de pertenencia. RES GESTA, nro 55, Instituto de Historia – Fac. Der. y Cs. Ss. del Rosario – UCA. Rosario – Argentina. ISSN: 2525-0884

Cinti S (2017) Las Islas de la bahía Blanca/Los forjadores de su historia, 1ra Edición. Vacasagrada Ediciones, Bahía Blanca

Coleman AH (1948) Mi vida de ferroviario inglés en la Argentina: 1887–1948. Ediciones del autor, Bahía Blanca

Ferrera IM (2003) El polo petroquímico de Bahía Blanca como eje y estrategia del desarrollo en el corredor bioceánico Sur. Revista Tiempo y Espacio. Temuco, Chile. 10(13):63–80

González Uriarte M (1984) Características geomorfológicas de la porción continental que rodea la Bahía Blanca, Provincia de Buenos Aires. IX Congreso Geológico. Actas III. 556–576

González MA, Weiler N (1983) Ciclicidad de niveles marinos holocénicos en Bahía Blanca en delta del río Colorado (Provincia de Buenos Aires), en base a edades Carbono-14. Actas de la

Oscilación del Nivel Medio del Mar durante el último Hemiciclo De glacial de la Argentina. 69–88. Mar del Plata

Gutiérrez Téllez B, Schillizzi R (2002) Asociaciones de Diatomeas en Paleoambientes Cuaternarios de la Costa Sur de la Provincia de Buenos Aires. Argentina Pesquisas em Geociências 29:59e70

INDEC (2010) Censo Nacional de Población, Hogares y Viviendas 2010 – Datos provistos por la Agencia de Inversión y Comercio Exterior del Municipio de Bahía Blanca

London SA (2017) Methodological Note about the Analysis of the Economic Growth and Environment Ecology & Environmental Sciences vol. 2

Melo W (2004) Origen del estuario de Bahía Blanca. En: Piccolo MC, Hoffmeyer MS (eds) Ecosistema del Estuario de Bahía Blanca. Instituto Argentino de Oceanografía. Sapienza Ed, pp 21–27

Melo WD, Carbone ME (2012) Regionalización ambiental del Estuario de Bahía Blanca. IX Jornadas Nacionales de Geografía Física de Bahía Blanca, Bahía Blanca

Melo WD, Limbozzi F (2008) Geomorphology, hidrological systems and land use of Bahía Blanca Estuary region. In: Neves R, Baretta J, Mateus M (eds) Perspectives on integrated coastal zone management in South America. IST Press. Scientific Publishers, Lisboa, pp 317–331

Melo WD, Piccolo MC (2006) Cartografía del estuario de Bahía Blanca. XXIII Reunión Científica Asociación Argentina de Geofísicos y Geodesias

Melo WD, Cuadrado D, Gómez E (1997) Influencia de las condiciones ambientales en la planificación territorial de Puerto Rosales. I^{ras} Jornadas Nacionales de Geografía Física 1996

Melo WD, Schillizzi R, Perillo GME et al (2003) Influencia del área continental pampeana en la evolución morfológica del estuario de Bahía Blanca. AAS Revista 10(1):39–52

Melo WD, Piccolo MC, Perillo GME (2005) Propuesta de desarrollo turístico en playas del estuario de Bahía Blanca: Faro El Rincón y Baliza Monte Hermoso. VII Jornadas Nacionales y Primer Simposio Internacional de Investigación-Acción en Turismo

Perillo GM, Piccolo MC (1990) Physical characteristics of the Bahía Blanca Estuary (Argentina). Estuar Coast Shelf Sci 34:303–347

Perillo GME, Piccolo MC (1999a) Geomorphological and physical characteristics of the Bahía Blanca Estuary, Argentina. In: Piccolo MC, Pino-Quivira M (eds) Perillo GME. Estuaries of South America. Environmental Science. Springer, Berlin/Heidelberg, pp 195–216

Perillo GM, Piccolo MC (1999b) Geomorphology and physical characteristics of the Bahía Blanca Estuary (Argentina). In: Piccolo MC, Pino-Quivira M (eds) Perillo GME. Estuaries of South America. Environmental Science. Springer, Berlin/Heidelberg, 223 p

Piccolo MC, Perillo GME, Melo WD (2008) The Bahía Blanca Estuary: an integrated overview of its geomorphology and dynamics. In: Neves R, Baretta J, Mateus M (eds) Perspectives on integrated coastal management in South America. IST Press, Lisboa, pp 231–240

Prieto AR (1996) Late Quaternary Vegetational and Climatic Changes in the Pampa Grassland of Argentina. Quat Res 45:73–88

Spagnuolo J, Farinati E, Aliotta S (2000) Tafonomía de *Brachidontes rodriguezi* (Bivalvia) de cordones costeros holocenos de Argentina. Profil. Institut für Geologie und Paläontologie, Universität Stuttgart, Band 18, CD- ROOM (76)

Spalletti LA, Isla FI (2003) Características y Evolución del delta del río Colorado ("Colú-Leuvú"), Provincia de Buenos Aires, República Argentina. Rev Asoc Argent Sedimentol 10(1):23–37

Chapter 3
Physical Oceanography of the Bahía Blanca Estuary

Gerardo M. E. Perillo and M. Cintia Piccolo

3.1 Introduction

Through history, estuaries have played a significant role in the evolution of coastal civilizations as they have fulfilled many essential roles that accelerated the establishment and economic advance of the ports and cities located on their shores (Perillo 1995a). Most estuaries are associated with rivers that allow a connection with the hinterland, especially those that have navigational capability. A low percentage of the major estuaries in the world does not have significant rivers providing freshwater and sediment input. The Bahía Blanca Estuary (Fig. 2.1; Chap. 2) is an exception as the present-day freshwater input is low and concentrated mostly at the head and middle reach of the Canal Principal.

The Bahía Blanca Estuary has a very complex evolution during the Late Pleistocene-Early Holocene (for detailed descriptions, see Chapters 1 and 2). The estuary today was the northern portion of a very large (over 300 km wide) delta formed by the combination of the Colorado and Negro Rivers (Melo et al. 2003, 2013), but mainly dominated by the former. The formation of the delta corresponds to a significant input of sediments from the Colorado River through several distributaries as well as other rivers that are no longer active. The excess sediment provided during about 10,000 years generated a delta which its actual extension into the inner shelf is today a matter of discussion since there is no clear geomorphologic and sedimentological indication of its outer border within the El Rincón area (Fig. 1.2; Chap. 1). Upon the diminishing discharges of the Colorado River, including the disappearance of many of its distributaries (Melo 2004), sediment input reduced sensibly, being presently basically negligible.

G. M. E. Perillo (✉) · M. C. Piccolo
Instituto Argentino de Oceanografía, IADO (Universidad Nacional del Sur-CONICET), Bahía Blanca, Argentina

Departamento de Geología, Universidad Nacional del Sur, Bahía Blanca, Argentina

S. M. Fiori, P. D. Pratolongo (eds.), *The Bahía Blanca Estuary*,
https://doi.org/10.1007/978-3-030-66486-2_3

On the other hand, mean sea level (msl) rose at extraordinary rates (about 9 mm/year; Gómez and Perillo 1995) from 8000 to 6000 years ago, reaching the highest level so far measured in the world of 7 m above present. Therefore, what is today the estuary was a shallow sea which penetrated along the Vidriera Salt Flat up to the Chasicó Lake (Melo 2004; Melo et al. 2013). Sea level dropped at a rate of the order of 4 mm/year (Gómez and Perillo 1995) until about the Little Ice Age (1400–1700 year AC) where msl bounced back and started to rise again. Nevertheless, Gómez et al. (2006) have clear proofs (correlated with similar data from Brazil) that about 2400 year BP msl dropped about 3 m below present. Still, the actual story of the origin of the estuary needs to unfold further. In the last 1000 years, the conditions have changed dramatically. Since there is no further sediment input, added to the advance of the msl of the order of 1.6 mm/year (Lanfredi et al. 1988, 1998), the Bahía Blanca Estuary has become a marine-dominated environment being in a significant erosional stage (Perillo 1995b).

To understand the dynamic processes that occur in the estuary is essential to have a clear image of the geomorphology of the system since all processes are controlled by the shape of the channels and the distribution of the intertidal and island areas. Therefore, the aim of this chapter is to provide an integral review of the geomorphologic, climatic, and dynamic processes acting on the estuary.

3.2 Geomorphologic Features of the Bahía Blanca Estuary

The geomorphology of the Bahía Blanca Estuary has been described many times already (i.e., Perillo and Piccolo 1991, 1999; Perillo et al. 2001; Melo 2004). Therefore, for the present case, only its general structure will be discussed with the addition of specific features that appear as the most relevant, but also those that have the higher impact on the dynamic processes. The Bahía Blanca Estuary is a complex network of tidal channels, with a set of major ones (Principal, Bermejo, Bahía Falsa, Bahía Verde, and Caleta Brightman) which have a general NW-SE orientation. All these major channels are remnants of the original distributaries of the Colorado River.

Although at first sight, the Bahía Blanca Estuary appears with little or no changes along the time, when analyzed in detail, there are significant modifications in the migration of major channels (Ginsberg and Perillo 1999; Ginsberg et al. 2009), as well as significant changes that occur among the various marshes and tidal flats. After the msl retreated about 1500 years ago, the intertidal and supratidal areas were dominated by *Sarcocornia* marshes. However, the erosional processes undermined these marshes lowering the level of the surface making them more prone to have a longer hydroperiod (meaning passing from inundations about 40–70 times per year to about 700 times per year). Most of the *Sarcocornia* marshes were replaced by tidal flats which themselves are, in part, populated by *Spartina alterniflora* marshes, plants that are more adaptable to longer periods of inundation (Pratolongo et al. 2013).

Nevertheless, as one travels along the tidal channels of the estuary, the original strata of the delta deposits are outcropping along the flanks. Although, there are clear signs of erosion in the inner channels and along the island coasts. It is common to find parts of the islands with their typical continental vegetation deposited, after being eroded, along the banks of the channels being latter redistributed by the tidal currents.

The sediment eroded, mostly from the inner portion of the estuary, is transported toward the mouth and deposited on the lower tidal flats. Therefore, there is a sediment compaction gradient from the head to the mouth of the estuary. At the head, where most of the material is still the original from the delta, the level of compaction is higher. However, toward the mouth, most of the sediment that was eroded and was not transported outside the estuary by the ebb currents is deposited on the low flats. These are generating layers of unconsolidated sediments, which, if affected by strong currents or waves, may be easily resuspended. The places where these sediments are deposited tend to be natural traps, meaning that the hydrodynamic conditions make them adequate to concentrate the material.

One of the most interesting things observed in the Bahía Blanca Estuary, especially in the outer portions of the estuary, is that the *Spartina* marshes develop along the tidal channels (Fig. 3.1a) whereas tidal flats are located farther away from the channel (Perillo 2019). The normal distribution in all coastal wetlands of the world is precisely the opposite (Allen 2000; Hopkinson et al. 2018; Pratolongo et al. 2019). Typically, one expects a standard ramp from the channel inland where the lower levels are dominated by unvegetated (considering vascular plants only) areas. Along the channel, usually the bare deposits of the tidal flat appear and, as we move further up, the stability of the sediments and the slightly lower hydroperiod allows for the settlement of the pioneer plants and further development of the marsh structure. The presence of plants (specifically their leaves and stems) increases frictional drag, thus reducing the turbulence generated by waves and currents and helping the deposition of the sediments being transported by those processes (Leonard and Luther 1995; Pratolongo et al. 2019).

In the case of the outer portions of the Bahía Blanca Estuary, the mechanism is inverse likely due to the possible presence of remnants of older delta channel levees. These levees, being slightly higher than the adjacent flats, are better suited to allow the settlement of plants right at the border of the flat. Once the pioneer plants settle, they start working in reducing the turbulence at the moment water flows over the bankfull levels inducing sediment deposition. Together with the sediment, nutrients are also deposited, which are employed by the plants for further growth, reducing still more the sediment transported farther inland. Therefore, it becomes a virtuous circle, higher plants reduce more the turbulence, generating more sedimentation and nutrient concentration, which then helps increase the size of the plants. *Spartina* plants grow from rhizomes in a radial structure (Fig. 3.1b). As these circles grow, they coalesce and then form a large patch that spreads along the border of the channel (Perillo 2019).

Tidal flats and marshes of the Bahía Blanca Estuary are dominated by a complex network of tidal courses (sensu Perillo 2009) and depressions (Revollo Sarmiento

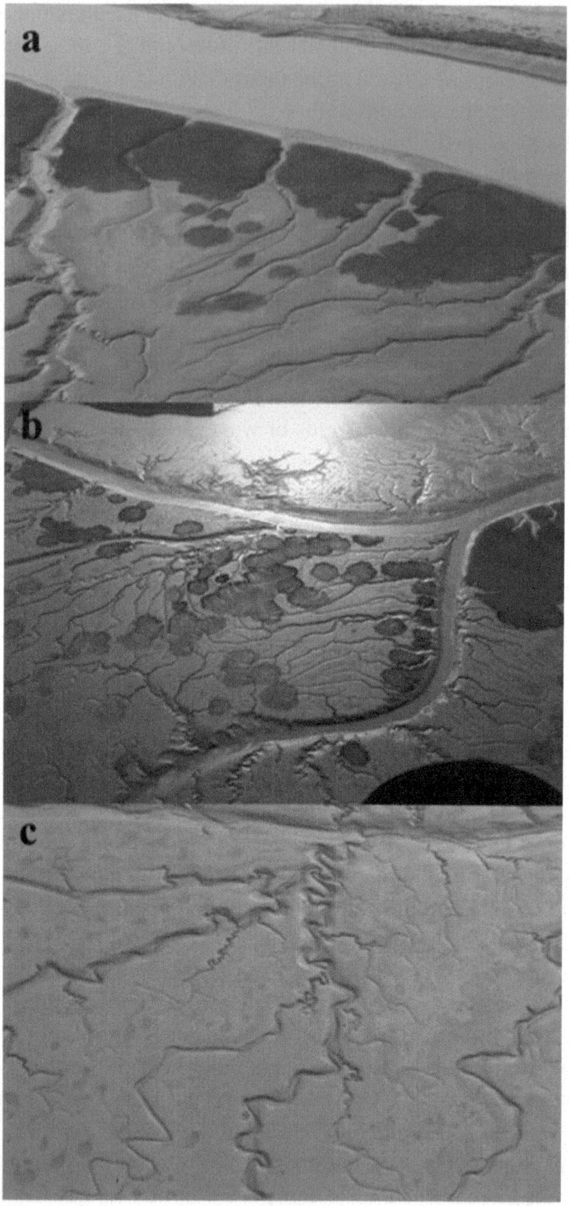

Fig. 3.1 Examples of the characteristics of the typical wetlands and tidal courses found in the Bahía Blanca Estuary. (**a**) Development of *Spartina* marsh along the border of a tidal channel while the tidal flats are farther away, (**b**) complex tidal course pattern on a tidal flat being colonized by circular patches of *Spartina*, (**c**) example of a tidal creek with higher sinuosity near the mouth plus some rectangular meanders probably due to the influence of the underlying deltaic sediments. (Photos by Gerardo M. Perillo)

et al. 2016, 2020; Perillo 2019). Courses vary from incipient rills and groves along the flanks of larger courses to extensive creeks and channels. They were described in detail by Perillo and Cuadrado (1991), Angeles et al. (2004), Ginsberg and Perillo (2004), among others; however, Perillo (2009, 2019) provided an analysis of the various types of drainage systems as well as the particular shape of the courses. The latter requires further analysis, though, because most of the courses change their meandering pattern from the head to the mouth.

Most commonly, as also occurs in rivers, courses tend to have higher sinuosity in the low relief sectors of the basin. In the case of the Bahía Blanca Estuary occurs the contrary. The low relief part of the tidal flats presents creeks with low sinuosity meanders, but as they approach their mouth on the steeper side of larger creeks or channels, the degree of sinuosity increases significantly (Fig. 3.1c). A possible explanation is that meandering in these creeks does not develop, as usually occur in a river system, as a mechanism to lengthen the channel to retain more water under the bankfull level, but a process known as instability propagation (Perillo 2019). The initial meander occurs at the mouth of the creek due to the formation of a bar or spit due to the sediment transported along the flank of the main channel. This deviation of the mouth creates a damming of the ebbing water which is forced to erode the course upstream generating a new meander. In some cases, these meanders are activated by the formation of erosional cusps (Ginsberg and Perillo 1990; Zhao et al. 2019). Each new meander generates a modification of the circulation pattern within the creek that propagates headward.

On the other hand, the Bahía Blanca Estuary is worldwide unique due to the number and diversity of tidal depressions (Perillo 2019) found mostly on the tidal flats (Fig. 3.2). They range from ponds to pools with circular, elliptic, or complex shapes. Based on the analysis of Google Earth images of 14 sites selected from the estuary, the pond density varies from 6 per 100 m^2 to 25 per 100 m^2 (Revollo Sarmiento et al. 2020), whereas the size range varies from 0.6 to 173 m^2. Depressions fulfill a significant role in both the ecology and geomorphology of the tidal flats and marshes. In the former, due to the level of water retention, they provide the conditions for certain species of worms and algae to establish colonies. In the latter case, depressions are many times related to the process of formation of tidal courses.

The Bahía Blanca Estuary is also characterized by the presence of scour holes that only were previously reported for North Bay (USA) (Kjerfve et al. 1979). The most significant scour holes are located at the intersection of tributary channels like

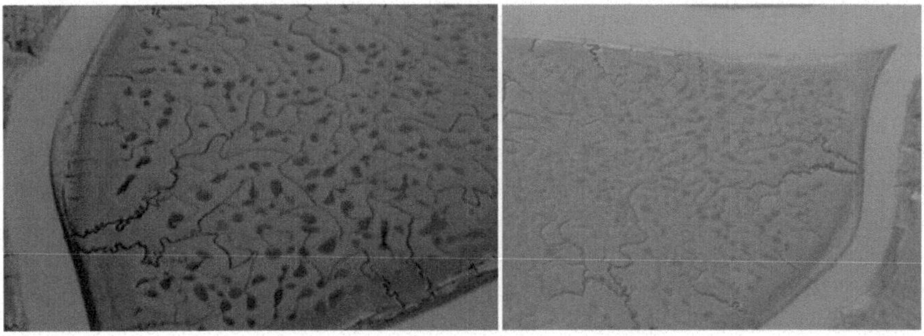

Fig. 3.2 Examples of the extensive distributions of tidal depressions, some of them connected by small grooves and creeks. Their interconnection may develop into a tidal creek. (Photos by Gerardo M. Perillo)

those found at the Tres Brazas, Cabeza de Buey (Ginsberg and Perillo 1999), or Maldonado channels (Pierini et al. 2005). Nevertheless, there is a particular scour hole that is unique for the Bahía Blanca Estuary found at the center of a meander (Mavo Manstretta et al. 2018; Perillo 2019) which is developed by the formation of a particular eddy due to interaction of the tidal currents with the specific geomorphology of the channel.

Large bedforms in the deeper portions of the Principal Channel are another unique feature of the Bahía Blanca Estuary. Aliotta and Perillo (1987) described the first-ever large dunes (up to 6 m high) that generate flow separation at the crest due to the high slope of the lee side (in some dunes they are 30° (average 11°) near the angle of repose of the sediment). The particular characteristic of these dunes is that they are solitary bedforms as defined by Perillo and Ludwick (1984) since they are separated by a flat bed of fine sediment (silt/clay). The average migration rate was calculated in 33 m/year which was later confirmed by a newer study (Minor Salvatierra et al. 2015) of the same bedform field reaching an average migration of 43 m/year. Although this is the major dune field, there are others found both on the middle reach (Cuadrado et al. 2010) and in the outer portion of the estuary (Gómez and Perillo 1992).

Even though the extensive area that the Bahía Blanca Estuary (2300 km^2) has, most of it has not been studied from both its geomorphology and dynamics. On the other hand, about 90% of the publications on these subjects are concentrated in Canal Principal's issues. The Principal Channel (Fig. 2.1; Chap. 2) has a total length of 60 km and varies in width from about 3–4 km at the mouth (22 m depth) to 200 m at the head (3 m depth); both depth and width increase almost an order of magnitude from head to mouth (Perillo and Piccolo 1991). Like other major channels (bays) that flow toward the inner shelf, the Canal Principal is partly closed by a modified ebb delta (Cuadrado and Perillo 1997). The channel cross section is steep on the sides, with a U-shaped bottom having a small asymmetry to the right. Upstream of Puerto Galvan (Fig. 2.2; Chap. 2), the channel narrows and becomes more V-shaped with the asymmetry following the meandering pattern headward (Gómez et al. 1996).

At the confluence with the Principal Channel, the funnel-shaped mouth of tributary channels is turned seaward, due to the ebb dominance in the Principal Channel. Most ebb deltas at the mouth of major channels have changed (Gómez and Perillo 1992; Cuadrado and Perillo 1997a, b); however, the original delta shape and southward orientation of the ebb channel and associated shoals in the Principal Channel are still preserved despite strong marine dynamics. This is due to its stable connection to the northern shore and its position on top of a sill (Chasicó Formation), which has reduced sediment transport and served as an anchor for the ebb delta (Cuadrado and Perillo 1997a, b). The mobility of delta shoals depends on the approach angle of tidal currents (Perillo and Cuadrado 1991), while 3D dunes in the Canal Principal and tributary channels form due to high current velocities and geomorphologic "traps" which favor sedimentation (Aliotta and Perillo 1987; Gómez et al. 1996).

Except for few bedforms in flood-dominated channels, all 3D dunes and shoals in the Bahía Blanca Estuary have ebb dominance. Since sediment runoff from rivers is virtually absent and ebb delta characteristics impede sediment input from the shelf, the high concentration of suspended sediments in the estuary is due to erosion of tidal flats and island shores (Ginsberg and Perillo 1990). The southern coast of the Principal Channel across the port of Ingeniero White has retreated up to 50 m between 1980 and 1986 and 1.5 million m^3 were exported from an 8 km stretch in the mid-reach of the channel (Perillo and Sequeira 1989). However, the dredging of the port and navigation channel to a nominal depth of 45′ (13.5 m) in 1989–1990 affected most of the tidal flats located along large portions of the southern coast of the Principal Channel. The tidal flats have become *Spartina* marshes in about 3 years after the dredging. At present, there are indications that the *Spartina* marshes are now in an incipient erosional stage, probably due to lack of sediment input.

Because most sediments in the estuary are silts and clays (Gelós et al. 1988), strong currents and short slack water intervals impede their deposition in channels and on the tidal flats, while short, locally generated waves erode old sediments and prevent any accumulation of new ones. These conditions explain the erosional stage of most of the estuary and the prevalence of sediments from the deltaic deposition period. Furthermore, the biological and physical interactions in the system are rather crucial in the development of tidal creeks (Perillo and Iribarne 2003) and play a role in the erosional processes. For instance, Minkoff et al. (2005) and Minkoff (2007) have demonstrated that crabs and plants acting together were responsible for eroding over 2200 m^3 of sediments of a 270 ha high marsh in 45 years.

3.3 Climatology

The Bahía Blanca Estuary is located in a temperate climate zone in the southeast of Buenos Aires Province. Mid-latitude westerly winds and the influence of the Subtropical South Atlantic High dominate the typical weather pattern of the region. Climate variability and air-water interaction processes determine the main characteristics of the estuarine water parameters and dynamics. The area is windy with prevailing winds direction from NNW, NW, and N with a mean velocity of 24 km/h during 40% of the year and gusts over 100 km/h. The wind is high during the spring and summer and diminishes in the fall. The highest average speed occurred in the winter of 1986 with 82 km/h. The highest mean record was 70.3 km/h in 1980 and the lowest 28.2 km/h in 1964. A study of 50 years (1960–2014) of meteorological variables on the estuary coast shows the significance of the climate variability in different decades. The maximum average wind speeds were observed in 1970–1980 with mean values (62 km/h) that exceeded the average (50.3 km/h). The 1980–1990 decade had an average of 51.4 km/h. After that decade, the average speed decreased registering speeds of 45.5 km/h in 1990–2000 and 47.2 km/h in 2000–2010. From 2003 to the present, this parameter evidenced a decrease in the annual maximum average speeds (Ferrelli 2016; Ferrelli et al. 2019).

The mean annual temperature varies from 14 to 20 °C, with an average annual temperature of 15.5 °C. The lowest value corresponded to 1964 with 14 °C and the highest to 2014 with 16.6 °C. Periods in which the air temperature was lower than the average (15.3 °C) occurred in 1960–1966, 1971–1979, and 2005–2007, while the warm ones were 1980–1985, 1993–2001, 2008–2009, and 2011–2014. Due to the interaction processes between water and the atmosphere, an increase in the average air temperature causes an increase in the water temperature. The occurrence of El Niño events approximately coincided with air and water temperatures above the mean values, while the manifestation of La Niña events in the Bahía Blanca Estuary caused air and water temperatures below the mean values. During the last 50 years (1960–2014), the air temperature of the area increased 0.9 °C, the maximum wind speed decreased 7.5 km/h, and the relative humidity presented two different cycles, marking wet and dry periods. The annual mean rainfall is 654 mm.

Like temperature, precipitation in the estuary showed notable differences on a seasonal scale. Summer is the rainiest season with an average of 206.2 mm, presenting a maximum of 405 mm in 1985 and a minimum of 56.5 mm in 1972. The spring average is 198.4 mm. The maximum was recorded in 1976 with 496 mm and the minimum in 1991 with 83.7 mm. Autumn presents an average value of 138.1 mm, ranging between 364.2 and 30.7 mm in 1982 and 1988, respectively. The winter average value is 96.3 mm with a maximum of 238.5 mm in 1989 and a minimum of 11.5 mm in 1995. The dry and wet years were related to the Oceanic Niño Index (ONI) since they occurred together with weak and moderate El Niño and La Niña years, while the normal ones generally coincided with weak Niño or Niña periods. The climate of the region presents a marked seasonality, and the climate variability has an essential influence of the estuary dynamics, because wind, among others, affects tides, waves, and water circulation in general.

3.4 Physical Oceanography of the Estuary

As indicated, the Bahía Blanca Estuary receives very little input of freshwater, but it is enough to generate a thermohaline circulation (Piccolo and Perillo 1990). The two major tributaries enter the estuary from the northern shore. The Sauce Chico River (a drainage area of 1600 km^2) discharges into the Principal Channel about 3 km downstream from the head of the estuary, and the Napostá Grande Creek (a drainage area of 1240 km^2) reaches the estuary about 1 km downstream of the port of Ingeniero White (Fig. 2.2; Chap. 2). Both tributaries behave similarly in spring and summer during maximum mean rainfall, but they are out of phase in autumn when the Sauce Chico River has a secondary mode (Piccolo and Perillo 1990).

Although mean annual runoff flows of the Sauce Chico River and the Napostá Grande Creek are low (1.5–1.9 and 0.5–0.9 m^3/s, respectively), runoff from the Sauce Chico River may peak between 10 and 50 m^3/s, with a recorded maximum of 106 m^3/s in 1977, although other unpublished sources indicate a maximum of 200 m^3/s. Besides these significant freshwater inputs, the inflows from other, smaller tributaries into the estuary are intermittent and only significant during periods of high local precipitation. However, the most substantial input of freshwater, nutrients, and contaminants is provided by the sewage discharges from Bahía Blanca, Punta Alta, and Ingeniero White cities. These sewage discharges are located at the Maldonado Channel and Napostá Creek (for Bahía Blanca City), Puerto Rosales (Punta Alta City) (Figs. 2.2 and 2.4; Chap. 2). Only in the last year, all three discharges have treatment plants, otherwise, before the sewage had only some minor primary treatment which implied a high level of pollution mostly around the discharge places (see Chap. 4, "Bahía Blanca Estuary: A Chemical Oceanographic Approach"). However, the high level of turbulence and mixing due to the tidal dynamics allows a dramatic dilution of the pollutants within a few kilometers from the discharges.

When a flash flood in either of both tributaries occurs, the plume of freshwater only remains within a few tens of kilometers from the mouth of the rivers. In those conditions, the water column shows a significant stratification that in places may reach a salt wedge structure. However, rarely are more than partly mixed conditions. These conditions are exceptional since most of the time, the estuary behaves as vertically homogeneous (Perillo et al. 2001).

Mean annual (13 °C), summer (21.6 °C), and winter (8.5 °C) surface water temperatures (Fig. 3.3a) in the Principal Channel are always slightly higher at the head

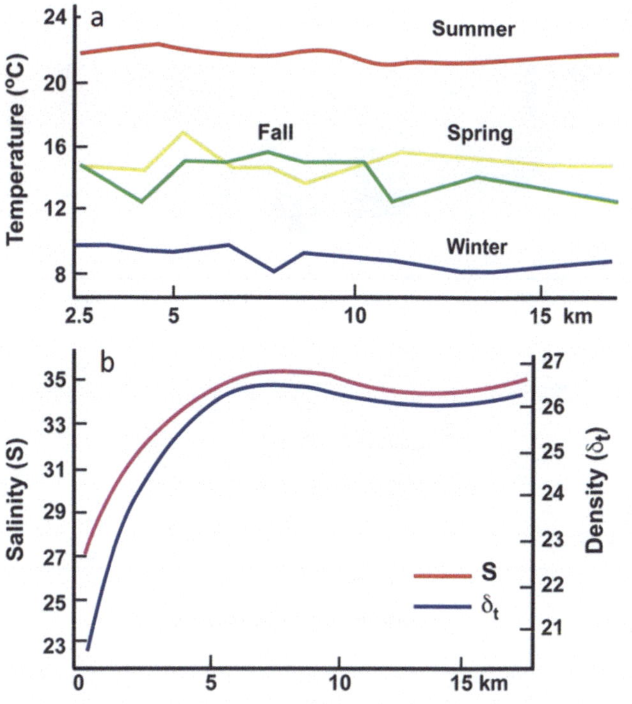

Fig. 3.3 Average historical distribution of (**a**) temperature and (**b**) salinity and density along the Principal Channel of the Bahía Blanca Estuary. Position 0 km corresponds to the mouth of the Sauce Chico River. (Modified from Piccolo and Perillo 1990)

of the estuary (Piccolo et al. 1987; Piccolo and Perillo 1990; Perillo et al. 2001), while mean surface salinity increases exponentially from the head to mid-reaches of the estuary (Fig. 3.3b). Longitudinal temperature distributions vary between rainy periods in spring/summer and low runoff in winter, when the vertical thermal structure of the estuary is homogeneous, and there are longitudinal variations. The data provided in Fig. 3.4 correspond to average values taken from historical measurements made between 1967 and 1986 along the Principal Channel (Perillo et al. 1987). However, recent monitoring made along the middle reach of the channel has shown similar temperature values but marked an increase in the salinity data varying from 33.8 in winter to 37.1 in summer (IADO 2016), but, in some cases, values reached up to 45.3 in the fall of 2018 basically along the whole channel (IADO 2018).

Because of its intricate geomorphology, the dynamic processes acting in the estuary are also complex. The main driver is the tide, but the wind plays a significant role that

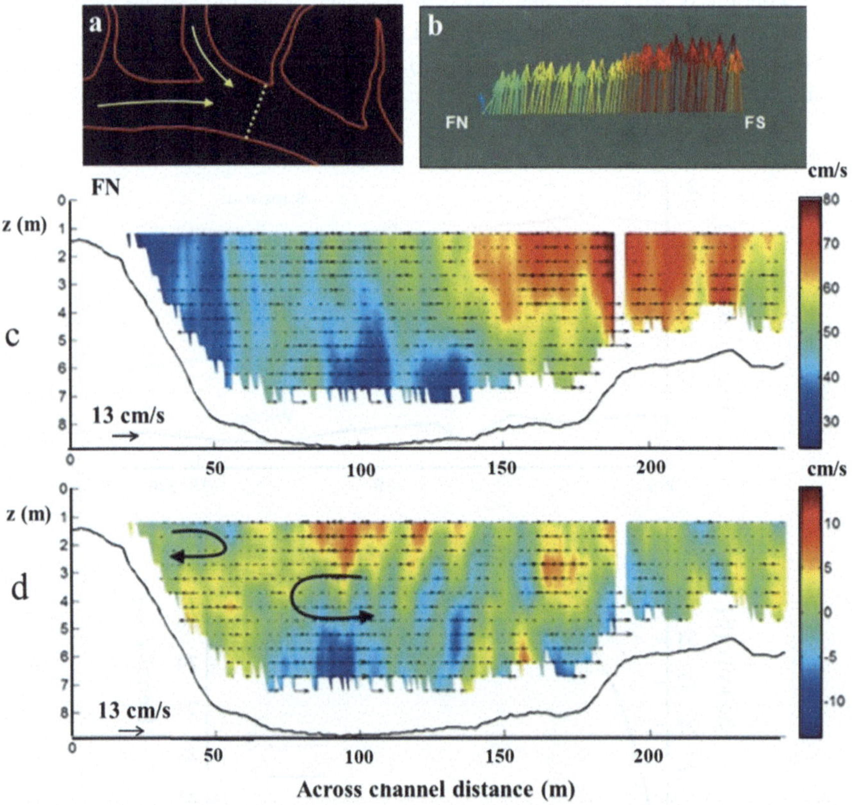

Fig. 3.4 Examples of cross-section velocity profiles measured at different places of the inner reach of the Principal Channel showing the effects of the tributaries, adjacent tidal flats, as well as the dynamics along a large channel meander. (**a**) Location of the cross section, (**b**) velocity vectors along the cross section, (**c**) amplitude and direction of cross-section velocity vectors during flood conditions, (**d**) amplitude and direction of cross-section velocity vectors during ebb conditions. Note the changes in speed across the channel due to the geomorphology. (Modified from López Gregori et al. 2017)

cannot be discounted in the circulation of the waters. In the present section, we propose to provide a simplified view of the main processes and the interaction among them.

The estuary is dominated by a quasi-stationary tidal wave that cooscillates with the tidal wave that propagates from south to north along the Argentinian continental shelf (Palma et al. 2004). Due to how the wave propagates, the water penetrates first through the Bahía Falsa and Bahía Verde channels and later through the Principal Channel (Perillo and Piccolo 1991). The ebbing condition is reverse, moving out first through the southern channels and ending with the outflow from the Principal Channel. The complexity of this situation arises because all these main channels are interconnected by the complex network of channels and creeks that transfer water from one primary channel to another. Furthermore, when the tide level covers the flats and marshes, except for the islands, all the estuary is covered by water with a mean depth of the order of 0.5–1 m (Perillo et al. 2001). Therefore, water that may have initially entered the estuary through one of the southern bays, it may ebb by one or more of the northern channels. In several channels that connect two major channels (i.e., El Embudo), the direction of the flow may change up to four times during the same tidal cycle depending on which major channel is generating the most considerable influence in the circulation.

The complexity of this circulation has a significant correlate because it makes impossible to establish with adequate precision the residence time of the water, but still worse, of any possible substance that is introduced in the estuary. Numerical models of the estuary so far (Pierini et al. 2008a, b) failed in defining this situation because we still lack detailed geomorphology of the estuary which is the essential element to have a well-behaved model.

The Bahía Blanca Estuary is a mesotidal, semidiurnal system with tidal ranges that were measured only along the Principal Channel with mean values ranging from 2.2 m at the Oceanographic Tower (OT) (Fig. 2.4; Chap. 2) to 3.5 m in Puerto Belgrano, 3.8 m in the Ingeniero White port reaching a maximum of 4 m at the head in Villarino Viejo (Perillo and Piccolo 1991). Although along all the estuary the tide behaves as pure semidiurnal, at the OT the tide is mixed semidiurnal (Perillo and Piccolo 1991). The relationship between the convergence of the channel and the friction effect of the geomorphology results in a hypersynchronous estuary (based on the classification by Le Floch 1961) as the tidal range and the intensity of the tidal currents increase headward.

The comparison of the amplitude of the main harmonics at the three most important tidal stations between the data back in 1991 with information gathered from 2008 to 2017 shows basically no changes. However, when the phases are compared, they show variations of up to 40° (Table 3.1). Although there is no specific correlation between the changes in geomorphology and the possible shift in the phases of the tide, as a first approximation we assume that the changes in this variable may have occurred due to the dredging of the Principal Channel as well as the significant deposition of the dredged material on top of the tidal flats and marshes along the Principal Channel.

As indicated, the estuary is fully dominated by the tides and behaves as a lower, marine estuary as all of it is affected by marine processes. The middle and upper estuaries are concentrated within the last 5 km of the Sauce Chico River, at the head of the estuary (Piccolo and Perillo 1990; Perillo et al. 2001).

Table 3.1 Changes in the amplitude and phase of the main harmonic components between the data informed by Perillo and Piccolo (1991) (91) and the values estimated in 2017 (17) by Blanco Monroy (in preparation)

Station	Harmonic	IW	PB	OT
Zo		2.32	2.43	2.93
M_2	A(91)	1.63	1.46	1.09
	Ø(91)	271	266	233
	A(17)	1.56	1.41	1.05
	Ø(17)	244	236	213
S_2	A(91)	0.22	0.18	0.17
	Ø(91)	48	44	356
	A(17)	0.2	0.20	0.17
	Ø(17)	5	356	334
O_1	A(91)	0.13	0.14	0.15
	Ø(91)	43	51	19
	A(17)	0.14	0.16	0.15
	Ø(17)	30	25	12
K_1	A(91)	0.19	0.17	0.20
	Ø(91)	110	81	85
	A(17)	0.18	0.19	0.18
	Ø(17)	87	81	73
N_2	A(91)	0.24	0.12	0.17
	Ø(91)	172	149	159
	A(17)	0.2	0.17	0.12
	Ø(17)	163	154	134

IW Ingeniero White, *PB* Puerto Belgrano, *OT* Oceanographic Tower

Winds in the Bahía Blanca Estuary play a significant role in all aspects of the estuarine dynamics. As indicated, wind speeds can be very high blowing from rather particular directions, specifically N and NW, SE and SW. All these directions have immediate repercussions regarding their relationship with the orientation of the largest channels (mostly NW-SE). Winds generate waves, storm surges, and subtidal sea-level variations in the estuary. Wind waves (about 5–10 cm high and 1–3 m long) occur in channels and on tidal flats when covered by water. The incoming tide, together with N and NW winds, forms interaction waves which are steep and up to 1.5 m high and 10–20 m long. However, the main effect of the predominant NW and N winds is advancing the time of low water, delaying the time of high water, and modifying predicted water levels. Significant deviations between predicted astronomical tides and actual tide levels occur at the Ingeniero White port (−4.01 and 2.39 m) and the Oceanographic Tower (−1.51 and 1.87 m; Perillo and Piccolo 1991). The maximum negative values coincide with NW winds and maximum positive values with winds from the SW. At the same stations, the low-frequency sea-level response to wind indicates that time scales of more than 10 days prevail while short-scale energy peaks correspond to approximately 3-day periods (Piccolo and Perillo 1989).

Perillo and Sequeira (1989) proposed three types of waves for the middle reach of the Bahía Blanca Estuary (although they can be observed in any of the other indicated environments): (1) formed by wind action, (2) formed by the interaction of wind and tidal currents, and (3) formed by ship wakes. Wind waves are originated by the continuous action of the wind over any fetch large enough to develop them. If the wind blows parallel to channels, short, very steep waves are the most common ones. These low period waves have little effect on the channel bottom, but they can be very effective in resuspending sediments on tidal flats and eroding channel banks. It is common to observe along channel borders a turbid, relatively narrow band of water derived by this wave activity. The continuous impinging of these very steep waves upon the channel flanks may also be a decisive factor in the sediment fatigue and resuspension of any material deposited in the tidal flats during high tide slack which may result in a coarse sediment but thin laminae within the sediment structure of the flats (Ginsberg and Perillo 1990). In places where strong winds blow parallel to large channels, as is common along the Principal Channel of the Bahía Blanca Estuary, very steep waves, up to 1.5 m high and with wavelengths of the order of 10–30 m, are formed by the interaction of the incoming tide and the wind.

The presence of these waves sometimes makes the navigation of small crafts and fishing vessels, which are active in the area, very difficult. Once generated, these waves propagate against the flanks, contributing to their mechanical erosion. These waves disappear as soon as the wind or tide changes directions. Waves are not only a natural phenomenon in estuaries with active harbors, wake action against the channel coast and machine vibration may also induce fatigue in the interparticle bonding (Ginsberg and Perillo 1990), thus increasing the erosion potential of the material especially if the channels are relatively narrow.

Tidal currents in the estuary, in synchronicity with the geomorphology, are also rather complicated because of the interaction of the channels with the adjacent tidal flats and marshes. During the tidal stages in which the water level is below the bankfull, the water flow is similar to what may be observed in a regular river with the difference of its reversibility. However, when the water level floods the adjacent flats and marshes, there is a tendency to reduce the velocity while the water spreads out over the large flat areas. This is also because the tide must work against the slope, although small, still large enough to extend the time to complete the inundation. As the conditions reverse, the return flow is accelerated by two reasons: first is the geomorphologic slope in the direction of the flow, but also because there is a marked increase in the energy slope between the water on the flats and that in the main channel.

An example of the tidal circulation in the Principal Channel that probes the previous ideas is the study made by López Gregori et al. (2017) at the inner portion of the channel. Along two successive tidal cycles (over 27 h) measured twice in spring and neap conditions, in 7 cross sections a total of 430 transects (there is no previous reference of so many transects elsewhere). The idea was to analyze both the circulation along the channel but also to define the meandering effects as well as the influence of tributary channels upon the Principal Channel. As expected, ebb currents were significantly stronger than their flood counterparts, even though the duration of the flood was about 1.5 h longer than the ebb. The current direction was, for the most part, parallel to the flanks; however, in transects adjacent to tributary channels, we observed a higher level of variability (Fig. 3.4c). Employing the velocity profiles in the thalweg, they estimated the average direction vector for the flood and ebb currents with a difference between them of 176°.

We detected a progressive variation of the velocity along the width of the transect, being this phenomenon more pronounced during the ebb (Fig. 3.5a). In the first case, the depth-averaged velocity analysis showed a lateral reduction of this parameter from the southern flank to the north flank, estimated at an average value of the order of 0.60 m/s. In contrast, we observed an inversion for transect II in the location of the zones of maximum and minimum velocity with respect to the former. The mean difference was 0.52 m/s, and the core of higher velocities was characterized by a larger lateral extension. During the flood, the lateral variations of the

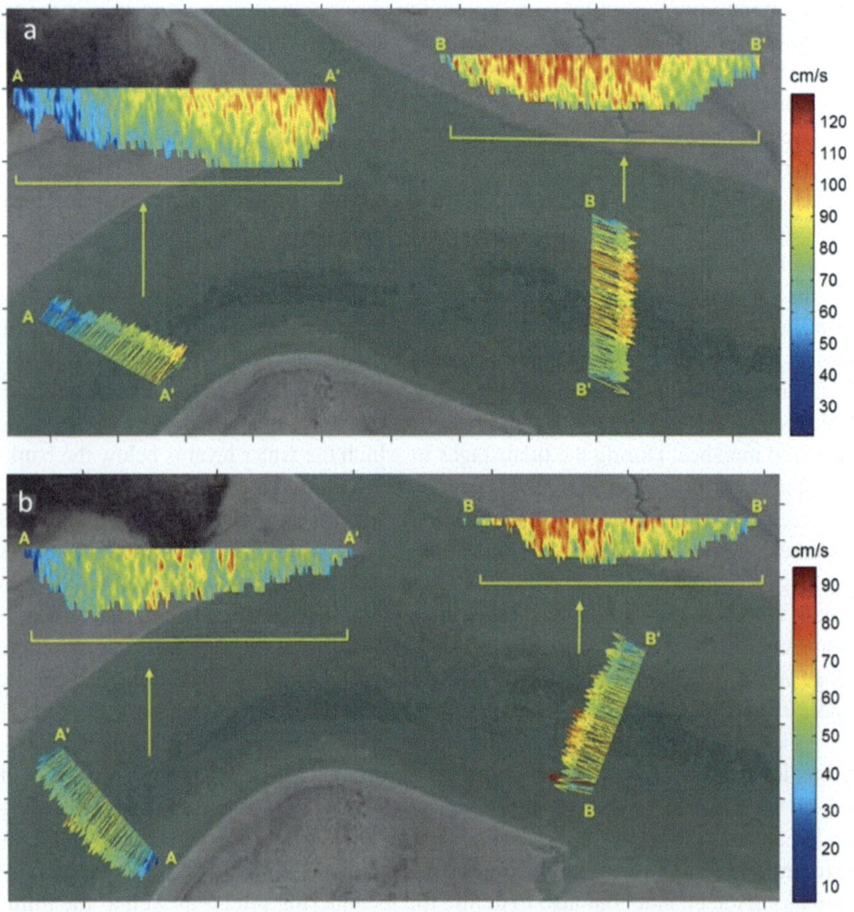

Fig. 3.5 Examples of cross-section velocity profiles measured at different places along a channel meander in the inner reach of the Principal Channel showing the effects of the adjacent tidal flats as well as the dynamics along a large channel meander: (**a**) ebb conditions, (**b**) flood conditions. (Modified from López Gregori et al. 2017)

velocity for both transects were diffusive, being possible to observe a field of speeds defined by a higher intensity in the vicinity of the southern slope (Fig. 3.5b).

During the whole tidal cycle, on the northern portion of the transects, we observed significant variability in the magnitude and direction of the current. For instance, mouthward the Bahía del Pejerrey Channel, they detected a strong velocity gradient. While in the center of the channel, maximum depth-averaged velocities were between 0.80 and 0.90 m/s southward, and between 0.15 and 0.37 m/s northward, respectively (Fig. 3.5). This effect was more pronounced during mid ebb, disappearing as the cycle approached low-water slack. Similar results were observed by Payares Peña et al. (2018) along a pronounced meander where a major scour hole was observed.

Velocity profiles clearly show the effects of acceleration and deceleration (Sassi 2009) at the surface levels (Fig. 3.6). Nevertheless, wind influence plays an essential role in the velocity profile by actually generating two boundary layers: (1) a bottom boundary layer principally characterized by the logarithmic velocity law, and (2) at the surface, the wind shear stress produces a boundary layer (Fig. 3.7). The momentum transfer to the lower levels results in a logarithmic defect velocity profile (Sassi and Perillo 2006).

3.5 Summary

The Bahía Blanca Estuary is one of the most complex estuaries in the world and, besides more than 50 years of intensive studies, one of the less known from the oceanographic point of view. The complex geomorphology, the logistics required to obtain adequate data from areas south of the Principal Channel, makes the task rather complicated and particularly expensive both in time and budget. Therefore,

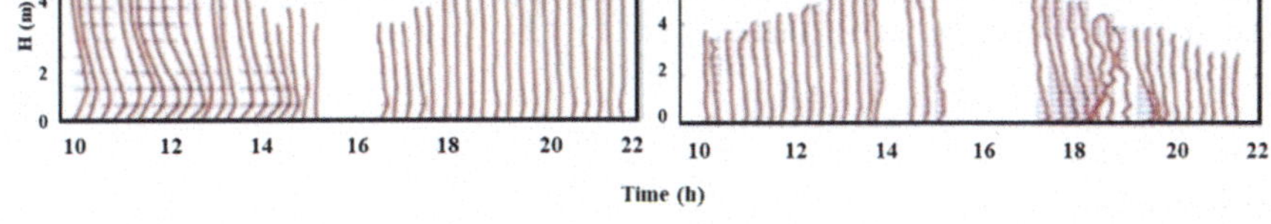

Fig. 3.6 Examples of velocity profiles along a tidal cycle of the longitudinal and transversal components of the flow on the Bahía del Pejerrey Channel. These profiles are purely tidal with no wind influence. In all cases (**a–d**), the effects of acceleration and deceleration of the flow are observed in particular during the times around slack water. (Modified from Sassi and Perillo 2006; Sassi 2009)

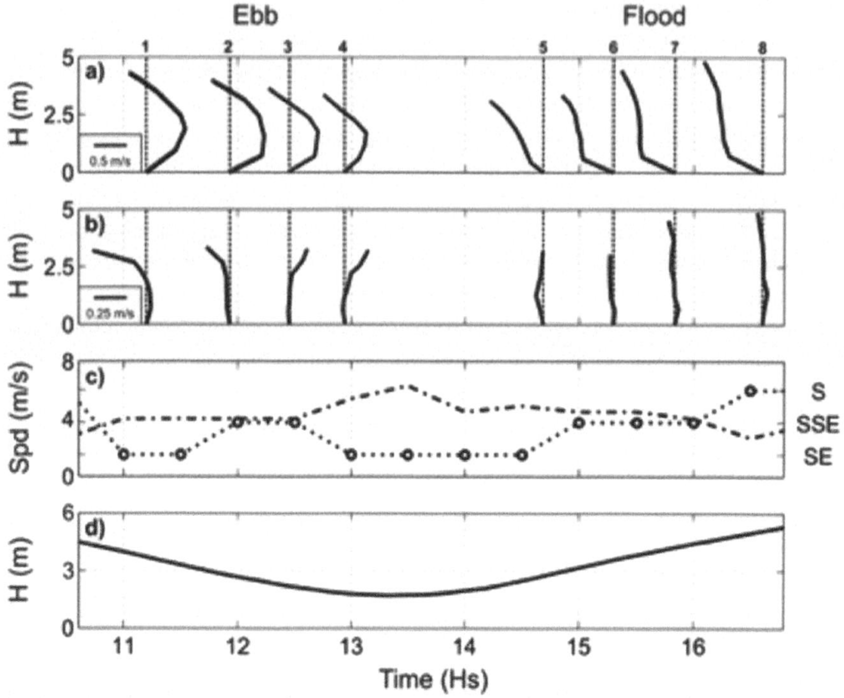

Fig. 3.7 Examples of velocity profiles along a tidal cycle for the longitudinal and transversal components of the flow in the Principal Channel near the mouth of the Maldonado Channel but strongly affected by wind. In all cases (**a–d**), the effects of acceleration and deceleration of the flow are observed in particular during the times around slack water. (Modified from Sassi and Perillo 2006; Sassi 2009)

most researchers concentrated their effort on the Principal Channel and the areas adjacent to use a pilot zone that may represent the rest of the estuary.

However, this may not be true. The Principal Channel has both geomorphologic and dynamic conditions that are unique within the context of the whole estuary. None of the other channels has the same structure neither they are being effected by anthropic activities. Rather the contrary, channels like Bermejo or Bahía Falsa and Bahía Verde are quite pristine and require to be approached in a different way as we have done for the Principal Channel.

References

Aliotta S, Perillo GME (1987) A sand wave field in the entrance to the Bahía Blanca Estuary, Argentina. Mar Geol 76(1):1–14

Allen J (2000) Morphodynamics of Holocene salt marshes: a review sketch from the Atlantic and Southern North Sea coasts of Europe. Quat Sci Rev 19:1155–1231

Angeles GR, Perillo GME, Piccolo MC, Pierini JO (2004) Fractal analysis of tidal channels in the Bahía Blanca Estuary (Argentina). Geomorphology 57:263–274
Cuadrado DG, Perillo GME (1997) Principal component analysis applied to geomorphologic evolution. Est Coast Shelf Sci 44(4):411–419
Cuadrado DG, Perillo GME (1997a) Migration of intertidal sand banks at the entrance of the Bahia Blanca Estuary, Argentina. J Coast Res 13:139–147
Cuadrado DG, Perillo GME (1997b) Principal component analysis applied to geomorphologic evolution. Est Coast Shelf Sci 44:411–419
Cuadrado DG, Gómez EA, Pierini JO (2010) Sand transport on an estuarine submarine dune field. Geomorphology 121:257–265
Ferrelli F (2016) Análisis del clima local y microlocal de la ciudad de Bahía Blanca. PhD dissertation (unpublished). Universidad Nacional del Sur
Ferrelli F, Brendel AS, Aliaga VS et al (2019) Climate regionalization and trends based on daily temperature and precipitation extremes in the south of the Pampas (Argentina). Cuad Investn Geogr 45:393–416
Gelós EM, Spagnuolo JO, Lizasoain GO (1988) Mineralogía y caracterización granulométrica de sedimentos actuales de la plataforma Argentina entre los paralelos de 39° y 43° de latitud sur y del Golfo San Matías. Rev Asoc Geol Arg 43:63–79
Ginsberg SS, Perillo GME (1990) Channel bank recession in the Bahía Blanca Estuary, Argentina. J Coast Res 6(4):999–1010
Ginsberg SS, Perillo GME (1999) Deep scour holes at the confluence of tidal channels in the Bahia Blanca Estuary, Argentina. Mar Geol 160:171–182
Ginsberg SS, Perillo GME (2004) Characteristics of tidal channels in a mesotidal estuary of Argentina. J Coast Res 20:489–497
Ginsberg SS, Aliotta S, Lizasoain GO (2009) Morphodynamics and seismostratigraphy of a deep hole at tidal channel confluence. Geomorphology 104:53–261
Gómez EA, Perillo GME (1992) Largo Bank: a shoreface-connected linear shoal at the Bahía Blanca Estuary entrance, Argentina. Mar Geol 104:193–204
Gómez EA, Perillo GME (1995) Sediment outcrops underneath shoreface-connected sand ridges, outer Bahía Blanca estuary, Argentina. Quat South Am Antar Pen 9(3):27–42
Gómez EA, Ginsberg SS, Perillo GME (1996) Geomorfología y sedimentología de la zona interior del Canal Principal del Estuario de Bahía Blanca. Rev Asoc Arg Sediment 3(2):55–61
Gómez EA, Martinez DE, Borel CM et al (2006) Negative sea-level oscillation in Bahía Blanca Estuary related to a global climatic change around 2,650 yr. B.P. J of Coast Res SI 39 (Proceedings of the 8th International Coastal Symposium), 181–185
Hopkinson CS, Wolanski E, Brinson MM, Cahoon DR, Perillo GME (2018) Coastal wetlands: a synthesis. In: GME P, Wolanski E, Cahoon DR, Hopkinson CS (eds) Coastal wetlands: an integrated ecosystem approach. Elsevier, Amsterdam, pp 1–75
IADO (2016) Programa de monitoreo de la calidad ambiental de la zona interior del estuario de Bahía Blanca. Informe Final, Instituto Argentino de Oceanografía, 228 pp. Available in: http://www.bahiablanca.gov.ar/cte/index.html
IADO (2018) Programa de monitoreo de la calidad ambiental de la zona interior del estuario de Bahía Blanca. Informe Final, Instituto Argentino de Oceanografía, 364 pp. Available in: http://www.bahiablanca.gov.ar/cte/index.html
Kjerfve B, Shao C-C, Stapor FW (1979) Formation of deep scour holes at the junction of tidal creeks: a hypothesis. Mar Geol 33:M9–M14
Lanfredi NW, D'Onofrio EE, Mazio CC (1988) Variations of the mean sea level in the Southwest Atlantic Ocean. Cont Shelf Res 8:1211–1220
Lanfredi NW, Pouza JL, D'Onofrio EE (1998) Sea-level rise and related potential hazards on the Argentine coast. J Coast Res 14:47–60
Le Floch J (1961) Propagation de la marée dans l'estuaire de la Seine et en Seine Maritime. PhD dissertation, Université de Paris, 507 pp (unpublished)

Leonard LA, Luther ME (1995) Flow hydrodynamics in tidal marsh canopies. Limnol Oceanogr 40:1474–1484
López Gregori M, Alvarez LI, Perillo GME (2017) Circulation pattern in a tidal channel influenced by tributary channels and adjacent tidal flats. Rio Accoustics 2017. https://ieeexplore.ieee.org/document/8349706
Mavo Manstretta GM, Payares Peña NB, Vitale AJ et al (2018) Banco sedimentario como agente modelador de corrientes y transporte de sedimentos en un canal de marea del estuario de Bahía Blanca, Argentina. XVI Reunión Argentina de Sedimentología, Gral. Roca (abstract)
Melo WD, Schillizzi R, Perillo GME et al (2003) Influencia del área continental pampeana sobre el origen y la morfología del estuario de Bahía Blanca. Rev Asoc Arg Sediment 10:65–72
Melo WD (2004) Cuenca de Recepción del Estuario de Bahía Blanca. PhD. Dissertation (unpublished) Departamento de Geografía, Universidad Nacional del Sur
Melo WD, Perillo GME, Perillo MM et al (2013) Late Pleistocene-Holocene deltas in the southern Buenos Aires Province, Argentina. In: Young G, GME P (eds) Deltas: landforms, ecosystems and human activities, IAHS Publication 358. IAHS Press, Wallingford, pp 187–195
Minkoff DR (2007) Geomorfología y dinámica de canales de mareas en ambientes intermareales. PhD dissertation Departamento de Ingeniería, Universidad Nacional del Sur (unpublished)
Minkoff DR, Escapa CM, Ferramola FE et al (2005) Erosive processes due to physical - biological interactions based in a cellular automata model. Lat Am J Sediment Basin Anal 12:25–34
Minor Salvatierra M, Aliotta S, Ginsberg SS (2015) Morphology and dynamics of large subtidal dunes in Bahia Blanca estuary, Argentina. Geomorphology 246:168–177
Palma ED, Matano RP, Piola AR (2004) A numerical study of the southwestern Atlantic shelf circulation: Barotropic response to tidal and wind forcing. J Geophys Res 109:C08014. https://doi.org/10.1029/2004JC002315
Payares Peña NB, Mavo Manstretta GM, Arena M et al (2018) Características hidrodinámicas de un canal de marea del estuario de Bahía Blanca, Argentina. XXVIII Congreso Latinoamericano de Hidráulica. Buenos Aires (extended abstract)
Perillo GME (1995a) Geomorphology and sedimentology of estuaries: an introduction. In: GME P (ed) Geomorphology and sedimentology of estuaries. Elsevier, Amsterdam, pp 1–16
Perillo GME (1995b) Definition and geomorphologic classifications of estuaries. In: Perillo GME (ed) Geomorphology and sedimentology of estuaries. Elsevier, Amsterdam, pp 17–47
Perillo GME (2009) Tidal courses: classification, origin and functionality. En: Perillo GME, Wolanski E, Cahoon DR, y Brinson MM (eds) Coastal wetlands: an integrated ecosystem approach. Elsevier, Amsterdam. 185–210
Perillo GME (2019) Geomorphology of tidal courses and depressions. In: Perillo GME, Wolanski E, Cahoon DR, y Hopkinson CS (eds) Coastal wetlands: an integrated ecosystem approach. Elsevier, Amsterdam. 221–261
Perillo GME, Cuadrado DG (1991) Geomorphologic evolution of El Toro Channel, Bahía Blanca Estuary (Argentina) prior its dredging. Mar Geol 97(3/4):405–412
Perillo GME, Iribarne OO (2003) New mechanisms studied for creek formation in tidal flats: from crabs to tidal channels. EOS Am Geophys Union Trans 84:1–5
Perillo GME, Ludwick JC (1984) Geomorphology of a sand wave in lower Chesapeake Bay, Virginia, USA. Geo-Mar Lett 4(2):110–114
Perillo GME, Piccolo MC (1991) Tidal response in the Bahía Blanca Estuary. J Coast Res 7:437–449
Perillo GME, Piccolo MC (1999) Geomorphologic and physical characteristics of the Bahía Blanca Estuary. Argentina. En: Perillo GME, Piccolo MC, y Pino Quivira M (eds) Estuaries of South America: their geomorphology and dynamics. Environmental Science Series, Springer-Verlag, Berlín 195–216
Perillo GME, Sequeira ME (1989) Geomorphologic and sediment transport characteristics of the middle reach of the Bahía Blanca Estuary, Argentina. Journal of Geophysical Research-Oceans 94(C10):14351–14362
Perillo GME, Arango JM, Piccolo MC (1987) Parámetros físicos del estuario de Bahía Blanca. Período 1967–1986. Instituto Argentino de Oceanografia Contribución Técnica, 250 pp

Perillo GME, Piccolo MC, Parodi E et al (2001) The Bahía Blanca estuary, Argentina. In: Seeliger U, Kjerfve B (eds) Coastal marine ecosystems of Latin America. Springler, Berlín, pp 205–217
Piccolo MC, Perillo GME (1989) Subtidal sea level response to atmospheric forcing in Bahía. Proceedings of the 3rd International Congress on Southern Hemisphere Meteorology and Oceanography, Buenos Aires, pp 323–324
Piccolo MC, Perillo GME, Arango JM (1987) Hidrografía del estuario de Bahía Blanca, Argentina. Revista Geofisica 26:75–89
Piccolo MC, Perillo GME (1990) Physical characteristics of the Bahía Blanca estuary (Argentina). Est Coast Shelf Sci 31:303–317
Pierini JO, Perillo GME, Carbone ME et al (2005) Residual flow structure at a scour hole in Bahía Blanca Estuary, Argentina. J Coast Res 21:784–796
Pierini JO, Marcovecchio J, Campuzano F, Perillo GME (2008a) Evolution of salinity and temperature in Bahía Blanca Estuary, Argentina. En: Neves R, Baretta J, Mateus M (eds) Perspectives on integrated coastal zone management in South America. IST Press, Lisbon. 501–509
Pierini JO, Campuzano F, Marcovecchio J, Perillo GME (2008b) The application of MOHID to assess the potential effect of sewage discharge system at Bahía Blanca Estuary (Argentina). In: Neves R, Baretta J, Mateus M (eds) Perspectives on integrated coastal zone management in South America. IST Press, Lisbon. 511–518
Pratolongo P, Mazzon C, Zapperi G et al (2013) Land cover changes in tidal salt marshes of the Bahía Blanca estuary (Argentina) during the past 40 years. Est Coast Shelf Sci 113:23–31
Pratolongo P, Leonardi N, Kirby JR et al (2019) Temperate coastal wetlands: morphology, sediment processes, and plant communities. In: Perillo GME, Wolanski E, Cahoon DR et al (eds) Coastal wetlands: an integrated ecosystem approach. Elsevier, Amsterdam, pp 105–152
Revollo Sarmiento GN, Cipolletti MP, Perillo MM et al (2016) Methodology for classification of geographical features with remote sensing images: application to tidal flats. Geomorphology 257:10–22
Revollo Sarmiento GN, Perillo GME, Delrieux CA (2020) Morphological characterization of ponds and tidal courses in coastal wetlands using rjr. Est Coast Shelf Sci (accepted)
Sassi MG (2009) Circulación y transporte de sedimentos en estuarios. PhD dissertation. Departamento de Física, Universidad del Centro de la Provincia de Buenos Aires (unpublished)
Sassi M, Perillo GME (2006) Influencia del viento en el perfil de velocidades de un canal de mareas. VI Jornadas de Ciencias del Mar, Puerto Madryn (abstract)
Zhao K, Gong Z, Xu F et al (2019) The role of collapsed bank soil on tidal channel evolution: a process-based model involving bank collapse and sediment dynamics. Water Resour Res. https://doi.org/10.1029/2019WR025514

Chapter 4
Bahía Blanca Estuary: A Chemical Oceanographic Approach

Jorge E. Marcovecchio, Ana L. Oliva, Noelia S. La Colla, Andrés H. Arias, Sandra E. Botté, Pía Simonetti, Analía V. Serra, Vanesa L. Negrin, Ana C. Ronda, and Claudia E. Domini

J. E. Marcovecchio (✉)
Instituto Argentino de Oceanografía IADO (Universidad Nacional del Sur-CONICET), Bahía Blanca, Argentina

Facultad Regional Bahía Blanca (UTN-BHI), Universidad Tecnológica Nacional, Bahía Blanca, Argentina

Facultad de Ingeniería, Universidad FASTA (FI-UFASTA), Mar del Plata, Argentina

Academia Nacional de Ciencias Exactas, Físicas y Naturales (ANCEFN), Ciudad Autónoma de Buenos Aires, Argentina
e-mail: jorgemar@iado-conicet.gob.ar

A. L. Oliva · N. S. La Colla · P. Simonetti · A. V. Serra
Instituto Argentino de Oceanografía IADO (Universidad Nacional del Sur-CONICET), Bahía Blanca, Argentina

A. H. Arias
Instituto Argentino de Oceanografía IADO (Universidad Nacional del Sur-CONICET), Bahía Blanca, Argentina

Departamento de Química, Universidad Nacional del Sur (DQ-UNS), Bahía Blanca, Argentina

S. E. Botté · V. L. Negrin · A. C. Ronda
Instituto Argentino de Oceanografía IADO (Universidad Nacional del Sur-CONICET), Bahía Blanca, Argentina

Departamento de Biología, Bioquímica y Farmacia, Universidad Nacional del Sur (DBByF-UNS), Bahía Blanca, Argentina

C. E. Domini
Departamento de Química, Universidad Nacional del Sur (DQ-UNS), Bahía Blanca, Argentina

S. M. Fiori, P. D. Pratolongo (eds.), *The Bahía Blanca Estuary*,
https://doi.org/10.1007/978-3-030-66486-2_4

4.1 The Chemical Scenario Within the Estuary

Aquatic environments are fully characterized by a set of parameters and chemical processes that make up the natural scenario where organisms can properly develop (Odum 2014). The mentioned parameters include temperature, salinity, pH, dissolved oxygen, inorganic nutrients, organic matter, other dissolved gases, and photosynthetic pigments, among others (Roy et al. 2011). In addition, these parameters can show numerous processes that take place within the corresponding biogeochemical cycles (i.e., seasonal variations, adsorption/desorption, alternative circulation between water, sediments, organisms, and atmosphere driven by physicochemical control) (Yakushev and Newton 2013). The interaction of these parameters and processes determines the scenario where the whole biological processes will occur, also including population movements along the different spatial gradients due to physiological problems (Hester and Harrison 2007). Moreover, the chemical scenario results in an essential framework to carry out the biological production within the system, including the most transcendental biological processes for the environment such as photosynthesis and respiration (Yosim and Fry 2015), and the consequent transference of energy to higher trophic levels (Xu et al. 2011). Finally, it is important to remark that this chemical scenario fully and continuously interact with other components within the environment (i.e., hydrological, geomorphological, or climatic systems) to allow the development of biodiversity over time (Grimm et al. 2013).

Two sets of chemical and/or physicochemical parameters could be mentioned to build up the abovementioned scenario: (i) structural parameters and (ii) eco-physiological parameters (Marcovecchio and Freije 2013). The first group provides the conditions under which biological processes develop at the stage considered and includes parameters such as temperature, salinity, and pH/alkalinity, among others. Meanwhile, the second group indicates the production capacity of the system under study and includes inorganic nutrients, photosynthetic pigments, and organic matter, among others. The integration of both groups provides an image of the health of the ecosystem as well as the functioning of its biological components (Balvanera et al. 2006).

Bahía Blanca Estuary has been the object of a large amount of environmental studies along the last 40 years, including most of the scientific disciplines (e.g., chemistry, biology, geology, physics, hydrology, meteorology, among others). Within this framework, a large time-series database on physicochemical parameters has been developed at the inner part of the estuary (Puerto Cuatreros and Ingeniero White Port; Fig. 4.1) which allows to bring a nice picture of the condition and functioning outline of this area. It is also a very useful tool to diagnose the environmental condition of the system considering that the mentioned inner zone of the estuary includes not only the discharges of the main rivers from the region but also of the largest anthropic activities such as cities, industries, and harbor locations, with their impacts on the estuary. In addition, much information has also been

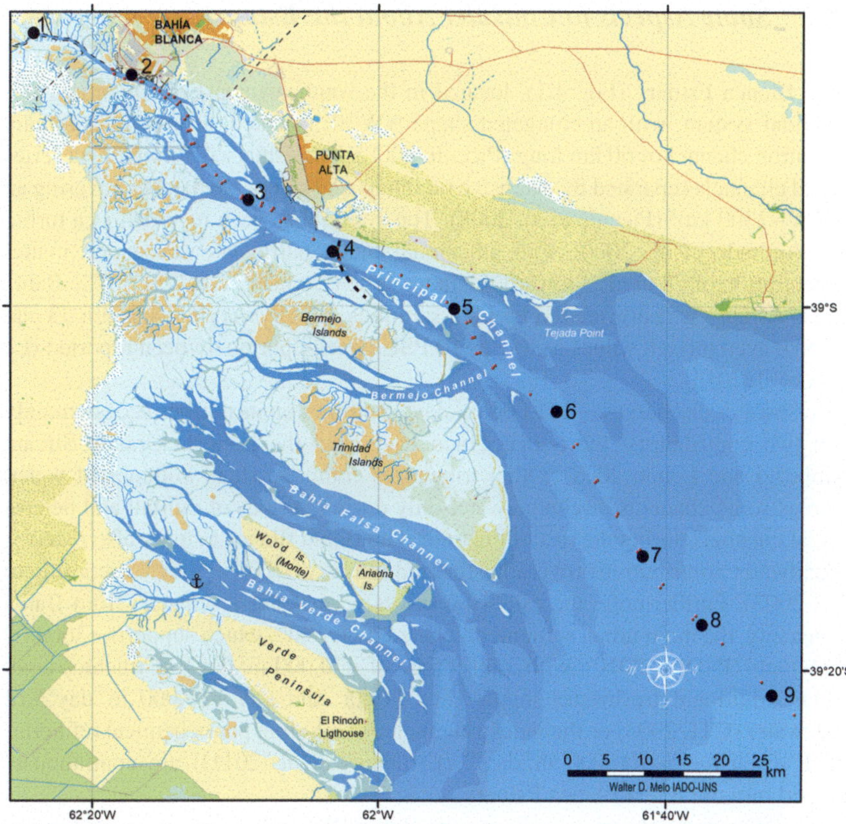

Fig. 4.1 Location of sampling stations for chemical studies within Bahía Blanca Estuary. (Map by Walter D. Melo)

recorded through cruises along the Principal Channel of the estuary, as well as on the external areas within the estuary (Fig. 4.1).

The present chapter has two main goals: (1) to summarize the available information on the physicochemical conditions of the Bahía Blanca Estuary, including the corresponding range of values of the analyzed parameters (temperature, salinity, turbidity, dissolved oxygen, inorganic nutrients (of N, P, and Si), particulate organic matter, and photosynthetic pigments (chlorophyll *a* and phaeopigments), identifying seasonal variations, as well as the influence of external sources to the estuarine balance, and (2) to present a brief summary of the state of contamination within the estuary, including records of trace metals, hydrocarbons, pesticides, and microplastics in both abiotic (estuarine water, sediments) and biological (fish species) compartments.

4.1.1 Some Aspects to Consider About the Estuary

Bahía Blanca Estuary (Fig. 4.1), located in the southeast coast of Argentina, is a mesotidal system, with an elongated shape NW-SE oriented, and includes a main navigation channel of 60 km long (Piccolo 2008). The estuary is formed by a series of tidal channels separated by extensive tidal flats and saltmarsh patches covering an area of ~2300 km^2 (Piccolo et al. 2008). The system has been regarded as a turbid one (Andrade et al. 2000) with an annual mean particulate suspended matter concentration of 78 $mg.L^{-1}$ toward the inner zone (Guinder et al. 2009). The Bahía Blanca Estuary is characterized by high salinities, usually varying between 33 and 40 PSU, even though values as low as 10 PSU have opportunely been reported (La Colla et al. 2015).

Freshwater discharges (~241,000 $m^3.day^{-1}$) inflows on its northern coast, mainly from two contributors: the Sauce Chico River and the Napostá Grande Stream (Limbozzi and Leitào 2008). Other tributaries comprise small volumes of water, whereas some channels discharge water only during periods of rainfall and behave as tidal channels during the rest of the time. The Bahía Blanca Estuary also receives groundwater contributions (overall estimated inflow of 2000 $m^3.day^{-1}$ according to CTE 2003). Furthermore, the estuary receives wastewater discharges from Bahía Blanca city, the main urban settlement located on its side, which supports ~300,000 inhabitants (INDEC 2010), with a discharge of ~70,000 $m^3.day^{-1}$ (Cifuentes et al. 2011). Additional freshwater inputs (i.e., Punta Alta city ~19,000 $m^3.day^{-1}$ of wastewater (CTE 2003) or the industrial nucleus including petrochemical refineries which discharges ~106,000 $m^3.day^{-1}$ (Cifuentes et al. 2011)) are continuously dumped into the Bahía Blanca Estuary.

The cities and industries enclosing the Bahía Blanca Estuary are in continuous development (expansion and production). By 2002, the industrial area surrounding the petrochemical center embraced only 9 industries, while in 2012, it included more than 135 (Sznaiberg 2012). The harbor area modifies the coastal environment through the corresponding maintenance dredging activities, the infill material from the dredged areas, and the modification of the coastlines. So, this set of anthropogenic stressors result in the deterioration of the Bahía Blanca estuarine environment.

4.2 Main Chemical Features Within the Bahía Blanca Estuary

A huge amount of information on the chemical aspects of the Bahía Blanca Estuary has been collected, processed, and published in the last five decades, and it has allowed identifying both spatial and temporal variations and distribution trends, as well as the main biogeochemical cycles which drive the observed performance.

4.2.1 Temperature

Estuarine water temperature was one of the usually measured parameters within the Bahía Blanca Estuary, and the analysis of the obtained data (achieved between 1974 and 2019) has shown a very stable behavior along the Principal Channel, from the head of the estuary and up to its mouth (Fig. 4.2) (Marcovecchio and Freije 2004). The corresponding statistical analysis has indicated that non-significant differences have been observed between the considered areas of the estuary (inner, middle, and outer), allowing to sustain that estuarine temperature is governed by both the ocean water temperature as well as the air one, which regulate its variations within the estuary (Marcovecchio et al. 2010a). Unlike this, freshwater discharges have not shown to be a significant regulator of the temperature even on the area close to the outlets (Marcovecchio 2001). Furthermore, water temperature measured in the ports of Puerto Cuatreros and Ingeniero White (at the inner estuarine area) have demonstrated to be strongly regulated by the air temperature from the nearby region, and their corresponding curves of measured values have shown quite similar trend in their distribution of values (Fig. 4.3) (Freije and Marcovecchio 2004).

During the herein analyzed period (1974–2019), the range of temperature values recorded within the estuary water column has oscillated between 4.1 °C (on June 1983) and 27.8 °C (on January 2011). The distribution of temperature values has followed a sinusoidal curve, which indicates the occurrence of a thermic cycle opportunely indicated as characteristic of the estuarine conditions (James et al. 2013; Ralston et al. 2014). This trend in coastal water temperatures is also reflected in estuarine conditions as well as in the biological processes occurring (Barletta and Valença Dantas 2016).

In addition, it has been observed a smooth but permanent rise in water temperature within the estuary along the last 40 years (Table 4.1). This fact fully agrees with a general trend described for numerous coastal and estuarine systems all over the world (i.e., Nguyen et al. 2011; Sunda and Cai 2012; Wetz and Yoskowitz 2013), which indicates the global increase that temperature is showing within different environments from the Earth (Hansen et al. 2010; IPCC 2018).

4.2.2 Salinity

The Bahía Blanca Estuary has been characterized as a particular estuarine environment, considering it is not associated to the outflow of big river (Piccolo and Perillo 1999). Consequently, the salinity distribution pattern does not present a sharp gradient along the estuary, which is one of the main characteristics usually applied to describe traditional estuaries (Nguyen et al. 2012; Cloern et al. 2017; Bowman 2018). Nevertheless, a clear variation in salinity values could be observed at the inner estuary, with ranges from 17.9 PSU and up to 41.3 PSU opportunely

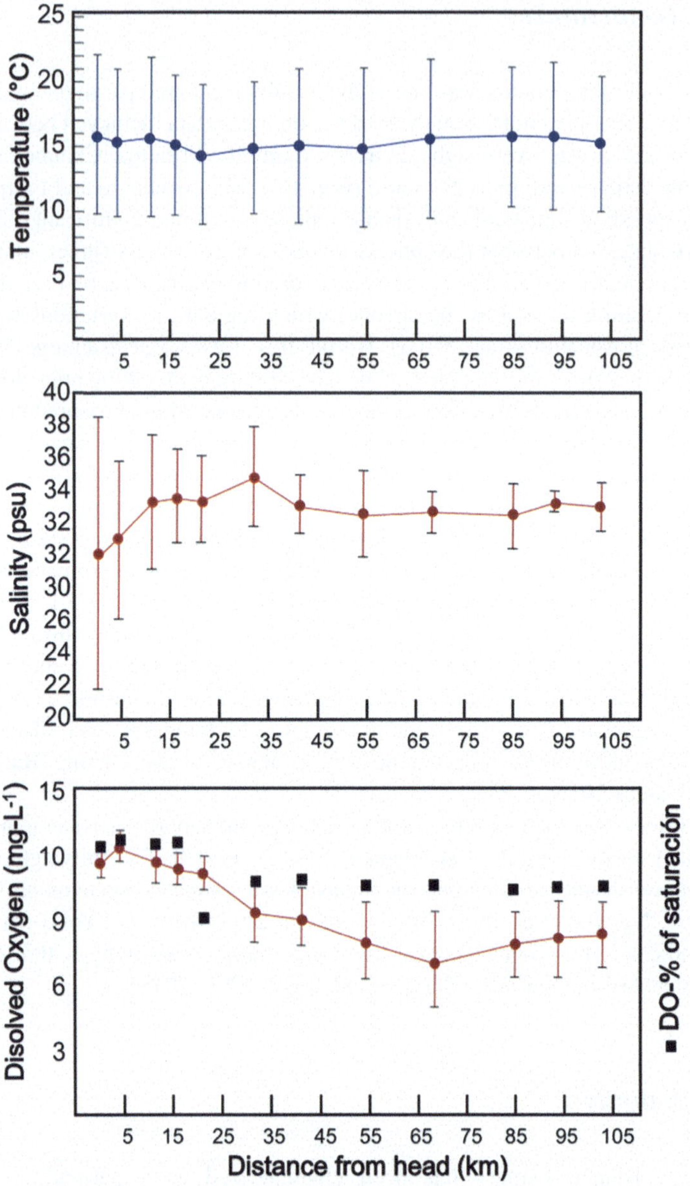

Fig. 4.2 Distribution of physico-chemical parameters values along the Principal Channel within Bahía Blanca Estuary. Temperature. Salinity. Dissolved oxygen and % of saturation

recorded (Freije and Marcovecchio 2004), and allowing to characterize the Bahía Blanca Estuary as an estuarine environment (Perillo et al. 2001).

An important fact is that the estuary behaves as "hypersaline" within its inner area (Marcovecchio and Freije 2004) at almost every summer transforms it in a

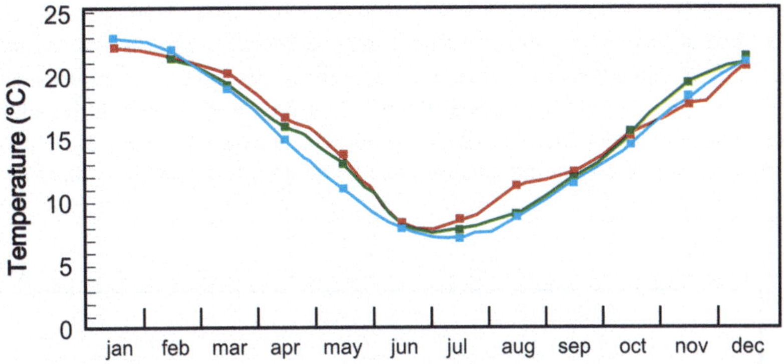

Fig. 4.3 Annual cycle of temperature in two fixed sampling stations (Ingeniero White and Puerto Cuatreros) located in the inner area within the Bahía Blanca Estuary

Table 4.1 Decadal variation of physico-chemical parameters' average concentrations within Bahía Blanca estuarine water

Analyzed parameters	Decadal averages			
	1974–1984	1984–1994	1994–2004	2004–2014
Temperature (°C)	13.8 ± 5.8	14.2 ± 5.5	14.4 ± 5.5	15.7 ± 5.5
Salinity (psu)	31.79 ± 3.66	33.58 ± 3.50	32.65 ± 3.60	27.12 ± 12.06
Dissolved oxygen (mg.L^{-1})	7.7 ± 1.5	–	7.9 ± 1.8	7.4 ± 1.6
pH	–	–	8,2 ± 0.48	8.2 ± 0.32

Mean value ± standard deviation

negative one during that period – in the terms as defined by de Silva Samarasinghe and Lennon 1987, Nunez Vaz et al. 1990, or Nunez Vaz 2012, among others, allowing an inward flux. Furthermore, it should be highlighted that salinity has shown a very stable behavior along the Principal Channel of the Bahía Blanca Estuary, with its maximum variations in the inner area within the estuary (Fig. 4.2).

Finally, the observed distribution of salinity values fully agrees with previous reports by other authors on different estuarine systems (i.e., Telesh and Khlebovich 2010; Whitfield et al. 2012; Telesh et al. 2013).

4.2.3 pH

pH measurements have presented a consistent distribution trend of values along the estuary, with nearly constant levels within the different sampling stations. In addition, it has been observed that the range of measured values has shown differences, usually linked to seasonal changes and related to biological processes (Feely et al. 2010). In this way, highest pH values have been recorded after the occurrence of the large phytoplankton blooms (winter and summer), reaching up to levels of ~9 (Fig. 4.4a) (Popovich and Marcovecchio 2008; Marcovecchio et al. 2010a). In the analysis of the long-time series at Puerto Cuatreros and Ingeniero White Ports, it can be observed that, even though the distribution trend is the same at both sites, the measured values at Puerto Cuatreros Port seemed to be slightly higher than those from Ingeniero White Port. The maximum values at both sites have been identified on summer (January) and winter (August), at the same time the

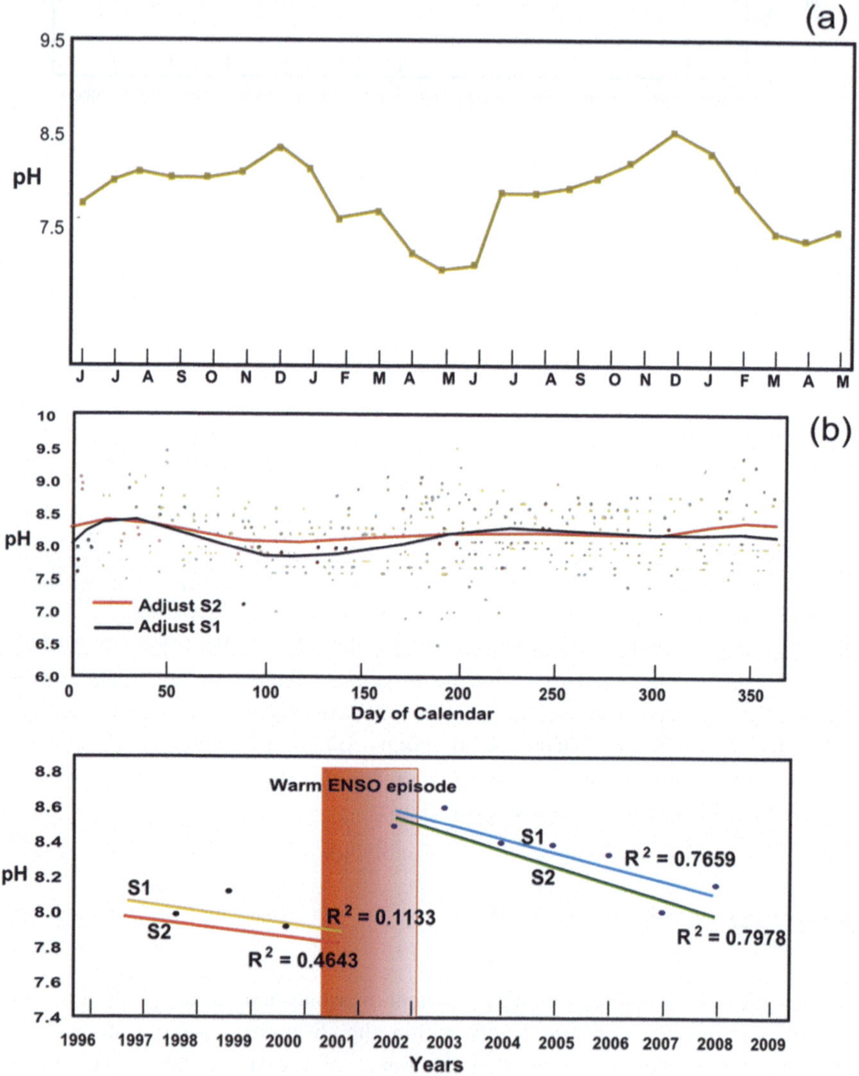

Fig. 4.4 Distribution and variation of pH values within Bahía Blanca Estuary along the studied period. (a) pH temporal variation for the 2004–2006 period, (b) pH distribution values along the 1996–2009 period

main phytoplankton blooms took place (Fig. 4.4a). This kind of distribution pattern as well as the importance of this parameter values on the framework of biological processes development have been opportunely highlighted by several authors on different estuarine systems (Nixon et al. 2015; Villafañe et al. 2015).

Moreover, it must be highlighted that along the assessed period (1974–2019) the interannual variability of pH was quite stable, even though the results show that during 2001 and 2002 the occurrence of the climatic event "El Niño Southern Oscillation" produced the highest rainfall of the decade in line with the higher pH values. In particular, during those years and taking into account the previous condition of acid-base balance, there was a significant change in the acidification of water toward its alkalization. In fact, during that period, the pH increased significantly more than 0.5 units, doubling the average standard deviation in less than a year (Fig. 4.4b) (Arias et al. 2012). Thereafter, pH values gradually returned to general acidification trend.

4.2.4 *Turbidity*

Turbidity concentrations have demonstrated to decrease from the inner area of the estuary seaward, considering that near the head it ranges between 50 and 300 ntu, while seaward it decreases to less than 200 ntu in the middle area of the estuary, and at the open ocean observed values close to Bahía Blanca's mouth are lower than 30 ntu. In this sense, two consistent processes adding suspended sediments into the estuarine water column must be considered: (i) the main land sources are located at the inner area (i.e., streams, rivers, sewage outfalls, harbors), and (ii) the increasing depth from the head to the mouth of the estuary which generates higher sediment resuspension (Cuadrado et al. 1994; Perillo et al. 2005). In addition, the analysis of Puerto Cuatreros and Ingeniero White Port long-time data series has shown a slightly lower mean turbidity values at Ingeniero White Port compared to Puerto Cuatreros, even in both cases the maximum values have been recorded during winter. This distribution trend agreed with the reports from other authors regarding different estuaries at other latitudes, and not only from field work (Yang et al. 2014; Watts et al. 2017) but also from remote sensing analyses (Chen et al. 2007; Garaba and Zielinski 2015).

4.2.5 *Dissolved Oxygen*

The distribution of dissolved oxygen (DO) within the Bahía Blanca Estuary has shown enough high values so as to support a valuable biological production, with average levels close to 7 $mg.L^{-1}$ and reaching up to approximately 13 $mg.L^{-1}$ during the highest productive periods (winter and late summer) (Fig. 4.2). Furthermore, and even considering that the highest concentrations of dissolved oxygen were always recorded at the inner area of the estuary, the spatial distribution trend of this parameter has shown very stable values along the whole estuary (Fig. 4.2), presenting not any level associated to hypoxia or anoxia (Gillanders et al. 2015; Breitburg et al. 2015). This fact, together with the recognized higher productivity ability of the system within the inner area, allows to sustain that both the mixture processes occurring inside the estuary due to different drivers (tide, currents, waves, winds) as well as the input of highly oxygenated seawater incoming from the continental shelf tend to homogenize the DO levels within the system (Borja et al. 2015; Harris et al. 2015). This observation has also agreed with the corresponding distribution of the oxygen saturation percentage, which has been homogeneous along the estuary, and presented the highest values at the inner region, with an average value close to 85%, and the lowest at the region where the sewage outfall from Bahía Blanca city discharges into the estuary (Fig. 4.2). Finally, it can be mentioned that this percentage occasionally surpasses the maximum theoretical value (100%) during the phytoplankton blooms at the inner zone (Popovich and Marcovecchio 2008).

4.2.6 Ecophysiological Parameters: Inorganic Nutrients–Organic Matter–Chlorophyll a

The Bahía Blanca Estuary has been recognized as a nutrient-enriched environment, which usually maintains high levels of these inorganic compounds along most of the year (Freije and Marcovecchio 2004). The available studies developed since 1974 have consistently shown a general nutrients distribution trend, with their maximum concentrations within the inner area of the estuary and slightly decreasing toward the mouth, showing different concentrations but a similar distribution model with no complete depletions anywhere in the studied environment (Fig. 4.5). Thus, the oxidized nitrogen species have behaved as follows: the mean levels of nitrate and nitrite have varied from ~10μM and ~ 2μM, respectively, at the inner area down to ~0.1μM for both close to the estuary's mouth. It is important to highlight that the levels of nitrate at the inner zone are usually no less than 4–5μM, reaching up to 10–12μM, with the exception of the phytoplankton blooms' periods (late winter and early spring: August–September), when the concentration of this nutrient can be close to depletion (Popovich et al. 2008). On the other hand, a very high stock of ammonium is usually available within the system, mainly also in the inner area, with mean values of ~20–25μM and reaching up to peaks of ~100μM (Fig. 4.5) (Popovich et al. 2008). This is a quite important aspect, considering that this nitrogenated compound has never been completely depleted, and so it represents a permanent potential stock of nitrogen for the estuary's biological processes.

In addition, it should be highlighted that – within the above-described nitrogen compounds' distribution trend – these nutrients are not acting as limiting elements of biological production in the terms defined opportunely by Moss et al. (2013) and Kennish (2019).

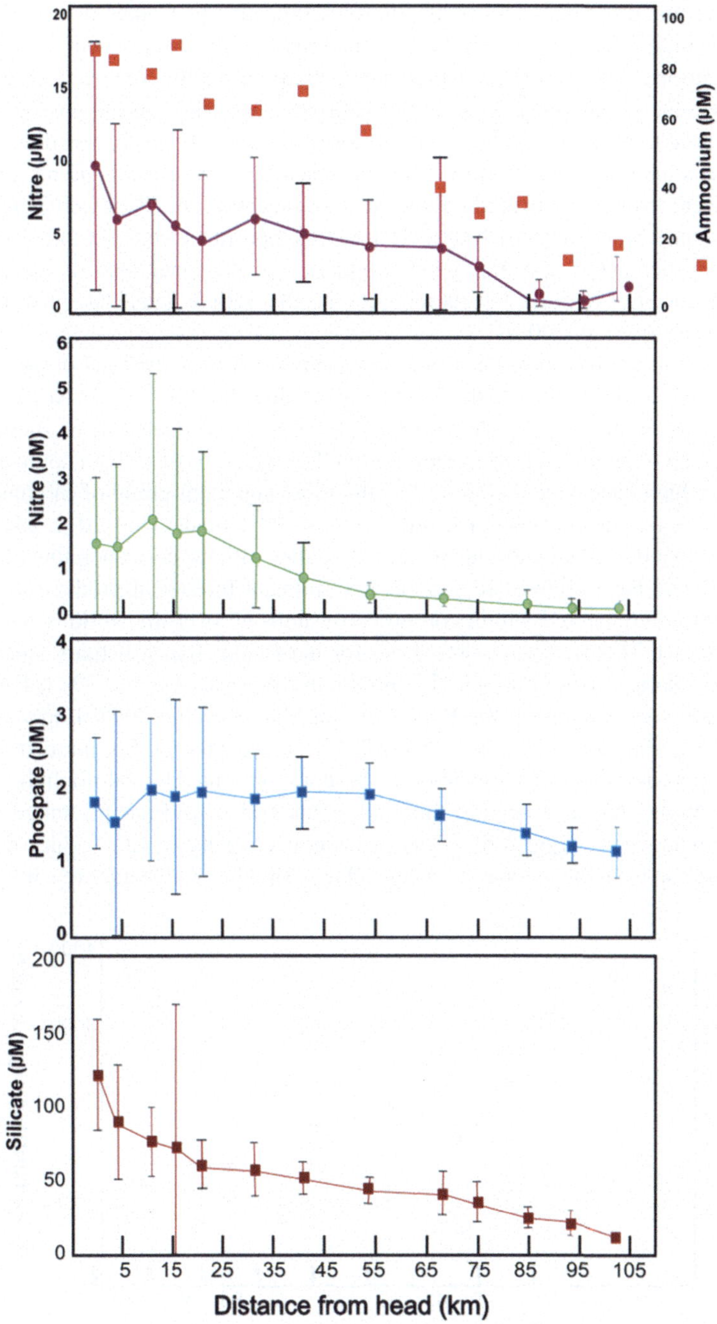

Fig. 4.5 Distribution of inorganic nutrients values along the Principal Channel within Bahía Blanca Estuary

Unlike the nitrogenated nutrients, dissolved reactive phosphate distribution has shown a slight decrease from the head to the mouth of the estuary, with quite stable concentrations and variations which are usually not fully depleted (Fig. 4.5). Nevertheless, at the inner area of the estuary where the largest phytoplankton blooms used to occur, phosphate concentrations occasionally decreased down to ~0, but these situations have been quickly surpassed, and the corresponding levels of this nutrient strongly recovered up to its usual concentrations (Popovich et al. 2008; Popovich and Marcovecchio 2008). Despite this, phosphate could eventually act as a limiting element of the biological production at the inner area of the estuary although for brief periods, but its recovery up to baseline levels has proven to be very fast (Freije et al. 2008).

Finally, silicate has shown a strong decreasing trend from the head of the estuary (mean value ~120μM) down to the mouth (mean value ~8μM) (Fig. 4.5). Nevertheless, these values of silicate have demonstrated to be fully adequated as to support the development of phytoplankton blooms within the estuary, most of them dominated by diatoms which have been recognized as the most important group of Si consumers within this environment (Popovich and Gayoso 1999; Guinder et al. 2012, 2013).

The above-described nutrients scenario together with the corresponding physico-chemical one have allowed to support a significant biological production which deserves to be remarked within coastal ecosystems all over the world (Cloern and Jassby 2010). This fact was pointed out for the Bahía Blanca Estuary since long time ago, characterizing it as a highly productive one (i.e., Gayoso 1983, 1998a, b, 1999; Freije and Gayoso 1988; Popovich and Marcovecchio 2008; López Abbate et al. 2015; Guinder et al. 2009, 2013, 2015, among others). So, the distribution trend of photosynthesizer pigments (i.e., herein represented by chlorophyll *a* mean concentrations) has also been evaluated for this system and has presented a very consistent tendency (Fig. 4.6). Thus, a clear decreasing trend of chlorophyll *a* concentration from the inner area (mean value ~10μg.L^{-1} with values reaching up to

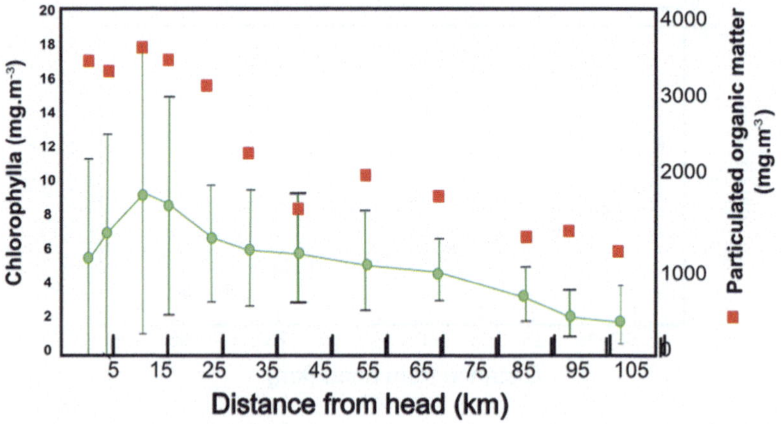

Fig. 4.6 Distribution of chlorophyll *a* and particulate organic matter values along the Main Navigation Channel within Bahía Blanca Estuary

~20μg.L^{-1}) down to the outer one (mean value ~2μg.L^{-1}) has been observed (Fig. 4.6). However, it is important to highlight that the chlorophyll *a* levels within the estuary have never been null, which indicates that the system is a permanently productive estuary, and its lower values are similar to those usually recorded on coastal marine waters from the Argentine Sea even during bloom periods (Rivas et al. 2006; Romero et al. 2006; Olguín and Alder 2011; Paparazzo et al. 2017). In addition, during the three decades from the 1970s to 1990s, chlorophyll *a* values of approximately 55μg.L^{-1} have been recorded at different years (Popovich and Marcovecchio 2008); this situation was modified after the big event *El Niño* from 2001, when structural conditions of the system changed (Arias et al. 2012). As a consequence of the new environmental scenario, the previously described trend of biological production was preserved both in time and in characteristics, but the usual dominant species on the periodical blooms (the diatom *Thalassiosira curviseriata* after Gayoso 1998 or Popovich and Gayoso 1999) was replaced by *Thalassiosira minima*, which is significantly smaller and consequently includes less chlorophyll molecules than the first one (Guinder et al. 2012, 2017); so, from 2001 and up to the present, maximum chlorophyll values as recorded within the estuary have not surpassed ~12μg.L^{-1} (Guinder et al. 2017).

Simultaneously, with the above-described pigments distribution and concentrations, the values of net primary productivity (NPP) determined at the inner area of the estuary have reached up to ~300 mgC.m^{-3}.h^{-1} (Fernández et al. 2014), which could be mentioned between the highest records reviewed in the international literature (Sheaves et al. 2015). These high productivity levels determine the occurrence of high concentrations of organic matter (OM) within the system (mean values ~2000 mg C.m^{-3}all along the Principal Channel of the estuary, and with top levels reaching up to 3680 mg C.m^{-3}; Fig. 4.6); furthermore, the maximum OM levels agree with the peaks of chlorophyll *a*, as well as with the depletion of nitrate, nitrite, silicate, and dissolved oxygen, indicating that most of the determined particulate organic matter (POM) is originated through biological production. In fact, the analysis of stable isotopes signature has demonstrated that Bahía Blanca Estuary phytoplankton was the main contributor of organic matter to the sediments, followed by sewage and microbial mats within the system (La Colla et al. 2014). In addition, different OM sources occur within the system, including sewage outfall discharges, rivers, and streams, among others, which could significantly modify the available OM stock for the system (Marcovecchio et al. 2008).

4.3 Studies on Pollutants Occurrence, Levels, and Distribution Within Bahía Blanca Estuary

Bahía Blanca Estuary is an excellent study case of potentially toxic substances occurrence within the environment as well as related pollution processes, because it is a large transitional environment with a great human activity within its inner area, including non-adequate use of soils (including unplanned urbanization), untreated domestic or industrial sewage discharge, harbor activities, and/or incorrect solid waste disposal, between others (Marcovecchio et al. 2008). This environment has been particularly studied since the 1970s, including its water physico-chemical parameters, associated biological processes, and pollutants occurrence. These studies allow to characterize the potential effects on the system, as well as to recognize its response ability. The information included in the next paragraphs presents a brief overview on the occurrence, levels, and distribution trends of the main pollutants determined within Bahía Blanca Estuary (i.e., heavy metals, hydrocarbons, pesticides, or microplastics) in both the abiotic and the biological compartments. The identified trends are analyzed within a historical viewpoint, which allows pointing out evolutive processes on the estuarine environmental quality. This information is largely useful to carry out concrete monitoring and management programs within the estuary, and the main information recorded along the last decades is synthesized in the next paragraphs.

4.3.1 Heavy Metals

Studies on heavy metals at Bahía Blanca Estuary (i.e., Cd, Cr, Cu, Fe, Pb, Zn) started on the early 1980s and included data on estuarine water, sediments, suspended particulate matter (SPM), and biota. This information has been obtained applying internationally standardized methodologies (i.e., samples' wet acid mineralization followed by atomic absorption spectroscopy (AAS) or induced coupled plasma with optical resolution (ICP-OES)) opportunely described and compiled in different previous papers (i.e., Marcovecchio and Ferrer 2005; De Marco et al. 2006; Botté et al. 2010a, b; Marcovecchio et al. 2014). Most of these results were obtained within the inner area of the estuary, even additional information exists on other regions along the system (Fig. 4.1). Analytical quality (AQ) of the developed analysis was checked against internationally certified reference materials, provided by the National Institute for Environmental Studies (NIES) from Tsukuba (Japan) (mussel and pepperbush tissues, marine and estuarine sediments), as well as by the National Institute of Standards and Technology (NIST) from Boulder, Colorado (USA) (mussel tissue, marine sediments). Statistical comparisons were developed using analysis of variance (ANOVA), mean values assessment (Tukey's test), correlation analysis, and single linear regression analysis (Sokal and Rohlf 1995).

The main obtained results are summarized in Table 4.2, which shows interesting trends within their distribution. So, dissolved metals have presented a spread range of values from non-detectable ones (meaning lower than the applied analytical method detection limit) which have been more of the analyzed samples along the considered four decades, and up to relatively high concentrations (i.e., 6.8μg $Cd.L^{-1}$, 48.6μg $Cr.L^{-1}$, or 19.6μg $Pb.L^{-1}$) which have been eventually recorded within the Bahía Blanca Estuary (Table 4.2). It is important to highlight that the dissolved fraction of trace metals indicates the recent entry of these elements into the system taking into account their ephemeral persistency within the aquatic environment (Kalnejais et al. 2010; Kent and Vikesland 2016). In this sense, the occurrence of eventual inputs of dissolved metals into the estuary could be supported, although it must be exhaustively clarified that these incomes should not be continuous and that the possibility of the existence of metals released from those deposited in sediments due to changes in the physico-chemical conditions that govern them must also be considered (Namieśnik and Rabajczyk 2010; de Souza Machado et al. 2016). However, everything seems to indicate that the occurrence of the first mentioned situation is much more likely than the second one (Gautam et al. 2014).

Table 4.2 Trace metals (Cd, Cu, Cr, Fe, Hg, Ni, Pb, Zn) in estuarine water (μg.L^{-1}), surface sediments (μg.g^{-1}, dry weight) and fish muscle (μg.g^{-1}, wet weight) from Bahía Blanca Estuary

		Range of values (maximum–minimum)			
Analyzed metals		1980–1989	1990–1999	2000–2009	2010–2019
Cadmium (Cd)	Dissolved (μg.L^{-1})	n.d.–1.86	n.d.–1.79	n.d.–1.21	n.d.–6.8
	Sediments (μg.g^{-1}, d.w.)	n.d.–2.36	n.d.–3.17	n.d.–2.20	n.d.–0.94
	Fish tissues (μg.g^{-1}, w.w.)	n.d.–0.46	n.d.–0.54	n.d.–0.66	n.d.–0.02
Copper (Cu)	Dissolved (μg.L^{-1})	–	–	0.54–16.1	0.31–9.7
	Sediments (μg.g^{-1}, d.w.)	3.86–29.9	4.77–31.3	2.46–25.9	8.7–34.1
	Fish tissues (μg.g^{-1}, w.w.)	n.d.–2.66	n.d.–2.16	n.d.–4.47	n.d.–1.70
Chromium (Cr)	Dissolved (μg.L^{-1})	–	2.04–29.4	0.75–21.1	n.d.–48.6
	Sediments (μg.g^{-1}, d.w.)	2.9–16.3	3.5–18.8	1.27–21.06	7.2–14.7
	Fish tissues (μg.g^{-1}, w.w.)	n.d.–3.39	n.d.–3.84	n.d.–6.67	n.d.–2.10
Iron (Fe)[a]	Dissolved (μg.L^{-1})	–	–	0.01–38.6	n.d.–62.5
	Sediments (mg.g^{-1}, d.w.)	5.9–51.5	6.1–43.3	4.7–36.6	5.9–82.2
	Fish tissues (μg.g^{-1}, w.w.)	–	–	–	–
Mercury (Hg)[b]	Dissolved (ng.L^{-1})	n.d.	n.d.–1.11	n.d.–2.07	n.d.–1.96
	Sediments (ng.g^{-1}, d.w.)	10–1670	10–720	n.d.–200	n.d.–220
	Fish tissues (ng.g^{-1}, w.w.)	220–3970	200–640	90–120	n.d.–100
Nickel (Ni)	Dissolved (μg.L^{-1})	–	–	n.d.–9.5	n.d.–1.35
	Sediments (μg.g^{-1}, d.w.)	–	–	1.75–26.1	4.5–14.4
	Fish tissues (μg.g^{-1}, w.w.)	–	–	n.d.–2.08	n.d.–0.66
Lead (Pb)	Dissolved (μg.L^{-1})	n.d.–10.4	n.d.–6.55	n.d.–15.7	n.d.–19.6
	Sediments (μg.g^{-1}, d.w.)	4.76–31.05	3.87–26.65	n.d.–42.71	4.3–19.3
	Fish tissues (μg.g^{-1}, w.w.)	n.d.–3.02	n.d.–3.83	n.d.–5.94	n.d.–1.70
Zinc (Zn)	Dissolved (μg.L^{-1})	n.d.–50.6	n.d.–73.6	n.d.–71.3	n.d.–54.7
	Sediments (μg.g^{-1}, d.w.)	19.6–103.7	21.75–706.1	14.2–98.5	21.4–82
	Fish tissues (μg.g^{-1}, w.w.)	n.d.–11.2	0.2–7.13	n.d.–8.09	0.9–30.1

After Marcovecchio (1988), Marcovecchio et al. (1986, 1988a, b, c), Botté et al. (2008, 2010a, b), IADO (1997, 1999, 2002, 2006, 2008, 2009, 2010, 2012, 2014, 2016, 2018), Marcovecchio and Ferrer (2005), Marcovecchio et al. (2010b, 2016), Simonetti et al. (2017), Buzzi and Marcovecchio (2018) and La Colla et al. (2018a, b, 2019)

Range of values (minimum–maximum) along the 1980–2019 period

n.d.: values below the detection limit of the applied analytical method

[a]Values in (mg.g^{-1}); [b]Values in (ng.g^{-1})

A totally different trend was observed in the distribution of trace metals in sediments of the estuary. In this case, most of the metals studied were permanently detected in that compartment, and only a few exceptions (i.e., Cd) showed values below the corresponding detection limits (Table 4.2). Although the range of concentrations of metals measured in sediments of the estuary has been very wide and varied, the corresponding average values can be classified as intermediate ones (Cairns Jr. 2009), indicating the existence of anthropogenic sources that input them into the system but without reaching critical levels for organisms living there (Sindermann 2006). These sediments' contamination *status* due to their trace metal content was checked against opportunely reported Geo-accumulation Index (I*geo*) and Enrichment Factor (EF) within Bahía Blanca Estuary (Fernández Severini et al. 2018, following Förstner et al. 1990) as well as comparison with different Sediment Quality Guidelines (i.e., Canadian, European Union, FAO, among others)(Roach 2005; Hübner et al. 2009). In this latter sense, it must be considered that these Quality Guides must be generated for particular environments or regions, and therefore are not universalizable (Kwok et al. 2013). Thus, the most convenient is to generate own guides for each environment or region evaluated (Förstner et al. 1990).

Finally, when trace metals distribution within fishes from Bahía Blanca Estuary was considered, a variable trend has been observed but showing that most of studied metals seemed to be present within the analyzed tissues (Table 4.2). In all cases the obtained results include numerous *non-detectable* values, which means that the corresponding environmental exposure to this element is far to critical levels (Jezierska et al. 2009; Govind and Madhuri 2014). In any case, the levels of metals determined in the tissues of the analyzed fish demonstrate the existence of exposure to these compounds (eventual or periodic), as well as the corresponding biological accumulation processes opportunely described for this environment (Marcovecchio et al. 1986, 1988a, b; De Marco et al. 2006; La Colla et al. 2018b, 2019; among others). An interesting point to be considered has been the decrease in Hg concentrations in fish tissues that occurred from the 1980s to the present, and which meant a significant environmental recovery of these species (Marcovecchio et al. 2001; De Marco et al. 2006).

4.3.2 Hydrocarbons, Including Polycyclic Aromatic Hydrocarbons (PAHs)

PAHs can be originated from three possible sources: petrogenic, pyrolytic and natural. Although PAHs can be originated naturally, anthropogenic activities are generally considered the major source of PAHs release into the environment. There are important differences in the chemical composition of PAHs mixtures depending on the sources of emission. In general, pyrolytic PAHs present dominance of high molecular mass compounds (corresponding to four to six rings compounds), while petrogenic PAHs present dominance of two/three rings compounds. Moreover, pyrolysis process produces PAHs associated to soot carbon. A quite stronger association and persists during is deposited in aquatic system affecting partitioning and bioavailability of PAHs.

The Bahía Blanca Estuary has a large history of addressing sources and origin of this type of hydrocarbons, since Arias et al. (2011), followed by Arias et al. (2009) and Oliva et al. (2015), between others. PAHs can be introduced into the Bahía Blanca Estuary by different ways: for instance, spillage of fossil fuels, ship traffic, atmospheric depositions, urban runoff, and municipal and industrial wastewater discharge. From these, the atmospheric transport is the most important pathway for their environmental distribution. Particle-bound PAHs can be transported long distances, from the emission source as far as remote areas where they are removed from the atmosphere through precipitation and dry deposition. Recently, Orazi et al. tracked the sources and distribution of atmospheric PAHs around the Bahía Blanca Estuary (Orazi et al. 2020). Results showed a range from 27.97 to 1052.99 ng. m^{-3} and from 52.40 to 2118.34 ng. g^{-1} d.w. for air and soil samples, respectively. The highest air-PAHs levels were registered in Bahía Blanca city (1052.99 ng. m^{-3}, d.w.), confirming the hypothesis of a significant PAHs source impacting Bahía Blanca Estuary through the air. Atmospheric PAHs sources are dominated by pyrolytic origin: high density of traffic, coal combustion, residential emissions, and industrial activities were tagged as the principal diffuse sources. Finally, PAHs have been shown to bioaccumulate in the BBE biota, i.e., mussels and pelagic fish (Arias et al. 2010; Oliva et al. 2015). Figure 4.7 shows the source apportionment of PAHs in biota at the Bahía Blanca Estuary.

4.3.3 Pesticides

When organochlorine pesticides (OCPs) enter the ocean, they tend to accumulate in biota and bottom sediments. Smaller organisms incorporate them primarily through their respiratory surfaces while larger animals do so through food intake (UNEP 2002). Due to their fat solubility and high persistence to biological degradation, they are biomagnified along the trophic web (UNEP 2002; Roche et al. 2009; Lailson-Brito et al. 2010); so the higher trophic levels, such as marine mammals, top predatory fish or seabirds, usually show the highest OCPs concentrations. The biota could be affected to toxic effects caused by short- or long-term exposure to these pesticides. These effects include reproductive damage (Gross et al. 2002), endocrine disruption, immune suppression, and cancer, among others (Bergman et al. 2012; Menzies et al. 2013).

In South America, most of the OCPs were intensively used and produced, between 1950 and 1990, except Endosulfan, Dicofol, and Methoxychlor, which were used until recently. During their use, a large part of the OCPs entered the continental environment and subsequently into the sea, and although all the South American countries banned or restricted their use and production, detectable and even harmful environmental concentrations are likely to remain present in the seas

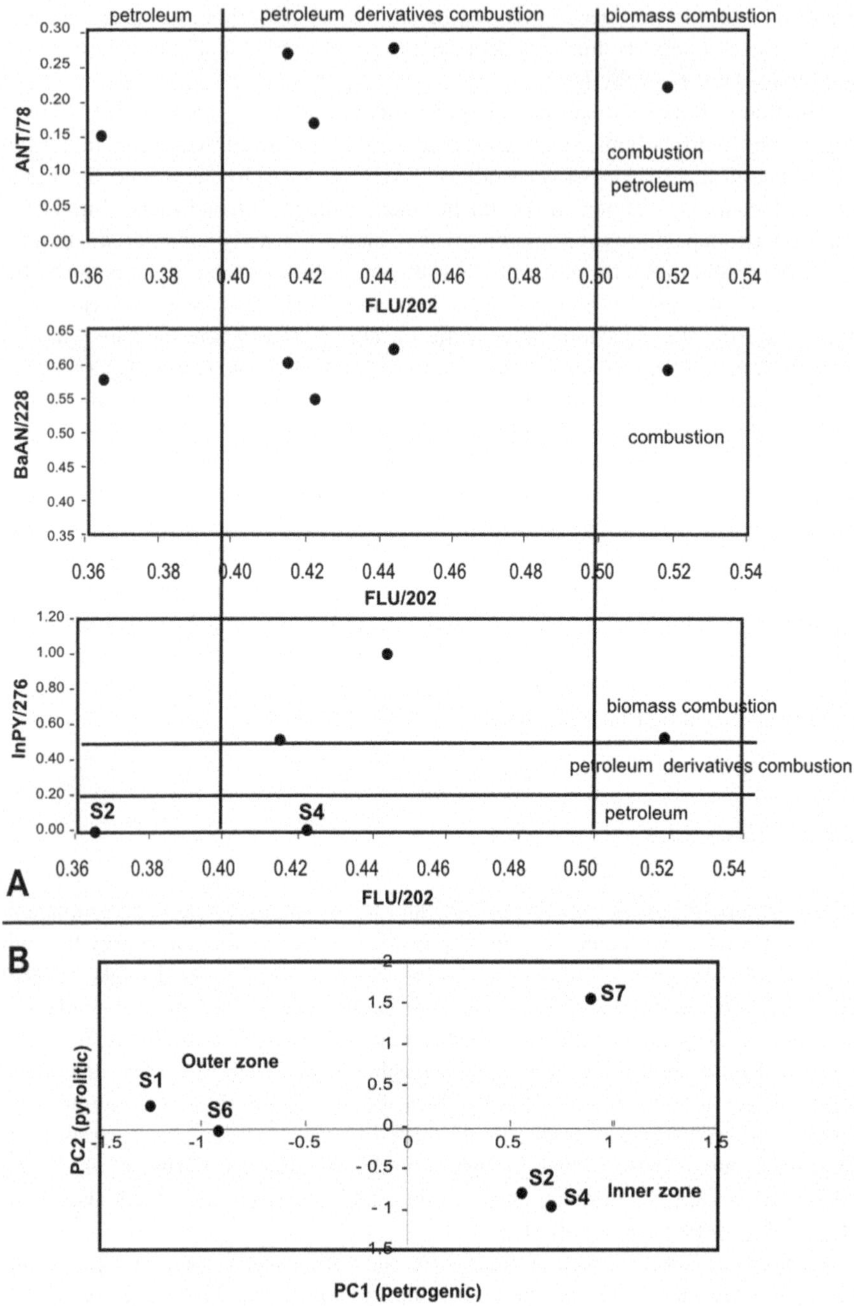

Fig. 4.7 Source origin plot indicating main sources at the Bahía Blanca Estuary biota. (**a**) Mussel's IP/276, BaA/228, and An/178 source ratio plot vs Fl/202 ratio, showing different origin zones. (**b**) PCA score plot of biota samples. (After Arias et al. 2010)

and oceans for several years due to the intrinsic characteristics and environmental dynamics of the OCPs (UNEP 2002; Iwata et al. 1993; Wania et al. 2006).

The Bahía Blanca Estuary is located in one of the most important agriculture regions of Argentina. Large amounts of organochlorine compounds (OCs) are known to be used in areas located close to the estuary basin (Renella and Quirós 2000), and the abundant water and sediments loads of Sauce Chico River, Maldonado River, Napostá Grande Stream, and other tributaries may play an important role on the transport and distribution of OCs. In spite of this, in the history of the Bahía Blanca Estuary, apart from Sericano and Pucci (1984), there was no OC pesticides information or previous studies on this topic until 2010 (Arias et al. 2010). From then, frequent monitoring studies could set average values and OCs entries to the estuary. It has been demonstrated that the estuary sediments are functioning as a sink for persistent OC pesticides (or their metabolites) that are either in use or have been recently banned for agricultural purposes, pointing out an extensive use of OCs in the catchments of the Bahía Blanca Estuary in the past and a present use of some, also revealing a declining trend in the environmental burden of persistent pesticides. The main OC access way to the Bahía Blanca Estuary suggested to function by land runoff and subsequent transport to the estuary by draining tributaries and/or evaporation and atmospheric wet deposition (Arias et al. 2010; Tombesi et al. 2018; Gironés et al. 2020).

DDT and HCH residues are present, but in the low range of levels in comparison to other worldwide locations as well as to the world coastal sediment concentrations (Fig. 4.8). Unlike this, $\sum$OCs average is in the medium range, indicating that the Bahía Blanca Estuary is not grossly polluted by HCHs or DDTs and that a significant proportion of OCs pollution is provided by other pesticides (i.e., Endosulfan).

4.3.4 Microplastics

Microplastics (MPs) have passed from being considered emerging pollutants to be recognized as an emerged threat, with the urgent need to better assess their distribution in the marine environment, as well as the ecotoxicological and ecological risks that they have (Avio et al. 2017). Although MPs have been considered as vectors for other pollutants (Rochman 2015; Hartmann et al. 2017), these particles have also been determined themselves as Persistent Organic Pollutants (POP) in the Convention on POPs (UNEP 2001) due to their characteristics of persistence, bioaccumulation, long-range transport, and adverse effects. The study of the presence and distribution of MPs in marine environments has exponentially increased in recent years; however, there are very few studies in the Bahía Blanca Estuary. The first evidence was recently published in the gastrointestinal tract of the marine commercial fish *Micropogonias furnieri* (Arias et al. 2019). The investigations of this work focused on two areas of the Bahía Blanca Estuary, finding more particles in the specimens from the internal zone than in the middle zone of the estuary. Although the abundance of ingested MPs could reflect variations in the quantity and

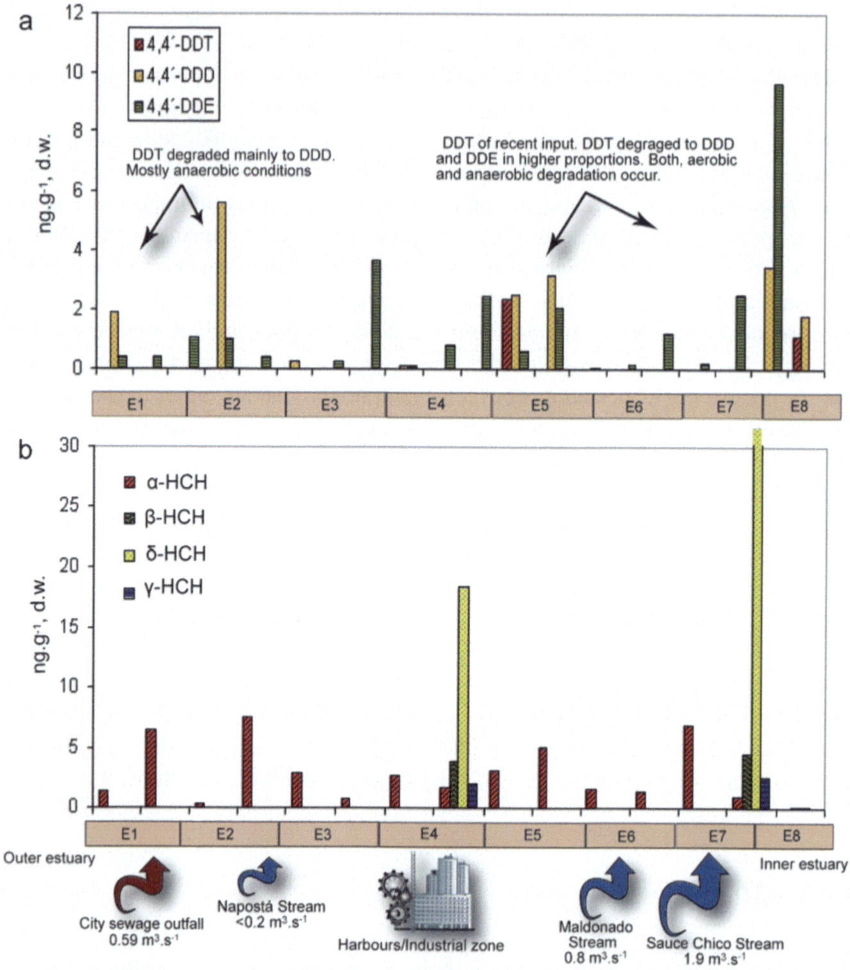

Fig. 4.8 (**a**) Example of DDT and degradative isomers spatial distribution at the Bahía Blanca Estuary. (**b**) HCH's spatial distribution over the sampled stations. Schematically, main sewage, industrial, and freshwater inputs are indicated

type of food consumed between individuals of the same species and between diverse species, the fact that the same species and length class was analyzed pointed out a different human anthropic pressure along the Bahía Blanca Estuary with respect to this pollutant. The second work published for this area was in the same year and demonstrated, for the first time in the Bahía Blanca Estuary, the existence of MPs in bottom sediments and surface seawater, and it was determined at the outermost area of the estuary (Ronda et al. 2019). Consistently with previous wide world environmental studies, MPs levels detected in Bahía Blanca Estuary were comparable to other marine environments, consisted mainly by fibers and the most abundant size was less than 1 mm (Rochman 2015; Gago et al. 2018). The fact that it has been evidenced MPs contamination in several matrices along the Bahía Blanca Estuary (organisms, sediment, and seawater) argues the need to continue conducting research that not only evaluates the presence and distribution of these particles but also the ecotoxicological and ecological effects that they could have on the Bahía Blanca Estuary ecosystem.

4.4 Brief Comment on the Chemical Functioning of the Estuary

Bahía Blanca Estuary is a very large system, whose functioning is clearly characterized by several processes that on the whole determine the success of the biological productivity occurring there.

The distribution of the structural parameters within the system is very stable, mainly in terms of temperature, pH, and turbidity. Salinity presents a relative stability, even though significant variations occur at the inner area, which could alternatively be increased or decreased according to the season.

The estuary is usually highly nutrient enriched, and levels of compounds of nitrogen (basically ammonium) are always available, even though the concentrations of oxidized nitrogen compounds (NO_2^- and NO_3^-) and phosphorus are eventually used down to close-depletion during the periodical phytoplankton blooms. Nevertheless, the recovery of these consumed nutrients through mineralization of organic matter is extremely fast. In addition, a large stock of silicate is usually available in the estuary, mainly at the inner area, which is completely adequate to support the biological demand within the system.

Concerning the biological primary production, the most important annual period is the late winter – early spring – when the highest phytoplankton blooms have historically occurred. It is developed because during this time the nutrients (N, P, and Si) are largely available, and both the temperature and light intensity are sufficiently low (~5 to 7 °C, and 400–700μE m^{-2} s^{-1}, respectively; after Popovich et al. 2008) as required by the dominant diatom species responsible for the mentioned bloom. Thus, very high levels of chlorophyll _a_ were detected during this phenomenon (with values reaching up to 55 $mg.m^{-3}$), representing densities of ~13 × 10^6 cells L^{-1} or net primary productivities of ~300 $mgC.m^{-3}.h^{-1}$.

The very high amounts of organic matter generated by these biological processes ensure the regenerated nutrients production, through mineralization processes occurring within the estuary (Freije et al. 2008). The obtained results seemed to indicate that a predominant liberation of ammonium was observed within the estuarine sediments at the inner estuary, even significant amounts of oxidized nitrogen compounds (NO_2^- and NO_3^-) were eventually also produced (Popovich et al. 2008). In addition, these are the first nitrogenated nutrients to be consumed during the phytoplankton bloom, and just when both NO_2^- and NO_3^- were depleted, the NH_4^+ started to be consumed (Popovich et al. 2008).

On the other hand, different kind of pollutants have eventually been recorded within Bahía Blanca Estuary, mainly in its inner area. Trace metals have been determined in water, sediments, and fish tissues within the estuary, even their concentrations were not within recognized critical values. Several of them have been recorded in the dissolved phase meaning that presumably sources of these elements are linked to the estuary, being able to discharge them both continuously or eventually. Sediments within Bahía Blanca Estuary have demonstrated to function as sink of trace metals, accumulating them at different rates but along the whole system. This fact, joint to corresponding physico-chemical conditions, fully governs the probabilities of trace metals input into Bahía Blanca Estuary biota.

As expected to a heavily anthropized environment, a widespread state of PAHs pollution has been shown to occur at the Bahía Blanca Estuary. Levels range from negligible (rural) to high in several *hotspots*, including harbor/industrial facilities (sediments) and dense urban environments (air).

DDT and HCH residues are present, but in the low range of levels in comparison to other worldwide locations as well as to the world coastal sediment concentrations. On the opposite, $\sum$OCs average is in the medium range, indicating that the Bahía Blanca Estuary is not grossly polluted by HCHs or DDTs and that a significant proportion of OCs pollution is provided by other pesticides (i.e., Endosulfan).

Microplastic pollution has been assessed in several matrices along the Bahía Blanca Estuary (organisms, sediment, and seawater) arguing the need to perform further research evaluating their ecotoxicological and ecological effects.

Bahía Blanca Estuary's productive cycle, regulated through bio-geochemical joint processes, has well functioned for a long time (at least during the last 40 years, when these studies started at the earlier 1970s). During early 2000 a strong modification on the composition of the phytoplankton bloom which characterizes Bahía Blanca Estuary was identified; nevertheless, the system continued to harmoniously work without showing significant changes in its bio-geochemical condition. Even though this environment is not strongly stressed by pollutants, different kinds of potentially toxic compounds have been repeatedly determined, even at low concentrations. Consequently, this is a nice scenario to control the evolution and progress of the mentioned estuarine chemical processes, as well as to monitor the potential occurrence of changes within the identified trends.

References

Andrade S, Pucci AE, Marcovecchio JE (2000) Cadmium concentrations in the Bahía Blanca estuary (Argentina). Potential effects of dissolved cadmium on the diatom *Thalassiosira curviseriata*. Oceanologia 42:505–520

Arias AH (2008) Comportamiento de los hidrocarburos aromáticos policíclicos (PAHs) en aguas, sedimentos y organismos de la zona interna del estuario de Bahía Blanca. Doctoral Thesis, Universidad Nacional del Sur (UNS). Bahía Blanca (Argentina), 178 pp

Arias AH, Spetter CV, Freije RH et al (2009) Polycyclic aromatic hydrocarbons in water, mussels (*Brachidontes* sp., *Tagelus* sp.) and fish (*Odontesthes* sp.) from Bahía Blanca Estuary, Argentina. Estuar Coast Mar Sci 85(1):67–81. https://doi.org/10.1016/j.ecss.2009.06.008

Arias AH, Vázquez-Botello A, Tombesi N et al (2010) Presence, distribution, and origins of polycyclic aromatic hydrocarbons (PAHs) in sediments from Bahía Blanca estuary, Argentina. Environ Monit Assess 160(1–4):301–314. https://doi.org/10.1007/s10661-008-0696-5

Arias AH, Pereyra MT, Marcovecchio JE (2011) Multi-year monitoring of estuarine sediments as ultimate sink for DDT, HCH, and other organochlorinated pesticides in Argentina. Environ Monit Assess 172(1–4):17–32. https://doi.org/10.1007/s10661-010-1315-9

Arias AH, Piccolo MC, Spetter CV et al (2012) Lessons from multi-decadal oceanographic monitoring at an estuarine ecosystem in Argentina. Int J Environ Res 6(1):219–234. https://doi.org/10.22059/IJER.2011.488

Arias AH, Ronda A, Oliva AL et al (2019) Evidence of microplastic ingestion by fish from the Bahía Blanca estuary in Argentina, South America. B Environ Contam Tox 102(6):750–756. https://doi.org/10.1007/s00128-019-02604-2

Avio CG, Gorbi S, Regoli F (2017) Plastics and microplastics in the oceans: from emerging pollutants to emerged threat. Mar Environ Res 128:2–11. https://doi.org/10.1016/j.marenvres.2016.05.012

Balvanera P, Pfisterer AB, Buchmann N et al (2006) Quantifying the evidence for biodiversity effects on ecosystem functioning and services. Ecol Lett 9:1146–1156. https://doi.org/10.1111/j.1461-0248.2006.00963.x

Barletta M, Valença Dantas D (2016) Environmental gradients. In: Kennish MJ (ed) Encyclopedia of estuaries. Springer, Heidelberg, pp 237–241. 778 pp. ISBN:978-94-017-8800-7

Bergman A, Heindel JJ, Jobling S et al (eds) (2012) State of the science of endocrine disrupting chemicals. UNEP (United Nations Environment Programme)-WHO (World Health Organization), Geneve. 213 pp. ISBN:978-92-807-3274-0

Borja A, Dauer DM, Elliott M et al (2015) Medium- and long-term recovery of estuarine and coastal ecosystems: patterns, rates and restoration effectiveness. Estuaries Coast 33:1249–1260. https://doi.org/10.1007/s12237-010-9347-5

Botté SE, Freije RH, Marcovecchio JE (2008) Dissolved heavy metal (Cd, Pb, Cr, Ni) concentrations in surface water and porewater from Bahia Blanca estuary tidal flats. B Environ Contam Tox 79(4):415–421. https://doi.org/10.1007/s00128-007-9231-6

Botté SE, Delucchi F, Freije RH et al (2010a) Chapter 11: Cadmium and organotin pollution in an estuarine environment from Argentina: and overview. In: El Nemr A (ed) Impact, monitoring and management of environmental pollution. Nova Publishers, New York, pp 263–283. ISBN:978-1-60876-487-7

Botté SE, Freije RH, Marcovecchio JE (2010b) Distribution of several heavy metals in tidal flats sediments within Bahía Blanca Estuary (Argentina). Water Air Soil Pollut 210:371–388. https://doi.org/10.1007/s11270-009-0260-0

Bowman MJ (2018) Chapter 7: Estuarine fronts. In: Kjerfve B (ed) Hydrodynamics of estuaries, Vol.1: Estuarine physics. CRC Press – Taylor and Francis Group, Boca Ratón, pp 85–131. 227 pp. ISBN:0-8493-4370-4

Breitburg DL, Hondorp D, Audemard C et al (2015) Landscape-level variation in disease susceptibility related to shallow-water hypoxia. PLoS One 10(2):e0116223. https://doi.org/10.1371/journal.pone.0116223

Buzzi NS, Marcovecchio JE (2018) Heavy metal concentrations in sediments and in mussels from Argentinean coastal environments, South America. Environ Earth Sci 77:321–333. https://doi.org/10.1007/s12665-018-7496-1

Cairns J Jr (2009) Chapter 1: The influence of contaminated sediments on sustainable use of the planet. In: Kassim TA, Barceló D (eds) Contaminated sediments. Springer, Heidelberg, pp 1–19. 193 pp. ISBN: 978-3-540-88013-4

Chen Z, Hu C, Muller-Karger F (2007) Monitoring turbidity in Tampa Bay using MODIS/Aqua 250-m imagery. Remote Sens Environ 109:207–220. https://doi.org/10.1016/j.rse.2006.12.019

Cifuentes O, Escudero D, Medus S et al (2011) Estudio de la dinámica (espacial y temporal) de los efluentes líquidos industriales y urbanos en la zona del Polo Petroquímico y área portuaria de Bahía Blanca. En: Puliefito E (ed) Contaminación atmosférica e hídrica en Argentina. (EdUTecNe), Buenos Aires, pp 539–545

Cloern JE, Jassby AD (2010) Patterns and scales of phytoplankton variability in estuarine-coastal ecosystems. Estuar Coasts 33:230–241. https://doi.org/10.1007/s12237-009-9195-3
Cloern JE, Jassby AD, Tara S et al (2017) Ecosystem variability along the estuarine salinity gradient: examples from long-term study of San Francisco Bay. Limnol Oceanogr 62:272–291. https://doi.org/10.1002/lno.10537
CTE (Comité Técnico Ejecutivo – Municipalidad de Bahía Blanca) (2003) Programa integral de monitoreo Bahía Blanca. http://www.bahiablanca.gov.ar/areas-de-gobierno/medioambiente/comite-tecnicoejecutivo/
Cuadrado D, Perillo GME, Marcos A (1994) Análisis preliminar del transporte del sedimento en suspension en Puerto Rosales. V Reunión Argentina de Sedimentología:229–234
De Marco SD, Botté SE, Marcovecchio JE (2006) Mercury distribution in abiotic and biological compartments within several estuarine systems from Argentina: 1980–2005 period. Chemosphere 65(2):213–233. https://doi.org/10.1016/j.chemosphere.2006.02.059
de Silva Samarasinghe JR, Lennon GW (1987) Hypersalinity, flushing and transient salt-wedges in a tidal gulf: an inverse estuary. Estuar Coast Shelf Sci 24:483–498. https://doi.org/10.1016/0272-7714(87)90129-6
de Souza Machado AA, Spencer K, Kloas W et al (2016) Metal fate and effects in estuaries: a review and conceptual model for better understanding of toxicity. Sci Total Environ 541:268–281. https://doi.org/10.1016/j.scitotenv.2015.09.045
Feely RA, Alin SR, Newton J et al (2010) The combined effects of ocean acidification, mixing, and respiration on pH and carbonate saturation in an urbanized estuary. Estuar Coast Shelf Sci 88:442–449. https://doi.org/10.1016/j.ecss.2010.05.004
Fernández Severini MD, Carbone ME, Villagran DM et al (2018) Toxic metals in a highly urbanized industry-impacted estuary (Bahía Blanca Estuary, Argentina): spatio-temporal analysis based on GIS. Environ Earth Sci 77(10):393–416. https://doi.org/10.1007/s12665-018-7565-5
Fernández EM, Garzón CJE, Martínez AM et al (2014) Estimación de la productividad primaria del fitoplancton en la zona interna del estuario de Bahía Blanca. In: Marcovecchio JE, Botté SE, Freije RH (eds) Procesos Geoquímicos Superficiales en Iberoamérica. Sociedad Iberoamericana de Física y Química Ambiental (SIFyQA), Salamanca, pp 261–278. ISBN:978-84-937437-6-5
Förstner U, Ahlf W, Calmano W et al (1990) Part II, Chapter 18: Sediments criteria development. In: Heling D, Rothe P, Förstner U et al (eds) Sediments and environmental geochemistry. Springer, Berlin, pp 311–338. 381 pp. ISBN-13:978-3-642-75099-1
Freije RH, Gayoso AM (1988) Producción primaria del estuario de Bahía Blanca. Informes UNESCO, Ciencias del Mar 47:112–114
Freije RH, Marcovecchio JE (2004) Chapter 8: Oceanografía química del estuario de Bahía Blanca. En: Piccolo MC, Hoffmeyer MS (eds) El ecosistema del estuario de Bahía Blanca. Instituto Argentino de Oceanografía (IADO – CONICET/UNS), Bahía Blanca, 69–78. ISBN:987-9281-96-9
Freije RH, Spetter CV, Marcovecchio JE et al (2008) Chapter 23: Water chemistry and nutrients of the Bahía Blanca Estuary. In: Neves R, Baretta J, Mateus M (eds) Perspectives on Integrated Coastal Zone Management in South America. Part B: From shallow water to the deep fjord: the study sites. IST Scientific Publishers, Lisbon, pp 243–256. ISBN:978-972-8469-74-0
Gago J, Carretero O, Filgueiras AV, Viñas L (2018) Synthetic microfibers in the marine environment: A review on their occurrence in seawater and sediments. Marine pollution bulletin, 127:365–376
Garaba SP, Zielinski O (2015) An assessment of water quality monitoring tools in an estuarine system. Remote Sens Environ 109(2):207–220. https://doi.org/10.1016/j.rse.2006.12.019
Gautam RK, Sharma SK, Suresh Mahiya S et al (2014) Chapter 1: Contamination of heavy metals in aquatic media: transport, toxicity and technologies for remediation. In: Sharma SK (ed) Heavy metals in water – presence, removal and safety. The Royal Society of Chemistry, Cambridge, pp 1–24. 380 pp. ISBN:978-1-84973-885-9
Gayoso AM (1983) Estudio de Fitoplancton del Estuario de Bahía Blanco:(Pcia. de Buenos Aires, Argentina). Zona interna. Puerto Cuatreros. Studia oecológica (2):73–88

Gayoso AM (1998) Long-term phytoplankton studies in the Bahías Blanca estuary, Argentina. ICES Journal of Marine Sciences 55:655–660

Gayoso AM (1998a) Estudio de Fitoplancton del Estuario de Bahía Blanco (Pcia. de Buenos Aires, Argentina). Zona interna: Puerto Cuatreros. Stud Oecol 2(2):73–88. ISSN:0211-4623

Gayoso AM (1998b) Long-term phytoplankton studies in the Bahía Blanca Estuary, Argentina. ICES J Mar Sci 55:655–660. https://doi.org/10.1006/jmsc.1998.0375

Gayoso AM (1999) Seasonal succession patterns of phytoplankton in the Bahía Blanca Estuary (Argentina). Bot Mar 42(4):367–375. https://doi.org/10.1515/bot.1999.042

Gillanders BM, Elsdon TS, Halliday IA et al (2015) Potential effects of climate change on Australian estuaries and fish utilising estuaries: a review. Mar Freshw Res 62:1115–1131. https://doi.org/10.1071/MF11047

Gironés L, Oliva AL, Marcovecchio JE et al (2020) Spatial distribution and ecological risk assessment of residual organochlorine pesticides (OCPs) in South American marine environments. Curr Environ Health Rep:1–14. https://doi.org/10.1007/s40572-020-00272-7

Govind P, Madhuri S (2014) Heavy metals causing toxicity in animals and fishes. Res J Anim Vet Fish Sci 2(2):17–23. ISSN:2320-6535

Grimm NB, Stuart Chapin FS, Bierwagen B et al (2013) The impacts of climate change on ecosystem structure and function. Front Ecol Environ 11(9):474–482. https://doi.org/10.1890/120282

Gross TS, Arnold BS, Sepúlveda MS et al (2002) Chapter 39: Endocrine disrupting chemicals and endocrine active agents. In: Hoffman DJ, Rattner B, Burton GAJ et al (eds) Handbook of ecotoxicology. CRC Press, Boca Raton, pp 1033–1098. 1315 pp. ISBN:1-56670-546-0

Guinder VA, Popovich CA, Perillo GME (2009) Particulate suspended matter concentrations in the Bahía Blanca estuary, Argentina: implication for the development of phytoplankton blooms. Estuar Coast Shelf Sci 85(1):157–165. https://doi.org/10.1016/j.ecss.2009.05.022

Guinder VA, Molinero JC, Popovich CA et al (2012) Dominance of the planktonic diatom *Thalassiosira minima* in recent summers in the Bahía Blanca Estuary, Argentina. J Plankton Res 34(11):995–1000. https://doi.org/10.1093/plankt/fbs060

Guinder VA, Popovich CA, Molinero JC et al (2013) Phytoplankton summer bloom dynamics in the Bahía Blanca Estuary in relation to changing environmental conditions. Cont Shelf Res 52:150–158. https://doi.org/10.1016/j.csr.2012.11.010

Guinder VA, López-Abbate MC, Berasategui AA et al (2015) Influence of the winter phytoplankton bloom on the settled material in a temperate shallow estuary. Oceanologia 57:50–60. https://doi.org/10.1016/j.oceano.2014.10.002

Guinder VA, Molinero JC, López Abbate MC et al (2017) Phenological changes of blooming diatoms promoted by compound bottom-up and top-down controls. Estuar Coast 40:95–104. https://doi.org/10.1007/s12237-016-0134-9

Hansen J, Ruedy R, Sato M et al (2010) Global surface temperature change. Rev Geophys 48:RG4004. https://doi.org/10.1029/2010RG000345

Harris LA, Hodgkins CLS, Day MC et al (2015) Optimizing recovery of eutrophic estuaries: impact of estratification and re-aeration on nutrient and dissolved oxygen dynamics. Ecol Eng 75:470–483. https://doi.org/10.1016/j.ecoleng.2014.11.028

Hartmann NB, Rist S, Bodin J et al (2017) Microplastics as vectors for environmental contaminants: exploring sorption, desorption, and transfer to biota. Integr Environ Asses 13(3):488–493. https://doi.org/10.1002/ieam.1904

Hester RE, Harrison RM (eds) (2007) Biodiversity under threat. Royal Society of Chemistry, Cambridge. 291 pp. ISBN:978-0-85404-251-7

Hübner R, Astin KB, Herbert RJH (2009) Comparison of sediment quality guidelines (SQGs) for the assessment of metal contamination in marine and estuarine environments. J Environ Monit 11:713–722. https://doi.org/10.1039/b818593j

IADO (1999) Programa de monitoreo de la calidad ambiental de la zona interior del estuario de Bahía Blanca. Informe Final, Instituto Argentino de Oceanografía, 25 + 68 pp. Available in: http://www.bahiablanca.gov.ar/cte/index.html

IADO (2002) Programa de monitoreo de la calidad ambiental de la zona interior del estuario de Bahía Blanca. Informe Final, Instituto Argentino de Oceanografía: 76 pp. Available in.: http://www.bahiablanca.gov.ar/cte/index.html

IADO (2006) Programa de monitoreo de la calidad ambiental de la zona interior del estuario de Bahía Blanca. Informe Final, Instituto Argentino de Oceanografía, 82 pp. Available in: http://www.bahiablanca.gov.ar/cte/index.html

IADO (2008) Programa de monitoreo de la calidad ambiental de la zona interior del estuario de Bahía Blanca. Informe Final, Instituto Argentino de Oceanografía, 107 pp. Available in: http://www.bahiablanca.gov.ar/cte/index.html

IADO (2009) Programa de monitoreo de la calidad ambiental de la zona interior del estuario de Bahía Blanca. Informe Final, Instituto Argentino de Oceanografía, 103 pp. Available in: http://www.bahiablanca.gov.ar/cte/index.html

IADO (2010) Programa de monitoreo de la calidad ambiental de la zona interior del estuario de Bahía Blanca. Adenda al Informe Final 2009, Instituto Argentino de Oceanografía, 92 pp. Available in: http://www.bahiablanca.gov.ar/cte/index.html

IADO (2012) Programa de monitoreo de la calidad ambiental de la zona interior del estuario de Bahía Blanca. Informe Final, Instituto Argentino de Oceanografía, 122 pp. Available in: http://www.bahiablanca.gov.ar/cte/index.html

IADO (2014) Programa de monitoreo de la calidad ambiental de la zona interior del estuario de Bahía Blanca. Informe Final, Instituto Argentino de Oceanografía, 242 pp. Available in: http://www.bahiablanca.gov.ar/cte/index.html

IADO (2016) Programa de monitoreo de la calidad ambiental de la zona interior del estuario de Bahía Blanca. Informe Final, Instituto Argentino de Oceanografía, 228 pp. Available in: http://www.bahiablanca.gov.ar/cte/index.html

IADO (2018) Programa de monitoreo de la calidad ambiental de la zona interior del estuario de Bahía Blanca. Informe Final, Instituto Argentino de Oceanografía, 364 pp. Available in: http://www.bahiablanca.gov.ar/cte/index.html

IADO (Instituto Argentino de Oceanografía, CONICET/UNS) (1997) Programa de monitoreo de la calidad ambiental de la zona interior del estuario de Bahía Blanca. Informe Final, Instituto Argentino de Oceanografía, 65 pp. Available in: http://www.bahiablanca.gov.ar/cte/index.html

INDEC (2010) Instituto Nacional de Estadística y Censos, Argentina. http://www.indec.gov.ar

IPCC (Intergovernmental Panel on Climate Change) (2018) Global warming of 1.5°C: an IPCC special report on the impacts of global warming of 1.5°C above pre-industrial levels and related global greenhouse gas emission pathways, in the context of strengthening the global response to the threat of climate change, sustainable development, and efforts to eradicate poverty. Masson-Delmotte V, Zhai P, Pörtner H-O et al (eds) (in press)

Iwata H, Tanabe S, Sakai N et al (1993) Distribution of persistent organochlorines in the oceanic air and surface seawater and the role of ocean on their global transport and fate. Environ Sci Technol 27:1080–1098. https://doi.org/10.1021/es00043a007

James NC, van Niekerk L, Whitfield AK, Potts WM, Götz A, Paterson AW (2013) Effects of climate change on South African estuaries and associated fish species. Clim Res 57:233–248. https://doi.org/10.3354/cr01178

Jezierska B, Ługowska K, Witeska M (2009) The effects of heavy metals on embryonic development of fish (a review). Fish Physiol Biochem 35:625–640. https://doi.org/10.1007/s10695-008-9284-4

Kalnejais LH, Martin WR, Bothner MH (2010) The release of dissolved nutrients and metals from coastal sediments due to resuspension. Mar Chem 121(1–4):224–235. https://doi.org/10.1016/j.marchem.2010.05.002

Kennish MJ (ed) (2019) Ecology of estuaries. Vol 2: Biological aspects. CRC Press – Taylor & Francis Publ.Co, Boca Raton. 276 pp. ISBN:0-8493-5892-2

Kent RD, Vikesland PJ (2016) Dissolution and persistence of copper-based nanomaterials in undersaturated solutions with respect to cupric solid phases. Environ Sci Technol 50:6772–6781. https://doi.org/10.1021/acs.est.5bo4719

Kwok KWH, Batley GE, Wenning RJ et al (2013) Sediment quality guidelines: challenges and opportunities for improving sediment management. Environ Sci Pollut R 21(1):17–27. https://doi.org/10.1007/s11356-013-1778-7

La Colla N, Franco M, Serra A et al (2014) Saltmarshes: an approach to organic matter contribution in sediments through isotopic signature. In: III Reunión Argentina de Geoquímica de la Superficie (III RAGSU), Mar del Plata (Pcia. de Buenos Aires, Argentina), 02-05/DIC/2014. ISBN:978-987-544-598-7

La Colla NS, Negrin VL, Marcovecchio JE et al (2015) Dissolved and particulate metals dynamics in a human impacted estuary from the SW Atlantic. Estuar Coast Shelf Sci 166:45–55. https://doi.org/10.1016/j.ecss.2015.05.009

La Colla NS, Botté SE, Negrin VL et al (2018a) Influence of human-induced pressures on dissolved and particulate Pb, Cr, Zn and Ni concentrations in the Bahía Blanca estuary, South America. Environ Monit Assess 190:532–547. https://doi.org/10.1007/s10661-018-6930-x

La Colla N, Botté SE, Marcovecchio JE (2018b) Metals in coastal zones impacted with urban and industrial wastes: insights on the metal accumulation pattern in fish species. J Mar Syst 181:53–62. https://doi.org/10.1016/j.jmarsys.2018.01.012

La Colla NS, Botté SE, Marcovecchio JE (2019) Mercury cycling and bioaccumulation in a changing coastal system: from water to aquatic organisms. Mar Pollut Bull 140:40–50. https://doi.org/10.1016/j.marpolbul.2018.12.051

Lailson-Brito J, Dorneles PR, Azevedo-Silva CE et al (2010) High organochlorine accumulation in blubber of Guiana dolphin, *Sotalia guianensis*, from Brazilian coast and its use to establish geographical differences among populations. Environ Pollut 158:1800–1808. https://doi.org/10.1016/j.envpol.2009.11.002

Limbozzi F, Leitào TE (2008) Chapter 31: Characterization of Bahía Blanca main existing pressures and their effects on the state indicators for surface and groundwater quality. In: Neves R, Baretta J, Mateus M (eds) Perspectives on integrated coastal zone management in South America. IST Scientific Publishers, Lisbon, pp 333–350. ISBN:978-972-8469-74-0

López Abbate MC, Molinero JC, Guinder VA et al (2015) Microplankton dynamics under heavy anthropogenic pressure. The case of the Bahía Blanca Estuary, southwestern Atlantic Ocean. Mar Pollut Bull 95(1):305–314. https://doi.org/10.1016/j.marpolbul.2015.03.026

Marcovecchio JE (1988) Estudio comparativo de la distribución de los metales traza mercurio, cadmio y zinc en organismos de dos zonas estuariales de Argentina: Bahía Blanca y desembocadura del Rio de la Plata. Doctoral thesis, Mar del Plata National University, 212 pp

Marcovecchio JE (2001) Impact of freshwater on coastal estuaries: South America Atlantic littoral. In: Buddemeier RW, Crossland C, Maxwell B et al (eds) *LOICZ/UNEP* Regional Shyntesis Programme: Australasia-Asia, The Americas, Africa-Europe. Summary Report and Compendium. LOICZ Reports & Studies N°22, 77 pp. ISSN:1383-4304

Marcovecchio JE, Ferrer LD (2005) Distribution and geochemical partitioning of heavy metals in sediments of the Bahía Blanca estuary, Argentina. J Coast Res 21(4):826–834. https://doi.org/10.2112/014-NIS.1

Marcovecchio JE, Freije RH (2004) Efectos de la intervención antrópica sobre sistemas marinos costeros: el estuario de Bahía Blanca, vol 56. Anales de la Academia Nacional de Ciencias Exactas, Físicas y Naturales (ANCEFN), Buenos Aires, pp 115–132. ISSN:0365-1185

Marcovecchio JE, Freije RH (eds) (2013) Procesos Químicos en Estuarios, Editorial de la Univ Tecnológica Nacional (EdUTecNe), Buenos Aires (Argentina). E-book. ISBN:978-987-1896-16-5

Marcovecchio JE, Moreno VJ, Pérez A (1986) Bio-magnification of total mercury in Bahia Blanca shark. Mar Pollut Bull 17(6):276–278. https://doi.org/10.1016/0025-326X(86)90064-0

Marcovecchio JE, Moreno VJ, Pérez A (1988a) The sole, *Paralichthys* sp. as an indicator species of heavy metal pollution in the Bahía Blanca estuary, Argentina. Sci Total Environ 75(2–3):191–200. https://doi.org/10.1016/0048-9697(88)90032-0

Marcovecchio JE, Moreno VJ, Pérez A (1988b) Determination of some heavy metal baselines in the biota of Bahía Blanca, Argentina. Sci Total Environ 75(2–3):181–190. https://doi.org/10.1016/0048-9697(88)90031-9

Marcovecchio JE, Moreno VJ, Pérez A (1988c) Total mercury contents in marine organisms of the Bahía Blanca estuary trophic web. In: Seeliger U, Lacerda LD, Patchineelam SR (eds) Metals in coastal environments of Latin America. Springer-Verlag, Heidelberg, pp 122–129

Marcovecchio JE, Andrade JS, Ferrer LD et al (2001) Mercury distribution in estuarine environments from Argentina: the detoxification and recovery of salt-marshes after 15 years. Wetland Ecol Manag 9(4):317–322. https://doi.org/10.1023/A:1011860618461
Marcovecchio JE, Botté SE, Delucchi F et al (2008) Chapter 28: Pollution processes in Bahía Blanca estuarine environment. In: Neves R, Baretta J, Mateus M (eds) Perspectives on Integrated Coastal Zone Management in South America. Part B: From shallow water to the deep fjord: the study sites. IST Scientific Publishers, Lisbon, pp 303–316. ISBN:978-972-8469-74-0
Marcovecchio JE, Freije RH, Popovich CA, Botté SE et al (2010a) Long-term observational system for oceanographic studies within Bahía Blanca estuary (Argentina): state of the art and perspectives. AQUASHIFT: Life in Warming Waters, Christian Albrecht's University, Kiel
Marcovecchio JE, Botté SE, Fernández Severini MD et al (2010b) Geochemical control of heavy metals concentrations and distribution within Bahía Blanca estuary (Argentina). Aquat Geochem 16(2):251–266. https://doi.org/10.1007/s10498-009-9076-1
Marcovecchio JE, Botté SE, Domini CE et al (2014) Chapter 15: Heavy metals, major metals, trace elements. In: Nollet L, de Geelder LSP (eds) Handbook of water analysis, 3rd edn. CRC Press, Taylor & Francis Group LCC, Boca Ratón, pp 385–433. 979 pp. ISBN:978-1-4398-8966-4
Marcovecchio JE, Botté SE, Fernández Severini MD (2016) Distribution and behavior of zinc in estuarine environments: an overview on Bahía Blanca estuary (Argentina). Environ Earth Sci 75:1168–1184. https://doi.org/10.1007/s12665-016-5942-5
Menzies R, Quinete NS, Gardinali P et al (2013) Baseline occurrence of organochlorine pesticides and other xenobiotics in the marine environment: Caribbean and Pacific collections. Mar Pollut Bull 70:289–295. https://doi.org/10.1016/j.marpolbul.2013.03.003
Moss B, Jeppesen E, Søndergaard M et al (2013) Nitrogen, macrophytes, shallow lakes and nutrient limitation: resolution of a current controversy? Hydrobiologia 710:3–21. https://doi.org/10.1007/s10750-012-1033-0
Namieśnik J, Rabajczyk A (2010) The speciation and physicochemical forms of metals in surface waters and sediments. Chem Spec Bioavailab 22(1):1–24. https://doi.org/10.3184/095422910X12632119406391
Nguyen KDT, Morley SA, Lai C-H et al (2011) Upper temperature limits of tropical marine ectotherms: global warming implications. PLoS One 6(12):e29340. https://doi.org/10.1371/journal.pone.0029340
Nguyen DH, Umeyama M, ShintaniT (2012) Importance of geometric characteristics for salinity distribution in convergent estuaries. J Hydrol 448–449:1–13. https://doi.org/10.1016/j.jhydrol.2011.10.044
Nixon SW, Oczkowski AJ, Pilson MEQ et al (2015) On the response of pH to inorganic nutrient enrichment in well-mixed coastal marine waters. Estuar Coast 38:232–241. https://doi.org/10.1007/s12237-014-9805-6
Nunes Vaz RA (2012) The salinity response of an inverse estuary to climate change and desalination. Estuar Coast Mar Sci 98:49–59. https://doi.org/10.1016/j.ecss.2011.11.023
Nunes Vaz RA, Lennon GW, Bowers DG (1990) Physical behaviour of a large, negative or inverse estuary. Cont Shelf Res 10(3):277–304. https://doi.org/10.1016/0278-4343(90)90023-F
Odum EP (2014) The strategy of ecosystem development. In: Ndubisi FO (ed) The ecological design and planning reader. Island Press, Washington, DC, pp 203–216. 625 pp. ISBN:978-1-61091-489-5
Olguín HF, Alder VA (2011) Species composition and biogeography of diatoms in Antarctic and Subantarctic (Argentine shelf) waters (37–76°S). Deep-Sea Res II 58:139–152. https://doi.org/10.1016/j.dsr2.2010.09.031
Oliva AL, Quintas PY, La Colla NS et al (2015) Distribution, sources, and potential ecotoxicological risk of polycyclic aromatic hydrocarbons in surface sediments from Bahía Blanca Estuary, Argentina. Arch Environ Contam Tox 69(2):163–172. https://doi.org/10.1007/s00244-015-0169-0
Orazi MM, Arias AH, Oliva AL et al (2020) Characterization of atmospheric and soil polycyclic aromatic hydrocarbons and evaluation of air-soil relationship in the Southwest

of Buenos Aires province (Argentina). Chemosphere 240:124847. https://doi.org/10.1016/j.chemosphere.2019.124847

Paparazzo FE, Williams GN, Pisoni JP et al (2017) Linking phytoplankton nitrogen uptake, macronutrients and chlorophyll-a in SW Atlantic waters: the case of the Gulf of San Jorge, Argentina. J Mar Syst 172:43–50. https://doi.org/10.1016/j.jmarsys.2017.02.007

Perillo GME, Piccolo MC, Parodi E et al (2001) Chapter 14: The Bahía Blanca Estuary, Argentina. In: Seeliger U, Kjerfve B (eds) Coastal Marine Ecosystems of Latin America, Ecological studies, vol 144. Springer, Heidelberg, pp 205–217. https://doi.org/10.1007/978-3-662-04482-7_15

Perillo GME, Pierini JO, Pérez DE, Piccolo MC (2005) Suspended sediment fluxes in the middle reach of the Bahia Blanca Estuary, Argentina. In: DM FG, Knight J (eds) High resolution morphodynamics and sedimentary evolution of estuaries. Springer, Heidelberg, pp 101–114. ISBN:13:978-1-4020-3296-7

Piccolo MC (2008) Chapter 23: Climatological features of the Bahía Blanca Estuary. In: Neves R, Baretta J, Mateus M (eds) Perspectives on integrated coastal zone management in South America. IST Scientific Publishers, Lisbon, pp 233–242. ISBN:978-972-8469-74-0

Piccolo MC, Perillo GME (1999) The Argentina estuaries: a review. In: GME P, Piccolo MC, Pino-Quivira M (eds) Estuaries of South America: their geomorphology and dynamics, vol 235. Springer, Heidelberg, pp 101–132. ISBN:13 978-3-642-64269-2

Piccolo MC, Perillo GME, Melo WD (2008) Chapter 22: The Bahía Blanca estuary: an integrated overview of its geomorphology and dynamics. In: Neves R, Baretta J, Mateus M (eds) Perspectives on integrated coastal zone management in South America. IST Scientific Publishers, Lisbon, pp 221–232. ISBN:978-972-8469-74-0

Popovich CA, Gayoso AM (1999) Effect of irradiance and temperature on the growth rate of *Thalassiosira curviseriata* Takano (Bacillariophyceae), a diatom bloom in Bahía Blanca estuary (Argentina). J Plankton Res 21(6):1101–1110. https://doi.org/10.1093/plankt/21.6.1101

Popovich CA, Marcovecchio JE (2008) Spatial variability of phytoplankton and environmental factors in a temperate estuary of South América (Atlantic Coast, Argentina). Cont Shelf Res 28:236–244. https://doi.org/10.1016/j.csr.2007.08.001

Popovich CA, Spetter CV, Marcovecchio JE et al (2008) Dissolved nutrients availability during winter diatom bloom in a turbid and shallow estuary (Bahía Blanca, Argentina). J Coast Res 24(1):95–102. https://doi.org/10.2112/06-0656.1

Ralston DK, Keafer BA, Brosnahan ML et al (2014) Temperature dependence of an estuarine harmful algal bloom: resolving interannual variability in bloom dynamics using a degree-day approach. Limnol Oceanogr 59(4):1112–1126. https://doi.org/10.4319/lo.2014.59.4.1112

Renella A, Quirós R (2000) Relevamiento del uso de pesticidas agrícolas en la alta cuenca del río Salado (Provincia de Buenos Aires). Proyecto UBA-Municipio de Junín (CS 2826/99). Área de Sistemas de Producción Acuática, Departamento de Producción Animal, Facultad de Agronomía, Universidad de Buenos Aires, 106 pp

Rivas AL, Dogliotti AI, Gagliardini DA (2006) Seasonal variability in satellite-measured surface chlorophyll in the Patagonian Shelf. Cont Shelf Res 26:703–720. https://doi.org/10.1016/j.csr.2006.01.013

Roach AC (2005) Assessment of metals in sediments from Lake Macquarie, New South Wales, Australia, using normalisation models and sediment quality guidelines. Mar Environ Res 59:453–472. https://doi.org/10.1016/j.marenvres.2004.07.002

Roche H, Vollaire Y, Persic A et al (2009) Organochlorines in the Vaccare's lagoon trophic web (Biosphere Reserve of Camargue, France). Environ Pollut 157:2493–2506. https://doi.org/10.1016/j.envpol.2009.03.016

Rochman CM (2015) Part II, Chapter 5: The complex mixture, fate and toxicity of chemicals associated with plastic debris in the marine environment. In: Bergmann M, Gutow L, Klages M (eds) Marine anthropogenic litter. Springer, Heidelberg, pp 117–140. 456 pp. ISBN:978-3-319-16509-7

Romero SI, Piola AR, Charo M et al (2006) Chlorophyll-a variability off Patagonia based on SeaWiFS data. J Geophys Res 111:C05021. https://doi.org/10.1029/2005JC003244

Ronda AC, Arias AH, Oliva AL et al (2019) Synthetic microfibers in marine sediments and surface seawater from the Argentinean continental shelf and a marine protected area. Mar Pollut Bull 149:110618. https://doi.org/10.1016/j.marpolbul.2019.110618

Roy S, Llewellyn C, Skarstad Egeland E et al (eds) (2011) Phytoplankton pigments: characterization, chemotaxonomy and applications in oceanography. Cambridge University Press, Cambridge. 892 pp. ISBN:978-1-107-00066-7

Sericano J, Pucci AE (1984) Chlorinated hydrocarbons in the seawater and surface sediments of Blanca Bay, Argentina. Estuar Coast Shelf Sci 19(1):27–51. https://doi.org/10.1016/0272-7714(84)90051-9

Sheaves M, Baker R, Nagelkerken I et al (2015) True value of estuarine and coastal nurseries for fish: incorporating complexity and dynamics. Estuar Coast 38:401–414. https://doi.org/10.1007/s12237-014-9846-x

Simonetti P, Botté SE, Marcovecchio JE (2017) Occurrence and spatial distribution of metals in intertidal sediments of a temperate estuarine system (Bahía Blanca, Argentina). Environ Earth Sci 76:636–647. https://doi.org/10.1007/s12665-017-6975-0

Sindermann CJ (ed) (2006) Coastal pollution – effects on living resources and humans. CRC Press, Taylor & Francis Group LCC, Boca Ratón. 309 pp. ISBN:10:0-8493-9677-8

Sokal RR, Rohlf FJ (eds) (1995) Biometry: the principles and practice of statistics in biological research. WH Freeman & Co, New York. 487 pp. ISBN:13:978-0716724117

Sunda WG, Cai W-J (2012) Eutrophication induced CO_2-acidification of subsurface coastal waters: interactive effects of temperature, salinity, and atmospheric pCO_2. Environ Sci Technol 46:10651–10659. https://doi.org/10.1021/es300626f

Sznaiberg L (2012) Parques Industriales: Luz verde para producir futuro. Revista Informe Industrial N° 233. http://www.informeindustrial.com.ar/

Telesh IV, Khlebovich VV (2010) Principal processes within the estuarine salinity gradient: a review. Mar Pollut Bull 61:149–155. https://doi.org/10.1016/j.ecss.2013.10.013

Telesh IV, Schubert H, Skarlato S (2013) Life in the salinity gradient: discovering mechanisms behind a new biodiversity pattern. Estuar Coast Mar Sci 135:317–327. https://doi.org/10.1016/j.ecss.2013.10.013

Tombesi N, Pozo K, Arias A et al (2018) Records of organochlorine pesticides in soils and sediments on the southwest of Buenos Aires Province, Argentina. Environ Earth Sci 77(11):403–414. https://doi.org/10.1007/s12665-018-7582-4

UNEP (2001) GEO América Latina y el Caribe. Perspectivas del medio ambiente. PNUMA, San José (Costa Rica), 144 pp

UNEP (United Nations Environment Programme) (2002) Regionally based assessment of persistent toxic substances – Eastern and Western South America. UNEP Regional Report. http://www.chem.unep.ch/pts

Villafañe VE, Valiñas MS, Cabrerizo MJ et al (2015) Physio-ecological responses of Patagonian coastal marine phytoplankton in a scenario of global change: role of acidification, nutrients and solar UVR. Mar Chem 177:411–420. https://doi.org/10.1016/j.marchem.2015.02.01

Wania F, Breivik K, Persson NJ et al (2006) CoZMo-POP 2 – a fugacity-based dynamic multi-compartmental mass balance model of the fate of persistent organic pollutants. Environ Model Softw 21:868–884. https://doi.org/10.1016/j.envsoft.2005.04.003

Watts MJ, Mitra S, Marriott AL et al (2017) Source, distribution and ecotoxicological assessment of multielements in superficial sediments of a tropical turbid estuarine environment: a multivariate approach. Mar Pollut Bull 115(1–2):130–140. https://doi.org/10.1016/j.marpolbul.2016.11.057

Wetz MS, Yoskowitz DW (2013) An "extreme" future for estuaries? Effects of extreme climatic events on estuarine water quality and ecology. Mar Pollut Bull 69:7–18. https://doi.org/10.1016/j.marpolbul.2013.01.020

Whitfield AK, Elliott M, Basset A et al (2012) Paradigms in estuarine ecology: a review of the Remane diagram with a suggested revised model for estuaries. Estuar Coast Shelf Sci 97:78–90. https://doi.org/10.1016/j.ecss.2011.11.026

Xu S, Chen Z, Li S et al (2011) Modeling trophic structure and energy flows in a coastal artificial ecosystem using mass-balance Ecopath model. Estuar Coast 34:351–363. https://doi.org/10.1007/s12237-010-9323-0

Yakushev EV, Newton A (2013) Chapter 1: Introduction: redox interfaces in marine waters. In: Yakushev EV (ed) Chemical structure of pelagic redox interfaces: observation and modelling. Springer, Heidelberg, pp 1–12. 292 pp. ISBN:978-3-642-32124-5

Yang Y-P, Li Y, Sun Z-H et al (2014) Suspended sediment load in the turbidity maximum zone at the Yangtze River Estuary: the trends and causes. J Geogr Sci 24(1):129–142. https://doi.org/10.1007/s11442-014-1077-3

Yosim AE, Fry RC (2015) Chapter 1: Systems biology in toxicology and environmental health. In: Fry RC (ed) Systems biology in toxicology and environmental health. Elsevier Sci. Publ, Amsterdam, pp 1–10. 273 pp. ISBN:978-0-12-801564-3

Chapter 5
Plankton Ecology and Biodiversity in the Bahía Blanca Estuary

Anabela A. Berasategui, M. Sofía Dutto, Celeste López-Abbate, and Valeria A. Guinder

5.1 Introduction

Estuaries support high plankton biomass that often exceeds those found over the adjacent continental shelves due to the input of organic carbon and nutrients from land. The species that comprise this high biomass vary according to the geographical location of the estuary and typically show minimum diversity at high latitudes, which gradually increases until maximum values at 15–20° and decreases at the equator. This productivity is one of the key reasons that has attracted humans to populate estuarine shorelines throughout history. In addition to nutrient and organic matter availability, estuaries provide a great number of habitat types for plankton communities mainly defined by physical gradients and geomorphology. Spatial gradients in estuaries pose unique short-term (tidal cycles, freshwater input) and long-term (water movement, chemical cycling, and physical structure) environmental variability that forces plankton to display wide adaptive responses in order to survive (Day et al. 2013).

Adaptive responses of planktonic communities in estuaries consist on several behavioral and physiological trade-offs to cope with environmental changes, mainly caused by freshwater input and the tide dynamic. In this sense, the intertidal region is subjected to extreme temperatures, water level fluctuations, drying conditions, and salinity changes (Elliott et al. 2015; Smyth and Elliott 2016). Salinity is largely the key physical factor that regulates the spatial distribution and structure of estuarine phytoplankton and zooplankton. In this context, estuarine species are classified according to their tolerance to salinity as *stenohaline* and *euryhaline*. The first one encompasses those species which can tolerate only a narrow range of salinity and are generally found near the mouth of the estuary. The *euryhaline* species are those that tolerate salinity fluctuations within a wider range and can thus penetrate further

---

A. A. Berasategui (✉) · M. S. Dutto · C. López-Abbate · V. A. Guinder
Instituto Argentino de Oceanografía (IADO-CONICET/UNS), Consejo Nacional de Investigaciones Científicas y Técnicas (CONICET), Bahía Blanca, Argentina

S. M. Fiori, P. D. Pratolongo (eds.), *The Bahía Blanca Estuary*,
https://doi.org/10.1007/978-3-030-66486-2_5

up the estuary. Protists are adapted to either low or high salinity values and display species-specific distribution patterns based on salinity gradients (Lancelot and Muylaert 2011). Similarly, most estuarine metazoan zooplankton are osmoconformers, and each taxon is associated with a specific range of salinity that is within their physiological tolerance. The pattern of faunal distribution along estuarine gradients usually displays the highest number of species in the more saline waters near the mouth, and hence richness tends to decrease with decreasing salinity toward the riverine area (Whitfield et al. 2012). In this sense, the planktonic species from the Bahía Blanca Estuary have been adapted for generations to high salinity values above those in adjacent open sea, which is an uncommon feature in estuaries. For instance, the average salinity values for the inner and middle zones of the Bahía Blanca Estuary are around 33, and display hypersalinity (values between 41 and 43) during hot and dry summers. High evaporation rates and restricted water circulation in the shallower inner reach can lead to an inverse salinity distribution along the estuary showing decreasing salinity from the head toward the mouth. Moreover, hypersalinity is a consequence of the low mean annual runoff of the rivers associated with the Bahía Blanca Estuary and the residual circulation pattern in the inner zone. This particularity implies that other environmental factors than salinity gradients come into play in the distribution of species, such as tides, wind, nutrient loading, and human disturbance.

In general, species diversity increases from highly disturbed habitats such as the Principal Channel of navigation, toward less impacted secondary channels at the mouth of the estuary. Tides and wind are the main energy source that contribute to concentrate higher plankton abundance toward the head of the Bahía Blanca Estuary during the ebb (Guinder et al. 2009; Menéndez et al. 2012; Chazarreta et al. 2015; López-Abbate et al. 2019b). The effect of the tidal cycle on short-term variations in the mesozooplankton community has been reported in this estuary by Menéndez et al. (2012) and Chazarreta et al. (2015). These authors reported that the abundance of key mesozooplankton species was greater near the bottom during most of the tidal cycles and also suggested a lateral movement of key copepod species to areas of decreased flushing, such as channel margins. These authors also pointed that this vertical abundance pattern was not observed at low tide, possibly because of the shallowness of the ecosystem. In fact, at ebb tide, when fast currents occur, higher abundances of zooplankton were observed near the bottom, which can be interpreted as a retention mechanism of organisms within the estuary.

In relation to the continental inputs to the Bahía Blanca Estuary, the largest freshwater and nutrient inputs are those provided by the sewage discharges of Bahía Blanca, Punta Alta, and Ingeniero White cities. There is a subtle balance between beneficial nutrient enrichment and overfertilization, which can seriously accelerate primary production, bacterial respiration, and eutrophication. Hence, if elevated nitrogenous and/or phosphorous loadings occur as a result of anthropic perturbations, the supply rate of silicate may usually become limiting, leading to changes in the phytoplankton community composition (Day et al. 2013). In this sense, the Bahía Blanca Estuary is usually nutrient-enriched showing a large natural stock of silicate and a high level of nitrogenous compounds, which are adequate to support phytoplankton demand of the system, dominated by diatoms (Popovich et al. 2008; Guinder et al. 2010). Nitrogen-rich effluents, however, produce a decrease on microzooplankton probably due to the toxic effect of ammonium (López-Abbate et al. 2019a).

Another factor that drives plankton distribution is the light availability, which plays a key role in the control of biomass-specific productivity in turbid and nutrient-rich coastal systems like the Bahía Blanca Estuary (Popovich et al. 2008; Guinder et al. 2009). The reduction in the underwater light availability affects primary production by phytoplankton, and in the inner shallow area of the estuary, light penetration depends upon the wind intensity and direction that erodes the margins of the tidal flats and saltmarshes and releases soft sediments into the water column (López-Abbate et al. 2019b). In agreement, the decrease of suspended particulate matter concentration in winter with a concomitant increase in the penetration of solar radiation seems to be one of the main drivers for the development of the phytoplankton winter bloom in the Bahía Blanca Estuary, ascribed as the main biomass event of the annual cycle (Guinder et al. 2009). When production is reduced by turbidity, highly nutritive phytoplankton may be replaced by detrital material in the plankton food web of the estuary (Diodato and Hoffmeyer 2008; Dutto et al. 2014).

It is worth noting that organic detritus plays a key role in this turbid ecosystem where saltmarshes are crucial contributors to the organic fraction of suspended particulate matter (Negrin et al. 2011, 2013). Phytoplankton, microzooplankton, and detritus have an important role in the natural diet of mesozooplankton. For example, experimental data revealed that key copepods consumers exert higher filtration rates on microzooplankton than on phytoplankton; however in late spring detritus is a greater contributor to their natural diet (Diodato and Hoffmeyer 2008). In fact, the seasonal fatty acid pattern of the mesozooplankton indicates different feeding strategies over the year, suggesting an active feeding mode on protistan plankton during summer and a more terrestrially derived compounds diet in winter (Dutto et al. 2014). High environmental tolerance along with feeding plasticity allows plankton communities to survive under the wide-ranging conditions of the Bahía Blanca Estuary. Growing human disturbance, however, is pushing plankton to their adaptive edge, and profound ecological responses are beginning to become evident.

5.2 Phytoplankton and Microzooplankton Biodiversity and Seasonality

The phytoplankton in the Bahía Blanca Estuary has been studied on a biweekly basis over more than 30 years (1978–2010), with a monthly or more sporadic sampling in the last decade (2011–2019). The high frequency monitoring allowed the characterization of the phenology and biodiversity of the phytoplankton blooms in relation with changing environmental conditions (Gayoso 1981, 1989, 1998; Popovich and Gayoso 1999; Popovich et al. 2008; Guinder et al. 2010, 2013, 2016).

Similarly, the microzooplankton community (phagotrophic protists with cell size between 20 and 200 μm, mainly tintinnids and oligotrichs) have been extensively studied by light microscopy in the estuary during discontinuous periods from 1986 to 2011 (Pettigrosso 2003; Pettigrosso and Popovich 2009; Barría de Cao et al. 2005, 2011; López-Abbate et al. 2015, 2019a, b). It is worth noting that collection and processing of samples have been consistent throughout the study period, including the identification and quantification of species >5 μm in cell size under light microscopy. This high resolution plankton survey has been performed by the same group of specialists, ensuring that changes in the community structure and/or composition are due to environmental dynamics and not attributable to analytical processes. Together with the sustained observations of protistan plankton, in situ sea surface temperature, salinity, chlorophyll *a* and dissolved inorganic nutrients (silicates, DIN, and phosphate) have been measured.

Overall, the long-term monitoring raised key aspects of the phytoplankton phenology and biodiversity in the Bahía Blanca Estuary: (1) dominance of diatoms all year-round, followed by dinoflagellates and nanoflagellates, (2) recurrence of a late winter-early spring bloom, regarded as the main yearly biomass event, and (3) occurrence of a summer bloom of lower magnitude and duration and with higher interannual variability. Dominance of diatoms is a common feature of eutrophic, turbid, and vertically mixed estuaries (e.g., Cloern and Dufford 2005; Guinder et al. 2009) and adjacent coastal areas (Garibotti et al. 2011; Guinder et al. 2018). Together with dinoflagellates, both phytoplankton groups are rich in long-chain essential fatty acids, what confer them high nutritional quality to support secondary production (Winder et al. 2017). Concerning microzooplankton, tintinnids dominate the biomass of phagotrophic protists during most of the annual cycle, but especially during summer, while oligotrichs do not show a clear recurrent pattern. The ability of oligotrichs to retain plastids from prey confers them with certain independence from environmental conditions. As a result, oligotrichs may take advantage of eventual favorable conditions, regardless of seasons and prey abundance. Overall, the total biomass of microzooplankton shows maximum values during summer, with the dominance of tintinnids and oligotrichs, and minimum during winter, when the community shifts toward the dominance of rotifers, phagotrophic dinoflagellates, and mixotrophic oligotrichs (Barría de Cao et al. 2011; Pettigrosso and Popovich 2009; Pettigrosso et al. 2016). This pattern discloses the seasonal coupling and the tight trophic interconnections between both phototrophic and phagotrophic protistan communities. Accordingly, microzooplankton are often the preferred prey of mesozooplankton, in spite of their lower concentration compared to that of phytoplankton, due to specific characteristics such as body size, swimming mode, and nutritional quality (Stoecker and Capuzzo 1990). In fact, tintinnids represent the main energy resource of the dominant copepod in the estuary, *Acartia tonsa* Dana, during its productive season (summer) (Diodato and Hoffmeyer 2008). The high carbon transfer efficiency from phytoplankton and microzooplankton to higher trophic levels in the Bahía Blanca Estuary highlights the key role of protistan plankton in producing harvestable fish and in providing foraging and nursery services for several permanent and migratory animal species (López Cazorla 2007; Hoffmeyer et al. 2009b; Marrari et al. 2013; Fiori et al. 2016).

5.2.1 *Observed Changes in Protistan Plankton*

The more than 40 years of data records revealed a decrease in the mean annual levels of chlorophyll *a*, attributed to a gradual shift in the seasonal patterns of phytoplankton biomass. The annual biomass cycle changed from unimodal to bimodal due to a decrease in the winter-early spring bloom and an intensification of the summer bloom (Winder and Cloern 2010; López-Abbate et al. 2017) (Fig. 5.1). These phenological changes have been driven by compound effects of multiple biotic and abiotic factors, i.e., warming, changes in wind patterns, increase in water turbidity, precipitation and nutrient inputs, invasion of species, and shifts in grazing pressure (Guinder et al. 2010, 2012, 2016; López-Abbate et al. 2017). The regimen shift in the plankton realm has become evident since the early 2000s, when the typical blooming species began to decrease and to be replaced, likely causing irreversible changes in the composition of the phytoplankton blooms (Fig. 5.1).

The onset of the late winter-early spring bloom is around June and commonly lasts until October. Chlorophyll *a* over the 1978–present period for the winter-early spring bloom reached maxima of 54 μg l^{-1} in 1980, 44 μg l^{-1} in 2002, and 25 μg l^{-1} in 2007, displaying a long-term decline at an yearly rate of 1% (López-Abbate et al. 2017). For more than two decades, until the early 2000s, the most abundant and diverse taxa of the winter-spring bloom were the diatom genera *Thalassiosira* and *Chaetoceros*, with *T. curviseriata* Takano as the key component of this biomass event. Over more than 20 years, *T. curviseriata* has been the dominant blooming diatom in winter-early spring, reaching up to 80–90% of the total phytoplankton abundance, representing almost monospecific winter blooms in some years, like in 1991–1993 when it reached maximal abundances of 2.8×10^6 and 12.7×10^6 cells l^{-1} (Popovich and Gayoso 1999). *T. curviseriata* was a key prey for the copepod *Eurytemora americana* Williams which was introduced in the estuary in the late 1980s via ballast waters (Hoffmeyer 2004; Berasategui et al. 2009; Guinder et al. 2016). *T. curviseriata* has decreased toward recent years, with maxima not surpassing the 1.4×10^5 cells l^{-1} in the late 2000s, and not registered in the samples in the last winter-spring periods. Other co-dominant blooming *Thalassiosira* species were *T. anguste-lineata* Fryxell and Hasle, *T. pacifica* Gran and Angst, *T. rotula* Meunier, *T. eccentrica* Ehrenberg Cleve, *T. hibernalis* Gayoso, *T. hendeyi* Hasle and Fryxell, and *T. minima* Gaarder. Among the most frequent *Chaetoceros* species were *C. debilis* Cleve, *C. diadema* (Ehrenberg) Gran, *C. similis* Cleve, and a small *Chaetoceros* sp. (ca. 8 μm in diameter) characterized by delicate setae. Other frequent species during the winter bloom but also commonly present all-year round were the diatoms *Skeletonema costatum* (Greville) Cleve, *Asterionellopsis glacialis* (Castracane) Round, *Ditylum brightwellii* (West) Grunow, *Leptocylindrus minimus* Gran, *Guinardia delicatula* (Cleve) Hasle, *Cerataulina pelagica* (Cleve) Hendey,

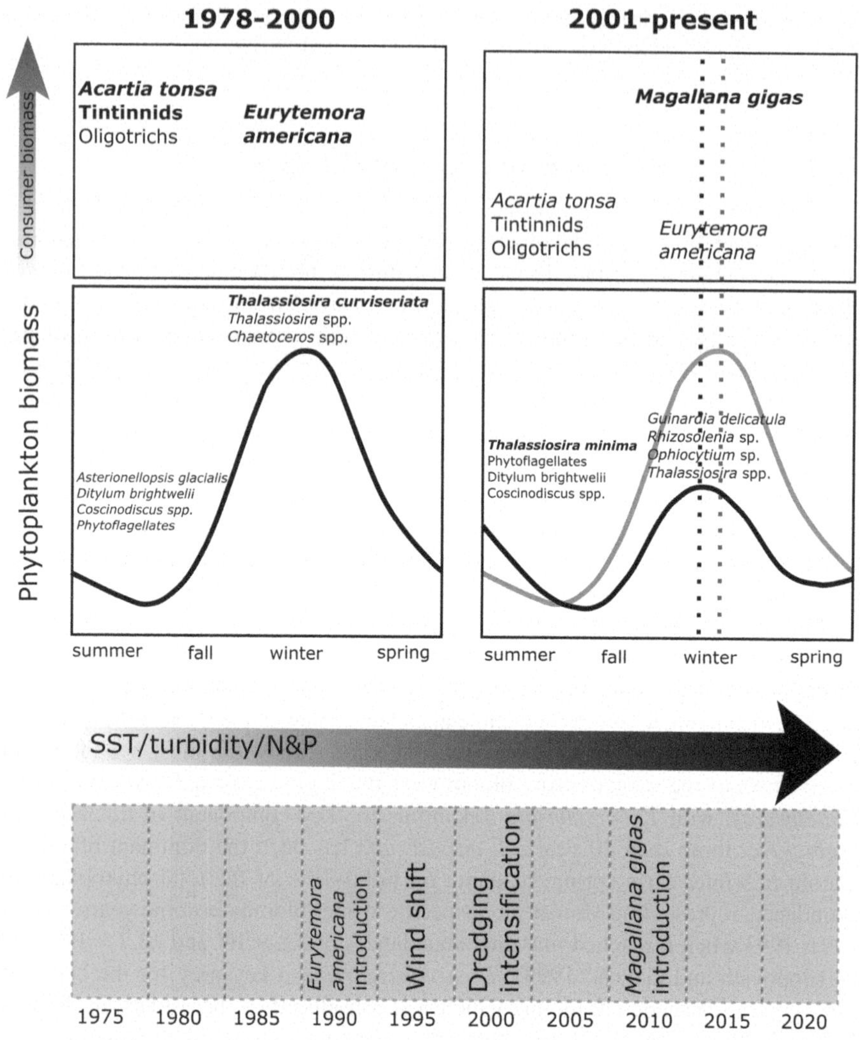

Fig. 5.1 Schematic representation of the regimen shift in phytoplankton in relation to the multiple interactive effects of environmental and biological drivers. During the time series (1978–present) a positive trend of SST, turbidity, nitrogen, and phosphorus was recorded, along with major environmental impacts which are enumerated in the lower gray bar. As a consequence, the annual cycle of phytoplankton biomass changed from unimodal (winter bloom) to bimodal (winter and summer blooms), while in recent years the winter bloom occurred ca. 4 weeks earlier (dashed lines). Dominant blooming species also shifted in recent years, denoting profound community-level reorganization. Parallel to phytoplankton changes, a shift in the abundance and composition of consumer's communities was also observed and is illustrated in the upper panels

Thalassiosira hendeyi and *Paralia sulcata* (Ehrenberg) Cleve. These species display seasonal and interannual variability and tend to bloom during summertime (December–February). Small phytoflagellates (10–20 μm) are present along the year with maxima during summer. The dinoflagellates *Scrippsiella trochoidea* (Stein) Loeblich III are important in late-spring and early summer, as well as the heterotrophic dinoflagellates of the genera *Protoperidinium* Bergh. During summer, unidentified species of gymnodinians and Cryptophyceae are conspicuous. The Xantophyceae *Ophiocytium* sp. is an important component of the phytoplankton in the Bahía Blanca Estuary, increasing its abundance in the last decade, together with the large diatom *Rhizosolenia* sp. in the spring season. In particular, the abundance of the small diatom *Thalassiosira minima* (5–15 μm) has risen over the last summers, likely related to the resuspension of bottom sediments driven by winds and dredging operations in the estuary, and the consequent release and germination of resting cysts of this planktonic species in the water column (Guinder et al. 2012).

Tintinnids in the Bahía Blanca Estuary are broadly dominated by species with agglutinated lorica, consistent with the occurrence of this group within neritic areas (Dolan 2012). *Tintinnidium balechi* Barría de Cao represents the dominant species during most of the annual cycle, followed by several species from the genus *Tintinnopsis* (*T. brasiliensis* Kofoid and Campbell, *T. gracilis* Kofoid and Campbell, *T. parva* Merkle, *T. baltica* Brandt, *T. beroidea* Stein, *T. levigata* Kofoid and Campbell) and *Codonellopsis lusitanica* Jörgensen (Barría de Cao et al. 2005; López-Abbate et al. 2019b). While tintinnids were in most cases identified to the species level, oligotrichs were in some cases identified to the genus level and in other cases were counted as a whole given that species identification requires complex staining and in vivo observation. In spite of these limitations, several species have been identified as recurrent within the estuary, such as *Strombidium capitatum* (Leegaard) Kahl, *S. emergens* (Leegaard) Kahl, *S. acutum* Leegaard, *S. dalum* Lynn, Montagnes and Small, *Strombidinopsis elongata* Song and Bradbury, *Strobilidium epacrum* Lynn and Montagnes, *Lohmanniella oviformis* Leegaard, *Laboea strobila* Lohmann, and *Tontonia appendiculariformis* Fauré-Fremiet.

In the last decades (1986–2011), tintinnids have revealed an interannual trend characterized by a yearly rate of decline of 2.8% (López-Abbate et al. 2019b). In the same way, oligotrichs declined at a yearly rate of 1.6% in the period 1994–2011, but with no significant trend. Although no clear species replacement was evident in both tintinnids and oligotrichs, observations suggest a trend toward the dominance of tintinnid species with wider oral diameter. Oral diameter is proportional to prey encounter rate, denoting a competitive advantage against species with narrower oral area constrained to a smaller prey size spectrum. These profound changes have been ascribed to the multiple effects of climate and human activities and represent the tipping point of long-term impacts on mesozooplankton and planktivorous fish dynamics.

5.2.2 Environmental Drivers of Long-Term Changes in Phytoplankton and Microzooplankton

The Bahía Blanca Estuary is characterized by the intense interaction between land and sea given that half of the estuarine area consists of extensive low-slope tidal flats densely fragmented by tidal courses (Perillo 2009). The bottom of the estuarine basin is composed by a massive deposit of fine sediments which relates to an ancient mouth of a river that was partially buried with marine and tidal plain sediments during the Holocene marine transgression (Pratolongo et al. 2017). The grain size in the lower mudflats influenced by tides ranges between very coarse to fine, with a modal value of 32 μm. The continuous influence of the mesotidal regime and the effect of high energy wind waves promote the mobilization of the softer sediments from the bottom toward the water column, generating high values of suspended sediments (>50 mg l^{-1}) within a size range between 1 and 50 μm (mode of 10 μm) (Cuadrado et al. 2005; Guinder et al. 2015; Zapperi et al. 2017).

Water turbidity is an important factor that determines the ecology of suspended populations in estuaries. Suspended sediments interfere with phytoplankton light harvesting by reducing light penetration and potentially excluding light-limited species. In addition, a high concentration of inedible suspended particles within the prey size range of phagotrophic protists complicates the prey uptake ultimately reducing population growth (Boenigk and Novarino 2004). In spite of the limitations posed by turbidity, plankton species in the Bahía Blanca Estuary have developed a series of adaptations to prosper under turbid conditions. For instance, *T. curviseriata*, the keystone species dominating the winter phytoplankton bloom until the early 2000s, and *T. minima* dominating in recent summers are well adapted to thrive light under low intensities (Popovich and Gayoso 1999; Guinder et al. 2012). Similarly, phagotrophic protists, broadly dominated by tintinnids with agglutinated lorica, on one hand take advantage on high sediments availability to build their lorica and get some protection against predation and on the other hand, avoid mechanical cell damage produced by suspended minerals (Dolan 2012).

The annual variability of water turbidity is mainly affected by precipitation which enhances sediment runoff of continental origin and also by the biological disturbance of tidal flats. During summer, turbidity reaches maximum values due to the intense activity of crabs that excavate large burrows and remove up to 5 kg m^{-2} d^{-1} of sediments (Zapperi et al. 2017). In contrast, water turbidity reaches minimum values in winter when benthic communities are less active and precipitation is low (Guinder et al. 2009). In the last three decades, however, water turbidity has revealed a significant trend beyond natural variability in the inner area of the estuary. This trend has been attributed to several factors. One of the primary factors is related to the shift of wind patterns toward low intensity but highly persistent NW winds in the last 25 years (López-Abbate et al. 2017). NW winds run parallel to the estuarine main channels and produce high energy wind waves that enhance the erosion of the extensive tidal flats (Perillo and Sequeira 1989) and produce the mobilization of soft sediments toward the water column. Shift on wind patterns has been documented in the region as a result of the southward displacement of the South Atlantic subtropical anticyclone and the persistent anomalies of the Southern Annular Mode (Dragani et al. 2010). Dredging is an additional factor that produces shear stress and destabilizes bottom sediments. Commercial harbor activities in the inner estuary have grown gradually in the last decades, which implied the intensification of sediment removal to allow the navigation of large ships. Deepening and straightening of waterway began in the late 1950s, while an important deepening occurred in 1999–2000 and produced an 11-fold increase of monthly sediment extraction to maintain navigation conditions (López-Abbate et al. 2017). Shear stress of tidal flats is also produced by the present rate of sea level rise (Lanfredi et al. 1988). Lateral erosion of estuarine margins produced a 33% loss of saltmarsh area since 1967 (Pratolongo et al. 2013). Marsh boundary erosion likely stimulated the continuous transport of mud deposits to the adjacent water column and contributed with the increase on water turbidity.

The long-term trend of water turbidity implied important consequences on planktonic communities in the Bahía Blanca Estuary. Although estuarine plankton is well adapted to turbid conditions, natural populations follow a typical response curve with a threshold value beyond their tolerance range. When this threshold is exceeded and individuals are given no enough time to adapt, they can be locally excluded (Boyd and Brown 2015). Indeed, the concentration of chlorophyll *a* (as a proxy of phytoplankton), oligotrichs, tintinnids, and the copepod *A. tonsa* experienced a significant decrease over the last 30 years (López-Abbate et al. 2019b). Phytoplankton and tintinnids were more vulnerable to the increase on water turbidity. Long-term patterns of phytoplankton showed a drastic erosion of seasonal peaks and a shift from the typical unimodal productivity pattern, i.e., the dominance of a winter bloom, toward a bimodal regime with moderate productivity events in summer and winter (Fig. 5.1). The weakening of the winter bloom produced a loss of synchrony between tintinnids and primary productivity. In addition to the indirect effect of turbidity, tintinnids were directly affected by the higher concentration of suspended sediments since the size range of such sediments overlaps precisely with the prey size spectrum of most tintinnids in the Bahía Blanca Estuary. This suggests that the decreasing ratio between sediments and phytoplankton forced tintinnids to an additional energy expenditure to sort inedible particles. Oligotrichs, with their ability to act both as phototrophs and phagotrophs (i.e., mixotrophs), were less sensitive to water turbidity probably as a result of the exclusion of light-limited phytoplankton and filter feeders (López-Abbate et al. 2019b). The negative trend of tintinnids impacted on the populations of their main predator, *Acartia tonsa*. This copepod dominates mesozooplankton during most of the annual cycle and reaches maximum abundances in summer. During this period, adults prey preferentially upon tintinnids, which represents nearly 78% of total particle's filtration (Diodato and Hoffmeyer 2008). Although many other factors may have contributed to the long-term trend of *A. tonsa* (e.g., competitive exclusion due to the introduction of invasive species, vulnerability to heavy metal, and nutrient pollution), sustained observations revealed that the drop on tintinnid's concentration produced a negative impact on the numerical abundance of this species (López-Abbate et al. 2019b).

The long-term decline of chlorophyll *a* was opposed to the positive trend documented in the continental shelf (Marrari et al. 2017). In the Bahía Blanca Estuary, the trend was attributed to an ensemble of global and local-scale factors with a time-varying dominance that corresponded to the changing estuarine conditions (Guinder et al. 2010; López-Abbate et al. 2017). At the beginning of the time series (1978–1993), phytoplankton was driven by nutrients concentration and water temperature. However, after 1993, the environmental influence shifted toward the dominance of turbidity and wind velocity, whereas climate signals (SAM and ENSO) boosted indirect effects through its influence on precipitation, wind, water temperature, and turbidity. In recent years, the notable dominance of larvae of *Magallana gigas* Thunberg in the plankton of the estuary has sharply contributed to the phytoplankton clearance, related to the fast growth and voracious feeding of this invasive species (Fiori et al. 2016). This oyster was introduced in the Argentine coast (40°S) in 1981 for aquaculture, and documented for the first time in the Bahía Blanca Estuary in 2010 (Dos Santos and Fiori 2010).

Long-term changes in the Bahía Blanca Estuary denote the prominent role of wind patterns on water turbidity within estuaries with internal sources of sediments. Hence, wind modifications driven by climate change, along with other erosive processes such as sea level rise, have the potential to profoundly impact the growth, phenology, and synchronization of estuarine plankton. Moreover, the introduction and expansion of invasive species, a growing threat of global warming, lead to significant impacts on the ecosystem structure, functioning, and biodiversity. Under the future scenario of climate change and urban development, natural plankton communities will thereby face unprecedented vulnerability, which urges us to develop effective tools to mitigate ecological impacts.

Among the most widespread effects of human settlements in coastal areas is the shift on nutrient balance. The Bahía Blanca Estuary receives excess nitrogen and phosphorus from point-source sewage effluents. On average, the effluents transport four times the concentration of nitrogen and three times the concentration of phosphorus in estuarine waters. In particular, the concentration of ammonium (NH^{4+}) within the sewage plume (ca. 1700 m) frequently exceeds the tolerance threshold of plankton (ca. 100 μM). Sustained observations revealed that when exposed to severe eutrophication, both phytoplankton and microzooplankton are negatively affected by a significant reduction on population biomass (López-Abbate et al. 2015). This reduction on plankton biomass may be partly explained by the dilution of suspended populations by the continuous inflow of sewage. In fact, sewage effluent contributes with significant amounts of freshwater into the estuary, which can represent up to 23% of the total freshwater input. In spite of the direct effect on plankton biomass, excess NH^{4+} also inhibits the growth of phytoplankton and the uptake of NO^{3-} (Glibert et al. 2014). Although nutrient loading does not produced the expected stimulation on phytoplankton growth, growth inhibition does not occur in response to excess NH^{4+} in the Bahía Blanca Estuary (López-Abbate et al. 2016; López-Abbate et al. 2019a). Instead, nutrient addition promoted a shift on species composition by the increased dominance of nanoflagellates. Severe eutrophication reduced the grazing rate of microzooplankton on nanoplankton and of nanoflagellates on bacteria (López-Abbate et al. 2019a). The lower reactivity potential to control newly produced biomass under severe eutrophication restricted the ability of microzooplankton to control phytoplankton blooms (López-Abbate et al. 2016). From a biogeochemical perspective, an inefficient use of primary producer's biomass decreases the overall carbon transfer across the food web (Schmoker et al. 2016) and produces a positive feedback with sewage effluents by an extra organic matter accumulation. Environmental quality degradation associated with the dumping of urban effluents can thereby undermine the compensatory capacity of microbial communities and reduce their potential to offset nutrient imbalance (Box 5.1).

Box 5.1: Parasitic Outbreak Alleviates Competition and Temporarily Benefits Thalassiosira curviseriata Population

Changes at the phytoplankton community level emerge from multiple interactive effects of biotic and abiotic stressors at the population level. The species-specific responses to compound environmental drivers are non-linear and depend on individual life history and interspecific connections (Litchman and Klausmeier 2008). It is well known that the high functional diversity of microbial plankton confers plasticity to their response against natural variability (Worden et al. 2015). Still, sustained impacts such as warming and coastal eutrophication gradually alter the phytoplankton yield (López-Abate et al. 2017). Likewise, sporadic impacts such as extreme weather events and parasitic outbreaks drive significant shifts in the community structure and species composition (Guinder et al. 2017). Eventually, these changes at the base of the pelagic food web can trigger substantial alterations in the ecosystem functioning.

In the Bahía Blanca Estuary, the long-term phytoplankton monitoring allowed the identification of the key blooming species *Thalassiosira curviseriata* Takano as a case study to assess the compound effects of multiple environmental drivers on the phenology of blooms. *T. curviseriata* is a small centric diatom (6–21 μm in diameter) arranged in curved chains of several cells. This species is eurythermal (5–20 °C) and euryhaline (20–40) and well adapted to growth at relatively low underwater light intensity (Guinder et al. 2009). Specimens isolated from the turbid inner zone of the estuary showed inhibited growth at ~150 μmol m^{-2} s^{-1} (Popovich and Gayoso 1999). As described in previous sections, *T. curviseriata* has been the dominant blooming diatom in winter-early spring in the Bahía Blanca Estuary for more than 20 years (1978–2003), but eventually its abundance has notably decreased over the last years along with changes in its phenology (Guinder et al. 2016).

The winter phytoplankton bloom portrays the main biomass event in the annual cycle, commonly composed by the same cluster of species from the genera *Thalassiosira* and *Chaetoceros*. Nevertheless, in the last decade the winter bloom has exhibited a species turnover toward the co-occurrence of diatoms in detriment of the formerly dominant *T. curviseriata*. The abundance decline of *T. curviseriata* was responsible for the conspicuous biomass drop of the phytoplankton bloom in recent years, concurrently with an earlier peak of ca. 4 weeks (Guinder et al. 2010; López-Abbate et al. 2017). The relevance of this species within the phytoplankton community allowed to compare dominant environmental factors acting at community level (indirectly estimated through chlorophyll *a*) versus the species level. The main drivers of the species' decreasing trend were identified from the characterization of its ecological niche and from structural equation modelling (Fig. 5.2): precipitation, water temperature, salinity, turbidity, dissolved inorganic nutrients, and grazing pressure exerted by the invasive copepod *Eurytemora americana* Williams (Guinder et al. 2016). The multiple interplay of these drivers has direct and indirect effects on its population, affecting the competition for resources and the predator-prey interactions. The rising temperature and drier conditions in the region over the former years (1990–2008) permeated the pelagic environment leading to warmer and saltier conditions in recent winters, which together with changes in nutrient ratios (i.e., rise in nitrite, nitrate, and phosphate) (Guinder et al. 2010; López-Abbate et al. 2015, 2016) impaired the growth of *T. curviseriata*. Moreover, the increase in the population size of the copepod *E. americana* since its introduction in the estuary in the late 1980s (Hoffmeyer 2004; Berasategui et al. 2009, 2013) has caused a temporal niche differentiation of these two species. The small diatom has shifted its seasonal distribution toward higher annual ranges of temperature and salinity to avoid seasonal overlapping with its consumer (Guinder et al. 2016).

Interestingly, in winter 2012 during routinary field research, a severe and ephemeral parasitic infection on co-dominant blooming diatoms was observed, concurrent with the re-emergence of *T. curviseriata* in the plankton (Guinder et al. 2017). The outbreak was caused by the parasitic nanoflagellate *Pirsonia* sp., with a high host-specificity on *Thalassiosira pacifica* and *Chaetoceros diadema* (more than 40% of the cells were infected). The infection resulted in a severe reduction of their populations and allowed the development of the fast-growing opportunistic species *T. curviseriata*, able to exploit open niches. The changes in species composition comprised a restructuring of the community by cell sizes and shapes (Guinder et al. 2017). The ephemeral parasitic outbreak took place during a period of extreme precipitation in the area (fourfold higher than in the same period in the previous decade 2000–2011) that dropped the salinity from 36.5 to 30.0 in 2 weeks. The abrupt change in salinity was likely responsible of the observed osmotic collapse of blooming diatoms, manifested by the breakout of cell's cytoplasm. Eventually, these extreme changes might overcome the physiological plasticity of phytoplankton cells and affect their osmotic regulation. In addition, the sudden extreme rains might result in unbalanced nutrient ratios and increased turbidity. These changes in the pelagic environment and the massive parasitism on blooming diatoms seemed to release species competition and allowed the development of the confined population of *T. curviseriata*, as a compensatory response to the severe reduction in the bloom biomass, by more than 80% (Guinder et al. 2017).

Commonly, the ephemeral nature of parasitic episodes and the rapidly re-establishment of the phytoplankton community after the infection make their detection difficult. Nevertheless, the high frequency, species-specific survey in the Bahía Blanca Estuary allowed the discrimination of changes in the community structure due to environmental stress from natural succession of phytoplankton species. Protists' parasitism appears to be a prominent vector of phytoplankton bloom collapse and species turnover, besides compound effects of hydroclimatic drivers, nutrients, and grazers. Other studies have documented that mass host mortalities caused by specific parasitism have enhanced species diversity and trophic chain complexity (Salomon et al. 2009; Dunne et al. 2013; Gsell et al. 2013). Overall, these findings highlight the importance of addressing the proximal ecology of the dominant species to understand structural modifications in the microbial plankton and their repercussion in higher trophic levels.

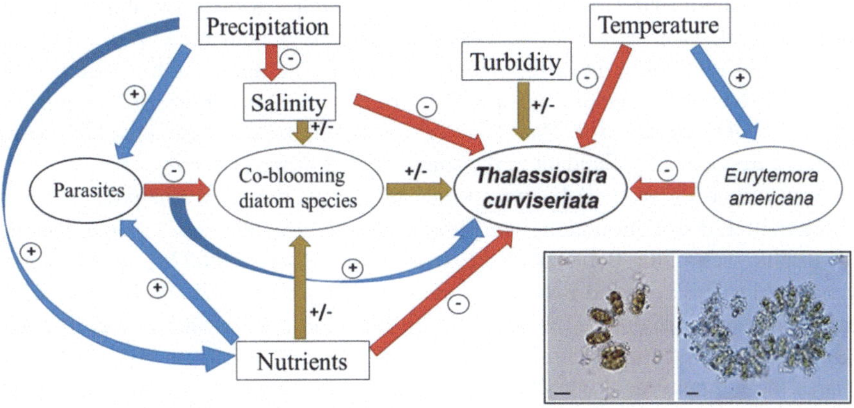

Fig. 5.2 Main biotic and abiotic drivers and their compound direct and/or indirect effects on the abundance and phenology of the diatom *Thalassiosira curviseriata* assessed in the Bahía Blanca Estuary, based on high frequency, long-term field observations. Scale bar in the photos of *T. curviseriata* chains: 10 μm

5.3 An Overview of Metazoan Zooplankton Community from the Inner Zone of the Bahía Blanca Estuary

Zooplankton in aquatic ecosystems are the functional link between primary producers and higher trophic levels, mainly the size fraction between 0.2 and 20 mm (mesozooplankton) (Turner 2004; Sieburth et al. 1978). These organisms rapidly respond to changes in the environment and, therefore, are valuable bioindicators of ecosystem changes (Silva et al. 2004; Chang et al. 2009; Uriarte and Villate 2004).

Zooplankton composition and dynamics in the Bahía Blanca Estuary have been studied for more than four decades (reviewed by Hoffmeyer 2007; Hoffmeyer and Mianzan 2007; Hoffmeyer and Cervellini 2007; Berasategui et al. 2019). Neritic zooplankton is well adapted to large seasonal variability in temperature, salinity, turbidity, and nutrient inputs. It is known that there is a positive response of the mesozooplankton abundance from the Bahía Blanca Estuary to salinity, which indicates that this community appears to be well acclimated to the commonly high salinity values in the estuary (average annual salinity around 33, with maximum values of 43) (Berasategui et al. 2016; Berasategui et al. 2019). The study of Guerrero et al. (1976) was pioneer on mesozooplankton from the inner zone of the Bahía Blanca Estuary, and reported a maximal abundance of 8.10^5 ind. m^{-3}. Subsequently, the meso- and macrozooplankton (>20 mm, according to the classification by Sieburth et al. 1978) were extensively studied by Hoffmeyer and colleagues (Piccolo and Hoffmeyer 2007), where high seasonal variability in the abundance of the meso- and macrozooplankton was documented. These authors recorded 34 taxa in the mesozooplankton fraction. The holoplankton was mostly represented by the calanoid copepods *Acartia tonsa* Dana, *Paracalanus parvus* Claus, *Labidocera fluviatilis* Dahl F, and *Calanoides carinatus* Krøyer, which were the most frequent and abundant in the study area (Hoffmeyer 2007).

Acartia tonsa is the best-known species among small marine copepods. It has a cosmopolitan distribution along temperate coastal areas and plays a pivotal role in the trophic web in the Bahía Blanca Estuary (Lopez Cazorla 2007; Lopez Cazorla et al. 2011). This copepod is found in the water column throughout the year, with maximal abundance during the summer season and minimal abundance during the winter season (Sabatini 1989; Hoffmeyer 1994, 2004; Berasategui et al. 2016). It shares the temporal ecological niche with the exotic copepod *Eurytemora americana* Williams, during the austral winter-spring (Hoffmeyer 2004; Hoffmeyer et al. 2009a).

Within the holoplankton, Hoffmeyer (2007) also reported the presence of harpacticoid copepods (i.e., *Euterpina acutifrons* Dana, *Tisbe varians* Scott T., Robertsonia *propinqua* Scott T., and *Heterolaophonte* sp. Lang). This group is dominated by species adapted to live in close association with bottom layers; however, intense resuspension processes lead to a constant transfer of these organisms from the bottom toward the water column. Several species from gelatinous zooplankton (i.e., *Mnemiopsis leidyi* Mayer, *Turritopsis nutricula* McCrady, and *Obelia* sp. Péron and Lesueur) and the mysids *Neomysis americana* Smith are also common inhabitants within estuarine waters, especially during the warm season (Hoffmeyer and Minazan 2007). Regarding meroplankton, the presence of cirripedia, mollusca, bryozoa, polichaeta, ascidiacea, cnidaria, and decapods larvae were reported, being the decapoda larvae the most abundant group during the warm season with 31 reported taxa (Hoffmeyer and Cervellini 2007). The highest abundances in this group were attributed to the zoeas of the crab *Neohelice granulata* Dana and shrimp larvae of *Peisos petrunkevitchi* Burkenroad and *Artemesia longinaris* Spence Bate.

Subsequent surveys during the years 2009–2010 showed a restructuring of the mesozooplankton community toward the dominance of typical estuarine species and a decrease in total mesozooplankton abundance, in comparison to those recorded by Hoffmeyer (2007) during the years 1990–1991 (Berasategui et al. 2019). The mesozooplankton community in 2009–2010 was composed mainly of *A. tonsa* (annual mean 695.48 ind m^{-3}), *N. granulata* (annual mean of 81.97 ind m^{-3}), *B. glandula* (annual mean of 28.89 ind m^{-3}), *A. amphitrite* (annual mean of 13.35 ind m^{-3}), *P. parvus* (annual mean of 9.63 ind m^{-3}), *E. americana* (annual mean of 3.91 ind m^{-3}), *E. acutifrons* (annual mean of 3.62 ind m^{-3}), and *Spionidae* spp. (annual mean of 1.15 ind m^{-3}). Berasategui et al. (2019) also documented a decrease in the abundance of copepod species typical from the adjacent shelf area, such as *L. fluviatilis* and *C. carinatus* (Boltovskoy 1981; Bradford Grieve et al. 1999), and a shift in the distribution of decapod larvae of *Pachycheles laevidactylus* Ortmann 1892, *Alpheus puapeba* Christoffersen 1979, and *Cyrtograpsus altimanus* Rathbun 1914, probably related to the joint effects of dredging activities in the main channel and the tidal dynamics of the ecosystem.

In more recent years (2014–2018), mesozooplankton denoted similar annual dynamics (Table 5.1) to that reported by Hoffmeyer (2007) and Berasategui et al. (2019) in previous years. Overall, mesozooplankton showed the greatest abundances during the warm months which ranged between 11,704.67 and 557.84 ind. m^{-3}. During the warm season *A. tonsa* and the decapod zoeas of Grapsidae dominated (Fig. 5.3a), whereas in winter, a decrease in the abundance of *E. americana* (Fig. 5.3b) was notorious in relation to population densities reported in previous years (Berasategui et al. 2009, 2012, 2019). These changes in the mesozooplankton community in the estuary are closely interconnected with the long-term modifications in the microbial plankton phenology, and composition, mentioned in Sects. 5.1 and 5.2 (Fig. 5.1).

The inner zone of the Bahía Blanca Estuary is highly susceptible to biological invasions due to a poor seawater exchange and its exposure to intense maritime traffic and periodic dredging. These characteristics have probably favored the effective introduction (via ballast water) and dispersion of exotic species already mentioned by Hoffmeyer et al. (2004), such as *B. glandula* and *E. americana*, and more recently, the oyster *Magallana gigas* (Fig. 5.3d) Thunberg (Dos Santos and Fiori 2010; Chazarreta et al. 2015) as well as the fiddler crab *Leptuca* (=*Uca*) *uruguayensis* Nobili (Truchet et al. 2019). Indeed the planktonic larvae of both species are increasingly frequent in zooplankton samples.

Regarding gelatinous fauna, the Bahía Blanca Estuary seems to encompass suitable environmental conditions for the development of this kind of organisms (Dutto

Table 5.1 Zooplankton taxa in the Bahía Blanca Estuary

Functional groups	Species	Abund.
Hydrozoa	*Eucheilota ventricularis* (McCrady, 1859)	*
Hydrozoa	*Gossea brachymera* (Bigelow, 1909)	*
Hydrozoa	*Liriope tetraphylla* (Chamisso & Eysenhardt, 1821)	**
Hydrozoa	*Turritopsis nutricula* (McCrady, 1857)	*
Hydrozoa	*Obelia* spp. (Péron & Lesueur, 1810)	**
Hydrozoa	*Olindias sambaquiensis* Müller (1861)	**
Hydrozoa	*Actinula (L)*	*
Scyphozoa	*Chrysaora lactea* (Eschscholtz, 1829)	*
Ctenophora	*Mnemiopsis leidyi* (A. Agassiz, 1865)	***
Ctenophora	*Beroe ovata* (Bruguière, 1789)	*
Ctenophora	*Pleurobrachia pileus* (O. F. Müller, 1776)	*
Mollusca	*Gastropoda and Bilvalvia undet.(LV)*	*
Mollusca	*Crassostrea gigas* (Thunberg, 1793)	***
Polychaeta	Spionidae (Grube, 1850)	**
Polychaeta	Syllidae (Grube, 1850)	**
Polychaeta	Polynoidae (Kinberg, 1856)	**
Polychaeta	Onuphidae (Kinberg, 1865)	**
Polychaeta	Aphroditidae (Malmgren, 1867)	*
Chaetognatha	*Parasagitta friderici* (Ritter-Záhony, 1911)	*
Cladocera	*Ceriodaphnia* (Dana, 1853)	**
Cladocera	*Bosmina longirostris* (O.F. Müller, 1785)	*
Cladocera	*Daphnia* (O.F. Müller, 1785)	*
Mysida	*Neomysis americana* (S.I. Smith, 1873)	**
Mysida	*Arthromysis magellanica* (Cunningham, 1871)	*
Amphipoda	*Corophium* sp. (Latreille, 1806)	*
Amphipoda	Corophium, Latreille, 1806 Caprella, Lamarck 1801	*
Decapoda	*Cyrtograpsus angulatus* (Dana, 1851)	**
Decapoda	*Cyrtograpsus altimanus* (Rathbun, 1914)	**
Decapoda	*Neohelice granulata* (Dana, 1851)	***
Decapoda	*Pagurixus* sp. (Melin, 1939)	*
Decapoda	*Alpheus puapeba* (Christoffersen, 1979)	*
Decapoda	*Corystoides chilensis* (Lucas in H. Milne Edwards & Lucas, 1844)	*
Decapoda	*Artemesia longinaris* (Spence Bate, 1888)	**
Decapoda	*Pinnixa patagoniensis* (Rathbun, 1918)	*
Decapoda	*Pachycheles laevidactylus* (Ortmann, 1892)	*
Decapoda	*Leptuca uruguayensis* (Nobili, 1901)	*
Decapoda	*Peisos petrunkevitchi* (Burkenroad, 1945)	***
Sessilia	*Balanus glandula* (Darwin, 1854)	***
Sessilia	*Amphibalanus amphitrite* (Darwin, 1854)	***
Copepoda	*Acartia tonsa* (Dana, 1849)	****
Copepoda	*Calanoides carinatus* (Krøyer, 1849)	*
Copepoda	*Euterpina acutifrons* (Dana, 1847)	***

(continued)

Table 5.1 (continued)

Copepoda	*Labidocera fluviatilis* (Dahl F., 1894)	**
Copepoda	*Monstrilla helgolandica* (Claus, 1863)	*
Copepoda	*Eurytemora americana* (Williams, 1906)	**
Copepoda	*Monstrilla* sp. (Dana, 1849)	*
Copepoda	*Oithona nana* (Giesbrecht, 1893)	*
Copepoda	*Paracalanus parvus* (Claus, 1863)	**
Copepoda	*Delavalia* aff. *palustris* (Brady, 1868)	*
Copepoda	*Microarthridion* aff. *littorale* (Poppe, 1881)	**
Copepoda	*Halicyclops* affi. *crassicornis* (Herbst, 1955)	*
Copepoda	*Nannopus* aff. *palustris* (Brady, 1880)	*
Copepoda	*Dactylopusia tisboides* (Claus, 1863)	*
Copepoda	*Tisbe* sp. (Lilljeborg, 1853)	*
Copepoda	*Longipedia* (Claus, 1862)	**
Copepoda	*Ectinosomatidae* (Sars G.O., 1903)	*
Crustacea	*Isopoda undefined*	*
Crustacea	*Stomatopoda undefined*	*
Bryozoa	*Amathia* sp. (Lamouroux, 1812)	*
Bryozoa	Cyphonauta (L)	*
Anthozoa	Cerinula (L)	*
Echinodermata	*Ophiuroidea* (L)	*
Tunicata	Ascidiacea	***
Ichthyoplankton	*Brevoortia aurea* (Spix & Agassiz, 1829)	**
Ichthyoplankton	*Ramnogaster arcuata* (Jenyns, 1842)	*
Ichthyoplankton	Sciaenidae (Cuvier, 1829)	*
Ichthyoplankton	Atherinidae (Risso, 1827)	*
Chordata	Oikopleuridae (Lohmann, 1915)	*

Abundance data are from 2014 to 2018 (Puerto Cuatreros and Puerto Rosales). Categories of mean abundance: * less than 10 ind. m^{-3}, ** 10–100 ind. m^{-3}, *** 100–1000 ind. m^{-3}, and **** more than 1000 ind. m^{-3}

et al. 2017). Gelatinous zooplankton is a polyphyletic assemblage of disparate organisms with a gelatin-like consistency (Haddock, 2004). The most common gelatinous species are those included into the phyla Cnidaria and Ctenophora (i.e., medusae and ctenophores, commonly grouped as "jellyfish"). These organisms are important modulators of marine ecosystems due to their ability to consume large quantities of zooplanktonic prey including fish larvae and eggs and their "blooming" nature that potentiates their ecological roles (Boero 2013).

In the temperate Southwestern Atlantic Ocean, investigations on jellyfish have been intensified in recent years. Studies on the abundance, species composition, biodiversity patterns, invasions, and ecological traits of medusae and ctenophores have been carried out at different sites along the Argentine Sea and revealed the biological relevance of these species within pelagic ecosystems (e.g., Genzano et al. 2006, 2008a, b, 2009a, b; Rodriguez et al. 2007, 2012, 2014, 2017; Schiariti et al. 2008, 2012, 2014; Diaz Briz et al. 2012; Dutto et al. 2019a). As already mentioned,

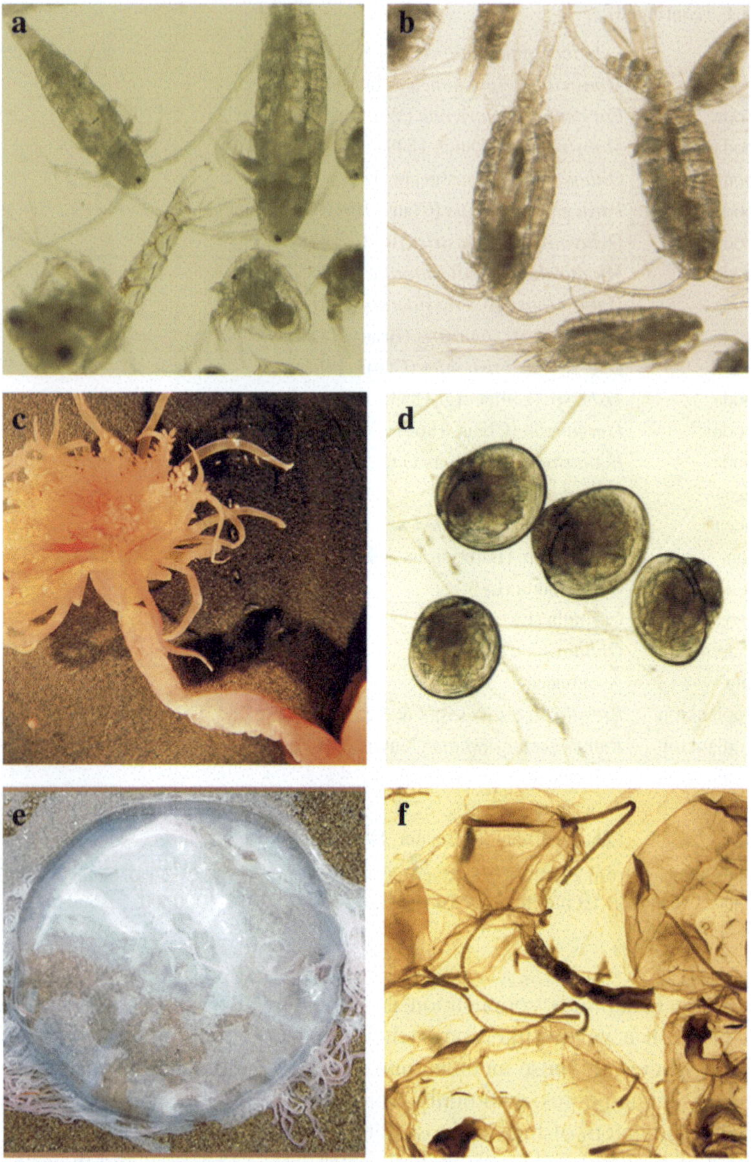

Fig. 5.3 Zooplankton key species from the Bahía Blanca Estuary. (**a**) Late-spring mesozooplankton with the copepod *Acartia tonsa*, Grapsidae zoea, and nauplii of *Amphibalanus amphitrite*. (**b**) Mesozooplankton in winter dominated by the copepod *Eurytemora americana*, (**c**) hydroid of *Corymorpha januarii*, (**d**) *Magallana gigas*, (**e**) *Olindias sambaquiensis*, and (**f**) *Liriope tetraphylla*. (Photos by Anabela Berasategui (**a**, **b**, and **d**) and Sofía Dutto (**c**, **e**, and **f**))

one of the geographic zones that seems to enclose suitable environmental conditions for the development of gelatinous zooplankton communities is the Bahía Blanca Estuary and the adjacent inner shelf (Dutto et al. 2017, 2019a). Species composition and ecological studies on medusae and ctenophores in this area concentrated in the 1980s (Mianzan and Sabatini 1985; Zamponi and Mianzan 1985; Mianzan 1986a, b, 1989a, b, c; Hoffmeyer 1990), but they were discontinued. Currently, the study of gelatinous zooplankton in the estuarine region has recommenced. Twenty-three hydromedusae, 3 scyphomedusae, and 3 ctenophores species have been reported for the area, including some related to economic and sanitary concerns, such as the cases of *Chrysaora lactea* Eschscholtz (Scyphozoa), the hydromedusae *Olindias sambaquiensis* Müller (Fig. 5.3e), and *Liriope tetraphylla* Chamisso (Fig. 5.3f) and Eysenhardt (Dutto et al. 2017). The recent records of high abundances of *L. tetraphylla* in the region of the Bahía Blanca Estuary (Dutto et al. 2019a) deserve further investigation and monitoring in order to assess its population dynamic and ecological role in this coastal ecosystem. The proliferation of *L. tetraphylla* may be boosted by the high productivity of the ecosystem and also promoted by the traits and plasticity of this holoplanktonic species. Likewise, hydroids are also commonly found in this area (Genzano et al. 2009a, b). Recently, a population of the solitary and seasonal polyp of *Corymorpha januarii* was described from a small tidal channel in the inner area of the Bahía Blanca Estuary, and one of the first approaches to their feeding ecology by gut content analysis was provided (Dutto et al. 2019b). Polyps of *C. januarii* show a variable diet composed mainly of organic matter and zooplanktonic prey. The stomach content analysis reveals that these polyps probably select copepods, mysids, and other zooplanktonic prey of lower swimming capacity (e.g., barnacles larvae and benthic invertebrates, and fish eggs) (Boxes 5.2 and 5.3).

Box 5.2: Eurytemora americana as Bioindicator of Water Quality

Contaminants associated with anthropic activities such as pesticides, heavy metals, and hydrocarbon derivatives produce negative effects on the metabolism of various marine organisms, generating oxidative damage and stress, as well as a reduction on reproduction and survival (Regoli et al. 2002; Suderman and Marcus 2002; Hack et al. 2008; Tsangaris et al. 2010). In this sense, copepods respond fast to changes in the environment driven by compound effects of global scale hydro-climate variability and anthropogenic pressure at local/regional scale. Their fast response to environmental conditions along with their short life cycles and fast rate of reproduction constitutes valuable properties that motivate their use as bioindicators (Parmar et al. 2016; Battuello et al. 2017). These crustaceans have a crucial role in the trophic transfer of contaminants in the marine food web, since they tend to accumulate pollutants from their prey (Wang 2002; Parmar et al. 2016). It is known that in copepods, pesticides, aromatic polycyclic hydrocarbons, and heavy metals lead to disruptive effects at the endocrine level and affect growth and reproductive rates (Marcus 2004).

The Bahía Blanca Estuary is subject to high anthropogenic impact derived from industrial, port, and urban development, which affects the dynamic and composition of plankton and seston (Marcovecchio et al. 2008; Hoffmeyer et al. 2008). It is also a highly productive ecosystem and constitutes an important nursery area for birds, crustaceans, and fish of commercial interest, such as *Micropogonias furnieri* (Desmarest, 1823), *Cynoscion guatucupa* (Cuvier, 1830), *Odontesthes argentinensis* (Valenciennes, 1835) and *Engraulis anchoita* (Hubbs and Marini, 1935) (Lopez Cazorla 2007; Marrari et al. 2013). Key estuarine copepods (*E. americana* and *A. tonsa*) seem well adapted to severe eutrophication (Biancalana and Torres 2011; Biancalana et al. 2012; Dutto et al. 2014); however the apparent adaptation of these species to highly eutrophic conditions may be threatened by rising effluent and pollutant disposal derived from intense human activities (Fernández Severini et al. 2017).

The exotic calanoid copepod *E. americana* (Fig. 5.2b) was introduced in the ecosystem via ballast water translocation through commercial ships coming from the Northern Hemisphere (Hoffmeyer 1994), and has successfully established in the planktonic food web, with maximum abundance from winter to spring (Hoffmeyer et al. 2009b; Berasategui et al. 2019). This species displays two distinct types of reproductive behavior depending on the prevailing environmental conditions (Berasategui et al. 2009, 2012). During the growing season, females produce subitaneous eggs, which hatch as population continues to rise in the plankton. When the population peak begins to decrease after spring, diapause eggs are produced. Thereafter, the species disappears from the pelagic habitat and recruits in banks of diapause eggs in the bottom sediments, until optimal environmental conditions of subsequent years trigger their hatching (Berasategui et al. 2013).

In order to assess the adaptive capacity of copepods exposed to sewage and industrial pollution, Berasategui et al. (2018) evaluated the reproductive performance of *E. americana* through laboratory experiments under different levels of contamination. Experimental data revealed that egg production, number of nauplii, number of fecal pellets, survival, and fertility state of females responded negatively to increasing concentrations of heavy metals (cadmium, lead, copper, zinc, and chrome), dissolved ammonium and phosphate, and turbidity. The results showed that bioavailable contaminants from the dissolved phase of sewage effluent reduced *E. americana* fertility in a 53.09% (egg production 12.4 ± 2.9 egg/female.clutch), while the water from the bottom of the sewage discharge site was undoubtedly lethal for this species (Fig. 5.4a). When exposed to the particulate phase of sewage effluents, only 40% of the females showed regeneration of their gonads for a second egg laying (Fig. 5.4b), while the dissolved phase effluent lead to null gonad regeneration for a second egg laying. The current rate of environmental change in the Bahía Blanca Estuary highlights that empirical studies on the physiological responses of copepods to anthropogenic impacts are of urgent and critical concern for planning resource and ecosystem management.

Box 5.3: *Olindias sambaquiensis*, a Stinging Jellyfish from the Southwestern Atlantic

Olindias sambaquiensis (Limnomedusa: Olindiidae, see Fig. 5.3e) is a conspicuous hydromedusan species endemic to the coastal waters of southern Brazil, Uruguay, and Argentina. Its distribution ranges approximately from 23°S to 42°S, where this jellyfish displays outbreaks of variable magnitude and duration (Chiaverano et al. 2004; Resgalla Jr. et al. 2019). This species has a meroplanktonic life cycle, which includes small and primitive colonial polyps (asexual phase), only known through laboratory cultivation and never found in nature. Its free-living medusa (sexual phase) may reach 10 cm in diameter and shows diel vertical migration habits staying at the bottom during the day being, presumably, nocturnal (Chiaverano 2001; Ale et al. 2007). This species is responsible for a large number of stings to bathers in Argentina and Uruguay during summer and also during winter and autumn in southern Brazil (Mianzan and Ramírez 1996; Resgalla Jr. et al. 2011). *Olindias sambaquiensis* presents three different types of tentacles, and, as an adult, it may have more than 400 tentacles, all of them full of stinging cells or cnidocytes. Its whole body may also be covered by cnidocytes. The venom of this jellyfish is very complex and may include 29 putative toxins similar to venom proteins from diverse phyla, and also two novel cytolysins (Weston et al. 2013; Haddad Jr. et al. 2014). It can produce persistent skin pain and envenomation resulting in erythema, severe dermatitis, and systemic symptoms (Kokelj et al. 1993; Mianzan and Ramírez 1996; Mianzan et al. 2001; Mosovich and Young 2012). The poisoning by this jellyfish is characterized as moderate to severe, and can lead to cardiovascular complications including cardiopulmonary arrest (Haddad Jr. et al. 2014). Despite its sanitary importance, the population dynamic in those regions where this jellyfish is common and problematic is far from being understood. Its life cycle is poorly known because of the small size of the polyp that makes it difficult to find and, therefore, to study in nature, and the difficulty in generating and maintaining the polyps under laboratory conditions (Zamponi and Facal 1987; Resgalla Jr. et al. 2019). However, some congeners have been successfully cultured shedding light on the development of *Olindias* spp. and the environmental factors influencing their population dynamics (Patry et al. 2014; Toshino et al. 2019). Considering the results of these investigations, the dynamic of *O. sambaquiensis* is presumably tightly modulated by water temperature. Winter temperatures may have an important role in the maintenance and reproduction of polyps, whereas the liberation and growth of the medusa phase may be linked to rising temperatures (Patry et al. 2014; Resgalla Jr. et al. 2019). Although this hypothesis needs to be confirmed, it is clear that the environment, particularly the temperature, has a crucial role in the development of the life cycle phases of *O. sambaquiensis*, and therefore, in the occurrence and the magnitude of bloom events. Water temperature would be acting at life history level, whereas other factors, such as winds, seem to have a physical role favoring the presence or the absence at coastal waters of those jellyfish already liberated. In southern Brazil, for instance, southerly winds, which may promote the convergence of waters toward the coast accumulating the jellyfish in the beach zone, are associated with high number of stinging events (Resgalla Jr. et al. 2005, 2011). In the Bahía Blanca region, in contrast, offshore winds (northerly winds) trigger coastal upwellings, which produce the emergence of the bottom layer, favoring the concentration of *Olindias* jellyfish at coastal waters (Mianzan and Zamponi 1988; Brendel et al. 2017). In fact, the absence of the jellyfish in the beach zone of the Bahía Blanca region in recent years would be explained, in part, by the significant decrease of offshore wind speed (Brendel et al. 2017). The orientation of the coast, along with other features, such as the depth and bottom geomorphology, may explain the distinct patterns observed in the Southwestern Atlantic waters in relation to the coastal occurrence of *O. sambaquiensis*. Although progress has been made in the knowledge of *O. sambaquiensis* (see Resgalla Jr. et al. 2019), much remains to be understood, particularly, in relation to the life history of this species and the role of environmental factors in the modulation of the population dynamic. The comprehension of the life cycle and the life history of *O. sambaquiensis* is essential to answer how the environment can modulate the timing and the magnitude of blooms. To optimize the cultivation and the maintenance of the species, which will enable the experimentation, is therefore essential.

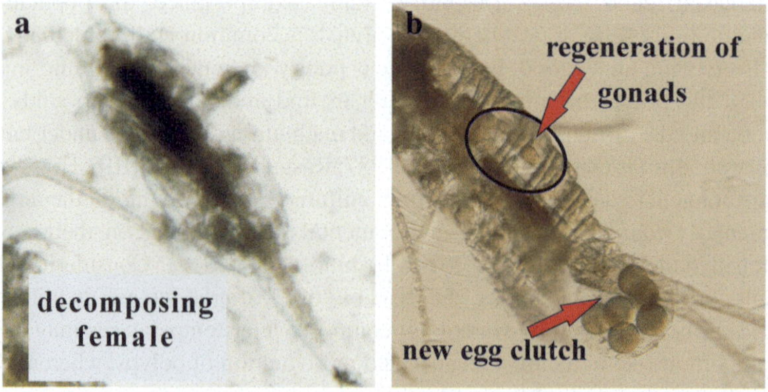

Fig. 5.4 Sewage effluent effects on fecundity in *Eurytemora americana*. Experimental response in females' culture. (**a**) Specimen exposed to water from the bottom of the sewage discharge, (**b**) specimen exposed to the particulate phase of the sewage effluents. (Photos by Anabela Berasategui)

References

Ale E, Imazu M, Oliveira OMP et al (2007) Ocorrência de *Olindias sambaquiensis* (cnidaria: hydrozoa) em hábito demersal ao redor da ilha de são sebastião (sp, Brasil). XII Congresso Latino-Americano de Ciências do Mar – XII COLACMAR Florianópolis

Barría de Cao MS, Beigt D, Piccolo MC (2005) Temporal variability of diversity and biomass of tintinnids (Ciliophora) in a southwestern Atlantic temperate estuary. J Plankton Res 27:1103–1111

Barría de Cao MS, Piccolo MC, Perillo GME (2011) Biomasss and microzooplankton seasonal assemblages in the Bahía Blanca estuary, Argentinean coast. J Mar Biol Assoc UK 91(5):953–959

Battuello M, Mussat Sartora R, Briziob P et al (2017) The influence of feeding strategies on trace element bioaccumulation in copepods (Calanoida). Ecol Indic 74:311–320

Berasategui AA, Hoffmeyer MS, Biancalana F et al (2009) Temporal variations in abundance and fecundity of the invading copepod *Eurytemora americana* in Bahía Blanca estuary during an unusual year. Estuar Coast Shelf Sci 85:82–88

Berasategui AA, Hoffmeyer MS, Dutto MS et al (2012) Seasonal variation in the egg morphology of the copepod *Eurytemora americana* and its relationship to reproductive strategy in a temperate estuary in Argentina. ICES J Mar Sci 69(3):380–388

Berasategui AA, Dutto MS, Chazarreta (2013) Seasonal occurrence and hatching success of benthic eggs of calanoid copepods in the Bahía Blanca Estuary, Argentina. Mar Biol Res 9(10):1018–1028

Berasategui AA, Fernandez-Severini MD, Menendez MC (2016) Reproductive trade-off of the copepod *Acartia tonsa* in a hypersaline estuary of Southwestern Atlantic. Temporal variations in the morphology of eggs. Mar Biol Res 12(8):817–829

Berasategui AA, Biancalana F, Fricke A et al (2018) The impact of sewage effluents on the fecundity and survival of *Eurytemora americana* in a eutrophic estuary of Argentina. Estuar Coast Shelf Sci 211:208–216

Berasategui AA, López Abbate MC, D'Agostino V et al (2019) Mesozooplankton structure and seasonal dynamics in three coastal systems of Argentina: Bahía Blanca Estuary, Pirámide Bay and Ushuaia Bay. In: Hoffmeyer M, Sabatini M, Brandini FP et al (eds) Plankton ecology of the Southwestern Atlantic from the subtropical to the Subantarctic Realm. Springer, Cham, p 327

Biancalana F, Torres AI (2011) Variation of mesozooplankton composition in a eutrophicated semienclosed system (Encerrada Bay, Tierra del fuego, Argentina). Braz J Oceangr 59(2):195–199

Biancalana F, Menendez MC, Berasategui AA et al (2012) Sewage pollution effects on mesozooplankton structure in a shallow temperate estuary. Environ Monit Assess 184:3901–3913

Boenigk J, Novarino G (2004) Effect of suspended clay on the feeding and growth of bacterivorous flagellates and ciliates. Aquat Microb Ecol 34:181–192

Boero F (2013) Review of jellyfish blooms in the Mediterranean and Black Sea. General Fisheries Commission for the Mediterranean FAO, Rome

Boltovskoy D (Ed) (1981) Atlas del zooplancton del Atlántico Sudoccidental y métodos de trabajo con el zooplancton Marino. Publicacion especial INIDEP, Mar del Plata, Argentina, pp 938

Boyd PW, Brown CJ (2015) Modes of interactions between environmental drivers and marine biota. Front Mar Sci 2

Bradford-Grieve JM, Markhaseva EL, Rocha CEF et al (1999) Copepoda. In Boltovskoy D (Ed). South Atlantic Zooplankton. Backhuy publishers, Leinden, The Netherlands pp 869–1098

Brendel AS, Dutto MS, Menéndez MC et al (2017) Wind pattern change along a period of coastal occurrence variation of a stinging medusa on a SW Atlantic beach. Anuário do Instituto de Geociencias UFRJ 40:303–315

Chang KH, Doi H, Nishibe Y et al (2009) Spatial and temporal distribution of zooplankton communities of coastal marine waters receiving different human activities (fish and pearl oyster farmings). The Open Mar Biol J. https://doi.org/10.2174/18744508090301008

Chazarreta CJ, Hoffmeyer MS, Cuadrado DG (2015) Tidal effects on short-term mesozooplankton distribution in small channels of a temperate-turbid estuary, Southwestern Atlantic. Braz J Oceanogr 63(2):83–92

Chiaverano L (2001) Historia de vida de *Olindias sambaquiensis* (Limnomedusae, Olindiidae) durante su fase sexual en la zona de El Rincón (Buenos Aires, Argentina): Estructura de tallas, crecimiento, desarrollo e influencia ambiental en sus agregaciones. PhD thesis, Universidad Nacional de Mar del Plata, Argentina, p 70

Chiaverano L, Mianzan H, Ramírez F (2004) Gonad development and somatic growth patterns of *Olindias sambaquiensis* (Limnomedusae, Olindiidae). Hydrobiology 530(531):373–381

Cloern JE, Dufford R (2005) Phytoplankton community ecology: principles applied in San Francisco Bay. Mar Ecol Progr Ser 285:11–28

Cuadrado DG, Gómez EA, Ginsberg SS (2005) Tidal and longshore sediment transport associated to a coastal structure. Estuar Coast Shelf S 62:291–300

Day JW, Crump BC, Kemp WM et al (eds) (2013) Estuarine ecology. Wiley-Blackwell. Wiley, Hoboken

Diaz Briz L, Martorelli S, Genzano G et al (2012) Parasitism (Trematoda, Digenea) in medusae from the southwestern Atlantic Ocean: medusa hosts, parasite prevalences, and ecological implications. Hydrobiologia 690:215–226

Diodato SL, Hoffmeyer MS (2008) Contribution of planktonic and detritic fractions to the natural diet of mesozooplankton in Bahía Blanca Estuary. Hydrobiologia 614:83–90

Dolan JR (2012) Introduction to tintinnids. In: Dolan JR, Montagnes DJ, Agatha S, Coats DW, Stoecker D (eds) The biology and ecology of tintinnid ciliates. Wiley, p 16

Dos Santos EP, Fiori SM (2010) Primer registro sobre la presencia de *Crassostrea gigas* en el estuario de Bahía Blanca (Argentina). Commun Soc Malacol Urug 9:245–252

Dragani WC, Martin PBC, Simionato G et al (2010) Are wind wave heights increasing in southeastern south American continental shelf between 32°S and 40°S? Cont Shelf Res 30:481–490

Dunne JA, Lafferty KD, Dobson AP et al (2013) Parasites affect food web structure primarily through increased diversity and complexity. PLoS Biol. https://doi.org/10.1371/JOURNAL.PBIO.1001579

Dutto MS, Kopprio GA, Hoffmeyer MS et al (2014) Planktonic trophic interactions in a human impacted estuary of Argentina: A fatty acid marker approach. J Plankton Res. https://doi.org/10.1093/plankt/fbu012

Dutto MS, Genzano GN, Schiariti A et al (2017) Medusae and ctenophores from the Bahía Blanca Estuary and neighboring inner shelf (Southwest Atlantic Ocean, Argentina). Mar Biodivers Rec 2017:10–14

Dutto MS, Chazarreta CJ, Rodriguez CS et al (2019a) Macroscale abundance patterns of hydromedusae in the temperate Southwestern Atlantic (27°–56°S). PLoS One 14(6):e0217628

Dutto MS, Carcedo MC, Nahuelhual EG et al (2019b) Trophic ecology of a corymorphid hydroid population in the Bahia Blanca Estuary, Southwestern Atlantic. Reg Stud Mar Sci 31:100746

Elliott M, Borja Á, McQuatters-Gollop A et al (2015) Force majeure: will climate change affect our ability to attain Good Environmental Status for marine biodiversity? Mar Pollut Bull 95:7–27

Fernández Severini DM, Villagran D, Biancalana F et al (2017) Heavy metal concentrations found in seston and microplankton from an impacted temperate shallow estuary along the southwestern Atlantic Ocean. J Coast Res 33(5):1196–1209

Fiori SM, Pratolongo PD, Zalba SM et al (2016) Spatially explicit risk assessment for coastal invaders under different management scenarios. Mar Biol 163:245. https://doi.org/10.1007/s00227-016-3017-5

Garibotti IA, Ferrario ME, Almandoz GO et al (2011) Seasonal diatom cycle in Anegada Bay, El Rincón estuarine system, Argentina. Diatom Res 26(2):227–241

Gayoso AM (1981) Estudio de las diatomeas del estuario de Bahía Blanca. Doctoral dissertation, Universidad Nacional de La Plata, Argentina, p 100

Gayoso AM (1989) Species of the diatom genus *Thalassiosira* from the coastal zone of the South Atlantic (Argentina). Bot Mar 32:331–337

Gayoso AM (1998) Long-term phytoplankton studies in the Bahía Blanca Estuary, Argentina. ICES J Mar Sci 55:655–660

Genzano GN, Mianzan HW, Acha EM et al (2006) First record of the invasive medusa *Blackfordia virginica* (Hydrozoa: Leptomedusae) in the Río de la Plata estuary, Argentina-Uruguay. Rev Chil Hist Nat 79:257–261

Genzano GN, Mianzan H, Bouillon J (2008a) Hydromedusae (Cnidaria: Hydrozoa) from the temperate southwestern Atlantic Ocean: a review. Zootaxa 1750:1–18

Genzano GN, Mianzan HW, Diaz BL et al (2008b) On the occurrence of *Obelia* medusa bloom and empirical evidence of an unusual *Obelia* and *Amphisbetia hydroids* shoreline massive accumulations. Lat Am J Aquat Res 36:301–307

Genzano GN, Giberto D, Schejter L et al (2009a) Hydroid assemblages from the Southwestern Atlantic Ocean (34–42° S). Mar Ecol 30:33–46

Genzano GN, Rodriguez C, Pastorino G et al (2009b) The hydroid and medusa of *Corymorpha januarii* in temperate waters of the Southwestern Atlantic Ocean. Bull Mar Sci 84:229–235

Glibert PM, Wilkerson FP, Dugdale RC et al (2014) Phytoplankton communities from San Francisco Bay Delta respond differently to oxidized and reduced nitrogen substrates-even under conditions that would otherwise suggest nitrogen sufficiency. Front Mar Sci 1:1–16

Gsell AS, de Senerpont Domis LN, Verhoeven KJF (2013) Chytrid epidemics may increase genetic diversity of a diatom spring-bloom. ISME J 7:2057–2059

Guerrero MA, Izquierdo M, Canelo S (1976) Observaciones mensuales sobre algunos aspecto biológicos de la ría de Bahía Blanca entre pto Ing. White y Pto. Cuatreros. Contr Cient IADO 36:1–7

Guinder VA, Popovich CA, Perillo GME (2009) Particulate suspended matter concentrations in the Bahía Blanca Estuary, Argentina: implication for the development of phytoplankton blooms. Estuar Coast Shelf S 85:157–165

Guinder VA, Popovich CA, Molinero JC et al (2010) Long-term changes in phytoplankton phenology and community structure in the Bahía Blanca Estuary, Argentina. Mar Biol 157:2703–2716

Guinder VA, Molinero JC, Popovich CA et al (2012) Dominance of the planktonic diatom *Thalassiosira minima* in recent summers in the Bahia Blanca Estuary, Argentina. J Plankton Res 34:995–1000

Guinder VA, Popovich CA, Molinero JC et al (2013) Phytoplankton summer bloom dynamics in the Bahía Blanca Estuary in relation to changing environmental conditions. Cont Shelf Res 52:150–158

Guinder VA, López-Abbate MC, Berasategui AA et al (2015) Influence of the winter phytoplankton bloom on the settled material in a temperate shallow estuary. Oceanologia 57:50–60

Guinder VA, Molinero JC, López Abbate MC (2016) Phenological changes of blooming diatoms promoted by compound bottom-up and top-down controls. Estuar Coast 40:95–104. https://doi.org/10.1007/s12237-016-0134-9

Guinder VA, Carcedo MC, Buzzi N et al (2017) Ephemeral parasitism on blooming diatoms in a temperate estuary. Mar Fresh Res 69(1):128–133

Guinder VA, Tillman U, Krock B et al (2018) Plankton multiproxy analyses in the Northern Patagonian Shelf, Argentina: community structure, phycotoxins and characterization of Alexandrium strains. Front Mar Sci 5:394

Hack LA, Tremblay LA, Wratten SD et al (2008) Toxicity of estuarine sediments using a full life-cycle bioassay with the marine copepod *Robertsonia propinqua*. Ecotox Environ Safe 70:469–474

Haddad V Jr, Zara F, Marangoni S et al (2014) Identification of two novel cytolysins from the hydrozoan *Olindias sambaquiensis* (Cnidaria). J Venom Anim Toxins Incl Trop Dis 20:10. https://doi.org/10.1186/1678-9199-20-10

Haddock SH (2004) A golden age of gelata: past and future research on planktonic ctenophores and cnidarians. Hydrobiologia 530(13):549–556

Hoffmeyer MS (1990) Algunas observaciones sobre la alimentación de *Mnemiopsis mccradyi* Mayer (Ctenophora-Lobata). Iheringia Ser Zool 70:55–65

Hoffmeyer MS (1994) Seasonal succession of Copepoda in the Bahía Blanca Estuary. In: Ferrari and Bradley (eds) Ecology and morphology of copepods, developments in hydrobiology, hydrobiology 292(293): 303–308
Hoffmeyer M (2004) Decadal change in zooplankton seasonal succession in the Bahía Blanca estuary, Argentina, following the introduction of two zooplankton species. J Plankton Res 26(2):181–189
Hoffmeyer MS (2007) Mesozooplancton. In Piccolo MC. Hoffmeyer MS (eds). Ecosistema del Estuario de Bahía Blanca. FUNS press, pp 133–141
Hoffmeyer MS, Cervellini PM (2007) Zooplancton temporario. In: Piccolo MC, Hoffmeyer MS (eds) Ecosistema del Estuario de Bahía Blanca, 3rd edn. EdiUNS, Bahía Blanca, p 153
Hoffmeyer MS, Mianzan HW (2007) Macrozooplankton del estuario y aguas costeras adyacentes. In: Piccolo MC, Hoffmeyer MS (eds) Ecosistema del Estuario de Bahía Blanca, 3rd edn. EdiUNS, Bahía Blanca, p 143
Hoffmeyer MS, Fernandez-Severini MD, Menendez MC et al (2008) Composition and dynamics of mesozooplankton assemblages in the Bahía Blanca Estuary. In: Neves R, Baretta J, Mateus M (eds) Perspectives on integrated coastal zonemanagement in South America. ISTPress: Scientific Publishers, Lisboa, p 303
Hoffmeyer MS, Menendez MC, Biancalana F et al (2009a) Ichthyoplankton spatial pattern in the inner shelf off Bahía Blanca Estuary, SW Atlantic Ocean. Estuar Coast Shelf S 84:383–392
Hoffmeyer MS, Berasategui AA, Beigt D et al (2009b) Environmental regulation of the estuarine copepods *Acartia tonsa* and *Eurytemora americana* during coexistence period. J Mar Biol Assoc U K 89:355–361
Kokelj F, Mianzan H, Avian M et al (1993) Dermatitis due to *Olindias sambaquiensis*: a case report. Cutis 51:339–342
Lancelot C, Muylaert K (2011) Trends in estuarine phytoplankton. Ecology:5–15
Lanfredi NW, D'Onofrio EE, Mazio CA (1988) Variations of the mean sea level in the Southwest Atlantic Ocean. Cont Shelf Res 8:1211–1220
Litchman E, Klausmeier CA (2008) Trait-based community ecology of phytoplankton. 599. Annu Rev Ecol Evol S. 39:615–639
Lopez Cazorla A (2007) Peces. In: Piccolo MC, Hoffmeyer MS (eds) Ecosistema del Estuario de Bahía Blanca, 3rd edn. EdiUNS, Bahía Blanca, p 191
Lopez Cazorla A, Pettigrosso RE, Tejera L et al (2011) Diet and food selection by *Ramnogaster arcuata* (Osteichthyes, Clupeidae). J Fish Biol 78(7):2052–2066
López-Abbate MC, Molinero JC, Guinder VA et al (2015) Microplankton dynamics under heavy anthropogenic pressure. The case of the Bahia Blanca Estuary, Southwestern Atlantic Ocean. Mar Pollut Bull 9(1):305–314
López-Abbate MC, Barría de Cao MS, Pettigrosso RE (2016) Seasonal changes in microzooplankton feeding behavior under varying eutrophication level in the Bahía Blanca estuary (SW Atlantic Ocean). J Exp Mar Biol Ecol 481:25–33
López-Abbate MC, Molinero JC, Guinder VA et al (2017) Time-varying environmental control of phytoplankton in a changing estuarine system. Sci Total Environ 609:1390–1400
López-Abbate MC, Molinero JC, Barría de Cao MS et al (2019a) Eutrophication disrupts summer trophic links in an estuarine microbial food web. Food Webs 20:00121. https://doi.org/10.1016/j.fooweb.2019.e00121
López-Abbate MC, Molinero JC, Perillo GM et al (2019b) Long-term changes on estuarine ciliates linked with modifications on wind patterns and water turbidity. Mar Environ Res 144:46–55
Marcovecchio JE, Botte SE, Delucchi F et al (2008) Pollution processes in Bahía Blanca Estuary environment. In: Neves R, Baretta J, Mateus M (eds) Perspectives on integrated coastal zone management in South America. IST Press, Lisboa, p 301
Marcus N (2004) An overview of the impacts of eutrophication and chemical pollutants on copepods of the coastal zone. Zool Stud 43(2):211–217

Marrari M, Signorini S, McClain CR et al (2013) Reproductive success of the Argentine anchovy, *Engraulis anchoita*, in relation to environmental variability at a mid-shelf front 875 (Southwestern Atlantic Ocean). Fish Oceanogr 22:247–261

Marrari M, Piola AR, Valla D (2017) Variability and 20-year trends in satellite-derived surface chlorophyll concentrations in large marine ecosystems around South and Western Central America. Front Mar Sci 4:372

Menéndez MC, Piccolo MC, Hoffmeyer MS (2012) Short-term variability on mesozooplankton community in a shallow mixed estuary (Bahía Blanca, Argentina): influence of tidal cycles and local winds. Estuar Coast Shelf Sci 112:11–22

Mianzan HW (1986a) Estudio sistemático y bioecológico de algunas medusas Scyphozoa de la región subantártica. PHD thesis, Universidad Nacional de La Plata, La Plata

Mianzan HW (1986b) *Beroe ovata* en aguas de la Bahía Blanca, Argentina (Ctenophora). Spheniscus 2:29–32

Mianzan HW (1989a) Las medusas Scyphozoa de la Bahía Blanca. Boletim do Instituto Oceanografico São Paulo 37:29–32

Mianzan HW (1989b) Sistemática y zoogeografía de Scyphomedusae en aguas neríticas argentinas. Inv Mar CICIMAR 4:15–34

Mianzan HW (1989c) Distribución de *Olindias sambaquiesi*s Müller, 1861 (Hydrozoa, Limnomedusae) en el Atlántico Sudoccidental. Iheringia Sér Zoo 69:155–157

Mianzan HW, Ramírez F (1996) *Olindias sambaquiensis* stings in the South West Atlantic. In: Williamson JAH, Fenner PJ et al (eds) Venomous and poisonous marine animals: a medical and biological handbook. Univ. New South Wales Press, Brisbane, pp 206–208 and 301

Mianzan HW, Sabatini M (1985) Estudio preliminar sobre distribución y abundancia de *Mnemiopsis maccradyi* en el estuario de Bahía Blanca (Ctenophora). Spheniscus 1:53–68

Mianzan HW, Zamponi MO (1988) Estudio bioecológico de *Olindias sambaquiensis* Muller, 1861 (Limnomedusae, Olindiidae) en el área de Monte Hermoso, II. Factores meteorológicos que influyen en su aparición. Iheringia 2:63–68

Mianzan HW, Fenner PJ, Cornelius PFS et al (2001) Vinegar as a disarming agent to prevent further discharge of the nematosyst of the stinging hydromedusa *Olindias sambaquiensis*. Cutis 68:45–48

Mosovich JH, Young P (2012) Picadura de medusa *Olindias sambaquiensis*. Análisis de 49 casos. Medicina 72:380–388

Negrin VL, Spetter CV, Asteasuain RO et al (2011) Influence of flooding and vegetation on carbon, nitrogen, and phosphorus dynamics in the pore water of a *Spartina alterniflora* salt marsh. J Environ Sci 23(2):212–221

Negrin VL, Spetter CV, Guinder VA et al (2013) The role of *Sarcocornia perennis* and tidal flooding on sediment biogeochemistry in a South American wetland. Mar Biol Res 9(7):703–715

Parmar TK, Rawtani D, Agrawal YK (2016) Bioindicators: the natural indicator of environmental pollution. Front Life Sci 9:110–118

Patry W, Knowles T, Christianson I, Howard M (2014) The hydroid and early medusa stage of *Olindias formosus* (Cnidaria, Hydrozoa, Limnomedusae). J Mar Biol Assoc UK 94:1409–1415

Perillo GME (2009) Tidal courses: classification, origin and functionality. In: Perillo GME, Wolanski E, Cahoon DR et al (eds) Coastal wetlands: an integrated ecosystem approach. Elsevier, Amsterdam, p 185

Perillo GME, Sequeira ME (1989) Geomorphologic and sediment transport characteristics of the middle reach of the Bahia Blanca estuary (Argentina). J Geophys Res Oceans 94:14351–14362

Pettigrosso RE (2003) Planktonic ciliates Choreotrichida and Strombidiida from the inner zone of the Bahía Blanca estuary, Argentina. Iheringia 93:117–126

Pettigrosso RE, Popovich CA (2009) Phytoplankton-aloricate ciliate community in the Bahía Blanca Estuary (Argentina): seasonal patterns and trophic groups. Braz J Oceanogr 57(3):215–227

Pettigrosso RE, Garcia MD, Uibrig RA et al (2016) Mixotrophic ciliate dynamics in two zones of a temperate and highly turbid estuary in South America, Argentina. Ecol Austral 26:107–119

Piccolo MC, Hoffmeyer MS (2007) Ecosistema del Estuario de Bahía Blanca, 3rd edn. EdiUNS, Bahía Blanca, p 233
Popovich CA, Gayoso AM (1999) Effect of irradiance and temperature on the growth rate of *Thalassiosira curviseriata* Takano (Bacillariophyceae), a bloom diatom in Bahía Blanca estuary (Argentina). J Plankton Res 21:1101–1110
Popovich CA, Spetter CV, Marcovecchio JE et al (2008) Dissolved nutrient availability during winter diatom bloom in a turbid and shallow estuary (Bahía Blanca, Argentina). J Coast Res 24:95–102
Pratolongo P, Mazzon C, Zapperi G et al (2013) Land cover changes in tidal salt marshes of the Bahía Blanca estuary (Argentina) during the past 40 years. Estuar Coast Shelf S 133:23–31
Pratolongo P, Piovan MJ, Cuadrado DG et al (2017) Coastal landscape evolution on the western margin of the Bahía Blanca Estuary (Argentina) mirrors a non-uniform sea-level fall after the mid-Holocene highstand. Geo-Mar Lett 37:373–384
Regoli F, Gorbi S, Frenzilli G et al (2002) Oxidative stress in ecotoxicology: from the analysis of individual antioxidants to a more integrated approach. Mar Environ Res 54:419–423
Resgalla C Jr, Goncalves VC, Klein AHF (2005) The occurrence of jellyfish stings on the Santa Catarina coast, southern Brazil. Brazil. Braz J Oceanogr 53:183–186
Resgalla C Jr, Rosseto AL, Haddad JV (2011) Report of an outbreak of stings caused by *Olindias sambaquiensis* Muller, 1861 (Cnidaria: Hydrozoa) in southern Brazil. Braz J Oceanogr 59:391–396
Resgalla C Jr, Petri L, Teodoro da Silva BG et al (2019) Outbreaks, coexistence, and life cycle of jellyfish species in relation to abiotic and biological factors along a South American coast. Hydrobiologia 839:87–102
Rodriguez C, Genzano G, Mianzan H (2007) First record of *Eutonina scintillans* Bigelow, 1909 (Hydrozoa: Leptomedusae: Eirenidae) in temperate waters of the southwestern Atlantic Ocean. Investig Mar 35:135–138
Rodriguez C, Miranda TP, Marques AC et al (2012) The genus *Hybocodon* (Cnidaria, Hydrozoa) in the southwestern Atlantic Ocean, with a revision of *Hybocodon* species recorded in the area. Zootaxa 3523:39–48
Rodriguez CS, Pujol MG, Mianzan HW et al (2014) First record of the invasive stinging medusa *Gonionemus vertens* in the southern hemisphere (Mar del Plata, Argentina). Lat Am J Aquat Res 42:653–657
Rodriguez CS, Marques AC, Mianzan HW et al (2017) Environment and life cycles influence distribution patterns of hydromedusae in austral South America. Mar Biol Res 13:659–676
Sabatini ME (1989) Ciclo anual del copépodo *Acartia tonsa* Dana 1849 en la zona interna de Bahía Blanca (Provincia de Buenos Aires). Sci Mar 53:847–856
Salomon PS, Granéli E, Neves MHCB, Rodriguez EG (2009) Infection by Amoebophrya spp. parasitoids of dinoflagellates in a tropical marine coastal area. Aquat Microb Ecol 55:143–153
Schiariti A, Kawahara M, Uye SI et al (2008) Life cycle of the jellyfish *Lychnorhiza lucerna* (Scyphozoa: Rhizostomeae). Mar Biol 156:1–12
Schiariti A, Sal Moyano MP, Giberto DA et al (2012) First record of the association between *Lychnorhiza lucerna* (Scyphozoa, Rhizostomeae) and *Cyrtograpsus affinis* (Decapoda, Varunidae). Lat Am J Aquat Res 40:1090–1093
Schiariti A, Morandini AC, Jarms C et al (2014) Asexual reproduction strategies and blooming potential in Scyphozoa. Mar Ecol Prog Ser 510:241–253
Schmoker C, Russo F, Drillet G et al (2016) Effects of eutrophication on the planktonic food web dynamics of marine coastal ecosystems: the case study of two tropical inlets. Mar Environ Res 119:176–188
Sieburth JMCN, Smetacek V, Lenz J (1978) Pelagic ecosystem structure: heterotrophic compartments of the plankton and their relationship to plankton size fractions. Limnol Oceanogr 23:1256–1263
Silva AP, Neumann-Leitão S, Schwamborn R et al (2004) Mesozooplankton of an impacted bay in North Eastern Brazil. Braz Arch Biol Tech 47(3):485–493

Smyth K, Elliott M (2016) Effects of changing salinity on the ecology of the marine environment. In: Solan M, Whiteley NM (eds) Stressors in the marine environment. Oxford University Press Oxford University Press, London

Stoecker DK, Capuzzo JM (1990) Predation on Protozoa: its importance to zooplankton. J Plankton Res 12:891–908

Suderman BL, Marcus NH (2002) The effects of Orimulsion and Fuel Oil #6 on the hatching success of copepod resting eggs in the seabed of Tampa Bay, Florida. Environ Pollut 120:787–795

Toshino S, Tanimoto M, Minemizu R (2019) *Olindias deigo* sp, a new species (Hydrozoa, Trachylinae, Limnomedusae) from the Ryukyu Archipelago, southern Japan. Zoo Keys 900:1–21

Truchet MD, Buzzi NS, Carcedo C et al (2019) First record of the fiddler crab *Leptuca (Uca) uruguayensis* in the Bahía Blanca estuary (Buenos Aires, Argentina) with comments on its biology in South America. Regional Studies in Marine. https://doi.org/10.1016/j.rsma.2019.100539

Tsangaris C, Kormasb K, Strogyloudia E et al (2010) Multiple biomarkers of pollution effects in caged mussels on the Greek coastline. Comp Biochem Phys 151(3):369–378

Turner JT (2004) The importance of small planktonic copepods and their roles in pelagic marine food webs. Zool Stud 43:255–266

Uriarte I, Villate F (2004) Effects of pollution on zooplankton abundance and distribution in two estuaries of the Basque coast (Bay of Biscay). Mar Pollut Bull 49:220–228

Wang WX (2002) Interactions of trace metals and different marine food chains. Mar Ecol Prog Ser 243:295–309

Weston AJ, Chung R, Dunla WC, Morandini AC et al (2013) Proteomic characterisation of toxins isolated from nematocysts of the South Atlantic jellyfish *Olindias sambaquiensis*. Toxicon 71:11–17

Whitfield AK, Elliott M, Basset A et al (2012) Paradigms in estuarine ecology –a review of the Remane diagram with a suggested revised model for estuaries. Estuar Coast Shelf S 97:78–90

Winder M, Cloern JE (2010) The annual cycles of phytoplankton biomass. Philos Trans R Soc B. https://doi.org/10.1098/rstb.2010.0125

Winder M, Carstensen J, Galloway AWE et al (2017) The land-sea interface: a source of high-quality phytoplankton to support secondary production. Limnol Oceanogr 62:S258–S271. https://doi.org/10.1002/lno.10650

Worden AZ, Follows MJ, Giovannoni SJ et al (2015) Rethinking the marine carbon cycle: factoring in the multifarious lifestyles of microbes. Science 347(6223):1257594. https://doi.org/10.1126/science.1257594

Zamponi M, Facal O (1987) Estudio Bioecológico de *Olindias sambaquiensis* Muller, 1861 en el área de Monte Hermoso. l. Ciclo de vida. (Limnomedusae, Olindidae). Neotropica 33:119–126

Zamponi MO, Mianzan HW (1985) La mecánica de captura y alimentación de *Olindias sambaquiensis* Müller, 1861 (Limnomedusae) en el medio natural y en condiciones experimentales. Hist Nat 5:269–278

Zapperi G, Piovan MJ, Pratolongo P (2017) Community structure and spatial zonation of benthic macrofauna in mudflats of the Bahía Blanca Estuary, Argentina. J Coast Res 34(2):318–327

Chapter 6
Biology and Ecology of the Benthic Algae

M. Emilia Croce, M. Cecilia Gauna, Carolina Fernández, Ailen M. Poza, and Elisa R. Parodi

6.1 What Are Algae?

Algae form a heterogeneous group of organisms that range from single microscopic cells to giant seaweeds belonging to diverse evolutionary lineages. The term algae has no formal taxonomic standing; it is routinely used to designate a polyphyletic, non-cohesive, and artificial group of eukaryotes and prokaryotic photosynthetic organisms. Algae include two main groups, according to their body size, macroscopic algae (macroalgae) and a highly diversified group of microorganisms known as microalgae (Barsanti and Gualtieri 2006). The profound diversity of forms, sizes, ecological niches, levels of organization, photosynthetic pigments, storage products, structural polysaccharides, and life histories reflects the separate evolutionary origins of this diverse group.

By definition, algae are considered photoautotrophs; they use sunlight and CO_2 to produce carbohydrates and ATP, depending entirely upon their photosynthetic

M. E. Croce (✉) · M. C. Gauna
Instituto Argentino de Oceanografía, IADO (Universidad Nacional del Sur-CONICET), Bahía Blanca, Argentina

Departamento de Biología, Bioquímica y Farmacia, Universidad Nacional del Sur, Bahía Blanca, Argentina
e-mail: ecroce@criba.edu.ar

C. Fernández
Instituto Argentino de Oceanografía, IADO (Universidad Nacional del Sur-CONICET), Bahía Blanca, Argentina

Centro de Emprendedorismo y Desarrollo Territorial Sostenible (CEDETS), Universidad Provincial del Sudoeste (UPSO) – Comisión de Investigaciones Científicas de la Provincia de Buenos Aires (CIC), Bahía Blanca, Argentina

A. M. Poza · E. R. Parodi
Instituto Argentino de Oceanografía, IADO (Universidad Nacional del Sur-CONICET), Bahía Blanca, Argentina

S. M. Fiori, P. D. Pratolongo (eds.), *The Bahía Blanca Estuary*,
https://doi.org/10.1007/978-3-030-66486-2_6

apparatus for metabolic processes. However, most algal lineages include colorless heterotrophic species that obtain organic carbon from the environment, either by taking up dissolved substances or by engulfing bacteria and other cells. Some species of algae are called mixotrophs, as they combine both photoautotrophy and heterotrophy as nutritional strategies. There is a gradient among these three strategies; thus, the algae can be classified as obligate heterotrophic, facultative mixotrophic (e.g., some dinoflagellates), obligate phototrophic (e.g., *Dinobryon* Ehrenberg species), and obligate mixotrophic (e.g., *Euglena* Ehrenberg species) (Graham et al. 2016).

6.1.1 Habitats Colonized by Algae

Algae are mainly aquatic organisms, except for subaerial species that live exposed to the atmosphere and rely on liquid or vapor to carry out their metabolic functions. Aquatic algae display a broad range of tolerance to pH, temperature, turbidity, and concentrations of O_2 and CO_2, being able to colonize almost every aquatic environment; however, obligate photoautotrophic species are limited to shallow areas because of the rapid attenuation of light with depth. Aquatic algae can be found suspended in water bodies (planktonic algae), even under ice in polar areas, or attached to substrates or within sediments (benthic algae).

Benthic algae, either macroscopic or microscopic, can grow on stones (epilithic), on mud or sand (epipelic), on the thallus of other algae or plants (epiphytic), or on animals (epizoic). The term microphytobenthos is used to designate the community composed of eukaryotic microalgae, mainly diatoms, and cyanobacteria, that live on illuminated surfaces of a wide variety of aquatic habitats, ranging from tidal flats and marshes to submerged aquatic vegetation beds and subtidal sediments (Macintyre et al. 1996; Miller et al. 1996). The term periphyton is also used to designate this group of algae; however, it is more frequently used to refer to freshwater benthic microalgae. Even though microphytobenthos is less conspicuous than macroalgae or vascular plants, it contributes significantly to total primary productivity in coastal areas, and in many shallow aquatic systems, the biomass of the microphytobenthic community far exceeds that of phytoplankton (Pinckney and Zingmark 1993; Daehnick et al. 1992). The distribution of benthic microscopic algae varies extensively from place to place depending on the presence of suitable substrata, water depth, light availability, and physical perturbation (Sand-Jensen and Borum 1991). If strongly attenuated irradiances reach the sea bottom because of deep and/or turbid water, then benthic primary producers will be absent. In contrast, if high irradiances reach the bottom, as in shallow transparent waters, then the primary production of benthic microalgae can be dominant (Stevenson 1996).

Microphytobenthic algae occur in marine and freshwater habitats, while benthic macroalgae mainly habit marine environments. Large marine macroalgae are usually called *seaweeds*. Macroalgae that live in estuaries and marine coasts are usually classified according to the levels of the coast at which they occur. Supralittoral macroalgae are those that grow above the high-tide level, within the reach of waves and spray; intertidal macroalgae are those that grow on shores exposed to tidal cycles; and sublittoral macroalgae are those that grow in the benthic environment from the extreme low-water level to around 200 m deep (Graham et al. 2016). Their distribution along the coastal gradient is related with their photosynthetic capacities (red and brown macroalgae have accessory pigments that allow them to photosynthesize in regions where the light is attenuated) and their tolerance to salinity and desiccation. The influence of these factors on macroalgal assemblages will be described in Sect. 6.3.1.

6.1.2 Forms and Body Types of Algae

Many microalgal species occur solitary as unicellular individuals with or without flagella, hence motile or non-motile, in a variety of shapes. Other algae exist as aggregates of several single cells held together loosely or in a highly organized fashion, the colony. In these types of aggregates, the cell number is indefinite, growth occurs by cell division of its components, and each cell can survive on its own. When the number and arrangement of cells are determined at the time of origin and remain constant during the lifespan of the individual colony, this is termed coenobium (e.g., *Volvox* Linnaeus). Solitary and colonial species propelled by flagella are referred to as flagellates; however, not all flagellates include algal species.

Another common type of algal body is the filament, formed by a series of cells arranged end to end, with adjacent cells sharing a common cross wall, and forming chains where the daughter cells are connected to each other by their end wall. Filaments may be formed by one, two, or several rows of cells, and they may be unbranched as in *Oscillatoria* Vaucher ex Gomont (Cyanophyceae) or *Ulothrix* Kützing (Ulvophyceae), have false branching as in *Tolypothrix* Kützing ex É. Bornet and C. Flahault (Cyanophyceae), or have true branching as in *Cladophora* Kützing (Ulvophyceae) (Barsanti and Gualtieri 2006).

Macroalgal body is called *thallus*; it may be constructed by branched or unbranched filaments, coenocytic siphons (siphonous thallus), parenchyma, or pseudoparenchyma. Coenocytic thalli are composed of one giant multinucleate cell, lacking internal cell walls, except for septa that occur in reproductive states. Examples of this morphology can be found in the family Bryopsidales such as *Bryopsis* J.V. Lamouroux and *Codium* Stackhouse. Parenchymatous thallus is composed of tissues with a three-dimensional array of cells often connected by intercellular connections; this tissue organization is found in *Ulva* Linnaeus (Ulvophyceae) and many of the brown algae such as kelps. Pseudoparenchymatous algae are made up of a loose or close aggregation of numerous, intertwined, branched filaments that collectively form the thallus, held together by mucilages, especially in red algae such as *Gelidium* J.V. Lamouroux (Florideophyceae). Thallus construction is entirely based on a filamentous construction with little or no internal cell differentiation (Graham et al. 2016).

6.2 Classification of Algae

6.2.1 Taxonomic Classification

Among algal systematists, a wide range of classification schemes have been proposed (Bold and Wynne 1978; South and Whittick 1987; Margulis et al. 1990; Van den Hoek et al. 1995; Graham and Wilcox 2000). Considering the polyphyletic nature of algae as a group of organisms, algal classification is somewhat difficult to cope with the traditional taxonomic systems. However, it is still useful to represent the general characters and levels of organization, despite the fact that taxonomic opinion may change as new information accumulates.

The algae are traditionally classified into lineages or phyla, according to numerous characteristics: the photosynthetic pigments; the chemical nature of the energy storage product; the organization of the thylakoid membranes and other features of the chloroplasts; the composition and structure of the cell wall; the number, arrangement, and ultrastructure of the flagella; and the life cycle. Cyanobacteria make up a particular phylum of the domain Eubacteria, whereas eukaryotic algae are classified into more than 10 phyla of the domain Eukarya; these are Chlorophyta (green algae), Rhodophyta (red algae), Glaucophyta (glaucophytes), Cryptophyta (cryptomonads), Dinophyta (dinoflagellates), Bacillariophyta (diatoms), Phaeophyta (brown algae), Haptophyta (haptophytes), Chlorarachniophyta (chlorarachniophytes), and Euglenophyta (euglenoids). Molecular studies aiming to assess the internal genetic coherence of nuclear genes and ribosomal RNA are increasingly unraveling the relationships among these major lineages (Graham et al. 2016).

Cyanophyta (also known as blue-green algae, cyanobacteria, or cyanophytes) are unique algae, as they are prokaryotes. This is the most ancient algal lineage, with fossils dating back almost 3000 million years. The blue-green algae exhibit a variety of forms and are the most widely distributed of algal groups. Their cellular organization can be unicellular, branched or unbranched filamentous, or unspecialized colonial. The cyanophytes are distributed in marine and freshwater environments, occasionally forming blooms in eutrophic waters. They are an important component of benthic systems forming mats on soil, mudflats, and hot springs but are less conspicuous in soils along reef margins. Cyanobacteria can be found as symbiotic organisms in diatoms, ferns, lichens, cycads, sponges, algae, and other systems (Stewart et al. 1983; Costa and Lindblad 2002; Charpy et al. 2012; Rikkinen 2013).

Glaucophyta algae can be unicellular flagellates or form colonies in freshwater habitats, although some species can be found in soil samples. They are dorsoventrally constructed and have blue-green photosynthetic plastids. Glaucophytes are of particular importance in evolutionary studies because their plastids differ from those of other eukaryotic algae and resemble cyanobacteria in some way (Barsanti and Gualtieri 2006).

Red algae, or formally, Rhodophyta, occur as single cells, individual filaments, aggregations of filaments, or sheets of cells. Photosynthetic pigments are present in the red plastids of most species, except in certain parasitic genera. They are unusual among eukaryotes because of the lack of flagella in any stages of their life cycle and the presence of accessory phycobiliproteins organized in phycobilisomes. Although a few studies have noticed subtle motility of red algal propagules (Rosenvinge 1927; Geitler 1944; Lin et al. 1975; Hill et al. 1980; Pickett-Heaps et al. 2001), it is generally accepted that they are non-motile, especially compared with brown and green algae. The dispersal and settlement of red algal propagules are strongly dependent on the abiotic factors of the benthic environment (Ogata 1953; Luther 1976; Harlin and Lindbergh 1977). In most species, cytokinesis is incomplete; thus, the daughter cells are separated by a proteinaceous plug that fills the junction between cells (pit connection), which successively becomes a plug. Most red algae have sexual life cycles which usually involve alternation of three generations. Red algae are especially diverse and abundant in tropical and subtropical marine waters, but they are also present in freshwater and terrestrial environments. Red algae can be classified under two main classes, the Bangiophyceae that retain morphological characters that are found in the ancestral pool of red algae, ranging from single cells to multicellular filaments or sheetlike thalli, and the Florideophyceae that include morphologically complex red algae and are widely considered to be a derived, monophyletic group. One of the most striking features of the Florideophyceae is that they are the most diverse algal group regarding the thallus construction.

The phylum Phaeophyta is defined by one particular feature, that is, the presence of two different flagella in the cells. Flagellate cells are thus termed heterokont, possessing a long mastigonemate flagellum, and a short smooth one that points backward along the cell. This division includes several classes, Phaeophyceae, Bacillariophyceae, Cryptophyceae, Dictyochophyceae, Haptophyceae, and Xanthophyceae, of which the first two are the most relevant in coastal environments. The class Phaeophyceae (brown algae) is almost exclusively marine and is dominant in temperate waters. They include more than 250 genera, ranging from microscopic filaments to giant kelps that can reach 80 m long. Some species of Phaeophyceae display the greatest complex organization of tissues and cells among algae. Many have photosynthetic blades, as well as specialized blade-bearing stipes, holdfasts, and specialized conductive cells. The class Bacillariophyceae, integrated by organisms informally called diatoms, represents the most common algal group. They are very abundant and thus significant primary producers responsible for an estimated 20% of global carbon fixation. Diatoms dominate the phytoplankton of cold, nutrient-rich waters, such as upwelling areas of the oceans and recently circulated lake waters. Diatoms are the most significant producers of biogenic silica, dominating the marine silicon cycle. It is estimated that over 30 million km^2 of ocean floor is covered with sedimentary deposits of diatom frustules (Harris et al. 2006). These organisms are also important components of the benthic estuarine environment forming the microphytobenthic community.

The members of the phylum Dinophyta (dinoflagellates) are typically unicellular flagellates. Dinoflagellates have two flagella with independent beating patterns, conferring a characteristic rotatory swimming motion. Flagella are apically inserted in a region close to the midpoint of the ventral side of the cell. They are common components of the freshwater and marine habitats (Barsanti and Gualtieri 2006).

The phylum Euglenophyta includes mostly unicellular widely distributed flagellates; predominantly occupants of interfaces, such as the air-water and sediment-water boundaries. They are especially abundant in highly eutrophic environments and are known to be tolerant to extreme conditions of desiccation, low pH, and heat (Walne and Kivic 1990).

Chlorarachniophyta are amoeboid, coccoid or flagellate cells with secondary green plastids (Hibberd and Norris 1984). They are phototrophic and phagotrophic, engulfing bacteria, flagellates, and eukaryotic algae. The name chlorarachniophytes refers to the green color of their plastids and the spider shape of the cells. Chlorarachniophytes occur in temperate and tropical marine waters, growing among sand grains, on mud, in tide pools, on seaweeds, or in plankton.

The phylum Chlorophyta (green algae) displays a large range of somatic differentiation varying from flagellates to complex multicellular thalli differentiated into macroscopic structures that resemble organs. The different levels of thallus organization (unicellular, colonial, filamentous, siphonous, and parenchymatous) have served as the basis of their classification. Chlorophytes include at least nine lineages of early-diverging, unicellular prasinophytes (Lemieux et al. 2014) and a more-derived, monophyletic assemblage known as the "core Chlorophyta" (Fučikova et al. 2014). Traditionally, the chlorophytes were divided into the classes Ulvophyceae (ulvophyceans), Trebouxiophyceae (trebouxiophyceans), and Chlorophyceae (chlorophyceans).

The taxonomic treatment that will be used throughout this chapter for the algae occurring in the Bahía Blanca Estuary originates from morphological identifications of recently collected specimens or from local coastal reports. For this reason, some algae will be mentioned to the genus level, while others to the species level.

6.2.2 *Morpho-functional Classification*

In every biological community, each species occupies a unique ecological niche and commonly there are groups of species that utilize the resources in similar ways. That is, there are species that may be geographically distant and evolutionarily distinct, but have similar ecological functions in the ecosystem and thus occupy similar adaptive zones. As mentioned before, benthic algae can be single cells or large seaweeds with internal structural complexity. This variability and diversity can be simplified by classifying algae into functional-morphological categories. In oceanic systems, this classification is used to describe macroalgal assemblages (Vanderklift and Lavery 2000; Konar and Iken 2009), to address physiological questions (Littler and Arnold 1982; Johansson and Snoeijs 2002), and to examine the impact of several types of disturbance on benthic communities (Dethier 1981). The recognition of such aggregates by ecologists enables them to understand and predict the outcome of interspecific interactions and to interpret patterns in community structure without studying individual species (Steneck and Watling 1982). Littler and Littler (1980) and Steneck and Watling (1982) proposed models for algae, where the overall form of the thallus was hypothesized to predict aspects of its physiology and ecology, and resistance to consumers. The general models hypothesize that the type of growth and mineralization of algae dictates relative rates of primary productivity, growth rate, competitive ability, resistance to consumption by grazers, resistance to physical disturbance, tolerance to physiological stress, and successional stage, and that all of these functions are correlated with each other.

The model proposed by Littler and Littler (1984) considers algal form groups: sheet, filamentous, coarsely branched, thick leathery, jointed calcareous, and crustose. Steneck and Watling (1982) later created a slightly different ranking that included filamentous, foliose, corticated, leathery, articulated calcareous, and crustose forms. A more recent model of classification, proposed by Steneck and Dethier (1994), is based on the productivity and susceptibility to grazing, incorporating algal groups such as microalgae, filamentous, crustose, foliose, corticated foliose, corticated macrophyte, leathery macrophyte, and articulated calcareous (Table 6.1). Balata et al. (2011) proposed a new expanded classification of morphological functional groups based on several characteristics of the thallus like structure, growth form, branching pattern, and also on the taxonomic affinities of the alga (Table 6.2).

Table 6.1 Classification of algae into morpho-functional groups

Functional groups	External morphology	Internal anatomy	Texture	Representatives
Microalgae	Single cell			Diatoms – Cyanobacteria
Filamentous	Delicately branched	Uniseriate, multiseriate, or lightly corticated	Soft, flexible	*Gelidium, Giffordia, Cladophora, Ectocarpus, Acrochaetium, Chaetomorpha*
Sheetlike	Thin tubular or flattened	Uncorticated, one-several cells thick	Soft, flexible	*Colpomenia, Dictyota, Ulva, Porphyra*
Coarsely branched	Coarsely branched, upright, terete	Corticated	Fleshy-wiry	*Codium, Colpomenia*
Thick leathery	Thick blades and branches	Differentiated, heavily corticated, thick walled	Leathery, rubbery	kelps
Jointed calcareous	Articulated, calcareous, upright	Calcified genicula, flexible intergenicula	Stony	*Corallina*
Crustose	Epilithic, prostrate, encrusting	Calcified or uncalcified parallel cell rows	Stony or tough	Ralfsioid crust

Adapted from Littler and Littler (1984) and Steneck and Dethier (1994)

Table 6.2 Classification of macroalgae from the Bahía Blanca Estuary into morpho-functional groups

Functional group	Description	Taxa
Chlorophyta		
1. Filamentous uniseriate	Uniseriate filaments, either branched or not	*Cladophora surera*, *Chaetomorpha linum*
2. Bladelike	Thalli forming blades one to few layers of cells thick	*Ulva lactuca*, *Ulva prolifera*, *Ulva intestinalis*, *Ulva flexuosa*
3. Tubular	Macroscopic thalli consisting of one-layered hollow tubes	*Blidingia marginata*, *Blidingia minima*
4. Siphonous with thin separate filaments	Algae consisting of branched, single-celled filaments	*Bryopsis plumosa*
Phaeophyta		
1. Filamentous uniseriate	Uniseriate branched filaments	*Ectocarpus siliculosus*, *Hincksia hincksiae*
2. Compressed with bladelike habit	Corticated algae producing expanded blades with erect habit	*Punctaria latifolia*, *Planosiphon nakamurae*
Rhodophyta		
1. Filamentous uniseriate and pluriseriate with erect thallus	Thin filaments (up to 2 mm wide), forming bushy branched thalli devoid of (or with limited) prostrate parts	*Polysiphonia abscissa*, *Polysiphonia morrowii*, *Ceramium diaphanum*, *Acrochaetium* sp.
2. Bladelike	Thalli forming blades one to few layers of cells thick	*Pyropia* sp.
3. Larger-sized corticated	Branched corticated algae with erect habit and comparatively thick axes (more than 2 mm wide)	*Gelidium pusillum*, *Gelidium crinale*, *Gracilaria verrucosa*
4. Smaller-sized corticated	Branched corticated algae with erect habit and relatively thin axes (up to 2 mm wide)	*Neosiphonia harveyi*

According to Balata et al. (2011)

6.3 Factors Affecting Algal Assemblages in Estuarine Environments

6.3.1 *Abiotic Factors*

Estuarial environments are fluctuating habitats, characterized by a mixture of freshwater and seawater which originates unstable conditions; hence, the communities that develop in estuaries are well adapted to this habitat. Physical and chemical factors of the estuarine environment clearly influence the functional morphology and behavior of the benthic communities. For example, in intertidal flats community structure is regulated by sediment particle size, oxygen concentration, salinity, and temperature, to which the species may respond differently (Coull 1999; Schweiger et al. 2008). Periodic inundation and exposure to air is harsh in these environments. Light intensity, salinity, pH, nutrient levels, grazing, and sediment stability limit the productivity of algae (Kennish 2017). In estuaries, high biogeochemical rates account for a relatively low number of species (Costanza et al. 1993); this is because only a limited number of species are adapted to salinity variation (Webb et al. 1997).

Microphytobenthic and macroalgal communities are regulated by light, temperature, and nutrient availability, which are the major parameters controlling their reproduction (Denis et al. 2012). Seaweeds disperse either as free-floating macroscopic forms or as microscopic propagules (e.g., spores, gametes, and zygotes). The spatial patterns of seaweed in estuaries are contingent upon the dispersal capabilities of the populations; hence, tidal currents play a crucial role in these habitats as they influence the dispersal, settlement, and attachment of macroalgae (Kennish 2017).

6.3.2 *Biotic Interactions*

Community structure of estuarine environments is influenced by various biotic interactions, involving grazing and competition (Buffan-Dubau and Carman 2000). The trophic dynamics of estuaries are complex because this environment is occupied by different types of primary producers such as phytoplankton, salt marsh plants, submersed seagrass, and benthic algae. Unlike the open sea, where practically all phytoplankton is consumed alive, in estuaries, primary producers are not heavily grazed, but die and begin to decompose before being consumed, contributing to the detritus food web (Omstedt et al. 2014). Algae generate abundant organic detritus constituting an important food source available all year round; they also can provide new habitats where some species can find a refuge from predators (Valiela et al. 1997; Raffaelli et al. 1998). Thus, organic matter from benthic macroalgae together with phytoplankton may support both benthic and pelagic food webs in the intertidal and adjacent subtidal areas of shallow bay systems (Kang et al. 2003). Macroalgal thallus tends to have higher nitrogen and phosphorus contents than seagrass tissues (Atkinson and Smith 1983; Duarte 1995). When macroalgal decomposition occurs, the derived nutrients accumulate and increase the nutritional value of the sediment, stimulating the growth of other primary producers, microalgae and bacteria (Rossi 2006), therefore promoting the increase in abundance and biomass of grazers and deposit feeders (Hull 1987; Ford et al. 1999; Rossi and Underwood 2002). In the case of organic matter transported through the water column, this is used by filter feeders such as oysters, clams, and mussels; while the organic matter that accumulates in the sediment is later used by deposit feeders such as worms, amphipods, and a many other small organisms (Day et al. 2013).

In estuarine soft sediments, macroalgae are often spatially distributed forming a mosaic of patches of different species interspersed with bare substratum (Berezina et al. 2007). Several studies have detected that benthic macroalgal communities support higher abundances of both epibenthos and infauna than do other comparable unvegetated bottoms (Summerson and Peterson 1984; Irlandi and Peterson 1991). The effects of macroalgae on benthic organisms are probably density dependent, because at moderate densities, macroalgae create a more heterogeneous environment (Thiel and Watling 1998). This favors the increase of subsurface deposit feeders and the abundance of epifaunal species, which find abundant shelter and food in the new algal substratum (Raffaelli et al. 1998). At high densities, macroalgal blooms alter the natural balance between production and decomposition of organic matter and can have dramatic effects on the local fauna (Valiela et al. 1997). The extent of these effects depends on the composition of the dominant faunal assemblages at the bloom onset, the magnitude and nature (drifting or steady) of the bloom, the season, and, to a lesser extent, the type of bloom-forming algae (Raffaelli et al. 1998).

Isaksson and Pihl (1992) observed that when algae cover approximately 30–50% of the surface in soft-bottom marine ecosystems, the abundance of epibenthic fauna associated with vegetation increases, whereas with algal coverages of 90%, the abundance of mobile epibenthic fauna declines. The decomposition of large deposits of macroalgae can affect the availability of oxygen in the sediment and cause episodes of anoxia and sulfide production, accentuated by reduced water flow under the algae and increased sedimentation rate (Nedergaard et al. 2002). In these situations, the abundance and diversity of the assemblages can decrease because animals can be displaced or killed or because of an alteration between predator and prey relationships (Renaud et al. 1999; Kelaher and Levinton 2003). In areas with fluctuating oxygen levels, mobile predators can temporarily leave hypoxic areas and then return when oxygen rises to capture the infaunal invertebrates that emerged during low oxygen (Nestlerode and Diaz 1998).

Infauna from shallow soft bottoms, for example burrowing adult bivalves, respond to high biomass of bloom-forming algae by migrating on the surface. Juvenile bivalves, on the other hand, can reach high abundance among algal strands and filaments, while highly mobile species, such as the gastropod *Hydrobia* W. Hartmann, 1821, often reach high densities in algal mats (Raffaelli et al. 1998). In contrast, Franz and Friedman (2002) observed that the blooms of *Ulva lactuca* Linnaeus drastically reduced the abundance of epibenthic copepods, attributing that reduction to the anoxic conditions registered within the algal mats. In drifting algae, Norkko et al. (2000) found high epifaunal abundance on the macroalgae *Ectocarpus siliculosus* (Dillwyn) Lyngbye and *Pilayella littoralis* (Linnaeus) Kjellman, with intermediate oxygen levels within the drifting seaweed, and only found hypoxic conditions at the algal-sediment interface. In the case of the mobile infauna, filamentous algae provide a refuge from physical stresses. However, for sedentary infauna, the cover by laminar algae forms a barrier between the animals and the oxygenated water column generating an adverse effect on survivorship (Perkins and Abbott 1972; Bach and Josselyn 1979; Dauer and Conner 1980). The physical structure provided by algae serves as a refuge and reduces the predation rate of mobile epibenthic species in soft-sediment communities (Lenanton et al. 1982; Robertson and Lenanton 1984; Marx and Herrnkind 1985; Wilson et al. 1990; Smith and Herrnkind 1992). For instance, the increased cover of *Ulva* and *Cladophora* reduces the predation rates of the crustaceans *Crangon crangon* (Linnaeus, 1758) and *Carcinus maenas* (Linnaeus, 1758) by the cod *Gadus morhua* (Linnaeus, 1758)

(Isaksson et al. 1994; Pihl et al. 1995). The juveniles of many fish species also find shelter and abundant food between macroalgal communities, and, after reaching a certain size, they swim to deeper waters (Coles et al. 1993; Ross and Moser 1995; Rooker et al. 1998). Seagrass meadows thus play an important role as "nurseries" for numerous fish species (Heck et al. 2003).

In summary, anoxic conditions have an overall negative effect on the abundance of the faunal community in spite of all the other possible concomitant positive effects of the bloom (i.e., enhanced refuge, food, and structure for recruitment) (Sagasti et al. 2001). The relative tolerance of organisms to hypoxic conditions generated by algae may determine the increase or decrease of predation rates. For example, predation on larval fish by jellyfish can increase during hypoxic episodes, because jellyfish tolerate hypoxia, but larval fish are unable to evade predators in low oxygen. Concurrently, predation on larval fish by adult fish decreases, because the adult fish are intolerant to hypoxia (Breitburg et al. 1994).

On the other hand, it is important to mention the impact of herbivores on the biomass of benthic algae (i.e., top-down control). Microphytobenthic algae are affected by grazing (Hillebrand et al. 2000). Several studies report that the extent of herbivory on benthic algae is highly variable (Cebrian et al. 1998; Cebrian 1999, 2002). In some cases, herbivores may remove a large percentage of the production of microphytobenthic communities (Nicotri 1977; Baird and Ulanowicz 1993), but in other instances, herbivory only represents a minor loss for the community (Admiraal et al. 1983; Montagna 1984). In spite of the large variability found within each community type, microphytobenthos and macroalgae tend to lose higher percentages of primary production due to grazing than do seagrass communities (Bennett et al. 1999; Blanchard et al. 2001), influencing the dynamics of producers biomass and nutrient recycling.

6.4 Adaptations Mechanisms of the Benthic Algae

Estuaries are habitats with high variability; hence, the communities that live in them are frequently disturbed. Some species of algae rapidly recolonize the substrates after a disturbance; a new generation can appear within a few weeks, and several generations may develop in one year. Such a fast succession of generations allows the adaptation of algal species that are able to tolerate the abrupt changes that occur in estuarine conditions (Larsen and Sand-Jensen 2006). These changes favor the development of fast-growing, short-lived, and morphologically simple algae, like phytoplankton, epiphytic algae, and ephemeral macroalgae, whereas benthic macrophytes such as kelp do not occur (Pedersen and Borum 1996).

Macroalgal species tend to exhibit vertical patterns of distribution, from the upper to the lower tidal levels, originating different zonation patterns. This is because different species have different adaptive responses to the physical, chemical, and biotic conditions. The lack of hard substrates, the low light penetration, the variations of salinity, pollution, competition, and grazing are factors that explain the reduced seaweed richness of estuaries and the occurrence of some annual species (Druehl 1967; Larsen and Sand-Jensen 2006). Although estuaries are stressful environments to most marine algae, a few adapted species can flourish. For example, eutrophic estuaries promote the proliferation of opportunistic and tolerant macroalgae like *Ulva*, *Chaetomorpha* Kützing, *Cladophora*, *Monostroma* Thuret, *Ceramium* Roth, *Gracilaria* Greville, *Porphyra* C. Agardh, *Pyropia* J. Agardh, *Ectocarpus* Lyngbye, and *Pilayella* Bory, which can bloom into large proportions (Morand and Merceron 2005; Scanlan et al. 2007).

Wilkinson (1981) exposed three fundamental points to explain macroalgal distribution patterns in estuaries: (1) The colonization occurs almost wholly by marine species, with freshwater ones abundant only in the uppermost reaches of the estuary. (2) In the uppermost regions of estuaries, there is a reduction in species number and diversity due to the decrease of red algae (Rhodophyta) and then of brown algae (Phaeophyta). Green algae (Chlorophyta) and blue-green algae (Cyanophyta) do not necessarily become more numerous, in terms of species, but they constitute a much greater proportion of the algal community because they extend further inland. (3) Colonization of the mid-reaches is dominated by brackish water algae.

The algae that live in estuaries generally proliferate under a broad range of temperatures, irradiances, salinities, and nutrient conditions (Martins et al. 1999; Taylor et al. 2001; Sousa et al. 2007; Choi et al. 2010; Kim et al. 2011). For example, the growth rate of *Ulva* increases at low salinity and high nutrient levels in laboratory culture conditions (Taylor et al. 2001). *Ulva* has physiological adaptations to grow as free-floating mats. Some species of *Ulva* have partially hollow branched tubular thalli filled with air generated through photosynthesis, which under unfavorable conditions produce thalli that can float for 2 or 3 months favoring survival (Kim et al. 2011).

The filamentous algae are well adapted to overcome the low light intensities of estuarine waters. They have thin photosynthetic tissues, with high contents of pigments per cell volume that reduce the trajectory of light through the tissue. As they have high area/volume ratio, they use efficiently the light and nutrients per biomass unit (Kirk 1994; Niklas 1992; Nielsen and Sand-Jensen 1990; Duarte 1995; Duarte et al. 1995).

Estuarine filamentous green algae with heterotrichous growth such as *Cladophora* present densely pigmented assimilatory cells that penetrate upward through the mud and the covering mats that contrast greatly with the weakly pigmented cells of the prostrate system. This morphological characteristic constitutes an ecological adaptation to burial by soft sediment, as it helps retain moisture during low tides (Boedeker and Hansen 2010).

Cladophorales algae have other strategies for protection against desiccation and fluctuating salinities, for example, the formation of a thick gelatinous cover produced by swelling of the outer cell wall layers (Wille 1909), endophytic habit (Polderman 1976), and the presence of hematochrome/oil droplets (Wille 1909).

Chaetomorpha has been mainly studied for its ecological role as a possible regulator of nutrient availability in estuarine habitats (Krause-Jensen et al. 1996, 1999; Menendez 2005) and for its ability to tolerate a wide range of salinities. The identification of ascorbate oxidase activity in *Chaetomorpha linum* (O.F. Müller) Kützing suggests a novel mechanism of adaptation to increased salinity, because this enzyme could have a role in salt stress by catalyzing intracellular production of water, which could mitigate the stress (Caputo et al. 2010).

Algae that are unicellular, colonial, or have a thallus with 1–4 cell layers can change their pigment concentration, affecting light absorbance as a mechanism of adaptation to irradiance variations (Agusti et al. 1994).

In most estuaries, microphytobenthic communities are composed of a mixture of different taxa that can form conspicuous biogenic structures on intertidal and supratidal sediments commonly called biofilms and microbial mats, the latter being among the oldest ecological structures on Earth (van Gemerden 1993). The main difference between biofilms and mats is that the first one is formed by a single layer of microorganisms embedded in an organic matrix, while microbial mats are vertically stratified microbial communities dominated by cyanobacteria. Microbial mats are regarded as advanced stages of biofilms (Noffke 2010). Biofilms and microbial mats result from the growth and activity of microphytobenthic organisms that trap sediment particles and bind them in extracellular polymeric substances (EPS) produced by the organisms (Margulis et al. 1980; Krumbein 1994). Although many microorganisms are capable of secreting EPS, phototrophs are especially important since they produce them *de novo* through CO_2 fixation, while chemotrophic organisms form EPS by converting other organic compounds, which may be limited (Stal 2006).

Extracellular polymeric substances are a complex mixture of high-molecular-weight polymers consisting of 90% or more of polysaccharides (Hoagland et al. 1993). They allow the locomotion of microorganisms and also provide protection against changes in salinity, temperature, UV radiation, and desiccation (Decho 2000). At the same time, they generate high cohesion in the sediment since they form an adherent cover on the particles (de Winder et al. 1999; Decho 2000). Microphytobenthos develops mainly in marine sediments of the intertidal and supratidal region because the extreme and strongly fluctuating environmental conditions that prevail in this region exclude, or at least reduce the abundance of grazers (Stal 2006). In open coastal environments, coarse sand is deposited over high-energy areas, becoming unsuitable for the development of microbial mats or biofilms because of the damage caused by the abrasion of sand particles induced by the action of wind and waves (Eckman et al. 2008). On the other hand, silt and clay sediments are deposited in low-energy areas such as estuaries and deltas; these sediments are characterized by low light penetration and high sedimentation rate, which difficult the phototactic migration response of cyanobacteria. In addition, this type of sediment usually contains large amounts of nutrients, conditions under which the cyanobacteria are outcompeted by opportunistic organisms, mainly diatoms, which have high growth rates under high nutrient conditions. This is why cyanobacteria prefer sediments of fine to medium sand as a substrate for the formation of microbial mats, while in silt-clay sediments, biofilms of diatoms predominate (Watermann et al. 1999).

In extreme conditions, microphytobenthic organisms are able to regulate their photosynthesis to avoid photoinhibition (Cartaxana et al. 2011), thereby maintaining relatively high abundance in the sediment. For example, epipelic life forms are able to migrate vertically in the sediment to position themselves in favorable light conditions, and diatoms can adapt their photosynthetic apparatus efficiently to the light conditions in a few minutes (Glud et al. 2002). In these ways, physiological photoinhibition is avoided at high light levels of solar radiation. However, the microgradients of sediment grain sizes and organic particle sizes, percentages of organics, porosity, light attenuation, and oxygen govern vertical microalgal distribution patterns (Kennish 2017).

6.5 The Benthic Algae of the Bahía Blanca Estuary

Among the benthic algae that occur in the Bahía Blanca Estuary, microscopic epipelic algae are the most studied. The first studies of the microphytobenthos appeared in the 1980s (i.e. Cicerone 1987; Farías 1988) which initially depicted the diversity of the microscopic algae of this region. These preliminary studies were subsequently followed by qualitative studies conducted on specific coasts of the estuary (Parodi and Barria 2003). The most recent studies focus on ecological aspects (Da Rodda 2004) or were motivated by the anthropic impact caused by the expansion of the industrial area (Pizani 2009).

Thanks to these studies, there is a fair knowledge of the diversity of microalgal taxa that can be found in many of the coastal regions of the estuary. The relevance that diatoms and Cyanophyceae have in the process of sediment stabilization and their contribution to the formation of microbial mats has led the phycological studies in this region. These topics will be discussed later in the chapter and in Box 6.1.

A different scenario stands for the macroscopic algae (including their microscopic life stages), to which less attention has been paid. There are fair explanations for this situation. The Bahía Blanca Estuary is located in a coastal region considered to be poor in macroalgal flora (Liuzzi et al. 2011). From a biogeographic point of view, the Bahía Blanca Estuary is located in a transition zone between the Argentinean and the Magellanic biogeographic provinces that is characterized by a reduction of macroalgal biodiversity (Balech and Ehrlich 2008). This scarcity of macroscopic algae is a consequence of two factors that are related with the characteristics of the sediment and its dynamics: (1) in the majority of the coasts, the dominant fraction of the sediment is composed of small-sized particles, which may prevent the settlement of the macroalgal reproductive cells and/or compromise the successful attachment of the microscopic life stages of macroalgae (Parodi 2004); and (2) the turbid waters resulting from the suspended fine sediment reduce the penetration of light into the lower layers of the water column, limiting the depth at which the macroalgae are able to photosynthesize. These combined conditions restrict the occurrence of the microscopic and macroscopic thalli to the hard substrata submerged in shallow waters or exposed to air during low tide, where the reproductive cells, either motile or non-motile, can fix and access to light intensities that allow them to photosynthesize. Examples of appropriate substrates are the consolidated sediments and any artificial or natural solid object set on the coast. Despite that the turbid waters significantly reduce the penetration of light in the water, the shallow tidal flats that span in some sectors of the Bahía Blanca Estuary allow the development of small intertidal communities of macroalgae, where several species can thrive.

As a consequence of these unfavorable conditions, the macroalgal assemblages of the Bahía Blanca Estuary are less conspicuous and go unnoticed, unlike the assemblages of other coasts of Argentina. The estuarial coasts have high sedimentation rates; therefore, some macroalgae are usually buried under the sediment. This adds to the fact that the actual diversity of macroalgal species is usually obscured by the existence of cryptic species. As the worldwide trend suggests, it is expected that the number of macroalgal taxa in the Bahía Blanca Estuary increases as the coastal industrialization advances, providing available artificial substrate, increasing the nutrient loads in the water, and promoting the ingression of alien marine species through ballast water.

6.5.1 Diversity and Composition of Soft-Bottom Microalgal Assemblages

Many studies about different microphytobenthic communities have been carried out in the Bahía Blanca Estuary to examine the biodiversity, structure, and dynamics of this estuarine benthic ecosystem (Cicerone 1987; Parodi and Barría de Cao 2003; Da Rodda and Parodi 2005; Pizani 2009). A particular feature of the tidal flats of the Bahía Blanca Estuary is that they are colonized by extensive microbial mats (Cuadrado and Pizani 2007; Cuadrado et al. 2011, 2012). They have been widely studied to explore the relationship between microphytobenthos, sediment, and physical factors, such as irradiance, temperature, sedimentation rate, and wave height (Cuadrado et al. 2012, 2013; Pan et al. 2013a). Most of these studies have been conducted in Puerto Rosales and, to a lesser extent, in Puerto Cuatreros, Villarino Viejo, and Almirante Brown locations (Figs. 2.2, 2.3, and 2.4; Chap. 2).

Even though the surface of the mudflats of the Bahía Blanca Estuary is often apparently devoid of vegetation, the richness of microphytobenthic algae is high since a total of 144 taxa have been recorded by different authors (Parodi and Barría de Cao 2003; Da Rodda 2004; Da Rodda and Parodi 2005; Pizani 2009; Fernández et al. 2018). Diatoms are the dominant group with 109 taxa, whereas 34 taxa of Cyanobacteria and only 1 taxon of Euglenophyta have been registered. Regarding diatoms, *Nitzschia* Hassall and *Navicula* Bory are the best represented genera, with 27 and 13 taxa, respectively. Figure 6.1 shows some common benthic microalgae found in the Bahía Blanca Estuary.

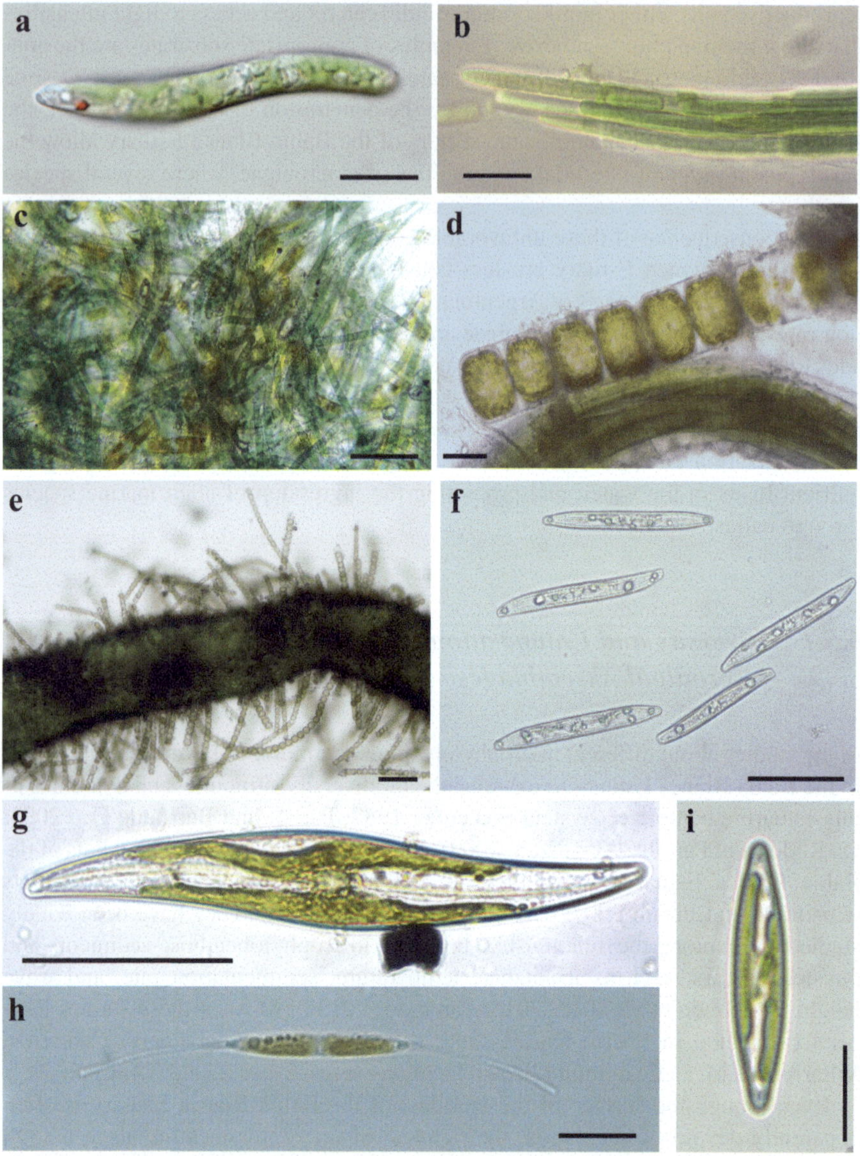

Fig. 6.1 Common benthic microalgae of the Bahía Blanca Estuary, (**a**) *Euglena* sp. (scale bar: 6 µm), (**b**) trichomes of *Coleofasciculus chthonoplastes* (scale bar: 20 µm), (**c**) biofilm of diatoms and filamentous cyanobacteria developing on hard substrate (scale bar: 50 µm), (**d**) centric diatom *Melosira* and filamentous cyanobacteria (scale bar: 6 µm), (**e**) centric diatoms on *Blidingia* sp. (scale bar: 60 µm), (**f**) *Nitzschia clausii* (scale bar: 50 µm), (**g**) *Gyrosigma spencerii* (scale bar: 50 µm), (**h**) *Cylindrotheca closterium* (scale bar: 6 µm), (**i**) *Navicula* sp. (scale bar: 6 µm). (Photos by (**a**) Natalia Pizzani, (**b**, **d**, **f**, **g**, **h**, and **i**) Carolina Fernandez, (**c** and **e**) M. Emilia Croce)

The unique characteristics of estuarine environments allow the coexistence of freshwater, estuarine, and marine species; according to this, the diatom species found in the microphytobenthos of the Bahía Blanca Estuary are mostly holo-euryhaline forms, which can tolerate large changes in the salinity of water from hypotonic to hypertonic. On the other hand, freshwater species are also present, which arrive from the discharges of freshwater tributaries (Pizani 2009). In addition, some of the species found in surface sediments are typically planktonic forms, namely, species of *Triceratium* Ehrenberg, *Podosira stelligera* (Bailey) A. Mann, *Planothidium delicatulum* (Kützing) Round and Bukhtiyarova (reported as *Achnanthes delicatula* (Kützing) Grunow), *Luticola mutica* (Kützing) D.G. Mann, *Craticula halophila* (Grunow) D.G. Mann (cited as *Navicula halophila* (Grunow) Cleve), *Navicula salinicola* Hustedt (reported as *Navicula incertata* Lange-Bertalot), *Gyrosigma attenuatum* (Kützing) Rabenhorst, *Paralia sulcata* (Ehrenberg) Cleve, and *Petrodictyon gemma* (Ehrenberg) D.G. Mann (cited as *Surirella gemma* (Ehrenberg) Kützing), or are species that are usually found in both planktonic and benthic habitats, such as *Entomoneis alata* (Ehrenberg) Ehrenberg, *Cylindrotheca closterium* (Ehrenberg) Reimann and J.C. Lewin, *Gyrosigma fasciola* (Ehrenberg) J.W. Griffith and Henfrey, *Nitzschia sigma* (Kützing) W. Smith, and *Petrodictyon gemma* (Ehrenberg) D.G. Mann (Parodi and Barría de Cao 2003; Da Rodda and Parodi 2005; Pizani 2009). This is because the surface of the sediment in the tidal mudflat is constantly subjected to the action of waves and tidal currents, which results in the resuspension of benthic individuals, as well as in the incorporation of planktonic species by sedimentation.

The microphytobenthic community of Puerto Rosales consists mainly of filamentous cyanobacteria and small diatoms. Among filamentous cyanobacteria, *Coleofasciculus chthonoplastes* (Thuret ex Gomont) M. Siegesmund, J.R. Johansen, and T. Friedl (reported as *Microcoleus chthonoplastes* Thuret ex Gomont) is dominant, whereas species of *Oscillatoria* and *Arthrospira* Sitzenberger ex Gomont are less abundant. *C. chthonoplastes* typically has many trichomes within a common sheath threaded into a spiral arrangement; the resulting mesh of interweaving cyanobacterial filaments together with the microbially secreted EPS traps the sand grains and significantly increases the cohesiveness of sediments (Stal et al. 1985). In that sense, the dominance of *C. chthonoplastes* is indicative of well-developed microbial mat, presenting an elevated resistance to erosion, and a protective cover to the underlying sediments. The microbial mats in which this cyanobacterium is dominant were termed "epibenthic mats" by Noffke (2010) and are typically found in the supratidal zone.

The diatoms recorded in the Bahía Blanca Estuary are mainly small pennate diatoms; the genera *Nitzschia*, *Navicula*, *Diploneis* Ehrenberg ex Cleve, and *Amphora* Ehrenberg ex Kützing are quantitatively dominant. Other larger diatoms of the genera *Pleurosigma* W. Smith, *Gyrosigma* Hassall, and *Cylindrotheca* Rabenhorst are also present, but they are less frequent. Among central diatoms, the genera *Thalassiosira* Cleve and *Coscinodiscus* Ehrenberg and the species *Cyclotella meneghiniana* Kützing and *Paralia sulcata* have been mentioned by different authors (Pizani 2009; Pan et al. 2013a, b). The dominance of small diatoms is attributed to the fact that small cells have higher growth and nutrient uptake rates than bigger cells since they have higher surface/volume ratio, which allows small cells to outcompete bigger cells when they are subjected to frequent physical disturbances (Williams 1964; Snoeijs et al. 2002).

Box 6.1: Ecosystem Engineers in the Bahía Blanca Estuary: The Crab Neohelice granulata

Tidal currents are responsible for sediment transport, and waves produce either sediment deposition or erosion. These factors interact with the biological components of the coast determining whether deposition or erosion is the dominant process in a specific site. These complex processes ultimately determine the type and abundance of organisms in the sediment (Blanchard et al. 2000; Dyer et al. 2000).

Bioturbation is defined as the biological reworking of sediments and soils through animal activities like feeding and burrowing that generates changes in chemical gradients and relocates resources and microorganisms. Such sediment restructuring also promotes physical alterations, by changing the balance of material transported, and affects the structure and functioning of assemblages (Meysman et al. 2006; Kristensen et al. 2012). Organisms that directly or indirectly modify the physical environment and regulate the availability of resources for other species are called ecosystem engineers (Statzner et al. 2000; Gutiérrez et al. 2003; Berkenbusch and Rowden 2003; Cardinale et al. 2004).

Neohelice granulata (Dana, 1851) (= *Chasmagnathus granulata*) is an estuarine crab that excavates semipermanent burrows generating extensive burrowing beds which cover up to 80% of the intertidal areas of SW Atlantic estuaries and bays (Botto et al. 2006; Iribarne et al. 2005). *N. granulata* distributes from the northeastern coast of Patagonia, Argentina (42°25′S, 64°36′W), to Río de Janeiro, Brazil (22°57′S, 42°50′W) (Spivak 2010). In the muddy salt marshes of the Bahía Blanca Estuary, an association between the crab and the halophyte plant *Sarcocornia perennis* (Miller) A. J. Scott was described by Perillo and Iribarne (2003). This association has a particular configuration, where the plants form a ring surrounding a non-vegetated salt pan densely excavated by the crab. These ring-shaped configurations are 1.5–8 m in diameter and have high water retention at the inner part. Such ring-shaped configuration of the halophyte vegetation is the macroscopic evidence of the changes in salinity, humidity, and hardness that occur in the sediment and which are a consequence of the plant-crab interaction (Escapa et al. 2007).

The composition and structure of the microphytobenthic assemblage differ considerably among the different environments composing the rings. A thin diatom biofilm, characterized by high abundance of diatoms and reduced abundance of cyanobacteria, is observed in the inner part of the rings, associated with small grain sediment, while well-developed microbial mats, characterized by the presence of *Coleofasciculus chthonoplastes*, are present in the outer region of the rings. Both structures show differences in the EPS matrix, since the biofilms dominated by diatoms are less developed and more irregular than those formed in the presence of cyanobacteria (Fig. 6.5a). In cyanobacterial mats, EPS are found as an embedded continuous matrix, whereas in diatom biofilms, they develop as a web of spongelike fibrils with void spaces (Fernández et al. 2018) (Fig. 6.5b). Such differences in the microphytobenthic assemblages are the result of changes in the physical properties of sediments caused by the bioturbation of the burrowing crab.

A common consequence of the activity of burrowing deposit feeders is the increase in softness and water content of the sediment. Also, the burrows contribute to the accumulation of fine particles since the sediment with a high percentage of clay accumulates at the burrow tunnel during high tide (Davis 1993; Botto and Iribarne 2000). This physical mixture would restrain the formation of large and well-developed microbial mats on the superficial sediments of the salt marsh with abundant crabs. Then, the death of the plants in the inner part of the rings can also be associated with permanent modifications in the development of the microphytobenthos succession resulting from the biodisturbing action of crabs (Fernández et al. 2018).

Given the important role of microphytobenthos in recycling of nutrients, biofiltration, and sediment stabilization in coastal ecosystems, the study of the distribution patterns of micro- and macroorganisms provides valuable information for the formulation of integrated management plans, aiming to reduce the ecosystem erosion and contamination.

6.5.2 Hard Substrate Available for Algal Settlement

In the Bahía Blanca Estuary, the hard substrate available for the settlement of algae comprises natural and artificial structures (Fig. 6.2). The hard substrate of natural origin consists of rocks made of consolidated fine sedimentary particles (henceforth named *outcrops*) and the mollusk shells (oysters, mussels, snails, and barnacles). The outcrops are made up of compacted sand and clay material giving rise to a relatively hard substrate (Spalleti 1980; Aliotta and Lizasoain 2004). These outcrops are common in the southeastern coasts of the estuary like Villa del Mar (Fig. 2.4; Chap. 2). These substrates are called outcrops because they appear interrupting the large tidal flats or sand beaches. In the Bahía Blanca Estuary, macroalgal assemblages inhabit the depressions (tidal pools) that form on these outcrops, which remain filled with water during low tide (Fig. 6.2b).

The main artificial hard substrate consists of wooden docks, concrete barriers, metal ladders, floating platforms and bridges, mooring ropes, sewer pipes, buoys,

Fig. 6.2 Artificial and natural substrata of the Bahía Blanca Estuary colonized by benthic algae, (**a**) general view of a bloom of green macroalgae on a salt marsh in Villa del Mar, (**b**) macroalgal assemblage in a tidal pool located in the outcrops of Villa del Mar, (**c**) macroalgae attached to mollusk shells, (**d**) macroalgae attached to a floating platform in CNBB, (**e**) green and red macroalgae growing on concrete rocks in the upper intertidal zone of Puerto Rosales, (**f**) detail of *Blidingia* sp. on a concrete rock in Puerto Rosales, (**g**) green macroalgae on a wheel in Villa del Mar, (**h**) green macroalgae covering the surface of a sewer pipe in Club Náutico Bahía Blanca (CNBB). (Photos by M. Emilia Croce)

and rubber wheels. All these surfaces appear in the different coasts of the Bahía Blanca Estuary and are colonized by micro- and macroalgae. Although the hard substrate is scarce in the Bahía Blanca Estuary, the industrial growth in the region has modified the environment in such a way that the surfaces available for algal attachment have increased in number by the artificial structures constructed by humans.

Although microbial mats develop conspicuous biogenic structures on soft sediments, the biofilms composed of diatoms and filamentous cyanobacteria can also be found covering hard natural or artificial substrates in the intertidal zone (Fig. 6.1c). In contrast to microalgae, macroscopic algae are always found attached to hard substrates (except for the drifting species). This is because the main condition for a macroalga to colonize a substrate is that the substrate is relatively stable, for the algae to remain attached and to avoid being flushed away by waves or currents. Consequently, macroscopic algae are virtually able to occupy any type of substrate as long as the surface is suitable to attach for the microscopic propagules (spores or gametes), or a fragment of the thallus, in the case of vegetative propagation (Amsler et al. 1992; Fletcher and Callow 1992). Because of the different tolerances of each species of macroalgae to desiccation stress (Dromgoole 1980; Davison and Pearson 1996), the success to colonize a stable substrate greatly depends on the location of the substrate with respect to the coastline (Lobban and Harrison 1994). The establishment and persistence of benthic algal populations depend on the reproductive performance of the species but also on the successful survival of their propagules. Any factor that influences the recruitment, the settlement, and/or the post-settlement of algal propagules becomes an important factor determining the establishment, dynamics, and structure of algal communities (Wahl and Hoppe 2002). For example, substratum microtopography has been recognized as one of the major factors structuring marine benthic communities (Emson and Faller-Fristch 1976; Woodin 1978; Menge et al. 1983; Brault and Bourget 1985; Bergeron and Bourget 1986). In the case of artificial substrate, the type of material that constitutes the substrate also influences the colonization and survival of the macroalgal species. For example, substrates that retain water like ropes are favorable for species that are less tolerant to desiccation (Nienhuis 1969).

In general, marine macroalgae have a wide capacity of dispersal through a variety of forms, from unicellular to multicellular propagules, either sexual or asexual (Santelices 1990; Norton 1992). Due to their potential for colonization of new habitats, any modification of the habitat can quickly lead to changes in the macroalgal diversity of the coast. The factors associated with the expansion of the industrial area, such as the increased availability of substrate, the introduction of exotic species through ballast water, and the increase of nutrient loads into the water, are promoting changes in the diversity and distribution of macroalgae in the Bahía Blanca Estuary.

6.5.3 Diversity and Composition of Macroalgal Assemblages

The richness of macroalgae in the Bahía Blanca Estuary is low compared with other coasts of Argentina (Miloslavich et al. 2011). According to the literature, a total of 19 macroalgal taxa have been recorded on the coasts of this estuary (Perillo et al. 2001; Parodi 2004; Bremec et al. 2004; Croce et al. 2015; Hoshino et al. 2020; Koller 2021). Eighteen of those taxa were found in the most recent surveys from 2015 to 2019, together with three new taxa, *Blidingia marginata* (J. Agardh) P.J.L. Dangeard ex Bliding, *Blidingia minima* (Nägeli ex Kützing) Kylin, and a species of *Pyropia* J. Agardh. The complete list of 22 taxa is shown in Table 6.2. Red and green macroalgae are the dominant groups with nine taxa each. According to the classification of functional groups proposed by Balata et al. (2011), six morpho-functional groups are recognized; the majority of the taxa belong to the categories bladelike and filamentous uniseriate and pluriseriate with erect thallus.

Although the turbid waters in this region limit the growth of macroalgae because of the low light penetration through the water column, some tolerant species such as opportunistic species with rapid growth rates and small turf-like forms flourish in these habitats. In terms of biomass, the most abundant red macroalgae are *Polysiphonia abscissa* J.D. Hooker and Harvey, *Polysiphonia morrowii* Harvey, and *Ceramium diaphanum* (Lightfoot) Roth. These three filamentous species grow in the tidal pools of the outcrops in Villa del Mar, where they are present almost all year round, although their bushy thalli are larger in winter (Fig. 6.3a, c). Species of *Polysiphonia* Greville also live attached to floating platforms near the surface in the recreational harbor of the Bahía Blanca Nautic Club (CNBB), where they are continuously submerged. The presence of *Gracilaria verrucosa* (Hudson) Papenfuss in the tidal flats of Villa del Mar is reported in the literature (Parodi 2004); however, this species has not been found in recent surveys. The macroscopic thalli of *Pyropia* are abundant during the winter on the hard substrates of the upper intertidal of Puerto Rosales. *Pyropia* grows in the mooring ropes that are frequently exposed to air during low tide, on the wooden pillars that support the docks, and on concrete rocks (Fig. 6.3d). In general, the distribution of benthic red macroalgae that inhabit the Bahía Blanca Estuary is limited to the intertidal flats and salt marshes of Villa del Mar; except for *Pyropia*, they have not been recorded in other coasts with hard substrate available. The lack of mobility of the propagules may restrict the spread of red macroalgae in this estuary. Two species of *Gelidium* J.V. Lamouroux grow exclusively in the tidal pools of the outcrops located in the tidal flats of Villa del Mar. They form dense mats on the consolidated sediment or grow as epiphytes on mollusk shells (Fig. 6.3b). *Gelidium* species are perennial in these coasts and reproduce sexually all year round. They also reproduce and spread vegetatively, due to their ability for regeneration and reattachment from fragments by the formation of rhizoids. This behavior has been observed during culture experiments (unpublished data) and is reported for other species of *Gelidium* as well (Santelices and Varela 1994; Titlyanov and Titlyanova 2006; Otaiza et al. 2018).

Fig. 6.3 Common benthic red macroalgae of the Bahía Blanca Estuary, (**a**) *Polysiphonia morrowii*, detail of the thallus and axes (scale bar: 500 µm), (**b**) *Gelidium pusillum*, general aspect of the thallus (scale bar: 1 cm), (**c**) *Ceramium diaphanum*, general aspect of the thallus (scale bar: 1 cm), (**d**) *Pyropia* sp.: thalli in nature (scale bar: 1 cm). (Photos by (**a**, **b**, and **d**) M. Emilia Croce, (**c**) Ailen Poza)

The green algae are the second dominant group, represented by nine taxa (Table 6.2). *Ulva* Linnaeus is the most conspicuous genus, represented by four species that reach high biomass in winter (Fig. 6.4c). They cover large surfaces of the intertidal flats and tidal pools in Villa del Mar. They also colonize submerged objects and cover the concrete walls and rocks located in the upper intertidal zone of the harbors. Green algae of the family Ulvophyceae are the most common group of macroalgae in estuarine habitats. This is because they withstand desiccation occasioned by the strong irradiance and wind during low tide and tolerate a wide range of salinities. The presence of the genus *Enteromorpha* Link was reported by Perillo et al. (2001), Parodi (2004), and Bremec et al. (2004), although this genus has been now merged into *Ulva* (Hayden et al. 2012). Isolated thalli of *Cladophora surera*

Fig. 6.4 Common benthic green and brown macroalgae of the Bahía Blanca Estuary, (**a**) *Blidingia marginata*, detail of the thallus (scale bar: 60 µm), (**b**) *Bryopsis plumosa*, general aspect of the thallus (scale bar: 2 cm), (**c**) *Ulva lactuca*, general aspect of the thallus (scale bar: 1 cm), (**d**) *Punctaria latifolia* (scale bar: 2 cm). (Photos by (**a**, **c**, and **d**) Ailen Poza, (**b**) M. Emilia Croce)

E.R. Parodi and E.J. Cáceres are usually found in tidal pools or on the concrete structures, but are most frequently found as an epiphyte on other macroalgae. *C. surera* and *Ulva flexuosa* Wulfen (= *Enteromorpha flexuosa* (Wulfen) J. Agardh) are tolerant to salinity changes and more related to freshwater environments (Parodi 2004). The coenocytic algae *Bryopsis plumosa* (Hudson) C. Agardh forms dense bushes during winter in the upper tidal pools of the outcrops of Villa del Mar (Fig. 6.4b). *Chaetomorpha linum* grows as an epiphyte on other macroalgae, but it also occurs on the shaded regions of the concrete walls of the harbor of the CNBB. Green macroalgae such as *Blidingia minima* and *B. marginata* grow usually on concrete rock in the upper intertidal zone where they can reach high densities (Fig. 6.4a).

The brown algae are less frequent in this habitat; they are represented only by four taxa, the Ectocarpaceae *Ectocarpus siliculosus* and *Hincksia hincksiae* (Harvey) P.C. Silva, the Chordariaceae *Punctaria latifolia* Greville, and the Scytosiphonaceae *Planosiphon nakamurae* M. Hoshino, M.E. Croce, Hanyuda, and Kogame (Table 6.2). The Ectocarpaceae have been almost exclusively found as epiphytes on other macroalgae. Macroscopic thalli of *P. nakamurae* can be found in winter, growing on the walls and concrete blocks in the harbor of CNBB that are completely exposed during low tide. *P. latifolia* occurs in the salt marshes dominated by *Spartina* Schreb., located in Villa del Mar. The stems of this vascular plant offer a temporary substrate for the attachment during the winter where the macroscopic thalli of *P. latifolia* (Fig. 6.4d) attach sparsely (Parodi 2004, as *Punctaria latifolia* var. *crouanii*).

Among the macroalgae that live in the Bahía Blanca Estuary, three of them are considered exotic (or alien): *Polysiphonia morrowii*; *Neosiphonia harveyi* (Bailey) M.-S. Kim, H.-G. Choi, Guiry, and G.W. Saunders; and *Planosiphon nakamurae*. These species have been registered in other coasts of the Patagonian region as well. *P. morrowii* is present in Puerto Madryn (Raffo et al. 2014), Bahía Anegada (Croce and Parodi 2014), and Las Grutas (personal observation). There are different hypotheses of the introduction of *P. morrowii*. The vectors may have been ballast water (Hewitt et al. 2007) or other exotic marine organisms such as the Pacific oyster in Bahía Anegada or the macroalgae *Undaria pinnatifida* (Harvey) Suringar, that have been introduced in the 1980s and the 1990s (Verlaque 2001; Kim et al. 2004; Geoffroy et al. 2012). There is no evidence about the initial site of their introduction, but it may have dispersed along the South Atlantic coast by shipping among different harbors. *N. harveyi* is also present in the southern coasts of Argentina (Raffo et al. 2014), and it is catalogued as introduced, given that it is worldwide known as an invasive species. *Planosiphon nakamurae* was recently found in the coasts of the South Atlantic Ocean. Its identity was confirmed by molecular tools, which also evidenced its relationship with a Japanese haplotype (Hoshino et al. 2020).

The exotic macroalgae are characterized by fast vegetative growth and dispersal, usually by fragmentation, by a rapid completion of the life cycle, and wide tolerances to environmental variables (Nyberg and Wallentinus 2005). For several reasons, exotic macroalgae are common in estuarine environments. First, because estuaries are areas of entrance to the new environment, usually through ballast water that is released by the international vessels. Second, the strong environmental variations, mainly salinity changes and desiccation, are more easily tolerated by exotic species. And third, because estuaries are regions occupied by human settlements, and consequently, they constitute eutrophic environments where the exotic species with fast-growing capabilities can exploit outcompeting with the native species.

6.6 Epiphytic Algae

Epibiosis is defined as a nonsymbiotic facultative association between an epibiont (an organism that lives attached to a living surface) and a basibiont (a substrate organism or host) (Wahl 1989). This phenomenon is common in aquatic environments and among macroalgae (epiphytes), which may live attached to animals, vascular plants, or other macroalgae.

In the Bahía Blanca Estuary, there are natural substrates such as the leaves and stems of halophyte plants, and the thalli of the macroalgae themselves, where epiphytic macroalgae and microalgae can attach. For example, *P. latifolia* grows attached to the stems of *Spartina*. Although the population is relatively ephemeral, this epiphytic association is persistent throughout the years, suggesting that it is well established. The most frequent taxa that live attached to other macroalgae are the filamentous red and green macroalgae *Ceramium* Roth, *Acrochaetium* Nägeli, *Cladophora*, and *Ulva* sp. (Koller 2021); however, the diatoms *Melosira* C. Agardh, *Achnanthes* Bory, and *Cocconeis* Ehrenberg frequently appear attached to the surfaces of the macroalgae thallus, solitary or in chains. *Blidingia* species are sometimes covered by epiphytic diatoms; the most frequent is *Melosira* that forms dense brown tufts of long chains (Fig. 6.1e).

Epiphytic associations among macroalgae and benthic fauna are also common in the Bahía Blanca Estuary. *Polysiphonia* and *Gelidium* species are epiphytes on two dominant bivalves, the native mussel *Brachidontes rodriguezii* (d'Orbigny, 1842) and the exotic Pacific oyster *Magallana gigas* (Thunberg, 1793) (= *Crassostrea gigas* Thunberg, 1793). Succession and ecological studies in other coasts of the northern Patagonian region have shown that *M. gigas* offers a suitable substrate for the settlement of native and exotic macroalgae (Borges 2006; Croce and Parodi 2012) in areas where no suitable hard substrate is available. *Gelidium* lives also attached to the gastropod *Crepidula aculeata* (Gmelin, 1791). This association is very frequent suggesting that the symbiosis may be beneficial to both species. The macroalgae may benefit from nutrient loadings produced by the gastropod beneath, and the gastropod may enhance the dispersal of the macroalgae by carrying them from one place to another. From the perspective of the mollusk, the benefit may be related to the avoidance of predators, since the macroalgae cover the shell completely.

6.7 Mat-Forming Macroalgae

The term "turf" is widely used in marine ecology to identify a layer of short and densely branched algae that is several millimeters to a few centimeters tall (Connell et al. 2014). Several macroalgal species fall under this definition, although the more common are small-sized species of red algae representatives of the orders Ceramiales, Gelidiales, and Corallinales. A more specific term, the word "mat", identifies a small group of algae defined as short and densely branched and formed by prostrate and erect axes and which grow entangled into a thick mass (Hay 1981). These groups of algae, either turf-forming or mat-forming species, are relevant to the dynamics of benthic ecosystems (Airoldi et al. 1995; Bulleri and Benedetti-Cecchi 2006; Gorman and Connell 2009). One reason for that is that these cushions trap large amounts of sediment particles in relation with their small size, influencing the transportation of energy and organic matter in intertidal marine environments (Airoldi 2003). It is known from several studies that the macroalgae that form this type of mats have an important influence on the nutrient dynamics. Besides their role in the dynamics of the mentioned abiotic factors, they are also key components of biotic assemblages as they provide an excellent niche for little crustaceans, polychaetes, annelids, mollusks, and other small invertebrates (Prathep et al. 2003). Through many years of surveys in the intertidal regions of the Bahía Blanca Estuary, it has been noticed that the populations of the mat-forming Gelidiales may be key components of the benthic communities in this coast. The vegetative growth of these algae has been studied in culture conditions, and preliminary results showed that these algae have a high capacity of regeneration by producing numerous branches, rhizoids, and rhizoidal filaments, and they can also withstand a high load of epiphytes for long periods of time (unpublished data).

Box 6.2: Potential Use of Native Estuarine Macroalgae for Biomitigation

Marine macroalgae are used in a variety of domestic and industrial processes. The more ancestral use is as fresh food (Lee et al. 2017), but due to their varying intrinsic characteristics and chemical composition, they can also be used for the production of a variety of algal products, for example, fertilizers (Selvam and Sivakumar 2014), biogas by anaerobic digestion (Hinks et al. 2013), phycocolloids (Chan and Matanjun 2017), and polymers that can be incorporated into conventional plastic formulations to develop biodegradable plastics (Freile-Pelegrín et al. 2007). Macroalgae are also a source of bioactive compounds (polysaccharides, proteins, lipids, polyphenols, carotenoids, and vitamins). These phytochemicals have different functional groups including carboxyl, hydroxyl, phosphate, and amine that can bind pollutants (Areco and dos Santos 2010, Sanjeewa et al. 2016). The presence of sulfated polysaccharides in the cell wall of macroalgae, mainly in their fibrous matrix and intercellular spaces, is the main reason for their high capacity to bind contaminants. In fact, hydroxyl, sulfate, and carboxyl groups of the polysaccharide chains are strong ion exchangers; therefore, they are the important sites of complexation of metal cations (Vasconcelos and Leal 2001). Biosorption is one of the most promising remediation technologies for aquatic areas that are polluted with heavy metal ions (Gupta et al. 2015). The major advantages of biosorption using the macroalgal biomass for wastewater treatment are the low cost and investment needed, the simple design and easy operation, and the use of nontoxic substances. Hence, recently, the interest in using seaweed as biomitigators or for bioremediation of marine environments is increasing (Kim et al. 2017). For this reason, macroalgae can be exploited as a resource at the same time as it is used for ecological services.

Along the eastern coast of the Bahía Blanca Estuary, there are several human settlements. From the inner part of the estuary to the mouth, we find the towns of Villarino Viejo, General Daniel Cerri, Bahía Blanca, and Punta Alta. Of all these emplacements, the city of Bahía Blanca has the largest demographic growth due to the settlement of a fertilizer industry, a large petrochemical pole, and thermoelectric plants, as well as the expansion of the harbor. As a consequence of this economic development, the estuary is the receptor of waste discharges from industrial origin (oil derivatives, pesticides, heavy metals, etc.) as well as untreated domestic sewage, which have generated problems of contamination (Marcovecchio et al. 2010).

The macroalgal assemblages that inhabit the Bahía Blanca Estuary are potentially useful for implementing methodologies of pollutants remotion and eutrophication management. The most promising candidate for biomitigation is the green alga *Ulva*, which is particularly useful in the biosorption of heavy metals and other compounds due to its high surface area, relatively simple structure, and uniform distribution of binding sites (Sari and Tuzen 2008; Turner et al. 2007).

The multiple functions and uses of seaweeds discussed above would promote the cultivation of seaweeds to obtain high-quality raw materials for different applications. Biosorption by seaweeds is a promising method that utilizes efficiently the naturally existing raw material. It is noticeable that very few studies have used real wastewater for the treatment and most of the experiments have used simulated wastewater. Therefore, it is recommended that future studies consider the use of real wastewater, especially in impacted environments such as the Bahía Blanca Estuary.

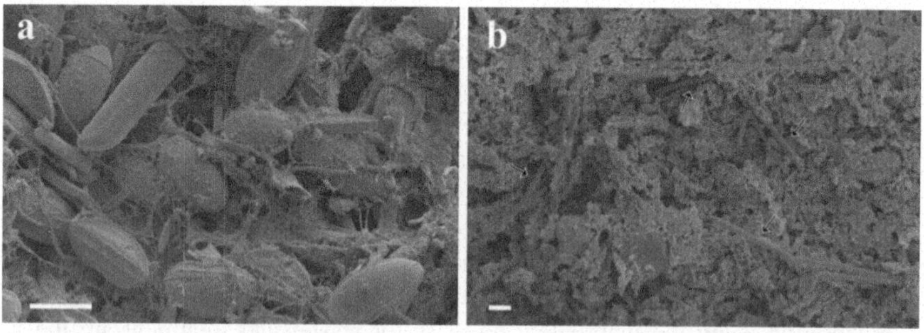

Fig. 6.5 Structure of microphytobenthic assemblages, (**a**) scanning electron photomicrograph of biofilms dominated by pennate diatoms embedded in a compacted matrix of EPS (scale bar: 20 μm), (**b**) microbial mat dominated by filamentous cyanobacteria (arrows) (scale bar: 6 μm). (Photos by Constanza Da-Rodda)

Glossary

Amoeboid	Type of cell organization that lacks a cell wall and the protoplasm undergoes frequent changes in shape.
Blade	(= lamina) Flattened structure that is somewhat leaflike.
Bloom	Massive or conspicuous growth of algae, usually a large percentage of the total cells belong to one or a few species.
Coccoid	Simple cell type that is spherical, subspherical, or rod-shaped.
Cytokinesis	The process by which one cell physically divides into two cells.
Epilithic	Organisms living attached to the surfaces of rocks and stones.
Epipelic	Organisms living attached to the surfaces of mud or sand.
Epiphytic	Organisms living on the surfaces of plants or algae, using it for support but not for nutrition.
Epizoic	Organisms living on the surfaces of animals.
Eukaryotic	Organisms composed of cells with membrane-bounded nucleus. Most contain cells with a complex organization, with microtubules, membrane-bounded organelles, and chromatin organized into more than a single chromosome.
Eutrophic	Waters that contain relatively high levels of dissolved nutrients (e.g., nitrate, phosphate); typically exhibit high levels of primary productivity.
Flagellum	Long, threadlike organelle that projects out of the cell and functions in motility. In eukaryotic cells, they consist of a 9 doublet + 2 central singlet array of microtubules.
Frustule	In diatoms, the silica cell wall or test, composed of two valves.
Holdfast	A cell or multicellular structure that functions in attachment to a substrate.
Kelp	Large phaeophytes that are members of the Laminariales.
Mastigonemate	Stiff, lateral hairs borne by a flagellum, consisting of a base, a tubular shaft, and several terminal hairs.
Monophyletic	Evolutionary term referring to a trait or group of organisms that evolved directly from a common ancestor.
Phagotrophic	Mode of nutrition referring to heterotrophic protoctists or tissue cells that ingest solid food particles by phagocytosis.
Phycobiliproteins	Complex of phycobilins with protein found in cyanobacteria, rhodophytes, glaucocystophytes, and some cryptophytes.
Phycobilisome	Cellular structure containing phycobilin pigments and arranged as protrusions on the surface of the thylakoids of cyanobacteria, rhodophytes, and glaucocystophytes, but within the thylakoids (between membranous stacks) in the plastids of cryptophytes.

Photosynthetic pigments	Pigments are chemical compounds present in algae by which the energy of sunlight is captured for photosynthesis.
Plastids	Generic term for photosynthetic organelles present in algae. Plastids contain the enzymes and pigments for photosynthesis, ribosomes, and other structures.
Polyphyletic	Evolutionary term referring to a trait or group of organisms that derived from more than one common evolutionary ancestor or ancestral group.
Prokaryotic	Typically unicellular organism lacking a distinct nucleus and membrane-bound organelles.
Propagules	Generative structure, any unicellular or multicellular, capable of survival, dissemination, and further growth.
Somatic	General term referring to the body (soma) of an organism. The part not involved in reproduction or germination.
Spore	Reproductive cell (motile or not) that originates by mitosis (mitospore) or meiosis (meiospore).
Stipe	Stalk, stemlike portion of the algal thallus.
Symbiotic	Physical association between organisms of different species.
Thallus	The body of an alga that is not differentiated into roots, stalks, and leaves as in other plants.
Thylakoid	Flat saclike structure formed by membranes that is present in the cytoplasm of cyanobacterial cells and in plastids of eukaryotic algae and plants.

References

Admiraal W, Bouwman LA, Hoekstra L et al (1983) Qualitative and quantitative interactions between microphytobenthos and herbivorous meiofauna on a brackish intertidal mudflat. Int Rev Ges Hydrobiol 68:175–191

Agusti S, Enriques S, Frost-Christensen S et al (1994) Light harvesting among photosynthetic organisms. Funct Ecol 8:273–279

Airoldi L (2003) The effects of sedimentation on rocky coast assemblages. Oceanogr Mar Biol 41:161–236

Airoldi L, Rindi F, Cinelli F (1995) Structure, seasonal dynamics and reproductive phenology of a filamentous turf assemblage on a sediment influenced, rocky subtidal shore. Bot Mar 38:227–237

Aliotta S, Lizasoain GO (2004) Tipos de fondos y su caracterización geológica por métodos sismoacústicos. In: Piccolo MC, Hoffmeyer M (eds) Ecosistema del estuario de Bahía Blanca. Instituto Argentino de Oceanografía, Bahía Blanca, pp 51–59

Amsler CD, Reed DC, Neushul M (1992) The microclimate inhabited by macroalgal propagules. J Br Phycol 27:253–270

Areco MM, dos Santos A (2010) Copper, zinc, cadmium and lead biosorption by *Gymnogongrus torulosus*. Thermodynamics and kinetics studies. Colloid Surface B 81:620–628

Atkinson MJ, Smith SV (1983) C:N:P ratios of benthic marine plants. Limnol Oceanogr 28:568–574

Bach SD, Josselyn MN (1979) Production and biomass of *Cladophora prolifera* (Chlorophyta, Cladophorales) in Bermuda. Bot Mar 22:163–168

Baird D, Ulanowicz RE (1993) Comparative study on the trophic structure, cycling and ecosystem properties of four tidal estuaries. Mar Ecol Prog Ser 99:221–237

Balata D, Piazzi L, Rindi F (2011) Testing a new classification of morphological functional groups of marine macroalgae for the detection of responses to stress. Mar Biol 158:2459–2469

Balech E, Ehrlich MD (2008) Esquema biogeográfico del Mar Argentino. Rev Invest Des Pesq 19:45–75

Barsanti L, Gualtieri P (2006) Algae: anatomy, biochemistry, and biotechnology. Taylor & Francis, Boca Raton

Bennett A, Bianchi TS, Means JC et al (1999) The effects of polycyclic aromatic hydrocarbon contamination and grazing on the abundance and composition of microphytobenthos in salt marsh sediments. J Exp Mar Biol Ecol 242:1–20

Berezina N, Tsiplenkina IG, Pankova ES et al (2007) Dynamics of invertebrate communities in stony littoral of the Neva Estuary (Baltic Sea) under macroalgal blooms. Transit Waters Bull 1:49–60

Bergeron P, Bourget E (1986) Shore topography and spatial partitioning of crevice refuges by sessile epibenthos in an ice disturbed environment. Mar Ecol Prog Ser 28:129–145

Berkenbusch K, Rowden AA (2003) Ecosystem engineering-moving away from "just-so" stories. N Z J Ecol 27:67–73

Blanchard GF, Paterson DM, Stal LJ et al (2000) The effect of geomorphological structures on potential biostabilisation by microphytobenthos on intertidal mudflats. Cont Shelf Res 20:1243–1256

Blanchard GF, Guarini J-M, Orvain F et al (2001) Dynamic behaviour of benthic microalgal biomass in intertidal mudflats. J Exp Mar Biol Ecol 264:85–60

Boedeker C, Hansen GI (2010) Nuclear rDNA sequences of *Wittrockiella amphibia* (Collins) comb. nov. (Cladophorales, Chlorophyta) and morphological characterization of the mat-like growth form. Botanica Marina 53(4):351–356

Bold HC, Wynne MJ (1978) Introduction to the algae: structure and reproduction. Prentice-Hall, Inc., Englewood Cliffs

Borges ME (2006) Ecología de las ostras en ambientes del sur bonaerense: cultivo y manejo de sus poblaciones. PhD thesis, Universidad Nacional del Sur, Bahía Blanca, p 247

Botto F, Iribarne O (2000) Contrasting effects of two burrowing crabs (*Chasmagnathus granulata* and *Uca uruguayensis*) on sediment composition and transport in estuarine environments. Estuar Coast Shelf Sci 51:141–151

Botto F, Iribarne O, Gutiérrez J et al (2006) Ecological importance of passive deposition of organic matter into burrows of the SW Atlantic crab *Chasmagnathus granulatus*. Mar Ecol Prog Ser 312:201–226

Brault S, Bourget E (1985) Structural changes in an estuarine subtidal epibenthic community: biotic and physical causes. Mar Ecol Prog Ser 21:63–73

Breitburg DL, Steinberg N, DuBeau S et al (1994) Effects of low dissolved oxygen on predation on estuarine fish larvae. Mar Ecol Prog Ser 64:235–246

Bremec CS, Martinez DE, Elias R (2004) Asociaciones bentónicas de fondos duros y comunidades incrustantes. In: Piccolo MC, Hoffmeyer M (eds) Ecosistema del estuario de Bahía Blanca. Instituto Argentino de Oceanografía, Bahia Blanca, pp 171–178

Buffan-Dubau E, Carman KR (2000) Diel feeding behaviour of meiofauna and their relationships with microalgal resources. Limnol Oceanogr 2:381–395

Bulleri F, Benedetti-Cecchi L (2006) Mechanisms of recovery and resilience of different components of mosaics of habitats on shallow rocky reefs. Oecologia 149:482–492

Caputo E, Ceglie V, Lippolis M et al (2010) Identification of a NaCl-induced ascorbate oxidase activity in *Chaetomorpha linum* suggests a novel mechanism of adaptation to increased salinity. Environ Exp Bot 69:63–67

Cardinale BJ, Gelmann ER, Palmer MA (2004) Net spinning caddisflies as stream ecosystem engineers: the influence of Hydropsyche on benthic substrate stability. Funct Ecol 18:381–387

Cartaxana P, Ruivo M, Hubas C et al (2011) Physiological versus behavioral photoprotection in intertidal epipelic and epipsammic benthic diatom communities. J Exp Mar Biol Ecol 405:120–127

Cebrian J (1999) Patterns in the fate of production in plant communities. Am Nat 154:449–468

Cebrian J (2002) Variability and control of carbon consumption, export and accumulation in marine communities. Limnol Oceanogr 47:11–22

Cebrian J, Williams M, McClelland J et al (1998) The dependence of heterotrophic consumption and C accumulation on autotrophic nutrient content in ecosystems. Ecol Lett 1:165–170

Chan PT, Matanjun P (2017) Chemical composition and physicochemical properties of tropical red seaweed, *Gracilaria changii*. Food Chem 221:302–336

Charpy L, Casareto BE, Langlade MJ et al (2012) Cyanobacteria in coral reef ecosystems: a review. J Mar Biol 2012:1–9

Choi TS, Kang EJ, Kim JH, Kim KY (2010) Effect of salinity on growth and nutrient uptake of *Ulva pertusa* (Chlorophyta) from an eelgrass bed. Algae 25:17–26

Cicerone D (1987) Estimación de la biomasa de diatomeas bentónicas en áreas mesolitorales de la porción interna del Estuario de Bahía Blanca, a partir del análisis de pigmentos fotosintéticos. Lic thesis. Universidad Nacional del Sur, Bahía Blanca

Coles RG, Lee Long WJ, Watson RA (1993) Distribution of seagrasses, and their fish and penaeid prawn communities, in Cairns Harbour, a tropical estuary, Northern Queensland, Australia. Aust J Mar Fresh Res 44:193–126

Connell SD, Foster MS, Airoldi L (2014) What are algal turfs? Towards a better description of turfs. Mar Ecol Prog Ser 495:299–307

Costa JL, Lindblad P (2002) Cyanobacteria in symbiosis with cycads. In: Rai AN, Bergman B, Rasmussen U (eds) Cyanobacteria in symbiosis. Springer, Dordrecht

Costanza R, Kemp WM, Boynton WR (1993) Predictability, scale, and biodiversity in coastal and estuarine ecosystems-implications for management. Ambio 22:88–96

Coull BC (1999) Role of meiofauna in estuarine soft-bottom habitats. Aust J Ecol 24:327–343

Croce ME, Parodi ER (2012) Seasonal dynamic of macroalgae in intertidal pools formed by beds of *Crassostrea gigas* (Mollusca, Bivalvia) on the north Patagonian Atlantic coast. Bot Mar 55:49–58

Croce ME, Parodi ER (2014) The Japanese alga *Polysiphonia morrowii* (Rhodomelaceae, Rhodophyta) on the South Atlantic Ocean: first report of an invasive macroalga inhabiting oyster reefs. Helgol Mar Res 68(2):241–252

Croce ME, Gauna MC, Fernández C et al (2015) Intertidal seaweeds from North Atlantic Patagonian coasts, Argentina. Check List 11(5):1739

Cuadrado DG, Pizani NV (2007) Identification of microbially induced sedimentary structures over a tidal flat. Latin Am J Sedimentol Basin Anal 14:65–116

Cuadrado DG, Carmona NB, Bournod CN (2011) Biostabilization of sediments by microbial mats in a temperate siliciclastic tidal flat, Bahía Blanca estuary (Argentina). Sediment Geol 237:95–61

Cuadrado DG, Carmona NB, Bournod CN (2012) Mineral precipitation on modern siliciclastic tidal flats colonized by microbial mats. Sediment Geol 271–272:58–66

Cuadrado DG, Bournod CN, Pan J et al (2013) Microbially-induced sedimentary structures (MISS) as record of storm action in supratidal modern estuarine setting. Sediment Geol 296:1–8

Da Rodda C (2004) Biodiversidad y distribución espacial de las comunidades microalgales bentónicas de la marisma "El Cangrejal" en el Estuario de Bahía Blanca. Lic Thesis, Universidad Nacional del Sur, Bahía Blanca, p 51

Da Rodda C, Parodi ER (2005) Cyanophyceae Epipélicas de la Marisma El Cangrejal en el estuario de Bahía Blanca (Buenos Aires, Argentina). Bol Soc Argent Bot 40:157–168
Daehnick AE, Sullivan MJ, Moncreiff CA (1992) Primary production of the sand microflora in seagrass beds of Mississippi sound. Bot Mar 35:131–139
Dauer DM, Conner WG (1980) Effects of moderate sewage input on benthic polychaete populations. Estuar Coast Mar Sci 6:335–346
Davis WR (1993) The role of bioturbation in sediment resuspension and its interaction with physical shearing. J Exp Mar Biol Ecol 171:187–200
Davison IR, Pearson GA (1996) Stress tolerance in intertidal seaweeds. J Phycol 32:197–211
Day JW, Yáñez-Arancibia A, Kemp WM et al (2013) Introduction to estuarine ecology. In: Day JW, Crump BC, Kemp WM et al (eds) Estuarine ecology, 2nd edn. Wiley, Hoboken, pp 1–18
De Winder B, Staats N, Stal LJ et al (1999) Carbohydrate secretion by phototrophic communities in tidal sediments. J Sea Res 42:131–146
Decho AW (2000) Exopolymer microdomains as a structuring agent for heterogeneity within microbial biofilms. In: Riding RE, Awramik SM (eds) Microbial sediments. Springer, Berlin, pp 1–9
Denis L, Gevaert F, Spilmont N (2012) In situ intertidal variability of microphytobenthic production in a macrotidal temperate estuarine system. J Soils Sediments:1517–1529
Dethier M (1981) Heteromorphic algal life histories: the seasonal pattern and response to herbivory of the brown crust *Ralfsia californica*. Oecologia 49:333–339
Dromgoole FI (1980) Desiccation resistance of intertidal and subtidal algae. Bot Mar 23:149–159
Druehl LD (1967) Vertical distribution of some benthic marine algae in a British Columbia inlet, as related to some environmental factors. J Fish Res Bd Can 24:33–46
Duarte CM (1995) Submerged aquatic vegetation in relation to different nutrient regimes. Ophelia 41:87–11
Duarte CM, Sand-Jensen K, Nielsen SL et al (1995) Comparative functional ecology. Trends Ecol Evol 6:418–421
Dyer KR, Christie MC, Wright EW (2000) The classification of intertidal mudflats. Cont Shelf Res 20:639–660
Eckman JE, Andres MS, Marinelli RL et al (2008) Wave and sediment dynamics along a shallow subtidal sandy beach inhabited by modern stromatolites. Geobiology 6:21–32
Emson RH, Faller-Fristch RJ (1976) An experimental investigation into the effect crevice availability on abundance and size-structure in a population of *Littorina rudis* (Manton) (Gastropoda; Prosobranchia). J Exp Mar Biol Ecol 23:285–297
Escapa M, Minkoff DR, Perillo GM et al (2007) Direct and indirect effects of burrowing crab *Chasmagnathus granulatus* activities on erosion of Southwest Atlantic Sarcocornia dominated marshes. Limnol Oceanogr 52:2340–2349
Farías L (1988) Estudio del microfitobentos en áreas mesolitorales de la porción interna del Estuario de Bahía Blanca (Puerto Cuatreros). Seminario Lic. Oceanografía. Universidad Nacional del Sur, Bahia Blanca
Fernández C, Da-Rodda C, Gauna MC et al (2018) The role of plant-crab interaction in structuring microphytobenthic communities in a shallow temperate estuary. Braz J Oceanogr 66:307–314
Fletcher RL, Callow ME (1992) The settlement, attachment and establishment of marine algal spores. J Br Phycol 27:303–329
Ford RB, Thrush SF, Probert PK (1999) Macrobenthic colonisation of disturbances on an intertidal sandflat, the influence of season and buried algae. Mar Ecol Prog Ser 191:163–174
Franz DR, Friedman I (2002) Effects of macroalgal mat (*Ulva lactuca*) on estuarine sand flat copepods: an experimental study. J Exp Mar Biol Ecol 271:209–226
Freile-Pelegrín Y, Madera-Santana T, Robledo D et al (2007) Degradation of agar films in a humid tropical climate: thermal, mechanical, morphological and structural changes. Polym Degrad Stab 92:244–252

Fučikova K, Lewis PO, Lewis LA (2014) Widespread desert affiliation of trebouxiophycean algae (Trebouxiophyceae, Chlorophyta) including discovery of three new desert genera. Phycol Res 62:294–305

Geitler L (1944) Furchungsteilung, simultane Mehrfachteilung, Lokomotion, Plasmoptse und Ókologie der Bangiacee *Porphyridium cruentum*. Flora 37:300–333

Geoffroy A, Le Gall L, Destombe C (2012) Cryptic introduction of the red alga *Polysiphonia morrowii* Harvey (Rhodomelaceae, Rhodophyta) in the North Atlantic Ocean highlighted by a DNA barcoding approach. Aquat Bot 60:67–71

Glud RN, Kühl M, Wenzhöfer F et al (2002) Benthic diatoms of a high Arctic fjord (Young Sound, NE Greenland): importance for ecosystem primary production. Mar Ecol Prog Ser 238:15–29

Gorman D, Connell SD (2009) Recovering subtidal forests on human-dominated landscapes. J Appl Ecol 46:1258–1265

Graham LE, Wilcox LW (2000) Algae. Prentice Hall, Upper Saddle River

Graham LE, Graham JM, Wilcox LW et al (2016) Algae, 3rd edn. LJLM Press, LLC, Madison

Gupta VK, Nayak A, Agarwal S (2015) Bioadsorbents for remediation of heavy metals: current status and their future prospects. Environ Eng Res 20:1–18

Gutiérrez JL, Jones CG, Strayer DL et al (2003) Mollusks as ecosystem engineers: the role of shell production in aquatic habitats. Oikos 61:79–90

Harlin MM, Lindbergh JM (1977) Selection of substrata by seaweed: optimal surface relief. Mar Biol 40:33–40

Harris MA, Cumming BF, Smol JP (2006) Assessment of recent environmental changes in New Brunswick (Canada) lakes based on paleolimnological shifts in diatoms species assemblages. Can J Botany 84:151–163

Hay ME (1981) The functional morphology of turf-forming seaweeds: persistence in stressful marine habitats. Ecology 62:739–750

Hayden H, Blomster J, Maggs CA et al (2012) Linnaeus was right all along: *Ulva* and *Enteromorpha* are not distinct genera. Eur J Phycol 38:277–294

Heck KL Jr, Hays G, Orth RJ (2003) Critical evaluation of the nursery role hypothesis for seagrass meadows. Mar Ecol Prog Ser 253:123136

Hewitt CL, Campbell ML, Schaffelke B (2007) Introductions of seaweeds: accidental transfer pathways and mechanisms. Botanica Mar 50:326–337

Hibberd DJ, Norris RE (1984) Cytology and ultrastructure of *Chlorarachnion reptans* (Chlorarachniophyta) *division nova*, Chlorarachniophyceae, (*Clasis nova*). J Phycol 20:36–330

Hill SA, Towill LR, Sommerfeld MR (1980) Photomovement responses of *Porphyridium purpureum*. J Phycol 16:444–448

Hillebrand HB, Worm B, Lotze HK (2000) Marine microbenthic community structure regulated by nitrogen loading and grazing pressure. Mar Ecol Prog Ser 204:27–38

Hinks J, Edwards S, Sallis PJ et al (2013) The steady state anaerobic digestion of *Laminaria hyperborea*. Effect of hydraulic residence on biogas production and bacterial community composition. Bioresour Technol 143:221–230

Hoagland KD, Rosowski JR, Gretz MR et al (1993) Diatom extracellular polymeric substances: function, fine structure, chemistry, and physiology. J Phycol 29:537–566

Hoshino M, Croce ME, Hanyuda T, Kogame K (2020) Species delimitation of *Planosiphon gracilis* morphospecies (Scytosiphonaceae, Phaeophyceae) from Japan and the description of *Pl. nakamurae* sp. nov. Phycologia 59:116–126

Hull SC (1987) Macroalgal mats and species abundance, a field experiment. Estuar Coast Shelf Sci 25:519–532

Iribarne O, Bruschetti M, Escapa M et al (2005) Small and large-scale effect of the SW Atlantic burrowing crab on habitat use by migratory shorebirds. J Exp Mar Biol Ecol 315:87–61

Irlandi EA, Peterson CH (1991) Modification of animal habitat by large plants mechanisms by which seagrasses influence clam growth. Oecologia 87:307–318

Isaksson I, Pihl L (1992) Structural changes in benthic macrovegetation and associated epibenthic faunal communities. Neth J Sea Res 30:131–140

Isaksson I, Pihl L, van Montfrans J (1994) Eutrophication-related changes in macro vegetation and foraging of young cod (*Gadus morhua* L.): a mesocosm experiment. J Exp Mar Biol Ecol 177:203–217

Johansson G, Snoeijs P (2002) Macroalgal photosynthetic responses to light in relation to thallus morphology and depth zonation. Mar Ecol Prog Ser 244:63–72

Kang CK, Kim JB, Lee KS et al (2003) Trophic importance of benthic microalgae to macrozoobenthos in coastal bay systems in Korea: dual stable C and N isotope analyses. Mar Ecol Prog Ser 259:79–92

Kelaher BP, Levinton JS (2003) Variation in detrital enrichment causes spatio-temporal variation in soft-sediment assemblages. Mar Ecol Prog Ser 261:8597

Kennish M (2017) Ecology of estuaries: volume 2: biological aspects, 1st edn. CRC Press, Boca Raton

Kim MS, Yang CE, Mansilla A et al (2004) Recent introduction of *Polysiphonia morrowii* (Ceramiales, Rhodophyta) to Punta Arenas, Chile. Bot Mar 47:389–394

Kim JH, Kang EJ, Park MG et al (2011) Effects of temperature and irradiance on photosynthesis and growth of a green-tide-forming species (*Ulva linza*) in the Yellow Sea. J Appl Phycol 23:421–432

Kim JK, Yarish C, Hwang EK et al (2017) Seaweed aquaculture: cultivation technologies, challenges and its ecosystem services. Algae 32:1–13

Kirk JTO (1994) Light composition and photosynthesis in aquatic communities. Cambridge University Press, Cambridge

Koller L (2021) Cultivo de una especie de *Gelidium* (Gelidiales, Rhodophyta) del Estuario de Bahía Blanca, Thesis, Universidad Nacional del Sur, Bahía Blanca

Konar B, Iken K (2009) Influence of taxonomic resolution and morphological functional groups in multivariate analyses of macroalgal assemblages. Phycologia 48:24–31

Krause-Jensen D, McGlathery K, Rysgaard S et al (1996) Production within dense mats of the filamentous macroalga *Chaetomorpha linum* in relation to light and nutrient availability. Mar Ecol Prog Ser 134:207–216

Krause-Jensen D, Christensen PB, Rysgaard S (1999) Oxygen and nutrient dynamics within mats of the filamentous macroalga *Chaetomorpha linum*. Estuaries 22:31–38

Kristensen E, Penha-Lopes G, Delefosse M et al (2012) What is bioturbation? The need for a precise definition for fauna in aquatic sciences. Mar Ecol Prog Ser 446:285–302

Krumbein WE (1994) The year of the slime. In: Krumbein WE, Paterson D, Stal LJ (eds) Biostabilization of sediments. Bibliotheks und Informationssystem der Carl von Ossietzky Universitat, Oldenburg, pp 1–7

Larsen A, Sand-Jensen K (2006) Salt tolerance and distribution of estuarine benthic macroalgae in the Kattegat–Baltic Sea area. Phycologia 45:13–23

Lee WK, Lim YY, Leow ATC et al (2017) Factors affecting yield and gelling properties of agar. J Appl Phycol 29:1527–1540

Lemieux C, Otis C, Turmel M (2014) Six newly sequenced chloroplast genomes from prasinophyte green algae provide insights into the relationship among prasinophyte lineages and the diversity of streamlined genome architecture in picoplanktonic species. BMC Genomics 15:857

Lenanton RCJ, Robertson AI, Hansen JA (1982) Nearshore accumulations of detached macrophytes as nursery areas for fish. Mar Ecol Prog Ser 9:51–57

Lin HP, Sommerfeld MR, Swafford JR (1975) Light and electron microscope observations on motile cells of *Porphyridium purpureum* (Rhodophyta). J Phycol 11:452–457

Littler MM, Arnold KE (1982) Primary productivity of marine macroalgal functional-form groups from southwestern North America. J Phycol 18:307–311

Littler MM, Littler DS (1980) The evolution of thallus form and survival strategies in benthic marine macroalgae: field and laboratory tests of a functional forma model. Am Nat 116:25–44

Littler MM, Littler DS (1984) Relationships between macroalgal funtional form groups and substrata stability in a subtropical rocky-intertidal system. J Exp Mar Biol Ecol 74(1):13–34

Liuzzi MG, Lopez Gappa J, Piriz ML (2011) Latitudinal gradients in macroalgal biodiversity in the Southwest Atlantic between 36 and 55°S. Hidrobiologia 673(1):205–214

Lobban C, Harrison PJ (1994) Seaweed ecology and physiology. Cambridge University Press, New York

Luther G (1976) Bewuchsuntersuchungen auf Natursteinsubstraten im Gezeitenbereich des Nordsylter Wattenmeeres: Algen. Helgolander wiss Meeresunters 28:318–351

Macintyre HL, Geider RJ, Miller DC (1996) Microphytobenthos: the ecological role of the "secret garden" of unvegetated, shallow-water marine habitats. I. Distribution, abundance and primary production. Estuaries 19:186–201

Marcovecchio JE, Botté SE, Fernandez Severini MD et al (2010) Geochemical control of heavy metal concentrations and distribution within Bahía Blanca Estuary (Argentina). Aquat Geochem 16(2–3):251–266. https://doi.org/10.1007/s10498-009-9076-1

Margulis L, Barghoorn ES, Ashendorf D et al (1980) The microbial community in the layered sediments at Laguna Figueroa, Baja California, Mexico: does it have Precambrian analogues? Precambrian Res 11:93–123

Margulis L, Corliss JO, Melkonian M et al (eds) (1990) Handbook of Protoctista. Jones and Bartlett Publishers, Boston

Martins I, Oliveira JM, Flindt MR et al (1999) The effect of salinity on the growth rate of the macroalgae *Enteromorpha intestinalis* (Chlorophyta) in the Mondego estuary (West Portugal). Acta Oecol 20:259–265

Marx JM, Herrnkind WF (1985) Macroalgae (Rhodophyta: *Laurencia* spp.) as habitat for young juvenile spiny lobsters, *Panulirus argus*. Bull Mar Sri 36:423–431

Menendez M (2005) Effect of nutrient pulses on photosynthesis of *Chaetomorpha linum* from a shallow Mediterranean coastal lagoon. Aquat Bot 82:181–192

Menge BA, Ashkenas LR, Matson A (1983) Use of artificial holes in studying community development in cryptic marine habitats in a tropical rocky intertidal region. Mar Biol 77:128–141

Meysman FJR, Middelburg JJ, Heip CHR (2006) Bioturbation: a fresh look at Darwin's last idea. Trends Ecol Evol 21:688–695

Miller DC, Geider RJ, Macintyre HL (1996) Microphytobenthos: the ecological role of the "secret garden" of unvegetated, shallow-water marine habitats. II Role in sediment stability and shallow-water food webs. Estuaries 19:202–212

Miloslavich P, Klein E, Diaz JM et al (2011) Marine biodiversity in the Atlantic and Pacific coasts of South America: knowledge and gaps. PLoS One 6(1):e14631

Montagna PA (1984) In situ measurement of meiobenthic grazing rates on sediment bacteria and edaphic diatoms. Mar Ecol Prog Ser 18:119–130

Morand P, Merceron M (2005) Coastal eutrophication and excessive growth of macroalgae. In: Pandalai SG (ed) Recent research developments in environmental biology. Research Signpost, Trivandrum, pp 395–449

Nedergaard RI, Risgaard-Peterson N, Finster K (2002) The importance of sulfate reduction associated with *Ulva lactuca* thalli during decomposition: a mesocosm experiment. J Exp Mar Biol Ecol 275:15–29

Nestlerode JA, Diaz RJ (1998) Effects of periodic environmental hypoxia on predation of a tethered polychaete, *Glycera americana*: implications for trophic dynamics. Mar Ecol Prog Ser 172:185–195

Nicotri ME (1977) Grazing effects of four marine intertidal herbivores on the microflora. Ecology 58:620–632

Nielsen SL, Sand-Jensen K (1990) Allometric scaling of maximal photosynthetic growth rate to surface/volume ratio. Limnol Oceanogr 251:77–180

Nienhuis PH (1969) The significance of the substratum for intertidal algal growth on the artificial rocky shore of the Netherlands. Int Revue Ges Hydrobiol 54(2):207–215

Niklas KJ (1992) Plant Allometry: the scaling of form and process. University of Chicago Press, Chicago

Noffke N (2010) Microbial mats in sandy deposits from the Archean era to today. Springer, Berlin

Norkko JE, Bonsdorff E, Norkko A (2000) Drifting algal mats as an alternative habitat for benthic invertebrates. Species specific responses to a transient resource. J Exp Mar Biol Ecol 248:79–64

Norton TA (1992) Dispersal by macroalgae. Br Phycol J 27:293–301

Nyberg CD, Wallentinus I (2005) Can species traits be used to predict marine macroalgal introductions? Biol Invasions 7(2):265–279

Ogata E (1953) Some experiments on the settlement of spores of red algae. Jpn J Ecol 3:128

Omstedt A, Humborg C, Pempkowiak J et al (2014) Biogeochemical control of the coupled CO_2–O_2 system of the Baltic Sea: a review of the results of Baltic-C. Ambio 43:49–59

Otaiza RD, Rodriguez CY, Caceres JH et al (2018) Fragmentation of thalli and secondary attachment of fragments of the agarophyte *Gelidium lingulatum* (Rhodophyta, Gelidiales). J Appl Phycol 30:1921–1931

Pan J, Bournod CN, Cuadrado DG et al (2013a) Interaction between estuarine microphytobenthos and physical forcings: the role of atmospheric and sedimentary factors. Int J Geosci 4:352–361

Pan J, Bournod CN, Pizani NV et al (2013b) Characterization of microbial mats from a siliciclastic tidal flat (Bahía Blanca estuary, Argentina). Geomicrobiol J 30:665–674

Parodi ER (2004) Marismas y algas bentónicas. In: Piccolo MC, Hoffmeyer M (eds) Ecosistema del estuario de Bahía Blanca. Instituto Argentino de Oceanografía, Bahia Blanca, pp 61–67

Parodi ER, Barría de Cao S (2003) Benthic microalgal communities in the inner part of the Bahía Blanca estuary (Argentina): a preliminary qualitative study. Oceanol Acta 25:279–284

Pedersen M, Borum J (1996) Nutrient control of algal growth in estuarine waters. Nutrient limitation and the importance of nitrogen requirements and nitrogen storage among phytoplankton and species of macroalgae. Mar Ecol Prog Ser 142:261–272

Perillo GME, Iribarne O (2003) Processes of tidal channels develop in salt and freshwater marshes. Earth Surf Proc Land 28:1473–1482

Perillo GME, Piccolo MC, Parodi ER et al (2001) Bahia Blanca Estuary ecosystem: a review. In: Seeliger U, Kjerfve B (eds) Coastal marine ecosystems of Latin America. Springler Verlag, Berlín, pp 205–217

Perkins EJ, Abbott OJ (1972) Nutrient enrichment and sand flat fauna. Mar Pollut Bull 3:70–72

Pickett-Heaps J, West J, Wilson S et al (2001) Time-lapse videomicroscopy of cell (spore) movement in red algae. Eur J Phycol 36(1):9–22

Pihl L, Isaksoon I, Wennhage H et al (1995) Recent increases of filamentous algae in shallow Swedish Bays: effects on the community structure of epibenthic fauna and fish. Neth J Aquat Ecol 29:349–358

Pinckney JL, Zingmark RG (1993) Modelling the annual production of intertidal benthic microalgae in estuarine ecosystems. J Phycol 29:396–407

Pizani N (2009) Valorización de las interacciones microfitobentos-sedimentos en planicies de marea impactadas por el dragado hidráulico. Ph.D Thesis, Universidad Nacional del Sur, Bahía Blanca

Polderman PJG (1976) *Wittrockiella paradoxa* Wille (Cladophoraceae) in n.w. European saltmarshes. Hydrobiol Bull 6:98–63

Prathep A, Marrs R, Norton T (2003) Spatial and temporal variations in sediment accumulation in an algal turf and their impact on associated fauna. Mar Biol 142(2):381–390

Raffaelli DG, Raven JA, Poole LJ (1998) Ecological impact of green macroalgal blooms. Oceanogr Mar Biol Annu Rev 36:97–125

Raffo MP, Geoffroy A, Destombe C et al (2014) First record of the invasive red alga *Polysiphonia morrowii* Harvey (Rhodomelaceae, Rhodophyta) on the Patagonian shores of the Southwestern Atlantic. Botanica Marina 57(1):21–26

Renaud PE, Syster DA, Ambrose WGJ (1999) Recruitment patterns of continental shelf benthos of North Carolina, USA, effects of sediment enrichment and impact on community structure. J Exp Mar Biol Ecol 237:89–66

Rikkinen J (2013) Molecular studies on cyanobacterial in lichen symbioses. Mycokeys 6:3–32

Robertson AI, Lenanton RCJ (1984) Fish community structure and food chain dynamics in the surf-zone of sandy beaches: The role of detached macrophyte detritus. J Exp Mar Biol Ecol 84:265–283
Rooker JR, Holt SA, Soto MA et al (1998) Post settlement patterns of habitat use by Sciaenid fishes in subtropical seagrass meadows. Estuaries 21:318–327
Rosenvinge LK (1927) On mobility in the reproductive cells of the Rhodophyceae. Bot Tidsskrift 40:72–80
Ross SW, Moser ML (1995) Life history of juvenile gag, *Mycteroperca microlepis*, in North Carolina estuaries. Bull Mar Sci 56:222–237
Rossi F (2006) Small-scale burial of macroalgal detritus in marine sediments: effects of *Ulva* spp. on the spatial distribution of macrofauna assemblages. J Exp Mar Biol Ecol 332:84–89
Rossi F, Underwood AJ (2002) Small-scale disturbance and increased nutrients as influences on intertidal macrobenthic assemblages experimental burial of wrack in different intertidal environments. Mar Ecol Prog Ser 241:29–39
Sagasti A, Schaffner LC, Duffy JE (2001) Effects of periodic hypoxia on mortality, feeding and predation in an estuarine epifaunal community. J Exp Mar Biol Ecol 258:257–283
Sand-Jensen K, Borum J (1991) Interactions among phytoplankton, periphyton, and macrophytes in temperate freshwaters and estuaries. Aquat Bot 41:137–175
Sanjeewa KKA, Kim EA, Son KT et al (2016) Bioactive properties and potentials cosmeceutical applications of phlorotannins isolated from brown seaweeds: a review. J Photochem Photobiol B 162:60–65
Santelices B (1990) Patterns of organization of intertidal and shallow subtidal vegetation in wave exposed habitats of Central Chile. Hidrobiologia 192(1):35–57
Santelices B, Varela D (1994) Abiotic control of reattachment in *Gelidium chilense* (Montagne) Santelices & Montalva (Gelidiales; Rhodophyta). J Exp Mar Biol Ecol 177:145–155
Sarı A, Tuzen M (2008) Biosorption of Pb (II) and Cd (II) from aqueous solution using green alga (*Ulva lactuca*) biomass. J Hazard Mater 152(1):302–308
Scanlan CM, Foden J, Wells E et al (2007) The monitoring of opportunistic macroalgal blooms for the water framework directive. Mar Poll Bull 55:162–171
Schweiger O, Settele J, Kudrna O et al (2008) Climate change can cause spatial mismatch of trophically interacting species. Ecol 89:3472–3479
Selvam GG, Sivakumar K (2014) Influence of seaweed extract as an organic fertilizer on the growth and yield of *Arachis hypogea* L. and their elemental composition using SEM-Energy Dispersive Spectroscopic analysis. Asian Pac J Reprod 3:18–22
Smith KN, Herrnkind WF (1992) Predation on early juvenile spiny lobsters *Panulirus argus* (Latreille): influence of size and shelter. J Exp Mar Biol Ecol 157:3–18
Snoeijs P, Busse S, Potapova M (2002) The importance of diatom cell size in community analysis. J Phycol 38:265–272
Sousa AI, Martins I, Lilleb AI et al (2007) Influence of salinity, nutrients and light on the germination and growth of *Enteromorpha* sp. spores. J Exp Mar Biol Ecol 341:142–150
South GR, Whittick A (1987) Introduction to phycology. Blackwell, Oxford
Spalleti L (1980) Paleoambientes sedimentarios en secuencias silicoclásticas. Asoc Geológ Argent 8:1–175
Spivak ED (2010) The crab *Neohelice* (=*Chasmagnathus granulata*): an emergent animal model from emergent countries. Helgoland Mar Res 64:149–154
Stal LJ (2006) Microphytobenthos as a biogeomorphological force in intertidal sediment stabilization. Ecol Eng 36:236–245
Stal LJ, van Gemerden H, Krumbein WE (1985) Structure and development of a benthic marine microbial mat. FEMS Microbiol Ecol 31:111–125
Statzner B, Fièvet E, Champagne JY et al (2000) Crayfish as geomorphic agents and ecosystem engineers: biological behavior affects sand and gravel erosion in experimental streams. Limnol Oceanogr 45:630–640

Steneck RL, Dethier MN (1994) A functional group approach to the structure of algal-dominated communities. Oikos 69:476–498

Steneck RL, Watling L (1982) Feeding capabilities and limitation of herbivorous mollusks: a functional group approach. Mar Biol 68:299–319

Stevenson RJ (1996) An introduction to algal ecology in freshwater benthic habitats. In: Stevenson RJ, Bothwell ML, Lowe RL (eds) Algal ecology: freshwater benthic ecosystems. Academic, San Diego, pp 3–30

Stewart WDP, Rowell P, Rai AN (1983) Cyanobacteria-eukaryotic plant symbioses. Ann Inst Pasteur Microbiol 134(1):205–228

Summerson HC, Peterson CH (1984) The role of predation in organizing benthic communities of a temperate-zone seagrass bed. Mar Ecol Prog Ser 15:63–77

Taylor R, Fletcher RL, Raven JA (2001) Preliminary studies on the growth of selected 'green tide' algae in laboratory culture: effects of irradiance, temperature, salinity and nutrients on growth rate. Bot Mar 44:327–336

Thiel M, Watling L (1998) Effects of green algal mats on infaunal colonization of a New England mud flat–long-lasting but highly localized effects. Hydrobiologia 375:177–189

Titlyanov EA, Titlyanova TV (2006) Production of plantlets of the red alga of *Gelidium* genus (Rhodophyta) from thallus fragments. Russ J Mar Biol 32(5):284–288

Turner A, Lewis MS, Shams L et al (2007) Uptake of platinum group elements by the marine macroalga, *Ulva lactuca*. Mar Chem 65(3–4):271–280

Valiela I, McClelland J, Hauxwell J et al (1997) Macroalgal blooms in shallow estuaries, controls and ecophysiological and ecosystem consequences. Limnol Oceanogr 42:165–1118

Van den Hoek C, Mann DG, Jahns HM (1995) Algae – an introduction to phycology. Cambridge University Press, Cambridge

van Gemerden H (1993) Microbial mats: a joint venture. Mar Geol 113:3–25

Vanderklift MA, Lavery PS (2000) Patchiness in assemblages of epiphytic macroalgae on *Posidonia coriacea* at a hierarchy of spatial scales. Mar Ecol Prog Ser 192:127–135

Vasconcelos MTSD, Leal MFC (2001) Seasonal variability in the kinetics of Cu, Pb, Cd and Hg accumulation by macroalgae. Mar Chem 74:65–85

Verlaque M (2001) Checklist of the macroalgae of Thau Lagoon (Herault, France), a hot spot of marine species introduction in Europe. Oceanol Acta 24:29–49

Wahl M (1989) Marine epibiosis. I. Fouling and antifouling: some basic aspects. Mar Ecol Prog Ser 58:175–189

Wahl M, Hoppe K (2002) Interactions between substratum rugosity, colonization density and periwinkle grazing efficiency. Mar Ecol Prog Ser 225:239–249

Walne PL, Kivic PA (1990) Euglenida. In: Margulis L, Corlissis JO, Melkonian M et al (eds) Handbook of Protoctista. Jones & Bartlett Publishers, Boston, pp 270–287

Watermann F, Hillebrand H, Gerdes G et al (1999) Competition between benthic cyanobacteria and diatoms as influenced by different grain sizes and temperatures. Mar Ecol Progr Ser 187:77–87

Webb P, Wooldridge T, Schlacher T (1997) Osmoregulation and spatial distribution in four species of mysid shrimps. Comp Biochem Physio A 117:427–431

Wilkinson M (1981) Survival strategies of attached algae in estuaries. In: Jones NV, Wolff WJ (eds) Feeding and survival strategies of estuarine organisms. Springer, Boston, pp 29–38

Wille N (1909) Algologische Notizen XV. Über *Wittrockiella* nov. gen. Nytt Mag Naturvidensk 47:209–225

Williams RB (1964) Division rates of salt marsh diatoms in relation to salinity and cell size. Ecology 45:877–880

Wilson KA, Able KW, Heck JKL (1990) Predation rates on juvenile blue crabs in estuarine nursery habitats: evidence for the importance of macroalgae (*Ulva lactuca*). Mar Ecol Prog Ser 58:243–251

Woodin SA (1978) Refuges, disturbance, and community structure: a marine soft-bottom example. Ecology 59:274–284

Chapter 7
The Intertidal Meiobenthos of the Bahía Blanca Estuary

Verónica N. Bulnes, Agustín G. Menechella, Kevin A. Rucci, and Michel Sciberras

7.1 Introduction

The meiofauna is present in every aquatic environment, from freshwater to marine habitats, in all kinds of sediments from muddy soft to coarse gravel, even associated to surface structures like vegetation, sea ice and biogenetic structures, like corals and worm tubes. Most meiofaunal organisms live in the interstitial spaces between the sediment particles, especially near the surface, but they may be also found living commensally in animal tubes, hydrozoan colonies and mollusc pallial cavities. In marine habitats, their representatives inhabit from high beach sediments to the deepest depths of the sea.

The organisms gathered in the meiofaunal realm have been known since the eighteenth century, but it was not until 1942 that Mare defined the term

V. N. Bulnes (✉)
Instituto de Ciencias Biológicas y Biomédicas del Sur, INBIOSUR (Universidad Nacional del Sur, CONICET), Bahía Blanca, Argentina

Zoología de Invertebrados I. Departamento de Biología, Bioquímica y Farmacia. Universidad Nacional del Sur, Bahía Blanca, Argentina
e-mail: vebulnes@criba.edu.ar

A. G. Menechella
Zoología de Invertebrados I. Departamento de Biología, Bioquímica y Farmacia. Universidad Nacional del Sur, Bahía Blanca, Argentina

Instituto Argentino de Oceanografía, IADO (Universidad Nacional del Sur, CONICET), Bahía Blanca, Argentina

K. A. Rucci
Instituto de Investigaciones Biológicas y Tecnológicas, Universidad Nacional de Córdoba, CONICET, IIBYT, Córdoba, Argentina

M. Sciberras
Zoología de Invertebrados I. Departamento de Biología, Bioquímica y Farmacia. Universidad Nacional del Sur, Bahía Blanca, Argentina

S. M. Fiori, P. D. Pratolongo (eds.), *The Bahía Blanca Estuary*,
https://doi.org/10.1007/978-3-030-66486-2_7

meiobenthos, as the animals smaller than macrofauna but larger than microfauna. Almost all along the twentieth century, the study of these organisms was centred on describing and classifying them, and in the 1970s, specialists showed a growing interest in the ecology of these animals. From that moment on, defining meiofauna for quantitative studies required the establishment of practical limits relevant to the size and not always agreeing with taxonomic classifications. The formal size boundaries of meiofauna were operationally defined, based on the standardized mesh width of sieves with 500 μm (1000 μm) as upper and 44 μm (63 μm) as lower limits (Giere 2009). All benthic organisms that pass through the coarse sieve but are retained by the finer sieve during processing are considered meiobenthos. Since this chapter will deal only with benthic marine metazoan phyla, the term meiobenthos will be used as a synonym of meiofauna.

After 20 years of quantitative research, the specialist hypothesized that some marine taxa may have developed singular meiofaunal traits over evolutionary time and what we define today as meiofauna is not only a definite size category but also a separate biological and ecological unit. The most important characteristics are interpreted as consequences of the miniaturization of the body regarding the interstitial habitus, and they are mostly related to reproductive aspects, like the completely benthic direct life cycle, the short generation times and the semelparity. They are usually motile forms and feed by seeking food particles in a highly discriminate manner, and their dispersion phase take place only as adults. Opposite to the meiofaunal traits, the macrofauna is characterized by a planktonic larval development and dispersal, with long generation times, iteroparity, and either sedentary or motile, they seem to feed more unselectively on particles of food (Warwick et al. 2006).

From the 36 recognized animal phyla, 22 are considered meiofaunal taxa. Most of them constitute the 'permanent meiofauna', since their representatives remain within the meiofaunal size limits throughout their life. The rest of them are part of the 'temporary meiofauna' and include the taxa with immature stages that fall within the meiofaunal size range, but when they achieve sexual maturity, they reach macrofaunal dimensions.

The meiofauna consume a wide spectrum of food sources, including microalgae, eukaryotic microbes, small metazoans, bacteria and organic detritus, although at species level they often show a specialized diet. The meiofaunal activity modifies a series of physical, chemical and biological sediment properties, positively and negatively affecting various ecosystem services including sediment stabilization, biochemical cycling, waste removal and food web dynamics, at various spatial and temporal scales. This evidence is still scattered, but the interest is growing and it is expected that unravelling and quantifying the role of the meiofauna will support future environmental management policies (Schratzberger and Ingels 2017).

Although ubiquitous in marine habitats, the meiofaunal taxa are not evenly distributed. On a large scale, the highest abundances are recorded in estuarial environments, while the lowest are usually found at the deep sea. On a smaller scale, for example, on a beach slope, the meiofauna distributes patchily. The causes of such distributions are difficult to define, but benthologists agree that primarily the sediment particle size, the temperature and the salinity are the physical factors limiting the abundance and species composition (Coull 1999). The grain size is the key factor since it directly determines spatial and structural conditions and indirectly determines the physical and chemical conditions of the sediment. In muddy estuarine sediments and intertidal flats, the meiofauna is restricted to the narrow 2–3 cm from the surface, where it faces highly variable and severe conditions. The freshwater supply varies seasonally, while the seawater supply does it twice a day. Additionally, the temperature, the intensity of light and the exposition vary in accordance with the season of the year, the tides, the time of the day and the air temperature. On top of that, the existence of anthropogenic pressures like a high local population density, the presence of harbours and dredging activities limits the ecosystem availability for the meiobenthic organisms.

The specialist estimates there is a world average of 10^6 meiofaunal organisms per square metre in every uncontaminated estuarine sediment, with a biomass of 0.75–2 mg per square metre (Coull 1999), which is low compared to other benthic components. On the other hand, the throughput of carbon can be orders of magnitude higher than its standing stock, which makes the meiofauna the main responsible for the production and flow of energy (rather than its storage) in the ecosystem (Schratzberger and Ingels 2017).

In sediments, the nematodes are usually the most abundant taxon comprising 60–90% of the total fauna and copepods are typically second at 10–40%. Occasionally, a group other than nematodes may predominate (e.g. turbellarians) or copepods are not second in abundance (Coull 1999). Meiofauna can be locally and temporally very abundant, particularly after a large spawning or a settling event. The diversity and composition of the nondominant taxa vary depending on sediment particle size, exposure, etc.

In South America, there is a serious gap in the knowledge of the marine microscopic animals. The biodiversity of meiofaunal taxa is heavily underestimated and even less is known about the interstitial community. There are a few isolated examples in central Chile and Brazil (Rodriguez et al. 2001; Albuquerque et al. 2007; Dupuy et al. 2015) where most meiofaunal organisms are gathered in higher taxa like phylum, class and order, without further discrimination.

In Argentina, the meiofaunal biodiversity is scarcely known, and even in some comprehensive contributions of the Argentinean marine fauna (Boschi and Cosseau 2004; Calcagno 2014), they remain neglected. Except for the free-living nematodes, there is a lack of information about the interstitial marine biodiversity, both spatial and temporal distribution, as well as its role in the ecosystems, especially in estuarine environments, like the Bahía Blanca Estuary.

With extensive tidal plains that are discovered at low tide, a salinity continuous gradient and a predominance of muddy to sandy sediment with silt and clay, the Bahía Blanca Estuary houses a benthic community with high taxonomic diversity. When it comes to the knowledge of the estuarine communities of the southeast Buenos Aires province, there are some long-term studies centred in the macrofaunal component of the benthos (Bremec et al. 2007; Elías et al. 2007), overlooking the meiofaunal component, probably due to the small size of the meiobenthic organisms, the time-consuming sample sorting, as well as the extreme complexity of their taxonomy. It was not until 2007 when the study of the meiofaunal taxa begun with the taxonomic contributions on microturbellarians (Bulnes 2007), nematodes (Martelli 2010; Villares et al. 2013) and harpacticoid copepods (Sciberras 2018; Sciberras et al. 2014, 2018, 2021). These contributions revealed many species new to science suggesting that the contribution of meiofauna to the biodiversity of the Bahía Blanca Estuary remains underestimated.

7.2 Nematoda

The Nematoda constitute a highly diverse group of animals that have been extensively studied, especially due to the economical and medical importance associated to the parasitic representatives. The phylum comprises about 25,000 described species, including the parasitic taxa, although this represents a small percentage of the estimated biodiversity. The marine roundworms include 4000–5000 species, and they are considered the most diverse and widespread group of nematodes, occurring from shallow shores to the abyss. They are permanent meiobenthic inhabitants, and at a local scale, they are highly abundant. A typical surface area meiocore (10 cm^2) may contain about 60–90% of the total faunal diversity, followed by harpacticoid copepods (10–40%) and other less represented taxa like Platyhelminthes, Gastrotricha, etc. (Schmidt-Rhaesa 2014).

The nematodes' life cycle is strictly bound to the substrate. They are easily recognized in the sediment samples, since they move undulating in the dorsoventral plane, pushing against a substratum, such as a sand grain or any submerged surface. They are non-segmented, cylindrical, threadlike worms, with a simple body plan and morphologically highly homogeneous among species (Higgins and Thiel 1988).

For years, the taxonomy of the nematodes has been established on the external characteristics, shape and cuticularization of the buccal cavity. Nevertheless, the recent discovery that many morphospecies are in fact complexes of several genetically distinct species that are hard or impossible to discriminate based on morphology has forced specialists to integrate both morphological and molecular information to obtain a more accurate biodiversity status in the ecosystems (Schmidt-Rhaesa 2014).

Meiobenthic marine nematodes are small-sized animals ranging between 0.5 and 3 mm long. Nematodes inhabiting sandy habitats tend to be more slender, whereas nematodes from muddy habitats are generally more robust, probably because the former move through the interstitial apertures, while the latter borrow through the sediments.

Despite their high diversity and abundance, their total biomass is much lower than that of bacteria and macrobenthos, and there is a still ongoing discussion whether nematodes significantly contribute to ecosystem processes. Nevertheless, there is no doubt that nematodes contribute to particular ecosystem functions, such as the decomposition of organic matter and the stabilization of intertidal muds.

In Argentina, the marine Nematoda of the Magellanic zoogeographical region (42° to 55° South) have been extensively studied over 40 years by Catalina Pastor de Ward and collaborators (Pastor de Ward 2001, 2003, 2004; Pastor de Ward and Lo Russo 2007; Harguinteguy et al. 2012; Pastor de Ward et al. 2014, 2015a, b; Villares et al. 2015, 2016; Martelli et al. 2017). On the other hand, the biodiversity of nematodes from the Argentinean zoogeographical region (20° to 42° South), including the Bahía Blanca Estuary, is still poorly known. During December 2012, Martelli sampled in three sites along the northern coast of the Bahía Blanca Estuary and recorded over 60 different genera (Table 7.3). Until today, only 13 taxa have been identified to species level (see species list), two of them exclusive from the estuary, *Campylaimus bonariensis* Villares, Martelli, Lo Russo and Pastor, 2013, and *Campylaimus arcuatus* Villares, Martelli, Lo Russo and Pastor, 2013 (Table 7.2). The fact that 50% of the species registered by Pastor and collaborators in the adjoining zoogeographical region were new to science led us think the biodiversity of Nematoda in the Bahía Blanca Estuary is heavily underestimated.

The community structure, biodiversity and distribution of meiobenthic nematodes along the intertidal mudflats in the Bahía Blanca Estuary are a result of the combination of environmental variables, such as the mean grain size, the proportion of silt and clay, the food availability, the eutrophication degree and the presence of deposited heavy metals. Their abundance is related to the organic matter concentration, and their role in the regulation of biogeochemical cycling of nutrients is not completely understood, but their ubiquity in the sediments of the Bahía Blanca Estuary suggest they may function as mediators associated with the strength of energy flux through trophic webs.

The meiofaunal taxa are distributed in small patches (Giere 2009). These patterns are also observed in nematodes, and for that reason, parameters like the monthly abundance in one site, even with a high number of samples, are not good enough to describe the community. The results from systematic sampling during 2013 and 2014 show sharp differences in the standard deviation of each sample (Table 7.1), hiding the seasonal differences, and revealing the study of this community must be complemented with other parameters.

Aside from the abundances and according to the environmental variables and the distance to the estuary's mouth, the nematode communities can be clustered in three different groups: the inner estuary group, the mid-portion estuary group and the outer estuary group. These differences are partially explained by their feeding habits. Following Wieser's (1953) classification, nematodes from the Bahía Blanca Estuary can be divided into four different feeding types. The first two types are deposit feeders – the selective deposit feeders, devoid of a developed buccal cavity; and the nonselective deposit feeders, with a conical buccal cavity but without any armature. Both types are exploiting the particulate organic matter. A third type are the epistratum feeders, possessing a buccal cavity armed with denticles and other small sclerotized structures, predominantly grazing over surface diatoms and other microalgae. The fourth type gather the omnivores and predators, and they possess spacious mouth openings and buccal cavities with well-developed teeth and other cuticularized structures.

Table 7.1 Monthly abundance of Nematoda and Harpacticoida in Arroyo Pareja beach (Bahía Blanca Estuary), based on 10 samples each month (n = 10)

	Nematoda		Harpacticoida	
	Mean ± SD	[min – max]	Mean ± SD	[min – max]
April 2013	734 ± 269.79	[448–1347]	232.2 ± 50.87	[172–300]
May 2013	535.9 ± 283	[12–933]	41.2 ± 30	[5–88]
August 2013	556.8 ± 295.77	[963–92]	28.8 ± 21.46	[8–78]
October 2013	612.9 ± 160.85	[359–903]	15.6 ± 9.7	[3–21]
December 2013	899.66 ± 443.82	[7–1370]	14.88 ± 11.37	[1–38]
January 2014	192 ± 230.91	[1–762]	48.4 ± 54.48	[4–184]
February 2014	502.4 ± 165.84	[212–660]	169.2 ± 71.69	[77–286]

The data are presented as mean ± standard deviation and [minimum–maximum] range (Sciberras, unpublished data)

Table 7.2 List of recorded species of Nematoda in the Bahía Blanca Estuary

List of recorded species
Campylaimus arcuatus Villares, Martelli, Lo Russo and Pastor, 2013
Campylaimus bonariensis Villares, Martelli, Lo Russo and Pastor, 2013
Daptonema laxum (Wieser, 1956)
Enoploides vectis Gerlach, 1957
Molgolaimus typicus Furstenberg and Vincx, 1992
Nudora besnardi (Gerlach, 1956)
Paraethmolaimus dahli (Gerlach, 1953)
Paramonohystera parabutschlii (Timm, 1961)
Prochromadorella ditlevseni (de Man, 1922)
Sabatieria wieseri Platt, 1985
Setosabatieria hilarula (de Man, 1922)
Thalassoalaimus macrosmaticus Wieser, 1953
Thalassomonhystera parva (Bastian, 1865)

Adapted from Martelli (2013)

The communities inhabiting the inner and mid portion of the estuary are strongly dominated by genera belonging to the nonselective deposit feeders (Fig. 7.1a, b). In these microambients, a low oxygen level, small grain size and high organic matter concentrations coexist. This combination of factors seems to reduce the survival possibilities of many nematodes, resulting in communities characterized by a low diversity and richness, and only two genera, *Paraethmolaimus* Jensen, 1994, and *Terschellingia* de Man, 1888, dominated the biodiversity of the inner estuary portion.

Table 7.3 List of recorded genera of Nematoda in the Bahía Blanca Estuary

List of recorded Nematoda genera	
Adoncholaimus Filipjev, 1918	*Monoposthia* de Man, 1889
Aegialoalaimus de Man, 1907	*Neochromadora* Micoletzky, 1924
Anticyathus Cobb, 1920	*Odontophora* Bütschli, 1874
Araeolaimus de Man, 1888	*Oncholaimellus* de Man, 1890
Calyptronema Marion, 1870	*Oncholaimus* Dujardin, 1845
Campylaimus Cobb, 1920	*Oxystomina* Filipjev, 1918
Ceramonema Cobb, 1920	*Paracyatholaimoides* Gerlach, 1953
Chromadora Bastian, 1865	*Paracyatholaimus* Micoletzky, 1922
Chromadorina Filipjev, 1918	*Paraethmolaimus* Jensen, 1994
Cobbia de Man, 1907	*Linhomoeus* Bastian, 1865
Comesoma Bastian, 1865	*Paramonhystera* Steiner, 1916
Cyartonema Cobb, 1920	*Polysigma* Cobb, 1920
Daptonema Cobb, 1920	*Prochromadorella* Micoletzky, 1924
Diplolaimella Allgén, 1929	*Parachromadorita* Blome, 1974
Diplopeltula Gerlach, 1950	*Promonhystera* Wieser, 1956
Enoplolaimus de Man, 1893	*Pselionema* Cobb, 1933
Halalaimus de Man, 1888	*Pseudocella* Filipjev, 1927
Halichoanolaimus de Man, 1886	*Pseudosteineria* Wieser, 1956
Hopperia Vitiello, 1969	*Richtersia* Steiner, 1916
Laimella Cobb, 1920	*Sabatieria* Rouville, 1903
Leptolaimus de Man, 1876	*Setosabatieria* Platt, 1985
Linhystera Juario, 1974	*Siphonolaimus* de Man, 1893
Marylynnia (Hopper, 1972)	*Spilophorella* Filipjev, 1917
Mesacanthion Filipjev, 1927	*Spirinia* Gerlach, 1963
Metachromadora Filipjev, 1918	*Steineria* Micoletzky, 1922
Metalinhomoeus de Man, 1907	*Synonchiella* Cobb, 1933
Metoncholaimus Filipjev, 1918	*Terschellingia* de Man, 1888
Microlaimus de Man, 1880	*Thalassoalaimus* de Man, 1893
Molgolaimus Ditlevsen, 1921	*Thoracostomopsis* Ditlevsen, 1918
Monhystera Bastian, 1865	*Viscosia* de Man, 1890

Adapted from Martelli (2013)

On the other hand, towards the estuary's mouth, the sediments grain size and therefore the permeability rises and the animals do not have to face the low tensions of oxygen present in the inner portion of the estuary. Although this portion is still dominated by the nonselective deposit feeders (Fig 7.1c), in particular the genus *Setosabatieria* Platt, 1985, there is a growing presence of epistratum feeders, followed by the remaining feeding types. The increase in the abundance and diversity of nematodes suggests the presence of a wider range of food resources in this area.

To see a complete list of recorded species and genera, please check Tables 7.2 and 7.3 (Martelli 2013).

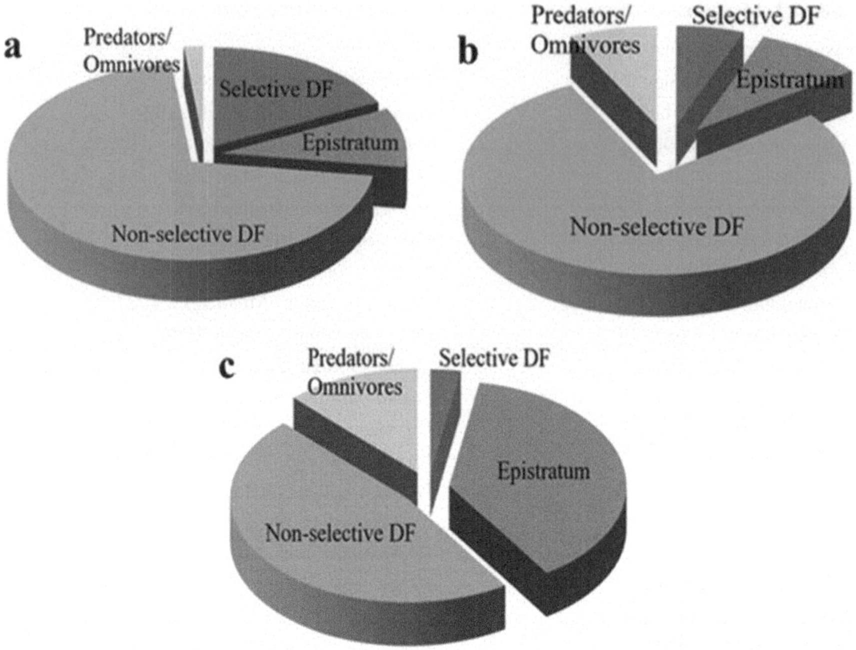

Fig. 7.1 Nematoda community based on their feeding habits in the different regions of the Bahía Blanca Estuary: (**a**) inner zone, (**b**) middle zone, (**c**) external zone. Adapted from Martelli (2013)

7.3 Interstitial Copepods

The subclass Copepoda possesses more than 12,500 described species. It is a relatively small group of arthropods, but it is incredibly abundant and probably constitutes the main group of animals in terms of the number of individuals. In marine sediments, the copepods are usually the second most abundant taxon in the meiofauna after the nematodes, and they can turn out the dominant group as the particle size of the sediment becomes coarser. Between the copepods, the order Harpacticoida assembles most of their benthic representatives, although some members of the order Cyclopoida have also colonized this environment. Among these benthic organisms, those who have successfully exploited the interstitial space between the sediment particles are usually characterized by their small size and elongate worm-like bodies (Huys et al. 1996; Boxshall and Halsey 2004).

Typically, marine interstitial sediment communities consist of around 30 species of copepods. Usually most individuals belong from three to five species, revealing a high degree of dominance. Most species are present throughout the year, and their abundances often show seasonal cyclical changes. Often, copepod abundance is high from the end of the spring to the beginning of the autumn (warm period) and remains low along the winter (Table 7.1). Nonetheless, some contributions have reported abundance peaks during the autumn and over the winter (Feller 1980; Coull and Dudley 1985; Davidovich and Chepurnov 1991). Along 1 year, some of the species display simple life cycles, with a single abundance peak, while others undergo more complex life cycles, with multiple abundance peaks. In general, before an abundance peak, a reproduction peak occurs, where there is an increasing number of females carrying eggs. When these ovigerous females are present 3–6 months of the year, the species is considered to exhibit a restricted reproductive period. On the other side of the spectrum, this period is considered prolonged, if the egg-carrying females remain present over 3 months, but disappear after 12 months. Eventually, some species exhibit a continuous reproductive period, and the ovigerous females are present all year long. Even when there are a few examples of species displaying continuous reproductive behaviour, most meiobenthic copepod species have either restricted or prolonged reproductive periods (Coull and Dudley 1985).

The study of the copepods has been attached to the study of the meiofauna since the beginning. The best-known fauna is that of the northern hemisphere. Most contributions regarding the ecology of meiobenthic harpacticoids describe the results obtained in Europe and North America. This knowledge continues to grow since they are now considered trustworthy bioindicators of pollution (along with the nematodes) (see Box 7.1).

Box 7.1: Invisible Yet Powerful: Meiofauna as a Bioindicator

Since its birth, meiobenthology has been a neglected discipline, since it deals with tiny and almost invisible animals. Unfortunately, a defined trend took over: the human mind seemed to be focused only on the study of 'big animals', such as large invertebrates or the well-known vertebrates. For a long time, meiofaunal animals were not very interesting, not glamorous, and not dangerous, and they look all the same – simply unimpressive tiny animals. In recent years, molecular studies brought new interest to these old known phyla. Meiofaunal taxa are now considered cornerstones in the explanation of animal phylogeny and world awarded champions when it comes to explaining the molecular machinery of the metazoan development (model organisms of three Nobel-Prize-rewarded investigations).

Meiobenthic animals are interstitial and live between the sediment grains. The absence of dispersal phases, like planktonic larvae, keep them their entire life, and even for generations, in the near vicinity. These features are shared with all meiofaunal taxa, and they have been a key factor why these organisms gained ecological importance in the last decade.

As a consequence of the increase in human activities, marine environments have been strongly affected. Pipeline inputs, domestic wastes, regular port and harbour dredging, oil spills and shipping and pollutants derived from these sources, many of which are lethal or have long-term deleterious effects, are trapped in marine sediments where they accumulate over long periods. This hidden danger affects all organisms in the ecosystem, and in front the line, the invisible and neglected meiofauna.

Since the 1970s, the benthos, which are the animals inhabiting the sediments, have been identified as a suitable ecological group for biomonitoring the effects of pollution, especially the macrobenthos. Since the specialist realized the meiofauna represent a separate biologically and ecologically defined group of animals, rather than an arbitrarily defined size range of benthic invertebrates, the scientist has been increasingly using the meiofauna as a bioindicator of the environmental quality on coastal marine systems.

What are the features that make these animals promising candidates for biomonitoring?

- They are ubiquitous.
- They are found in a high number of species and individuals (rich and abundant), even in small volume samples.
- Some taxa display a high sensitivity to several pollution agents.
- They are characterized by rapid life cycles and thus rapid generation turnover and lack larval dispersion.
- They demand lower costs related to field sampling.

This combination of traits provides the scientist with rapid and reproducible information regarding marine pollutants and their effects on ecosystem quality. Their application spectrum is high including all known methods applied in pollution studies:

- Field studies and computer-based models
- Toxicity essays based on living organisms, obtaining uptake rates or determining sublethal reactions to deterioration or recovering over generations
- Micro- and mesocosm experiments, which allows the obtention of 'realistic' results
- Analytical, histological and genetical approaches applied to single animals

The meiofauna is diverse, and finding a suitable model organism for these kinds of studies is difficult. For that matter, environmental surveys have been using the most abundant and richest taxa. The nematodes and harpacticoid copepods come in first line, followed by other phyla like foraminifers, etc.

For example, some researchers, based on their abundance, have proposed the nematode to copepod ratio (N:C) as a fast, easy and reliable tool for monitoring the effect of organic matter enrichment, a common marine contaminant. Since the nematodes seem more resistant to the environmental stress induced by the increasing dissolved organic matter content, a higher ratio in polluted areas is expected.

Another parameter used to evaluate the effects of pollutants is the index of trophic diversity (ITD) of nematodes. This index is based on the proportion of the different feeding groups (see Sect. 7.2). In a given area, any environmental disturbance will affect the food supply. When the ITD grows, it means a single trophic group dominates the community, which is the expected outcome, associated with increased stress related to the changing nature or proportion of the food items. However, the biomonitoring with meiofaunal taxa seems a promising tool and at the same time is at serious risk, particularly since the number of scientists dealing with their complex taxonomy is sinking every year.

Despite the available amount of literature worldwide that discloses the relevance of copepods as components of the meiobenthos, in Argentina, only a few ecological studies have includes them. During a period of 15 years in the late 1970s, Rosa Pallares made the first contributions to the taxonomic knowledge of this group in the south of Argentina, surveying the Ría Deseado (47°45′S), the Staten Island (54°46′S) and Tierra del Fuego (˃54°48′S), including scattered data on its ecology (e.g. Pallares 1968a, b, 1969, 1975, 1982). The state knowledge of meiobenthic copepods from the Bahía Blanca Estuary is even scarcer. Sciberras and collaborators described three new species since 2014 (Sciberras et al. 2014, 2017, 2021). A more comprehensive study of their ecology is available in her PhD thesis (Sciberras 2018), which provides the first descriptive analysis of the seasonal variation of the abundance, diversity and reproductive period of the meiobenthic copepod community in Arroyo Pareja beach, a locality placed on the northern coast of the Bahía Blanca Estuary (Fig. 2.4; Chap. 2).

The study conducted by Sciberras (2018) revealed the presence of 13 harpacticoid species, five of them displaying temporal and numerical dominance. They were present ten out of 12 months and displayed the highest abundances. *Halectinosoma parejae* Sciberras, Huys, Bulnes and Cazzaniga, 2017 (Ectinosomatidae), was the dominant species during most part of the year and constitutes about the 68% of the total of harpacticoid community in Arroyo Pareja. Regarding the abundances of the species, *Halectinosoma parejae* was followed by one species of the genus *Delavalia* (Miraciidae), after that *Nannopus* sp. (Nannopodidae) (Sciberras et al. 2021, in press), followed by a still undetermined species of the family Miraciidae and finally by *Quinquelaophonte aestuarii* Sciberras, Bulnes and Cazzaniga, 2014 (Laophontidae). All the abovementioned species together attained the 94,06% of the harpacticoid community in this locality. The remaining 6% was represented by other eight species, always in a very low proportion (Fig. 7.2).

Only one species of the genus *Delavalia* may be found present throughout the whole year; nevertheless, its density remains always low (Fig. 7.3c). *Halectinosoma parejae* (Fig. 7.3b) was the dominant species throughout the summer (December–February) and autumn (March–May). In July, at the beginning of the winter, the

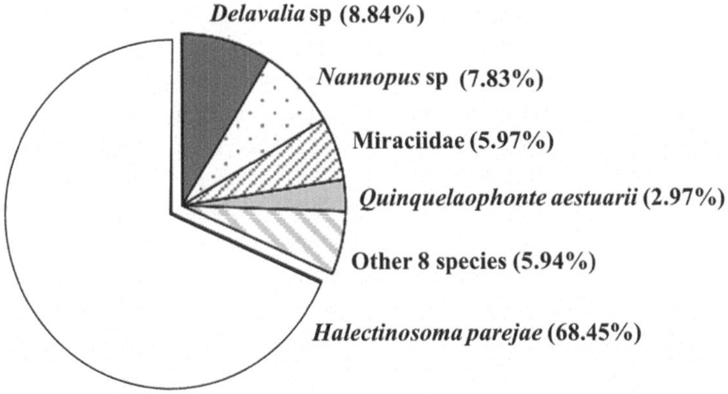

Fig. 7.2 Total composition of benthic copepods of the Bahía Blanca Estuary

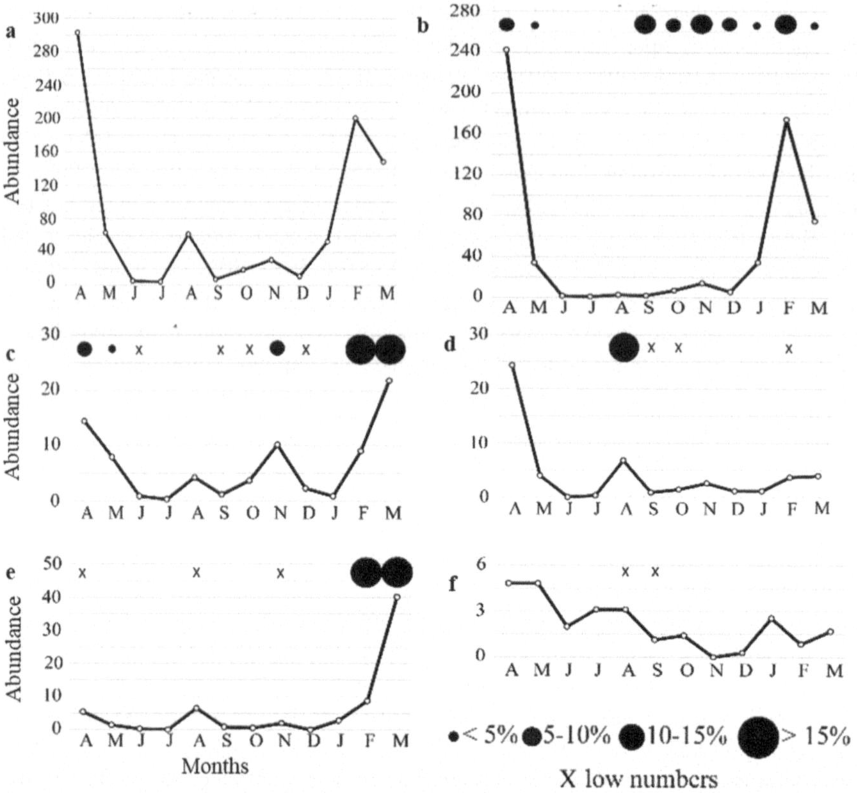

Fig. 7.3 Benthic copepods of the Bahía Blanca Estuary. (**a**) Abundance of the harpacticoid community from Arroyo Pareja (13 putative species), (**b**) abundance of *Halectinosoma parejae* Sciberras, Huys, Bulnes and Cazzaniga, 2017, (**c**) abundance of *Delavalia* sp., (**d**) abundance of species of Miraciidae, (**e**) abundance of *Nannopus* sp., (**f**) abundance of *Quinquelaophonte aestuarii* Sciberras, Bulnes and Cazzaniga, 2014

adult harpacticoids disappeared almost completely. Towards the end of the coldest season and the beginning of the spring (August–October), the species less represented won importance, until the end of the spring, when the abundance of *Halectinosoma parejae* regained its dominancy. There was a strong seasonality in terms of density in the harpacticoid community from Arroyo Pareja. The highest abundances were registered in summer and autumn, and reached their deepest point during August, in the middle of the winter (Fig. 7.3a). The seasonal changes of the whole community were determined by the changes in the population of *Halectinosoma parejae*. This species was more abundant in summer, although there was a small peak of abundance in winter. The ovigerous females (females carrying eggs) of *Halectinosoma parejae* were present nine out of 12 months, recording their highest abundances during the spring, just before the population abundance peak of February–April (Fig. 7.3b). In terms of abundance, *Delavalia* sp. followed second (Fig. 7.2), showing a similar seasonal pattern to *Halectinosoma parejae*, with the most important peak in March and a small peak in July (Fig. 7.3c). Similar to *Halectinosoma parejae*, the ovigerous females are present 9 months of the year, being more abundant at the end of the summer (February–March).

The species belonging to the genus *Nannopus*, which represents around 8% of the harpacticoid community (Fig. 7.2), has low densities during most part of the year, with an abrupt peak towards the end of the summer (February–March), a period when most part of ovigerous females are registered (Fig. 7.3e).

An undetermined species belonging to the family Miraciidae (Fig. 7.2), representing around 6% of the harpacticoid community, shows a similar pattern to *Nannopus* sp. since it is found in low densities during most part of the year but it shows an abrupt peak in autumn (April), although most part of ovigerous females are registered during the small abundance peak during the winter (August) (Fig. 7.3d).

Regarding *Quinquelaophonte aestuarii* (Fig. 7.2), just a few data can be pointed out since it represents just 3% of the harpacticoid community and it is present in very low densities during the whole year. The samples never contained more than 17 specimens per 30 cubic centimetres, even in the months of highest abundance; thus, only the presence was registered during the end of the winter to the beginning of the spring (August–September) (Fig. 7.3f).

According to these data, the number of harpacticoid species in Arroyo Pareja agrees with the expected number for marine interstitial environments. Besides, the harpacticoid community follows the high dominance model since *Halectinosoma parejae* is the dominant species and, together with just a few more species, they represent more than 90% of the total harpacticoid community. The temporal pattern of higher abundances in the warm period is verified, although a lower abundance peak occurs in winter (August), followed by the disappearance of practically all harpacticoid adults in June and July.

The absence of *Halectinosoma parejae* in the winter samples suggests that this species has an annual life cycle. The peaks of abundance of three of the less abundant species in August (*Delavalia* sp., *Nannopus* sp. and the undetermined species belonging to the family Miracidae) could be related to the absence of the dominant species (*Halectinosoma parejae*) or with the development of an algal bloom, characteristic of this environment (Freije and Gayoso 1988; Gayoso 1998; Popovich et al. 2008). Although these blooms have been extensively studied associated to the phytoplankton of the estuary, the sudden availability of nutrients that causes it probably has a similar effect in microfitobenthos.

The four most abundant species appear to have different reproductive periods. *Halectinosoma parejae* ovigerous females can be found in all the months that adults are present. *Delavalia* sp. and *Nannopus* sp. also have prolonged reproductive periods, but with a much higher proportion of ovigerous females in summer, while the unidentified species of Miraciidae seems to concentrate its reproduction in the winter.

The meiobenthic copepod community of the estuary is diverse, and its study adds another dimension to the community dynamics in a complex and changing environment that remains still little explored. Although the data exposed in this section arise from short-term standardized surveys on a single locality of the Bahía Blanca Estuary (Sciberras 2018), it constitutes the basis for planning the next steps regarding this well-represented but little-known zoological group.

7.4 The Turbellarians

For over 140 years the turbellarians have been important research objects. They display an extraordinary regeneration power and they are considered one of the early-most spiralian Eubilateria lineages. For a long time, the turbellarians were gathered in a class within the Platyhelminthes. In the last two decades, with the development and use of advanced molecular studies to unravel the metazoan evolutionary relationships, strong evidence is supporting that the turbellarians are not only a paraphyletic group of Platyhelminthes, but they also gather more than one phylum of metazoan animals.

In this chapter, the phylum Acoelomorpha Ehlers, 1985, comprising the clade Acoela Uljanin, 1870, and Nemertodermatida Karling, 1970, as well as some traditional Platyhelminthes families, will be considered turbellarians, as it is usually considered in the scientific literature.

According to their size, turbellarians can be divided into two major groups, with no taxonomic association: the macroturbellarians or large worms, including specimens with a length between 1 cm and 30 cm, and the microturbellarians, which usually range in length from 10 μm to 300 μm. Marine triclads and polyclads constitute the macroturbellarians, which generally occur on hard seafloor, under stones, or associated with macrophytes, algae or gravel. The microturbellarians usually occur on sandy or muddy bottoms, in sheltered tide flats, in the lower beach slope and in the swash zone (Giere 2009).

There are more than 3000 described species of turbellarians, most of them interstitial microturbellarians inhabiting the marine littoral. They are small-sized bilaterian acoelomate worms, oval to round in cross section, extremely flexible that combine muscular and ciliary action to achieve movement, with a ventrally positioned mouth, without an anus and usually with an epidermal dual-gland adhesive system, which make them well adapted to the interstitial life. Additionally, they display direct benthic development, dispersal only as adults, short generation times and semelparity.

Their diversity is high, but their small size, as well as their taxonomic complexity, usually excludes them from ecological surveys. The determination always requires an assortment of separation methods, the study of live material under light microscope in delicate squash preparations, time-consuming sectioning and interpretation of histological slides and, finally, comparison with, in some cases, rather incomplete or imprecise descriptions often hidden in obscure journals rarely accessible in our country.

The density and biomass of turbellarians are usually higher in sandy habitats than in muddy bottoms. In sand, their diversity can be of the same magnitude as that of nematodes, and turbellarian biomass can be even higher.

In the meiofaunal community, the majority of the turbellarian species are considered to be predators of small zoobenthos, representing the top consumers, where they usually play the role, that in muddy sediments are carried out by the macrofaunal predators (Martens and Schockaert 1986). Based on gut analysis and observations on feeding, there are also grazers feeding on bacteria and microalgae. Occasionally, diatoms are found in the gut of predators, but it remains unclear whether they entered the gut via prey animals or they show some grade of mixotrophic behaviour (Reise 1984).

The turbellarians are not beneficiated by the sediment destabilization, and they usually avoid living in funnels and mounds, but they inhabit the vicinity of burrows, especially near the feeding pocket of other borrowing taxa, like annelids (Reise 1985).

The first monographic treatises on turbellarians are almost 130 years old, and since then, a lot has been achieved in numerous aspects including taxonomy, systematic, biogeography, evolutionary development, physiology and ecology. Nevertheless, most of the studies have been carried out in the northern hemisphere (Rieger 1998). In South America, most studies were done on Brazilian turbellarians between 1940 and 1960, and in the Galapagos Islands, in the 1970s. In Argentina, most of the microturbellarians surveys have been carried out by Dr. Damborenea and Dr. Brusa in La Plata Museum, focusing their interest mainly on freshwater species (Brusa 2006a, b; Damborenea et al. 2007; Brusa et al. 2008).

Since 2004 Bulnes has taken over the study of the meiobenthic marine turbellarian community. The first studies were carried out in two localities of the northern coast of the Bahía Blanca Estuary: Arroyo Pareja beach and Baterías beach. The first locality represents a highly anthropized area, where a number of industrial activities as well as seasonal touristic and sport activities are performed. On the other hand, Baterías beach shows a low anthropogenic influence, since it is located in a restricted area of the Naval Base Puerto Belgrano (Fig. 2.4; Chap. 2). Between November 2004 and February 2006, a systematic survey on these localities revealed some ecological information of the Bahía Blanca Estuary's community of meiobenthic turbellarians. This community displayed a great degree of spatial, temporal and abundance diversity. The sampling was aimed to examine the first five superficial centimetres of the surf zone, and in the intertidal and supralittoral zone, the sediments from the surface, down to the water table depth (McLachlan and Defeo 2018).

During this survey, 3278 specimens were extracted and taxonomically determined. The results revealed the presence of over 28 species of interstitial free-living Platyhelminthes and Acoelomorpha (Bulnes 2007).

Although the turbellarians are present all year long in both localities, two maximum abundance peaks were registered during the autumn and the spring in April and May and October. The lowest abundances occur in summer (January and February) and at the end of the winter (August and September) (Fig. 7.4a).

To see a complete list of recorded species and genera, please check Tables 7.4 and 7.5.

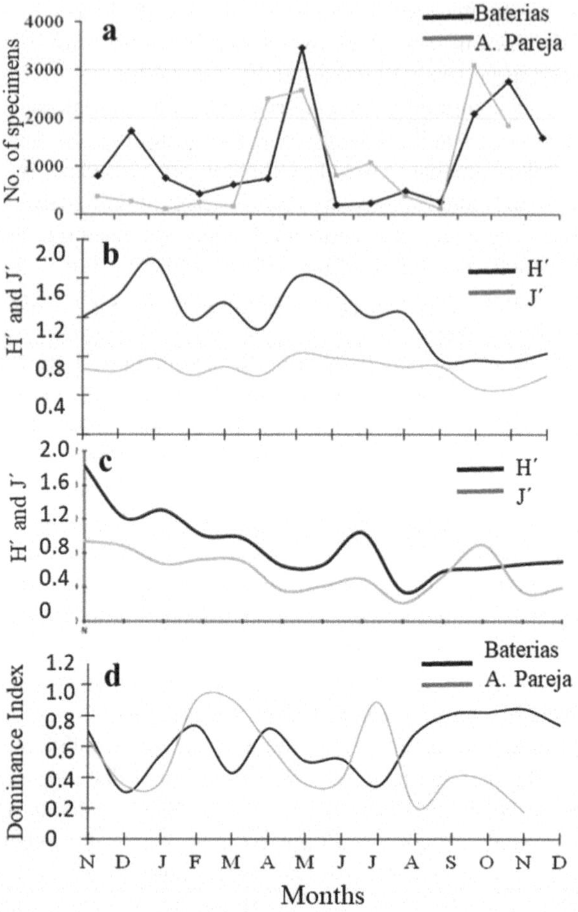

Fig. 7.4 Turbellarians of the Bahía Blanca Estuary. (**a**) Total abundance in Baterías and Arroyo Pareja beaches, (**b**) Shannon (H′) and Pielou (J′) index in Baterías beach, (**c**) Shannon (H′) and Pielou (J′) index in Arroyo Pareja beach, (**d**) dominance index variation in Baterías and Arroyo Pareja beaches

Most of the species were distributed independently from each other, and they were present for at least 3 months of the year displaying gradual taxon substitution. In Baterías beach, the highest diversity index (H′) was found in January (ten species), during one of the lowest abundance peaks, while in September the lowest diversity index (three species) agreed with a low abundance peak. The highest Pielou equity indexes (J′) were detected in February, April, October and November, whereas the lowest in May, June, July and September (Fig. 7.4b). In Arroyo Pareja beach, the highest diversity indexes occurred in November, January and June (eight species) and reached the lowest diversity index (two species) in July. The highest values of the Pielou equity index were detected in September and November,

Table 7.4 List of turbellarian species registered in the Bahía Blanca Estuary

Acoelomorpha		
		Philocelis robrochai Hooge and Rocha, 2006
Platyhelminthes	Proseriata	
		Vannuccia talea Marcus, 1954
	Macrostomorpha	
		Karlingia lutheri (Marcus, 1948)
		Myozona evelinae Marcus, 1949
		Archimacrostomum brasiliensis (Marcus, 1952)
	Dalytyphloplanida	
		Kalyla gabriellae Marcus, 1951
	Kalyptorhynchia	
		Cheliplana uruguayensis Van Steenkiste, Volonterio, Schockaert and Artois, 2008

Table 7.5 List of turbellarian genera registered in the Bahía Blanca Estuary

Acoelomorpha	
	Deuterogonaria Dörjes, 1968
	Kuma Marcus, 1950
	Symsagittifera Kostenko and Mamkaev, 1990
Platyhelminthes	
	Macrostomum Schmidt, 1848
	Microstomum Schmidt, 1848
	Promonotus Beklemischev, 1927
	Meidiama Marcus, 1946
	Carcharodorhynchus Meixner, 1938
	Cheliplana de Beauchamp, 1927
	Gnathorhynchus Meixner, 1929

whereas the lowest in April and July (Fig. 7.4c). Although Baterías beach showed higher dominance indexes than Arroyo Pareja beach, both showed the highest diversity in autumn and spring, and the lowest were in winter and summer.

The calculated dominance indexes (Fig. 7.4d) suggest that the species founded favourable conditions, especially regarding the food availability, leading to a fast growth in the abundance of fewer species that exploit rapidly the resources, probably multiplying asexual reproduction cycles, producing a high number of gametes or even achieving sexual maturity in a short period (Heitkamp 1988). On the other hand, the high diversity values calculated in the summer and the winter may be a consequence of the fast development of highly specialized species, present only in low numbers, but adapted to explore this temporarily impoverished environment.

In Baterías beach, 50% of the specimens belong to the clade Proseriata, followed by the Acoela, the Macrostomorpha, the Kalyptorhynchia and the Neodalyellida (Fig. 7.5a), while in Arroyo Pareja beach, 75% of the animals were representatives

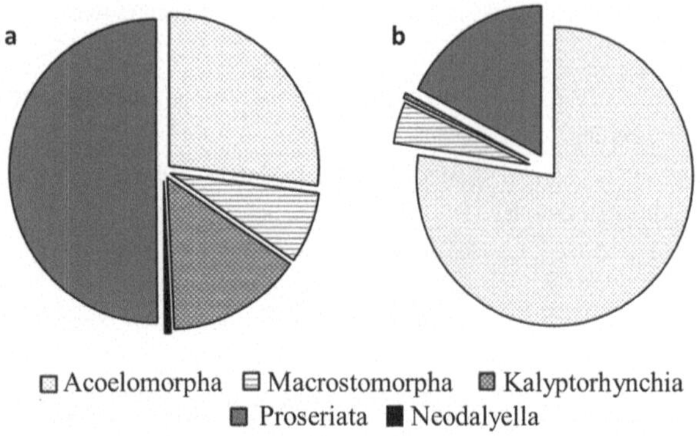

Fig. 7.5 Turbellarians of the Bahía Blanca Estuary. (**a**) Relative abundances of turbellarian groups in Baterías beach, (**b**) in Arroyo Pareja beach

of the clade Acoela, followed by the Proseriata, the Macrostomorpha and the Kalyptorhynchia (Fig. 7.5b).

The dominancy values remain high, since *Deuterogonaria* sp. represented the 50.24% of the collected specimens, being completely dominant in Arroyo Pareja beach during March and April, although absent in July and August (Fig. 7.6a). Regarding the abundance, a new species of *Symsagittifera* and a species belonging to the genus *Promonotus* followed the abovementioned acoel. Both *Symsagittifera* sp. and *Promonotus* sp. showed high abundance peaks in different months. The 14.3% of all the turbellarians were *Symsagittifera* sp., while 13.44% were *Promonotus* sp. The Proseriata *Promonotus* displayed an abundance peak in autumn, like *Deuterogonaria* sp., while *Symsagittifera* sp. was more abundant in spring (October), exhibiting one of the examples of the already mentioned gradual taxon substitution (Fig. 7.6b). In Baterías beach, the dominancy values of *Deuterogonaria* sp. (23.02%) and *Meidiama* sp. (26.5%) were alike, although lower than in Arroyo Pareja beach. Moreover, *Meidiama* sp. was absent during the months when *Deuterogonaria* sp. was dominant, and *Deuterogonaria* was absent during the period when *Meidiama* sp. was dominant, suggesting there is some degree of competition between these two species (Fig. 7.6c).

In general, the Acoela were dominant in Arroyo Pareja beach. This may be directly connected to the fact that Arroyo Pareja beach sediments are a mixture of fine-grain sand with a high content of clay, since mudflats and shallow tidal channels surround this locality, where the benthic microalgae proliferate. This combination of small-sized interstitial spaces and the abundance of diatoms as a food source may condition the development of the small-sized Acoela over the other larger turbellarian taxa. In Baterías, the beach profile is slightly more exposed to the wave action, and thus, the clay content is lower and the interstitial water currents are stronger. Here, the development of the turbellarians with adhesive dual glands for fixation to the substrate seems favoured. The high energy of this environment conditions the proliferation of diatoms, and the interstitial microhabitats are dominated by the predacious Proseriata and Kalyptorhynchia (Wellner and Reise 1989).

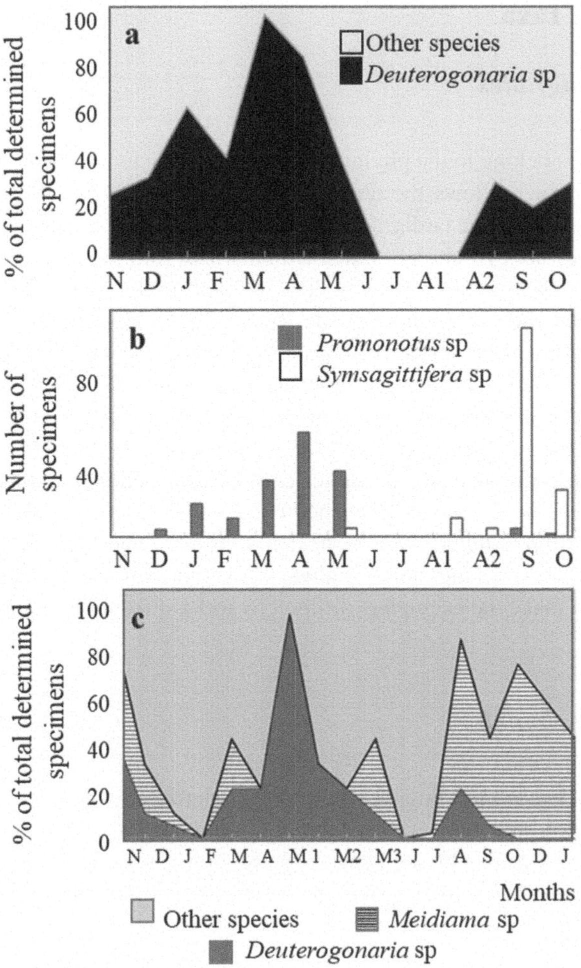

Fig. 7.6 Turbellarians of the Bahía Blanca Estuary. (**a**) Relative abundance of *Deuterogonaria* sp. in Arroyo Pareja beach, (**b**) *Promonotus* sp. and *Symsagittifera* sp. abundances in Arroyo Pareja beach, (**c**) relative abundance of *Deuterogonaria* sp. and *Meidiama* sp. in Baterías beach

The direct observation of diatoms in the gut content of Macrostomorpha and Proseriata suggests that although these Rhabditophora usually predate on copepods, nematodes and other turbellarians, they are also able to exploit different food sources, especially when the preys are scarce. Although the reproductive behaviour of the turbellarian agrees with the generalization made for meiobenthic organisms, the seasonal low selectivity seems to collide with the generalization made by Warwick and collaborators, when they characterized the meiofaunal taxa as selective feeders, being temporarily opportunistic, under given conditions (Warwick et al. 2006).

7.5 Other Taxa

7.5.1 *Tardigrades*

The water bears belong to the phylum Tardigrada. They live associated with a coating of water, which allows the animals to move and achieve all their metabolic requirements. The marine tardigrade may be found from the tide line up to the abyssal depths. They are cylindrical-bodied animals with four pair of limbs and possess numerous sensory structures, usually concentrated on the head, a complete digestive system and they are gonochoristic. Since 2015, Menechella and collaborators described three new species of marine tardigrades, all from a city in the direct vicinity of the Bahía Blanca Estuary (Menechella et al. 2015, 2017).

In Arroyo Pareja beach, Bulnes (unpublished data) found tardigrades in the samples with a higher sand content, from March to June and always in low numbers, not higher than five individuals per 25 cubic centimetres of sediment. The most frequent species is *Batillipes amblypyge* Menechella, Bulnes and Cazzaniga, 2017 (Fig. 7.6a); and rarely *Batillipes lingularum* Menechella, Bulnes and Cazzaniga, 2017, was present. The latter is smaller and it is always covered with some debris on its dorsal surface, which make the exemplars difficult to sort out from the samples.

7.5.2 *Kinorhynchs*

The Kinorhyncha or mud dragons are marine microscopic animals, usually more diverse in muddy sediments, rather than in sandy habitats. They have a short neck and a trunk divided into 11 segments, usually with several sensory structures like spines, sensory and glandular spots. The head displays numerous sensory appendages, the scalids, usually protected by a set of flattened appendages, the placids, which combine to form a movable protective structure of the head.

In the Bahía Blanca Estuary, only two species have been registered. In Arroyo Pareja beach, in sediments with higher sand content, Rucci and colaborators found *Franciscideres* cf. *kalenesos* Dal Zotto, Di Domenico, Garrafoni and Sørensen, 2013 (Rucci et al. 2020), a species with some unique adaptations to life in sandy habitats. In areas dominated by finer sediments, another species belonging to the clade Allomalorhagida has been collected, but they have been accidental observations, since the sampling and separation of kinorhynchs for taxonomic studies require specific techniques.

7.5.3 *Annelids*

The interstitial life seems to have evolved many times in different taxonomic groups, especially between the polychaetes. In many microscopic annelids, the segmental parapodia, the cirri as well as head appendages are reduced or even lacking. Many species seem to have retained some paedomorphic features associated with their miniaturization (Higgins and Thiel 1988).

In the Arroyo Pareja beach, a species of *Polygordius* Schneider, 1868, has been observed, especially in the spring months (Fig. 7.7d). Several other archiannelid species have been observed, but there is still a lack of taxonomic information regarding this group.

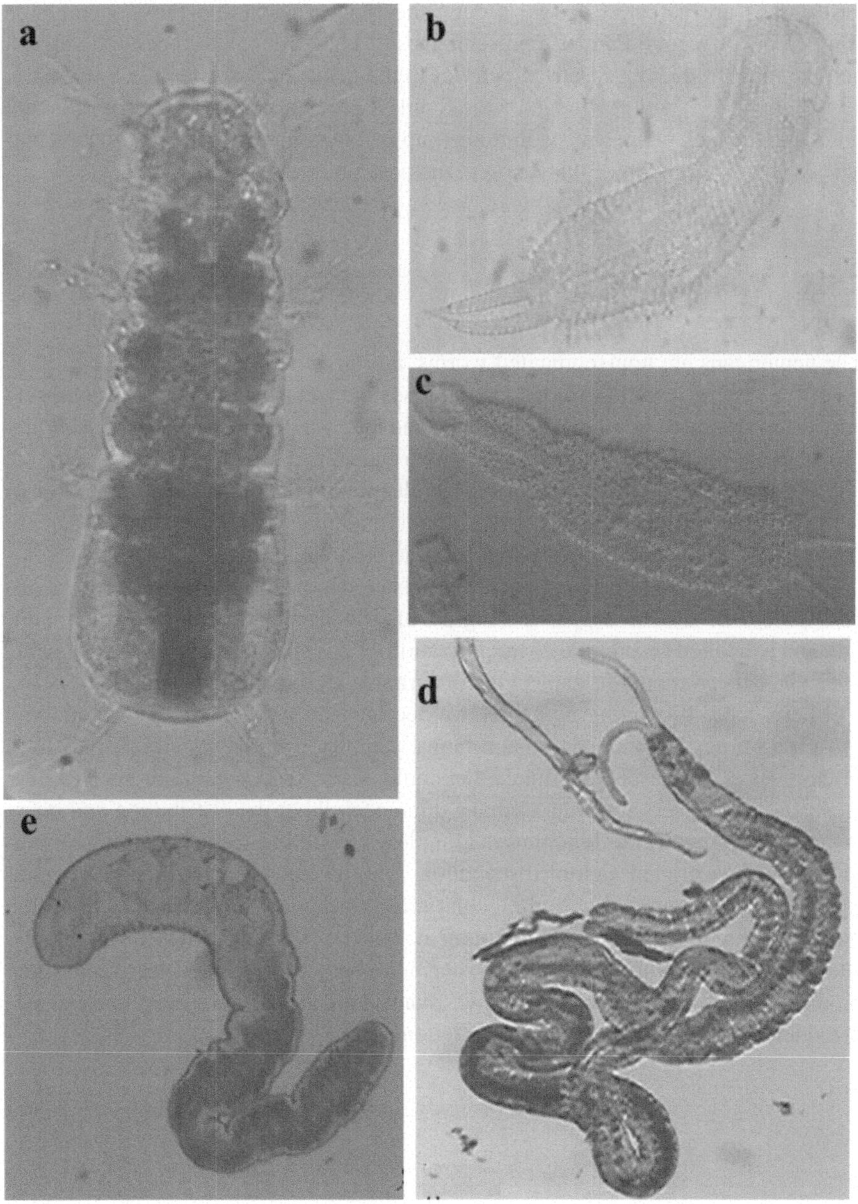

Fig. 7.7 Meiofaunal taxa of the Bahía Blanca Estuary. (**a**) *Batillipes amblypyge* Menechella, Bulnes and Cazzaniga, 2017, (**b**) gastrotrich Chaetonotoidea, (**c**) *Pseudostomella* sp., (**d**) *Polygordius* sp., (**e**) *Ototyphlonemertes* sp. (Photos by Verónica N. Bulnes)

7.5.4 Gastrotrichs

They are small, aquatic and interstitial worms. They are characterized by the dense ciliature on their ventral surface, as well as an anterior terminal mouth and a distal pair of adhesive tubes. Their identification depends on the characterization of the ornamentation of the body, like scales, plates and spines covering the trunk. The gastrotrichs are gathered in two orders: the Macrodasyda and the Chaetonotida (Fig. 7.7b). In Arroyo Pareja, representatives of both orders have been recognized, and the most frequent gastrotrich belongs to the genus *Pseudostomella* Swedmark, 1956 (Fig. 7.7). The knowledge of the South American marine gastrotrichs is still poor, and the lack of specialists in this group in Argentina will not allow us to characterize the community of this taxon in the near future.

7.5.5 Nemerteans

The nemerteans are non-segmented worms, a protrusible proboscis, with complete digestive system, a well-developed nervous system and a closed blood vascular system. They are found in all marine habitats and they are considered voracious predators, making them top consumers in the meiofaunal food chain.

Like with many other animals, the collection, separation and preparation of nemerteans require specific methods, still not implemented. Nevertheless, the methods used until now allowed us to obtain some preliminary data.

In the Bahía Blanca Estuary, we have registered a species belonging to the genus *Ototyphlonemertes* Diesing, 1863 (Fig. 7.7e). This genus is considered cosmopolitan, and a detailed taxonomic characterization of this species may provide valuable data, since there are major gaps in the geographic records of this genus.

Microscopic free-living animals have been considered somehow cosmopolitan for a long time. In contrast to macrofauna, samples from new studied areas have often revealed microscopic animals that could be ascribed to familiar species. It is clear we possess very few data to make inferences about the distribution and many cosmopolitan species could represent complexes of sibling species, each with probably a much-restricted area of distribution. Only intensive sampling and morphological studies combined with molecular techniques may throw some light on the biogeography of the meiofauna (Artois et al. 2011). In the Bahía Blanca Estuary, the study of the meiofauna started a few years ago, but although the description of the biodiversity will take some more time, many meiobenthic taxa have been recognized and await for a proper characterization.

References

Albuquerque EF, Brandão Pintol AP, d'Alcântara de Queiroz Perez A et al (2007) Spatial and temporal changes in intertidal meiofauna on a sandy beach of South America. Braz J Oceanogr 55(2):121–131. https://doi.org/10.1590/S1679-87592007000200005

Artois T, Fontaneto D, Hummon WD et al (2011) Ubiquity of microscopic animals? Evidence from the morphological approach in species identification. In: Fontaneto D (ed) Biogeography of Microscopic Organisms. Is everything small everywhere? The systematic association special 79:244–283

Boschi EE, Cousseau MB (eds) (2004). La vida entre mareas: vegetales y animales de las costas de Mar del Plata, Argentina. Publicaciones Especiales INIDEP, Argentina

Boxshall GA, Halsey SH (2004) An introduction to copepod diversity, vol I. Ray Society, London

Bremec CS, Martínez DE, Elías R (2007) Asociaciones bentónicas de fondos duros y comunidades incrustantes. In: Piccolo MC, Hoffmeyer M (eds) Ecosistema del estuario de Bahía Blanca. EDIUNS, Bahía Blanca

Brusa F (2006a) First record of Prolecithophora (Platyhelminthes) from Argentina: *Plagiostomum cilioejaculator* sp. n. Zootaxa 1223:47–53. https://doi.org/10.11646/zootaxa.1223.1.4

Brusa F (2006b) Macrostomida (Platyhelminthes: Rhabditophora) from Argentina, with descriptions of two new species of *Macrostomum* and of stylet ultrastructure. Zool Sci 23:853–862. https://doi.org/10.2108/zsj.23.853

Brusa F, Damborenea MC, Noreña C (2008) Dalyellioida (Platyhelminthes, Rhabdocoela) from the Rio de la Plata estuary in Argentina, with the description of two new species of Gieysztoria. Zootaxa 1861:1–16. https://doi.org/10.11646/zootaxa.1861.1.1

Bulnes VN (2007) Microturbelarios (Acoelomorpha y Platyhelminthes) litorales del estuario de Bahía Blanca. PhD Dissertation. Universidad Nacional del Sur, Argentina

Calcagno JA (ed) (2014) Los invertebrados marinos. Fundación de Historia Natural Félix de Azara, Argentina

Coull BC (1999) Role of meiofauna in estuarine soft-bottom habitats. Austral Ecol 24(4):327–343. https://doi.org/10.1046/j.1442-9993.1999.00979.x

Coull BC, Dudley BW (1985) Dynamics of meiobenthic copepod populations: a long-term study (1973–1983). Mar Ecol Prog Ser 24:219–229

Damborenea MC, Brusa F, Noreña C (2007) New Dalyelliidae (Platyhelminthes, Rhabditophora) from Buenos Aires province, Argentina, and their stylet ultrastructure. Zool Sci 24:803–810. https://doi.org/10.2108/zsj.24.803

Davidovich NA, Chepurnov V (1991) Periodical components of harpacticoids abundance dynamics in upper Black-sea sublittoral near Karadag. Okeanologiya 31(1):140–145

Dupuy C, Nguyen TH, Mizrahi D et al (2015) Structure and functional characteristics of the meiofauna community in highly unstable intertidal mud banks in Suriname and French Guiana (North Atlantic coast of South America). Cont Shelf Res 110:39–47. https://doi.org/10.1016/j.csr.2015.09.019

Elías R, Iribarne O, Bremec CS et al (2007) Comunidades bentónicas de fondos blandos. In: Piccolo MC, Hoffmeyer M (eds) Ecosistema del estuario de Bahía Blanca. EDIUNS, Bahía Blanca, pp 179–190

Feller RJ (1980) Quantitative cohort analysis of a sand-dwelling meiobenthic harpacticoid copepod. Est Coast Mar Sci 11(4):459–476. https://doi.org/10.1016/S0302-3524(80)80068-5

Freije RH, Gayoso AM (1988) Producción primaria del estuario de Bahía Blanca. Informes UNESCO, Ciencias del Mar, Uruguay 7:112–114

Gayoso AM (1998) Long-term phytoplankton studies in the Bahía Blanca Estuary, Argentina. ICES J Mar Sci 55:655–660

Giere O (2009) Meiobenthology. The microscopic motile fauna of aquatic sediments, 2nd edn. Springer, Berlin

Harguinteguy CA, Cofré MN, Pastor de Ward CT (2012) Change in the meiofaunal community structure of sandy beaches of the Nuevo Gulf (Chubut, Argentina). Pap Avulsos Zool 52(34):411–422. https://doi.org/10.1590/S0031-10492012021400001

Heitkamp U (1988) Life-cycles of microturbellarians of pools and their strategies of adaptation to their habitats. Prog Zool 36:449–455

Higgins RP, Thiel H (1988) Introduction to the Study of Meiofauna. Smithsonian Institution Press, Washington, DC

Huys R, Gee JM, Moore CG, Hamond R (1996) Marine and brackish water harpacticoid copepods. Part 1. Keys and notes for identification of the species, Synopses of the British Fauna (new series). Academic Press, Linnean Society of London, London

Martelli A (2010) Estudio taxonómico preliminar de nematodos de vida libre del estuario de Bahía Blanca. Degree thesis. Universidad Nacional del Sur, Argentina

Martelli A (2013) Technical report. Comisión de Investigaciones Científicas de la Provincia de Buenos Aires. La Plata, Argentina

Martelli A, Lo Russo V, Villares G et al (2017) Two new species of free-living marine nematodes of the family Oxystominidae Chitwood, 1935 (Enoplida) with a review of the genus *Thalassoalaimus* de Man, 1893 from the Argentine coast. Zootaxa 4250(4):347–357. https://doi.org/10.11646/zootaxa.4250.4.5

Martens P, Schockaert ER (1986) The importance of turbellarians in the marine meiobenthos: a review. Hydrobiologia 132:295–303. https://doi.org/10.1007/BF00046263

McLachlan A, Defeo O (2018) The ecology of sandy shores, 3rd edn. Academic Press, London

Menechella AG, Bulnes VN, Cazzaniga NJ (2015) A new Batillipedidae (Tardigrada, Arthrotardigrada) from Argentina. Zootaxa 4032(3):339–344. https://doi.org/10.11646/zootaxa.4032.3.11

Menechella AG, Bulnes VN, Cazzaniga NJ (2017) Two new species of *Batillipes* (Tardigrada, Arthrotardigrada, Batillipedidae) from the Argentinean Atlantic coast, and a key to all known species. Mar Biodivers 48:239–247. https://doi.org/10.1007/s12526-017-0640-4

Pallares RE (1968a) Copépodos marinos de la Ría Deseado (Santa Cruz, Argentina). Contribución sistemática-ecológica I. Servicio de Hidrografía Naval H-1024:11–25

Pallares RE (1968b) Copépodos marinos de la Ría Deseado (Santa Cruz, Argentina). Contribución sistemática-ecológica II. Physis 27(75):245–262

Pallares RE (1969) El género *Scutellidium* en la Ría Deseado (Crustacea, Copepoda). Physis 29(78):51–72

Pallares RE (1975) Copépodos marinos de la Ría Deseado (Santa Cruz, Argentina). Contribución sistemática-ecológica IV. Physis 34(89):213–227

Pallares RE (1982) Copépodos harpacticoides marinos de Tierra del Fuego (Argentina). IV. Bahía Thetis. Centro de Investigación de Biología Marina (CIBIMA). Contribución Científica 186:1–39

Pastor de Ward CT (2001) A new nematode from West Patagonian coast, *Biarmifer madrynensis* sp. n. with redefinition of the genus *Biarmifer* Wieser 1954 (Nematoda, Cyatholaimidae). Bull Inst R Sci Nat Belg Biol 71:139–149

Pastor de Ward CT (2003) Two new species of *Sabatieria* (Nematoda, Comesomatidae) from Golfo Nuevo, Chubut (Argentina). Zootaxa 172:1–12. https://doi.org/10.11646/zootaxa.172.1.1

Pastor de Ward CT (2004) New species of *Hopperia* (Nematoda, Comesomatidae) and *Metachromadora* (Nematoda, Desmodoridae) from Patagonia, Chubut, Argentina. Zootaxa 542:1–7. https://doi.org/10.11646/zootaxa.542.1.1

Pastor de Ward CT, Lo Russo V (2007) A review of the genus *Richtersia* (Nematoda: Selachinematidae): new species from Golfo San José and Golfo San Matías, Chubut (Argentina). J Mar Biol Assoc UK 87(5):1153–1160. https://doi.org/10.1017/S0025315407056755

Pastor de Ward CT, Lo Russo V, Villares G (2014) Two new species of *Prasaveljevia* Wieser, 1953 (Thoracostomopsidae, Nematoda) from Argentinean coasts (Chubut, Argentina). Zootaxa 4027(4):551–564. https://doi.org/10.11646/zootaxa.3999.4.2

Pastor de Ward CT, Lo Russo V, Villares G (2015a) *Patagonema iubatum* gen. Nov. and sp. nov. (Nematoda, Oncholaimidae, Octonchinae) from Patagonia sandy beach (Argentina). J Mar Biol Assoc UK 95(1):101–109. https://doi.org/10.1017/S0025315414000708

Pastor de Ward CT, Lo Russo V, Villares G et al (2015b) Free-living marine nematodes from San Julián Bay (Santa Cruz, Argentina). ZooKeys 489:1–133. https://doi.org/10.3897/zookeys.489.7311

Popovich CA, Guinder VA, Pettigrosso RE (2008) Composition and dynamics of phytoplankton and aloricate ciliate communities in the Bahía Blanca estuary. In: Neves R, Baretta JW, Mateus M (eds) Perspectives on integrated coastal zone management in South America. IST Press, Lisboa

Reise K (1984) Free-living Platyhelminthes (Turbellaria) of a marine sand flat: an ecological study. Microfauna Mar 1:1–62

Reise K (1985) Tidal flat ecology. An experimental approach to species interactions, Ecological studies 54. Springer, Berlin/New York

Rieger RM (1998) 100 years of research on 'Turbellaria'. Hydrobiol 383:1–27. https://doi.org/10.1023/A:1003423025252

Rodriguez JG, Lopez J, Jaramillo E (2001) Community structure of the intertidal meiofauna along a gradient of morphodynamic sandy beach types in southern Chile. Rev Chil Hist Nat 74:885–897. https://doi.org/10.4067/S0716-078X2001000400015

Rucci KA, Neuhaus B, Bulnes VN et al (2020) New record of the soft-bodied genus Franciscideres (Kinorhyncha) from Argentina, with notes on its movement and morphological variation. Zootaxa 4780(1):107–131. https://doi.org/10.11646/zootaxa.4780.1.5

Schmidt-Rhaesa A (ed) (2014) Handbook of zoology. Gastrotricha, Cycloneuralia and Gnathifera. Volume 2 Nematoda. De Gruyter, Berlin/Boston

Schratzberger M, Ingels J (2017) Meiofauna matters: the roles of meiofauna in benthic ecosystems. J Exp Mar Biol Ecol 502:12–25. https://doi.org/10.1016/j.jembe.2017.01.007

Sciberras M (2018) Biodiversidad de copépodos harpacticoideos bentónicos en un área del estuario de Bahía Blanca. PhD dissertation. Universidad Nacional del Sur, Argentina. http://repositoriodigital.uns.edu.ar/handle/123456789/4460

Sciberras M, Bulnes VN, Cazzaniga NJ (2014) A new species of *Quinquelaophonte* (Copepoda: Harpacticoida) from Argentina. Zoologia (Curitiba) 31:496–502. https://doi.org/10.1590/S1984-46702014000500010

Sciberras M, Huys R, Bulnes VN et al (2017) A new species of *Halectinosoma* Vervoort, 1962 (Copepoda: Harpacticoida) from Argentina, including a key to species with unusual leg armature patterns, notes on wrongly assigned taxa and an updated key to ectinosomatid genera. Mar Biodivers 48:407–422. https://doi.org/10.1007/s1252601708060

Sciberras M, Cazzaniga NJ, Huys R (2021) Description of a new species of *Nannopus* Brady, 1880 (Copepoda: Harpacticoida) from Argentinean waters, including an updated key to species. Zootaxa. (in press)

Villares G, Martelli A, Lo Russo V et al (2013) Three new species and one new record of *Campylaimus* (Diplopeltidae, Nematoda) from argentine coasts (Buenos Aires and Santa Cruz, Argentina). Zootaxa 3613(1):083–096. https://doi.org/10.11646/zootaxa.3613.1.4

Villares G, Lo Russo V, Pastor CT (2015) A new species of *Pontonema* (Oncholaimidae, Nematoda) and a redescription of *Pontonema incisum* Wieser, 1953 from Santa Cruz and Chubut Provinces, Argentina. Zootaxa 4058(3):417–428. https://doi.org/10.11646/zootaxa.4058.3.8

Villares G, Lo Russo V, Pastor de Ward CT et al (2016) Free-living marine nematodes from San Antonio Bay (Río Negro, Argentina). ZooKeys 574:43–55. https://doi.org/10.3897/zookeys.574.7222

Warwick RM, Dashfield SL, Somerfield PJ (2006) The integral structure of a benthic infaunal assemblage. J Exp Mar Biol Ecol 330:12–18. https://doi.org/10.1016/j.jembe.2005.12.013

Wellner G, Reise K (1989) Plathelminth assemblages from an exposed and a sheltered Sandy Beach of the Island of Sylt in the North Sea. Microfauna Mar 5:277–294
Wieser W (1953) Die Beziehung zwischen Mundhöhlengestalt, Ernährungsweise und Vorkommen bei freilebenden marinen Nematoden. Ark Zool 2:439–484

Chapter 8
The Intertidal Soft-Bottom Macrobenthic Invertebrates

M. Cecilia Carcedo, Sabrina Angeletti, Georgina Zapperi, Eder P. Dos Santos, and Sandra M. Fiori

8.1 The Macrobenthic Invertebrates and Their Distribution in Intertidal Areas of the Estuary

Sheltered coasts of bays and estuaries are depositional environments with fine-grained sediments and gently sloping intertidal areas, commonly known as tidal flats. These environments form in those places where tides and tidal currents dominate over other hydrodynamic forces (Klein 1985). Tidal flats are inhabited by a diverse assemblage of invertebrates, ranging from microscopic organisms within interstitial spaces of sediment particles to large forms such as crabs and shrimps. Specifically, those invertebrates larger than 500 μm and living associated with the bottom are called macrobenthic invertebrates. They constitute a diverse group of animals in coastal areas, in terms of its number of species, habitat preferences and feeding habits. Based on their position relative to the water/sediment interface, the macrobenthic invertebrates can be classified into three categories: (1) the endobenthos, that is, animals living buried within the sediment; (2) the epibenthos, that is, animals living on top of the sediment surface; and (3) the hyperbenthos, that is, animals living in the water layer close to the sea bed, but also found within the

M. C. Carcedo (✉) · S. M. Fiori
Instituto Argentino de Oceanografía IADO, (Universidad Nacional del Sur-CONICET), Bahía Blanca, Argentina

Departamento de Biología, Bioquímica y Farmacia, Universidad Nacional del Sur, Bahía Blanca, Argentina
e-mail: ccarcedo@iado-conicet.gob.ar

S. Angeletti
Instituto de Ciencias Biológicas y Biomédicas del Sur, Universidad Nacional del Sur, CONICET, INBIOSUR, Bahía Blanca, Argentina

G. Zapperi · E. P. Dos Santos
Instituto Argentino de Oceanografía IADO, (Universidad Nacional del Sur-CONICET), Bahía Blanca, Argentina

S. M. Fiori, P. D. Pratolongo (eds.), *The Bahía Blanca Estuary*,
https://doi.org/10.1007/978-3-030-66486-2_8

sediments due to the close proximity to the bottom (Pearson and Rosenberg 1987; Mees and Jones 1997). All these organisms, including slow-moving, relatively mobile and burrowing forms, have a range of effects on the physical structure and dynamics of the system, contributing as architects of their own habitat (Hansell 2005).

In relation to the feeding habits, protected areas such as estuarine tidal flats are typically inhabited by deposit feeders, while suspension feeders tend to be the most abundant group in high-energy environments like exposed sandy beaches (Snelgrove 1999). Dense deposit-feeding macrobenthic populations found on tidal flats substantially contribute to nutrient cycling by affecting sediment transport processes through bioturbation activities (Meysman et al. 2006; Van Colen et al. 2010). Moreover, these populations represent important trophic linkages through forage provision for fishes, birds and other epibenthic invertebrates, on one hand, and by feeding on benthic algae and bacteria, on the other hand (Van Oevelen et al. 2006). Reise (1985) showed that large carnivorous fishes and birds had little effect on the benthic community in sand flats (protected beaches, nearly estuarine environments). However, small epibenthic predators such as crabs, shrimps and juvenile fishes had significant effects on the small macrofauna and on juveniles of the larger macrofauna. Through this study, Reise (1985) suggested a conceptual hierarchy of biological processes influencing community structure, pointing out the importance of biological interactions on tidal flats (Defeo and McLachlan 2005; McLachlan and Defeo 2018). It is considered that in these environments, the lower limit of the intertidal area is mainly controlled by biotic factors such as the presence of competitors or predators, whereas the upper limit is mainly governed by abiotic processes such as immersion period, salinity and desiccation (Paterson et al. 2019).

The sheltered coast of the Bahía Blanca Estuary gives rise to an extensive, gently sloping and low-energy intertidal area, characterized by salt marshes and unvegetated tidal plains. Both environments are inhabited by a diverse community of macrobenthic invertebrates, with a great representation of deposit feeders, being polychaetes, crustaceans and molluscs the most represented groups. Early studies conducted in the Bahía Blanca Estuary by Elías (1985) and Elías and Bremec (1986) described for the first time the macroinfaunal associations of intertidal areas, specifically in four points covering the inner and mid sections of the Principal Channel (Fig. 2.1; Chap. 2). In the inner area, Puerto Cuatreros (Figs. 2.1 and 2.2; Chap. 2) was characterized by an intertidal area with salt marshes and mudflats where the benthic macrofaunal community was dominated by the polychaete *Laeonereis acuta* (Fig. 8.1a) in the upper intertidal zone and *Leodamas verax* (Fig. 8.1b), *Scoletoma* (= *Lumbrineris*) *tetraura* and *Kinbergonuphis* sp. in the lower intertidal zone. Another site studied in the inner section of the estuary was Puerto Galván (Figs. 2.1 and 2.3; Chap. 2). This site was characterized by the polychaetes *Laeonereis acuta* and *Eteone* sp. occupying the lower and middle zones of the intertidal area, and sometimes replaced by the snail *Heleobia australis*, mainly in middle and higher elevations within the intertidal area. The crab *Cyrtograpsus altimanus* was distributed along the entire mesolittoral area.

In the mid section of the estuary, in Puerto Rosales (Figs. 2.1 and 2.4; Chap. 2), the community was dominated by the snail *Heleobia australis* along the entire intertidal

Fig. 8.1 Some of the species found in intertidal soft-bottom areas in the Bahía Blanca Estuary. The polychaetes (**a**) *Laeonereis acuta* and (**b**) *Leodamas verax*; the bivalves (**c**) *Tagelus plebeius* and (**d**) *Nucula semiornata*; the amphipods (**e**) *Monocorophium insidiosum* and (**f**) *Heterophoxus videns*. (Photos by M. Cecilia Carcedo (**a, b, d, e, f**) and © Guido & Philippe Poppe - www.conchology.be (**c**))

area, and sometimes it is replaced by *Laeonereis acuta* at middle and lower elevations. Other species found were the priapulid *Priapulus tuberculatospinosus*, the snail *Buccinanops deformis*, the polychaetes *Scoletoma tetraura* and *Ninoe* sp., the clam *Tagelus plebeius* (Fig. 8.1c) and the crab *Cyrtograpsus altimanus*. The lower intertidal level was characterized by several deposit feeder bivalves such as *Nucula semiornata* (Fig. 8.1d), *Malletia* sp. and *Pitar rostratus* and other organisms such as the cnidarian *Stylatula darwinii*. Finally, the fourth site studied was Los Pichones (38°56′46″S, 62°17′56″W), located far from industries and urban settlements and therefore considered less influenced by human activities. This site was characterized by a low intertidal area dominated by polychaetes such as *Kinbergonuphis* sp., *Leodamas verax*, *Ninoe* sp. and *Axiothella* sp. The middle and higher levels of the intertidal area were exclusively inhabited by the crab *Neohelice granulata*.

A recent study, carried out in Villarino Viejo, in the inner section of the Principal Channel (Figs. 2.1 and 2.2; Chap. 2), characterized the community structure and the spatial zonation of benthic macrofauna in intertidal mudflats. This study mentions the crab *Neohelice granulata* as the typical species representative of the upper levels, along with the polychaete *Polydora* sp. (Zapperi et al. 2017). Both species showed a clear seasonal pattern of higher crab densities during spring to early autumn and higher polychaete density during late autumn and winter. The same study also remarked the presence of the dominant polychaete *Laeonereis acuta* in both sections of the mudflat, with higher abundances in the lower zone. The polychaete *Eteone* sp. and the priapulid *Priapulus tuberculatospinosus*, both species mentioned for the intertidal areas 30 years earlier (Elías 1985), were also found in the lower mudflat.

The burrowing crab *N. granulata* is largely considered the most conspicuous benthic species in tidal flats from the southwestern Atlantic coast (Spivak 2010). It has a broad distribution in the Bahía Blanca Estuary, occurring on both salt marshes and mudflats. The term 'cangrejales', from the Spanish word 'cangrejo' (crab), is commonly used locally to collectively denote the entire mudflat, due to the high density of burrows made by these crabs, with records of up to 172 burrows/m^2 in unvegetated areas, and up to 88 burrows/m^2 in salt marshes (Angeletti 2017). Crab burrows have a diameter of up to 12 cm, and they can reach up to 150 cm deep into the sediment (Escapa 2007). In these environments, *N. granulata* represents an important species within the benthic macrofaunal community in relation to its ecological role and its high population density (see Box 8.1).

Box 8.1: The Burrowing Crab Neohelice granulata: A Key Species in South American Estuaries

Neohelice granulata (Dana, 1851) is a crab from the northeastern coast of Patagonia, Argentina (42°25′S, 64°36′W), to Rio de Janeiro, Brazil (22°57′S, 42°50′W) (Spivak 2010), with a characteristic burrowing behaviour (Fig. 8.3). It digs and keeps open semipermanent burrows, allowing it to adopt a semiterrestrial life. Its presence grants a special physiognomy to the regions where it is found, called 'cangrejal' in Argentina and Uruguay, and 'caranguejal' in Brazil (Spivak 1997). Due to their semiterrestrial habits, the species are active in spring and summer but remain inactive or hidden inside their burrows throughout most autumn and winter. Consequently, the time available for feeding and growth should be limited by cold periods (Bas et al. 2005; Angeletti and Cervellini 2015).

Regarding the controversy of the systematics of this species, it can be mentioned briefly that Rathbun (1918) redescribes the species under the name *Chasmagnathus granulata*, which was used during most of the twentieth century, despite the disagreement of the scientific community in the masculine gender of specific names. However, the name *Chasmagnathus granulatus* has also been used by some authors (Lozada et al. 1988). Sakai et al. (2006) reviewed and reclassified all species previously attributed to the genera *Helice* and *Chasmagnathus*, redefined both genera and restricted them for the species of the seas of East Asia. They also introduced a new genus for the species of South America (*Neohelice*), New Zealand (*Austrohelice*) and the tropical and subtropical Indo-Pacific (*Pseudohelice*). Therefore, the species is currently called *Neohelice granulata*, belonging to the Grapsoidea superfamily and Varunidae family (Schubart et al. 2002).

This species has a broad feeding spectrum. It is considered omnivorous as it feeds on both algae and plants as well as polychaetes and small molluscs. It is also a deposit feeder, since crabs ingest sediments with a mixture of organic and inorganic matter. Stomach content studies showed traces of plants in crabs inhabiting marshes, but sediments (along with polychaetes, diatoms and nematodes) in crabs inhabiting mudflats, thus indicating that crabs act as deposit feeders in mudflats and herbivores in salt marshes (Iribarne et al. 1997; Botto et al. 2005; Méndez Casariego et al. 2011b). They can ingest on the surface and eliminate faeces in depth or vice versa, thus producing an important vertical mixture of sediments and generating important consequences in the processing of organic matter and nutrient flows (Reise 2002; Alberti et al. 2011).

Neohelice granulata lives in burrows to protect itself from wave action, extreme temperatures and desiccation. The burrows also provide shelter from aerial and terrestrial predators during periods of low tide, and from aquatic predators at high tide, thus avoiding the stress of finding refuge or escaping predators. Moreover, burrows are the places where two important events in a crab's life occur, moulting and reproduction, and also where young recruits are safeguarded until they reach larger sizes (Milner et al. 2010; Sal-Moyano et al. 2012). When building and maintaining their burrows, crabs bring sediments to the surface, forming mounds near burrow entrances (Murray et al. 2002). This bioturbation affects sediment structure because the cohesive nature of the organic matrix is disrupted by this process. Apart from directly affecting sediment porosity and permeability, this process has ecological significance in the aeration of soils containing anoxic sediments and in the distribution of halophytes. Consequently, high densities of active burrowers can increase erosion rates and sediment mobility (Botto and Iribarne 2000).

As a result of this disturbance in the sediment, *Neohelice granulata* may directly or indirectly affect other species that inhabit the same environment. This interaction can be advantageous for some organisms by promoting their growth or being harmful. Organisms that are sensitive to disturbance or excavation can suffer high mortality due to abrasion or dispersion (Kneib 1991; Billick and Case 1994), and predators of the coastal environment can attack areas where the infauna is exposed (Auster and Crockett 1984). The affected species include halophytic plants such as *Spartina densiflora*, *Spartina alterniflora* and *Sarcocornia perennis* (Bortolus and Iribarne 1999; Perillo and Iribarne 2003; Minkoff et al. 2006; Escapa et al. 2007; Daleo and Iribarne 2009; Alberti et al. 2008, 2011, 2014); other decapod crabs such as *Cyrtograpsus angulatus* and *Leptuca uruguayensis* (Spivak et al. 1994, 1996; Botto and Iribarne 2000; César et al. 2005; Méndez Casariego et al. 2009); polychaetes such as *Laeonereis acuta* and *Heteromastus similis* (Palomo et al. 2003; Escapa et al. 2004b; Rosa and Bemvenuti 2005); clams such as *Tagelus plebeius* (Gutiérrez and Iribarne 1998; Lomovasky et al. 2006); birds such as *Calidris fuscicollis*, *Charadrius falklandicus*, *Pluvialis squatarola*, *Haematopus palliatus*, *Larus atlanticus* and *Tringa melanoleuca* (Botto et al. 1998; Palomo et al. 2003; Iribarne et al. 2005; Berón et al. 2011); and rodents such as *Akodon azarae* and *Oligoryzomys flavescens* (Canepuccia et al. 2008), among others.

Therefore, the presence of the burrowing crab *Neohelice granulata* has important implications for the functioning of the estuarine ecosystem, being of great importance in the study of biological and ecological problems. In addition, the species is being used simultaneously for toxicology, biochemistry, neurophysiology and neurobiology and behaviour studies, becoming one of the most studied crab species, and considered as an emergent animal model for biological research at a global scale (Spivak 2010). At a local scale, *N. granulata* was declared as an 'emblematic' species by the Honorable Deliberative Council of Bahía Blanca city, according to ordinance N° 12.671 (2004). Other species reached by this ordinance were the olrog's gull (*Larus atlanticus*), the silver or Franciscan dolphin (*Pontoporia blainvillei*), the sharks (*Carcharias taurus*, *Carcharhinus brachyurus*) and migratory birds. The main objective of the ordinance was to recognize the species and through them to be able to educate the community and foster local identity.

Fig. 8.2 Association between salt marshes and benthic macrofauna. (**a**) The burrowing crab *Neohelice granulata* in salt marshes of *Spartina alterniflora*, (**b**) burrow of *Neohelice granulata* in salt marshes of *Sarcocornia perennis*, (**c** and **d**) the hydrobiid snail *Heleobia australis*. (Photos by Sandra M. Fiori (**a**), Sabrina Angeletti (**b**) and Walter Nievas (**c**, **d**))

Fig. 8.3 Individuals of the burrowing crab *Neohelice granulate* around a burrow entrance in mudflats of the Bahía Blanca Estuary. (Photo by Sabrina Angeletti)

8.2 Associations Between Salt Marshes and Benthic Macrofauna

Salt marshes are highly productive systems, and therefore the biomass exported from these environments is often considered the basis of estuarine food webs (Chalmers et al. 1985; Day et al. 1989). They are recognized worldwide for the various ecosystem services they provide, including atmospheric carbon sequestration (Chmura et al. 2003), reduction of eutrophication through nutrient uptake (Sousa et al. 2008) and metal sequestration (Weis and Weis 2004). In addition, marsh plants interact with benthic nutrients, modifying the nitrogen and phosphorus cycles (Pedersen et al. 2004). Their own presence also reduces the water energy with the consequent resistance to surface erosion. These environmental changes can have, in many cases, strong effects on the macrobenthic community structure. Jones et al. (1994) define the ecosystem engineers as the organisms that directly or indirectly modulate the availability of resources (other than themselves) to other species, by causing physical state changes in biotic or abiotic materials. By doing so, they can modify, maintain and/or create habitats. A way the salt marshes can act over community is ameliorating harsh physical conditions and therefore providing a less stressful microenvironment, e.g. the release of oxygen with the consequent changes in the redox potential and the pH in the rhizosphere (Pedersen et al. 2004), the reduction of thermal stress, favouring the establishment of other organisms by shading (Callaway 1995; Bortolus et al. 2002) and the availability of refuge (Hovel et al. 2001; Lewis and Eby 2002) and food availability (Kneib 1984). Therefore, salt marsh plants can have large effects on species distributions in intertidal communities. In the Bahía Blanca Estuary, salt marshes are distributed along all the margins of channels and on all the islands. The dominant salt marsh plant is the smooth cordgrass *Spartina alterniflora*, a species recently reported as an invasive species for South America (Bortolus et al. 2015). Other important salt marsh plants in the area are *Spartina densiflora* and *Sarcocornia perennis*, both species typical of substrates with longer air exposure, which are therefore more restricted to the supralittoral areas.

8.2.1 The Interaction Between Salt Marsh Plants and the Burrowing Crab Neohelice granulata

The amelioration of harsh physical conditions by plants plays an important role in dictating the abundance and distribution patterns of the crab *Neohelice granulata*. In Mar Chiquita Lagoon, located in the north coast of Buenos Aires Province (Fig. 1.2; Chap. 1), a strong relationship between plant cover and the spatial distribution of *N. granulata* was registered in the high marsh, where the spatial distribution of these crabs strongly overlapped the distribution of both *Spartina densiflora* and *Sarcocornia perennis* (Fig. 8.2a, b). Plant cover drives the spatial distribution of the dominant macroinvertebrate through facilitation, by buffering crabs from harsh, stressful environmental conditions. However, this relationship was not observed in the low marsh, where crabs established in both vegetated and unvegetated areas. This suggests that environmental conditions are more stressful on bare than on vegetated substrates but only in the high marsh, where stressful temperature and dehydration are important factors controlling the establishment and permanence of crabs (Bortolus et al. 2002). *N. granulata* density decreases towards the upper intertidal zone, as noted by Bortolus and Iribarne (1999). However, the colonization of the high marsh by *N. granulata* is facilitated by the presence of plants. *Sarcocornia perennis* is a good colonizer of high and saline intertidal areas. It generates a shading area allowing the sediment to stay wetter, therefore softer and more likely to be excavated. It also dampens various environmental factors, such as desiccation and high surface temperatures (Bortolus et al. 2002; Angeletti 2012).

The herbivorous habit of *Neohelice granulata* in marshes can determine the lower limit of the distribution of *Spartina densiflora* (Alberti et al. 2010). In addition, burrows of *N. granulata* can increase nutrient circulation and promote the production of *S. densiflora* through increased oxygenation that facilitates the growth of a fungus that colonizes its roots (Daleo et al. 2007; Daleo and Iribarne 2009). Regarding the herbivorous habit of the burrowing crab, it has been shown that it preferentially consumes *S. densiflora* more than *Sarcocornia perennis* (Costa et al. 2003). A complex process generated by the biological interaction between *N. granulata* and *S. perennis* has been described by Perillo and Iribarne (2003) in the inner section of the Bahía Blanca Estuary. Due to the crab's burrowing activity, plants grow by generating ring-formed patches from 1.5 to 8 m in diameter. Plant clones concentrate at the edges, while the inner sector is dominated by crab burrows. This interaction facilitates the erosion of the surface, arising from a decreased soil resistance due to the presence of burrows, and induces the formation of salt ponds in salt marshes. As a result of continuous sediment removal by crabs, the ponds remain sunken and accumulate water (even at low tide). These circular shallow ponds are continuously expanding in size and can join each other, forming different configurations. Water trapped in ponds can drain to nearby areas or pre-existing tidal channels (Escapa 2007).

8.2.2 *The Joint Distribution of* Spartina alterniflora *and the Snail* Heleobia australis

Another species frequently found in salt marshes where cordgrasses such as *Spartina* spp. are the dominant plant is the hydrobioid snail *Heleobia australis* (d'Orbigny, 1835) (Fig. 8.2c, d). This snail is a generally dominant species in abundance and biomass constituting a food item of relevance for fishes and crustaceans. In turn, since it incorporates the organic matter present in the sediments through its diet, it is a fundamental link in the nutrient cycle through coastal food webs (Figueiredo-Barros et al. 2006). *Heleobia australis* has a broad geographical distribution from

Rio de Janeiro, Brazil (22°54′S), to San Antonio Oeste, northern Argentinean Patagonia (40°84′S) (Aguirre and Farinati 2000), and it is mostly associated to mud, high salinity and low-energy coastal areas (Canepuccia et al. 2007). In the Bahía Blanca Estuary, it represents the only hydrobioid snail, being one of the most abundant species of intertidal mudflats and salt marshes in this ecosystem (Elías et al. 2004; Carcedo and Fiori 2011).

At the microscale, its spatial distribution is determined by a set of complex behaviours that allow the active selection of areas with certain physical and chemical qualities. The adults of *H. australis* are capable of floating away from the sediment itself by creating a gas bubble inside its shell. This behaviour was observed by Little and Nix (1976) for *Hydrobia ulvae* and then by Echeverría et al. (2010) for *H. australis* as a dispersion strategy. Through this mechanism, adults can temporarily enter the water column, and be carried along by the tide- and wind-driven currents, being able to escape from excessively stressful situations in search of better environmental conditions. Indeed, within intertidal areas of the Bahía Blanca Estuary, individuals of *H. australis* can actively select vegetated areas of *Spartina alterniflora*, resulting in a joint distribution between both species (Canepuccia et al. 2007; Carcedo and Fiori 2011). The explanation of this distribution pattern could be related to different processes, and Canepuccia et al. (2007) evaluate different hypotheses. One of them was related to the capability of *S. alterniflora* stems to generate microhabitats that favour the establishment of species by reducing predation rates (Lewis and Eby 2002), but authors conclude that predation seems to be a weak force to explain the association between species. Another hypothesis was related to the feeding preferences of the snail, assuming that *S. alterniflora* indirectly provides food resources (as epiphytes) to the snails. Stable isotopic analysis shows that neither *S. alterniflora* nor its epiphytes directly contribute to the diet of *H. australis*, being the sediment its main food source. Therefore, the association was neither explained by this hypothesis. Canepuccia et al. (2007) finally suggest that the distribution and survival of *H. australis* in marshes could be regulated by the temperature and the dehydration stress (Hovel et al. 2001). This conclusion was based on the observations that *H. australis* established more in experimental shaded areas and also on the higher mortality registered under more intense sun irradiance in bare areas. Therefore, the reduction of temperature provided by the shading of *S. alterniflora* stems leads to the selection of these microhabitats by the snails.

Although *H. australis* typically inhabits soft bottoms, it is capable to colonize hard substrates and reefs. The hard substrates of the Bahía Blanca Estuary show, in most cases, intermediate densities of *H. australis* between salt marshes and tidal flats (Carcedo and Fiori 2011). Given the structural complexity of the rocks, these areas could give certain advantages for snails similar to what occurs in reefs of the Pacific oyster, *Magallana gigas* (ex *Crassostrea gigas*) in Anegada Bay, Argentina (Fig. 1.2; Chap. 1) (Escapa et al. 2004a), and in tubes of the polychaete *Ficopomatus enigmaticus* in Mar Chiquita (De Francesco and Isla 2003; Luppi and Bas 2002). The unvegetated tidal flats were the areas less selected by *H. australis*. Their high conductance, reflectivity and caloric capacity could make these areas reaching higher temperatures than those with vegetation cover or with shaded spaces within hard substrates.

The active selection of microenvironments, described for adults of *H. australis*, was also suggested for juveniles and recruits. Snails of different sizes are differentially distributed across microenvironments in intertidal areas of the Bahía Blanca Estuary (Carcedo and Fiori 2011). Salt marshes of *S. alterniflora* were the only microenvironments where individuals smaller than 2.5 mm were found. In addition, juveniles and sub-adult snails are also able to colonize the hard substrates, while adults are distributed in all microenvironments. A similar behaviour was observed in Mar Chiquita Lagoon by the crab *Cyrtograpsus angulatus*, whose adults and juveniles reach high densities in muddy substrates, while its megalops, recruits and small juveniles are almost exclusively found in reefs of the polychaete *F. enigmaticus*, due to their importance as refugees against predation and cannibalism (Spivak et al. 1994; Luppi and Bas 2002). Although the hypothesis of dehydration stress was not tested between snails of different sizes, the effect of higher temperatures on the snails could be exacerbated in juveniles and recruits, leading to the observed size distribution between microenvironments.

8.3 Processes Modulated by Local Macroinfauna

Benthic organisms have a direct relationship with the substrate, and their interactions with the bottom environments generate both stabilization and destabilization of sediments. Some benthic microalgae excrete extracellular polymeric substances and build an organic matrix that binds sediment grains forming the so-called biofilms, recognized for their stabilizing function (Decho 2000; Cuadrado et al. 2011). Macroinfauna, on the other hand, can promote the destabilization of cohesive sediments through the production of pellets and the formation and maintenance of different types of constructions, such as burrows and caves (Herman et al. 1999; Braeckman et al. 2011). In addition to causing a physical disturbance, local macroinfauna can change the characteristics of the sediment affecting the porosity, permeability, resistance to erosion and its mobility (Botto and Iribarne 2000; Escapa et al. 2008). Thus, both bio-stabilizing and bio-destabilizing organisms influence the properties of intertidal sediments, significantly affecting geomorphology (Widdows et al. 2000; Murray et al. 2002).

Macroinfaunal activity can have considerable effects on the structure of both terrestrial and marine sediments. Its importance was first demonstrated by Charles Darwin in 1881, who dedicated his latest scientific book to this subject: *The Formation of Vegetable Mould, Through the Action of Worms, with Observations on Their Habits*. The content of this book focused on sediment alterations caused by the activity of burrowing organisms, especially by worms, a process that was later called 'bioturbation' (Richter 1952). Darwin was the first person to realize that biological activity at the local level, carried out by small invertebrates, could have consequences on a landscape scale, due to its influence on processes that model geomorphology. Numerous studies of a variety of disciplines, including ecology, edaphology, hydrology, geomorphology and even archaeology, cite Darwin's book as the original reference.

Macroinfauna in marine ecosystems modify the environment at different spatial and temporal scales. Many of these modifications are initially at a microscale, but they are likely to have large-scale effects on benthic and pelagic seascapes (Meadows et al. 2012). Tidal flats and salt marshes are habitats of a great variety of macroinfauna, subject to abiotic and biotic processes that actively shape them and define their functions. These organisms, through their different ways of life and feeding strategies, modify the size of the particles and the chemistry of the soil (Cadée 2001; Méndez Casariego et al. 2011a).

8.3.1 The Bioturbation

In modern ecological theory, bioturbation is recognized as an archetypal example of ecosystem engineering, capable of modifying geochemical gradients, redistributing food resources, viruses, bacteria, a variety of larval stages and eggs. From an evolutionary perspective, it has been shown that bioturbation played a key role in the evolution of metazoan life at the end of the Precambrian era (Meysman et al. 2006). In recent decades, bioturbation has been rediscovered as an important factor in the evolution of the earth's surface, through its influence on soil formation, erosion and horizon stability (Dietrich and Perron 2006; Kristensen et al. 2012).

In sediments of estuarine environments, oxygen penetration and aerobic decomposition are limited to the first surface millimetres, and the deeper layers are characterized by anaerobic decomposition processes (Revsbech et al. 1980; Kostka et al. 2002). However, the bioturbation activity of the fauna and roots of plants can increase the depth of the sediment's oxic layer, favouring aerobic processes (Fanjul et al. 2015). In environments of low or intermediate energy, such as the intertidal areas of the Bahía Blanca Estuary, bioturbating macrobenthic fauna is the main agent for mixing sedimentary particles (Rhoads and Young 1970; Jones and Jago 1993; Botto and Iribarne 2000). It has been proven that the activity of macrobenthic invertebrates that build and maintain their burrows affects the oxygenation, geochemistry and transport of particulate materials in the sediment column, mixing the upper layers with deeper ones, in a few days (Botto et al. 2006; Fanjul et al. 2007).

The burrowing and semiterrestrial crab *Neohelice granulata* actively and constantly build its burrows in the intertidal zone during low tide. Crabs dig and keep open semipermanent burrows, allowing this invertebrate to adopt a semiterrestrial life. In the Bahía Blanca Estuary, burrows are tunnel-shaped, with ramifications and multiple entrances, which is directly associated with the type of substrate on which the burrows are built. However, there are differences in the structural morphology of burrows and the burrowing activities in contrasting microhabitats (e.g. salt marshes and mudflats), indicating that this species has an adaptive burrowing behaviour (Escapa et al. 2007; Angeletti et al. 2018a). Burrows from salt marshes have narrow entrances and deep tunnels, and burrows from mudflats present very narrow entrances and surface tunnels. Crabs would change the depth of their burrows in relation to water table depth, to reach groundwater levels. In this way, the burrows would contain water throughout the tide cycle, an adaptation to maintain the humidity in high marsh areas (Iribarne et al. 1997; Escapa et al. 2007). These multiple entrances may provide easy access to shelter, especially in unvegetated mudflats. In addition, there is evidence suggesting that a larger number of burrow entrances in intertidal mudflats increases the ability to entrap organic particles that may serve as food (Iribarne et al. 1997; Botto et al. 2006). In salt marshes, chambers were discovered inside tunnels, possibly used for reproductive purposes (Angeletti et al. 2018a).

Neohelice granulata is commonly considered as an ecosystem engineer due to its great burrowing activity and its potential effect on intertidal sediment characteristics. When building and maintaining their burrows, crabs bring sediments to the surface forming biogenic mounds near burrow entrances (Murray et al. 2002). This bioturbation affects sediment structure because the cohesive nature of the organic matrix is disrupted by this process. Besides affecting sediment porosity and permeability, this process has ecological significance in the aeration of soils containing anoxic sediments and in the distribution of halophytes. Consequently, high densities of active burrowers can increase erosion rates and sediment mobility (Botto and Iribarne 2000).

Studies developed in the Bahía Blanca Estuary showed that biogenic mounds removed by crabs did not have a specific orientation. No general pattern was observed in their position in relation to the mouth of the burrow or to the location relative to the subtidal zone (Angeletti 2017). Regarding the penetrability of sediments, i.e. the force necessary to penetrate the substrate, biogenic mounds had higher water content and penetrability than sediments not removed by crab's activity (Escapa et al. 2004b; Angeletti et al. 2018a). This trend has been demonstrated by controlled laboratory experiments, carried out by Escapa et al. (2007), where the inclusion of crabs modified the sediment penetrability. After the bioturbation activity, the average force required to penetrate the sediments was reduced to less than 20% of that of the control sediments. However, no differences were found in the organic matter content between biogenic mounds and control sediments, because remixing keeps organic matter homogeneously dispersed in the sedimentary column (Bortolus and Iribarne 1999; Gutiérrez et al. 2006). The activity of *Neohelice granulata* transports sediments with a large amount of silt and clay from deep layers to the surface, recovering these inorganic materials along with the associated organic matter, and making them available at the surface and the water column (Escapa et al. 2007; Angeletti et al. 2018a).

Bioturbation by *Neohelice granulata* interacts with local physical processes and hydrodynamic conditions and can affect a wide variety of sedimentological parameters. Biologically induced roughness modifies hydrodynamics above the sediment layer, which in turn affects erosion and resuspension (Boudreau and Jørgensen 2001). Even in coastal systems, which are traditionally seen as modelled only by the physical forces of currents and waves, hydraulic engineers have recognized that bioturbation is a crucial component in sedimentary dynamics models (Murray et al.

2002; Paarlberg et al. 2005). Several studies have linked biological data on bioturbation to adjust parameters in bioturbation models (Gilbert et al. 2003; Solan et al. 2004; Jarvis et al. 2010; Schiffers et al. 2011).

The suspended sediment contribution arising from the bioturbation activity of *N. granulata* in intertidal sites of the Bahía Blanca Estuary was analysed by Angeletti et al. (2018b) using several approaches, ranging from field experiments to numerical modelling. It was found that crabs from mudflats remove, trap and erode more sediment per unit area than crabs from salt marshes, due to their higher population density and the mobility of cohesive sediments. Results obtained through MOHID (water modelling system) simulations showed that the sediments in the inner section of the estuary were maintained in the water column much longer than sediments from the middle zone. This longer residence time in the inner area could be attributed to the geomorphological and hydrodynamic characteristics of this section of the estuary, where numerous tidal channels coexist and phenomena of 'retention' occur before entrance into the main channel. By contrast, in the middle section of the estuary, the sediments are affected by greater water depth and higher tidal current speeds. In addition, wind-generated waves can be a determining factor in the spatio-temporal evolution of the bioavailable sediment in the water column.

8.3.2 The Benthic-Pelagic Coupling

Benthic-pelagic coupling is the process which links bottom sediments and the water column in aquatic systems and reveals as the exchange of energy, mass or nutrients (Griffiths et al. 2017). This phenomenon plays a major role (along with terrigenous input and upwelling) in determining the production and biological structure of these aquatic habitats (e.g. Sommer 1989; Valiela 1995). Estuaries are commonly shallow ecosystems, where the occurrence of strong coupling of matter and energy between the water column and the sediment is enhanced. Their strong horizontal gradients of salinity, water current velocity, turbulence and turbidity affect the coupling between ecological and biogeochemical processes in the water column and the sediment (Burdige 2006; Griffiths et al. 2017). This interaction highly influences nutrient recycling and overall productivity of the ecosystem.

Processes in the benthic boundary layer are key components in the dynamics of coastal ecosystems (Chauvaud et al. 2000; Dale and Prego 2002; Marcus and Boero 1998). This is the layer in which the exchange of materials between sediments and the water column takes place and where biological communities may influence these fluxes through bioturbation, excretion and respiration, as well as organic deposition (Norkko et al. 2001). Benthic and pelagic communities, therefore, can significantly modify physical-chemical properties of sediments, changing redox conditions and enriching the sediments with organic matter.

The fate of the organic matter produced in the water column depends on many physical-chemical and biological factors, such as currents, salinity, temperature, algal density and aggregate formation (Alldredge and Jackson 1995; Kiørboe and Jackson 2001). However, the shallowness of the environment strengthens benthic-pelagic interactions. The sink of phytodetritus and deposition of pelagic material may increase the organic matter input into the benthic habitat, inducing microbial remineralization. In response, the bottom sediments will contribute dissolved nutrients to the water column, and it is estimated that 30–80% of the nitrogen required in shallow coastal waters is supplied by benthic release (Blackburn and Henriksen 1983; Boynton and Kemp 1985; Nixon 1981). The most important productivity event in the pelagic food webs is phytoplankton blooms (Legendre 1990) which also have a large influence on benthic nutrient release or regeneration (Hansen and Blackburn 1992; Jensen et al. 1990; Overnell et al. 1995), as well as in their fast consumption up to depletion (Boyer et al. 2009; Bricker et al. 2008).

Ecosystem dynamics in the Bahía Blanca Estuary is dominated by a winter-early spring phytoplankton bloom that commonly occurs in the shallow waters of the inner zone. Studies carried out along 30 years have demonstrated that the lower turbidity observed in late autumn-early winter is a repetitive pattern which has been previously associated with the start of the phytoplankton bloom (Guinder 2011; Guinder et al. 2009). In this shallow environment, the sink of phytoplankton cells may represent an important contribution of organic matter to the benthic communities, and nutrient regeneration in the surface sediments may contribute significantly to the pelagic environment. Except for isolated observations of material settlement during the blooming season, few studies were performed regarding benthic-pelagic coupling in the area. Guinder et al. (2015) have indirectly addressed the subject when they evaluated the exportation of particulate suspended matter to the benthos through the analysis of settled material inside sediment collectors. Their results showed that the sedimentation of viable and large diatoms may represent an important source of allochthonous carbon for the benthic community.

The evaluation of nutrient fluxes using microcosm experiments revealed important variations in flux magnitude and direction that were related to the phytoplankton dynamics (Zapperi et al. 2016). Moderate release of nitrite and nitrate ($NO_3^- + NO_2^-$) from the sediment to the water column was observed in the incubation experiments performed for the upper and lower zones of the intertidal mudflats. This flux towards the water column represents a vital input after depletion of these nitrogen (N) species during the post-phytoplankton bloom period. However, the most dramatic response registered in the incubation experiments was ammonium (NH_4^+) recovery. After the chlorophyll-a peak during the post-bloom stage, the lower mudflats showed extremely high rates of NH_4^+ regeneration, whereas the upper mudflats continued the uptake trend they showed during pre-bloom and bloom stages. Low macrofaunal densities were recorded in the upper mudflats during that same period, while lower mudflats showed high densities of the polychaete *Laeonereis acuta*. Breakdown of organic matter to produce NH_4^+ is a major mechanism of N remineralization (McGlathery et al. 2004), and the results suggest a key role of benthic fauna in N recycling, as well as in the rapid recovery of NH_4^+ levels in the water column.

In the Bahía Blanca Estuary, macrobenthic communities in the upper and lower mudflats show differences in terms of their species composition as well as their patterns of activity throughout the year. Some common species in the upper mudflats, like *Neohelice granulata* and *Polydora* sp., are virtually absent in lower mudflats, whereas frequent species in lower mudflats, like *Laeonereis acuta* and *Eteone* sp., appear at much lower abundances in upper mudflats (Zapperi et al. 2016). In the same study, it was observed that lower mudflats showed maximum faunal densities during the chlorophyll-a peak, right after the bloom, whereas the upper mudflats recovered macrobenthic activity several months after the phytoplankton bloom. Biological activity of benthic macrofauna is expected to have a large effect on benthic fluxes (Hughes et al. 2000; Michaud et al. 2006; Webb and Eyre 2004), and it is reported that bioturbation would enhance benthic fluxes up to one order of magnitude (Burdige 2006), which can be related to the NH_4^+ recovery patterns described above.

Another clear representation of the coupling, mediated by biota, is the variable abundance of benthic macrofauna through the year. *Neohelice granulata* showed a minimum activity in June, at the same time that *Laeonereis acuta* attained maximum density, in agreement with the lowest turbidity registered. Erosion rates in the study area are closely related to bioturbation activity by the burrowing crab *N. granulata* (Escapa et al. 2007). On the contrary, polychaetes might contribute to sediment stability, reducing susceptibility to physical resuspension (Mortimer et al. 1999). Studies have suggested that feeding activity of *L. acuta* promotes sediment stability by deposition of cohesive, organically enriched sediments on the surface (Palomo and Iribarne 2000). On the other hand, light limitation has been proposed as a major control of winter bloom development in the Bahía Blanca Estuary (Guinder et al. 2009) because of its elevated turbidity. Hence, sufficient light penetration in the water column, promoted by the turbidity decrease associated to reduced *N. granulata* activity, is a potential trigger for phytoplankton cell proliferation (Cloern 1987; Irigoien and Castel 1997; May et al. 2003).

Summarizing, benthic-pelagic coupling plays a key role in determining production and biological structure in the Bahía Blanca Estuary. Its importance is reflected in how benthic-pelagic coupling mediated by biological activity plays a significant role in creating the window of lower turbidity that allows a phytoplankton bloom, which sustains the high productivity of this ecosystem. As the bloom develops, dissolved nutrients in the water column are consumed, and organic matter is produced. As a counterpart, the local coupling between remineralization in the surface sediments and benthic flux to the water column as dissolved nutrients allows the recovery of nutrient levels and supports primary production in the forthcoming cycle.

8.4 New Registers of Benthic Macrofaunal Species During the Last Decade and Their Implications

Because of the vast surface of the intertidal area in the Bahía Blanca Estuary and the instability of its soft sediments, it is a largely unexplored environment. In addition, the great majority of the animals living in intertidal flats have a cryptic lifestyle, remaining hidden in the muddy bottoms inside their burrows in deeper layers of sediments and emerging only when necessary to feed or to perform other vital functions (e.g. crabs, polychaetes). Some species even spend their whole lifespan inside their burrows (e.g. pholadid clams), which are recognizable through small openings visible on the surface. Relatively little is known about the biology and ecology of benthic macrofaunal invertebrates in intertidal areas of the Bahía Blanca Estuary, and the recent discovery of newly established populations supports this assumption. The most important findings are developed in the next paragraphs together with a brief discussion of the feasible causes and consequences of their presence and the processes that could be implicated in the extension of their distribution ranges.

8.4.1 *Hidden in the Mud: The Exotic Piddock* **Barnea truncata**

The family Pholadidae, commonly known as piddocks and angel wings, are represented by bivalve species that build tubular burrows by using their shell to mechanically erode the substratum. The Atlantic Mud-piddock *Barnea truncata* that belongs to this family is restricted to Atlantic Ocean margins. In the western Atlantic, *B. truncata* is distributed from 24°S to 45.4°N, from southeastern Brazil (west of São Paulo) to northern USA (Maine). In the eastern Atlantic, it has been recorded from 15°N to 34°S latitude, from Senegal to South Africa (COSEWIC 2009; Hebda 2011). During 2010, individuals of *B. truncata* were found in the intertidal zone of the Bahía Blanca Estuary (Fiori et al. 2012), and this finding represents the first record of the species at more than 2000 km south from the edge of its native range in South America. The native species *Barnea lamellosa*, with a distribution limited to the coasts of Uruguay and Argentina (Turner 1954), was also registered in the same intertidal area where *B. truncata* was found, but in a lower number than its exotic congener.

As for many of the non-indigenous invasive species registered in the Bahía Blanca Estuary, the accidental introduction through ballast water from ships was proposed as the most probable entry route to the ecosystem. Fiori et al. (2012) indicate several assumptions that support this hypothesis; one was related to the location where the population of *B. truncata* was found, in the intertidal zone near Ingeniero White (Figs. 2.1 and 2.2; Chap. 2), a port with heavy maritime traffic. Another assumption was related to the long permanence of their larvae in the water column as indicated by Turner (1954) and Chanley (1965), and finally due to its absence in intermediate locations between its southern native range and the Bahía Blanca Estuary. However, Fiori et al. (2012) also suggest that the natural expansion of this species is a possibility that should be also considered. The lack of records in intermediate locations could be associated with the cryptic lifestyle of this species, which further restricts the chances of being discovered. This was the case at the northern limits of its distribution range, where a single population was detected in Nova Scotia, approximately 350 km north and east of the nearest record in Maine (Hebda 2011).

Invasive bivalve species usually become dominant in their communities, and therefore, bivalve activities such as shell production, filter feeding and bioturbation can markedly affect ecosystem structure and functioning. In the case of burrowing bivalves, their ecosystem effects include possible changes in grain size, organic matter and nutrient concentrations via bioturbation, as well as an increase in oxygen penetration to deep sediment layers because of deep burrowing (Sousa et al. 2009). In addition, the surfacing shells can cause changes in hydrodynamic conditions and an increase in substrate complexity by creating a diversity of new microhabitats (Gutiérrez et al. 2003), in the form of available substrates for epibionts as well as shelter to other organisms in the interstices between their shells. However, the specific effects of a non-indigenous species on the environment and in the local biota should be evaluated. Further studies are needed to understand the consequences of the presence of *B. truncata* in the intertidal area of the Bahía Blanca Estuary, especially in relation to its engineering activities.

8.4.2 Another Crab in the Estuary: The Fiddler Crab **Leptuca uruguayensis**

Fiddler crabs are widely distributed throughout the tropics and subtropics of the world, especially in South America, where they form dense populations in silty or silt-sandy ecosystems, like sheltered bays and estuaries (Spivak et al. 1991), where they construct burrows in the high levels of the intertidal area (Crane 1975). Adults of fiddler crabs are generally small, with only a few species exceeding 25 mm in carapace width. The most notable feature of these crabs is the sexual dimorphism of the chelae: adult males developed an enlarged cheliped, not present in females or in the early stages of males. This cheliped is developed during the period of sexual maturation, being useful for combat and territorial defence (Yamaguchi and Henmi 2008), and its red vibrant colouration, acquired during the reproductive season, is useful to define territory and to attract females (Crane 1975). It is the waving of this large chela that gives the popular name of 'fiddlers' to these crabs.

The southwestern Atlantic fiddler crab *Leptuca uruguayensis* (Nobili, 1901) is distributed from southern Brazil (33°S) to the northern coast of Argentina (38°S). It is restricted to the intertidal zones of sheltered bays and estuaries, which results in a puzzling distribution. One of the southernmost permanent population inhabits Samborombón Bay (Fig. 1.2; Chap. 1), in the southern limits of the Río de la Plata Estuary (35°30′–36°22′S), where it is the dominant intertidal species, reaching densities of up to 140 crabs per m^{-2} (Boschi 1964; Iribarne and Martínez 1999). Other populations are found in the northern coast of Buenos Aires Province at the Mar Chiquita Lagoon, at the Quequén Grande Estuary and at the Quequén Salado Estuary (Spivak et al. 1991; Boschi et al. 1992) (Fig. 1.2; Chap. 1), but with lower densities than those found in Samborombón Bay. In fact, a pattern of decreasing density with increasing latitude was proposed for the species, over a land distance of less than 300 km.

During February 2019, Truchet et al. (2019) reported for the first time the presence of *Leptuca uruguayensis* in the Bahía Blanca Estuary, in the upper intertidal zone of an unvegetated tidal flat in Puerto Rosales (Fig. 2.4; Chap. 2), at approximately 200 km of its southern record. The species was registered in a density of 144 ± 48 individuals/m^2, higher than those found in the Quequén Grande and Quequén Salado Estuaries and close to those reported in northern locations like the Samborombón Bay, where they reach 142 individuals/m^2 (César et al. 2005) and San Antonio Cape (the southern extreme of Samborombón Bay) where densities reach 133 to 207 individuals/m^2 (Torres Jorda and Roccatagliata 2002). However, these studies report populations that persist over extensive areas, unlike the population inhabiting the Bahía Blanca Estuary, which encompasses only in a small patchy distribution (Truchet et al. 2019).

Bogazzi et al. (2001) suggest that the southern distribution limit of *L. uruguayensis* is mainly due to wind patterns. They indicate that during the period of larval release and recruitment, the contribution of wind drift transport is less frequent coastwards from northern to southern locations. Given that the wind is the main force moving the upper layer of the water column and consequently crab larvae, authors expect a decrease in the chances of carrying megalopae to southern locations and that develop into adults. The presence of adults of *L. uruguayensis* in the Bahía Blanca Estuary indicates that crab larvae effectively arrived in the coast. The high density found in a small patchy distribution together with the scarce variation in carapace width suggests that these individuals may come from a batch of larvae that reached the coast at once (Truchet et al. 2019).

A possible explanation of the presence of *L. uruguayensis* in the Bahía Blanca Estuary could be the introduction of larvae into the estuary via ship ballast water (Truchet et al. 2019). The movement of commercial and transport ships and boats, exacerbated by the increased globalization of commerce, has facilitated the spread of species. The most studied cases are those related to invasive species; nevertheless, some cases correspond to native species that found new places to settle according to their biological requirements. Another possible explanation could be related to the natural shift in species distributions to higher latitudes. Predicted biological and ecological responses to rising temperatures in the oceans are expected to cause shifts in species distribution ranges (Hoegh-Guldberg and Bruno 2010; Burrows et al. 2014; García Molinos et al. 2015; Sunday et al. 2015). Particularly, in the southwestern Atlantic Ocean, warming-favoured species could benefit from future climate change scenarios and expand southwards their geographical distribution. Future studies will be necessary to analyse the permanence of this population in the Bahía Blanca Estuary, especially in relation to its viability and the possible expansion to other sites within the estuary.

8.4.3 The Ephemeral Polyp Corymorpha januarii

Dense populations of polyps have been observed by local fishermen for more than 20 years in the margin of muddy-bottom channels of the Bahía Blanca Estuary. This information remained unnoticed by local scientists until November 2013, when polyps collected in a small channel of the estuary were analysed for identification purposes (Dutto et al. 2019). The species was *Corymorpha januarii* Steenstrup, 1855 (Cnidaria: Hydrozoa), a corymorphid hydroid native from the southwestern Atlantic coasts. It lives in tropical and temperate shallow waters from Rio de Janeiro (Brazil) to Puerto Madryn (northern Argentinean Patagonia, 42°48′S), being endemic to this region (da Silveira and Migotto 1992; Genzano et al. 2009). The channel where the polyps were collected is called El Saco (Fig. 2.2; Chap. 2), a small and shallow NW-SE oriented channel (3 km long, 110 m wide and a maximum of 5 m depth in high tide) usually used for artisanal fishing, located in the inner section of the Bahía Blanca Estuary, south of the Principal Channel (Fig. 2.1; Chap. 2).

The *Corymorpha januarii* population observed along an area of 250 m long and 3 m wide was composed of an estimated number of 31,500 polyps (42 ± 5 polyps/m^2). Almost all visible polyps were lying on the mud, completely exposed to the sun and air during the ebb tide (3 h). The accompanying fauna consisted of an epibenthic species, the octocoral *Stylatula darwinii* as well as by endobenthic species such as the bivalves *Pitar rostrata*, *Malletia* sp. and *Nucula* sp.; the gastropod *Buccinanops deformis*; amphipods like *Monocorophium insidiosum* (Fig. 8.1e) and *Heterophoxus videns* (Fig. 8.1f); and several species of polychaetes, being *Leodamas verax* and *Axiothella* sp., the dominant ones (Dutto et al. 2019).

Polyps play an important role in energy transfer from the water column to the benthos (Gili and Coma 1998) because of their great abundances and high predation impact on a wide food spectrum, from detritus, diatoms and benthic microplankton to egg and fish larvae (e.g. Gili and Hughes 1995; Gili et al. 2008). Dutto et al. (2019) indicate that the most frequently trophic items consumed by *Corymorpha januarii* were particulate organic matter (POM) and certain harpacticoid copepod species, with *Microarthridion aff. littorale* being the most highly consumed species, followed by the calanoid copepod *Acartia tonsa* and the mysid *Neomysis americana*. Considering the total density of polyps, its consumption represented a daily carbon consumption of 75 mg C/m^2/day, and the daily mass-specific ingestion rate of polyps of *C. januarii* was 20.5% of the polyp biomass.

The study of intertidal hydroids has been largely neglected, particularly of those in southwestern Atlantic coasts (Genzano et al. 2017). Therefore, the fortuitous finding of this species substantially contributes to understanding the role of these populations as important biomass contributors in benthic communities, being an essential part of the 'animal forest' (Rossi et al. 2017, 2019) on coastal ecosystems.

8.5 Threats to Macrobenthic Species

The intertidal areas of estuaries represent important areas for recruitment, breeding, feeding and refugee of several species of fishes and birds, mainly due to their low wave energy. The Bahía Blanca Estuary is not an exception, as several species entry into the estuary for reproductive or feeding purposes and the great majority feed on macrobenthic invertebrates found in intertidal areas (Elías et al. 2004). These organisms tend to be highly vulnerable because they are permanently exposed to several toxic organic and inorganic pollutants found in seawater and sediments (Kennish 2002).

The economic and demographic development of coastal regions, with the consequent increase of urban, industrial and port settlements, generates great loads of waste, which are discharged into the water. This discharge of anthropogenic contaminants to the environment can have profound effects on natural ecosystems (Vitousek et al. 1997). In the Bahía Blanca Estuary, several studies reflect the anthropogenic impact to which macrofaunal communities are subject. Most studies evaluated the impacts of cloacal sewage discharge (Fiori et al. 2019), polycyclic aromatic hydrocarbons (Arias et al. 2009), heavy metals (Buzzi and Marcovecchio 2016; Simonetti et al. 2012, 2013, 2018) and microplastics (Villagrán et al. 2019). Some of them were conducted at the community level, but the vast majority of studies evaluate impacts on targeted populations as the razor clam *Tagelus plebeius* and the burrowing crab *Neohelice granulata*.

Several macrobenthic species found in intertidal areas of the estuary, such as the burrowing crab *Neohelice granulata*, are known to have important ecosystem effects, in relation to its capability to modify energy fluxes in the salt marshes. Crab bioturbation and herbivory, for example, may account for variability in the uptake of atmospheric CO_2 by salt marshes and can also be relevant to '*Blue Carbon*' studies (i.e. carbon that is sequestered and stored in coastal marine environments; Nellemann et al. 2009), having an impact on mitigation of climate change. Therefore, the preservation and restoration of salt marshes, along with the other vegetated coastal ecosystems (i.e. mangroves and seagrass meadows), represent an essential piece of the solution to global climate change. Through herbivory, crabs could potentially modulate the amount of carbon stored in salt marshes by reducing aboveground biomass. The effects mediated by bioturbation comprise a complex interplay between direct and indirect effects with both positive and negative outcomes in terms of carbon sequestration and storage (Martinetto et al. 2016).

Therefore, the preservation and restoration of these coastal ecosystems represent a win-win scenario for multiple objectives as biodiversity conservation and climate change mitigation. Some efforts have been made to conserve biodiversity in the Bahía Blanca Estuary. The Nature Reserve Bahía Blanca, Bahía Falsa and Bahía Verde (Fig. 2.1; Chap. 2) was created in 1998 with the primary purpose of safeguarding the marine ecosystem as a whole as well as to protect and preserve an important refuge and reproduction site for several species. In 2016, the Bahía Blanca Estuary was declared as 'site of regional importance' of the Western Hemisphere

Shorebird Reserve Network (WHSRN). This conservation strategy aims to protect the nesting, breeding and staging habitats of migratory shorebirds.

Another important threat to macrobenthic species in the Bahía Blanca Estuary is the introduction of exotic species which can lead to important changes in soft-bottom communities (see Box 8.2). While the introduction of invasive species to estuarine systems worldwide continues unabated, more data are needed to

Box 8.2: Islands of Heterogeneity and Complexity in the Intertidal Area: The Hard Substrates

Opposite to the soft-bottom environments, natural hard substrates occupy relatively small areas in the Argentine coast. Particularly, in the Bahía Blanca Estuary, the most important hard substrate areas are located in the northern and external coast, near Punta Alta city (Fig. 2.4; Chap. 2). Unfortunately, little attention has been paid to study the benthic macrofaunal community that lives in these habitats. These large rocky outcrops, called 'beachrocks', have relevant geological features, due to the fact that its genesis is linked to the coastal morphosedimentological evolution (Aliotta et al. 2009). One of the most important outcrops is in the coastal zone of Villa del Mar (Fig. 2.4; Chap. 2), in the upper intertidal zone. These rocks are dominated by the exotic barnacles *Balanus glandula* and *Amphibalanus amphitrite* (Fiori and Bieckzinsky 2008) and the Pacific oyster *Magallana gigas*, with higher densities in lower intertidal levels (Dos Santos and Fiori 2010).

Another important rocky outcrop found in intertidal areas of the Bahía Blanca Estuary is located near Puerto Rosales, distributed along the intermediate and lower levels of the intertidal zone. Here, individuals of *Magallana gigas* were registered ranging from 7.9 ± 4.7 individuals/m^2 (attached in scattered gravels and shells) to 82.4 ± 23.8 individuals/m^2 (forming oyster reefs) (Dos Santos, pers. comm.). Other species found inhabiting these hard substrates were the mussel *Brachidontes rodriguezii*, the snail *Heleobia australis* (Carcedo and Fiori 2011) and several invertebrate species reported for regional communities of intertidal mussel beds such as the polychaete *Alitta succinea* and the amphipod *Monocorophium insidiosum* (Dos Santos et al. 2018).

Organisms inhabiting natural hard substrates can colonize artificial structures like breakwaters, boats, rafts, buoys and seawater intakes. These artificial structures are widely distributed along the Bahía Blanca Estuary and represent the most extensive hard-bottom surfaces for the settlement of marine organisms. Therefore, the vast majority of studies of hard-bottom macrofauna in the Bahía Blanca Estuary have focused on these types of structures. The first studies were conducted during the 1970s, in the two main port areas: Ingeniero White (Fig. 2.3; Chap. 2), located in the middle zone of the estuary, and Naval Base Puerto Belgrano, in the external zone of the estuary (Fig. 2.4; Chap. 2). These studies covered the species identification, intertidal zonification, annual succession of subtidal communities and the main epibiotic phenomena (Bastida and Torti 1971; Bastida et al. 1974a, 1974b; Bastida and Lichtschein de Bastida 1978; Valentinuzzi de Santos 1971).

Valentinuzzi de Santos (1971) indicated that the intertidal fouling community was dominated by the balanid *Amphibalanus amphitrite*, the mussel *Brachidontes rodriguezii* and several species of polychaetes. During the 2000s, the dominance of the community was due to two balanid species, *A. amphitrite* and *Balanus glandula*, together with the bryozoan *Amathia* (= *Bowerbankia*) sp. (Bremec et al. 2004). More than 10 years later, the dominance of *B. glandula* increased over *A. amphitrite*, and a high abundance of the anemone *Diadumene lineata* (Roldán 2014) was also detected. Floating substrates such as rafts in the Port of Ingeniero White showed a significant increase in the biomass of the hydrozoan *Tubularia crocea* during the warm season, within a community dominated by passive and active filters of cosmopolitan distribution (Dos Santos et al. 2006). The presence of exotic and cosmopolitan species is the main feature of the subtidal hard-bottom community of the Bahía Blanca Estuary, like the polychaete *Hydroides dianthus* and the bryozoans *Bugulina flabellata*, *Bugula neritina*, *Bugulina simplex*, *Bugulina stolonifera* and *Cryptosula pallasiana* (Orensanz et al. 2002), included, among others, in a more extensive list of species (Bremec et al. 2004).

During the last few decades, a rapid proliferation of artificial hard structures (e.g. energy infrastructure, aquaculture, coastal defences) was observed in coastal settings, with the consequent replacement of natural, often sedimentary, environments with hard substrata (Firth et al. 2016). Artificial hard substrates play a particularly important role in the dynamics of marine invasions, being possibly the first substrates colonized by non-native species arriving in a bay or estuary. For example, in coastal North America, approximately 90% of the alien species inhabiting hard substrata have been reported from docks and marinas (Mineur et al. 2012). It may result partly from their predominance on developed shorelines, where vectors and supply of propagules are presumably concentrated (Ruiz et al. 2009). In addition, hard-bottom communities may facilitate initial invasions through greater availability of resources over those on natural substrata. These areas can, therefore, provide seed populations that persist and increase in abundance and, as a result, spread regionally to other environments, including natural substrata (Glasby et al. 2007; Ruiz et al. 2009).

A dispersion of the non-native balanid *Balanus glandula* from the Argentine Port of Mar del Plata to neighbouring rocky coasts has been observed since the 1970s (Spivak 2005), reaching southern Patagonian ports years later (Rico et al. 2001). *B. glandula* entered the Bahía Blanca Estuary in the 1980s, detected through the observation of its larvae in the local zooplankton community (Hoffmeyer 1983). Regarding the other non-native balanid, *Amphibalanus amphitrite*, it entered local port before 1970, as indicated by early records in intertidal areas of Port of Ingeniero White (Valentinuzzi de Santos 1971) and intertidal-subtidal areas of Naval Base Puerto Belgrano (Bastida and Torti 1971). In 2008, the anemone *Diadumene lineata* was registered in the Bahía Blanca Estuary, attached to stems and roots of *Spartina alterniflora*, in an association never registered before (Molina et al. 2008).

The more recent invasion registered in the Bahía Blanca Estuary was the oyster *Magallana* (= *Crassostrea*) *gigas* (Dos Santos and Fiori 2010) (Fig. 8.4), a species native in the northwestern Pacific Ocean and the Sea of Japan, and introduced illegally in Argentina in 1982 for aquaculture purposes (Escapa et al. 2004a; Borges 2005, 2006). Since then, it has spread both northwards until reaching the Bahía Blanca Estuary and southwards to the northern Patagonia Region, indicating its sustained expansion, mainly due to its capability to tolerate a wide range of environmental conditions (Shatkin et al. 1997).

Even though the natural and artificial hard substrates, i.e. the primary habitat for the settlement of the Pacific oyster, are confined to a few sectors in the Bahía Blanca Estuary, the species is able to settle on rocks, debris and shells from the lower intertidal zone to depths of 40 m. Once it settles on a shell or a small stone, clumps of oysters may merge to form dense aggregations and potentially a reef (Herbert et al. 2016), facilitating further settlement on soft-bottom substrates (Fiori et al. 2016). Thus, the soft-bottom environments of the estuary with presence of gravels and/or shells could be modified towards bivalve reef as a result of the settlement of the Pacific oyster on these substrates (Bravo 2013).

The establishment of oyster reefs on soft bottoms can promote structural changes with effects on infaunal invertebrates, such as an increase in the abundance of epifaunal organisms that use the reef as a refuge and a decrease in those that do not need shelters. A higher abundance of preys can also affect habitat selection by coastal birds (Escapa et al. 2004a). Due to the high adaptation and expansion capability of the species, a negative impact of *M. gigas* on environmental values and human health in the estuary should be expected. The impacts can be estimated through a recently developed risk index, which evaluates the effectiveness to manage the environmental matrix as the best option for preventing or reducing undesired effects of the oyster (Fiori et al. 2016). For example, the risk index could evaluate projections of direct control operations as eradication of adult oysters at specific locations or the impacts of specific management practices.

Important changes in the ecosystem are expected due to the establishment of *Magallana gigas*, both in the pelagic and benthic environments. The species has a high filtration rate, consuming suspended organisms, as well as detrital organic matter, being able to modify the plankton community. In the case of the benthic environment, oyster populations in high densities as those found on the Bahía Blanca Estuary have the potential to (1) reduce food availability for other native filter feeding, (2) modify populations due to changes in abundance of their planktonic life stages and (3) change richness and diversity of community due to the increased bottom complexity (Crooks 2002; Ruesink et al. 2005; Troost et al. 2009; Wilkie et al. 2013). These changes not only impact at a local level, but its successful establishment increases the chances of a possible expansion to other places vectorized by national and international ships that constantly circulate in the estuary (Dos Santos and Fiori 2010). In fact, studies have predicted the dispersion of the Pacific oyster along extensive areas of the Argentine coast (Carrasco and Baron 2010).

Fig. 8.4 The Pacific oyster *Magallana gigas* in artificial hard substrates of the Bahía Blanca Estuary. (Photo by M. Emilia Croce)

effectively assess their long-term environmental impacts (Kennish 2002). Furthermore, in the current scenario of global climate change, future studies should tend to analyse specific responses of organisms into global and integrative approaches. Whole-system perspectives are needed for understanding how this complex and dynamic ecosystem will respond to multiple anthropogenic drivers.

References

Aguirre ML, Farinati EA (2000) Aspectos sistemáticos, de distribución y paleoambientales de *Littoridina australis* (D'Orbigny 1835) (Mesogastropoda) en el Cuaternario marino de Argentina (Sudamérica). Geobios 33:569–597

Alberti J, Escapa M, Iribarne O, Silliman B, Bertness M (2008) Crab herbivory regulates plant facilitative and competitive processes in Argentinean marshes. Ecology 89(1):155–164

Alberti J, Casariego AM, Daleo P, Fanjul E, Silliman B, Bertness M, Iribarne O (2010) Abiotic stress mediates top-down and bottom-up control in a Southwestern Atlantic salt marsh. Oecologia 163(1):181–191

Alberti J, Cebrian J, Casariego AM, Canepuccia A, Escapa M, Iribarne O (2011) Effects of nutrient enrichment and crab herbivory on a SW Atlantic salt marsh productivity. J Exp Mar Biol Ecol 405(1):99–104

Alberti J, Daleo P, Fanjul E, Escapa M, Botto F, Iribarne O (2014) Can a single species challenge paradigms of salt marsh functioning? Estuar Coast 38(4):1178–1188

Aliotta S, Spagnuolo J, Farinati E (2009) Origen de una roca de playa en la región costera de Bahía Blanca, Argentina. Pesquisas em Geociências 36(1):107–116

Alldredge AL, Jackson GA (1995) Preface: aggregation in marine system. Deep Sea Res II Top Stud Ocean 42(1):1–7

Angeletti S (2012) Composición y estructura poblacional de *Neohelice granulata* (Varunidae) en una planicie de marea del estuario de Bahía Blanca (Villa del Mar). Tesis de Licenciatura en Ciencias Biológicas. Univ Nac del Sur, Argentina, p 50

Angeletti S (2017) Efecto bioturbador del cangrejo *Neohelice granulata* sobre la distribución y transporte de sedimento en ambientes intermareales próximos al límite sur de su distribución geográfica: Un estudio poblacional comparado. PhD thesis. Univ Nac del Sur, Argentina, p 180

Angeletti S, Cervellini PM (2015) Population structure of the burrowing crab *Neohelice granulata* (Brachyura, Varunidae) in a Southwestern Atlantic salt marsh. Lat Am J Aquat Res 43(3):539–547

Angeletti S, Cervellini PM, Lescano L (2018a) Burrowing activity of *Neohelice granulata* crab (Brachyura, Varunidae) in Southwest Atlantic intertidal areas. Cienc Mar 44(3):155–167

Angeletti S, Pierini JO, Cervellini PM (2018b) Suspension sediment contribution resulting from bioturbation in intertidal sites of SW Atlantic mesotidal estuary: data analysis and numerical modelling. Sci Mar 82(4):245–256

Arias A, Spetter C, Freije H, Marcovecchio J (2009) Polycyclic aromatic hydrocarbons (PAHs) distribution in water column, native mussels (*Brachydontes* sp and *Tagelus* sp) and fish (*Odontesthes* sp) from an industrialized south American estuary. Estuar Coast Shelf Sci 85:67–81

Auster PJ, Crockett LR (1984) Foraging tactics of several crustacean species for infaunal prey. J Shellfish Res 4:139–143

Bas C, Luppi T, Spivak E (2005) Population structure of the South American estuarine crab, *Chasmagnathus granulatus* (Brachyura: Varunidae) near the southern limit of its geographical distribution: comparison with northern populations. Hydrobiologia 537:217–228

Bastida R, Lichtschein de Bastida V (1978) Las incrustaciones biológicas de Puerto Belgrano. III Estudio de los procesos de epibiosis registrados sobre paneles acumulativos. CIDEPINT Anales 3:55–97

Bastida RO, Torti MR (1971) Estudio preliminar de las incrustaciones biológicas de Puerto Belgrano (Argentina). LEMIT Anales, Ser II 3:45–75

Bastida R, Spivak E, L'Hoste SG, Adabbo HE (1974a) Las incrustaciones biológicas de Puerto Belgrano. I Estudio de la fijación sobre paneles mensuales, período 1971/72. LEMIT Anales Ser II(274):97–165

Bastida R, Spivak E, L'Hoste SG, Adabbo HE (1974b) Las incrustaciones biológicas de Puerto Belgrano. II Estudio de los procesos de epibiosis registrados sobre paneles mensuales. LEMIT Anales Ser II(274):25–33

Berón MP, García GO, Luppi T, Favero M (2011) Age-related prey selectivity and foraging efficiency of Olrog's Gulls (*Larus atlanticus*) feeding on crabs in their nonbreeding grounds. Emu 111(2):172–178

Billick I, Case TJ (1994) Higher order interactions in ecological communities: what are they and how can they be detected? Ecology 75(6):1529–1543

Blackburn TH, Henriksen K (1983) Nitrogen cycling in different types of sediments from Danish waters. Limnol Oceanogr 28(3):477–493

Bogazzi E, Iribarne O, Guerrero R, Spivak E (2001) Wind pattern may explain the southern limit of distribution of a southwestern Atlantic fiddler crab. J Shellfish Res 59:483–500

Borges ME (2005) La ostra del Pacífico *Crassostrea gigas* (Thumberg, 1793) en la Bahía Anegada (Provincia de Buenos Aires). In: Penchaszadeh PE et al (eds) Invasores: Invertebrados exóticos en el Río de la Plata y región marina aledaña. EUDEBA, Buenos Aires, pp 311–367

Borges ME (2006) Ecología de las ostras en ambientes del sur bonaerense: cultivo y manejo de sus poblaciones. PhD thesis. Univ Nac del Sur, Argentina, p 247

Bortolus A, Iribarne OO (1999) Effects of the SW Atlantic burrowing crab *Chasmagnathus granulata* on a *Spartina* salt marsh. Mar Ecol Prog Ser 178:79–88

Bortolus A, Schwindt E, Iribarne OO (2002) Positive plant-animal interactions in the high marsh of an Argentinean coastal lagoon. Ecology 83:733–742

Bortolus A, Carlton JT, Schwindt E (2015) Reimagining South American coasts: unveiling the hidden invasion history of an iconic ecological engineer. Divers Distrib 21(11):1267–1283

Boschi EE (1964) Los crustáceos decápodos Brachyura del litoral bonaerense (R Argentina) In: Boletín del Instituto de Biología Marina, V6, pp 3–99

Boschi EE, Fischbach CE, Iorio MI (1992) Catálogo ilustrado de los crustáceos estomatópodos y decápodos marinos de Argentina. Frente Marítimo 10, 7(Sec. A):–94

Botto F, Iribarne OO (2000) Contrasting effects of two burrowing crabs (*Chasmagnathus granulata* and *Uca uruguayensis*) on sediment composition and transport in estuarine environments. Estuar CoastShelf Sci 51(2):141–151

Botto F, Iribarne OO, Martínez MM, Delhey K, Carrete M (1998) The effect of migratory shorebirds on the benthic species of three southwestern Atlantic Argentinean estuaries. Estuaries 21(4):700–709

Botto F, Valiela I, Iribarne O, Martinetto P, Alberti J (2005) Impact of burrowing crabs on C and N sources, control, and transformations in sediments and food webs of SW Atlantic estuaries. Mar Ecol Progr Ser 293:155–164

Botto F, Iribarne O, Gutierrez J, Bava J, Gagliardini A, Valiela I (2006) Ecological importance of passive deposition of organic matter into burrows of the SW Atlantic crab *Chasmagnathus granulatus*. Mar Ecol Progr Ser 312:201–210

Boudreau BP, Jørgensen BB (2001) The benthic boundary layer: transport processes and biogeochemistry. Oxford University Press, London, p 404

Boyer JN, Kelble CR, Ortner PB, Rudnick DT (2009) Phytoplankton bloom status: chlorophyll a biomass as an indicator of water quality condition in the southern estuaries of Florida, USA. Ecol Indic 9(6):S56–S67

Boynton WR, Kemp WM (1985) Nutrient regeneration and oxygen consumption by sediments along an estuarine salinity gradient. Mar Ecol Progr Ser 23(1):45–55

Braeckman U, Van Colen C, Soetaert K, Vincx M, Vanaverbeke J (2011) Contrasting macrobenthic activities differentially affect nematode density and diversity in a shallow subtidal marine sediment. Mar Ecol Progr Ser 422:179–191

Bravo ME (2013) Estado de la invasión de la ostra del Pacífico, *Crassostrea gigas*, en el estuario de Bahía Blanca. Tesis de Licenciatura en Ciencias Biológicas. Univ Nac del Sur, Bahía Blanca, p 40

Bremec CS, Martínez DE, Elías R (2004) Asociaciones bentónicas de fondos duros y comunidades incrustantes. In: Piccolo MC, Hoffmeyer MS (eds) Ecosistema del estuario de Bahía Blanca. Instituto Argentino de Oceanografía, Bahía Blanca, pp 171–178

Bricker SB, Longstaff B, Dennison W, Jones A, Boicourt K, Wicks C, Woerner J (2008) Effects of nutrient enrichment in the nation's estuaries: a decade of change. Harmful Algae 8(1):21–32

Burdige DJ (2006) Geochemistry of marine sediments. Princeton University Press, Princeton, p 630

Burrows MT, Schoeman DS, Richardson AJ, Molinos JG, Hoffmann A, Buckley LB et al (2014) Geographical limits to species-range shifts are suggested by climate velocity. Nature 507:492–495

Buzzi NS, Marcovecchio JE (2016) A baseline study of the metallothioneins induction and its reversibility in *Neohelice granulata* from the Bahía Blanca estuary (Argentina). Mar Poll Bull 112(1–2):452–458

Cadée GC (2001) Sediment dynamics by bioturbating organisms. In: Ecological comparisons of sedimentary shores. Springer, Berlin/Heidelberg, pp 127–148

Callaway RM (1995) Positive interactions among plants. Bot Rev 61:306–349

Canepuccia AD, Escapa M, Daleo P, Alberti J, Botto F, Iribarne OO (2007) Positive interactions of the smooth cordgrass *Spartina alterniflora* on the mud snail *Heleobia australis*, in southwestern Atlantic salt marshes. J Exp Mar Biol Ecol 353:180–190

Canepuccia AD, Fanjul MS, Fanjul E, Botto F, Iribarne OO (2008) The intertidal burrowing crab *Neohelice* (= *Chasmagnathus*) *granulata* positively affects foraging of rodents in south western Atlantic salt marshes. Estuar Coast 31(5):920–930

Carcedo MC, Fiori SM (2011) Patrones de distribución y abundancia de *Heleobia australis* (Caenogastropoda: Cochliopidae) en el estuario de Bahía Blanca, Argentina. In: Cazzaniga NJ (ed) El Género *Heleobia* (Caenogastropoda: Cochliopidae) en América del Sur. Amici Molluscarum (Núm Esp), pp 33–35

Carrasco MF, Baron PJ (2010) Analysis of the potential geographic range of the Pacific oyster *Crassostrea gigas* (Thunberg, 1793) based on surface seawater temperature satellite data and climate charts: the coast of South America as a study case. Biol Invasions 12:2597–2607

César II, Armendáriz LC, Becerra RV (2005) Bioecology of the fiddler crab *Uca uruguayensis* and the burrowing crab *Chasmagnathus granulatus* (Decapoda, Brachyura) in the Refugio de Vida Silvestre Bahía Samborombón, Argentina. Hydrobiologia 545(1):237–248

Chalmers GC, Wiegert RG, Wolff PL (1985) Carbon balance in a salt marsh: interactions of diffusive export, tidal deposition and rainfall-caused erosion. Estuar Coast Shelf Sci 21:757–771
Chanley PE (1965) Larval development of a boring clam, *Barnea truncata*. Chesap Sci 6:162–166
Chauvaud L, Jean F, Ragueneau O, Thouzeau G (2000) Long-term variation of the bay of Brest ecosystem: benthic-pelagic coupling revisited. Mar Ecol Progr Ser 200:35–48
Chmura GL, Anisfeld S, Cahoon D, Lynch J (2003) Global carbon sequestration in tidal, saline wetland soils. Global Biogeochem Cy 17:1–12
Cloern JE (1987) Turbidity as a control on phytoplankton biomass and productivity in estuaries. Cont Shelf Res 7(11):1367–1381
COSEWIC (2009) COSEWIC assessment and status report on the Atlantic Mud-piddoc *Barnea truncate* in Canada. Committee on the Status of Endangered Wildlife in Canada. Ottawa: vii + 42 p
Costa CS, Marangoni JC, Azevedo AM (2003) Plant zonation in irregularly flooded salt marshes: relative importance of stress tolerance and biological interactions. J Ecol 91(6):951–965
Crane J (1975) Fiddler crabs of the world, Ocypodidae: genus *Uca*. Princeton University Press, Princeton, p 736
Crooks JA (2002) Characterizing ecosystem-level consequences of biological invasions: the role of ecosystem engineers. Oikos 97:153–166
Cuadrado DG, Carmona NB, Bournod C (2011) Biostabilization of sediments by microbial mats in a temperate siliciclastic tidal flat, Bahia Blanca estuary (Argentina). Sediment Geol 237(1):95–101
da Silveira FL, Migotto AE (1992) Rediscovery of *Corymorpha januarii* Steenstrup, 1854 (Hydrozoa, Corymorphidae) on the southeastern and southern coasts of Brazil. Steenstrupia 18:81–89
Dale AW, Prego R (2002) Physico-biogeochemical controls on benthic-pelagic coupling of nutrient fluxes and recycling in a coastal upwelling system. Mar Ecol Progr Ser 235:15–28
Daleo P, Iribarne OO (2009) The burrowing crab *Neohelice granulata* affects the root strategies of the cordgrass *Spartina densiflora* in SW Atlantic salt marshes. J Exp Mar Biol Ecol 373(1):66–71
Daleo P, Fanjul E, Casariego AM, Silliman BR, Bertness MD, Iribarne OO (2007) Ecosystem engineers activate mycorrhizal mutualism in salt marshes. Ecol Lett 10(10):902–908
Day JW, Hall CAS Jr, Kemp WM, Yañez-Arancibia A (1989) Estuarine ecology. Wiley-Interscience Publication/Wiley, New York
De Francesco CG, Isla FI (2003) Distribution and abundance of hydrobiid snails in a mixed estuary and a coastal lagoon, Argentina. Estuaries 26:790–797
Decho AW (2000) Microbial biofilms in intertidal systems: an overview. Cont Shelf Res 20(10):1257–1273
Defeo O, McLachlan A (2005) Patterns, processes and regulatory mechanisms in sandy beach macrofauna: a multi-scale analysis. Mar Ecol Progr Ser 295:1–20
Dietrich WE, Perron JT (2006) The search for a topographic signature of life. Nature 439(7075):411–418
Dos Santos EP, Fiori SM (2010) Primer registro sobre la presencia de *Crassostrea gigas* (Thumberg, 1793) en el Estuario de Bahía Blanca (Argentina). Comunicaciones de la Sociedad Malacológica del Uruguay 93:245–252
Dos Santos EP, Piccolo CM, Parodi ER (2006) Desarrollo de la biomasa de una comunidad incrustante (*fouling*) relacionado con parámetros ambientales. VI Jornadas Nacionales de Ciencias del Mar, Puerto Madryn, Argentina, p 175
Dos Santos EP, Carcedo MC, Zotelo C, Fiori SM (2018) Effects of erosion-accretion processes on a biogenic reef formed by *Brachidontes rodriguezii* (Mollusca: Mytilidae) on a sandy beach along the South Atlantic coast. J Marine Syst 187:146–155
Dutto MS, Carcedo MC, Nahuelhual EG, Conte AF, Berasategui AA, Garcia MD et al (2019) Trophic ecology of a corymorphid hydroid population in the Bahía Blanca estuary, Southwestern Atlantic. Reg Stud Mar Sci 31:100746

Echeverría CA, Neves RAF, Pessoa LA, Paiva PC (2010) Spatial and temporal distribution of the gastropod *Heleobia australis* in an eutrophic estuarine system suggests a metapopulation dynamics. Nat Sci 2(8):860–867

Elías R (1985) Macrobentos del estuario de Bahía Blanca. I Mesolitoral. Spheniscus 1:1–33

Elías R, Bremec CS (1986) Macrobentos del área de la bahía Blanca (Provincia de Buenos Aires). II Relaciones entre las asociaciones de sustrato móvil. Spheniscus 3:51–52

Elías R, Iribarne OO, Bremec CS, Martínez DE (2004) Comunidades bentónicas de fondos blandos. In: Piccolo MC, Hoffmeyer MS (eds) Ecosistema del Estuario de Bahía Blanca. Instituto Argentino de Oceanografía, Bahía Blanca, pp 179–190

Escapa M (2007) Efecto de las interacciones biológicas en la erosión de marismas. PhD thesis, Univ Nac de Mar del Plata, Argentina, p 162

Escapa M, Iribarne OO, Navarro D (2004b) Effects of the intertidal burrowing crab *Chasmagnathus granulatus* on infaunal zonation patterns, tidal behavior and risk of mortality. Estuaries 27(1):120–131

Escapa M, Isacch JP, Daleo P, Alberti J, Iribarne OO, Borges M, Dos Santos EP, Galiardini D, Lasta M (2004a) The distribution and ecological effects of the invasive Pacific Oyster *Crassostrea gigas* (Thunberg, 1793) in Northern Patagonia. J Shellfish Res 23:765–772

Escapa M, Minkoff DR, Perillo GM, Iribarne OO (2007) Direct and indirect effects of burrowing crab *Chasmagnathus granulatus* activities on erosion of Southwest Atlantic *Sarcocornia*-dominated marshes. Limnol Oceanogr 52(6):2340–2349

Escapa M, Perillo GM, Iribarne OO (2008) Sediment dynamics modulated by burrowing crab activities in contrasting SW Atlantic intertidal habitats. Estuar Coast Shelf Sci 80(3):365–373

Fanjul E, Grela MA, Iribarne OO (2007) Effects of the dominant SW Atlantic intertidal burrowing crab *Chasmagnathus granulatus* on sediment chemistry and nutrient distribution. Mar Ecol Progr Ser 341:177–190

Fanjul E, Escapa M, Montemayor D, Addino M, Alvarez MF, Grela MA, Iribarne OO (2015) Effect of crab bioturbation on organic matter processing in south West Atlantic intertidal sediments. Jour Sea Res 95:206–216

Figueiredo-Barros MP, Leal JJF, Esteves F, De A, Rocha A, De M, Bozelli RL (2006) Life cycle, secondary production and nutrient stock in *Heleobia australis* (d'Orbigny 1835) (Gastropoda: Hydrobiidae) in a tropical coastal lagoon. Estuar Coast Shelf Sci 69:87–95

Fiori SM, Bieckzinsky F (2008) Cirripedios exóticos del estuario de Bahía Blanca. In: Cazzaniga NJ, Arelovich HM (eds) Ambientes y Recursos Naturales del Sudoeste Bonaerense: Producción, contaminación y conservación (Actas de las V jornadas Interdisciplinarias del Sudoeste Bonaerense). EdiUNS, Bahía Blanca, pp 421–431

Fiori SM, Simonetti P, Dos Santos EP (2012) First record of Atlantic mud piddock, *Barnea* (*Anchomasa*) *truncata* (Bivalvia, Pholadidae) in Argentina. Aquat Invasions 7(2):283–286

Fiori SM, Pratolongo PD, Zalba SM, Carbone ME, Bravo ME (2016) Spatially explicit risk assessment for coastal invaders under different management scenarios. Mar Biol 163(12):245

Fiori SM, Bravo ME, Elías R, Serra AV, Carcedo MC, Dos Santos EP, Botté SE (2019) Effects of sewage effluent on the subtidal macrobenthic assemblage in a urban estuary (Argentina). Ecol Austral (in press)

Firth LB, Knights AM, Bridger D, Evans AJ, Mieszkowska N, Moore PJ, O'Connor NE, Sheehan EV, Thompson RC, Hawkins SJ (2016) Ocean sprawl: challenges and opportunities for biodiversity management in a changing world. In: Oceanography and marine biology. CRC Press, New York, pp 201–278

García Molinos J, Halpern BS, Schoeman DS, Brown CJ, Kiessling W, Moore PJ et al (2015) Climate velocity and the future of global redistribution of marine biodiversity. Nat Clim Chang 61:83–88

Genzano GN, Mianzan H, Rodríguez C, Diaz Briz LM (2009) The hydroid and medusa of *Corymorpha januarii* in temperate waters of the Southwestern Atlantic Ocean. Bull Mar Sci 84:229–235

Genzano GN, Bremec CS, Díaz-Briz LM, Costella JH, Morandini AC, Miranda T, Marques AC (2017) Faunal assemblages of intertidal hydroids (Hydrozoa, Cnidaria) inhabiting salt marshes and intertidal outcrops from Argentinean Patagonia (SW Atlantic Ocean). Lat Am J Aquat Res 45:177–187

Gilbert F, Aller RC, Hulth S (2003) The influence of macrofaunal burrow spacing and diffusive scaling on sedimentary nitrification and denitrification: an experimental simulation and model approach. J Mar Res 61:101–125

Gili JM, Coma R (1998) Benthic suspension feeders: their paramount role in littoral marine food webs. Trends in Ecol Evol 13:316–321

Gili JM, Hughes RG (1995) Ecology of benthic hydroids. Oceanogr Mar Biol Annu Rev 33:351–422

Gili JM, Duró A, García-Valero J, Gasol JM, Rossi S (2008) Herbivory in small carnivores: benthic hydroids as an example. J Mar Biol Assoc UK 88:1541–1546

Glasby TM, Connell SD, Holloway MG, Hewitt CL (2007) Nonindigenous biota on artificial structures: could habitat creation facilitate biological invasions? Mar Biol 151:887–895

Griffiths JR, Kadin M, Nascimento FJ, Tamelander T, Törnroos A, Bonaglia S, Bonsdorff E, Brüchert V, Gårdmark A, Järnström M, Kotta J (2017) The importance of benthic–pelagic coupling for marine ecosystem functioning in a changing world. Glob Change Biol 23(6):2179–2196

Guinder VA (2011) Dinámica del fitoplancton en el estuario de Bahía Blanca y su relación con las variables ambientales en el marco del cambio climático. PhD thesis, Univ Nac del Sur, Argentina, p 153

Guinder VA, Popovich CA, Perillo GM (2009) Particulate suspended matter concentrations in the Bahía Blanca estuary, Argentina: implication for the development of phytoplankton blooms. Estuar Coast Shelf Sci 85(1):157–165

Guinder VA, López-Abbate MC, Berasategui AA, Negrin VL, Zapperi G, Pratolongo PD, Severini MDF, Popovich CA (2015) Influence of the winter phytoplankton bloom on the settled material in a temperate shallow estuary. Oceanologia 57(1):50–60

Gutiérrez JL, Iribarne OO (1998) The occurrence of juveniles of the grapsid crab *Chasmagnathus granulata* in siphon holes of the stout razor clam *Tagelus plebeius*. J Shellfish Res 17(4):925–930

Gutiérrez JL, Jones CG, Strayer DL et al (2003) Mollusks as ecosystem engineers: the role of shell production in aquatic habitats. Oikos 101:79–90

Gutiérrez JL, Jones CG, Groffman PM, Findlay SE, Iribarne OO, Ribeiro PD, Bruschetti CM (2006) The contribution of crab burrow excavation to carbon availability in surficial salt-marsh sediments. Ecosystems 9(4):647–658

Hansell MH (2005) Animal architecture. Oxford University Press, Oxford, p 343

Hansen LS, Blackburn TH (1992) Effect of algal bloom deposition on sediment respiration and fluxes. Mar Biol 112:147–152

Hebda A (2011) Information in support of a recovery potential assessment of Atlantic mudpiddock (*Barnea truncata*) in Canada. DFO Can Sci Advis Sec Res Doc 2010/117: vi + 30 p

Herbert RJ, Humphreys J, Davies CJ, Roberts C, Fletcher S, Crowe TP (2016) Ecological impacts of non-native Pacific oysters (*Crassostrea gigas*) and management measures for protected areas in Europe. Biodivers Conserv 25(14):2835–2865

Herman PMJ, Middelburg JJ, Van de Koppel J, Heip CHR (1999) Ecology of estuarine macrobenthos. Adv Ecol Res 29:195–240

Hoegh-Guldberg O, Bruno JF (2010) The impact of climate change on the world's marine ecosystems. Science 328:1523–1528

Hoffmeyer MS (1983) Zooplancton del área interna de la Bahía Blanca (Buenos Aires, Argentina). I-Composición faunística. Hist Nat, Corrientes, Argentina 3(8):73–94

Hovel KA, Bartholomew A, Lipcius RN (2001) Rapidly entrainable tidal vertical migrations in the salt marsh snail Littoraria irrorata. Estuaries 24:808–816

Hughes DJ, Atkinson RJA, Ansell AD (2000) A field test of the effects of megafaunal burrows on benthic chamber measurements of sediment-water solute fluxes. Mar Ecol Progr Ser 195:189–199

Iribarne OO, Martínez MM (1999) Predation on the southwestern Atlantic fiddler crab (*Uca uruguayensis*) by migratory shorebirds (*Pluvialis dominica, P. squatarola, Arenaria interpres* and *Numenius phaeopus*). Estuarine Res Fed 22:47–54

Iribarne OO, Bortolus A, Botto F (1997) Between-habitats differences in burrow characteristics and trophic modes in the southwestern Atlantic burrowing crab *Chasmagnathus granulatus*. Mar Ecol Progr Ser 155:132–145

Iribarne OO, Bruschetti M, Escapa M, Bava J, Botto F, Gutierrez J et al (2005) Small-and large-scale effect of the SW Atlantic burrowing crab *Chasmagnathus granulatus* on habitat use by migratory shorebirds. J Exp Mar Biol Ecol 315(1):87–101

Irigoien X, Castel J (1997) Light limitation and distribution of chlorophyll pigments in a highly turbid estuary: the Gironde (SW France). Estuar Coast Shelf Sci 44(4):507–517

Jarvis NJ, Taylor A, Larsbo M et al (2010) Modelling the effects of bioturbation on the re-distribution of ^{137}Cs in an undisturbed grassland soil. Eur J Soil Sci 61:24–34

Jensen MH, Lomstein E, Serrensen J (1990) Benthic NH_4^+ and NO_3^- flux following sedimentation of a spring phytoplankton bloom in Aarhus bight, Denmark. Mar Ecol Prog Ser 61:87–96

Jones S, Jago CF (1993) In situ assessment of modification of sediment properties by burrowing invertebrates. Mar Biol 115(1):133–142

Jones CG, Lawton JH, Shachak M (1994) Organisms as ecosystem engineers. In: Ecosystem management. Springer, New York, pp 130–147

Kennish MJ (2002) Environmental threats and environmental future of estuaries. Environ Conserv 29(1):78–107

Kiørboe T, Jackson GA (2001) Marine snow, organic solute plumes, and optimal chemosensory behavior of bacteria. Limnol Oceanogr 46(6):1309–1318

Klein G d V (1985) Intertidal flats and intertidal sand bodies. In: Davis RA (ed) Coastal sedimentary environments, 2nd edn. Springer, New York, pp 187–224

Kneib RT (1984) Patterns of invertebrate distribution and abundance in the intertidal salt marsh: causes and questions. Estuaries 2:392–412

Kneib RT (1991) Indirect effects in experimental studies of marine soft-sediment communities. Am Zool 31(6):874–885

Kostka JE, Roychoudhury A, Van Cappellen P (2002) Rates and controls of anaerobic microbial respiration across spatial and temporal gradients in saltmarsh sediments. Biogeochemistry 60(1):49–76

Kristensen E, Penha-Lopes G, Delefosse M, Valdemarsen T, Quintana CO, Banta GT (2012) What is bioturbation? The need for a precise definition for fauna in aquatic sciences. Mar Ecol Progr Ser 446:285–302

Legendre L (1990) The significance of microalgal blooms for fisheries and for the export of particulate organic carbon in oceans. J Plankton Res 12(4):681–699

Lewis DB, Eby LA (2002) Spatially heterogeneous refugia and predation risk in intertidal salt marshes. Oikos 96:119–129

Little C, Nix W (1976) The burrowing and floating behavior of the gastropod *Hydrobia ulvae*. Estuar Coast Mar Sci 4(5):537–544

Lomovasky BJ, Casariego AM, Brey T, Iribarne OO (2006) The effect of the SW Atlantic burrowing crab *Chasmagnathus granulatus* on the intertidal razor clam *Tagelus plebeius*. J Exp Mar Biol Ecol 337(1):19–29

Lozada M, Romano A, Maldonado H (1988) Effect of morphine and naloxone on a defensive response of the crab *Chasmagnathus granulatus*. Pharmacol Biochem Be 30(3):635–640

Luppi TA, Bas CC (2002) Rol de los arrecifes del poliqueto invasor *Ficopomatus enigmaticus* Fauvel 1923 (Polychaeta: Serpulidae) en el reclutamiento de *Cyrtograpsus angulatus* Dana, 1951 (Brachyura: Grapsidae). Cienc Mar 28(4):319–330

Marcus NH, Boero F (1998) Minireview: the importance of benthic-pelagic coupling and the forgotten role of life cycles in coastal aquatic systems. Limnol Oceanogr 43(5):763–768

Martinetto P, Montemayor DI, Alberti J, Costa CS, Iribarne OO (2016) Crab bioturbation and herbivory may account for variability in carbon sequestration and stocks in South West Atlantic salt marshes. Front Mar Sci 3:122

May CL, Koseff JR, Lucas LV, Cloern JE, Schoellhamer DH (2003) Effects of spatial and temporal variability of turbidity on phytoplankton blooms. Mar Ecol Progr Ser 254:111–128

McGlathery KJ, Sundback K, Anderson IC (2004) The importance of primary producers for benthic nitrogen and phosphorus cycling. In: Nielsen SL, Banta GT, Pedersen F (eds) Estuarine nutrient cycling: the influence of primary producers. Kluwer Academic, Dordrecht, pp 231–261

McLachlan A, Defeo O (2018) The ecology of sandy shores. Academic Press

Meadows PS, Meadows A, Murray JM (2012) Biological modifiers of marine benthic seascapes: their role as ecosystem engineers. Geomorphology 157:31–48

Mees J, Jones MB (1997) The hyperbenthos. Oceanogr Mar Biol 35:35–212

Méndez Casariego A, Alberti J, Luppi T, Iribarne OO (2009) Stage-dependent interactions between intertidal crabs: from facilitation to predation. J Mar Biol Assoc UK 89(4):781–788

Méndez Casariego A, Luppi T, Iribarne O, Daleo P (2011a) Increase of organic matter transport between marshes and tidal flats by the burrowing crab *Neohelice* (*Chasmagnathus*) *granulata* Dana in SW Atlantic salt marshes. J Exp Mar Biol Ecol 401(1–2):110–117

Méndez Casariego A, Alberti J, Luppi T, Daleo P, Iribarne OO (2011b) Habitat shifts and spatial distribution of the intertidal crab *Neohelice* (*Chasmagnathus*) *granulata* Dana. J Sea Res 66(2):87–94

Meysman FJR, Middelburg JJ, Heip CHR (2006) Bioturbation: a fresh look at Darwin's last idea. Trends Ecol Evol 21:688–695

Michaud E, Desrosiers G, Mermillod-Blondin F, Sundby B, Stora G (2006) The functional group approach to bioturbation: II. The effects of the *Macoma balthica* community on fluxes of nutrients and dissolved organic carbon across the sediment-water interface. J Exp Mar Biol Ecol 337(2):178–189

Milner RNC, Booksmythe I, Jennions MD, Backwell PRY (2010) The battle of the sexes? Territory acquisition and defence in male and female fiddler crabs. Anim Behav 79(3):735–738

Mineur F, Cook EJ, Minchin D, Bohn K, MacLeod A, Maggs CA (2012) Changing coasts: marine aliens and artificial structures. In: Oceanography and marine biology. CRC Press, New York, pp 198–243

Minkoff DR, Escapa M, Ferramola FE, Maraschín SD, Pierini JO, Perillo GM, Delrieux C (2006) Effects of crab–halophytic plant interactions on creek growth in a SW Atlantic salt marsh: a cellular automata model. Estuar Coast Shelf Sci 69(3):403–413

Molina LM, Valinas MS, Pratolongo PD, Elías R, Perillo GME (2008) First record of the sea anemone *Diadumene lineata* (Verrill 1871) associated to *Spartina alterniflora* roots and stems, in marshes at the Bahia Blanca estuary, Argentina. Biol Invasions 11:409–416

Mortimer RJG, Krom MD, Watson PG, Frickers PE, Davey JT, Clifton RJ (1999) Sediment–water exchange of nutrients in the intertidal zone of the Humber estuary, UK. Mar Poll Bull 37(3):261–279

Murray JM, Meadows A, Meadows PS (2002) Biogeomorphological implications of microscale interactions between sediment geotechnics and marine benthos: a review. Geomorphology 47(1):15–30

Nellemann C, Corcoran E, Duarte CM, Valdes L, DeYoung C, Fonseca L, Grimsditch G (2009) Blue Carbon. A Rapid Response Assessment. United Nations Environmental Programme, GRID-Arendal, Birkeland Trykkeri AS, Birkeland

Nixon SW (1981) Remineralization and nutrient cycling in coastal marine ecosystems. In: Neilson BJ, Cronin LE (eds) Estuaries and Nutrients. Humana Press, New York, pp 111–138

Norkko A, Hewitt JE, Thrush SF, Funnell GA (2001) Benthic-pelagic coupling and suspension feeding bivalves: linking site-specific sediment flux and biodeposition to benthic community structure. Limnol Oceanogr 46(8):2067–2072

Orensanz JM, Schwindt E, Pastorino G, Bortolus A, Casas G, Darrigran G, Elías R, Lopez Gappa JJ, Obenat S, Pascual MS, Penchaszadeh P, Piriz ML, Scarabino F, Spivak ED, Vallarino EA (2002) No longer the pristine confines of the world ocean: a survey of exotic marine species in the southwestern Atlantic. Biol Invasions 4:115–143

Overnell J, Edwards A, Grantham BE, Harvey SM, Jones JW, Leftley JW, Smallman DJ (1995) Sediment-water coupling and the fate of the spring phytoplankton bloom in Loch Linnhe, a Scottish fjordic sea-loch. Sediment processes and sediment-water coupling. Estuar Coast Shelf Sci 4:1–19

Paarlberg AJ, Knaapen MA, de Vries MB, Hulscher SJMH, Wang ZB (2005) Biological influences on morphology and bed composition of an intertidal flat. Estuar Coast Shelf Sci 64(4):577–590

Palomo G, Iribarne OO (2000) Sediment bioturbation by polychaete feeding may promote sediment stability. Bull Mar Sci 67(1):249–257

Palomo G, Botto F, Navarro D, Escapa M, Iribarne OO (2003) Does the presence of the SW Atlantic burrowing crab *Chasmagnathus granulatus* Dana affect predator–prey interactions between shorebirds and polychaetes? J Exp Mar Biol Ecol 290(2):211–228

Paterson DM, Fortune I, Aspden RJ, Black KS (2019) Intertidal Flats: Form and Function. In: GME P et al (eds) Coastal Wetlands. Elsevier, pp 383–406

Pearson TH, Rosenberg R (1987) Feast and famine: structuring factors in marine benthic communities. In: Gee JHR, Giller PS (eds) Organization of communities, past and present. Blackwell Scientific, Oxford, pp 373–395

Pedersen MF, Nielsen SL, Banta GT (2004) Interactions between vegetation and nutrient dynamics in coastal marine ecosystems: an introduction. In: Nielsen S, Banta G, Pedersen M (eds) Estuarine nutrient cycling: the influence of primary producers. Kluwer Academic Publishers, Dordrecht, pp 1–15

Perillo GME, Iribarne OO (2003) New mechanisms studied for creek formation in tidal flats: from crabs to tidal channels. EOS Trans Am Geophys Union 84(1):1–5

Rathbun MJ (1918) The grapsoid crabs of America. US Government Printing Office

Reise K (1985) Tidal flat ecology: an experimental approach to species interactions. Springer, Berlin

Reise K (2002) Sediment mediated species interactions in coastal waters. J Sea Res 48(2):127–141

Revsbech NP, Sorensen J, Blackburn TH, Lomholt JP (1980) Distribution of oxygen in marine sediments measured with microelectrodes. Limnol Oceanogr 25(3):403–411

Rhoads DC, Young DK (1970) The influence of deposit-feeding organisms on sediment stability and community trophic structure. J Mar Res 28:150–178

Richter R (1952) Fluidal-Textur in Sediment-Gesteinen und über Sedifluktion überhaupt. Notizbl Hess Landesamtes Bodenforsch Wiesb 6:67–68

Rico A, Lanas P, López Gappa J (2001) Temporal and spatial patterns in the recruitment of *Balanus glandula* and *Balanus laevis* (Crustacea, Cirripedia) in Comodoro Rivadavia harbor (Chubut, Argentina). Rev Mus Argentino Cienc Nat 3(2):175–179

Roldán JP (2014) Comunidades macrobentónicas incrustantes del sector portuario de Bahía Blanca. Tesis de Licenciatura en Ciencias Biológicas. Univ Nac del Sur, Bahía Blanca, p 33

Rosa LC, Bemvenuti CE (2005) Effects of the burrowing crab *Chasmagnathus granulata* (Dana) on meiofauna of estuarine intertidal habitats of Patos lagoon, southern Brazil. Braz Arch Biol Technol 48(2):267–274

Rossi S, Bramanti L, Gori A, Orejas C (2017) An overview of the animal forests of the world. In: Rossi S, Bramanti L, Gori A, Orejas C (eds) Marine animal forests. Springer, Cham, pp 1–26

Rossi S, Rizzo L, Duchêne JC (2019) Polyp expansion of passive suspension feeders: a red coral case study. Peer J Preprints 7:e27490v1

Ruesink JL, Lenihan HS, Trimble AC, Heiman KW, Micheli F, Byers JE, Kay MC (2005) Introduction of non-native oysters: ecosystem effects and restoration implications. Annu Rev Ecol Evol Syst 36:643–689

Ruiz GM, Freestone AL, Fofonoff PW, Simkanin C (2009) Habitat distribution and heterogeneity in marine invasion dynamics: the importance of hard substrate and artificial structure. In: Wahl M (ed) Marine hard bottom communities. Springer, Berlin/Heidelberg, pp 321–332

Sakai K, Türkay M, Yang SL (2006) Revision of the *Helice/Chasmagnathus* complex (Crustacea: Decapoda: Brachyura). Abh Senckenberg Naturforsch Ges 565:1–76

Sal-Moyano MP, Gavio MA, Luppi TA (2012) Mating system of the burrowing crab *Neohelice granulata* (Brachyura: Varunidae) in two contrasting environments: effect of burrow architecture. Mar Biol 159(7):1403–1416

Schiffers K, Teal LR, Travis JMJ et al (2011) An open source simulation model for soil and sediment bioturbation. PLoS One 6:e28028

Schubart CD, Cuesta JA, Felder DL (2002) Glyptograpsidae, a new brachyuran family from Central America: larval and adult morphology, and a molecular phylogeny of the Grapsoidea. J Crustacean Biol 22(1):28–44

Shatkin GS, Shumway SE, Hawes R (1997) Considerations regarding the possible introduction of the Pacific oyster (*Crassostrea gigas*) to the Gulf of Maine: a review of a global experience. J Shellfish Res 16:463–477

Simonetti P, Botté SE, Fiori SM, Marcovecchio JE (2012) Heavy-metal concentrations in soft tissues of the burrowing crab *Neohelice granulata* in Bahía Blanca estuary, Argentina. Arch Environ Contam Toxicol 62(2):243–253

Simonetti P, Botté SE, Fiori SM, Marcovecchio JE (2013) Burrowing crab (*Neohelice granulata*) as a potential bioindicator of heavy metals in the Bahía Blanca estuary, Argentina. Arch Environ Contam Toxicol 64(1):110–118

Simonetti P, Botté SE, Marcovecchio JE (2018) Heavy metal bioconcentration factors in the burrowing crab *Neohelice granulata* of a temperate ecosystem in South America: Bahía Blanca estuary, Argentina. Environ Sci Pollut R 25(34):34652–34660

Snelgrove PV (1999) Getting to the bottom of marine biodiversity: sedimentary habitats: ocean bottoms are the most widespread habitat on earth and support high biodiversity and key ecosystem services. Bioscience 49(2):129–138

Solan M, Wigham BD, Hudson IR et al (2004) In situ quantification of bioturbation using time lapse fluorescent sediment profile imaging (f SPI), luminophore tracers and model simulation. Mar Ecol Prog Ser 271:1–12

Sommer U (ed) (1989) Plankton ecology: succession in plankton communities. Springer, Berlin, p 369

Sousa AI, Lillebø AI, Caçador I, Pardal MÂ (2008) Contribution of *Spartina maritima* to the reduction of eutrophication in estuarine systems. Environ Pollut 156:628–635

Sousa R, Gutierrez JL, Aldrige DC (2009) Non-indigenous invasive bivalves as ecosystem engineers. Biol Invasions 11:2367–2385

Spivak ED (1997) Cangrejos estuariales del Atlántico sudoccidental (25°-41°S) (Crustacea: Decapoda: Brachyura). Investig Mar 25:105–120

Spivak E (2005) Los cirripedios litorales (Cirripedia, Thoracica, Balanomorpha) de la región del Río de la Plata y las costas marinas adyacentes. In: Penchaszadeh PE et al (eds) Invasores: Invertebrados exóticos en el Río de la Plata y región marina aledaña. EUDEBA, Buenos Aires, pp 251–309

Spivak ED (2010) The crab *Neohelice* (=*Chasmagnathus*) *granulata*: an emergent animal model from emergent countries. Helgol Mar Res 64(3):149–154

Spivak ED, Gavio MA, Navarro CE (1991) Life history and structure of the world's southernmost Uca population: *Uca uruguayensis* (Crustacea, Brachyura) in Mar Chiquita lagoon (Argentina). Bull Mar Sci 48:679–688

Spivak ED, Anger K, Luppi TA, Bas C, Ismael D (1994) Distribution and habitat preferences of two grapsid crab species in Mar Chiquita lagoon (province of Buenos Aires, Argentina). Helgol Meeresunters 48:59–78

Spivak ED, Anger K, Bas CC, Luppi TA, Ismael D (1996) Size structure, sex ratio, and breeding season in two intertidal grapsid crab species from Mar Chiquita lagoon, Argentina. Neritica 10:7–26

Sunday JM, Pecl GT, Frusher S, Hobday AJ, Hill N, Holbrook NJ et al (2015) Species traits and climate velocity explain geographic range shifts in an ocean-warming hotspot. Ecol Lett 18:944–953

Torres Jorda M, Roccatagliata D (2002) Population dynamics of *Leidya distorta* (Isopoda: Bopyridae) infesting the fiddler crab *Uca uruguayensis* at the Río de la Plata estuary, Argentina. J Crustacean Biol 22:719–727

Troost K, Gelderman E, Kamermans P, Smaal AC, Wolff W (2009) Effects of an increasing filter feeder stock on larval abundance in the Oosterschelde estuary (SW Netherlands). J Sea Res 61:153–164

Truchet DM, Buzzi NS, Carcedo MC, Marcovecchio JE (2019) First record of the fiddler crab *Leptuca* (= *Uca*) *uruguayensis* in the Bahía Blanca estuary (Buenos Aires, Argentina) with comments on its biology in South America. Reg Stud Mar Sci 27:100539

Turner RD (1954) The family Pholadidae in the Western Atlantic and the Eastern Pacific. Part I - Pholadinae. Johnsonia 3(33):1–63

Valentinuzzi De Santos S (1971) Estudio preliminar sobre las comunidades intercotidales del Pto de Ing White (Pcia de Buenos Aires). Physis 30(81):407–417

Valiela I (ed) (1995) Marine ecological processes, 2nd edn. Springer, New York, p 686

Van Colen C, De Backer A, Meulepas G, Van Der Wal D, Vincx M, Degraer S, Ysebaert T (2010) Diversity, trait displacements and shifts in assemblage structure of tidal flat deposit feeders along a gradient of hydrodynamic stress. Mar Ecol Progr Ser 406:79–89

Van Oevelen D, Soetaert K, Middelburg JJ, Herman PMJ, others (2006) Carbon flows through a benthic food web: integrating biomass, isotope and tracer data. J Mar Res 64:453–482

Villagrán DM, Truchet DM, Buzzi NS, Lopez ADF, Severini MDF (2019) A baseline study of microplastics in the burrowing crab (*Neohelice granulata*) from a temperate southwestern Atlantic estuary. Mar Poll Bull 110686

Vitousek PM, Mooney HA, Lubchenco J, Melillo JM (1997) Human domination of earth's ecosystems. Science 277:494–503

Webb AP, Eyre BD (2004) The effect of natural populations of the burrowing and grazing soldier crab (*Mictyris longicarpus*) on sediment irrigation, benthic metabolism and nitrogen fluxes. J Exp Mar Biol Ecol 309(1):1–19

Weis JS, Weis P (2004) Metal uptake, transport and release by wetland plants: implications for phytoremediation and restoration. Environ Int 30:685–700

Widdows J, Brown S, Brinsley MD, Salkeld PN, Elliott M (2000) Temporal changes in intertidal sediment erodability: influence of biological and climatic factors. Cont Shelf Res 20(10):1275–1289

Wilkie EM, Bishop MJ, O'Connor WA (2013) The density and spatial arrangement of the invasive oyster Crassostrea gigas determines its impact on settlement of native oyster larva. Ecol Evol 3(15):4851–4860

Yamaguchi T, Henmi Y (2008) Cheliped differentiation and sex ratio of the fiddler crab *Uca arcuata*. Crustacean Res 37:74–79

Zapperi G, Pratolongo P, Piovan MJ, Marcovecchio JE (2016) Benthic-pelagic coupling in an intertidal mudflat in the Bahía Blanca Estuary (SW Atlantic). J Coast Res 32(3):629–637

Zapperi G, Piovan MJ, Pratolongo P (2017) Community structure and spatial zonation of benthic macrofauna in mudflats of the Bahía Blanca estuary, Argentina. J Coast Res 34(2):318–327

Chapter 9
Taxonomic and Functional Assessment of Subtidal Macrobenthic Communities in the Bahía Blanca Estuary (Argentina)

M. Emilia Bravo, M. Cecilia Carcedo, Eder P. Dos Santos, and Sandra M. Fiori

9.1 Introduction

The structural and functional characteristics of the macrobenthic soft-bottom invertebrate communities are mainly controlled by properties of the habitat. This control results from interplay of species-specific relationships with the physical and chemical properties of the sediment and surrounding seawater (Nikora 2010; Breine et al. 2018). Thus, the biomechanical properties of the benthic fauna are related to the physical characteristics of the sediments (e.g., stability, shear strength, and texture) and the water (e.g., intensity and dynamics of flow) (Nikora 2010). These variables, together with the tolerance levels of each species to chemical characteristics (e.g., salinity, pH, organic matter content, dissolved oxygen), limit their distribution, affecting the ecological configuration of the benthic communities they integrate (Gray 2002; Teske and Wooldridge 2003; Breine et al. 2018). Other factors affecting the ecological characteristics of benthic communities are interspecific relationships. The relative importance of biotic versus abiotic factors in the structuring of communities is poorly understood and is still being discussed, especially for the subtidal

M. E. Bravo (✉)
Instituto de Geociencias Básicas, Aplicadas y Ambientales de Buenos Aires (IGEBA-UBA-CONICET), Departamento de Ciencias Geológicas, Facultad de Ciencias Exactas y Naturales, Universidad de Buenos Aires, Buenos Aires, Argentina
e-mail: mebravo@gl.fcen.uba.ar

M. C. Carcedo · S. M. Fiori
Instituto Argentino de Oceanografía, Universidad Nacional del Sur, CONICET, IADO, Bahía Blanca, Argentina

Departamento de Biología, Bioquímica y Farmacia, Universidad Nacional del Sur, Bahía Blanca, Argentina

E. P. Dos Santos
Instituto Argentino de Oceanografía, Universidad Nacional del Sur, CONICET, IADO, Bahía Blanca, Argentina

S. M. Fiori, P. D. Pratolongo (eds.), *The Bahía Blanca Estuary*,
https://doi.org/10.1007/978-3-030-66486-2_9

benthic communities (Wilson 1990). Altogether, benthic communities integrate the individual responses of each species to environmental properties in a dynamic way across space and time. Due to their relative immobility, they are chronically exposed to any environmental stressor, which makes them excellent indicators of environmental health (Dean 2008). Understanding their response to environmental disturbances provides an insight into the level of stress and resilience of the marine ecosystem (Ellis et al. 2017).

The characteristics of the environment can, in turn, be largely modified by the benthic subtidal organisms. In estuaries, the structure of benthic communities regulates ecosystem functioning through some biophysical processes such as bioturbation and bioirrigation and biogeochemical processes such as nutrient cycling (Van der Wal et al. 2017a). The benthic subtidal organisms are important secondary producers and prey items of various benthic-feeding fish species (Van der Wal et al. 2017a; Bremec and Giberto 2017). Therefore, knowledge of their composition and biological characteristics is important to establish the quality and quantity of food available for species of fishing or conservation interest (Bremec and Giberto 2017). At the same time, analyzing the ecosystem services in which benthic organisms are involved may be a useful proxy of the resilience capacity of the marine system (Dissanayake et al. 2018).

Human activities in estuarine environments often lead to large-scale changes ranging from habitat modification to functional changes. For example, high rates of sedimentation associated with anthropogenic activities can lead to loss of filter-feeding functional groups and replacement by deposit-feeders (Ellis et al. 2017). High-contaminant loads can lead to replacement by opportunistic fauna of smaller body size and therefore with smaller cave depths, as well as with less longevity (Dauer et al. 1992; Dean 2008). These changes lead to a lower biomass, oxygenation, and reworking of sediments, among other consequences. Taxonomic and functional changes in benthic communities can, in turn, lead to shifts in the ecosystem services. Understanding the ecological and functional role of benthic organisms is necessary for taking environmental management measures (Dissanayake et al. 2018).

In Argentina, the subtidal environments of coasts have mainly been studied in relation to the population dynamics of macrobenthic species with current or potential fishing value (Narvarte et al. 2007; Zaidman et al. 2016). For example, several studies have been carried out on currently exploited commercial species, such as the Tehuelche scallop *Aequipecten tehuelchus* (Soria et al. 2016), the spiral snail *Zidona dufresnei* (Giménez et al. 2005), the purple clam *Amiantis purpurata* (Morsán 2007), mussels *Mytilus edulis platensis* and *Aulacomya atra*, and the Patagonian octopus *Octopus tehuelchus* (Narvarte et al. 2007). Only a small number of population ecology studies have previously addressed non-commercial species, and these were selected because of their dominance in the communities (Schiariti et al. 2006; Güller and Zelaya 2017). Ecological studies of subtidal benthic communities have mainly been focused on the relationship between benthic community distribution and spatial heterogeneity in relation to sediment and salinity (Gilberto et al. 2004; Diez et al. 2009; Kaminsky et al. 2018).

The subtidal macrobenthic communities of the Bahía Blanca Estuary were first studied in the late 1980s (Elías 1987, 1992, 1995; Elías and Ieno 1993). These studies described the communities of subtidal macrobenthic invertebrates and their distribution patterns related to variations in water salinity and sediment grain size over the estuary. After a 20-year period, a renewed interest in the status of the subtidal macrobenthic communities was associated with the environmental monitoring program required by Bahía Blanca Municipal Government (see Chap. 19). These monitoring programs were intended to update the knowledge base of the subtidal communities in order to assess the status of the Bahía Blanca Estuary ecosystem in relation to anthropogenic pressures. In addition, recent studies had focused on the effect of wastewater discharge on benthic communities (Fiori et al. 2020). The sampling sites studied by Elías (1992) and Fiori (2016) are shown in Fig. 9.1. The subtidal seafloor of the Bahía Blanca Estuary has predominantly unconsolidated sediments. Although there are sectors of the seabed with other morphological features (sand dune fields, rocky outcrops, deep-holes, terraces), the seabed on the shores of the channels analyzed in this study are the most representative of gradual large-scale environmental variation (Fig. 9.1).

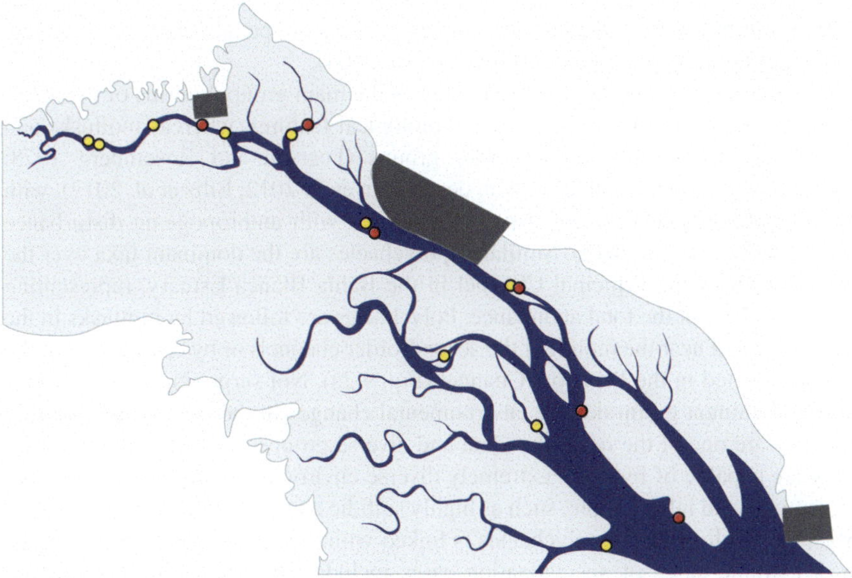

Fig. 9.1 Location of samples along the Principal Channel of the Bahía Blanca Estuary taken by Elías in 1987 (yellow dots) and Fiori in 2016 (red dots)

9.2 Patterns of Abundance and Biomass of the Main Taxonomic Groups

A central issue in community ecology is based on understanding the environmental drivers of changes in community structure across space and how these determine the distribution patterns of each species locally (Alves et al. 2020). In estuaries, the spatial variation of subtidal benthic communities is often related to salinity and sediment texture gradients (van der Linden et al. 2012; Lana et al. 2018; Alves et al. 2020). This reflects the response of each species and therefore the gradual change in community composition over a continuum (van der Linden et al. 2012; Lana et al. 2018; Alves et al. 2020). However, the relative importance of salinity with respect to grain size varies between temperate estuaries. For example, abundance and biomass in some northern hemisphere estuaries are related to salinity at the regional level, but to variation in grain size and/or hydrodynamic conditions at the local level (van der Linden et al. 2012; van der Wal et al. 2017a). On the other hand, in South African estuaries, abundance and biomass patterns have a weak relationship with the salinity gradient but a strong dependence on the sediment characteristics (Teske and Wooldridge 2003). In Argentina, the strong variation in salinity and turbidity along the Rio de la Plata Estuary is reflected in the appearance of a salt wedge, as well as a turbidity front leading to complex patterns of spatial variation of benthic communities along the gradient (Giberto et al. 2004).

In estuaries, polychaetes are often cited as the main group in terms of the abundance and diversity of both species and biological features, whereas mollusks and crustaceans are usually second-order groups (Pearson and Rosemberg 1978; Ysebaert et al. 2002; Dean 2008; Van der Linden et al. 2012; Ellis et al. 2017), with few exceptions, and they are usually associated with anthropogenic disturbance (Muniz and Venturini 2015). Similarly, polychaetes are the dominant taxa over the entire length of the Principal Channel in the Bahía Blanca Estuary, representing more than 70% of the total abundance. Polychaetes are followed by mollusks in the sectors located near the mouth of the second-order channels or by crustaceans at the stations located in the Principal Channel (Fig. 9.2a). Not surprisingly, polychaetes are the dominant group despite environmental changes in the estuary, considering that they are one of the most abundant and diverse groups in marine communities. They are capable of living in extremely diverse environments including some that are extreme and inhospitable, such as highly sulfidic anoxic sediments (Dean 2008). Also, the biodiversity of polychaetes is linked with diverse lifestyles. Some of the most extreme cases of specialization cases include the bone-eating polychaetes (Rouse et al. 2004) or those deriving energy from methane in deep-sea seeps and vents (Levin et al. 2017). Polychaete specialization includes different species with different adaptive strategies to pollutants, leading to specific responses to the different levels of pollution (Dean 2008; Ellis et al. 2017). Some of these specific responses to pollution are shared between polychaetes belonging to the same family, which is why several biological indices for assessing environmental quality are based on polychaetes (Dean 2008).

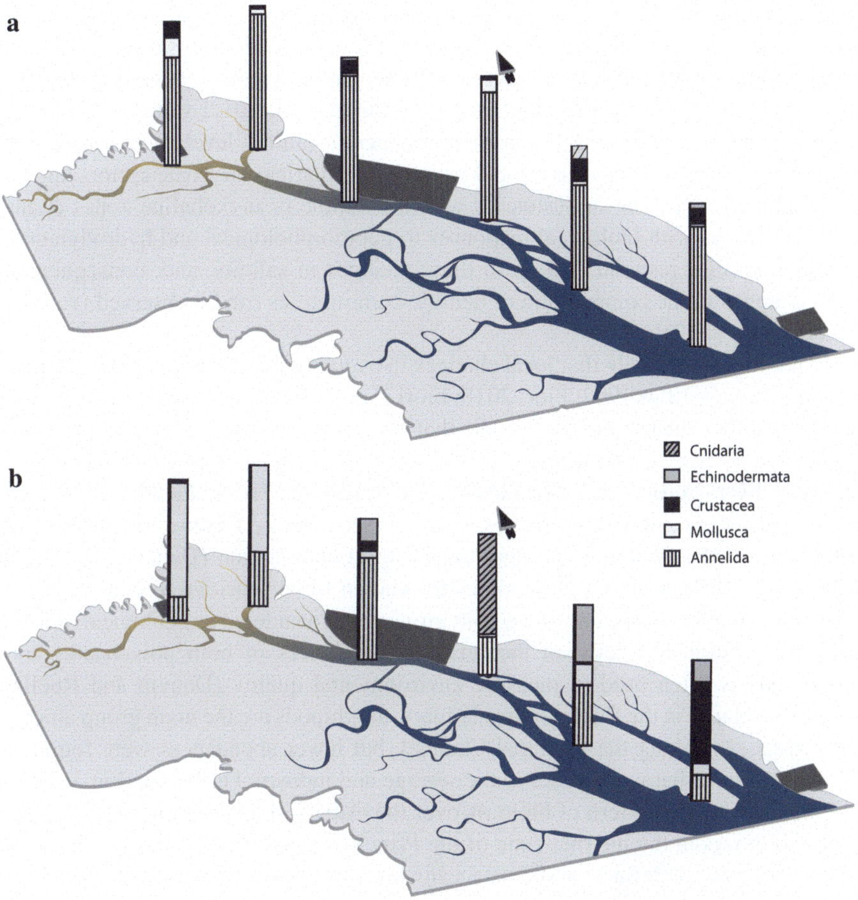

Fig. 9.2 Spatial variability of macrobenthic communities along the Principal Channel with detail of the relative contribution of the main taxa to (**a**) the total abundance and to (**b**) the total biomass

The relative contribution of secondary groups, like mollusks and crustaceans, differs between the internal and external areas of the Principal Channel. In general, mollusks are better represented in the inner zone of the Bahía Blanca Estuary, while crustaceans reach greater abundances in the external zone. However, this pattern associated with a linear environmental gradient is interrupted at the mouths of secondary channels. These local changes in benthic communities at the channel intersections may be related to local changes in environmental conditions (van der Wal et al. 2017a). The tidal channels and the Principal Channel are hydrodynamically articulated and their intersections may have particular configurations of the seabed, such as rocky outcrops, sand banks, and sand dunes (Ginsberg et al. 2009). The hydrodynamic and geomorphological complexity of the seafloor at the channel intersections can drastically change the benthic communities at a local scale, even leading to hot spots of diversity and/or biomass (van der Wal et al. 2017a). These local interruptions of the large-scale estuarine environmental gradient can also be found where the intersections of tidal channels lead to changes in salinity. This can be generated by the presence of rivers and streams with daily or seasonal changes in the amount of freshwater input van der Linden et al. 2012; van der Wal et al. 2017a; Lana et al. 2018). The seasonal changes in rainfall levels and consequent freshwater discharges from rivers and streams in estuaries can create sectors known as "micro-estuaries" or sub-estuaries within euhaline or myxohaline zones (Lana et al. 2018). As with biological responses to geomorphological and hydrodynamic changes, species respond rapidly to these changes in salinity, and, consequently, local changes in the composition of benthic communities can be observed (van der Linden et al. 2012).

In addition, it is likely that local changes in overall patterns are related to anthropogenic impact (Elias 1992; Fiori 2018; Fiori et al. 2020). Many studies on anthropically disturbed estuaries have found that the most stressful environments were dominated by annelids (Dauer et al. 1992; Dean 2008; Kristensen et al. 2014). This is a very diverse group, and a species-specific response to stress factors is expected (Ellis et al. 2017) and will be discussed further (see Sect. 9.5). Some polychaetes are often classified as tolerant or opportunistic to contamination (Dauer et al. 1992; Ellis et al. 2017), whereas amphipods are known to be sensitive to contaminated sediments. On this basis, and since both groups are considered as ecological indicators, the relationship between the relative abundances of both polychaetes and amphipods is often used to measure environmental quality (Dauvin and Ruellet 2007). Similarly, in the Bahía Blanca Estuary, amphipods are the main group among the crustaceans along the Principal Channel, but lower abundances were found at the two stations that were closer to the sewage and industrial effluents (Fig. 9.2a).

The distribution pattern of biomass over the estuary shows that mollusks have a higher biomass in the internal zone of the Principal Channel, whereas the biomass of polychaetes, cnidarians, and crustaceans increases from the middle to the external zone (Fig. 9.2b). Although crustaceans are relatively abundant, they do not represent an important contribution to the biomass due to their small body size (Fiori 2016). This is the opposite to the case of the cnidarians, because they are represented by the sea pen *Stylatula darwini* which is in low abundance, but it has a large body size and therefore contributes more to the total biomass. This also happens in the middle and outer zones of the Principal Channel, where the echinoderms, represented by ophiurans, contribute secondarily to the biomass in these zones (Fiori 2016). In the internal zone, the high dominance of mollusks is due to the bivalves *Corbula patagonica* and *Pitar rostratus* (Fiori 2016), which in previous studies were found to be dominant in the external zone (euhaline) in previous studies but were related to very fine sandy sediments in all cases (Elías 1992). Both species have also been mentioned as being abundant in relation to very fine sandy sediment on the adjacent continental shelf "El Rincón" indicating a wide tolerance to variation in salinity by these species (Bremec 1990). Grain size preference may also indicate that these species prefer bottom currents of low intensity, as expected for some suspension-feeders (e.g., van der Wal et al. 2017a).

9.3 Spatial Variations in Density, Biomass, Richness, and Diversity

The mean density of macrobenthic invertebrates was similar along the Principal Channel (Fig. 9.3a). The density is expressed here as a function of the sampling unit (Van Veen grab) with a surface area of 0.18 m^2. The mean densities were similar in the myxohaline and euhaline sectors, ranging from 83 individuals.$grab^{-1}$ at the innermost station to 104 individuals.$grab^{-1}$ at the outermost station. Interrupting this pattern, local changes with higher mean densities were found at the two stations that were closer to the sewage and industrial effluents (Fig. 9.3a). This increase in mean density associated with the sectors affected by sewage discharge is often attributed to strategies of rapid recruitment of opportunistic benthic macrofauna (Dauer et al. 1992).

The average biomass was similar along the Principal Channel with a slight decrease from the internal to the external zone, except for changes at stations located near to the port and sewage discharges where the total biomass was markedly lower (Fig. 9.3b). In both cases, these are possibly stations with marked environmental stress. The highest values of biomass were found at sites with the highest proportion of fine sediments (Fig. 9.3e). In the estuary, these characteristics are associated with environments with high rates of local deposition of sediments, nutrients, organic matter, and contaminants (Carbone et al. 2016). Moreover, the internal zone has a greater production of phytoplankton (Guinder et al. 2012) that could favor suspension-feeding organisms, such as the bivalve *Corbula patagonica*, the main contributor to the high biomass in the sector. Also, the high primary and secondary productivity of the inner zone gives it unique qualities for feeding and reproduction of several pelagic species, including species of commercial interest (Lopez Cazorla 2004). The area is also environmentally sensitive due to the combination of low water renewal rates and high anthropogenic pressure (Carbone et al. 2016). Considering the high benthic biomass, it is important to evaluate its ecological functions in relation to the retention and resuspension of pollutants (see Sect. 9.4).

The total species richness recorded in the Bahía Blanca Estuary was 155 species. This value is intermediate between that found in closed estuaries, even in tropical areas such as Brazil (Lana et al. 2018; Alves et al. 2020), and that of nearby open coastal areas. This coincides with the geographical characteristics of the estuary, which is semi-enclosed with little freshwater input concentrated in the interior zone, resulting in a salinity gradient that is much less marked than in other estuaries (Guinder et al. 2012). There are marked differences between myxohaline and euhaline zones (Fig. 9.3c). The myxohaline zone of the Bahía Blanca Estuary has the lowest species richness values, in ranges comparable to those found in other temperate area estuaries (van der Linden et al. 2012; Bremec and Giberto 2017). The greatest species richness is found in the external zone, where greater stability of physical-chemical parameters can lead to a more balanced state in the macrobenthic communities (van der Linden et al. 2012; van der Wal et al. 2017a). As for abundance and biomass, species richness presented minimum values at the stations affected by wastewater and industrial discharges (Fig. 9.3c). These stations also had

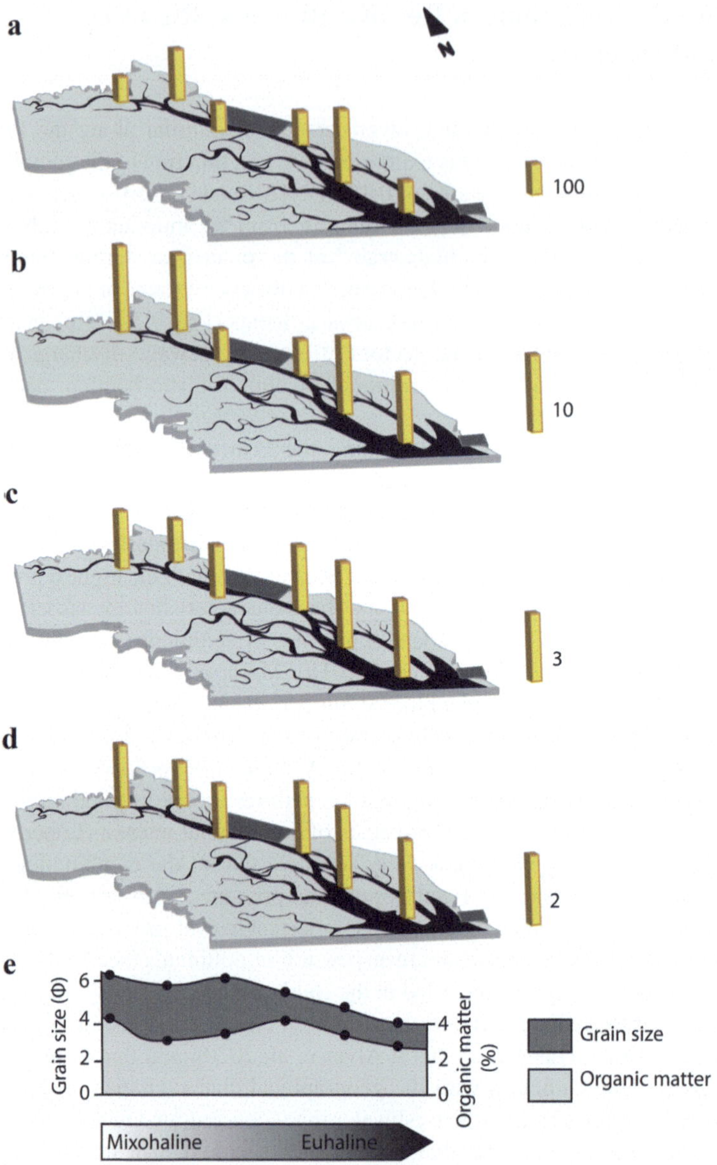

Fig. 9.3 Spatial variability of (**a**) mean density (ind.grab^{-1}), (**b**) mean biomass (g.grab^{-1}), (**c**) Margalef species richness, (**d**) Shannon-Wiener diversity index (H′) of benthic communities in relation to the variation of (**e**) grain size of sediments (phi units) and organic matter content (percentage) over the Principal Channel in the Bahía Blanca Estuary

the lowest species richness values of the three stations in the myxohaline zone, as well as those in the euhaline zone, so they may lack species with the physiological and behavioral adaptations necessary to overcome stress conditions, leading to lower species richness (Dauer et al. 1992; Ellis et al. 2017).

The soft-bottom subtidal areas of the Principal Channel in the Bahía Blanca Estuary were found to be inhabited by about 69 species according to Elias (1987), but the latest research listed about 155 different taxa (Fiori 2016, 2018; Bravo 2019). All the species listed to date for the area are found in Table 9.1, with information on the reference sources. Diversity by major taxa is also greater for the annelids represented by 58 taxa, followed by mollusks (40 taxa) and crustaceans (30 taxa) (Elias 1987; Fiori 2014, 2016, 2018). Other groups that constitute the benthic communities are the ophiuroids, pennatulaceans, priapulids, bryozoans, hydrozoans, pycnogonids, ostracods, nemerteans, and sponges (Elias 1987; Fiori 2014, 2016, 2018) (Fig. 9.4). The differences between the number of species found in the studies carried out by Elias (1987) and that of more recent studies (Fiori 2014, 2016, 2018) are due to the use of different mesh sizes during the separation of the organisms from the sediment. Elias (1987) used a 2 mm mesh, whereas Fiori (2014, 2016, 2018) used a 0.5 mm mesh.

In the inner zone (myxohaline), with muddy sediments and very fine sand, the species that dominate were the polychaetes *Nicolea* sp. and *Leodamas verax*. *Leodamas verax* is an ubiquitous and abundant species that can be found along the entire course of the Principal Channel and minor channels (Fiori 2016, 2018). Other polychaetes such as *Polydora* sp. and *Aphelocheta* sp. are important in the interior zone, decreasing towards the mouth of the estuary (Fiori 2016, 2018), while in the external zone (euhaline), *Axiothella* sp. and the bivalve *Corbula patagonica* dominate in the sandy mud sediments whose sand fraction is classified as fine to very fine (Elías 1987; Fiori 2014, 2016, 2018). Finally, the middle zone is characterized by *Terebellides totae*, associated to the sandy mud sediments (Elias 1987). Polychaetes also dominate in the stations affected by sewage and industrial discharges, and it is interesting that different families dominate in the different successional stages as will be discussed later (Sect. 9.6).

Although most of the changes are observed when evaluating polychaetes, when analyzing the secondary groups in abundance and diversity such as mollusks, they were represented by 40 species (20 Gastropoda, 18 Bivalvia, 1 Polyplacophora, 1 Scaphopoda, Table 9.1). The bivalves *Corbula patagonica*, *Nucula semiornata*, and *Pitar rostratus* were considered ubiquitous, mainly between 4 and 5 m depth. Slightly higher abundances of the gastropod *Buccinanops deformis* and the bivalves *Macoma uruguayensis* and *Tagelus plebeius* were found in shallower channels (0.50 cm depth, Fiori 2018). In the channels affected by wastewater discharges, mollusks represented by the bivalves *C. patagonica*, *N. semiornata*, *P. rostratus*, *Malletia cumingii*, and *Tagelus plebeius* and the gastropod *B. deformis* reached the same biomass as polychaetes (Fiori 2018; Fiori et al. 2020).

Species of the genus *Corbula* such as *Corbula gibba* have been classified as tolerant to a wide range of environmental disturbances such as coastal eutrophication, and successful recruitment has also been recorded in areas subject to mass

Table 9.1 List of the species described for soft-bottom subtidal areas of the Bahía Blanca Estuary

Species	Source
Annelida	
Aglaophamus sp.	Bremec et al. (1990)
Amphicteis sp.	Bremec et al. (1990)
Amphicteis gunneri	Elias et al. (2004)
Aphelocheta sp.	Bravo et al. (2018) and Fiori (2018)
Aricidea sp.	Bravo et al. (2018) and Fiori (2018)
Axiotella cf. *constricta*	Elias et al. (2001)
Axiotella sp.	Bremec et al. (1990), Elias et al. (2004) and Fiori (2018)
aff. Chone sp.	Bravo et al. (2018) and Fiori (2018)
Chone cf. *striata*	Elias et al. (2004)
Capitella capitata	Elias et al. (2001)
Cirratulidae	Elias et al. (2004) and Fiori (2018)
Dodecaceria sp.	Fiori (2018)
Dorvilleidae	Fiori (2018)
Drilonereis orensanzi	Bremec et al. (1990)
Eteone sp.	Elias et al. (2004), Bravo et al. (2018) and Fiori (2018)
Eusyllinae indet	Bravo et al. (2018) and Fiori (2018)
Flabelligeridae	Elias et al. (2004) and Fiori (2018)
Glycera americana	Elias et al. (2004) and Bravo et al. (2018)
Goniadidae	Elias et al. (2004)
Halosydna patagonica	Bravo et al. (2018)
Halosydnella australis	Elias et al. (2004)
Harmothoe sp.	Elias et al. (2004)
Harmothoinae indet	Bravo et al. (2018)
Kimbergonuphis sp.	Elias et al. (2004)
Kimbergonuphis tenuis	Fiori (2018)
Laeonereis acuta	Elias et al. (2004), Bravo et al. (2018) and Fiori (2018)
Leodamas verax	Elias et al. (2004), Bravo et al. (2018) and Fiori (2018)
Lepidonotinae	Fiori (2018)
Lumbrineriopsis mucronata	Bravo et al. (2018)
Lumbrineris tetraura	Elias et al. (2004), Bravo et al. (2018) and Fiori (2018)
Malacoceros sp.	Bravo et al. (2018) and Fiori (2018)
Melinna uruguayi	Bravo et al. (2018)
Myxicola infundibulum	Elias et al. (2004)
Neanthes sp.	Elias et al. (2004)
Nephtyidae	Elias et al. (2004)
Nereididae	Elias et al. (2004), Bravo et al. (2018) and Fiori (2018)

(continued)

Table 9.1 (continued)

Species	Source
Nicolea chilensis	Elias et al. (2004)
Ninoe brasiliensis	Bremec et al. (1990)
Ninoe falklandica	Elias et al. (2004)
Onuphis setosa	Bravo et al. (2018) and Fiori (2018)
Opheliidae	Elias et al. (2004) and Fiori (2018)
Paraprionospio sp.	Bremec et al. (1990)
Paraonidae	Elias et al. (2001), Bravo et al. (2018) and Fiori (2018)
Pectinaridae	Fiori (2018)
Phylo felix	Elias et al. (2004)
Polydora sp.	Bravo et al. (2018) and Fiori (2018)
Scolecolepides sp.	Elias et al. (2004)
Sabellariidae	Elias et al. (2004)
Sabellaria nanella	Fiori (2018)
Serpulidae	Fiori (2018)
Spionidae	Elias et al. (2004) and Fiori (2018)
Syllidae	Elias et al. (2004)
Syllis sp.	Bravo et al. (2018) and Fiori (2018)
aff. Telephus sp.	Fiori (2018)
Terebella plagiostoma (ex Telephus plagiostoma)	Elias et al. (2004) and Bravo et al. (2018)
Terebellides totae	Elias et al. (2004), Bravo et al. (2018) and Fiori (2018)
Travisia sp.	Bremec et al. (1990)
Tricobranchidae	Fiori (2018)
Arthropoda	
Artemesia longinaris	Fiori (2018)
Austinixa patagoniensis	Elias and Ieno (1993), Elias et al. (2004) and Fiori (2018)
Cyrtograpsus affinis	Fiori (2018)
Cyrtograpsus altimanus	Elias (1992), Elias and Ieno (1993), Elias et al. (2004), Bravo et al. (2018) and Fiori (2018)
Cyrtograpsus angulatus	Elias and Ieno (1993), Elias et al. (2004) and Fiori (2018)
Loxopagurus loxochelis	Bremec (1990)
Neohelice granulata	Elias and Ieno (1993) and Elias et al. (2004)
Pagurus criniticornis	Bremec (1990)), Bravo et al. (2018) and Fiori (2018)
Pagurus exilis	Bremec (1990)
Pagurus sp.	Elias (1992), Elias and Ieno (1993) and Elias et al. (2004)
Peisos petrunkevitchi	Bravo et al. (2018)
Pilumnus reticulatus	Elias et al. (2004) and Fiori (2018)
Pinnotheres sp.	Elias and Ieno (1993) and Elias et al. (2004)

(continued)

Table 9.1 (continued)

Species	Source
Pleoticus muelleri	Fiori (2018)
Arthromysis magellanica	Bravo et al. (2018) and Fiori (2018)
Neomysis americana	Bravo et al. (2018) and Fiori (2018)
Ampithoe sp.	Fiori (2018)
Caprella equilibra	Fiori (2018)
Heterophoxus sp.	Fiori (2018)
Heterophoxus videns	Fiori (2018)
Monocorophium insidiosum (= *Corophium* sp.)	Elias and Ieno (1993), Elias et al. (2004), Bravo et al. (2018) and Fiori (2018)
Anthuridae	Bremec (1990)
Rectarcturidae	Fiori (2018)
Cristaserolis marplatensis (= *Serolis marplatensis*)	Bremec (1990)
Idotea sp.	Fiori (2018)
Macrochiridothea sp.	Bremec (1990)
Sphaeroma serratum	Fiori (2018)
Synidotea marplatensis	Fiori (2018)
Thysanoserolis elliptica (= *Serolis elliptica*)	Fiori (2018)
Monokalliapseudes schubarti (= *Kalliapseudes schubarti*)	Elias et al. (2001) and Bremec et al. (2017)
Anoplodactylus sp.	Bravo et al. (2018) and Fiori (2018)
Mollusca	
Aclis sp. (*ex Pherusa* sp.)	Bremec (1990)
Adrana electa	Bremec (1990), Elias et al. (2004) and Fiori (2018)
Amiantis purpurata	Bremec (1990)
Anachis aff. Isabellei	Bravo et al. (2018)
Barnea sp.	Fiori (2018)
Bostrycapulus odites	Fiori (2018)
Brachidontes rodriguezii	Fiori (2018)
Buccinanops cochlidium (ex Buccinanops gradatus)	Bremec (1990)
Buccinanops deformis (ex Buccinanops globulosus)	Bremec (1990), Elias et al. (2004), Bravo et al. (2018) and Fiori (2018)
Buccinanops monilifer (ex Dorsanum monilifer)	Bremec (1990)
Carditamera plata	Elias et al. (2004) and Fiori (2018)
Chaetopleura angulata	Fiori (2018)
Columbellidae	Fiori (2018)
Corbula patagonica	Bremec (1990), Elias et al. (2004) and Fiori (2018)
Crepidula protea	Fiori (2018)
Crepidula unguiformis	Bremec et al. (1990)

(continued)

Table 9.1 (continued)

Species	Source
Duplicaria gemmulata (ex Terebra gemmulata)	Bremec et al. (1990)
Ennucula puelcha	Fiori (2018)
Entodesma patagonicum	Fiori (2018)
Epitonium georgettinum	Fiori (2018)
Eurytellina gibber	Bremec (1990) and Elias et al. (2004)
Glycera americana	Fiori (2018)
Heleobia australis	Elias et al. (2004) and Bravo et al. (2018)
Kellia suborbicularis	Fiori (2018)
Lyonsia alvarezii	Elias et al. (2004) and Fiori (2018)
Mactra marplatensis	Bremec (1990)
Macoma uruguayensis	Elias et al. (2004)
Malletia cumingii	Elias et al. (2004), Bravo et al. (2018) and Fiori (2018)
Mitrella moleculina (ex Anachis moleculina)	Bremec (1990)
Notocochlis isabelleana (ex Natica isabelleana)	Bremec (1990)
Nucula semiornata	Elias et al. (2004) and Fiori (2018)
Olivancillaria carcellesi	Bremec (1990)
Olivella puelcha (ex Olivella plata)	Bremec (1990) and Fiori (2018)
Olivella tehuelcha	Elias et al. (2004)
Ostrea stentina	Bravo et al. (2018) and Fiori (2018)
Parvanachis paessleri	Elias et al. (2004)
Periploma compressum	Elias et al. (2004)
Pitar rostratus	Bremec et al. (1990), Elias et al. (2004), Bravo et al. (2018) and Fiori (2018)
Pleurobranchaea inconspicua	Fiori (2018)
Polyschides tetraschistus (ex Cadulus quadridentatus)	Bremec (1990)
Marginella prunum	Bremec (1990)
Solen tehuelchus	Bremec (1990)
Sphenia fragilis	Elias et al. (2004) and Fiori (2018)
Thracia sp.	Elias et al. (2004)
Tagelus plebeius	Fiori (2018)
Tellina petitiana	Fiori (2018)
Transenpitar americana	Fiori (2018)
Echinodermata	
Amphioplus albidus	Bremec et al. (1990)
Amphiura crassipes	Bremec et al. (1990)
Amphiura eugeniae	Fiori (2018)
Encope emarginata	Bremec et al. (1990)
Ophioplocus januarii	Elias et al. (2004)

(continued)

Table 9.1 (continued)

Species	Source
Amphipholis squamata	Brogger pers. com.
Cnidaria	
Amphisbetia operculata	Fiori (2018)
Clytia sp.	Fiori (2018)
Laomedea (Obelia) sp.	Fiori (2018)
Obelia spp.	Bravo et al. (2018)
Obelia bicuspidata	Bremec et al. (1990)
Plumularia setacea	Bremec et al. (1990)
Stylatula darwini	Elias et al. (2004) and Bravo et al. (2018)
Symplectoscyphus sp.	Fiori (2018)
Nemertea	
Nemertino indet	Elias et al. (2004)
Priapulida	
Priapulus tuberculatospinosus	Elias et al. (2004)
Bryozoa	
Biflustra puelcha (ex Membranipora puelcha)	Bremec (1990)
Amathia imbricata	Bravo (2019)
Scruparia ambigua	Bravo (2019)
Anguinella palmata	Bravo et al. (2018) and Fiori (2018)
Bugula neritina	Bravo et al. (2018)
Bugulina simplex	Bravo et al. (2018)
Bugulina stolonifera	Bravo et al. (2018) and Fiori (2018)
Celleporella hyalina	Bremec et al. (1990)
Conopeum sp.	Bravo et al. (2018) and Fiori (2018))
Crisia sp.	Bravo et al. (2018) and Fiori (2018)
Electra monostachys	Bremec et al. (1990)
Membranipora sp.	Bremec (1990) and Fiori (2018)

mortalities of other species. The association of *Corbula patagonica* with other bivalves such as *Ennucula puelcha*, *Malletia cumingii*, and *Macoma uruguayensis* was found in the Bahía Blanca Estuary and in the coast of Uruguay (Scarabino et al. 2006), while *C. patagonica* was associated with the bivalves *Mactra marplatensis*, *Angulus gibber*, *Adrana electa*, *Solen tehuelchus*, and *Pitar rostratus* on the adjacent continental shelf (Bremec 1990). Special attention has been paid to the taxonomic identification of the Nudibranchia *Pleurobranchaea inconspicua* in the Bahía Blanca Estuary, due to the registration of the neurotoxin TTX (tetrodotoxin) in specimens of *Pleurobranchaea* sp. aff. *Maculata* in Puerto Quequén (Farias et al. 2015), a coastal site very close to the Bahía Blanca Estuary (Fig. 1.2; Chap. 1). Intake of only 1–2 mg of TTX can cause death in adult humans (Noguchi and Arakawa 2008). High levels of toxins have been found in specimens of all ages of *Pleurobranchaea maculata* (McNabb et al. 2010), but so far no TXT has been reported in specimens of *P. inconspicua*.

Fig. 9.4 Polychaetes found in subtidal soft-bottoms of the Bahía Blanca Estuary (**a**) *Terebellides totae* (**b**) *Leodamas verax* (**c**) *Corbula patagonica* (**d**) *Anachis* sp. (**e**) *Amphipholis squamata* (**f**) *Anoplodactylus* sp. (Photos by ME Bravo)

Crustaceans represent the third group in abundance and species richness, after polychaetes and mollusks, with 30 species distributed in 6 sites along the Principal Channel (Fiori 2018). The only species of crustaceans with a notable abundance was the amphipod *Corophium* sp. in shallow subtidal areas of the in the inner zone of the estuary, where it reached densities of up to 650 individuals/m^2 (Elias and Ieno 1993). This low representation of crustaceans in the Bahía Blanca Estuary contrasts with the adjacent continental shelf, "El Rincón." In this area, crustaceans are dominant in abundance, with tanaid (*Kalliapseudes schubarti*, sensu Elias et al. 2001; Bremec et al. 2017) being the dominant species (Bremec 1990), reaching a dominance of ~83%.

To date, only limited information is available on the temporal changes in the ecological patterns of the benthic subtidal communities in the Bahía Blanca Estuary, which includes some analyses of seasonal variation (Elias et al. 2004; Bravo 2019). The highest values of density, biomass, richness, and diversity of species are found during the spring (Elías et al. 2004; Bravo 2019), after the winter phytoplankton bloom (Guinder et al. 2015). As in other temperate estuaries (Chainho et al. 2006; Van der Linden et al. 2012; Quiroga et al. 2016), the maximum values of density, biomass, species richness, and diversity are usually found in spring and summer, after the blooms of primary producers (Chainho et al. 2006; Quiroga et al. 2016). We are currently at the beginning of baseline studies, conducted systematically since 2013 (see Chap. 19). The continuation of these studies will facilitate the evaluation of temporal variations in the medium term. Alternatively, sampling with methods comparable to those of the studies carried out by Elias (1987) would make it possible to evaluate some of the main changes that have occurred in recent decades.

9.4 Functional Ecology of Macrobenthic Species

Several studies focused on understanding the ecological functions of species and their distribution patterns along environmental gradients in estuaries (e.g., van der Linden et al. 2012; van der Wal et al. 2017a), mainly with coastal management and conservation aims. This concept was introduced in conservation biology by Ehrlich and Walker (1998) who proposed the protection of redundant functional groups in order to maximize ecosystem resilience. Quantitative analysis for assessing the functional redundancy may serve as a proxy of the resilience capacity of coastal ecosystems, under the current scenario of increasing anthropogenic and climatic pressures (Dissanayake et al. 2018). Different organization levels could be affected by the loss of species within functional groups, for example, their specific ability to adapt to environmental changes on different spatial scales and the productivity realized for their community as well as the ecosystems services in which they are involved (Ehrlich and Walker 1998; Wieters et al. 2012).

The abundance data of soft-bottom subtidal macrobenthic invertebrates from Fiori (2016) was classified in function of different biological traits in order to analyze their variability along the Principal Channel. Information on the biological traits of the different taxa has been sourced from previous analyses made in the inner zone of the Bahía Blanca Estuary by Bravo (2019), as well as from taxa-specific literature or syntheses and suitable databases, e.g., polytraits (Faulwetter et al. 2014). Biological traits included feeding modes considering the categories subsurface deposit-feeder, surface deposit-feeder, omnivore, carnivore, filter-feeder, herbivore, and bacterial grazer; body size classified as small (<1 cm), medium (1–5 cm), or large (>5 cm); motility classified as discretely motile, motile, or sessile; calcification classified as non-calcified, heavily calcified, or lightly calcified; and lifestyle classified as burrower, errant, tube builder, or attached (Table 9.2). The category "not assigned" was used for classification of taxa of low taxonomic

Table 9.2 Main taxa representing the biological traits with the categories to which they were assigned

Biological traits	Categories	Main taxa
Feeding mode	Subsurface deposit-feeder	*Leodamas verax*, Flabelligeridae undet., *Aricidea* sp., *Malletia cumingii*, *Ennucula puelcha*, *Nucula semiornata*, *Tellina petitiana*
	Surface deposit-feeder	*Onuphis setosa*, *Polydora* sp., *Malacoceros* sp., *Aphelochaeta* sp., *Axiothella* sp., *Terebellides totae*, *Terebella plagiostoma*, Dorvilleidae undet., *Melinna uruguayi*, *Heleobia australis*, *Adrana electa*, *Cyrtograpsus affinis*, *Cyrtograpsus altimanus*, *Serolis elliptica*, *Heterophoxus videns*, *Neomysis americana*, *Arthromysis magellanica*
	Omnivore	*Laeonereis acuta*, *Glycera americana*, *Lumbrineris tetraura*, *Ninoe brasiliensis*, *Lumbrineriopsis mucronata*, Lepidonotinae undet., *Buccinanops deformis*, *Pagurus criniticornis*, *Idotea* sp., *Monocorophium insidiosum*, *Anoplodactylus* sp., Nematoda undet.
	Carnivore	*Syllis* spp., *Epitonium georgettinum*, *Pleurobranchaea* sp., *Olivella* sp.
	Filter/suspension-feeder	*aff. Chone* sp., *Myxicola* sp., *Sabellaria nanella*, Serpulidae undet., *Corbula patagonica*, *Pitar rostratus*, *Stylatula darwini*, *Ophiura* sp.
	Herbivore	*Chaetopleura angulata*
	Bacterial grazer	*Eteone* sp., *Barnea* sp.
Body size	Small	*P. cornuta*, *Syllis* spp., *Aphelochaeta* sp., *Aricidea* sp., Dorvilleidae undet., *E. georgettinum*, *E. puelcha*, *N. semiornata*, *Eteone* sp., *H. australis*, *H. videns*, *M. insidiosum*, Nematoda undet.
	Medium	*L. verax*, *L. acuta*, *G. americana*, *O. setosa*, *L. tetraura*, *L. mucronata*, *N. brasiliensis*, *Malacoceros* sp., Flabelligeridae undet., *Axiothella* sp., *aff. Chone* sp., *T. totae*, *T. plagiostoma*, Lepidonotinae undet., *M. uruguayi*, *Myxicola* sp., *M. cumingii*, *T. petitiana*, *C. patagonica*, *B. deformis*, *P. rostratus*, *Barnea* sp., *A. electa*, *C. angulata*, *Olivella* sp., *C. affinis*, *C. altimanus*, *P. criniticornis*, *Idotea* sp., *Anoplodactylus* sp., *N. americana*, *A. magellanica*
	Large	*Pleurobranchaea* sp., *S. darwini*, *Ophiura* sp.
Motility	Discretely motile	*L. verax*, *Malacoceros* sp., *Aricidea* sp., *Aphelochaeta* sp., *Axiothella* sp., *E. puelcha*, *N. semiornata*, *M. cumingii*, *T. petitiana*, *C. patagonica*, *P. rostratus*, *Barnea sp.*, *A. electa*
	Motile	*L. acuta*, *G. americana*, *L. tetraura*, *L. mucronata*, *N. brasiliensis*, *Syllis* spp., *Eteone* sp., Lepidonotinae undet., *E. georgettinum*, *Pleurobranchaea* sp., *B. deformis*, *C. angulata*, *Olivella* sp., *H. australis*, *C. affinis*, *C. altimanus*, *P. criniticornis*, *Idotea* sp., *H. videns*, *Anoplodactylus* sp., *N. americana*, *A. americana*, Nemertea undet., Nematoda, *Ophiura* sp.
	Sessile	*P. cornuta*, *O. tenuis*, Flabelligeridae undet., *aff. Chone sp.*, *S. nanella*, *T. totae*, *T. plagiostoma*, Serpulidae undet., *M. insidiosum*, *S. darwini*

(continued)

Table 9.2 (continued)

Biological traits	Categories	Main taxa
Calcification	Non-calcified	*L. verax*, *L. acuta*, *G. americana*, *L. tetraura*, *L. mucronata*, *N. brasiliensis*, *Syllis* spp., *Eteone* sp., Lepidonotinae undet., *Malacoceros* sp., *M. uruguayi*, *Myxicola* sp., *Aricidea* sp., *Aphelochaeta* sp., *Axiothella* sp., *P. cornuta*, *O. tenuis*, Flabelligeridae undet., *aff. Chone sp.*, *S. nanella*, *T. totae*, *T. plagiostoma*, Serpulidae undet., Nemertea undet., Nematoda., *Pleurobranchaea* sp.
	Heavily calcified	*E. georgettinum*, *E. puelcha*, *N. semiornata*, *M. cumingii*, *T. petitiana*, *C. patagonica*, *P. rostratus*, *Barnea* sp., *A. electa*, *B. deformis*, *C. angulatus*, *Olivella* sp., *C. affinis*, *C. altimanus*, *Ophiura* sp.
	Lightly calcified	*P. criniticornis*, *Idotea* sp., *H. videns*, *M. insidiosum*, *Anoplodactylus* sp., *N. americana*, *A. magellanica*, *S. darwini*
Lifestyle	Burrower	*L. verax*, *G. americana*, *Malacoceros* sp., *L. tetraura*, *L. mucronata*, *N. brasiliensis*, *Aricidea* sp., *Aphelochaeta* sp., *E. puelcha*, *N. semiornata*, *C. patagonica*, *P. rostratus*, *Barnea* sp., Nematoda undet.
	Errant	*L. acuta*, *Syllis* sp., *Eteone* sp., Lepidonotinae undet., *E. georgettinum*, *Pleurobranchaea* sp., *B. deformis*, *C. angulata*, *Olivella* sp., *H. australis*, *C. affinis*, *C. altimanus*, *P. criniticornis*, *Idotea* sp., *H. videns*, *Ampithoe* sp., *Anoplodactylus* sp., *N. americana*, *A. magellanica*, Nemertea undet., *Ophiura* sp.
	Tube builder	*O. setosa*, *P. cornuta*, *M. uruguayi*, Flabelligeridae undet., *Axiothella* sp., aff. *Chone* sp., *Myxicola* sp., *S. nanella*, *T. totae*, *T. plagiostoma*, Serpulidae undet., *M. insidiosum*
	Attached	*B. rodriguezii*, *S. darwini*

resolution and/or a lack of taxa-specific literature. The percentage of a given biological trait category was calculated for each station along the Principal Channel, and the results are shown in Fig. 9.5.

The characteristics that were equally represented in the macrobenthic organisms along the Principal Channel were the deposit-feeding diet and the absence of calcified structures, while the remaining biological traits showed differences between the sites (Fig. 9.5). There were marked differences between the inner zone (stations 1, 2, and 3) and the outer zone (stations 4, 5, and 6). For example, the functional structure of the inner zone is dominated by small, discretely mobile, and burrowing organisms, whereas the relative importance of medium-sized, mobile organisms increases towards the outer zone (Fig. 9.5). The feeding guilds of benthic fauna from the Bahía Blanca Estuary are mainly represented by subsurface- and surface deposit-feeders, omnivores, filter/suspension-feeders, and carnivores. The greatest trophic diversity is found at stations 1 and 6, which are the innermost and the most external ones. Other biological traits (body size, motility, calcification, lifestyle) show greater diversity and functional evenness from the middle zone (stations 3 and 4) to the outer zone (stations 5 and 6).

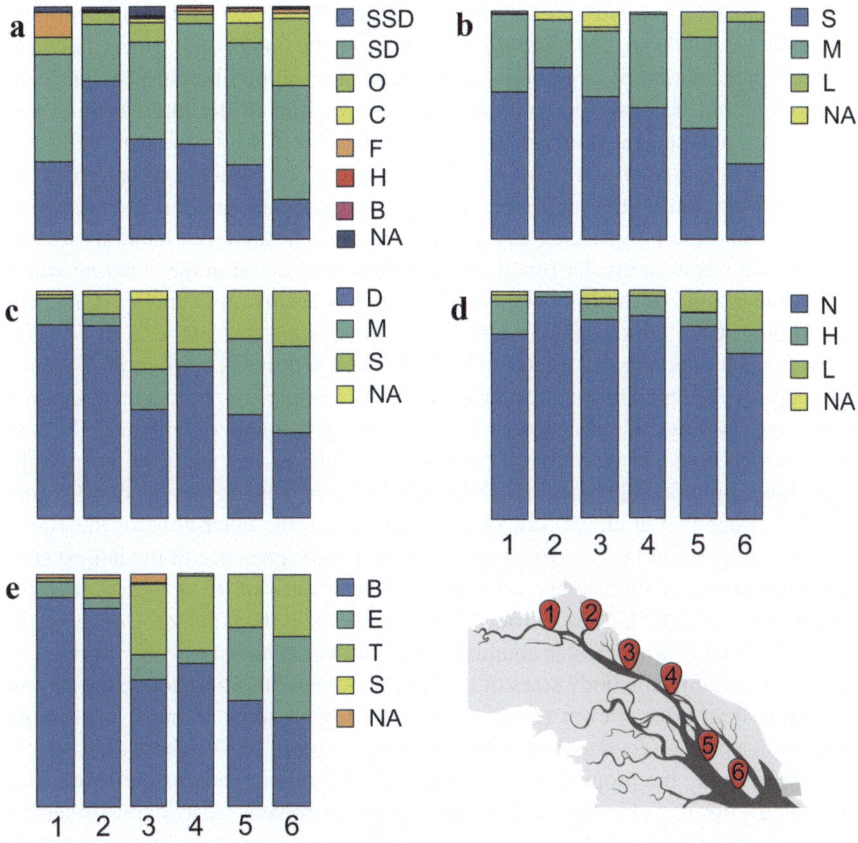

Fig. 9.5 Percentage of a given trait category at a given station, data from Fiori (2016). (**a**) Feeding mode: subsurface deposit-feeder (SSD); surface deposit-feeder (SD); omnivore (O); carnivore (C); filter feeder (F); herbivore (H); bacterial grazer (B); not assigned (NA). (**b**) Body size: small (S); medium (M); large (L); not assigned (NA). (**c**) Motility: discretely motile (D); motile (M); sessile (S); not assigned (NA). (**d**) Calcification: non-calcified (N); heavily calcified (H); lightly calcified (L); not assigned (NA). (**e**) Lifestyle: burrower (B); errant (E); tube builder (T); attached (S)

Although the mechanisms and quantification of the effects of environmental parameters, such as salinity, sediment grain size, organic matter content, and hydrodynamics, need to be better understood, it is already clear that they play an important role in setting up both the taxonomic and the functional composition of benthic communities. Some considerations below are still hypothetical and serve as illustrations of ecological implications, but they are already supported by several studies and may serve as potential research targets.

As well as the species-specific relationships of preference/tolerance with each of the environmental variables that generate changes in community structure in taxonomic terms, they also do so in functional terms. However, dominant diets did not vary significantly along the gradient even with the replacement of dominant species

(*Nicolea* sp. and *Leodamas verax* in the inner zone, *Terebellides totae* in the middle zone, and *Axiothella* sp. and *Corbula patagonica* in the outer zone). Nevertheless, a greater representation of large body-sized organisms with a filter/suspension feeding mode could be observed, such as *Stylatula darwini* in the internal zone and *Ophiura* sp. and several bivalves in the external zone. Also, in the external zone, there would be a greater representation of omnivorous and carnivorous organisms of low abundance but larger body size, like some crustaceans, such as *Cyrtograpsus altimanus*, and the polychaetes *Glycera americana*. On the other hand, as for the taxonomic analysis, a greater functional diversity is observed in the euhaline zone.

The inner zone of the Bahía Blanca Estuary has relative low energy (Ginsberg and Aliotta 2011) which may benefit the benthic fauna that are able to feed on highly concentrated organic matter in fine grain size sediments (surface and subsurface deposit-feeders, burrowing organisms), in comparison to other functional groups. At low flow rates, drag forces are reduced together with mixing rates (Nikora 2010), which coincide with a high content of organic matter and stability of substrate, favoring infaunal organisms characterized by burrowing and deposit-feeding habits (van der Wal et al. 2017a). On the other hand, the outer zone of the Bahía Blanca Estuary has relative high energy, reflected in the presence of the largest subtidal sand dunes of the estuary migrating at mean rates of 43 m year $^{-1}$ (Minor Salvatierra et al. 2015). These strong bottom currents in the outer zone of the estuary are associated with greater abundance of omnivores and carnivores, as well as the higher mobility and body sizes of the benthic fauna. Strong bottom currents can have large drag effects, even removing and destroying some animals, while high mixing rates provide food and an efflux of wastes (Nikora 2010). These drag effects may explain why the proportion of mobile and/or errant organisms increases with increasing current velocities relative to that of burrowers. Also deposit-feeding organisms are in low abundance in high-energy regions as the drag effects may flush them out (Nikora 2010; van der Wal et al. 2017a), whereas the high mixing rates can supply enough energy to sustain a higher diversity of feeding guilds (Nikora 2010; van der Wal et al. 2017a). However, the patterns observed only represent momentary conditions in the Bahía Blanca Estuary as hydrodynamic patterns in estuaries change constantly, associated with both natural and anthropogenic drivers, such as dredging activities (van der Wal et al. 2017a). It is expected that the distribution of the ecological patterns linked with hydrodynamic characteristics will change together with the morphology and hydrodynamics in the estuary.

Biological traits are proxies for understanding how organisms play crucial roles in marine ecosystems (Dissanayake et al. 2018). In estuaries, different ecosystem functions are regulated by biophysical processes mediated by benthic organisms such as bioturbation, bioirrigation, and biogeochemical processes (van der Wal et al. 2017a). Some macrobenthic organisms with biological traits such as burrowing habits, deposit-feeding strategies, or simply having highly calcified bodies may modify the physical structure of their own habitats, becoming ecosystem engineers (Kristensen et al. 2014; Tait et al. 2020). These organisms add environmental complexity, contributing disproportionally to ecosystem functions, and may buffer anthropogenic stressors (Xie et al. 2018; Tian et al. 2019; Tait et al. 2020). Rather than species richness per se, the presence, abundance, and biomass of ecosystem engineers are the main drivers of ecosystem function (Tait et al. 2020). In this sense, identification of the biological traits of benthic macrofauna involved in ecosystem functions (e.g., bioturbation, habitat-formers) may help to improve the environmental management of estuaries.

Burrowing and deposit-feeding are biological traits associated with bioturbation (Kristensen et al. 2014; Dissanayake et al. 2018). These biological traits are dominant in the muddy sediments in the inner zone of the Principal Channel of the Bahía Blanca Estuary and may change the oxygenation (redox potential), resuspension, cycling of organic matter, nutrients, and pollutants (Dean 2008). The changes in transport and biogeochemistry and, consequently, in the exchange of solutes between the water-sediment interphase are related to burrowing organisms through the construction and maintenance of burrows, as well as the ingestion and defecation of particles. Ventilation leads to rapid transport of solutes and is mediated by the water currents generated by the respiration and feeding of animals (Kristensen et al. 2014). Consequently, subsurface organic particles are transferred to the surface oxide sediments, and vice versa, accelerating the degradation of organic matter at the surface and favoring its resuspension by erosion (Kristensen et al. 2014). These functional groups are some of the first in the ecological succession (Dean 2008) to colonize environments enriched with organic matter, such as those found at sites affected by sewage discharge in the Bahía Blanca Estuary (Fiori et al. 2020).

Metal concentrations are positively correlated with fine grain size sediments in the Bahía Blanca Estuary (Marcovecchio et al. 2010; Fernández Severini et al. 2018). The inner zone of the estuary is particularly sensitive to the accumulation of pollutants, driven by a combination of high anthropogenic pressure and a low rate of water renewal. Although abiotic conditions are favorable for the precipitation of contaminants on the seabed, a high load of bioturbator organisms favors their resuspension into the water column (Kristensen et al. 2014; Xie et al. 2018; Tian et al. 2019). High rates of bioturbation carried out by burrowers and deposit-feeding benthic fauna in the inner sector of the estuary are expected to affect the dynamics of biogeochemical processes, as well as promote release, resuspension, and bioavailability of pollutant substances (Xie et al. 2018; Tian et al. 2019). In addition, as subtidal macrobenthos constitute key prey items for fish of commercial interest, such as *Micropogonias furnieri* and *Mustelus schmitti*, among others (Lopez Cazorla 2004), they may be driving the bioaccumulation of pollutants and even affecting human health (Kristensen et al. 2014; Barletta et al. 2019; Fiori et al. 2020).

Dominance of bioturbators observed at stations 1 and 2 along the Principal Channel can exclude some infaunal species due to changes in substrate conditions. On the one hand, they can make the substrate more unstable (enhancing erosion) and therefore exclude some infaunal species (such as discrete mobility or sessile fauna). On the other hand, the enhanced resuspension of particles may exclude fauna with filter/suspension feeding habits (Dean 2008; Kristensen et al. 2014). For example, some burrowing polychaetes, such as *Arenicola marina*, dramatically affect the species composition at the local scale through vigorous sediment mixing, strong disturbance, and competition for food (Kristensen et al. 2014). It is possible that similar effects are associated with the high load of bioturbator organisms at stations in the inner zone of the Bahía Blanca Estuary.

The resuspension of nutrients to the water column by burrowing species may be important in the inner zone of the Bahía Blanca Estuary, where otherwise low velocity bottom water currents promote their deposition (Carbone et al. 2016). The reverse state can be promoted by suspension-feeders that cause particle deposition and clear water (Kristensen et al. 2014). Burrowing and deposit-feeders are functional groups with higher tolerance to high loads of nutrients and metals than suspension feeders (Ellis et al. 2017). This may explain the low densities of suspension feeders in sectors with high food availability but that are polluted and or eutrophicated (Carbone et al. 2016) such as stations 2 and 4 in the Principal Channel.

The ecosystem engineering roles mentioned above are known as allogenic engineering due to physical alteration of the environment mechanically or chemically (Tait et al. 2020), whereas physical alteration of the environment by the bodies of benthic organisms, such as habitat building, is known as autogenic engineering (Tait et al. 2020). Heavily calcified and tube-building benthic organisms have biological traits typical of autogenic engineering organisms (Tait et al. 2020). These traits are mainly found in sandy sediments in the external part of the Bahía Blanca Estuary, and they enhance the settlement of sessile epifauna, such as bryozoans and hydrozoans. However, they are not represented by gregarious organisms, so this would not lead to any significant alterations in the three-dimensional configuration of the subtidal environment as may occur with some mollusks (Nikora 2010).

9.5 Acoustic Approach in Subtidal Benthic Habitats

Analysis of large-scale variation in the ecological patterns of macrobenthic communities would benefit of including acoustic classification of the seafloor. These acoustic techniques are widely used by marine geologists and geophysicists to map morphological features of the seabed with high spatial coverage in a short time (Anderson et al. 2008). Several studies find a good fit between seabed acoustic classification and ecological patterns of benthic communities and are therefore increasingly used for benthic habitat mapping (Anderson et al. 2008; van der Wal et al. 2017a, b; Mestdagh et al. 2020).

In the Bahía Blanca Estuary, acoustic anomalies derived from the presence of hydrocarbon deposits (shallow gas) on the seafloor were found to be associated with local changes in macrobenthic communities (see Box 9.1; Bravo et al. 2018, Bravo et al. 2020). It is expected that larger local changes in the ecological patterns than those described in this chapter were found associated with patchiness of morphosedimentary features of the estuarine seabed. Secondary channels may have a combination of muddy sediment patches alternating with sandy sediment patches with or without sand dunes of different dimensions (Ginsberg et al. 2009). It is possible that these patches behave as different benthic habitats because of their different textures and morphodynamics. In the estuary, sand dunes are usually small in the inner zone (Gómez et al. 1996) and in secondary channels (Ginsberg et al. 2009) and large in the external zone (Ginsberg and Aliotta 2011; Minor Salvatierra et al. 2015). These dunes provide good information about the dynamics of the bottom currents (Ginsberg and Aliotta 2011; Minor Salvatierra et al. 2015), also affecting the benthic communities of macroinvertebrates (Nikora 2010; Van der Wal et al. 2017b). Other sectors, mainly in dredged sectors and at the intersection of channels with intensified currents, may have patches with rocky outcrops hosting sessile benthic macrofauna (Aliotta and Lizasoain 2004; Ginsberg et al. 2009). There is also a wide range of bathymetric variations in the subtidal of the estuary, and benthic communities usually respond to this variable. It is common to observe deep-holes in areas of channel intersection, reaching relative relief of 2–17 m from the bottom of the channels (Ginsberg and Perillo 1999). Overall, this highlights the intersections of tidal channels as morphodynamically complex subtidal environments, with the potential to host biodiversity hot spots (Van der Wal et al. 2017a). At the same time, it highlights the potential impact of the deposition of dredging material at channel intersections, as occurred during 2013 at the mouths of secondary channels on the Principal Channel (Ginsberg et al. 2014).

9.6 Impacts and Conservation Issues

The impacts of human activities are cumulative and affect all parts of the oceans at different scales, but the greatest impacts have been found in coastal and continental shelf environments (Harris and Hughes 2012). Estuaries are among the most impacted coastal ecosystems, being affected by a wide range of anthropogenic activities as they serve as ports and waterways to urban and industrial centers (Saiz-Salinas and González-Oreja 2000). Kennish et al. (2014) identified a wide range of anthropogenic factors that degrade and damage estuary ecosystems in the face of climate change. These include nutrient loading and eutrophication, wastewater and organic waste, habitat loss and alteration, chemical pollutants, sediment particle inputs, overfishing, intensive aquaculture, introduced/invasive species, and altered hydrological regimes. Human activities can have a cumulative impact on the structure, function, and ecological health of estuaries (Kennish et al. 2014). Affected estuaries are expected to become less resilient to climate variability, most likely as a result of the loss of biological and functional diversity (Dolbeth et al. 2007).

In the Bahía Blanca Estuary, the main anthropogenic pressures are related to physical (dredging and deposition of dredging material), chemical (untreated urban and industrial effluent discharge), and biological (introduction of exotic species) impacts. Since benthic macroinvertebrates have a rapid and specific response to different types of stress, they are excellent indicators of environmental quality (Pearson and Rosenberg 1978; Guidetti et al. 2000; Hampel et al. 2009; Dauvin et al. 2012). Understanding the interaction of benthos and environmental disturbances could be useful for identifying areas of high environmental susceptibility by facilitating management (Muniz et al. 2013; Dissanayake et al. 2018). This is even more important considering the ecological role of the benthic fauna in relation to other biological groups in the pelagic fraction.

The Bahía Blanca Estuary is the most important deepwater port system in Argentina (see Chap. 2). This requires periodic maintenance of the Principal Channel by dredging and the deposition of dredging material in intertidal areas for the filling and expansion of the port complex (Ginsberg et al. 2014). These tasks have been carried out for more than 30 years, with the most important work done in 1989, 1991, and 2013. Both dredging and the deposition of dredging material have led to drastic changes in the morphology and sedimentology of the estuarine system (Ginsberg et al. 2014). Most of the 2013 dredging deposits were made at the mouth of secondary channels. This disrupts the natural hydrodynamics and balance between erosion and deposition in these channels that normally function as a hydrodynamic network (Ginsberg et al. 2014). Such changes are expected to alter the ecological patterns of large-scale benthic communities. At the local level, stabilization of dredging material deposits is associated with changes in the grain size of shallow subtidal sediments (higher representation of coarse sands, gravels, and poorer selection) related to lower specific richness and biomass, as well as changes in the species composition of macrobenthic communities (Fiori et al. 2020). They have also been associated with increased turbidity in the water column and higher concentrations of metals (chromium, nickel, iron, and lead; Fiori et al. 2020). In addition to physically altering the benthic habitat, dredging could promote resuspension and/or translocation of contaminating substances trapped on the seafloor (La Colla et al. 2018).

Through the monitoring program that evaluates the environmental quality of the estuary (see Chap. 19), pilot sampling has been carried out to learn about the ecological status of the subtidal macrobenthic communities in channels affected by industrial wastewater and petrochemical discharges (Fiori 2018). This study represented the first integrated approach to explore the impact of anthropogenic activities on the macrobenthic assemblages in the estuary. Preliminary results showed that the sediments of these channels contain metals, polycyclic aromatic hydrocarbons (PAH), hydrocarbon-degrading bacteria, and the fecal coliform bacteria *Escherichia coli* (Fiori 2018). As expected, biological descriptors of the benthic community in channels affected by anthropogenic effluents indicate that the community was impoverished in terms of abundance and diversity. In the channel where the industrial effluents are discharged, three species were numerically dominant: the polychaetes *Laeonereis acuta* and *Malacoceros* sp., and the clam *Tagelus plebeius* (Fiori 2018). These polychaetes are deposit-feeding organisms, and the clam is a suspension-feeding organism (Holland and Dean 1977). In addition, *T. plebeius* is a deep-digging species that inhabits permanent caves (up to 75 cm deep) that show vertical movements in each tidal cycle (Holland and Dean 1977). These species can affect the remobilization of some chemical compounds by their feeding mode or burrowing activity, increasing the release of contaminants from the sediment to the water column (Schaller 2014) or moving contaminants associated with suspended matter from the water column to the sediment (Klerks et al. 1997). These changes in the environmental distribution of contaminants will affect contaminant levels in other organisms (Klerks 2018).

The area closest to the wastewater effluent discharge, where sediments were anoxic with a high concentration of cadmium, lead, zinc, and copper, was dominated by *Laeonereis acuta* (Fiori et al. 2020). Polychaetes can respond to exposure to high concentrations of metals and toxins by forming metal granules, increasing the secretion of mucus or accumulation in specific areas of the body followed by exocytosis (Viarengo and Nott 1993; Gibbs et al. 2000; Berthet et al. 2003; Mouneyrac et al. 2003). According to toxicological tests, *L. acuta* exhibits some tolerance to cadmium exposure. This seems to be related to a high concentration of antioxidant enzymes that prevents damage caused by oxidative stress or formation of insoluble metal granules in its tissues for detoxification (Sandrini et al. 2006, 2008; Carricavur et al. 2018). Weis et al. (2017) showed that *L. acuta* is a species tolerant of synergisms of diffuse contaminants that occupies a niche space that is uncertain for other species more sensitive to contamination. The author found that the effects of contaminants on molecular levels occur earlier, which may encourage decision-making by managers and stakeholders in the use of preventive practices before damage occurs at the population or community level (Weis et al. 2017). Individuals of *L. acuta* in contaminated Brazilian estuaries were two to three times more exposed than those in undeveloped estuaries, suggesting that they may become more attractive prey to consumers and favor bottom-up transfer of pollution (biomagnification) along the estuarine food web (Weis et al. 2017).

Another threat to local biodiversity associated with port activities is the introduction of non-native species that can become invasive. Scientists and policymakers are increasingly considering invasive species as a major threat to the marine biodiversity in coastal environments (Occhipinti Ambrogi et al. 2011). The effects of invasion have been associated with decline in populations of indigenous species (Kappel 2005), changes in the food web (Oguz et al. 2008), and alterations to the structure and functioning of entire ecosystems (Vilà et al. 2011, Simberloff et al. 2013). The presence of non-native marine species in the Bahía Blanca Estuary has been documented from 1970 to the present, mainly associated with artificial hard substrate in the ports (bryozoans, Lichtschein and Bastida 1980; tunicates and crustaceans anemones, Molina et al. 2009; bivalves, Dos Santos and Fiori 2010). Many of these species became dominant in the biofouling communities, even replacing native biota. It has been proposed that accidental introduction through ship ballast water is the most likely route of entry of these species into the ecosystem. In contrast, the non-native species in the soft-bottom benthic communities of the Bahía Blanca Estuary are unknown, since the inventory is still being completed and only a fraction of the fauna has been identified at the species level (Fiori 2018). As already mentioned by other authors, the main source of uncertainty regarding biological invasions in the Southwest Atlantic is the poor documentation of marine biota (Orensanz et al. 2002).

The continental shelf adjacent to the Bahía Blanca Estuary, known as "El Rincón," concentrates demersal fisheries, and so it has been studied more intensely to characterize the feeding use of the seafloor by benthivorous fish of commercial interest (Bremec and Giberto 2017). However, in the Bahía Blanca Estuary, the capture of fish of commercial interest was historically carried out by artisanal fishermen, with little political-economic influence in comparison to that exerted by the petrochemical industry and port complex. This is reflected in the fact that both the fish stock and the artisanal fishermen are strongly threatened by the growing anthropogenic pressure in the area (Truchet et al. 2019; Speake et al. 2020). In turn, this threatens the natural, historical, and cultural heritage of the Bahía Blanca Estuary and associated urban centers (Speake et al. 2020). The combination of scientific research and the knowledge of artisanal fishermen could be the key for the correct development of management policies (Truchet et al. 2019; Speake et al. 2020). Research on the subtidal macrobenthic species could determine the quantity and quality of food that supports species of commercial interest, such as the whitemouth croaker (*Micropogonias furnieri*), narrownose shark (*Mustelus schmitti*), rays, and stripped weakfish (*Cynoscion guatucupa*) (Lopez Cazorla 2004; Bremec and Giberto 2017). This would not only allow the analysis of socioeconomic variables related to the quantity of fish stocks, but also it has health implications, especially considering that some of these species evidenced pollutant bioaccumulation of which the bioavailability is linked to their benthic preys (Oliva et al. 2017).

Assessing the dynamics of pollutants in relation to the macrobenthic community allows an ecosystem approach, which would include some ecosystem functions that are affected or used by humans. The results discussed in this chapter indicate that there are sectors within the estuary that are more sensitive than others in terms of the biological features of the species. The inner zone of the estuary would be particularly sensitive to pollution because of its low energy, shallow depth (high benthic-pelagic exchange), and dominance of bioturbating organisms. These organisms favor bioremediation of the environment but also the trophic transference and magnification of the pollutants. This would imply environmental and health sensitivity considering that this area is a feeding, spawning, and breeding area for species of conservation and fishing interest.

9.7 Closing Remarks

The sampling effort conducted over the last decade increased the number of species identified along the Principal Channel and smaller channels affected by human activities (see Chap. 19). Despite this progress, much research remains to be done to complete the inventory of the biodiversity of the subtidal macrobenthic communities. Most of the estuary remains unexplored, and some species may have not been discovered yet because they require further taxonomic research. For example, several species of jellyfish have been reported in the study area, but their polyps have never been found in benthic studies (Dutto et al. 2017). This also indicates that although many species in the Bahía Blanca Estuary are meroplanktonic, ecological research on plankton and benthos has been conducted separately. At this point, the analytical integration of the planktonic fraction would allow a holistic understanding of the ecological processes involving the subtidal macrobenthos, as with information regarding the marine geology of the area.

Box 9.1: Evaluation of the Effect of Shallow Gas on Benthic Communities in the Bahía Blanca Estuary

In several coastal ecosystems of the world, the increase in the sea level that occurred during Quaternary period generated the burial of continental organic matter by fine grain size marine sediments. This created an anaerobic environment where organic matter decomposition mediated by microorganisms formed methane gas. When methane gas concentrations exceeded the diffusion capacity, formed bubbles trapped in their interstitial space due to the low permeability of sediments. After a critic concentration (approximately 1%) and thanks to the difference in acoustic impedance generated by the bubbles, the gas deposits can be detected by seismo-acoustic methods, which are the most commonly used in shallow gas mapping.

Currently, the process by which vertical migration of methane gas deposits occurs remains not fully understood, but it is related with the relative permeability of sediment layers located above and the physical processes affecting them (e.g., erosion, deposition; Weschenfelder et al. 2016). This biogenic gas, cumulated at different distances to the seabed surface, can migrate until be in touch with the seabed surface or even escape through seabed to the water column and to the atmosphere. There is an increasing interest in understanding the roles played by the interaction of the seabed methane gas deposits with the surrounding environment (biotic and abiotic). This significance of biogenic shallow gas is associated with both their practical and immediate implications. For example, gas can be considered as a potential energy resource but also can impede engineering settlements by reducing shear strength of the seabed (Fleischer et al. 2001). Additionally, recent findings suggest that we are just beginning to understand its role in global carbon cycle and other biogeochemical cycles. For example, at local scale they are associated with significant alterations in diagenetic and redox processes in seabed sediments (Ramírez-Pérez et al. 2015; Ramírez-Pérez and De Blas 2017). At global scale some authors consider them as important sources of atmospheric methane emissions for climate change (Karisiddaiah and Veerayya 1994; Borges et al. 2016). However, due to logistical limitations, quantitative measurements of methane emissions from the seabed are scarce (García Gil et al. 2011). Research needs are becoming more important considering a possible increase in methane emissions from the seabed associated with rising water temperatures (Borges et al. 2016) and expansion of hypoxic areas which promote methanogenesis on buried organic carbon (Oppo et al. 2020) and can act in a synergetic way.

Marine methane gas can release to the water column from deep, thermogenic sources at continental slopes. These gas seeps can act as energy source, nursery, and refuge for local and surrounding benthic fauna (Grupe et al. 2015; Grey 2016; Levin et al. 2016). Moreover, the biota that lives associated with methane seepages is estimated to consume up to 80% of the total methane before it reaches the atmosphere (Boetius and Wenzhöfer 2013). Although the link between fauna and methane seems to be stronger at water depths greater than 200 m, in coastal systems, hard substrate derived from methane provides environmental complexity and suitability for sessile fauna (Jensen et al. 1992). Nevertheless, when gas is trapped in sediments, its ecological effect remains poorly understood. Most studies focused on more noticeable effects of thermogenic gas seeps, while few studies exploring the biological effect of shallow biogenic gas (e.g., Judd et al. 2002) have considered it as despicable. Nevertheless, its biological effect has received only isolated, sparse attention at shallow depths, and recent studies made in the Bahía Blanca Estuary indicate that these effects may had been underestimated (Bravo et al. 2018, Bravo et al. in press).

Biogenic shallow gas deposits have been found widely distributed in several coastal systems of South America, mainly in Brazil and Argentina (Weschenfelder et al. 2016). To date, in Argentina, its presence has been reported in the Río de La Plata Estuary, the San Matías Gulf, the Beagle Channel, and the Bahía Blanca Estuary (Weschenfelder et al. 2016). In the Bahía Blanca Estuary, gas deposits are located at different depths to the seabed covering extensive areas from the inner to the external zones of the Principal Channel, even extending its distribution to the neighboring continental shelf "El Rincón" (Andreoli 2018; Bravo et al. 2018). The gas deposits in the Bahía Blanca Estuary showed a tendency to be closer to the surface of the seabed or even to be in contact with it towards the innermost zone (Bravo et al. 2018). Wide sectors of the seabed surface of the subtidal area are characterized by gas-bearing sediments. Additionally, many of these sectors are in the area where the port complex is located (Bravo et al. 2018).

The wide coverage of gas-bearing sediments in the surface of the seabed in the Bahía Blanca Estuary gives an excellent opportunity to evaluate its environmental effects. The ecological effect of gas presence in soft-bottom seabed at shallow water depths was studied by Bravo et al. (2018) in the Bahía Blanca Estuary by comparing a gas site with a control site, both selected based on seismic evidences. These sites were adjacent to each other. Each sample location was selected based on the analysis of a seismic survey profile. This way, it was established with high precision that both sites only differed in gas presence, while sharing water depths (2–3 m below the reference level), and geomorphological and sedimentological features between them (Bravo et al. 2018; Bravo et al. in press).

In the Bahía Blanca Estuary, methane gas affects the physical and chemical properties of seabed sediments (Bravo et al. 2018; Bravo et al. in press). Despite sedimentological features commonly associated with gas when it is actively seeping, such as carbonate cement precipitation and local sediment lithification (e.g., Torry Bay, Scotland; Judd et al. 2002), were not observed in the Bahía Blanca Estuary, gas-bearing sediments were characterized by having lower shear strength and higher organic matter content than those from a control site (Bravo et al. 2018). The lower shear strength was coincident with findings from other estuarine environments such as in Chesapeake Bay (USA, Hill et al. 1992). Analyzing the seabed areas covered by gas-bearing sediments as a soft-bottom benthic habitat, they behave as an unstable substrate with more possibilities of resuspension (Bravo et al. 2018). Its organic matter enrichment is a common feature of gas-bearing sediments, characteristic shared with other estuaries and bays such as in Chesapeake Bay (USA, Hill et al. 1992), San Simón Bay (Ría de Vigo et al. 2011), and Skagerrak (Western Baltic Sea, Laier and Jensen 2007). Some authors had attributed the high amount of organic matter in the gas-bearing sediments to the use of methane as an alternative carbon source by the microbial consortium so that its organic carbon oxidative capacity was exceeded (Hill et al. 1992). This would imply that the vertical migration of buried methane, promoted by seawater warming, may add organic carbon on the coastal seafloor (Bravo et al. in press). Without considering the effect of methane per se, low shear strength and organic enrichment of gas-bearing sediments may change structural and functional configurations of faunistic groups associated to the soft-bottom seabed (Bravo 2019).

There were identified a total of 48 taxa inhabiting gas and control sites, which were represented differently (Czekanovski-Dice-Sorensen index = 0.52) in terms of abundance. The differences between sites were also observed in terms of Shannon diversity and evenness. Total abundance was lower, while diversity and evenness were higher at gas site than at control site. The dominant species in gas site, representing 66.7% of total abundance, was the polychaete *Aricidea* sp., while in control site, the amphipod *Monocorophium insidiosum* dominated the community, representing 72% of total abundance. In comparisons of biomass, cnidarians were dominant at gas site, while at control site annelids were the highest biomass group. Total biomass was higher at the gas site, and this was explained by the contribution of the sea pen *Stylatula darwini* which was absent at the control site. Until now the biology of this species has been poorly studied. It is interesting to carry out research on the possible preference for gas-bearing sediments by *S. darwini*, mostly considering that some other species of sea pens have been reported associated with gas seeps at continental slopes (Barrie et al. 2011; Bigham 2016; Palomino et al. 2016).

Shallow gas appears to affect abundance, composition, and distribution of benthic species in the Bahía Blanca Estuary. In general, the benthic community at gas site was dominated by polychaetes followed by crustaceans, whereas at control site this relationship was reversed. This could be due to a differential sensitivity of benthic organisms to the presence of gas in sediments. Previous studies attributed the macrobenthic community structure to organic matter enrichment as a result of sewage discharges even at sites distant from these disturbing sources (Elías 1987). Based on findings from Bravo et al. (2018), higher organic matter content at gas sites appears to be more related to methane than to sewage effluents.

Considering the expected fauna-sediment relations for the study area, characterized by sandy-mud sediments with high organic matter content, the dominant species should be deposit-feeders (Sanders 1958; Fauchald and Jumars 1979; Rhoads and Germano 1982). Nevertheless, this trend was only found at control site, while dominant feeding guilds at gas site were suspension-feeders, carnivores, and scavengers followed by surface deposit-feeders. At gas site dominates the Paraonidae *Aricidea* sp. whose feeding guild varies between species from deposit-feeder, carnivore, to suspension-feeder (Fauchald and Jumars 1979). In contrast, control site is dominated by a detritivore, *Monocorophium insidiosum* (Guerra-García et al. 2014). Some other authors had found organisms belonging to family Corophiidae avoiding gas, considering them as sensitive to sulfide-rich sediments (Meadows et al. 1981; Judd et al. 2002). As the methanogenesis process usually follows the hydrogen sulfide genesis (Judd 2004), it is expected that this gas could inhibit *M. insidiosum* in methane gas-bearing sediments in the Bahía Blanca Estuary.

Benthic communities at gas site showed a clearly different taxonomic structure, a markedly lower total abundance of organisms but higher total biomass. Some of the species from surroundig areas seems to avoid gas-bearing sediments, which may be related with high content of organic matter, low shear stress, and/or methane and hydrogen sulfide gas toxicity, whereas the species found in gas site may be obtaining benefits such as exclusion of competitors and/or energy source. For what we know, gas-bearing sediments in subtidal seabed affect taxonomic structure of benthic communities inhabiting them, but whether this leads to changes in their ecological functions should be analyzed in future studies. Thus, the presence of biogenic shallow gas on the seabed surface may behave as an important modelling factor for the benthic habitat, making it therefore necessary to take shallow gas distribution into account for researches on distribution of benthic communities. Understanding its ecological effects will be favored by future studies in other coastal sectors where shallow gas covers the surface of the seabed.

References

Aliotta S, Lizasoain G (2004) Tipos de fondos y su caracterización geológica por métodos sismoacústicos. In: Piccolo MC, Hoffmeyer MS (eds) Ecosistema del Estuario de Bahía Blanca. Instituto Argentino de Oceanografía, Bahía Blanca, Argentina, pp 51–59

Alves AT, Petsch DK, Barros F (2020) Drivers of benthic metacommunity structure along tropical estuaries. SC Rep 10(1):1–12

Anderson JT, Van Holliday D, Kloser R et al (2008) Acoustic seabed classification: current practice and future directions. ICES J Mar Sci 65(6):1004–1011

Andreoli A (2018) Análisis de los rasgos morfosedimentológicos de fondo en un sector de la Plataforma Continental al sur de la provincia de Buenos Aires. Tesis de Licenciatura, Universidad Nacional del Sur
Barrie JV, Cook S, Conway KW (2011) Cold seeps and benthic habitat on the Pacific margin of Canada. Cont Shelf Res 31(2):S85–S92
Barletta M, Lima AR, Costa MF (2019) Distribution, sources and consequences of nutrients, Persistent organic pollutants, metals and microplastics in South American estuaries. Sci Total Environ 651:1199–1218
Berthet B, Mouneyrac C, Amiard JC, Amiard-Triquet C, Berthelot Y, Le Hen A, Smith BD (2003) Accumulation and soluble binding of cadmium, copper, and zinc in the polychaete Hediste diversicolor from coastal sites with different trace metal bioavailabilities. Archives of Environmental Contamination and Toxicology 45(4):468–478
Bigham K (2016) Interpretation of the relationship between benthic fauna, geologic distributions, and methane seeps at Southern Hydrate Ridge, Oregon continental margin
Borges AV, Champenois W, Gypens N et al (2016) Massive marine methane emissions from near-shore shallow coastal areas. Sci Rep 6:27908
Boetius A, Wenzhöfer F (2013) Seafloor oxygen consumption fuelled by methane from cold seeps. Nat Geosci 6(9):725–734
Bravo ME (2019) Efecto de los depósitos de gas somero del fondo marino en las comunidades macrobentónicas del estuario de Bahía Blanca. Doctoral thesis, Universidad Nacional del Sur (UNS). Bahía Blanca (Argentina)
Bravo ME, Aliotta S, Fiori SM et al (2018) Distribution, vertical position and ecological implications of shallow gas in Bahía Blanca Estuary (Argentina). Estuar Coast Shelf Sci 202:222–231
Bravo ME, Fiori SM, Aliotta S et al (2020) Assessment of the impacts of shallow gas on taxonomic structure of benthic communities in Bahía Blanca Estuary (Argentina). Estuar Coast Shelf Sci. in press
Breine NT, De Backer A, Van Colen C et al (2018) Structural and functional diversity of soft-bottom macrobenthic communities in the Southern North Sea. Estuar Coast Shelf Sci 214:173–184
Bremec CS (1990) Macrobentos del área de Bahía Blanca (Argentina). distribución espacial de la fauna. Bol. Inst. Oceanogr 38(2):99–110
Bremec CS, Elías R, Giberto DA (2017) El Rincón. In: Bremec CS, Giberto DA (eds) Comunidades bentónicas en regiones de interés pesquero en Argentina. Instituto Nacional de Investigación y Desarrollo Pesquero, Mar del Plata, pp 40–46
Bremec CS, Giberto DA (2017) Comunidades bentónicas en regiones de interés pesquero en Argentina. Instituto Nacional de Investigación y Desarrollo Pesquero, Mar del Plata, 131 p
Carbone ME, Spetter CV, Marcovecchio JE (2016) Seasonal and spatial variability of macronutrients and Chlorophyll a based on GIS in the South American estuary (Bahía Blanca, Argentina). Environ Earth Sci 75(9):736
Carricavur AD, Boudet LC, Romero MB, Polizzi P, Marcovecchio JE, Gerpe M (2018) Toxicological responses of Laeonereis acuta (Polychaeta, Nereididae) after acute, subchronic and chronic exposure to cadmium. Ecotoxicology and environmental safety 149:217–224.
Chainho P, Costa JL, Chaves ML et al (2006) Seasonal and spatial patterns of distribution of subtidal benthic invertebrate communities in the Mondego River, Portugal—a poikilohaline estuary. In: Marine Biodiversity. Springer, Dordrecht, pp 59–74
Dauer DM, Rodi AJ, Ranasinghe JA (1992) Effects of low dissolved oxygen events on the macrobenthos of the lower Chesapeake Bay. Estuaries 15(3):384–391
Dauvin JC, Alizier S, Rolet C et al (2012). Response of different benthic indices to diverse human pressures. Ecol Indic 12(1):143–153
Dauvin JC, Ruellet T (2007) Polychaete/amphipod ratio revisited. Mar Pollut Bull 55(1–6):215–224
Dean HK (2008) The use of polychaetes (Annelida) as indicator species of marine pollution: a review. Rev Biol Trop 56(4):11–38

Diez MJ, Romero MC, Obenat S et al (2009) Distribución de Invertebrados Bentónicos en el Canal Beagle, Argentina. Distribution of Benthic Invertebrates in the Beagle Channel, Argentina. Anales del Instituto de la Patagonia 37(2):29–40

Dissanayake NG, Frid CL, Drylie TP et al (2018) Ecological functioning of mudflats: global analysis reveals both regional differences and widespread conservation of functioning. Mar Ecol Progr Ser 604:1–20

Dolbeth M, Cardoso PG, Ferreira SM et al (2007) Anthropogenic and natural disturbance effects on a macrobenthic estuarine community over a 10-year period. Mar Pollut Bull 54(5):576–585

Dos Santos EP, Fiori SM (2010) Primer registro sobre la presencia de Crassostrea gigas (Thumberg, 1793) en el Estuario de Bahía Blanca,(Argentina)

Dutto MS, Genzano GN, Schiariti A et al (2017) Medusae and ctenophores from the Bahía Blanca Estuary and neighboring inner shelf (Southwest Atlantic Ocean, Argentina). Mar Biodiver Rec 10(1):14

Ehrlich P, Walker B (1998) Rivets and redundancy. Bioscience 48(5):387–388

Elias R (1987) Estudio inventarial y ecológico del macrobentos de la Bahía Blanca. Doctoral thesis. Universidad Nacional de La Plata, Facultad de Ciencias Naturales y Museo. 262p

Elias R (1992) Quantitative benthic community structure in Blanca Bay and its relationship with organic enrichment. Mar Ecol 13(3):189–201

Elías R (1995) The subtidal macroinfauna from soft-bottom substrate of the Blanca Bay (Argentine). Thalassas 11:73–86

Elias R, Bremec CS, Giberto D et al (2001) Estudio faunístico de las comunidades bentónicas infaunales de El Rincón. Resultados de la campaña CC 14/00. Inf Téc Int DNI-INIDEP N° 8(/2001):11

Elias R, Ieno EN (1993) La comunidad de *Laeonereis acuta* Treadwell, 1923 (Polychaeta: Nereididae) en la región interna de la Bahía Blanca. Iheringia, Ser Zool 75:3–13

Elias R, Iribarne OO, Bremec CS et al (2004) Comunidades bentónicas de fondos blandos. In: Piccolo MC, Hoffmeyer MS (eds) Ecosistema del Estuario de Bahía Blanca. Instituto Argentino de Oceanografía, Bahía Blanca, Argentina, pp 179–190

Ellis JI, Clark D, Atalah J et al (2017) Multiple stressor effects on marine infauna: responses of estuarine taxa and functional traits to sedimentation, nutrient and metal loading. Sci Rep 7:12013. https://doi.org/10.1038/s41598-017-12323-5

Farías NE, Obenat S, Goya AB (2015) Outbreak of a neurotoxic side-gilled sea slug (Pleurobranchaea sp.) in Argentinian coasts. New Zeal J Zool 42(1):51–56

Fauchald K, Jumars PA (1979) The diet of worms: a study of polychaete feeding guilds. Oceanogr Mar Biol Annu Rev 17:193–284

Faulwetter MS, Markantonatou MV, Pavloudi MC et al (2014) Polytraits: a database on biological traits of marine polychaetes. Biodivers Data J 2:e1024

Fernández Severini MD, Carbone ME, Villagran DM et al (2018) Toxic metals in a highly urbanized industry-impacted estuary (Bahia Blanca Estuary, Argentina): spatio-temporal analysis based on GIS. Environ Earth Sci 77:393. https://doi.org/10.1007/s12665-018-7565-5

Fiori SM (2014) Capítulo IV Comunidades bentónicas. In: Programa de Monitoreo de la calidad ambiental del Estuario de Bahía Blanca. http://www.bahiablanca.gov.ar/areas-de-gobierno/medio-ambiente/comité-tecnico-ejecutivo/informes-medio-ambientales

Fiori SM (2016) Capítulo IV Comunidades bentónicas. In: Programa de Monitoreo de la calidad ambiental del Estuario de Bahía Blanca. http://www.bahiablanca.gov.ar/areas-de-gobierno/medio-ambiente/comité-tecnico-ejecutivo/informes-medio-ambientales

Fiori SM (2018) Capítulo IV Comunidades bentónicas. In: Programa de Monitoreo de la calidad ambiental del Estuario de Bahía Blanca. http://www.bahiablanca.gov.ar/areas-de-gobierno/medio-ambiente/comité-tecnico-ejecutivo/informes-medio-ambientales

Fiori SM, Bravo ME, Elias Ret AL (2020) Effects of sewage effluente on the subtidal mactobenthic assemblage in a urban estuary (Argentina). Ecol Austral 1(30):134–145. https://doi.org/10.25260/EA.20.30.1.0.954

Fleischer P, Orsi T, Richardson M et al (2001) Distribution of free gas in marine sediments: a global overview. Geo-Mar Lett 21(2):103–122

García-Gil S, de Blas E, Martínez-Carreño N et al (2011) Characterisation and preliminary quantification of the methane reservoir in a coastal sedimentary source: San Simón Bay, Ría de Vigo, NW Spain. Estuar Coast Shelf Sci 91(2):232–242

Gibbs PE, Burt GR, Pascoe PL, Llewellyn CA, Ryan KP (2000) Zinc, copper and chlorophyll-derivatives in the polychaete Owenia fusiformis. Journal of the Marine Biological Association of the United Kingdom 80(2):235–248

Giberto DA, Bremec CS, Acha E et al (2004) Large-scale spatial patterns of benthic assemblages in the SW Atlantic: the Rıo de la Plata estuary and adjacent shelf waters. Estuar Coast Shelf Sci 61(1):1–13

Gimenez J, Lasta M, Bigatti G (2005) Exploitation of the volute snail *Zidona dufresnei* in Argentine waters, Southwestern Atlantic Ocean. J Shellfish Res 24(4):1135–1140

Ginsberg SS, Aliotta S (2011) Sediment transport circulation pattern through Mesotidal channels system. In: Ginsberg SS (ed) Sediment Transport. InTech, Vienna, Austria, pp 275–294

Ginsberg SS, Aliotta S, Lizasoain GO (2009) Sistema interconectado de canales de marea del estuario de Bahía Blanca, Argentina: evaluación de la circulación de sedimento como carga de fondo por métodos acústicos. Lat Am J Aquat Res 7(2):231–245

Ginsberg SS, Aliotta S, Vecchi L et al (2014) Vinculación entre el canal de acceso a los puertos de Bahía Blanca y elsistema interconectado de canales de marea en relación a las tareas dedragado. XIV Reunión Argentina de Sedimentología, La Plata, Asamblea Argentina de Sedimentología, Asociación Argentina de Sedimentología

Ginsberg SS, Perillo G (1999) Deep-scour holes at tidal channel junctions, Bahia Blanca Estuary, Argentina. Mar Geol 160(1–2):171–182

Gómez EA, Ginsberg SS, Perillo GM (1996) Geomorfología y sedimentología de la zona interior del Canal Principal del Estuario de Bahía Blanca. Revista de la Asociación Argentina de Sedimentología 3(2):55–61

Gray JS (2002) Species richness of marine soft sediments. Mar Ecol Progr Ser 244:285–297

Grey J (2016) The incredible lightness of being methane-fuelled: stable isotopes reveal alternative energy pathways in aquatic ecosystems and beyond. Front Ecol Evol 4:8

Grupe BM, Krach ML, Pasulka AL et al (2015) Methane seep ecosystem functions and services from a recently discovered southern California seep. Mar Ecol 36:91–108

Guerra-García JM, De Figueroa JT, Navarro-Barranco C et al (2014) Dietary analysis of the marine Amphipoda (Crustacea: Peracarida) from the Iberian Peninsula. J Sea Res 85:508–517

Guidetti P, Modena M, La Mesa G, Vacchi M (2000) Composition, abundance and stratification of macrobenthos in the marine area impacted by tar aggregates derived from the Haven oil spill (Ligurian Sea, Italy). Marine Pollution Bulletin 40(12):1161–1166

Guinder VA, López-Abbate MC, Berasategui AA et al (2015) Influence of the winter phytoplankton bloom on the settled material in a temperate shallow estuary. Oceanologia 57(1):50–60

Guinder VA, Molinero JC, Popovich CA et al (2012) Dominance of the planktonic diatom Thalassiosira minima in recent summers in the Bahía Blanca Estuary, Argentina. J Plankton Res 34(11):995–1000

Güller M, Zelaya DG (2017) A hot-spot of biodiversity in Northern Patagonia, Argentina. Biodivers Conserv 26(14):3329–3342

Hampel H, Elliott M, Cattrijsse A (2009) Macrofaunal communities in the habitats of intertidal marshes along the salinity gradient of the Schelde estuary. Estuar Coast Shelf Sci 84(1):45–53

Harris PT, Hughes MG (2012) Predicted benthic disturbance regimes on the Australian continental shelf: a modelling approach. Mar Ecol Progr Ser 449:13–25

Hill JM, Halka JP, Conkwright R et al (1992) Distribution and effects of shallow gas on bulk estuarine sediment properties. Cont Shelf Res 12(10):1219–1229

Holland AF, Dean JM (1977) The biology of the stout razor clam Tagelus plebeius: I. Animal-sediment relationships, feeding mechanism, and community biology. Chesap Sci 18(1):58–66

Jensen P, Aagaard I, Burke RA Jr et al (1992) Bubbling reefs' in the Kattegat: submarine landscapes of carbonate-cemented rocks support a diverse ecosystem at methane seeps. Mar Ecol Progr Ser 83:102–112

Judd AG, Sim R, Kingston P et al (2002) Gas seepage on an intertidal site: Torry Bay, Firth of Forth, Scotland. Cont Shelf Res 22(16):2317–2331

Judd AG (2004) Natural seabed gas seeps as sources of atmospheric methane. Environ Geol 46(8):988–996

Kaminsky J, Varisco M, Fernández M et al (2018) Spatial analysis of benthic functional biodiversity in San Jorge Gulf, Argentina. Oceanography 31(4):104–112

Kappel CV (2005) Losing pieces of the puzzle: threats to marine, estuarine, and diadromous species. Frontiers in Ecology and the Environment 3(5):275–282

Karisiddaiah SM, Veerayya M (1994) Methane-bearing shallow gas-charged sediments in the eastern Arabian Sea: a probable source for greenhouse gas. Cont Shelf Res 14(12):1361–1370

Kennish MJ, Brush MJ, Moore KA (2014) Drivers of change in shallow coastal photic systems: an introduction to a special issue. Estuaries and coasts 37(1):3–19

Klerks PL, Leberg PL, Lance RF, McMillin DJ, Means JC (1997) Lack of development of pollutant-resistance or genetic differentiation in darter gobies (Gobionellus boleosoma) inhabiting a produced-water discharge site. Marine Environmental Research 44(4):377–395

Klerks PL, Kascak A, Cazan AM, Adhikary ND, Chistoserdov A, Shaik A, Louka FR (2018) Effects of the razor clam Tagelus plebeius on the fate of petroleum hydrocarbons: A mesocosm experiment. Archives of environmental contamination and toxicology 75(2):306–315

Kristensen E, Delefosse M, Quintana CO et al (2014) Influence of benthic macrofauna community shifts on ecosystem functioning in shallow estuaries. Front Mar Sci 1:41

La Colla NS, Botté SE, Negrin VL et al (2018) Influence of human-induced pressures on dissolved and particulate metal concentrations in a South American estuary. Environ Monit Assess 190(9):532

Laier T, Jensen JB (2007) Shallow gas depth-contour map of the Skagerrak-western Baltic Sea region. Geo-Mar Lett 27(2–4):127–141

Lana P, Christofoletti R, Gusmão JB Jr et al (2018) Benthic Estuarine Assemblages of the Southeastern Brazil Marine Ecoregion (SBME). In: Brazilian Estuaries. Springer, Cham, pp 117–175

Levin LA, Baco AR, Bowden DA et al (2016) Hydrothermal vents and methane seeps: rethinking the sphere of influence. Front Mar Sci 3:72

Levin LA, Mendoza GF, Grupe BM (2017) Methane seepage effects on biodiversity and biological traits of macrofauna inhabiting authigenic carbonates. Deep-Sea Res PT II 137:26–41

Lichtschein de Bastida V, Bastida R (1980) Los briozoos de las comunidades incrustantes de puertos argentinos

Lopez Cazorla A (2004) Peces. El Ecosistema del Estuario de Bahía Blanca. In: Piccolo MC, Hoffmeyer MS (eds) Ecosistema del Estuario de Bahía Blanca. Instituto Argentino de Oceanografía, Bahía Blanca, Argentina, pp 191–201

Marcovecchio J, Botté S, Severini MF et al (2010) Geochemical control of heavy metal concentrations and distribution within Bahia Blanca Estuary (Argentina). Aquat Geochem 16:251–266

McNabb P, Selwood AI, Munday R et al (2010) Detection of tetrodotoxin from the grey side-gilled sea slug *Pleurobranchaea maculata*, and associated dog neurotoxicosis on beaches adjacent to the Hauraki Gulf, Auckland, New Zealand. Toxicon 56:466–473

Meadows PS, Deans EA, Anderson JG (1981) Responses of *Corophium volutator* to sediment sulphide. J Mar Biol Assoc UK 61:739e748

Mestdagh S, Amiri-Simkooei A, van der Reijden KJ et al (2020) Linking the morphology and ecology of subtidal soft-bottom marine benthic habitats: a novel multiscale approach. Estuar Coast Shelf Sci 106687

Minor Salvatierra MS, Aliotta S, Ginsberg SS (2015) Morphology and dynamics of large subtidal dunes in Bahia Blanca estuary, Argentina. Geomorphology 246:168–177

Molina LM, Valinas MS, Pratolongo PD, Elias R, Perillo GM (2009) First record of the sea anemone Diadumene lineata (Verrill 1871) associated to Spartina alterniflora roots and stems, in marshes at the Bahia Blanca estuary, Argentina. Biological Invasions 11(2):409–416

Morsan E (2007) Spatial pattern, harvesting and management of the artisanal fishery for purple clam (Amiantis purpurata) in Patagonia (Argentina). Ocean & Coastal Management 50(5–6):481–497

Mouneyrac C, Mastain O, Amiard JC, Amiard-Triquet C, Beaunier P, Jeantet AY, Rainbow PS (2003) Trace-metal detoxification and tolerance of the estuarine worm Hediste diversicolor chronically exposed in their environment. Marine Biology 143(4):731–744

Muniz P, Lana P, Venturini N et al (2013) Un manual de protocolos para evaluar la contaminación marina por efluentes domésticos. No. 577.727 M3.131 p

Muniz P, Venturini N (2015) Macrobenthic communities in a temperate urban estuary of high dominance and low diversity: Montevideo Bay (Uruguay). CICIMAR Oceánides 30(1):9–20

Narvarte M, González R, Filippo P (2007) Artisanal mollusk fisheries in San Matías Gulf (Patagonia, Argentina): an appraisal of the factors contributing to unsustainability. Fish Res 87(1):68–76

Nikora V (2010) Hydrodynamics of aquatic ecosystems: an interface between ecology, biomechanics and environmental fluid mechanics. River Res Appl 26(4):367–384

Noguchi T, Arakawa O (2008) Tetrodotoxin – distribution and accumulation in aquatic organisms. and cases of human intoxication Mar Drugs 6:220–242

Oguz T, Fach B, Salihoglu B (2008) Invasion dynamics of the alien ctenophore Mnemiopsis leidyi and its impact on anchovy collapse in the Black Sea. Journal of plankton research 30(12):1385–1397

Oliva AL, La Colla NS, Arias AH et al (2017) Distribution and human health risk assessment of PAHs in four fish species from a SW Atlantic estuary. Environ Sci Pollut R 24(23):18979–18990

Occhipinti-Ambrogi A, Marchini A, Cantone G, Castelli A, Chimenz C, Cormaci M, Piraino S (2011) Alien species along the Italian coasts: an overview. Biological invasions 13(1):215–237

Oppo D, De Siena L, Kemp DB (2020) A record of seafloor methane seepage across the last 150 million years. Sci Rep 10(1):1–12

Orensanz JML, Schwindt E, Pastorino G, Bortolus A, Casas G, Darrigran G, Vallarino EA (2002) No longer the pristine confines of the world ocean: a survey of exotic marine species in the southwestern Atlantic. Biological Invasions 4(1):115–143

Palomino D, López-González N, Vázquez JT et al (2016) Multidisciplinary study of mud volcanoes and diapirs and their relationship to seepages and bottom currents in the Gulf of Cádiz continental slope (northeastern sector). Mar Geol 378:196–212

Pearson TH, Rosenberg R (1978) Macrobenthic succession in relation to organic enrichment and pollution of the marine environment. Oceanogr Mar Biol Ann Rev 16:229–311

Quiroga E, Ortiz P, González-Saldías R et al (2016) Seasonal benthic patterns in a glacial Patagonian fjord: the role of suspended sediment and terrestrial organic matter. Mar Ecol Prog Ser 561:31–50

Ramírez-Pérez AM, De Blas E, García-Gil S (2015) Redox processes in pore water of anoxic sediments with shallow gas. Sci Total Environ 538:317–326

Ramírez-Pérez AM, de Blas E (2017) Iron reactivity in anoxic sediments in the Ría de Vigo (NW Spain). Chemosphere 174:8–19

Rhoads DC, Germano JD (1982) Characterization of organism-sediment relations using sediment profile imaging: an efficient method of remote ecological monitoring of the seafloor (Remots™ System). Mar Ecol Prog Ser 8:115–128

Rouse GW, Goffredi SK, Vrijenhoek RC (2004) Osedax: bone-eating marine worms with dwarf males. Sci 305(5684):668–671

Saiz-Salinas JI, González-Oreja JA (2000) Stress in estuarine communities: lessons from the highly-impacted Bilbao estuary (Spain). J Aquat Ecosyst Stress Recovery 7(1):43–55

Sandrini JZ, Regoli F, Fattorini D, Notti A, Inácio AF, Linde-Arias AR, Monserrat JM (2006) Short-term responses to cadmium exposure in the estuarine polychaete Laeonereis acuta (Polychaeta,

Nereididae): Subcellular distribution and oxidative stress generation. Environmental Toxicology and Chemistry: An International Journal 25(5):1337–1344

Sandrini JZ, Lima JV, Regoli F, Fattorini D, Notti A, Marins LF, Monserrat JM (2008) Antioxidant responses in the nereidid Laeonereis acuta (Annelida, Polychaeta) after cadmium exposure. Ecotoxicology and environmental safety 70(1):115–120

Sanders HL (1958) Benthic studies in Buzzards Bay. I. Animal-sediment relationships. Limnol Oceanogr 3:245–258

Scarabino F, Zaffaroni JC, Clavijo C et al (2006) Bivalvos marinos y estuarinos de la costa uruguaya: faunística, distribución, taxonomía y conservación. En: Menafra R, Rodríguez-Gallego L, Scarabino F et al (eds) Bases para la conservación y el manejo de la costa uruguaya. Fauna Silvestre Uruguaya

Schaller J (2014) Bioturbation/bioirrigation by Chironomus plumosus as main factor controlling elemental remobilization from aquatic sediments?. Chemosphere 107:336–343

Schiariti A, Berasategui AD, Giberto DA et al (2006) Living in the front: *Neomysis americana* (Mysidacea) in the Río de la Plata estuary, Argentina-Uruguay. Mar Biol 149(3):483–489

Simberloff D, Martin JL, Genovesi P, Maris V, Wardle DA, Aronson J, Vila M (2013) Impacts of biological invasions: what's what and the way forward. Trends in ecology & evolution 28(1):58–66

Soria G, Orensanz JL, Morsán EM, Parma AM et al (2016) Scallops biology, fisheries, and management in Argentina. Dev Aquac Fish Sci 40:1019–1046

Speake MA, Carbone ME, Spetter CV (2020) Análisis del sistema socio-ecológico del estuario Bahía Blanca (Argentina) y su impacto en los servicios ecosistémicos y el bienestar humano

Tait LW, Lohrer AM, Townsend M et al (2020) Invasive ecosystem engineers threaten benthic nitrogen cycling by altering native infaunal and biofouling communities. Sci Rep 10(1):1–11

Teske PR, Wooldridge TH (2003) What limits the distribution of subtidal macrobenthos in permanently open and temporarily open/closed South African estuaries? Salinity vs. sediment particle size. Estuar Coast Shelf Sci 57(1–2):225–238

Tian S, Tong Y, Hou Y (2019) The effect of bioturbation by polychaete *Perinereis aibuhitensis* on release and distribution of buried hydrocarbon pollutants in coastal muddy sediment. Mar Pollut Bull 149:110487

Truchet DM, Noceti MB, Villagrán DM et al (2019) Fishers' ecological knowledge about marine pollution: what can FEK contribute to ecological and conservation studies of a Southwestern Atlantic Estuary? J Ethnobiol 39(4):584–606

van der Linden P, Patrício J, Marchini A et al (2012) A biological trait approach to assess the functional composition of subtidal benthic communities in an estuarine ecosystem. Ecol Indic 20:121–133

van der Wal D, Lambert GI, Ysebaert T et al (2017a) Hydrodynamic conditioning of diversity and functional traits in subtidal estuarine macrozoobenthic communities. Estuar Coast Shelf Sci 197:80–92

van der Wal D, Ysebaert T, Herman PM (2017b) Response of intertidal benthic macrofauna to migrating megaripples and hydrodynamics. Mar Ecol Progr Ser 585:17–30

Viarengo A, Nott JA (1993) Mechanisms of heavy metal cation homeostasis in marine invertebrates. Comparative Biochemistry and Physiology Part C: Comparative Pharmacology 104(3):355–372

Vilà M, Espinar JL, Hejda M, Hulme PE, Jarošík V, Maron JL, Pyšek P (2011) Ecological impacts of invasive alien plants: a meta-analysis of their effects on species, communities and ecosystems. Ecology letters 14(7):702–708

Weschenfelder J, Klein AH, Green AN et al (2016) The control of palaeo-topography in the preservation of shallow gas accumulation: examples from Brazil, Argentina and South Africa. Estuar Coast Shelf Sci 172:93–107

Weis WA, Soares CHL, de Quadros DPC, Scheneider M, Pagliosa PR (2017) Urbanization effects on different biological organization levels of an estuarine polychaete tolerant to pollution. Ecological Indicators 73:698–707

Wieters EA, McQuaid C, Palomo G et al (2012) Biogeographical boundaries, functional group structure and diversity of rocky shore communities along the Argentinean coast. PLoS One 7(11)

Wilson WH (1990) Competition and predation in marine soft-sediment communities. Annu Rev Ecol Evol S 21(1):221–241

Xie M, Wang N, Gaillard JF et al (2018) Interplay between flow and bioturbation enhances metal efflux from low-permeability sediments. J Hazard Mat 341:304–312

Ysebaert T, Herman PM (2002) Spatial and temporal variation in benthic macrofauna and relationships with environmental variables in an estuarine, intertidal soft-sediment environment. Marine Ecology Progress Series 244:105–124

Zaidman P, Kroeck MA, Van der Molen S, Williams G, Villalobos LG, Kissner EO, Morsan E (2016) Local scale variation in the reproductive pattern of the southern geoduck, Panopea abbreviata (Bivalvia: Hiatellidae), in Patagonia. Revista de biología marina y oceanografía 51(2):359–371

Chapter 10
Shrimps and Prawns

Patricia Marta Cervellini and Jorge Omar Pierini

10.1 Introduction

Marine shrimp are decapod crustaceans, distinguished from other crustaceans by its thoracic appendages. While the first three pairs (maxillipeds) are modified for feeding, the remaining five pairs (pereopods) are the walking legs, hence the name Decapoda or "ten-legs." They belong to the Suborder Dendrobranchiata, defined by branched gills, prominent hinges on the pleon, larvae hatching as nauplii or protozoeae, and the presence of a petasma in males (Tavares and Martin 2010; Tavares et al. 2009). In its adult stage, they present a body segmented into 19 somites or parts, divided into an anterior region or "head" covered by a rigid cephalothorax (the carapace) and a posterior region, the articulated abdomen (pleon), that represents the edible body part. The cephalothorax projects into a compressed tip or "face," with dorsal teeth, and, in some species, ventral teeth are also present.

The Suborder Dendrobranchiata is further divided into the Superfamilies Penaeoidea and Sergestoidea. The Superfamily Penaeoidea is characterized by the first three pairs of pereopods being chelate (ending in pincers) and of similar size and shape (Pérez Farfante 1988). This group includes a little more than 350 species of commercial interest; among them, about 100 species comprise most of the annual

P. M. Cervellini (✉)
Instituto de Ciencias Biológicas y Biomédicas del Sur, Universidad Nacional del Sur, CONICET, INBIOSUR, Bahía Blanca, Argentina

Departamento de Biología, Bioquímica y Farmacia. Universidad Nacional del Sur., Bahía Blanca, Argentina
e-mail: pcervell@uns.edu.ar

J. O. Pierini
Instituto Argentino de Oceanografía (IADO-CONICET-UNS), Bahía Blanca, Argentina

Comisión de Investigaciones Científicas de la Provincia de Buenos Aires (CIC), Bahía Blanca, Argentina

S. M. Fiori, P. D. Pratolongo (eds.), *The Bahía Blanca Estuary*,
https://doi.org/10.1007/978-3-030-66486-2_10

world shrimp catches (Fransen 2014). The Superfamily Sergestoidea can be separated from the penaeoids by the lack or reduction in size of the fourth and fifth pairs of pereopods and by having a smaller number of gills. These shrimps occur in commercial catches; however, because of their small size and soft bodies, they are of little or no economic value (Pérez Farfante 1988).

Penaeoid and sergestoid shrimps are widely distributed, occurring in marine, estuarine, and freshwater environments. Most marine species occupy shallow or moderately deep waters, but some are found at depths of almost 5700 m. Although many shrimps are pelagic, most are benthic, living on a large variety of bottoms including mud, sand, rock, fragments of shells, or mixtures of these materials. In addition, some species frequent coral reefs, and a few others live in association with other invertebrates (Pérez Farfante 1988; Fransen 2014). They are animals that feed on decaying particulate organic matter and various invertebrates that they capture with their multiple appendices located on the head and thorax. The legs in the abdomen (pleopods) are lined with fine silks that are used to swim, as well as removing bottom sediments to get buried, especially during daylight hours (Boschi 2004).

These shrimps are dioecious, and the external structures of the genital system are the major dimorphic features. The male has two pairs of modified abdominal appendages on the first and second abdominal segments (the petasma and appendix masculina) that deliver sperm to the female's external receptacle (the thelycum) located between the bases of the fifth walking legs (Bailey-Brock and Moss 1992). The thelycum can be closed (spermatic mass is internally placed on thelycum plates) or opened (spermatic mass is exposed on thelycum region) (Dall et al. 1990). They reproduce through copulating males and females, which release fertilized eggs that are left free in the water, with exceptions such as certain small shrimps from the Sergestoidea superfamily, which usually incubate eggs within their own bodies. The reproduction occurs in the sea (spring to autumn), and depending on the species, after fertilization the female stores the sperm (packs or spermatophore) in the seminal receptacles. The embryo develops inside the egg, and free larvae are born 12–24 h after fertilization, depending on the temperature (Scelzo 2016).

The planktonic larva, called nauplius (pl. nauplii), is characterized by having three pairs of appendices (antennules, antennas, and jaws) and an eye. Nauplius feed on their own reserves (yolk) until they transform into zoea larvae (protozoeae) that begin to feed on phytoplankton by filtration. Protozoeae are further transformed into mysis larvae of carnivorous habits, which acquire adult morphology during their evolution. There is a time span of approximately 20 days between the first and last mysis larvae stages, before reaching the postlarval stage that also spans 20 days. Postlarvae have adult morphology and modify their planktonic life form (life in the water column) to a benthic behavior (on the substrate), acquiring after a few days the appearance of adults and sexual maturity. The entire cycle lasts from 30 to 45 days. In the natural environment, postlarvae migrate from breeding areas to the coast and frequently enter estuaries or brackish lagoons to feed (Iorio et al. 1990; Scelzo 2016).

10.2 Shrimps Present in the Bahía Blanca Estuary

Because of their economic value and abundance in estuarine and coastal ecosystems, shrimps are widely studied, and the literature about them is extensive. In the next paragraphs, we will provide updated information on the biological traits and life cycles of three species of shrimps present in the Bahía Blanca Estuary, which were selected based on their ecological and economic relevance. Two of them belong to the Superfamily Penaeoidea, *Pleoticus muelleri* (Family Solenoceridae) and *Artemesia longinaris* (Family Penaeidae), both species targets of artisanal and industrial fisheries along the Argentine coast; the third species belongs to the Superfamily Sergestoidea, *Peisos petrunkevitchi* (Family Sergestidae), which represents an important link within local food webs.

10.2.1 The Argentine Red Shrimp Pleoticus muelleri

The Argentine red shrimp *Pleoticus muelleri* (Spence Bate, 1888) is a robust species characterized by a short rostrum, with a series of teeth in its superior margin (Boschi 1963) and a typically intense pink color in their exoskeleton (Fig 10.1a). This species is endemic to the southwestern Atlantic (Spivak et al. 2019), distributed

Fig 10.1 Species of shrimps exploited in Bahía Blanca Estuary (**a**) Argentine red shrimp *Pleoticus muelleri*; (**b**) Argentine stiletto shrimp *Artemesia longinaris*. (Photos by Sandra Fiori (a) and Juan Manuel Alvarez (b))

along South American coastal waters from Rio de Janeiro, Brazil, to Santa Cruz, Argentina (Angelescu and Boschi 1959; Boschi 1963; Boschi and Gavio 2005) and frequently found between 3 and 130 m depth (D'Incao 1999; Boschi et al. 1992). There is a marked sexual dimorphism in this species, as well as large variations in size and weight along its latitudinal range of occurrence.

In the Gulf of San Jorge, Argentina, the mean carapace length is 39 mm for males and 43 mm for females. Further north, in subtropical coasts of southeastern Brazil, mean carapace lengths are smaller (34.7 mm for females and 27.7 mm for males) (Castilho et al. 2008). The highest adult abundance occurs in the Gulf of San Jorge, in Patagonian waters, where there is the largest reproductive or *stock* population. Minor *stock* populations locate offshore the Bahía Blanca Estuary and in front of Mar del Plata (Buenos Aires Province) (Boschi 1986; Boschi and Selzo 1967, 1969, 1971; Cervellini and Mallo 1991; Iorio et al. 1996; Scelzo 2016).

The Argentine red shrimp thrives in water temperatures between 6°C and 23°C and salinities from 31.5 to 33.5. Because of these wide ranges, populations show different reproductive behaviors along its distributional range. While reproduction is a continuous process in tropical areas, it shows strong seasonal oscillations in higher latitudes (Boschi 1997; Costa et al. 2004; Castilho et al. 2008). In the Argentine Sea, the reproductive season begins in late spring, and spawning peaks in mid-summer, decreases during fall, and finally disappears in winter (Boschi 1986). This behavior is supported by the presence-absence of mature, impregnated females and larval stages in plankton samples. The spawning capacity of this species ranges from 129,000 to 477,000 eggs per female (Macchi et al. 1992), and larval densities rise up to 334 protozoa and 85 mysis per 100 m^3 (Iorio et al. 1996). Along the Argentine coast, larvae of *Pleoticus muelleri* move large distances, between 120 and 300 nautical miles (about 220 and 550 km, respectively), transported by coastal currents (Boschi 1989).

The Argentine red shrimp fishery is the main crustacean fishery in Argentina and one of the most important marine resources in the southwestern Atlantic. Most of the reported annual landings come from fishing grounds in the Patagonian region. However, there is still a substantial contribution of the artisanal fleet along Buenos Aires Province (Garza et al 2017)

10.2.2 The Argentine Stiletto Shrimp **Artemesia longinaris**

The Argentine stiletto shrimp *Artemesia longinaris* Spence Bate, 1888 is characterized by a long and sinuous rostrum (Boschi 1963). *A. longinaris* has a smaller size than *Pleoticus muelleri*, although there are size variations across its geographical range. In Buenos Aires Province, in Mar del Plata Port, males and females reach up to 24 and 29 mm, respectively (Castilho et al. 2008). Adults are transparent with pigmented spots, due to the presence of chromatophores (Gavio and Boschi 2016), allowing blending and going unnoticed on sandy bottoms (Fig 10.1b). *A. longinaris* is endemic to the southwestern Atlantic and distributed from Rio de Janeiro, Brazil, to Rawson, Chubut Province, Argentina (D´Incao 1999). The species is infra- and circalittoral (Boschi et al. 1992; D'Incao 1999; Carvalho-Batista et al. 2011), plentiful in shallow waters up to 20 m depth (Scelzo et al. 2002).

Populations of *A. longinaris* present phenotypic variations throughout the distribution range. The body size and the mean size at sexual maturity increase with the latitude, from São Paulo (Brazil) to Mar del Plata (Argentina), but decrease with latitude from Rio de Janeiro to São Paulo. There are also latitudinal differences in the reproductive period, which tends to be continuous in lower latitudes and seasonal in higher latitudes (Carvalho-Batista et al. 2014). These variations may result from reproductive adaptations to environmental factors, mainly temperature, nutrient supply, and subsequent plankton production. There are also intrinsic physiological limitations that constrain the reproductive behavior (Castilho et al. 2007). Despite this phenotypic plasticity, there is no genetic differentiation among *A. longinaris* populations through its entire geographical distribution. The genetic homogeneity is maintained by larval dispersal and their high migratory capacity, which ensures gene flow among populations (Carvalho-Batista et al. 2014).

Artemesia longinaris plays a critical role in trophic food webs of the southwestern Atlantic. This shrimp inhabits soft bottoms composed of sand, silt, and clay. Diet studies revealed that sand is the major component on stomach contents, and sand grains are covered by films of bacteria, which are the base of shrimp nutrition (Gavio and Boschi 2004). In turn, *A. longinaris* is a valuable food item for different species of fish and other invertebrates. Shrimps have higher caloric content than polychaetes, gastropods, and echinoderms (Thayer et al. 1973; Capitoli et al. 1994). Because of its high biomass and energetic quality, a large number of carnivorous fishes in the coastal waters of Buenos Aires Province prey upon these shrimps, which constitute their main food source (Boschi 1963; Gavio and Boschi 2016). In recent years, however, this species has become a common target of both artisanal and industrial fisheries. The increasing fishing pressure and the need for management strategies push knowledge about reproductive traits in this short-life cycle species.

10.2.3 The White Shrimp Peisos petrunkevitchi

The white shrimp *Peisos petrunkevitchi* Burkenroad, 1945 is a small shrimp species endemic to the southwestern Atlantic from Rio de Janeiro, Brazil, through Uruguay to Chubut, Argentina (Boschi et al. 1992; D'Incao and Martins 2000). Females reach a maximum of 45 mm, and the males being smaller than the females reach between 10 and 30 mm (Boschi and Cousseau 2004). White shrimps present a short rostrum, with two dorsal teeth (Costa et al. 2003). It is a typical species of coastal ecosystems, especially at larval stages.

The presence, distribution, and abundance of *P. petrunkevitchi* larvae in plankton samples were studied by Boschi and Scelzo (1969). The entire reproductive cycle takes place in coastal waters, at less than 30 nm from the shoreline. The spawning period lasts from October to December, as revealed by the appearance of eggs, nauplius, and protozoea in plankton samples (Mallo and Boschi 1982; Mallo and Cervellini 1988; Cervellini and Mallo 1991). Larval development from free eggs includes four nauplii, five elaphocaris (protozoea), one acanthosoma, and four mastigopus (postlarvae) stages. Larval development until the first postlarva takes 45–48 days at temperatures of 18 ± 2°C (Mallo 1986). All larval and postlarval stages of *P. petrunkevitchi* are active planktonic filter-feeders (Mallo and Boschi 1982).

White shrimps is an active planktonic filter at all stages of development (Mallo and Boschi 1982), feeding on diatoms and detritus. In turn, they represent an important food resource for other marine animals such as fish, birds, and other crustaceans (Omori 1974). Because of its high abundance and biomass, this is a key species in food webs of coastal ecosystems. It has also been described as prey for large whales, expanding the current knowledge on the possible trophic roles of this species (Bortolotto et al. 2016). In the Bahía Blanca Estuary, *P. petrunkevitchi* was reported as an important food item for juveniles of the striped weakfish *Cynoscion guatucupa* (Sardiña and López Cazorla 2005a), the whitemouth croaker *Micropogonias furnieri* (Sardiña and López Cazorla 2005b), and the Jenyns' sprat *Ramnogaster arcuata* (López Cazorla et al. 2011).

10.3 Migration of Shrimp Larvae and Postlarvae

Shrimps and prawns migrate from coastal to deep waters with trophic and reproductive purposes (Lindley 1987; Wehrtmann 1989; Morgan 1990, 1992). The mechanisms responsible for larval migration from the spawning areas towards estuaries and lagoons vary according to the larval location and developmental stage. However, the immigration of larvae through the estuarine inlets is regulated by the tidal currents, and these become the main mechanism for the recruitment of a variety of fishes and shrimps (Rothlisberg et al. 1995; Blanton et al. 1999; Forward and Tankersley 2001). Several aspects of migration process have been well documented. The number of larvae is generally greater during the flood tide than during the ebb tide (DeLancey et al. 1994; Rothlisberg et al. 1995; Burke et al. 1998; Jager and Mulder 1999). However, the number of larvae that migrate from coastal areas into estuaries aided by flood tide transport depends on the number that arrives at the inlet of the estuary (Blanton et al. 1999; Forward and Tankersley 2001). A second aspect is that many larvae of estuarine-dependent species concentrate outside the inlets in an optimal position to insure the best transport during the next flood cycle (Young and Carpenter 1977; Calderón–Pérez and Poli 1987; Poli and Calderón–Pérez 1987; Rothlisberg et al. 1995; Blanton et al. 1999; Condie et al. 1999).

The spatial and temporal distribution of shrimp larvae and postlarvae in the Bahía Blanca Estuary has been studied for more than two decades (Cervellini and Mallo 1991; Cervellini 1992, 2001). This study has been carried out obtaining samples of mesozooplankton monthly and during periods of active reproduction of the species, biweekly. During spring and summer (October and February), shrimps move offshore to the estuary of Monte Hermoso and migrate to deep waters 15 km from the coast, where salinity conditions are more stable (Fig. 10.2). The females spawn, outside the limits of the estuary, thousands of eggs which are fertilized by males. The eggs are derived in several larval stages and even change their way of feeding (from phyto- to zooplanktophagous). The first stages feed on microscopic algae (phytoplankton), but then they become zooplanktophagous (microcrustaceans) and ingest detritus, and the postlarvae grow until they become juveniles. That's when they enter the estuary and swarm in the many canals, where they find abundant food. The larvae and postlarvae stages of *Peisos petrunkevitchi* have been captured outside the estuary (Cervellini and Mallo 1991). From the results obtained from periodic sampling in the area, we can conclude that larvae and postlarvae stages of Argentine red shrimp *P. muelleri* and *A. longinaris* are exclusively marine; they do not migrate to coastal lagoons and/or estuaries as other penaeid families, thus being distinguished from most commercial subtropical and tropical shrimps.

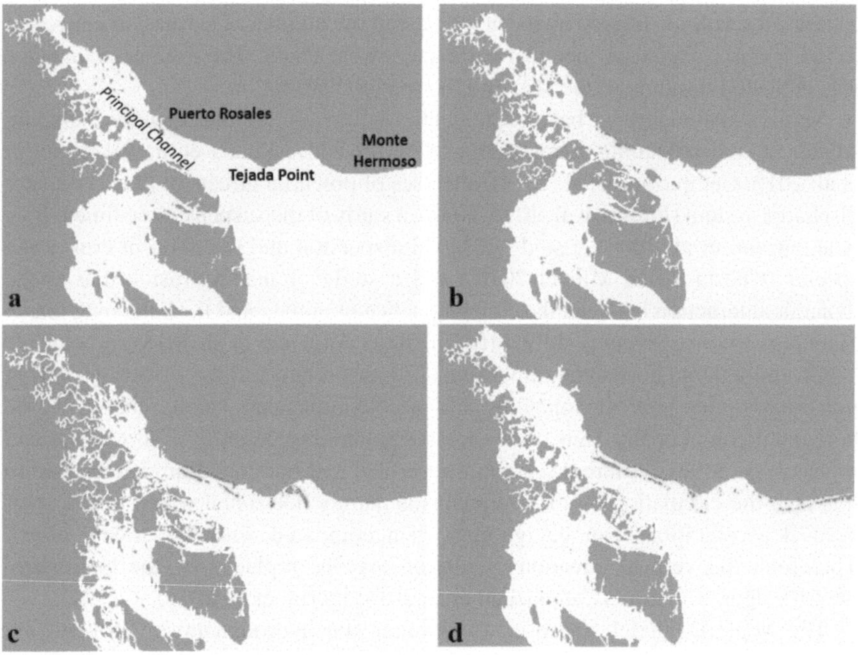

Fig. 10.2 Simulation of dispersal of Penaeidae larvae made with MOHDIS model. References: Localization of larvae in green (**a**) starting point 00:00 hs 05/09/2003, (**b**) 16:30 hs 05/09/2003, (**c**) 16:30 hs 06/09/2003, (**d**) 16:30 hs 07/09/2003

10.3.1 A Hydrodynamic Model

The spatial and temporal variability of physical, biological, and chemical processes observed at the Bahía Blanca Estuary suggests that a more detailed study of the processes controlling the dynamics of the area cannot be done based on field data alone (Campuzano et al. 2014; Angeletti et al. 2018; Pierini et al. 2019). Several measurement efforts would be necessary, and the associated costs would be high. A possible solution for the lack of data is the use of numerical models as sophisticated tools of interpolation and extrapolation of field data, both at spatial and temporal domains. These tools allow the study of different scenarios, forecasting the subsequent environmental answer as well as the determination of exchange fluxes between the estuary and the ocean.

Then to investigate the possible migration of larvae and postlarvae shrimp from penetrated, the estuary of Bahía Blanca was capture and the data obtained from abundance (n° ind/m3) was incorporate into a numerical model. Samples of plankton were collected at 13 stations located along the Principal Channel of the Bahía Blanca Estuary. Data were collected from the monthly survey with Coast Guard GC 75 of the Argentine Naval Prefecture, which covered the mouth of the estuary. The larvae were captured with a 50-cm-diameter plankton net that is 1.90 m long and has 224 mm pore opening. Horizontal and oblique horizontal trawls last 10 minutes at two knots. Samples were fixed with 4% formaldehyde. A flowmeter was used to estimate the volume filtered by the network, and the number of larvae was converted to ind/100 m3. Physical-chemical parameters were taken. The average abundance was indicated throughout the year and the spatial distribution.

Numerical models have been implemented in the Bahía Blanca Estuary, including studies of hydrodynamic characteristics (Pierini 2007; Rueda et al. 2013; Pierini et al. 2012; Campuzano et al. 2014), analyses of potential effects of the wastewater discharge system (Pierini et al. 2012, 2018), a study of the suspension sediment flow (Campuzano et al. 2008), a study of larval dispersion and retention of crustacean species (Cuesta 2010; Miguel 2010), and a study of marsh erosion due to the dynamic interaction between the crab Neohelice granulata and the halophyte marsh plant *Sarcocornia perennis* (Minkoff et al. 2006; Angeletti et al. 2018).

Given the large horizontal dimensions (70 km) relative to the vertical (10 m) in the estuary, the vertical velocities and accelerations are small relative to the horizontal components. Thus, the processes generating quantity of movement and transport occur at a different scale in the vertical and horizontal directions. Due to this fact, the circulation on these domains is mainly horizontal which implies that vertical accelerations can be ignored when compared with the gravity effect. Therefore, the vertical equation of motion may be replaced by the hydrostatic pressure approximation (Campuzano et al. 2014; Pierini et al. 2019).

The MOHID model was used to simulate the hydrodynamics 2D vertically averaged domain model with a horizontal resolution of 0.01° covering from coordinates −61.41W, −39.38S to the inner Bahía Blanca Estuary. The model was calibrated and validated by Campuzano et al. (2014). To adequately represent the study areas, we used a modified version of the high-density bathymetry (50 × 50 m grid) developed by Pierini (2007) with bathymetric data used to compose the model domain coming from two sources: the GEBCO digital atlas, a one-minute global bathymetric grid database, and data from the Bahía Blanca Port Management Consortium (CGPBB) with a waterline obtained from the evaluation of six sets of Landsat 5 TM and Landsat 7 ETM data resulting in a high-density bathymetry (Pierini 2007). Monthly average values of discharge from the main tributaries of the estuary (Sauce Chico River and Napostá Grande Creek) were based on monitored data (Pierini 2007; Campuzano et al. 2014). The tide data were provided by the Bahía Blanca Port Management Consortium. Wind intensity (vector module) and wind direction were provided by the CGPBB and by the private company Oiltanking Ebytem, located close to Villa del Mar (6 km) (Fig. 1.4, Chap. 1). The waves were considered a fetch-based wave generation model to predict the significant wave height and wave period inside the Bahía Blanca Estuary. The fetch wave model uses as input the fetch distances, water depth, wind velocity, and direction in each grid cell. The fetch distances were calculated for 16 directions. As wave propagation is not considered explicitly, this model is only adequate for areas where the ocean does not directly affect the waves. This is the case of our study area due to the geometry of the inlet channel, where only local wind waves are generated.

The MOHID model initially was calibrated and validated in the study area and incorporated with the information obtained at one of the sampling sites and later coupled to Lagrangian (particle-tracking) module, including larvae behavior. The MOHID model calculates the movement of particles that simulate larvae dynamics. In this case, it was implemented to adequately represent the larval migration of shrimp Penaeoidea. The model tracked the trajectories of larvae and was forced with tide, creek discharges (Napostá and Sauce Chico), and local wind conditions which occurred in April 2004, in order to capture a range of environmental variabilities experienced by larvae and postlarvae. We examined whether larval or postlarval behavior and tidal conditions could influence dispersal distance. The grid used in the model is 50 × 50 m. The number of particles released in Puerto Coronel Rosales zone was proportional to the density of larvae and the area of the grid size. Several generalizable results emerge from the present Lagrangian simulation model of larval dispersal. First, predictions of larvae dispersal can be developed from knowledge of the mean and fluctuating hydrodynamic components and wind conditions, coupled with basic information about larvae life history characteristics. Second, expected dispersal scales range from 1 to 100 km depending on these characteristics (Fig. 6.5). In the study and Lagrangian results shown that the larvae and postlarvae of shrimp Penaeoidea were recorded only at the mouth of the estuary of Bahía Blanca and distributed throughout north shore of the Bahía Blanca Estuary where the abiotic factors, mainly salinity, are more constant resembling those of the adjacent continental shelf (Cervellini 1988; Cervellini and Mallo 1991; Cervellini 1992; Perillo and Piccolo 1999; Delgado et al. 2017).

It is known that larval stages, particularly of penaeoid shrimp, require stable sea conditions, since small variations affect larval survival (García and Le Reste 1986). Larvae are found near the Faro buoy, at the height of Monte Hermoso, with certainty that the adults leave the estuary to spawn, where a very complex biological cycle begins and the new specimens enter the estuary as juveniles, to gain weight. Such habitat would offer shelter and abundance of food at early stages of the species. They eat microscopic organisms, such as plankton.

Lastly, future integrative research on dispersal patterns, ocean currents, and larval behaviors will be necessary to determine the relative contribution of oceanography and behavior to realized dispersal patterns.

10.4 Shrimp Fisheries in Argentina

The Argentine red shrimp, *Pleoticus muelleri*, is a target of coastal fisheries along its entire distributional range. In Brazil and Uruguay, the species is a resource of minor importance, but in Argentina it is the commercially most important crustacean, and its fishery is among the 25 major shrimp fisheries in the world (Poli and Calderon-Perez 1987). Based on biological and fisheries data, including reproduction areas and seasons, growth, recruitment, and fisheries concentration, three independent populations have been recognized in Argentine waters for *P. muelleri*: (1) Mar del Plata Port and neighboring areas up to Querandí Lighthouse (25 miles north from Mar del Plata Port), (2) Southern Buenos Aires Province including the Bahía Blanca Estuary, and (3) the Patagonian region (mainly Chubut Province). From the several past decades, landings of this species have been highly variable, but in the last years their catches have consistently increased (FAO 2016; SAGPyA 2017). The fluctuations in abundance of *P. muelleri* are related to changing environmental and oceanographic conditions, which could cause variable mortalities in early life stages. In consequence, there is considerable uncertainty in the sustainable yields (de la Garza et al. 2017). This species is currently captured up to the maximum advisable levels. Landings in 2016 were equivalent to 25% of the total landings of all the marine/continental species combined in Argentina (de la Garza et al. 2017).

Globally, bycatch, or the incidental catch of nontarget organisms, is a major issue in shrimp trawl fisheries, which may have detrimental effects on marine resources and wildlife (Gillet 2008; Kelleher 2008). More than a hundred species are caught by shrimp trawlers in the Patagonian region, and the Argentine hake, *Merluccius hubbsi*, is the most frequent and abundant species in these incidental catches (Bertuche 1999; Pettovello 1999; Gongora et al. 2009). Hake is the most important demersal species for the commercial fishing in Argentina (Cordo 2005). Incidental catches have been also recognized as a significant source of seabird mortality and one of the main threats to seabirds at sea (Croxall et al. 2012). The Patagonian shrimp fishery has a negative effect on populations of the Magellanic penguin, *Spheniscus magellanicus*, due to direct incidental mortality and also by overlapping between penguin diet (anchovy and hake) and bycatch species (Gandini et al. 1999; Marinao and Yorio 2014). Between 2010 and 2017, 70–100% of the red shrimp production was fished within San Jorge Gulf; currently, the area is closed for protection. The capture of Argentine red shrimp in the period between 2004 and 2016 is shown in Fig 10.3.

In recent years, the Argentine stiletto shrimp, *Artemesia longinaris*, has become an important target of the commercial fisheries as a consequence of the decline of more traditional and profitable marine shrimps. Artisanal fisheries occur along its entire distribution, and industrial fisheries are mainly concentrated in southern Brazil and Argentina (D'Incao et al. 2002). Catches are mostly commercialized in the internal market and show large seasonal and inter-annual fluctuations, leading to efforts through a commercially sustainable aquaculture, capable of providing a continuous supply. It is captured in appreciable concentrations over a narrow coastal strip that does not exceed 5 miles (depths between 2 and 30 m). Recent studies suggest that it is possible to consider *A. longinaris* as a metapopulation in which there are more stable subpopulations, with high densities throughout the year, such as those from southern Brazil to Argentina and from Macaé (Boschi 1969). These high-density subpopulations may serve as sources of new individuals for the less-stable sink populations, like those from São Paulo State. Source populations would be critical for the implementation of management strategies, such as the creation of protected areas and the implementation of off-season periods, directed to sustain fisheries along the entire range of distribution (Carvalho-Batista et al. 2014).

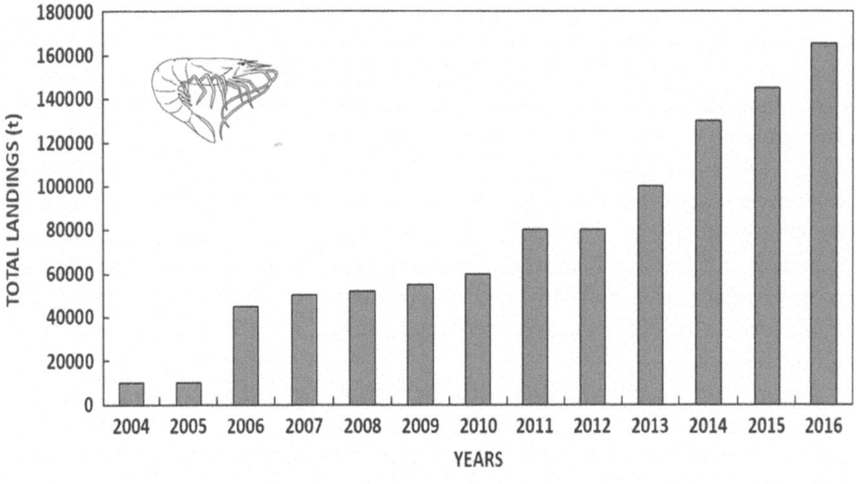

Fig 10.3 Evolution of landings registered for the Argentine red shrimp, *Pleoticus muelleri*, between 2004 and 2016 for all ports in Argentina. (Adapted Secretary of Fisheries and Aquaculture. Ministry of Agriculture, Livestock and Fisheries. Argentina)

10.4.1 Artisanal Shrimp Fisheries in the Bahía Blanca Estuary

In the southern coast of Buenos Aires Province, *Pleoticus muelleri* and *Artemesia longinaris* are target species of small-scale artisanal fisheries. Artisanal fisheries are defined as those who extract fishing resources for their commercialization, using small-scale, low-technology, and low-capital fishing practices (Defeo and Castilla 2005). Along the Bahía Blanca Estuary and Anegada Bay (Fig 1.2, Chap. 1), several sites were identified as fishing zones, but 75% of the total catches of both species come from three areas: the Principal Channel (45%) and Vieja Channel (11%) in the Bahía Blanca Estuary and Riacho Azul (14%) located in Anegada Bay (Suquele and Colautti 2005). The main landing ports are Ingeniero White and Coronel Rosales in the Bahía Blanca Estuary (Figs. 2.2 and 2.3, Chap. 2).

The artisanal fleet is composed of small boats (up to15 m length) locally known as *flota amarilla* (yellow fleet) with limited autonomy, and each boat may be accompanied by several smaller boats (Fig. 10.4). Instead of the typical bottom trawling, shrimp fishers in the Bahía Blanca Estuary make use of tidal currents. Fishing nets like those used for trawling are deployed at high tide or ebb tide, each net attached to a pair of anchors. Fishing operations are carried through complete flood tides or ebb tides, and shrimps, along with incidental fishes and crustaceans dragged by bottom tidal currents, get trapped into the net. Each fishing boat carries about 10 nets, along with 20 anchors, and the fishing maneuver is all manual labor carried by fishers on board the small accompanying boats (Bertuche et al. 1998).

Nets are revised twice during each tide to take the catch. The management and control of the nets are carried out by the canoes driven by an outboard motor or oars. The boats are generally in number of six. The catch of each net is transported to the main vessel, which remains fixed in one place throughout the fishing operation. On board, fishermen manually separate *P. muelleri* and *A. longinaris* from and accompanying fauna and select those shrimps that are commercial in size. In addition, the crustaceans that will be used for sale are processed (boiled, salted, or sulfided) or placed in a cool place to ensure they arrive in good condition at port. The choice of specimen preservation will depend on the remaining time and temperature for landing. In the Bahía Blanca Estuary, there are two fishing seasons: one in summer (January to May) and the other in winter (August to spring). In the former nets are located in the middle of water column; in the latter nets are anchored over the bottom.

Shrimp abundance in the Bahía Blanca Estuary has strong seasonal and inter-annual oscillations for both species (Cervellini and Piccolo 2007). The migratory breeder behavior would be related with seasonal landing fluctuations. In the past, outside the estuary, there was an important fishing activity of trawlers of greater magnitude and autonomy that come from northern Buenos Aires Province, such as Mar del Plata Port, that could impact negatively over the local fisheries. Since 2006 fishing with trawls is prohibited in El Rincón area, outside the Bahía Blanca Estuary; however, it has not been evaluated whether this management measure has favored

Fig. 10.4 (**a**) Artisanal shrimp fishing boat, (**b**) fishermen select the catches (**c**) bycatch. (Photo by: (a) Sandra Fiori, (b and c) Gabriela Blasina)

artisanal shrimp and prawn fisheries into the estuary. On the other hand, the inter-annual variability seems to be related with climatic conditions. It was observed that when the average annual temperature exceeded 16 °C, in the following year or in the same year, the catches of the resource decreased significantly (Fig. 10.5). The maximum catches in the Bahía Blanca Estuary are registered every 3–5 years (Cervellini and Piccolo 2007).

Since 1999, one of the rules that must be obeyed by those who practice artisanal or commercial fishing is the obligation to submit to the enforcement authority in the Delegation that has the Secretary of Fisheries Activities of the Province of Buenos Aires, in the Port of Ingeniero White; they must declare to the authority the catches made in each tide. When the information contained in the parts corresponding to a given period of time, provides valuable information, allow understand the dynamics of the fishery or implement management measures that tend to achieve sustainable exploitation of resources.

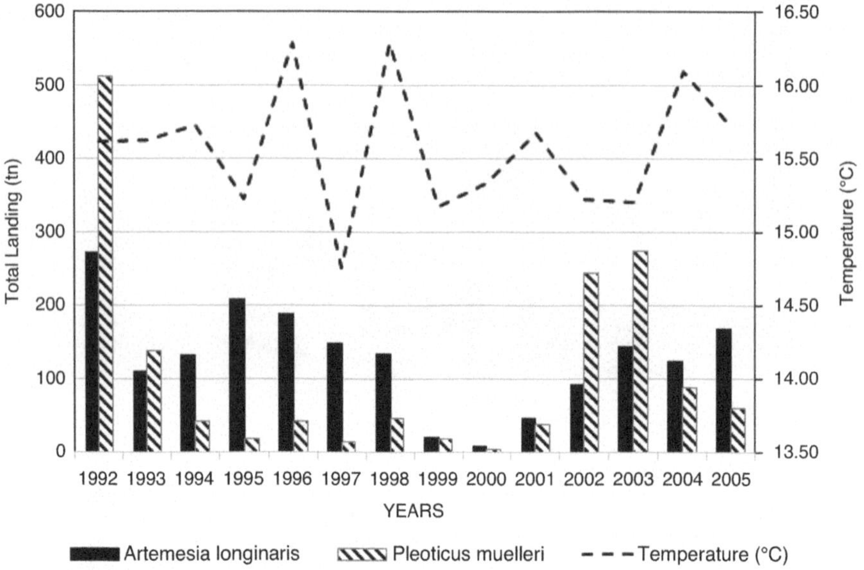

Fig. 10.5 Annual catches of *Pleoticus muelleri* and *Artemesia longinaris* and average annual temperature in the Bahía Blanca Estuary during 1992 and 2005. (Adapted from Cervellini and Piccolo 2007)

10.5 Conclusions

There are many factors that affect the presence of shrimp in the Bahía Blanca Estuary, both natural and anthropogenic. Both species have very different fluctuations in annual catch volumes. The dynamics of the fishery in the estuary is markedly seasonal and is related to the biological characteristics of the species. It is recommended to monitor the status of these populations taking into account two factors: the cycle study life and the dynamics of juveniles and adults. It is also recommended to control the fishing activities that take place in the outside area of the estuary. In this way, exploitation will be optimized through the creation of scientific-technical bases for sustainable development of the fisheries of these two resources so valuable to the community of Bahía Blanca. As we have anticipated in previous paragraphs, eggs and larval stages of crustaceans penaeoid are an important part of the zooplankton. The abundance is of great interest in the economy of the ecosystem and allows to know the fauna composition of the zooplankton, especially of those members that behave like meroplankton, the distribution and abundance of them becomes important within the plankton for being indicators of water bodies, for being food for another. Finally, we want to point out that one of the great advantages of using a mathematical model for integrated coastal zone management is that datasets collected in different periods and sampling intervals can be integrated into a single tool to reproduce and analyze the processes taking place in the water body. The hydrodynamic module of the MOHID model has been validated and calibrated for the entire estuary, but it is important to note that there is only oceanographic data about the Principal Channel and poor data acquisition for the rest of the estuary. From the performed simulations, it could be concluded that horizontal dispersion of larvae is only at the outer part (mouth) of the Bahía Blanca Estuary and has a significant importance over horizontal distribution throughout the northern shore of the estuary where currents and abiotic factors, mainly salinity, are more constant. This conclusion was obtained by model results, taking into account the sampling period of the larvae, where Lagrangian particles develop their dispersion like passive particles that move and extend over this coastal sector where they were also affected by the atmospheric (wind) and oceanographic parameters (currents, waves, etc.). The combination of data analysis and numerical modelling allows a more comprehensive conceptual model of the Bahía Blanca Estuary hydrodynamics that could aid in decision-making of local managers. One of the greatest advantages of using modelling tools in integrated coastal zone management is that datasets collected at different periods and sampling intervals could be consolidated in a tool to reproduce periodical phenomena, i.e., tides and currents, and to analyze different system descriptors such as tidal or current distribution, larval or postlarval spatial distribution, etc. Numerical modelling also allows the isolation and discrimination of single processes and can aid with data accuracy and consistency – important factors to consider if we want to analyze larval retention and nursery areas and in the future observe their influence in fishing. Multidisciplinary analysis to date has not been carried out in our study area which is vital for a correct coastal management.

References

Angelescu VA, Boschi EE (1959) Estudio biológico pesquero del langostino de Mar del Plata en conexión con la operación nivel medio. Servicio Hidrografía Naval, Público H-1017, 137 p

Angeletti S, Pierini JO, Cervellini PM (2018) Suspended sediment contribution resulting from bioturbation in intertidal sites of a SW Atlantic mesotidal estuary: data analysis and numerical modelling. Sci Mar 82(4):245–256

Bailey-Brock JH, Moss SM (1992) Penaeid taxonomy, biology and zoogeography. In: Fast AW, Lester LJ (eds) Marine shrimp culture: principles and practices. Developments in aquaculture and fisheries science, vol 23. Elsevier Science Publisher BV, Dordrecht

Bertuche D (1999) Sobre la captura incidental de la merluza en la pesquería del langostino patagónico Inf. Tec. INIDEP N° 76/99, 5pp

Bertuche D, Fischbach C, Boccanfuso J (1998) Síntesis de los resultados obtenidos en el monitoreo del recurso langostino realizado en el sector costero de la Pcia. De Bs. As, al sur de Puerto Ingeniero White. Inf. Téc. Inst. INIDEP N°129

Blanton J, Werner FE, Kapolnai A et al (1999) Wind-generated transport of fictitious passive larvae into shallow tidal estuaries. Fish Oceanogra 8(2):210–223

Bortolotto GA, Kolesnikovas CKM, Freire AS et al (2016) Young humpback whale *Megaptera novaeangliae* feeding in Santa Catarina coastal waters, Southern Brazil, and a ship strike report. Mar Biodivers Rec 9(1):29

Boschi EE (1963) Los camarones comerciales de la familia Penaeidae de la costa Atlántica de América del Sur. Clave para el reconocimiento de las especies y datos bioecológicos. Boletín del Instituto de Biología Marina (3):1–40

Boschi EE (1969) Estudio Biológico Pesquero del camarón *Artemesia longinaris* Bate de Mar del Plata. Bol Inst Biol Mar N° 18:1–4
Boschi EE (1986) La pesquería del langostino del litoral patagónico. Cuadernos de Redes 20:20–26
Boschi EE (1989) Biología pesquera del langostino del litoral patagónico de Argentina (*Pleoticus muelleri*). Ser Contrib INIDEP Mar del Plata 646:1–71
Boschi EE (1997) Las pesquerías de crustáceos decápodos en el litoral de la República Argentina. Invest Mar Valparaíso 25:19–40
Boschi EE (2004) Camarones y especies afines. In: Boschi EE, Cousseau MB (eds) La vida entre mareas, vegetales y animales de las costas de Mar del Plata, INIDEP, Argentina, pp 205–212
Boschi EE, Gavio MA (2005) On the distribution of decapod crustaceans from the Magellan Biogeographic Province and the Antarctic region. Sci Mar 69(S2):195–200
Boschi EE, Scelzo MA (1967) Campaña de Pesca Exploratorio Camaronera en el Litoral de la Provincia de Buenos Aires, 16-24 de Febrero de 1967, Proy Des Pesq (FAO), Publ N°2:1–22
Boschi EE, Scelzo MA (1969) Nuevas Campañas Exploratorias Camaroneras en el Litoral Argentino, con Referencias al Plancton de la Región. Proy Des Pesq (FAO), Publ N°16:1–31
Boschi EE, Scelzo MA (1971) Últimos resultados de las investigaciones sobre los peneidos comerciales en la Argentina (Marcaciones, Campañas camarones y cultivos). CARPAS/5/Doc Téc 4:1–17
Boschi EE, Cousseau MB (2004) La vida entre mareas: Vegetales y animales de las costas de Mar del Plata, Argentina, 1 ed. INIDEP, Mar del Plata
Boschi EE, Fischbach CE, Iorio MI (1992) Catálogo ilustrado de los crustáceos estomatópodos y decápodos marinos de Argentina
Burke JS, Ueno M, Tanaka Y et al (1998) The influence of environmental factors on early life history patterns of flounders. J Sea Res 40(1):19–32
Burkenroad MD (1945) A new sergestid shrimp (*Peisos petrunkevitchii*, n. gen., n. sp.), with remarks on its relationships. Trans Connecticut Acad Arts Sci 36:553–591
Calderón-Pérez JA, Poli CR (1987) A physical approach to the postlarval *Penaeus* immigration mechanism in a Mexican coastal lagoon (Crustacea: Decapoda: Penaeidae). Anales del Instituto de Ciencias del Mar y Limnología. Universidad Nacional Autónoma de México 14(1):147–156
Campuzano FJ, Pierini JO, Leitão PC (2008) Hydrodynamics and sediments in Bahia Blanca estuary: data analysis and modelling. In: Neves R, Baretta J, Mateus M (eds) Perspectives on integrated coastal zone Management in South America. IST Press, Lisbon, pp 483– 503
Campuzano FJ, Pierini JO, Leitão PC (2014) Characterization of the Bahía Blanca estuary by data analysis and numerical modelling. J Mar Syst 129:415–424. https://doi.org/10.1016/10.1016/j.jmarsys.2013.09.001
Capitoli R, Bager A, Ruffino M (1994) Contribuçao ao conhecimento das relaçoes tróficas bentônico-demersais nos fundos de pesca do camarao *Artemesia longinaris* Bate, na regiao da Barra da Lagoa dos Patos, RS, Brasil. Nauplius 2:53–74
Carvalho-Batista A, Simões SM, Lopes M et al (2011) Ecological distribution of the shrimp *Pleoticus muelleri* (Bate, 1888) and *Artemesia longinaris* Bate, 1888 (Decapoda, Penaeoidea) in the southeastern Brazilian littoral. Nauplius 19:135–143
Carvalho-Batista A, Negri M, Pileggi LG et al (2014) Inferring population connectivity across the range of distribution of the stiletto shrimp *Artemesia longinaris* Spence Bate, 1888 (Decapoda, Penaeidae) from DNA barcoding: implications for fishery management. ZooKeys 457:271
Castilho AL, Gavio MA, Costa RC et al (2007) Latitudinal variation in population structure and reproductive pattern of the endemic south american shrimp Artemesia longinaris (Decapoda: Penaeoidea). J Crustacean Biol 27(4):548–552
Castilho AL, Da Costa R, Fransozo A et al (2008) Reproduction and recruitment of the South American red shrimp, *Pleoticus muelleri* (Crustacea: Solenoceridae), from the southeastern coast of Brazil. Mar Biol Res 4(5):361–368. https://doi.org/10.1080/17451000802029536
Cervellini PM (1988) Las larvas y postlarvas de los crustáceos Decápoda en el estuario de Bahía Blanca. Variaciones estacionales y su relación con los factores ambientales. PhD thesis, Universidad Nacional del Sur, Bahía Blanca, Argentina, 246 pp

Cervellini PM (1992) Seasonal and spatial distribution of decapod Crustacea larvae in the Bahía Blanca estuary, Argentina. J Aqua Trop 7:35–46

Cervellini PM (2001) Variabilidad en la abundancia y retención de larvas de crustáceos decápodos en el estuario de Bahía Blanca, Provincia de Buenos Aires, Argentina. Invest Mar Valparaíso 29(2):25–33

Cervellini PM, Mallo J (1991) Distribución espacial de estadios larvales de Penaeidae en Bahía Blanca, Argentina. Biol Pesq 20:5–11

Cervellini PM, Piccolo MC (2007) Variación anual de la pesca del langostino y camarón en el estuario de Bahía Blanca. Geoacta 32:111–118

Condie SA, Lonerang NR, Die DJ (1999) Modeling the recruitment of tiger prawns *Penaeus esculentus* and *P. semisulcatus* to nursery grounds in the Gulf of Carpentaria, northern Australia: Implications for assessing stock-recruitment relationships. Mari Ecol Progr Series 178:55–68

Cordo H (2005) Evaluación del estado del efectivo sur de 41°S de la Merluza (*Merluccius hubbsi*) y estimación de la captura biológicamente aceptable correspondiente al año 2005. Instituto Nacional de Investigación y Desarrollo Pesquero (INIDEP), Mar del Plata

Costa RCD, Fransozo A, Melo GAS, Freire FADM (2003) Chave ilustrada para identificação dos camarões Dendrobranchiata do litoral norte do estado de São Paulo, Brasil. Biota Neotropica 3(1):1–12

Costa RC, Fransozo A, Pinheiro AP (2004) Ecological distribution of the shrimp *Pleoticus muelleri* (Bate, 1888) (Decapoda: Penaeoidea) in southeastern Brazil. Hydrobiologia 529(1-3):195–203

Croxall JP, Butchart SHM, Lascelles B et al (2012) Seabird conservation status, threats and priority actions: a global assessment. Bird Conservation International 22:1–34

Cuesta AC (2010) Composición, abundancia estacional y dispersión horizontal de especies de mesozooplancton en la parte media del estuario de Bahía Blanca (Puerto Rosales). Tesis de Licenciatura en Ciencias Biológicas de la Universidad Nacional del Sur, 50 pp

D'Incao F (1999) Subordem Dendrobranchiata (camarões marinhos). In: Buckup L, Bond Buckup G (eds) Os Crustaceos do Rio Grande do Sul. Editora da Universidade, Porto Alegre, pp 275–299

D'Incao F, Martins STS (2000) Brazilian species of the genera *Acetes* H. Milne Edwards, 1830 and *Peisos* Burkenroad, 1945 (Decapoda: Sergestidae). J Crustacean Biol 20(5):78–86

Dall W, Hill J, Rothlisberg PC et al (1990) The biology of Penaeidae. Adv Mar Biol 27

De la Garza J, Moriondo Danovaro P, Fernández, M (2017) An overview of the argentine red shrimp (*Pleoticus muelleri*, Decapoda, Solenoceridae) fishery in Argentina. Biology, fishing, management and ecological interactions. Mar del Plata: Instituto Nacional de Investigaciony Desarrollo Pesquero INIDEP. Contribution N° 2080

Defeo O, Castilla JC (2005) More than one bag for the world fishery crisis and keys for co-management successes in selected artisanal Latin American shellfisheries. Rev Fish Biol Fisher 15:265–283

DeLancey LB, Jenkins JE, Whitaker JD (1994) Result of long term, seasonal sampling for *Penaeus* postlarvae at Breach inlet, South Carolina. Fish Bull 92:633–640

Delgado AL, Ferrelli F, Piccolo MC et al (2017) Influence of Physical-Biological Variability and Legal Regulations on Fisheries Resources in the Southern Coastal Zone of Buenos Aires Province. Argentina Argentina Anuário do Instituto de Geociências – UFRJ 40:5

FAO (2016) El estado mundial de la pesca y la acuicultura 2016. Contribución a la seguridad alimentaria y la nutrición para todos. Roma, 224 p

Forward RB Jr, Tankersley RA (2001) Selective tidal stream transport of marine animals. Oceanogr Mar Biol. An Annual Review 39:305–353

Fransen CHJM (2014) Shrimps and prawns. The living marine resources of the Eastern Central Atlantic 1:37–196

Gandini PA, Frere E, Pettovello AD et al (1999) Inter-action between Magellanic penguins and shrimp fisheries in Patagonia, Argentina. Condor 101:783–789

García S, Le Reste L (1986) Ciclos vitales, dinámica, explotación y ordenación de las poblaciones de camarones peneidos costeros. FAO Doc. Téc. Pesca 203:180

Gavio MA, Boschi EE (2004) Biology of the shrimp *Artemesia longinaris* Bate, 1888 (Crustacea: Decapoda: Penaeidae) from Mar del Plata coast, Argentina. Nauplius 12(2):83–94

Gavio MA, Boschi EE (2016) Historia de vida del camarón de Mar del Plata *Artemesia longinaris* (Decapoda, Penaeoidea, Penaeidae). In: Boschi EE (ed) Los crustáceos de interés pesquero y otras especies relevantes en los ecosistemas marinos. El Mar Argentino y sus recursos pesqueros, 6. Instituto Nacional de Investigación y Desarrollo Pesquero INIDEP, Mar del Plata, pp 51–58

Gillet R (2008) Global study of shrimp fisheries. Fisheries technical paper no. 475. Food and Agriculture Organization of the United Nations, Rome, pp 331

Góngora ME, Bovcon ND, Cochia PD (2009) Ictiofauna capturada incidentalmente en la pesquería de langostino patagónico *Pleoticus muelleri* Bate, 1888. Rev Biol Mar Oceanogr 44(3):583–593

Incao, Fernando & Valentini, Helio & Rodrigues, Luiz. (2002). Avaliação da pesca de camarões nas regiões Sudeste e Sul do Brasil. Atlântica. 24. 103–116

Iorio MI, Scelzo MA, Boschi EE (1990) Larval and post-larval development of the shrimp *Pleoticus muelleri*, Bate, 1888 (Crustacea, Decapoda, Solenoceridae). Sci Mar 54(4):329–341

Iorio MI, Macchi GJ, Fischbach CE et al (1996) Estudios sobre la dinámica reproductiva del langostino *Pleoticus muelleri* en el área de Bahía Blanca (Provincia. De Buenos Aires, Argentina). Frente Marit 15:111–118

Jager Z, Mulder J (1999) Transport velocity of flounder larvae (*Platichthys flesus* L.) in the Dollard (Ems Estuary). Estuar Coast Shelf Sci 49(3):327–346

Kelleher K (2008) Descartes en la pesca de captura marina mundial. FAO Documento Técnico de Pesca 470:1–147

Lindley JA (1987) Continuous Plankton Records: the geographical distribution and seasonal cycles of decapod crustacean larvae and pelagic post-larvae in the north-eastern Atlantic Ocean and the North Sea, 1981-3. J Mar Biol Ass UK 67:145–167

López Cazorla A, Pettigrosso RE, Tejera L et al (2011) Diet and food selection by *Ramnogaster arcuata* (Osteichthyes, Clupeidae). J Fish Biol 78:2052–2060

Macchi GJ, Iorio MI, Christiansen HE (1992) Aspectos del desove y fecundidad del langostino *Pleoticus muelleri* (Bate, 1888) (Crustacea, Decapoda, Solenoceridae). Rev Biol Mar 27(1):43–58

Mallo JC (1986) Desarrollo larval y cultivo en laboratorio del camarón argentino *Peisos petrunkevitchi* Burkenroad, 1945 (Crustacea, Decapoda, Sergestidae). Biología Pesquera 15:316

Mallo JC, Boschi EE (1982) Contribución al conocimiento del ciclo vital del camarón *Peisos petrunkevitchi* de la región de Mar del Plata, Argentina (Crustacea, Decapoda, Sergestidae). Physis, Buenos Aires A 41:85–98

Mallo J, Cervellini PM (1988) Distribution and abundance of larvae and postlarvae of *Artemesia longinaris*, *Pleoticus muelleri* and *Peisos petrunkevitchi* (Crustacea, Decapoda; Penaeidae) in the coastal waters in the Bahía Blanca Bay, Argentina. Aquacult Trop 3:1–9

Marinao CJ, Yorio PM (2014) Fishery discards and incidental mortality of seabirds attending coastal shrimp trawlers at Isla Escondida, Patagonia, Argentina Wilson Ornithological Society 2011–2012

Miguel C (2010) Variación estacional de las larvas de Decápoda en la parte interna del estuario de Bahía Blanca (Puerto Cuatreros) y dispersión horizontal de las zoeas de *Neohelice granulata* (=*Chasmagnathus granulatus*) (Decápoda, Varunidae) en relación con la circulación del agua. Tesis de grado de la Universidad Nacional del Sur, 40 pp

Minkoff DR, Escapa M, Ferramola FE et al (2006) Effects of crab-halophytic plant interactions on creek growth in a S.W. Atlantic Salt Marsh: A cellular automata model. Estuar Coast Shelf Sci 69:403–413

Morgan SG (1990) Impact of planktivorous fishes of dispersal, hatching, and morphology of estuarine crab larvae. Ecology 71:1639–1652

Morgan SG (1992) Predation by planktonic and benthic invertebrates on larvae of estuarine crabs. J Exp Mar Biol Ecol 163:91–110

Omori M (1974) The biology of pelagic shrimps in the ocean. Adv Mar Biol 12:233–324

Pérez Farfante I (1988) Illustrated key to penaeoid shrimps of commerce in the Americas. NOAA Technical Report NMFS 64

Perillo GME, Piccolo MC (1999) Geomorphological, physical characteristics of Bahía Blanca estuary, Argentina. The Argentina estuaries: a review. In: Perillo GM, Piccolo MC, Pino Quivira M (eds) Estuaries of South America: their geomorphology and dynamics. Springer, Berlin, pp 195–216

Pettovello A (1999) By-catch in the Patagonian red shrimp (*Pleoticus muelleri*) fishery

Pierini JO (2007) Circulation and transport in coastal areas of the Bay estuary Blanca, PhD thesis, University of Buenos Aires, Buenos Aires, 225 pp. (in Spanish)

Pierini JO, Lovallo M, Telesca L et al (2012) Investigating prediction performance of an artificial neural network and a numerical model of the tidal signal at Puerto Belgrano, Bahía Blanca estuary (Argentina) Acta Geophysica 61(6) Dec 2013:1522–1537. https://doi.org/10.2478/s11600-012-0093-x

Pierini JO, Campuzano FJ, Leitão CP et al (2018) Atmospheric influence over the residence time in the Bahia Blanca Estuary, Argentina. Thalass Int J Mar Sci. https://doi.org/10.1007/s41208-018-0120-z

Pierini JO, Campuzano FJ, Leitão PC et al (2019) Atmospheric influence over the time residence in the Bahía Blanca estuary. Thalass Int J Mar Sci 35(1):275–286. https://doi.org/10.1007/s41208-018-0120-z

Poli CR, Calderón-Pérez JA (1987) Effect of hydrological changes at the mouth of the Baluarte River on the immigration of postlarvae from *Penaeus vannamei* (Boone) and *P. stylirostris* (Stimpson) to Huizache-Caimanero lagoon system, Sinaloa, Mexico (Crustacea: Seafood Watch 2018)

Rothlisberg PC, Church JA, Fandry B (1995) A mechanism for near-shore concentration and estuarine recruitment of post-larval *Penaeus plebejus* Hess (Decapoda, Penaeidae). Estuar Coast Shelf Sci 40(2):115–138

Rueda JG, Otero-Díaz LJ, Pierini JO (2013) Caracterización hidrodinámica en un estuario tropical de América del Sur con régimen microtidal mixto (Bahía de Cartagena, Colombia) Boletín científico. CIOH 31:159–174

SAGPyA (2017) Resolución 304-2017. Secretaría de Agricultura, Ganadería, Pesca y Alimentación. Sanidad Animal

Sardiña P, Lopez Cazorla A (2005a) Feeding habits of the juvenile striped weakfish, *Cynoscion guatucupa* Cuvier 1830, in Bahía Blanca estuary (Argentina): seasonal and ontogenetic changes. Hydrobiologia 532:23–30

Sardiña P, Lopez Cazorla A (2005b) Trophic ecology of the of whitemouth croaker, *Micropogonias furnieri* (Pisces: Sciaenidae), in south-western Atlantic waters. J Mar Biol Assoc UK 85:405–413

Scelzo MA (2016) Biología reproductiva del langostino y del camarón de las aguas marinas argentinas. In: Boschi EE (ed) Los crustáceos de interés pesquero y otras especies relevantes en los ecosistemas marinos. El Mar Argentino y sus recursos pesqueros, 6. Instituto Nacional de Investigación y Desarrollo Pesquero INIDEP, Mar del Plata, pp 71–88

Scelzo MA, Martínez Arca J, Lucero NM (2002) Diversity, density and biomass of the macrofauna component of the "shrimp-shrimp" fishing funds, in front of Mar del Plata, Argentina (1998–1999). National Institute for Fisheries Research and Development, Journal of Fisheries Research and Development 15:43–65

Spivak ED, Farías NE, Ocampo EH et al (2019) Annotated catalogue and bibliography of marine and estuarine shrimps, lobsters, crabs and their allies (Crustacea: Decapoda) of Argentina and Uruguay (Southwestern Atlantic Ocean). Frente Marítimo 26:1–164

Suquele P, Colautti D (2005) Pesca artesanal y comercial en los puertos de la Ría de Bahía Blanca, Buenos Aires. Análisis de las partes pesqueras correspondientes a los años 2000, 2001, 2002, 2003 y 2004. Dirección de Asuntos Agrícolas, 13 pp

Tavares C, Martin J (2010) Suborder Dendrobranchiata Bate, 1888. In: Schram FR, von Vaupel Klein JC, Forest J, Charmantier-Daures M (eds) Treatise on Zoology – Anatomy, Taxonomy, Biology – The Crustacea, Decapoda, Volume 9 Part A Eucarida: Euphausiacea, Amphionidacea, and Decapoda (partim), vol 9A, pp 99–164

Tavares C, Serejo C, Martin J (2009) A preliminary phylogenetic analysis of the Dendrobranchiata based on morphological characters. In: Martin JW, Crandall KA, Felder DL (eds) Decapod crustacean Phylogenetics. Crustacean Issues 18:255–273

Thayer GW, Schaaf WE, Angelovic JW et al (1973) Caloric measurements of some estuarine organisms. Fish B-NOAA 71(1):289–296

Wehrtmann IS (1989) Seasonal occurrence and abundance of caridean shrimp larvae at Helgoland. German Bight Helgoländer Meeresuntersuchungen 43(1):87–112

Young PC, Carpenter SM (1977) Recruitment of postlarval penaeid prawns to nursery areas in the Moreton Bay, Queensland. Mar Freshwater Res 28:745–773

Chapter 11
Ecology and Biology of Fish Assemblages

Juan Manuel Molina, Gabriela Blasina, and Andrea Lopez Cazorla

11.1 Estuaries as Key Habitat for Fish

Estuaries are important habitats for several species of fish, due to the high levels of primary production which supports highly diverse and abundant prey. Estuaries also provide nursery and feeding grounds for many species (Elliott et al. 2007). In these regions environmental gradients are large and generate a unique combination of biotic and abiotic factors (Day et al. 1989). The major biotic and abiotic factors which determine the distribution and abundance of fish in estuaries are shown in Fig. 11.1. These factors are not independent but interact directly and indirectly with the fish species that inhabit estuaries. Hydrographic conditions directly influence mouth condition, estuarine water temperature, salinity, turbidity, and dissolved oxygen concentrations and indirectly affect habitat diversity, productivity, fish recruitment, food availability, and competition (Whitfield 1999). The high productivity of estuaries has often been identified as the main reason why fish are attracted to these areas in such large numbers, which is explained by the large food webs these environments support. Food, especially detritus and benthic invertebrates, is abundant in most estuarine systems. However, the availability of a particular food type is likely to show marked fluctuations over time and space, particularly in response to environmental changes which characterize all types of systems on the subcontinent (Potter et al. 2015). Species that are broadly tolerant of biotic and abiotic variability are at a considerable advantage over those fish species that cannot survive such fluctuations, because the former are able to occupy a food-rich environment from which many potential competitors are excluded (Whitfield 1999).

Bahía Blanca Estuary thrives with fish life. This estuary is a complex ecosystem encompassing several islands, salt marshes, mudflats, and large tidal flats which

J. M. Molina (✉) · G. Blasina · A. L. Cazorla
Departamento de Biología, Bioquímica y Farmacia, Universidad Nacional del Sur, Bahía Blanca, Argentina

Instituto Argentino de Oceanografia (IADO-CONICET/UNS), Bahía Blanca, Argentina

S. M. Fiori, P. D. Pratolongo (eds.), *The Bahía Blanca Estuary*,
https://doi.org/10.1007/978-3-030-66486-2_11

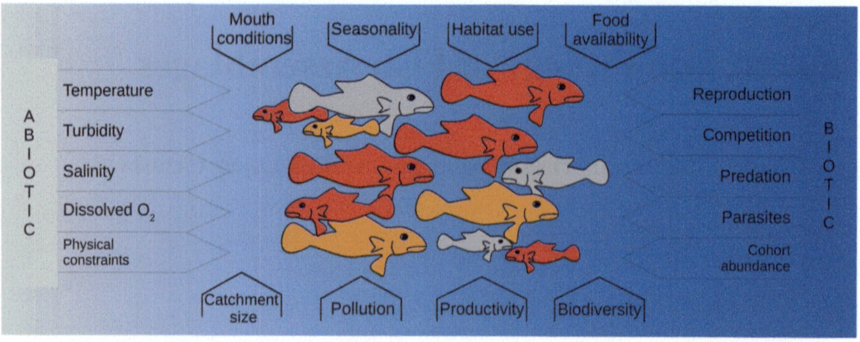

Fig. 11.1 Major abiotic (blue) and biotic (green) factors that influence the distribution and abundance of fish populations in estuaries. (Modified from Day et al. 1989)

generate marked environmental gradients thus promoting a high diversity of biological systems in a relatively small area (see Chaps. 1 and 3 of this book). This diverse ecotone provides several different types of habitats for fish. A world-scale review by Blaber (2002) states that the number of fish species in subtropical and tropical estuaries is much greater than in temperate regions: at least 100 species, with some reaching more than 200. Pasquaud et al. (2015) evaluated the latitude as potential factor determining fish species richness in estuaries, and they also identified a fish species richness distribution in relationship with latitude, with the lowest values of richness at 40° S. Yañez-Arancibia et al. (1980) recorded 121 species in tropical estuaries and lagoons of the Caribbean Sea. Atlantic areas appear relatively richness-poor regarding the worldwide average species richness of tropical estuaries (Simier et al. 2006). In the southwestern Atlantic temperate regions, the estuarine ichthyofauna is composed of ~110 species in dos Patos Lagoon (32° S), South Brazil (Chao et al. 1985); 119 fish species inhabit the Río de la Plata estuary (34° S) (Jaureguizar et al. 2016); and a total of 28 species were registered in Mar Chiquita coastal lagoon (37° S) in Argentina (González Castro et al. 2009). The number of fish species (32; Table 11.1) registered in Bahía Blanca Estuary (39° S) is consistent with the widely held view that latitude plays a critical role influencing richness, with tropical areas being more diverse than temperate ones.

More than 25 different mechanisms have been suggested for generating systematic latitudinal patterns in biodiversity, commonly emphasizing reasons as to why the tropics are highly specious. These include explanations based on chance, historical perturbation, environmental stability, habitat heterogeneity, productivity, and interspecific interactions. In these explanations, mean annual temperature is commonly used as a proxy of the energy of the system (Gaston 2000), i.e., the systems with a higher temperature would have higher energy, although this is not always the case. Among the different hypotheses that have been proposed in order to explain gradients in species richness, the species richness-energy relationship hypothesis received the greatest support, mostly from studies of terrestrial or freshwater organisms. For marine organism, a few studies addressed this framework, and their results are not consensual. In estuaries, the species richness-energy relationship has not

Table 11.1 Fish species presence in the inner, middle, and external zones of Bahía Blanca Estuary in the two periods of sampling: 1979–1983 and 2017–2018

	Inner zone		Middle zone		External zone
Species	1979–1983	2017–2018	1979–1983	2017–2018	1979–1983
Galeorhinus galeus					●
Mustelus schmitti		●	●	●	●
Notorhynchus cepedianus				●	●
Squatina guggenheim					●
Sympterygia acuta			●	●	●
Sympterygia bonapartii		●	●	●	●
Myliobatis goodei	●	●	●	●	●
Conger orbignianus		●	●	●	●
Brevoortia aurea	●	●	●	●	●
Ramnogaster arcuata	●	●	●	●	●
Anchoa marinii		●		●	
Lycengraulis sp.	●		●	●	●
Genidens barbus		●		●	
Macruronus magellanicus		●		●	
Porichthys porosissimus	●	●	●	●	●
Mugil liza	●		●		●
Odontesthes argentinensis	●	●	●	●	●
Acanthistius patachonicus			●		●
Dules auriga					●
Pomatomus saltatrix	●	●	●	●	●
Parona signata			●	●	●
Cynoscion guatucupa	●	●	●	●	●
Macrodon ancylodon			●		●
Menticirrhus americanus				●	●
Micropogonias furnieri	●	●	●	●	●
Pogonias cromis			●	●	
Umbrina canosai					●
Percophis brasiliensis					●
Stromateus brasiliensis	●	●	●	●	●
Paralichthys patagonicus					●
Paralichthys orbignyanus	●	●	●	●	●
Oncopterus darwinii					●
Symphurus jenynsi			●	●	●
Total species	12	16	21	23	29

been considered in former studies probably due to the complexity and the variability of these systems (Pasquaud et al. 2015).

Bahía Blanca Estuary is a shallow temperate estuarine system that produces hypersaline conditions, where the salt concentrations in the inner portion of the estuary often exceed those of the inner continental shelf (average salinity is 33 ppt even though values as low as 17.3 ppt and as high as 41.9 ppt have been registered;

Freije et al. 2008). The estuary can be divided in three zones, depending on the composition of fish species that make use of the environment: inner (from the head of estuary to Ingeniero White), middle (from Ingeniero White to Puerto Rosales), and external (from Puerto Rosales to the mouth of the estuary) (see references in Fig. 2.1; Chap. 2). The inner zone has abundant intertidal habitats that are harsh and variable, forcing fish to either perform tidal migrations or be exposed to the naturally variable abiotic conditions. This zone also gathers most of the freshwater outputs of the tributary river meaning salinity can vary greatly. The inner and middle zones are where most of the human activity takes place. The last zone is abundant in islands and channels of varying depth. Some of these islands and channels are populated by important salt marshes. This large mudflat contracts in low tide, forcing fish to the deeper channels. The changes in temperature in this area are the widest of the three zones. The external zone is not subjected to changes in environmental parameters as much as the other two and resembles oceanic conditions the most.

11.2 Distribution and Composition of Fish Assemblage

In 1979 Lopez Cazorla started surveying the fish species composition in Bahía Blanca Estuary. In these surveys, which were carried out until 1983, fishing was performed by the local artisanal fishermen, who used gear specific for their fishing needs, whether it is shrimp nets, channel closure nets, or gill nets (Fig. 11.2). These surveys covered the inner, middle, and external zones of Bahía Blanca Estuary. Between 2017 and 2018, our research team started conducting seasonal surveys on the estuary. Fishing was performed using two of the fishing gears used by fishermen (shrimp nets and channel closure nets, Fig. 11.2 a and b, respectively), albeit with lower fishing effort.

In this estuary, fishermen use these types of nets taking advantage of the strong flow of the tidal regime. Shrimp nets are usually set on high tide, and low tide flow forces fish into the net bag. Gill nets are placed in between tides, as fish moving in and out of the inner portions of the estuary are more likely to get entangled in them. Channel closure nets are placed on the inner channels on high tide, and as water level lowers during low tide, fish are pushed against the net, and the fishermen close the net and collect them. Demersal and demersal-benthic fish are most abundantly captured using these methods.

In the former surveys, 45 sampling trips were performed, while in the recent surveys 8 sampling trips were carried out. In both surveys fish were identified, and several measurements were taken (Table 11.1). Of the 32 species of fish found inhabiting the waters of Bahía Blanca Estuary, 7 were chondrichthyans and 25 osteichthyans.

From the initial surveys conducted in Bahía Blanca Estuary, summarized in Table 11.1, it is possible to conclude that there is a progression of species richness, from the inner zone (12 species) to the middle (21 species) and the external zones (29 species). The zoning pattern described at the start of this chapter can explain, up

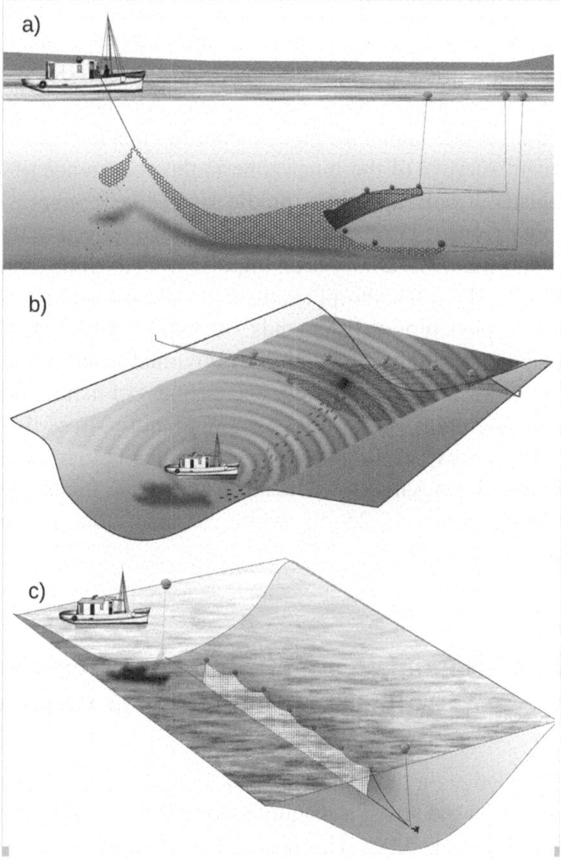

Fig. 11.2 Net types employed by artisanal fishermen of Bahía Blanca Estuary, also used for the scientific samplings. (**a**) Shrimp nets, (**b**) channel closure nets, and (**c**) gill nets. (Modified from Lopez Cazorla 2004)

to certain extent, these findings. The inner zone is challenging for fish given its wide range of water salinity. Anadromous fish, such as *Myliobatis goodei* and *Mugil liza*, have physiological adaptations that allow them to exploit the resources of this zone. Fish species with tolerance to changes in temperature and salinity can live or move in and out of the middle zone. Most chondrichthyans have less tolerance to abiotic stressors in comparison with bony fish; hence, they occur on the middle and external zones, with only *M. goodei* moving deeper into the inner zone. We have found exclusively marine fish as well, like *Dules auriga* or *Macruronus magellanicus*. The inner and middle zones of the estuary are common fishing grounds for the two study periods; hence, a comparison between both is only possible for those areas. Table 11.1 shows such comparison. In general, the more recent surveys found a greater number of species, especially considering those found in the inner zone, even though the fishing effort in 1979–1983 was considerably larger. In that period a total of 45 sampling trips were made, 14 in the inner zone (31.11%), 26 in the middle zone (57.78%), and 5 in the external zone (11.11%). In comparison, the 2017–2018 period was composed of 8 field samplings, 4 in the inner zone and 4 in the middle zone. Despite the comparatively low number of samples done in the second period, the increase in the number of species found is significant. Two species, in particular, appeared in the recent surveys that are noteworthy: *Genidens barbus* and *Macruronus magellanicus.* A notable absence in the recent samplings is the silverside mullet *Mugil liza.* Another fact increasing the number of species in the recent surveys of the inner zone is the occurrence of species that were previously registered only up to the middle zone, like *Conger orbignianus*, *M. schmitti*, and *S. bonapartii.* These species seem to have moved to the inner areas in recent years.

In estuaries, salinity is an important factor influencing fish abundance and distribution. The often abrupt changes in salinity common to estuaries can cause considerable physiological demands on fishes. Although fishes living in estuaries are adapted to salinity fluctuations, individual response to salinity stress varies by species and scale. Low diversity and richness of fishes observed in hypersaline systems have been attributed to the osmoregulatory stress that fishes withstand (Whitfield 2016). As expected, also in Bahía Blanca Estuary, richness values decrease as the amplitude of salinity fluctuation increases. Inner zone species were captured in salinity ranges of 25–36, while external zone species were captured in salinity ranges of 30–36.

11.3 Spatial, Seasonal, and Long-Term Variations in the Fish Assemblage of Bahía Blanca Estuary

Knowledge on the spatial and seasonal dynamics of fish species, especially in temperate coastal areas, is usually incomplete (García-Charton and Pérez-Ruzafa 2001; Topping et al. 2006). This lack of knowledge is an issue for effective and sustainable management of fish populations and their exploitation (McCormick and Choat 1987; Blyth-Skyrme et al. 2006). The seasonal data obtained in Bahía Blanca Estuary from the most recent set of surveys suggest that winter and spring contained the highest species richness, 20 and 17 species, respectively (Table 11.2). Conversely, autumn and summer had the lowest, 12 and 7, respectively. In the previous survey, however, richness was highest in autumn and summer (26 and 23 species, respectively), while it was lower in winter and spring (18 and 20 species, respectively). Some of the migratory species appear all year round in both periods, such as *C. guatucupa*, *M. furnieri*, and *M. schmitti*, although their peeks of abundance may vary according to the species. The occurrence of these species all year round was due to the presence of their juveniles, which remain in the estuary, while the adults migrate elsewhere. Resident species, like *Ramnogaster arcuata*, are present all year round.

For both periods, there is a clear differentiation between summer and autumn samples in the one hand and winter and spring samples on the other. While no clear pattern of species replacement between periods is evident in our results, these two season groups coincide with the periods of warmer and colder temperatures

Table 11.2 Fish species presence in autumn (A), winter (W), spring (Sp), and summer (Su) in Bahía Blanca Estuary in the two periods of sampling: 1979–1983 and 2017–2018

	1979–1983				2017–2018			
Species	A	W	Sp	Su	A	W	Sp	Su
Galeorhinus galeus	●			●	●			
Mustelus schmitti	●	●	●	●	●	●	●	●
Notorhynchus cepedianus		●	●	●		●		
Squatina guggenheim	●	●	●	●				
Sympterygia acuta	●	●		●		●		
Sympterygia bonapartii	●	●	●	●		●		●
Myliobatis goodei	●	●	●	●		●	●	
Conger orbignianus	●	●			●		●	
Brevoortia aurea	●	●	●	●	●	●	●	
Ramnogaster arcuata	●		●	●	●	●	●	●
Anchoa marinii					●	●	●	
Lycengraulis sp.	●		●	●	●		●	
Genidens barbus						●	●	
Macruronus magellanicus						●		
Porichthys porosissimus	●	●	●	●		●	●	●
Mugil liza	●	●	●	●				
Odontesthes argentinensis	●	●	●	●	●	●	●	
Acanthistius patachonicus	●	●		●				
Dules auriga	●			●				
Pomatomus saltatrix			●	●	●	●	●	
Parona signata	●	●	●	●		●	●	
Cynoscion guatucupa	●	●	●	●	●	●	●	●
Macrodon ancylodon	●	●		●	●			
Menticirrhus americanus		●				●		
Micropogonias furnieri	●	●	●	●	●	●	●	●
Pogonias cromis	●							●
Umbrina canosai		●						
Percophis brasiliensis	●						●	
Stromateus brasiliensis	●	●	●	●		●	●	
Paralichthys patagonicus	●							
Paralichthys orbignyanus	●	●	●	●		●	●	
Oncopterus darwinii	●							
Symphurus jenynsi	●		●	●		●		
Total	26	20	18	23	12	20	17	7

respectively, indicating that water temperature must play an important role in the seasonal species turnover. Being a shallow estuary means that the waters of Bahía Blanca Estuary heat up and cool off quicker than those of the open sea in front of it, as the heat retention within the estuary is poor. This desynchronized heating and cooling of the estuarine waters and the adjacent sea is important for species that employ an optimum temperature strategy to increase fitness, for example (Elisio et al. 2017). Many chondrichthyan species have been shown to exhibit this behavior, in which the individuals perform small-scale migrations to feed in high-temperature areas, and then return to colder areas to digest (Neer et al. 2007). This journey may involve swimming several kilometers and may take hours to complete. Seasonal dynamics such as this is thought to be responsible for the differences in species composition (Wonton 1992) and might be responsible for the species composition differences observed in Bahía Blanca Estuary.

Further spatial structure in the fish community of Bahía Blanca Estuary was found in the 1979–1983 surveys, using hierarchical classification procedures (Sneath and Sockal 1973) applied to the matrix of similarity indexes between sampling sites. Sampling sites close to shore were grouped together, some of them located on the internal zone and the southwestern quadrant of Bahía Blanca Estuary. A second separated cluster was formed by sites that were close to the main channel. Further subdivisions of these two groups had much smaller mean similarity indexes. The robustness of the discrimination between the two groups is supported by the total (100%) coincidence of the groups using the two similarity indexes. As explained at the first section of this chapter, depth plays an important role in structuring fish communities. Additionally, habitat resources such as food and shelter and environmental stressors vary greatly between close-shore intertidal environments and deeper subtidal zones. The main dredged portion of the inner zone of the navigation channel might represent an artificial deep subtidal environment, promoting species composition differences with the adjacent shallow intertidal shore habitats as described by Carbines and Cole (2009) in a similar estuary in New Zealand.

11.4 Species Spotlight: Biological Description of Some of the Fish Species of Bahía Blanca Estuary

11.4.1 Mustelus schmitti *(Springer, 1939)*

Locally called "gatuzo," the narrownose smoothhound *Mustelus schmitti* (Fig. 11.3a) is a small shark of the Triakidae family, attaining a maximum total length of 110 cm. This shark is endemic to the Southwest Atlantic Ocean, from the south of Brazil to the Argentinean Patagonia (22° S to 47°45' S), dwelling from coastal waters to up to 120 m of depth (Menni 1985). This shark is known to migrate seasonally in large numbers between wintering grounds in southern Brazil and summer grounds in Argentina (Vooren 1997) and also performs seasonal ingresses to estuaries, protected bays, and gulfs (Lopez Cazorla 2004; Chiaramonte and Pettovello 2000; Colautti et al. 2010). *M. schmitti* is one of the most studied sharks of Argentina: presently its reproduction, food habits, age, and growth and other of its biological processes have been described (Menni 1985; Chiaramonte and Pettovello 2000; Sidders et al. 2005; Segura and Milessi 2009; Colautti et al. 2010; Molina and Lopez

Fig. 11.3 Fish species spotlight of Bahía Blanca Estuary. (**a**) *Mustelus schmitti*, (**b**) *Myliobatis goodei*, (**c**) *Brevoortia aurea*, (**d**) *Ramnogaster arcuata*, (**e**) *Porichthys porosissimus*, (**f**) *Odontesthes argentinensis*, (**g**) *Cynoscion guatucupa*, (**h**) *Micropogonias furnieri*, (**i**) *Pogonias cromis*, and (**j**) *Paralichthys orbignyanus*. (Photos **a**, **b**, **c**, **f**, **g**, **h**, **i**, and **j** modified from Cousseau MB and Rosso JJ (2020) in Peces Argentina, edited by Vázquez Mazzini Editores, Ciudad de Buenos Aires-Argentina. Photos by Gabriela Blasina **d** and **e**)

Cazorla 2011; Molina et al. 2017). This species feeds mainly on crustaceans when close to the coast and on fish as it migrates to deeper waters. Seasonal and ontogenetic differences in diet composition have been described, with polychaetes being more important in the colder months of the year, while decapods become the main prey item in the warmer months (Molina and Lopez Cazorla 2011). Larger narrownose smoothhound sharks feed on larger crabs and fish, while neonates and juveniles prey on a greater variety of crustaceans and polychaetes. The species reaches sexual maturation at approximately 5 years old, with the males maturing faster than the females and also attaining less weight and length (Molina et al. 2017). The species migrates to the nursery areas of Bahía Blanca Estuary to give birth in spring-summer and then mates and leaves as temperature drops by the beginning of autumn. It gives birth to up to six pups, the size and number depending on the size and age of the female. The maximum age determined for this species is 20 years old (Molina et al. 2017).

The narrownose smoothhound is the most exploited elasmobranch species in Argentina, Brazil, and Uruguay, with an important percentage of the capture destined to exportation to England and China (Molina and Lopez Cazorla 2011). This shark is caught by both industrial and artisanal fishing fleets. The exploitation of this species throughout its distribution range led to recent declines in its populations despite maximum permitted catch regulations established by Argentina and Uruguay (Molina and Lopez Cazorla 2011). The narrownose smoothhound is currently considered endangered in these two countries by the IUCN (Massa et al. 2010).

11.4.2 **Myliobatis goodei** ***(Garman, 1885)***

The southern eagle ray, locally called "chucho," *Myliobatis goodei* (Fig. 11.3b) is a large stingray of the Myliobatidae family, reaching a meter in disc width. This species distributes from south California (35°N) to the south of Argentina (40°S). This chondrichthyan is diadromous, tolerating a wide range of salinity. It enters estuarine waters to give birth (Refi 1975). Two very similar species cohabit with *M. goodei*, *M. ridens* (Ruocco et al. 2012) and *M. freminvillii* (Aguiar et al. 2004).

M. goodei migrates to bays and estuaries during the warm months of spring and summer, to improve neonates and juveniles' access to food and shelter, providing effective protection against predators and optimum conditions for development (Castro 1993; Simpfendorfer and Mildward 1993). This migrating behavior of *M. goodei* was also reported by Molina and Lopez Cazorla (2015), Jaureguizar et al. (2003b), and Lopez Cazorla (1987) in Anegada Bay, Río de la Plata, and Bahía Blanca Estuary, respectively.

Molina and Lopez Cazorla (2015) inferred that the mating and spawning season for *M. goodei* may occur in summer, although the authors lack a complete yearly series of gonadosomatic index values (GSI) to accurately demonstrate this. The presence of mature males and pregnant females with highly developed embryos in the uterus in summer, and recently born pups, would strengthen this hypothesis and also imply that they became pregnant immediately after parturition (Hamlett 1999). *M. goodei* in the study area behave as generalist feeders, with a uniform diet composed mainly of bivalves. Trophic level of *M. goodei* in Anegada Bay (3.2) characterizes it as a secondary consumer (Molina and Lopez Cazorla 2015).

This species is captured as bycatch but retained and sold internally as well as exported in significant amounts. *M. goodei* is assigned as Data Deficient by the IUCN, given the possible population threats it faces (Stehmann 2009).

11.4.3 **Brevoortia aurea** *(Spix and Agassiz, 1829)*

Locally called "saraca," the Brazilian menhaden *Brevoortia aurea* (Fig. 11.3c) is a planktonic clupeid that reaches up to 40 cm in total length. *B. aurea* is a euryhaline fish, distributed from Salvador de Bahía in Brazil to San Matías Gulf (41°S) in Argentina. This species is abundant in estuarine waters. It has been estimated that it lives up to 12 years and the females reach larger sizes. Adults (larger than 20 cm TL) are captured inside Bahía Blanca Estuary in the spring and summer months, while only juveniles remain during autumn and winter (Lopez Cazorla 1985). While *B. aurea* is not targeted by any specific fishing fleet, it is an important bycatch component in coastal industrial and artisanal fisheries. The IUCN has evaluated the conservation status of this species as Least Concern with a stable population trend (Di Dario et al. 2017).

11.4.4 **Ramnogaster arcuata** *(Jenyns, 1842)*

Jenyns' sprat *Ramnogaster arcuata* (Fig. 11.3d), locally called "saraquita," is a pelagic fish species of the Clupeidae family. It has a wide distribution in coastal areas in the Southwestern Atlantic Ocean, from southern Brazil (estuary of the Patos Lagoon) to Tierra del Fuego in southern Argentina (Lopez Cazorla et al. 2011). It is confined to the external areas of rivers and coastal lagoon mouths, which are characterized by moderate salinity ranges. It is an estuarine-resident species that completes the entirety of its life cycle within estuaries (Garcia and Vieira 2001). The species reaches a total length of 130 mm and present a relatively short life span with a maximum age registered of 3 years. Sexual maturity is reached at 76 mm total length (1 year), and spawning season begins in spring (Lopez Cazorla and Sidorkewicj 2009). According to its trophic habits, it has been classified as a zooplanktivorous feeder, and its diet composition exhibited monthly variability in the main prey items consumed (Lopez Cazorla et al. 2011).

Ramnogaster arcuata is a main functional component of the ecosystem of Bahía Blanca Estuary, where it is not only one of the most abundant species but also one of the most commonly caught fish. It also represents a key food item for *C. guatucupa* and *P. orbignyanus*, two of the most economically important fish species in the area. The coastal habits and short life span of *R. arcuata* make it an excellent organism to be considered as a bio-indicator of aquatic environmental health (Lopez Cazorla and Sidorkewicj 2009; Ronda et al. 2019). The IUCN has evaluated the conservation status of this species as Least Concern (Di Dario et al. 2017).

11.4.5 Porichthys porosissimus *(Cuvier, 1829)*

Porichthys porosissimus (Fig. 11.3e) is a species of batrachoid known as Atlantic midshipman and locally as "lucerna." The reported maximum length for this species is 32 cm TL. It is an abundant species, caught as bycatch throughout its distribution range, from Río de Janeiro, Brazil, to Golfo de San Matías (41°S), Argentina. This species inhabits coastal waters from 30 to 200 m deep (Cousseau and Perrotta 2013). Other closely related species exhibit complex mating behaviors, which include nest building, sound- and bioluminescence-mediated courtship, and parental care (Tsujii et al. 1972). Photophores, a special type of skin cells that produce bioluminescence, are present in *P. porosissimus* and would likely play a homologous role in mating and communication between individuals. Mature individuals of this species croak when disturbed, meaning they are capable of producing sound, much like their northern counterparts.

Shrimp trawling and traps capture this species as bycatch, but only one study exists on the biology of this species (Vianna et al. 2000). In Bahía Blanca Estuary, it is captured mainly in the external zone, all year long. The largest adults were present during spring, while the smallest juveniles were found in summer. Size range was from 9 to 31 cm TL, and the most frequent sizes were 17–31 cm TL (Lopez Cazorla 1987). The IUCN has yet to evaluate the conservation status of this species.

11.4.6 Odontesthes argentinensis *(Valenciennes, 1835)*

Locally called "pejerrey," *Odontesthes argentinensis* (Fig. 11.3f) is a large-sized silverside that is widely distributed along the Atlantic Ocean coast between Sao Paulo, Brazil, and Rawson in Argentina (García 1987; Dyer 2000). It is a planktonic species that feeds on zooplankton, mainly crustaceans. This fish reaches 48 cm in total length, the males being larger than the females (Molina 2013). *O. argentinensis* inhabits shallow coastal waters, and juveniles have been found to be abundant in the surf zones of sand beaches. As for other atherinids, this fish shows a great phenotypic plasticity that allows its adaptation to different environments (Bamber and Henderson 1988) involving a wide range of salinities. This allows this species to inhabit estuaries and inshore waters, where it likely migrates during late spring and summer to reproduce (Cousseau and Perrotta 2013; Beheregaray and Levy 2000; Bemvenuti 2005). According to gonad ripening takes place between September and November with a peak in October, in the nearby location of Anegada Bay. Llompart et al. (2013) also describe age and growth of this species, which attains a maximum of 7 years, growing quickly in the first 2.

Odontesthes argentinensis in Bahía Blanca Estuary behave similarly, with the main captures being done in the external zone using gill nets and channel closure nets. Bahía Blanca Estuary represents a breeding area for this species, and spawning occurs from late August to November (Lopez Cazorla 2004). The commercial importance of this species is limited to the Argentinean market and is targeted by coastal artisanal fleets (Cousseau and Perrotta 2013). The IUCN has yet to evaluate the conservation status of this species.

11.4.7 Cynoscion guatucupa *(Cuvier 1829)*

The striped weakfish, *Cynoscion guatucupa* (Fig. 11.3g), is locally called "pescadilla de red." It is a pelagic fish species, which has a wide geographical distribution, extending from Río de Janeiro (22°S) in Brazil to San Matías Gulf (43°S) in Argentina (Cousseau and Perrotta 2013). This fish presents dietary shifts during ontogeny. It feeds from pelagic to benthic crustaceans on its early stages (mysids, sergestids, shrimps) and eventually increases progressively in ichthyophagy (mainly an increase consumption of alevins and young fish) as it develops into adulthood. It present dietary seasonal differences could be due to changes in abundance and availability of its prey species in the environment (Lopez Cazorla 1996; Sardiña and Lopez Cazorla 2005a). *C. guatucupa* performs seasonal migrations, swimming northwards between autumn and spring, leaving the fishing grounds of Uruguay and Argentina to move to the coast of Brazil, only to return to the south in summer (Villwock de Miranda and Haimovici 2007). Lopez Cazorla (1996) reports the influence of changes in temperature and salinity as triggers for the spawning movements of *C. guatucupa.* Spawning occurs outside of the estuary from spring to mid-autumn (Cassia 1986; Lopez Cazorla 2000), and the fry is pushed into the estuary by tidal movements. Small juveniles recruited from late spring move to deeper waters (25–50 m) in late autumn, when they reach a mean total length of 9.8 cm (age 0+). They remain there for the next 1–2 years before joining the adult stock's seasonal movements (Haimovici et al. 1996; Lopez Cazorla 2000; Sardiña and Lopez Cazorla 2005a). The total length of adult fish ranges from 34 to 63 cm, and the ages range from 3 to 23 years (Lopez Cazorla 2000; Ruarte and Sáez 2008). The IUCN has yet to evaluate the conservation status of this species.

11.4.8 Micropogonias furnieri *(Desmarest 1823)*

The whitemouth croaker *Micropogonias furnieri* (Fig. 11.3h), locally called "corvina rubia," is a demersal fish of the family Sciaenidae, inhabiting coastal waters up to 60 m deep. It is a euryhaline fish distributed widely in marine and estuarine systems of the eastern American coast from the Gulf of Mexico (20°20'N) to "El Rincón" (41°S) in Argentina (Carozza et al. 2004). The maximum recorded size for the species is 74 cm in TL, reaching sexual maturity at 33 cm of TL, which corresponds to 4 or 5 years of age. The reproductive period of *M. furnieri* is very long and extends from spring to summer (Macchi et al. 2003). Spawning occurs in highly saline coastal waters, and subsequently, larval *M. furnieri* enter coastal estuaries during winter months. In Bahía Blanca Estuary, species reproduction occurs in El Rincón area during spring, with the subsequent drift of eggs and larvae into estuary. Juveniles with sizes of 2 and 18 cm total length (Lt) remain inside the estuary from early summer to winter and then leave the region. At late spring, entrance to the estuary of individuals in the adult state with a size range from 30 to 72 cm Lt begins, and they remain in the area until autumn (Lopez Cazorla 2004). Young-of-year (YOY) and adult *M. furnieri* utilize estuarine habitats for feeding and growth (Jaureguizar et al. 2003a; Lopez Cazorla 2004). The species has been identified as a generalist feeder, and its stomach contents largely reflect seasonal changes in prey availability, meaning it has a broad dietary niche width (Mendoza-Carranza and Vieira 2008). Previous studies in estuarine habitats have documented ontogenetic changes in diet. YOY individuals rely heavily on polychaetes in their diets but also consume other food items such as chaetognaths, copepods, and amphipods. Evidence indicates that *M. furnieri* changes its feeding habits as it gets larger, relying more heavily on organisms such as mysids and fish. Adults have been described as opportunistic bottom-feeders that eat decapod crustaceans, such as crabs and shrimps, polychaetes, and, occasionally, small fishes (Sardiña and Lopez Cazorla 2005b; Giberto et al. 2007; Blasina et al. 2016).

Micropogonias furnieri is one of the most abundant demersal fishes of South American estuaries and an important component of artisanal and coastal industrial fisheries in Brazil, Uruguay, and Argentina (Carozza et al. 2004). *M. furnieri* fishery in the Río de la Plata estuary is mainly artisanal, with fish being caught mostly in winter in Samborombón Bay area of Argentina and during spring and summer in Santa Lucia area of Uruguay (Jaureguizar et al. 2003a, b). In Bahía Blanca Estuary, it was the second most important fish resource, captured in spring and summer. It reached 16% of the commercial landings between 1972 and 1992, although between 1994 and 1996 landings strongly decreased and values as low as 2% were reported (Lopez Cazorla 2004). The IUCN has evaluated the conservation status of *M. furnieri* as Least Concern, although a decreasing population trend is mentioned (Aguilera et al. 2015).

11.4.9 Pogonias cromis *(Linnaeus, 1766)*

The black drum *Pogonias cromis* (Fig. 11.3i), locally called "corvina negra," is a demersal coastal fish distributed along the western Atlantic Ocean from Massachusetts, USA, to south of Buenos Aires Province in Argentina. It is an estuarine-dependent species and the largest sciaenid found in the estuarine environments of Argentina. It reaches up to 120 cm in TL and is sexually mature at the end of the second year of life, at 28.5–33 cm of TL (Cousseau and Perrotta 2013). Spawning takes place in regions associated with estuaries, inside or outside of them, and mainly in spring. *P. cromis* juveniles live in estuarine areas as they can tolerate a wide range of salinities and water temperatures, eventually moving to offshore marine waters when they reach the adult stage. Adults are usually common in shallow coastal and estuarine waters and occasionally occur further from the coast (Macchi et al. 2002; Rubio et al. 2018).

Juvenile and adult *P. cromis* exploit a variety of benthic food resources and can use their strong pharyngeal teeth to crush the shells of mollusks and crustaceans (Blasina et al. 2016; Rubio et al. 2018). Gut content analyses have identified significant seasonal differences in the diet composition and trophic niche breadth. Because of this *P. cromis* has been classified as an opportunist predator (Blasina et al. 2010).

Pogonias cromis is the target of an important recreational and commercial fishery in the Gulf of Mexico, and it is commercially harvested in inshore waters of Samborombón Bay, a semi-enclosed region inside the Río de la Plata estuary in Argentina. Fishing effort occurs mainly between late winter and summer, but is especially high during spring, when *P. cromis* forms large schools in shallow waters. This behavior contributes to intensified commercial and recreational activity on this species (Macchi et al. 2002; Rubio et al. 2018). *P. cromis* is classified as Least Concern status by the IUCN, but a decreasing population trend is mentioned (Chao et al. 2015). In Bahía Blanca Estuary, in the period 1979–1983, this species was registered only in three opportunities, all in the same winter month (June) and in Ingeniero White (the middle zone) (Lopez Cazorla 1987). In 2017–2018 this species was captured also in the middle zone, but during summer (Table 11.1 and 11.2).

11.4.10 **Paralichthys orbignyanus** *(Jenyns, 1842)*

The flounder, locally called "lenguado," *Paralichthys orbignyanus* (Fig. 11.3j) is a commercially important species generally found in the shallow waters from Río de Janeiro (22° S) southwards in Brazil to San Matías Gulf (41° S) in Argentina (Cousseau and Perrotta 2013). Two other species of the genus *Paralichthys* live in Argentinean waters, *P. isosceles* and *P. patagonicus* (Lopez Cazorla 2005). *P. orbignyanus* is a typical benthonic fish with a wide temporal and spatial distribution in Bahía Blanca Estuary. Two other species of the genus *Paralichthys* live in Argentinean waters *P. isosceles* and *P. patagonicus* (Lopez Cazorla 2005).

The maximum ages recorded for both sexes corresponded to 6 years in males and 7 years in females, respectively, suggesting that *P. orbignyanus* is not a long-living species, and the females are longer and heavier than males (Lopez Cazorla 2005). Similar data have been published for *P. adspersus* by Escobar (1995), who aged this species and found they live a maximum of 6 years and reach up to 74 cm in total length. In addition, D'iaz de Astarloa and Munroe (1998) observed that the longest TL for *P. orbignyanus* was 61 cm for males and 103 cm for females, respectively, although they made no reference to age. Females are longer and heavier than males. Larger size in females could be indicative of a life history strategy supportive of increasing egg production (Masuda et al. 2000). The growth difference between females and males was also observed in *P. adspersus* females which exhibit a length significantly larger than males (Escobar 1995). The length growth registered in *P. orbignyanus* males and females in Bahía Blanca Estuary was significantly higher than that of *P. isosceles* reported by Fabré and Cousseau (1990).

Paralichthys orbignyanus has an active growth period in summer and interrupts its growth in winter. Spawning occurs in the period extending from November to January (spring-summer), and the eggs and larvae of this species are found in January and February (summer) in the area next to the estuary mouth (Lopez Cazorla 2005). This suggests that spawning occurs out of the estuary, as with other species of bony fish, such as *C. guatucupa* (Lopez Cazorla 1996, 2000). A similar behavior has been described for other Pleuronectidae (*Kareius bicoloratus*) which spawn off the coasts, at depths ranging from 20 to 50 m. Once larvae reach 10–15 mm in total length, they approach the coast, migrating to nursery grounds (Malloy et al. 1996). The IUCN has yet to evaluate the conservation status of this species.

11.5 Fish Habitat Uses

The life of fish in estuaries is conditioned by the abundance of food and variations in the abiotic parameters of the water (Elliott et al. 2007). Fish species found in estuaries use these systems in a variety of ways, and this usage can change at different life stages. Ecological characteristics of fish species found in estuaries can be divided into three main functional aspects: (1) the use fish make of the estuary during their life cycle, (2) reproductive characteristics, and (3) feeding preferences and strategies. Elliott et al. (2007) name the three functional groups as "estuarine use functional group," "reproductive mode functional group," and "feeding mode functional group" respectively.

11.6 Estuarine Use Functional Group

Many species spawn in marine waters and enter estuaries for variable periods, while others complete their life cycle within the estuary, and yet others employ the estuary as a feeding area (Potter et al. 2015). Thus, fish assemblages include estuarine-resident species, freshwater and marine species that typically use estuaries at a specific life stage, as well as migratory diadromous species (Elliott et al. 2007). Each of these categories is considered to contain two or more functional guilds that represent characteristics associated with the spawning, feeding, and/or refuge locations, which in some cases involve migratory movements between estuaries and other ecosystems (Whitfield 2016).

Guild approach categorization of fishes was proposed by Elliott et al. (2007) and refined by Potter et al. (2015). Two fish guilds are dominant in Bahía Blanca Estuary (Table 11.3): marine estuarine-opportunists and estuarine-residents; they are represented in estuaries by different life stages and are associated with different food chains (see Feeding mode functional group section). Marine-estuarine opportunists are predominantly juvenile fish making use of this ecosystem as a nursery area.

Table 11.3 Species frequency of occurrence in Bahía Blanca Estuary. *R* rare species (up to 33% of occurrence), *C* common species (34–66% of occurrence), *F* frequent species (more than 66% of occurrence)

Species	Resident	Migrant	Straggler
Galeorhinus galeus		R	
Mustelus schmitti		F	
Notorhynchus cepedianus		R	
Squatina guggenheim		R	
Sympterygia acuta	F		
Sympterygia bonapartii	F		
Myliobatis goodei		F	
Conger orbignianus		F	
Brevoortia aurea		F	
Ramnogaster arcuata	F		
Anchoa marinii	F		
Lycengraulis sp.	C		
Genidens barbus			R
Macruronus magellanicus			R
Porichthys porosissimus	F		
Mugil liza	C		
Odontesthes argentinensis	F		
Acanthistius patachonicus	R		
Dules auriga	R		
Pomatomus saltatrix		C	
Parona signata		C	
Cynoscion guatucupa		F	
Macrodon ancylodon		R	
Menticirrhus americanus		R	
Micropogonias furnieri		F	
Pogonias cromis		R	
Umbrina canosai		R	
Percophis brasiliensis		R	
Stromateus brasiliensis		R	
Paralichthys patagonicus	R		
Paralichthys orbignyanus	F		
Oncopterus darwinii	R		
Symphurus jenynsi	C		
Total species	14	17	2

These fish species regularly enter estuaries in substantial numbers but use, to varying degrees, coastal marine waters as alternative nursery areas. In small-scale studies, some authors have pointed out that estuary mouth width was the most important variable explaining a significant part of the variability in fish species richness (Nicolas et al. 2010; Pasquaud et al. 2015). Estuaries with large mouths can attract numerous brackish water species, as well as marine-estuarine opportunist fish species (Martinho et al. 2009; Vinagre et al. 2009). Estuarine-resident guilds are composed by species with populations in which the individuals complete their life cycle within the estuary. While a number of the marine estuarine-opportunist species have economic importance for the recreational and local artisanal fishermen, none of the small resident species, which are a highly productive component in this estuary, are utilized. In addition, a number of the marine straggler species are frequently registered. These species enter estuaries sporadically and in low numbers and are most common in zones where salinity typically does not decline far below approximately 33 ups. Due to hypersaline conditions in Bahía Blanca Estuary, no freshwater fish species has been registered (Table 11.3).

In Bahía Blanca Estuary, the biology and life history of each species condition the use they make out of the resources available. Resident fish species (i.e., estuarine-resident) here, for example, have a remarkable tolerance to environmental variations, while migrant species (i.e., marine estuarine-opportunists) exhibit behaviors that allow them to exploit the high productivity of the intertidal ecotone and leave the area when the conditions become unfavorable. The latter species have a comparatively low tolerance to shifts in abiotic variables.

11.6.1 Resident Species of Bahía Blanca Estuary

In Bahía Blanca Estuary, there are resident chondrichthyans and osteichthyans. Skates of the genus *Sympterygia* lay eggs all year round and are commonplace all along the waters of the estuary. The most ubiquitous species is *Sympterygia bonapartii*, the shortnose south Atlantic skate. This species lays eggs protected with a fibrous black capsule and with four tendrils. The eggs are placed around submerged vegetation or debris so that the tendril holds them in place.

The absence of natural hard substrate and the relative scarcity of submerged macroalgae offer little shelter for resident reef fish; however, species of genera *Dules* and *Acanthistius*, both associated with rocky bottoms and reefs, have been found to occur within the waters of the estuary. Soft sediments, however, are ideal for soles and flounders. In Bahía Blanca Estuary, there are four species of flounders and one of soles. Of these species, only the flounder *Paralichthys orbignyanus* occurs frequently and with considerable abundances, in so as to become a targeted species of the local artisanal fishermen.

The high turbidity of the estuary means that ambush predators like the southern Atlantic midshipman (*Porichthys porosissimus*) have no problem procuring food. Indeed this species is very abundant in the estuary, where it performs seasonal migrations. It enters from the external area of the estuary in spring and summer to mate and care for their young. This species is caught abundantly in shrimp nets from October to February during ebb tide, suggesting it uses the currents to move in and out of the inner part of the estuary on a daily basis. By autumn/winter it is already unlikely to fish any in the inner zone of the estuary. A possible explanation is the drop in water temperature which in the inner, and shallower, section of the estuary is much more intense and sharp. Changes in water salinity also offer another possible explanation, while it fluctuated sharply in the inner zone, values in the external and middle areas remained relatively constant (Lopez Cazorla 2004), so perhaps the distribution of this fish is due to a behavioral escape from the fluctuating conditions, of temperature and salinity, of the inner zone. Prey availability may also be a factor contributing to this species' distribution. Its main prey, the prawn *Peisos petrunkevitchi*, spawns by the end of winter, prompting an abundance peak during spring and summer. In the late summer months, spawners concentrate in the outer part of the estuary (Mallo and Cervellini 1988), representing a valuable protein and energy source for *P. porosissimus*, who might follow their seasonal movement patterns. Little else is known about this species, as there are no specific studies on it yet.

11.6.2 Migrant Species of Bahía Blanca Estuary

In Bahía Blanca Estuary, the increases in water temperature and salinity during the warmer months of spring are thought to trigger migratory movements of certain species of fish. However, little is known about the drivers of migration movements in the migrant species of Argentina. It is theorized that some species utilize the estuary as feeding ground, others as a nursery for their young, as mating area, or spawning waters. Regardless of the driver, several fish species shoal into the estuary in different times of the year and then leave.

The high productivity, the availability of refuge, and favorable conditions in spring and early summer in Bahía Blanca Estuary seem to be a reasonable explanation as to what draws migrating fish species to these waters. Sciaenids like *Cynoscion guatucupa* and *Micropogonias furnieri* migrate to the estuary and spawn before entering estuarine waters for feeding. Adults of *C. guatucupa* presented two abundance peaks in the estuary: one in early autumn and a more important peak in early spring. On the other hand, the higher abundance of *M. furnieri* adults into the estuary has been registered during spring and summer (Lopez Cazorla 2004). Given that nutrient load increases around winter, with a consequent increase in phytoplankton biomass, it is not surprising that these sciaenids have a bimodal spawning behavior. This behavior might favor the larvae produced in the early spawning event with more food availability, at the expense of lower growing temperatures. Larger juveniles of these species, preying on copepods and brachyuran larvae, are captured at the end of spring and throughout the summer, autumn, and winter, exploiting the zooplankton biomass explosion that follows the peaks of phytoplankton.

The abundant invertebrate assemblages also represent an outstanding food source for benthic predators like *M. furnieri*. This species uses the intertidal during high tide to prey on polychaetes and crabs among the "cangrejales." Local fishermen report they can see the nuzzling marks of *M. furnieri* in the mud of the "cangrejal" during low tide. Although this species has several specializations in its mouth to feed on hard-shelled prey, like crabs, it will opt for other benthonic prey if they are abundant, exhibiting an opportunistic feeding behavior.

Chondrichthyans also migrate into the estuary to give birth and lay eggs. Adults of the triakid *Mustelus schmitti* can be found in the waters of the estuary from the end of winter up to the beginning of summer. The juveniles can be found from summer to early winter. When this species enters the estuary, it shifts the diet to consume almost exclusively crabs. Females migrate into the inner parts of the estuary to give birth to 2–8 pups and then mate with the males. Stingrays of the genus *Myliobatis* also move into the estuary in spring to give birth and mate, utilizing the abundant populations of polychaetes and crabs to load up on energy reserves. They can be found up until April (autumn). Two species are known to occur here, *M. goodei* and *M. ridens*, which were thought to be one species until recently (Ruocco et al. 2012).

Large sharks like the seven-gill shark (*Notorynchus cepedianus*), copper shark (*Carcharhinus brachyurus*), and sand tiger shark (*Carcharias taurus*) also utilize estuarine waters as a nursery area during the warm months of spring and summer and also as a hunting ground for both fish and pinnipeds. Studies on these species in Bahía Blanca Estuary are lacking; the only available reference is the presence of *N. cepedianus* (Lopez Cazorla 1987).

11.6.3 Straggler Species

This category is represented by fish species that occur "accidentally" in estuaries; they generally occupy it for only very short periods of time and in limited areas. In Bahía Blanca Estuary, two species have been registered that come from other regions and that visit the estuary to use its favorable conditions opportunistically: *Genidens barbus* and *Macruronus magellanicus*. These findings represent the first record of both species of fish in Bahía Blanca Estuary. *G. barbus* is an anadromous species inhabiting estuaries and the marine continental shelf from Bahía in Brazil (17° 00′ S) to San Blas Bay in Argentina (40° 32′ S) (Avigliano and Volpedo 2015). On the other hand, *M. magellanicus* is distributed on intermediate platform of the Argentine Sea and in the gulfs of San Jorge and San Matías. On the platform its distribution is closely related to the Malvinas current, and it has been recorded at temperatures that varied from 3 to13° C (Cousseau and Perrotta 2013). Possible explanations for these irregular records are diverse; in some cases they could be due to atypical abiotic conditions, such as the extraordinary incursion of marine waters or the occurrence of adverse climatic conditions in the area outside the estuary, or it could even be due to an intrusion when following their preys.

11.7 Reproductive Mode Functional Group

The spawning features and the degree of parental care are required to define reproductive modes in fishes (DeMartini and Sikkel 2006; Elliott et al. 2007). Fish species are first divided into oviparous and viviparous, according to the maternal investment in individual offspring (DeMartini and Sikkel 2006). Viviparous species present internal fertilization and live-bearing of young with a broad range of post-fertilization provisioning, from no (strictly lecithotrophic viviparity) with live-bearing of young provisioned entirely by ovum yolk to extensive provisioning beyond the nutrition provided by ovum yolk (matrotrophic viviparity). On the other extreme are the oviparous species with lecithotrophic maternal provisioning (limited to the yolking of ovarian oocytes prior to fertilization) and external fertilization. Oviparous species are distinguished on the basis of their egg characteristics, mode of release, and the degree of parental care provided to eggs (Franco et al. 2008). These reproductive modes determine offspring survival; according to the optimization theory, parental care implies a greater investment on offspring (therefore, larger individual offspring) at the expense of the number of offspring in which it is performed. Within of viviparous species category, the most extreme example of parental care is matrotrophic viviparity (DeMartini and Sikkel 2006).

According to reproductive modes described in Jaureguizar et al. (2016), different reproductive strategies occur in estuarine and migrant assemblages of Bahía Blanca Estuary. The 78% of the species are oviparous (62·4% producing pelagic eggs, 6·2% eggs that settle on the substratum and adhesive eggs 9·4% and), followed by 15·6% of viviparous and 6·4% of ovoviviparous species. Of the 13 estuarine-resident species registered, most (61·5%) produce pelagic eggs, spawning within the estuary or in its influence area (e.g., *Ramnogaster arcuata*, *Oncopterus darwinii* and *Porichthys porosissimus*). Species that produce adhesive eggs that are able to attach to substrata and the vegetation are second in importance (23·5%; e.g., *Odontesthes argentinensis*), and finally two species (15%) were ovoviviparous (*Sympterygia acuta* and *S. bonapartii*). The reproductive strategy of marine migrant fish was similar to that of the estuarine-residents, as 58·8% of these species spawn pelagic eggs (e.g., *Micropogonias furnieri*, *Cynoscion guatucupa*, and *Brevoortia aurea*), followed by viviparous species (29·4%; e.g., *Mustelus schmitti* and *Myliobatis goodei*). There was only one ovoviviparous species whose male carry the eggs in their mouth (*Genidens barbus*).

11.8 Feeding Mode Functional Group

Trophic ecology studies seek to identify the feeding habits of species through the analysis of the major items consumed. Knowledge on the diets of species is one of the basic requirements for a closer examination of the relationships between organisms in a given ecosystem. A very close relationship exists between the quantity, quality, and availability of food and the distribution and abundance of consumer organisms (Dantas et al. 2013; Campos et al. 2015). The structures of fish assemblages that use the shallow areas of estuaries are strongly influenced by trophic relationships (Blasina et al. 2016). Knowledge on the structure of the trophic web allows the description of the energy flow in an ecosystem and the understanding of the ecological relationships among organisms (Dantas et al. 2013).

Although opportunism is a widely reported feeding strategy used by estuary-associated fish (Gerking 1994; Elliott et al. 2007), intrinsic factors such as morphological and behavioral constraints set the boundaries on what food items can be taken from the environment, thus affecting the individual's ability to obtain certain prey. Extrinsic interactions (of a species or an individual with both the environment and other community members) will also influence the diet of said individual (Elliott et al. 2002; Horn and Ferry-Graham 2006). For example, the foraging range of the fish will affect which prey are encountered and can be potentially included in the diet, while mouth adaptations and morphology will determine which among the potential prey are ultimately consumed. The trophic categories from literature were revised by Franco et al. (2008), and feeding mode functional groups were identified by combining information on predominant diet and feeding location. The trophic groups, indicating the main types of food exploited by fish within estuarine environments and the estuarine compartments (e.g., pelagic, benthic) where these resources are taken, are:

- Microbenthivores: feeds mainly on benthic, epibenthic, and hyperbenthic fauna, with prey size <1 cm.
- Macrobenthivores: feeds mainly on benthic, epibenthic, and hyperbenthic fauna, with prey size >1 cm.
- Planktivores: feeds predominantly on zooplankton and occasionally on phytoplankton in the water column, mainly by filter feeding.
- Hyperbenthivores/zooplanktivores: feeds just over the bottom, predominantly either on smaller mobile invertebrates living over the bottom or zooplankton; diverse prey capture mechanisms (ram, suction, or manipulation).
- Hyperbenthivores/piscivores: feeds just over the bottom, predominantly either on larger mobile invertebrates living over the bottom or fish; diverse prey capture mechanisms (ram, suction, or manipulation).
- Detritivores: feeds on all the small organisms in or on the surface layer of the substratum (e.g., benthic algae such as diatoms, microfauna, and, to a lesser extent, smaller meiofauna) and associated organic matter (usually of plant origin); ingests relatively large volumes of sand or mud (by suction mechanisms); digests the food material and passes out the inorganic particles.
- Herbivores: grazes predominantly on living macroalgal and macrophyte material.
- Omnivores: ingests both plant and animal material by feeding mainly on macrophytes, periphyton, epifauna, and filamentous algae.

11.8.1 Feeding Habits of the Fish of Bahía Blanca Estuary

With the objective to comparatively describe and analyze the trophic spectrum of the most common species and their interrelationship, Lopez Cazorla (1987) studied the stomach content of 1035 specimens belonging to 7 species of fish, between 1980 and 1982. The species studied were *Sympterygia bonapartii*, *Mustelus schmitti*, *Porichthys porosissimus*, *Odontesthes argentinensis*, *Micropogonias furnieri*, *Cynoscion guatucupa*, and *Paralichthys orbignyanus*. Her results indicate that *Neohelice granulata* is the most important food source for almost all the species studied. The remainder of the dietary items presented great differences in the proportions consumed by each of the species. *Sympterygia bonapartii* feeds mainly on benthic decapodic crustaceans, predominately peneids and brachyurans. The diet of *Mustelus schmitti* was found to be composed mainly of benthonic decapod crustaceans, polychaetes, and young fish. *Porichthys porosissimus* fed almost exclusively on *Peisos petrunkevitchi* with a small percentage of misidaceans. *Odontesthes argentinensis* showed a diet consisting principally of *Neohelice granulata*, gastropods, misidaceans, and amphipods. *Micropogonias furnieri* consumed *N. granulata* and *P. petrunkevitchi* as their most common prey. *Cynoscion guatucupa* presented a diet which consisted mainly of *Pleoticus muelleri*, *Artemesia longinaris*, *P. petrunkevitchi*, and young fish of the *Ramnogaster arcuata*, *Brevoortia aurea*, and *O. argentinensis*. *Paralichthys orbignyanus* preys on the following fish species: *O. argentinensis*, *B. aurea*, *R. arcuata*, *Parona signata*, *C. guatucupa*, *Pomatomus saltatrix*, *P. orbignyanus*, and *Lycengraulis olidus* (Fig. 11.4).

Crabs and polychaetes constitute the principal or most important food for benthonic and demersal fish, while Decapoda Natantia such as *A. longinaris*, *P. muelleri*, and *P. petrunkevitchi* were the principal food of demerso-planktonic fish. Although fish select certain types of prey, depending on their size and habitat, consumption of prey depends, above all, on the availability and the community structure of the prey.

In this regard, a brief description of the community structure of prey can help to interpret these findings. In Bahía Blanca Estuary, an adequate supply of food for the diverse life stages of the fish communities depends on a sequential abundance of progressively bigger prey, from autumn to late summer. Nutrient abundance increases by the end of summer, reaching a peak in the autumn months (between April and June) (see Chap. 3). This increase in the nutrient load produces a trophic cascade, triggering blooms in the planktonic fractions. The main phytoplankton bloom occurs between June (winter) and October (spring), followed by the mesozooplankton explosion in November (spring). November is the month when miscidaceans and brachyuran larvae, important prey items in the diet of several fish species of the estuary, register their maximum abundance. Teleost larvae of demersal species, like *Cynoscion guatucupa*, feed on small planktonic crustaceans like *Acartia tonsa* and then shift their diets to larger prey like miscidacean *Arthromysis magellanica*. Peneid crustaceans, like *Peisos petrunkevitchi*, *Artemesia longinaris*, and *Pleoticus muelleri*, occur mainly between January (summer) and June and are

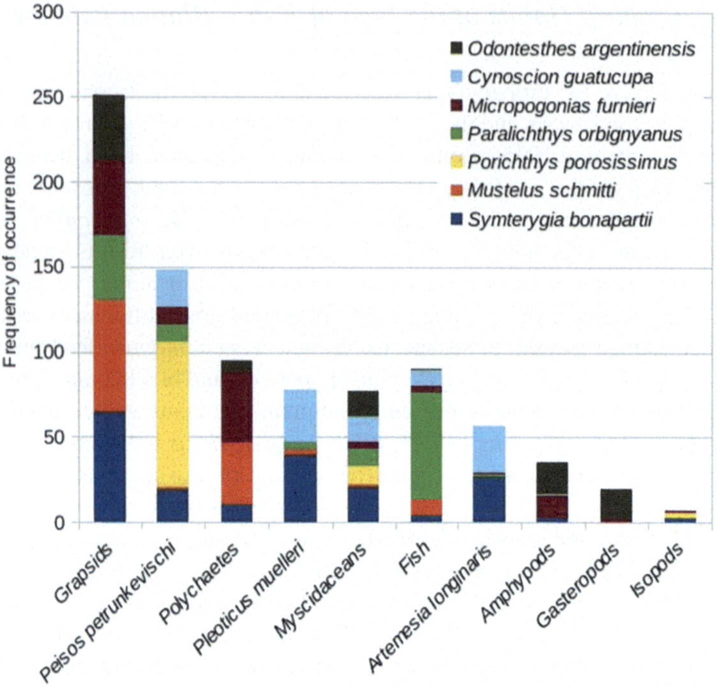

Fig. 11.4 Frequency of occurrence of major prey items in the stomach contents of the studied fish species of Bahía Blanca Estuary

the food of larger juveniles, which eventually prey on other fish. Pups of chondrichthyan are too large to feed on small planktonic prey, exploiting the abundance of larger crustaceans by the end of spring and mid-summer. Resident species like *Ramnogaster arcuata* breed in spring, when abundant food is available for the adults but also for the larvae and juveniles.

The benthic fish species that feed on reptant crustaceans and polychaetes do not experience such a pronounced seasonality as that of the plankton. The intertidal ecotone in Bahía Blanca Estuary is dominated by euryhaline vascular plants, like *Sarcocornia perennis* and *Spartina alterniflora*. These plants dominate the mud flats that cover most of the intertidal environment of the estuary. This habitat is characterized by fast-changing temperatures, oxygen-poor substrate, and lack of refuge. However, the association between the burrowing crab *Neohelice granulata* and the salt marsh plant *S. perennis* creates a unique habitat for other invertebrates, as it promotes sediment oxygenation and nutrient turnover (Parodi 2004). This particular type of habitat is called "cangrejal," in Spanish meaning "land of crabs." Polychaetes are particularly benefited, and their abundances in sites with *cangrejal* are significantly higher than in sites without *cangrejal*. This bolstered abundance of crabs and polychaetes might explain why these are the most predominant and important prey items for the fish species present in Bahía Blanca Estuary. *Cangrejal* sites could be essential for the thriving of the whole trophic network of the estuary (Elías et al. 2004).

Box 11.1 Artisanal Fishery in Bahía Blanca Estuary

It is widely recognized that small-scale fisheries play an important role in providing food and livelihood to people, contributing to poverty reduction and sustainable development in several places around the globe (FAO 2005). In particular, developing countries greatly benefit from this type of fishery, as they constitute the main source of both food and income for people living along the coast (Blaber et al. 2000). Artisanal fishermen communities develop an intricate relationship with the marine environment and the species that constitute their sustenance, which greatly aids the success of conservation and management practices.

The artisanal fishery in Bahía Blanca Estuary has been carried out since the beginning of 1900. Fishermen employ a combination of fishing gear throughout the year, consisting of shrimp nets, channel closures, and gill nets. Each type of net is employed to target a particular species or group of species and is performed in different times of the year to increase yield and reduce bycatch of unwanted species. Traditionally a family business, fishermen formed the *Cooperativa Pesquera Whitense* (White's Fishery Cooperative), a cooperative organization, between 1945 and 1999. The cooperative had a fish processing plant and handled the marketing of the fishermen catch, allowing a better income for the families of the fishermen and a regulatory frame for the activity. However, this organization closed in 1999 due to the collapse of the artisanal fisheries of several species (Lopez Cazorla et al. 2014).

The Collapse of a Fishery

Between 1972 and 1992, catches of *Cynoscion guatucupa* reached 50% of the total annual landings. However, at the end of the 1990s, catches dropped to 15%. In 2004 the Argentinean Government implemented fishing closures in El Rincón area, as a management measure to control the increasing landings and the decrease in biomass of *C. guatucupa* and several other commercial species (Carozza and Fernández Aráoz 2009). But the closure came too late to save the artisanal fishermen of Bahía Blanca Estuary; between 2000 and 2004, the artisanal fishery in the south of Buenos Aires Province collapsed.

In Lopez Cazorla et al. (2014), we explore the causes of the collapse of the fishery within the estuary. A cursory look at the reported landings suggested that the greatly increased fishing pressure from industrial vessels operating outside the estuary had depleted the stocks of *C. guatucupa*. The annual

(continued)

Box 11.1 (continued)

commercial landings of the Argentinean fleet reached 5000 t in the early 1970s. After that, landings increased sharply to 20,000–48,000 t in the decade between 1995 and 2004 (Villwock de Miranda and Haimovici 2007). From 1992 to 1998, the number of industrial vessels targeting striped weakfish at the northern continental shelf of Argentina doubled, and the amount of effort measured in fishing hours quadruplicated (Ruarte et al. 2000). For *C. guatucupa*, the first scientific results pinpointing the decrease in the yields of the artisanal fleet fishing this species in Bahía Blanca Estuary were presented by Lopez Cazorla (2004). Carozza et al. (2004) mentioned that since 2000, there was an increase in landings of several coastal species at El Rincón area, especially during the reproductive season of most of them. Additionally, Aubone et al. (2006) mentioned that from 1995 to 2006, biomass of *C. guatucupa* stocks south to the 39° S was severely depleted.

In Fig. 11.5, we plotted landings, effort, and yield (CPUE) for both fleets from 1992 to 2009. Landings in Bahía Blanca Estuary seem to increase steadily since 2004, whereas landings in El Rincón area increased to higher values between 1994 and 2002. Since 2001, landings dropped to lower values in El Rincón area, while they started to increase in Bahía Blanca Estuary.

Effort applied in Bahía Blanca Estuary decreased from 1992 to 2000; from 2001 up to 2004, it remained low and constant, and since 2005 a slight recovery was observed. Effort applied on El Rincón area increased in 1996–2001 period from the lower values of 1992–1995. From 2002, effort dropped considerably, remaining low until 2003, although a slightly positive trend can be observed until 2009. In Bahía Blanca Estuary, there was a steady drop in yield, from 1992 to 1994, and this decrease continues until 1998. On 1999, yield peaked at 265 kg/day and then dropped again until 2003. Starting on 2004, there was a net increase of the yield. Yields in El Rincón area increase since 1993, and up to 2001, around the same average values. Yield between 2002 and 2009 presented an important increase.

Our evidence suggests that the landings of the artisanal fleet operating in Bahía Blanca Estuary were affected by the increased fishing pressure exerted by the industrial fishing fleet of El Rincón area. Effective management of this fishery needs to be implemented to attain sustainability. While the fishing closures in El Rincón area provided certain respite for the weakfish populations, they are not enough to rebuild the stock of this species. *Cynoscion guatucupa* stocks need to be recovered before a sustainable fishery of this species can be implemented (Lopez Cazorla et al. 2014).

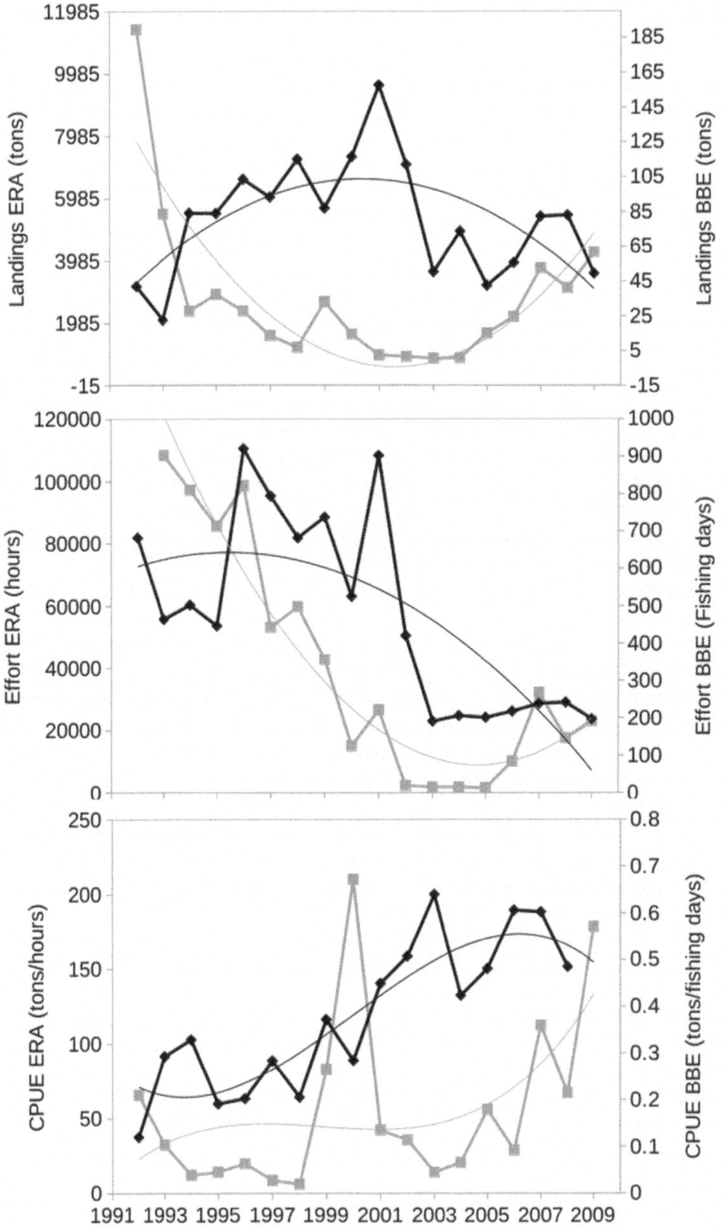

Fig. 11.5 Reported landings (**a**), effort (**b**), CPUE (**c**), and model estimates for Bahía Blanca Estuary (BBE) and El Rincón area (ERA) for *Cynoscion guatucupa* in 1992–2009. Data for Bahía Blanca Estuary is presented with boxes and a gray line for the model estimates. Data for El Rincón area is presented with triangles and a black line for the model estimates. (Modified from Lopez Cazorla et al. 2014)

References

Aguiar AA, Gallo V, Valentin JL (2004) Using the size independent discriminant analysis to distinguish the species of Myliobatis Cuvier (Batoidea: Myliobatidae) from Brazil. Zootaxa 464:1–7

Aguilera Socorro O, Fredou FL, Haimovici M et al (2015) The IUCN Red List of Threatened Species 2015: *Micropogonias furnieri*. https://doi.org/10.2305/IUCN.UK.2015-4.RLTS.T195076A49232972.en. Accessed 30 Aug 2019

Aubone A, Ruarte C, Di Marco E (2006) Un modelo matricial estructurado por estadios de tallas para la pescadilla de red (*Cynoscion guatucupa*) al sur de los 39°S, en el periodo 1995-2005. Inf Téc Ases y Transf DNI-INIDEP 20:16

Avigliano E, Volpedo AV (2015) New records of anadromous catfish *Genidens barbus* (Lacépéde, 1803) in the Paraná Delta (South America): evidence of extension in the migration corridor? Mar Biodivers Rec 8:e23. https://doi.org/10.1017/S175526721400147X

Bamber RN, Henderson PA (1988) Pre-adaptative plasticity in atherinids and the estuarine seat of teleost evolution. J Fish Biol 33:17–23

Beheregaray LB, Levy J (2000) Population genetics of the silverside *Odontesthes argentinensis* (Teleostei, Atherinopsidae): evidence for speciation in an estuary of southern Brazil. Copeia 2000:441–447

Bemvenuti MA (2005) Osteologia comparada entre as espécies de peixes-rei Odontesthes Everman & Kendal (Osteichthyes, Atherinpasidae) do sistema lagunar Patos-Mirim, no extremo sul do Brasil. Rev Bras Zool 22:293–305

Blaber SJM (2002) Fish in hot water: the challenges facing fish and fisheries research in tropical estuaries. J Fish Biol 61:1–20

Blaber SJM, Cyrus DP, Albaret JJ et al (2000) Effects of fishing on the structure and functioning of estuarine and nearshore ecosystems. ICES J Mar Sci 57:590–602

Blasina GE, Barbini SA, Díaz de Astarloa JM (2010) Trophic ecology of the black drum, *Pogonias cromis* (Sciaenidae) in Mar Chiquita coastal lagoon (Argentina). J Appl Ichthyol 26:528–534

Blasina GE, Molina JM, Lopez Cazorla A et al (2016) Relationship between morphology and trophic segregation in four closely related sympatric fish species (Teleostei, Sciaenidae). C R Biol 339:498–506

Blyth-Skyrme RE, Kaiser MJ, Hiddink JG et al (2006) Conservation benefits of temperate marine protected areas: variation among fish species. Conserv Biol 20:811–820

Campos D, Silva A, Sales N et al (2015) Trophic relationships among fish assemblages on a mudflat within a Brazilian Marine protected area. Braz J Oceanogr 63:429–442

Carbines G, Cole RG (2009) Using a remote drift underwater video (DUV) to examine dredge impacts on demersal fishes and benthic habitat complexity in Foveaux Strait, Southern New Zealand. Fish Res 96:230–237

Carozza C, Fernandez Araoz N (2009) Análisis de la actividad de la flota en el área de "El Rincón" dirigida al variado costero durante el período 2000-2008 y situación de los principales recursos pesqueros. Inf Téc INIDEP 23:p18

Carozza C, Lasta C, Ruarte C et al (2004) Corvina rubia (*Micropogonias furnieri*). In: Sánchez RP, Bezzi SI, Boschi EE (eds) El Mar Argentino y sus recursos pesqueros. Tomo 4: Los peces marinos de interés pesquero. Caracterización biológica y evaluación del estado de explotación. INIDEP, Mar del Plata, pp 255–270

Cassia MC (1986) Reproducción y fecundidad de la pescadilla de red (*Cynoscion striatus*). Publ Com Téc Mix Fr Mar 1:191–203

Castro JI (1993) The shark nursery of bulls bay, South Carolina, with a review of the shark nurseries of the southeastern coast of the United States. Environ Biol Fish 38:37–48

Chao LH, Pereira LE, Vieira JP (1985) Estuarine fish community of the dos Patos Lagoon, Brazil. A baseline study. In: Yañez-Arancibia A (ed) Fish community ecology in estuaries and coastal lagoons: towards an ecosystem integration. UNAM Press, Mexico, pp 429–450

Chao L, Vieira JP, Brick Peres M et al (2015) The IUCN Red List of Threatened Species 2015: *Pogonias cromis*. https://doi.org/10.2305/IUCN.UK.2015-2.RLTS.T193269A49230598.en. Accessed 30 Aug 2019

Chiaramonte GE, Pettovello AD (2000) The biology of *Mustelus schmitti* in southern Patagonia, Argentina. J Fish Biol 57:930–942

Colautti D, Baigun C, Lopez Cazorla A et al (2010) Population biology and fishery characteristics of Smoothhound *Mustelus schmitti* in Anegada Bay, Argentina. Fish Res 106:351–357

Cousseau MB, Perrotta RG (2013) Peces marinos de Argentina: biología, distribución, pesca. 4th. ed. Instituto Nacional de Investigación y Desarrollo Pesquero INIDEP, Mar del Plata

Cousseau MB, Pequeño G, Mabragaña E, Lucifora LO, Martínez P, Giussi A (2020) The Magellanic Province and its fish fauna (South America): Several provinces or one? J Biogeogr 47: 220– 234. https://doi.org/10.1111/jbi.13735

Dantas DV, Barletta M, Ramos JA et al (2013) Seasonal diet shifts and overlap between two sympatric catfishes in an estuarine nursery. Estuaries 36:237–256

Day JW, Hall CAS, Kemp WM, Yáñez-Arancibia A (1989) Estuarine Ecology. Wiley, New York

DeMartini EE, Sikkel PC (2006) Reproduction. In: Allen LG, Pondella DJ, Horn MH (eds) The ecology of marine fishes—California and adjacent waters. University of California Press, San Francisco, pp 483–452

Di Dario F, Williams JT, Palla H (2017) The IUCN Red List of Threatened Species 2017: *Ramnogaster arcuata*. https://doi.org/10.2305/IUCN.UK.2017-3.RLTS.T98822415A98886225.en. Accessed 30 Aug 2019

Díaz de Astarloa JM, Munroe TA (1998) Systematics, distribution and ecology of commercially important paralichthyid flounders occurring in Argentinean-Uruguayan waters (Paralichthys, Paralichthyidae): an overview. J Sea Res 39:1–9

Dyer BS (2000) Revisión sistemática de los pejerreyes de Chile (Teleostei, Atheriniformes). Est Oceanol 19:99–127

Elías R, Iribarne O, Bremec C et al (2004) Comunidades bentónicas de fondos blandos. In: Píccolo MC, Hoffmeyer MS (eds) El ecosistema del estuario de Bahía Blanca. Instituto Argentino de Oceanografía, Bahía Blanca, pp 179–190

Elisio M, Colonello JH, Cortés F et al (2017) Aggregations and reproductive events of the narrownose smooth-hound shark (*Mustelus schmitti*) in relation to temperature and depth in coastal waters of the South-Western Atlantic Ocean (38–42°S). Mar Freshw Res 68:732–742

Elliott M, Hemingway KL, Costello MJ et al (2002) Links between fish and other trophic levels. In: Elliott M, Hemingway K (eds) Fishes in estuaries. Wiley, New York, pp p124–p216

Elliott M, Whitfield AK, Potter IC et al (2007) The guild approach to categorizing estuarine fish assemblages: a global review. Fish Fish 8:241–268

Escobar BEA (1995) Dimorfismo sexual, crecimiento y fecundidad del lenguado común (*Paralichthys adspersus*) de la costa central del Perú. Thesis to obtain the Fishing Engineering degree, Facultad de Pesquería, Universidad Nacional Agraria La Molina Perú, p 63

Fabré NN, Cousseau MB (1990) Sobre la determinación de la edad y el crecimiento del lenguado *Paralichthys isosceles* aplicando retrocálculo. Rev Brasil Biol 50:345–354

FAO (2005) Increasing the contribution of small-scale fisheries to poverty alleviation and food security, FAO Technical Guidelines for Responsible Fisheries No 10. FAO, Rome, p 79

Franco A, Elliott M, Franzoi P, Torricelli P (2008) Life strategies of fishes in European estuaries: the functional guild approach. Mar Ecol Prog Ser 354:219–228

Freije RH, Spetter CV, Marcovecchio JE et al (2008) Water chemistry and nutrients of the Bahía Blanca Estuary. In: Neves R, Baretta J, Mateus M (eds) Perspectives on integrated coastal zone management in South America. IST Press, Lisboa, pp 241–254

García ML (1987) Contribución al conocimiento sistemático y biológico de los Atherinidae del Mar Argentino. PhD Thesis. Universidad Nacional de La Plata, La Plata

Garcia AM, Vieira JP (2001) O Aumento da diversidade de peixes no estuario da Lagoa dos Patos durante o episodio El Niño 1997–1998. Atlantica 23:133–152

García-Charton JA, Pérez-Ruzafa A (2001) Spatial pattern and the habitat structure of a Mediterranean rocky reef fish local assemblage. Mar Biol 138:917–934

Gaston KJ (2000) Global patterns in biodiversity. Nature 405:220–227

Gerking SD (1994) Feeding ecology of fish. Academic Press, California

Giberto DA, Bremec CS, Acha EM (2007) Feeding of the whitemouth croaker *Micropogonias furnieri* (Sciaenidae; Pisces) in the estuary of the Rio de la Plata and adjacent uruguayan coastal waters. Atlantica 29:75–84

González Castro M, Díaz de Astarloa JM, Cousseau MB et al (2009) Fish composition in a South-Western Atlantic temperate coastal lagoon: spatial–temporal variation and relationships with environmental variables. J Mar Biol Assoc U K 89:593–604

Haimovici M, Martins AS, Vieira PC (1996) Distribuicao e abundancia de peixes teleósteos demersais sobre a plataformacontinental do sul do Brasil. Rev Bras Biol 56:27–50

Hamlett WC (1999) Sharks, skates, and rays: the biology of elasmobranch fishes. The Johns Hopkins University Press, Maryland

Horn MH, Ferry-Graham LA (2006) Feeding mechanisms and trophic interactions. In: Allen LG, Pondella DJ, Horn MH (eds) The ecology of marine fishes: California and adjacent waters. University of California Press, Berkeley, pp 387–410

Jaureguizar A, Bava J, Carozza CR et al (2003a) Distribution of the whitemouth croaker (*Micropogonias furnieri*) in relation to environmental factors at the Río de la Plata estuary, South America. Mar Ecol Prog Ser 255:271–282

Jaureguizar A, Menni R, Bremec AC et al (2003b) Fish assemblage and environmental patterns in the Río de la Plata estuary. Estuar Coast Shelf Sci 56:921–933

Jaureguizar A, Solari A, Cortés F et al (2016) Fish diversity in the Río de la Plata and adjacent waters: an overview of environmental influences on its spatial and temporal structure. J Fish Biol 89:569–600

Llompart FM, Colautti DC, Maiztegui T et al (2013) Biological traits and growth patterns of pejerrey *Odontesthes argentinensis*. J Fish Biol 82:458–474

Lopez Cazorla A (1985) Edad, crecimiento y comportamiento migratorio de *Brevoortia aurea* (Agassiz, 1829) (Osteichthyes, Clupeidae) de Bahía Blanca (Argentina). Investig Pesq 49:297–314

Lopez Cazorla AC (1987) Contribución al conocimiento de la ictofauna marina del área de Bahía Blanca. PhD thesis, Universidad Nacional del Sur, Bahía Blanca

Lopez Cazorla A (1996) The food of *Cynoscion striatus* (Cuvier) (Pisces: Sciaenidae) in the Bahía Blanca area, Argentina. Fish Res 28:371–379

Lopez Cazorla A (2000) Age structure of the population of weakfish *Cynoscion guatucupa* (Cuvier) in the Bahıa Blanca waters, Argentina. Fish Res 46:279–286

Lopez Cazorla A (2004) Peces del estuario de Bahía Blanca. In: Píccolo M, Hoffmeyer M (eds) El ecosistema del estuario de Bahía Blanca. Instituto Argentina de Oceanografía, Bahía Blanca, pp 191–201

Lopez Cazorla A (2005) On the age and growth of flounder *Paralichthys orbignyanus* (Jenyns, 1842) in Bahía Blanca estuary, Argentina. Hydrobiologia 537:81–87

Lopez Cazorla AL, Sidorkewicj N (2009) Some biological parameters of Jenyns' sprat *Ramnogaster arcuata* (Pisces: Clupeidae) in South-Western Atlantic waters. Mar Biodivers Rec 2:127

Lopez Cazorla A, Pettigrosso R, Tejera L et al (2011) Diet and food selection by *Ramnogaster arcuata* (Osteichthyes, Clupeidae). J Fish Biol 78:2052–2066

Lopez Cazorla A, Molina JM, Ruarte C (2014) The artisanal fishery of Cynoscion guatucupa in Argentina: exploring the possible causes of the collapse in Bahía Blanca estuary. J Sea Res 88:29–35

Macchi GJ, Acha EM, Lasta CA (2002) Reproduction of black drum (*Pogonias cromis*) in the Rio de la Plata estuary, Argentina. Fish Res 59:83–92

Macchi GJ, Acha EM, Militelli MI (2003) Seasonal egg production of whitemouth croaker (*Micropogonias furnieri*) in the Río de la Plata estuary, Argentina–Uruguay. Fish Bull 101:332–342

Mallo P, Cervellini P (1988) Distribution and abundance of larvae and postlarvae of *Artemesia longinaris*, *Pleoticus muelleri* and *Peisos petrunkevitchi* (Crustacea: Decapoda: Penaeida) in the coastal waters of the Bahia Blanca Bay, Argentina. J Aquacult Trop 3:1–9

Malloy KD, Yamashita Y, Yamada H et al (1996) Spatial and temporal patterns of juvenile stone flounder *Kareius bicoloratus* growth rates during and after settlement. Mar Ecol Prog Ser 131:49–59

Martinho F, Dolbeth M, Viegas I et al (2009) Environmental effects on the recruitment variability of nursery species. Estuar Coast Shelf Sci 83:460–468

Massa A, Hozbor N, Chiaramonte GE et al (2010) Mustelus schmitti. The IUCN Red List of Threatened Species. https://doi.org/10.2305/IUCN.UK.2006.RLTS.T60203A12318268.en. Accessed 22 Oct 2019

Masuda Y, Ozawa T, Onoue O, Hamada T (2000) Age and growth of the flathead, *Platycephalus indicus*, from the coastal waters of West Kyushu, Japan. Fish Res 46:113–121

McCormick MI, Choat JH (1987) Estimating total abundance of a large temperate-reef fish using visual strip-transects. Mar Biol 96:469–478

Mendoza-Carranza M, Vieira J (2008) Whitemouth croaker *Micropogonias furnieri* (Desmarest, 1823) feeding strategies across four southern Brazilian estuaries. Aquat Ecol 42:83–93

Menni RC (1985) Distribución y biología de *Squalus acanthias*, *Mustelus schmitti* y *Galeorhinus vitaminicus* (Chondrichthyes) en el mar argentino en agosto-setiembre de 1978. Rev Mus La Plata 13:151–182

Molina JM (2013) La comunidad íctica de Bahía Anegada: estructura, composición, dinámica estacional y aspectos biológicos. PhD thesis, Universidad Nacional del Sur, Bahía Blanca

Molina JM, Cazorla AL (2015) Biology of *Myliobatis goodei* (Springer, 1939), a widely distributed eagle ray, caught in northern Patagonia. J Sea Res 95:106–114

Molina JM, Lopez Cazorla A (2011) Trophic ecology of *Mustelus schmitti* (Springer, 1939) in a nursery area of northern Patagonia. J Sea Res 65:381–389

Molina JM, Blasina GE, Lopez Cazorla AC (2017) Age and growth of the highly exploited narrownose smooth-hound (*Mustelus schmitti*) (Pisces: Elasmobranchii). Fish Bull 115:365–379

Neer JA, Rose KA, Cortés E (2007) Simulating the effects of temperature on individual and population growth of *Rhinoptera bonasus*: a coupled bioenergetics and matrix modeling approach. Mar Ecol Prog Ser 329:211–223

Nicolas D, Lobry J, Le Pape O, Boet P (2010) Functional diversity in European estuaries: relating the composition of fish assemblages to the abiotic environment. Estuar Coast Shelf Sci 88:329–338

Parodi E (2004) Marismas y algas bentónicas. In: Píccolo M, Hoffmeyer M (eds) El ecosistema del estuario de Bahía Blanca. Instituto Argentina de Oceanografía, Bahía Blanca, pp 101–107

Pasquaud S, Vasconcelos RP, França S et al (2015) Worldwide patterns of fish biodiversity in estuaries: effect of global vs. local factors. Estuar Coast Shelf Sci 154:122–128

Potter IC, Tweedley JR, Elliott M, Whitfield AK (2015) The ways in which fish use estuaries: a refinement and expansion of the guild approach. Fish Fish 16:230–239

Refi SM (1975) Myliobatidae y Dasyatidae del litoral bonaerense de la República Argentina y estudio comparado del mixopterigio (Chondrichthyes, Myliobatoidea). Physis 34:121–136

Ronda AC, Oliva AL, Arias AH et al (2019) Biomarker responses to polycyclic aromatic hydrocarbons in the native fish *Ramnogaster arcuata*, South America. Int J Environ Res 13:77–89

Ruarte CO, Sáez MB (2008) Estudio preliminar sobre la estructura de edades y el crecimiento de la pescadilla de red (*Cynoscion guatucupa*, Pisces, Sciaenidae) en el área sur de la Provincia de Buenos Aires. Rev Invest Des Pesq 19:37–44

Ruarte C, Lasta C, Carozza C (2000) Pescadilla de Red (*Cynoscion guatucupa*). In: Bezzi SI, Akselman R, Boschi EE (eds) Síntesis del estado de las pesquerías marítimas argentinas y de la Cuenca del Plata, Años 1997–1998, con la actualización de, vol 1999. INIDEP, Mar del Plata, pp 65–74

Rubio KS, Ajemian M, Stunz GW et al (2018) Dietary composition of black drum *Pogonias cromis* in a hypersaline estuary reflects water quality and prey availability. J Fish Biol 93:250–262

Ruocco NL, Lucifora LO, Díaz de Astarloa JM et al (2012) Morphology and DNA barcoding reveal a new species of eagle ray from the Southwestern Atlantic: *Myliobatis ridens* sp. nov. (Chondrichthyes: Myliobatiformes: Myliobatidae). Zool Stu 51:862–873

Sardiña P, Lopez Cazorla AC (2005a) Feeding habits of the juvenile striped weakfish, *Cynoscion guatucupa* Cuvier 1830, in Bahía Blanca estuary (Argentina): seasonal and ontogenetic changes. Hydrobiologia 532:23–38

Sardiña P, Lopez Cazorla A (2005b) Trophic ecology of the whitemouth croaker, *Micropogonias furnieri* (Pisces: Sciaenidae), in South-Western Atlantic waters. J Mar Biol Assoc U K 85:405–413

Segura AM, Milessi AC (2009) Biological and reproductive characteristics of the Patagonian smoothhound *Mustelus schmitti* (Chondrichthyes, Triakidae) as documented from an artisanal fishery in Uruguay. J Appl Ichthyol 25:78–82

Sidders MA, Tamini LL, Perez JE (2005) Reproductive biology of *Mustelus schmitti* springer, 1939 (Condrichtyes, Triakidae) in Puerto Quequen Buenos Aires Province. Rev Mus Argent Cienc Nat 7:89–101

Simier M, Laurent C, Ecoutin JM et al (2006) The Gambia River estuary: a reference point for estuarine fish assemblages studies in West Africa. Est Coast Shelf Sci 69:615–628

Simpfendorfer CA, Milward NE (1993) Utilisation of a tropical bay as a nursery area by sharks of the families Carcharhinidae and Sphyrnidae. Environ Biol Fish 37:337–345

Sneath PH, Sokal RR (1973) Numerical taxonomy: the principles and practice of numerical classification, 1st edn. W. H. Freeman, San Francisco

Stehmann M (2009) Myliobatis goodei. The IUCN Red List of Threatened Species. https://doi.org/10.2305/IUCN.UK.2009-2.RLTS.T161436A5423507.en. Accessed 22 Oct 2019

Topping DT, Lowe CG, Caselle JE (2006) Site fidelity and seasonal movement patterns of adult California sheephead *Semicossyphus pulcher* (Labridae): an acoustic monitoring study. Mar Ecol Prog Ser 326:257–267

Tsuji FI, Barnes AT, Case JF (1972) Bioluminescence in the marine teleost, *Porichthys notatus*, and its induction in a non-luminous form by *Cypridina luciferin* (Ostracoda). Nature 237:515–516

Vianna M, Tomas AR, Verani JR (2000) Aspects of the biology of the Atlantic Midshipman, *Porichthys porosissimus* (Teleostei, Batrachoididae): an important by-catch species of shrimp trawling off southern Brazil. Rev Brasil Oceanog 48:131–140

Villwock de Miranda L, Haimovici M (2007) Changes in the population structure, growth and mortality of striped weakfish *Cynoscion guatucupa* (Sciaenidae, Teleostei) of southern Brazil between 1976 and 2002. Hydrobiologia 589:69–78

Vinagre C, Santos FD, Cabral HN et al (2009) Impact of climate and hydrology on juvenile fish recruitment towards estuarine nursery grounds in the context of climate change. Estuar Coast Shelf Sci 85:479–486

Vooren CM (1997) Demersal elasmobranchs. In: Seeliger U, Odebrecht C, Castello JP (eds) Environment and biota of the Patos Lagoon Estuary. Springer-Verlag, Berlin, pp 141–146

Whitfield AK (1999) Ichthyofaunal assemblages in estuaries: a South African case study. Rev Fish Biol Fisher 9:151–186

Whitfield AK (2016) Biomass and productivity of fishes in estuaries: a South African case study. J Fish Biol 89:1917–1930

Wonton RJ (1992) Fish ecology. Blackie and Son, London

Yañez-Arancibia A, Linares AFA, Day JW (1980) Fish community structure and function in Terminos Lagoon, a tropical estuary in the southern Gulf of Mexico. In: Kennedy VS (ed) Estuaries perspectives. Academic Press, New York, pp 465–482

Chapter 12
Bahía Blanca Estuary and the Importance of Wetlands for the Conservation of Sea Turtles

Victoria Massola, Laura Prosdocimi, Cristina Suldrup, and Juan Facundo Sosa

Resumen
Las tortugas marinas, símbolo de longevidad, paciencia, fertilidad y libertad, son reptiles que aparecieron en la tierra hace unos 200 millones de años. Hay siete especies agrupadas en dos familias Cheloniidae y Dermochelyidae. Estos animales están presentes en todos los océanos y mares del mundo, excepto en las regiones polares. Las tortugas marinas son especies migratorias, que se mueve entre sus áreas de anidación y alimentación. Desde finales de primavera hasta principios de otoño es frecuente la aparición de tres especies de tortugas marinas en el litoral de la provincia de Buenos Aires, Argentina. Estas son la tortuga cabezona (*Caretta caretta*), la tortuga laúd (*Dermochelys coriacea*) y la tortuga verde (*Chelonia mydas*). El

V. Massola (✉)
Fundación para la Asistencia de Animales Marinos y Educación Ambiental FRAAM, Lokeren, Belgium

Reserva Natural Provincial Bahía Blanca – Organismo Provincial Para el Desarrollo Sostenible OPDS, Provincia de Buenos Aires, Bahía Blanca, Argentina

Programa Regional para la Investigación y Conservación de Tortugas Marinas en, Buenos Aires, Argentina

L. Prosdocimi
Programa Regional para la Investigación y Conservación de Tortugas Marinas en, Buenos Aires, Argentina

Dirección de Planificación y Gestión de Pesquerías, Subsecretaría de Pesca y Acuicultura, Ministerio de Agricultura, Ganadería y Pesca, Buenos Aires, Argentina

C. Suldrup · J. F. Sosa
Fundación para la Asistencia de Animales Marinos y Educación Ambiental FRAAM, Lokeren, Belgium

Programa Regional para la Investigación y Conservación de Tortugas Marinas en, Buenos Aires, Argentina

S. M. Fiori, P. D. Pratolongo (eds.), *The Bahía Blanca Estuary*,
https://doi.org/10.1007/978-3-030-66486-2_12

estuario de Bahía Blanca es un extenso humedal marino costero y es una importante área de alimentación, en el Atlántico sudoccidental, para la tortuga verde durante su etapa de desarrollo juvenil, principalmente debido a su dieta omnívora. En esta área costera es un importante centro urbano, industrial, portuario y de pesca artesanal. El objetivo de este capítulo es actualizar la información sobre el valor de los humedales del estuario de Bahía Blanca como una importante área de alimentación para las tortugas marinas, principalmente la especie de tortuga verde. Además, se presentarán algunos proyectos de investigación locales y los esfuerzos de conservación y educación ambiental que han estado en progreso desde 2003 hasta la actualidad.

Palabras Claves
Tortugas marinas, áreas de alimentación, captura incidental, especies en peligro de extinción

12.1 Introduction

Sea turtles belong to the reptile class; they are vertebrates that have been evolving since approximately 120 million years ago, a successful group of animals that has survived from periods of stable times to significant environmental change, adapting to marine life. There are seven species of sea turtles that are distributed in all the seas and oceans of the world, except in Antarctica. They swim in tropical, subtropical, and temperate waters. They are grouped into two families: Cheloniidae on one hand has six species whose shell has obvious shields – hawksbill turtle (*Eretmochelys imbricata*), green turtle (*Chelonia mydas*), loggerhead turtle (*Caretta caretta*), Kemp's ridley turtle (*Lepidochelys kempii*), olive ridley turtle (*Lepidochelys olivácea*), and flatback turtle (*Natator depressus*). Dermochelyidae on the other hand has a single species and shell without shields: leatherback turtle (*Dermochelys coriacea*). Throughout their life cycle, they occupy diverse ecosystems, moving from hundreds to thousands of kilometers between the nesting, reproduction, feeding, and developing areas, using beach, coastal zone, neritic, pelagic, and demersal environments. Some species use large "blue corridors" swimming across national and international coasts; therefore, they are considered highly migratory reptiles (Bolten 2003).

They play a highly important ecological role. In the food plot, they serve as prey and predators alike, contributing to the healthy maintenance of the structure and dynamics of the marine ecosystem. In adulthood, the green turtle becomes an herbivore, feeding on seaweed and seagrass. Through its feces, it contributes nitrogen to the environment, thus facilitating the regrowth of pastures and maintaining the replacement of nutrients. They are also considered indicators of the health of the sea and coasts (Gonzalez Carman et al. 2011).

In some ancestral cultures and folklore of many Asian and Western countries, they represent a symbol of longevity, calm, and patience – capacities and powers attributed to an animal considered mythical, sacred, and full of wisdom (Frazier 2003, 2005).

Despite being surrounded by so much spirituality, they have been intensely exploited since ancient times, as early as the sixteenth century, until late in the twentieth century for meat and egg consumption (Seminoff 2004; IUCN 2014), putting many populations of different species at severe risk (Broderick et al. 2001, 2006; Mc Clenachan et al. 2006).

On nesting beaches, where females spend a small part of their lives, they face real estate projects, looting of eggs, exploitation of meat, skin and shell trade by humans, attacks by domestic pets and other introduced animals, and presence of solid urban waste near nesting areas, as well as global warming that affects both sands where eggs mature and seawater. However, in recent years protection measures have been adopted on nesting beaches and other mitigation measures in the water, for example, to reduce interaction with fisheries, like the mandatory use of turtle-excluding devices (TED); the instruments are still insufficient for the protection and to ensure the viability of their populations.

Of the seven species of sea turtles, five can be found on the IUCN Red List of Threatened Species as either "endangered" or "critically endangered." The olive ridley turtle, hawksbill turtle, and leatherback turtle are listed as "critically endangered," the loggerhead and green turtle as "endangered," Kemp's ridley turtle as "vulnerable," and the flatback turtle as "insufficient data," which means that its conservation status is not clear due to the lack of data gathered (IUCN 2014; Bolten 2003).

Throughout their complex life cycle, postmodernity has exposed them to different threats including longlines, bycatch in coastal artisanal fisheries such as pelagic commercial fishing, boat accidents, oil spills, and discharge of other chemical substances, as well as solid urban waste pollution mainly plastic. In the last 100 years, threats of anthropic origin sea turtles must face have increased exponentially, compromising the viability of their population. For these reasons, the measures and actions taken for the protection and conservation of these wonderful marine reptiles must be applied cross-border based on the wide range of action and movement corridors used by sea turtles.

In the face of the unfavorable panorama of the Americas in the 1990s, a cooperation strategy was adopted between governmental and nongovernmental organizations and research groups aiming to coordinate conservation actions within the framework of a regional strategy. In 1996 in Caracas, Venezuela, the Inter-American Convention for the Protection and Conservation of Sea Turtles (CIT) was implemented – an intergovernmental treaty that promotes the protection, conservation, and recovery of sea turtle populations and the habitats they depend on. This was based on the most reliable data available and considered the environmental, socioeconomic, and cultural characteristics of the States Parties (ANNEX I – ACTA CFP No. 31/2018). The ILC was launched in 2001 as a space for dialogue and action in the Americas to achieve concerted management of the species (Marco Solano, Secretary Pro Tempore Inter-American Convention for the Protection and Conservation of Sea Turtles 2004).

12.2 Sea Turtles in Argentina

Until 2003 it was thought that the presence of sea turtles in Argentina was peculiar and accidental. There were few reports of beach-stranded animals with sanitary complications, accidentally bycaught by fishery boats, and in case of being dead, they are turned into local museum collections (Freiberg 1938; 1942, 1945; Gallardo 1977; Frazier 1984). This lack of information, plus the temperate coastal waters, leads to the belief that the presence of these marine reptiles was rare and that the few animals found here were disoriented or had been wrongly dragged by ocean currents (González Carman et al. 2011).

Since the conformation of the Programa Regional de Investigación y Conservación de Tortugas Marinas de Argentina (PRICTMA) (Regional Program of Investigation and Conservation of Argentine Sea Turtles) in 2003, protocols were established for systematic procedures to collect information about the presence of sea turtles in our country.

Different kinds of organizations began to work, led by this Program, by setting consensual objectives and strategies in order to optimize and enhance their technical and logistic capacities for the investigation and conservation of sea turtles.

At present, the PRICTMA is composed of nine organizations, distributed along the Argentine coast: the Ecoparque Aquarium in Buenos Aires; Peyú Project; Aquamarina; Sea World Foundation (at San Bernardo del Tuyú); ECOFAM Project; Mar del Plata Aquarium Foundation; Cooperative Association of the Multiple Use Natural Reserve Bahía Blanca, Bahía Falsa, Bahía Verde; the Marine and Fishery Biology Institute Alm. Storni; and Natural Patagonia Foundation (Fig. 12.1).

Thanks to the work strategies, of which the key point was the integration and cooperation between the different institutions in a logistic and academic level, some of the ruled objectives were reached in short time, in order to answer some basic questions about biology and about the different sea turtles visiting our waters. As a result of the working strategies, there also emerged two main tools: Marking Program and Satellite Monitoring Program for Sea Turtles.

In 2018, the National Action Plan for the Conservation of Sea Turtles (PAN Tortugas) was established, which has two programs: the National Action Program to reduce the interaction of sea turtles with fisheries in the Argentine Republic (Resolution of the Federal Fisheries Council No. 14/2018) and the National Action Program to reduce the interaction of sea turtles with marine debris in the Argentine Republic (Resolution of the Federal Environment Council No. 317/2015).

Within the framework of this plan, the Multiple Use Reserve and the FRAAM foundation are aligned on the objectives and conservation strategies for these marine reptiles, as well as for species of mammals, birds, and chondrichthyans.

Fig. 12.1 Institutions that are part of the Programa Regional de Investigación y Conservación de Tortugas Marinas de Argentina (PRICTMA). The map shows the distribution of the institutions: (1) Ecoparque; (2) Peyú Project; (3) Sea World Foundation; (4) Aquamarina; (5) Mar del Plata Aquarium Foundation; (6) Ecofam; (7) Multiple Use Reserve Bahía Blanca, Bahía Falsa, Bahía Verde (OPDS); (8) Almirante Storni Institute; (9) Natural Patagonia Foundation

12.3 Biology of the Species Found in Bahía Blanca

From the monitoring of the fishing localities and beach sectors, it was determined that, effectively, presence of sea turtles was not random; on the contrary, a stage in the life of three sea turtle species out of the seven species from the world happens in our sea. These species are the green turtle (*Chelonia mydas*), loggerhead turtle (*Caretta caretta*), and the leatherback turtle (*Dermochelys coriacea*). Recently, two specimens of hawksbill turtle (*Eretmochelys imbricata*) were found: the first one was by an accidental fishermen bycatch at San Bernardo's beach, and the second was found dead at Bahía Blanca's estuary, being the first records of this species in Argentina. Genetics studies made from them showed that they came from Brazilian nesting zones (Prosdocimi et al. 2014a, b).

Green turtle (*Chelonia mydas*): the green turtle is the biggest in the family of Cheloniidae (Pritchardand and Mortimer 1999). Generally, it shows an oval shell dorsoventrally flattened with five vertebral shields, four pairs of costal shields, and four pairs of inframarginal shields. It is characterized for having a couple frontal scales and four pairs of post-orbital scales. The upper jaw has a slightly denticulate border, while the inferior jaw has sharper denticulate border. Each fin shows one nail, though it may have two (Fig. 12.2) (Pritchardand and Mortimer 1999). Like the

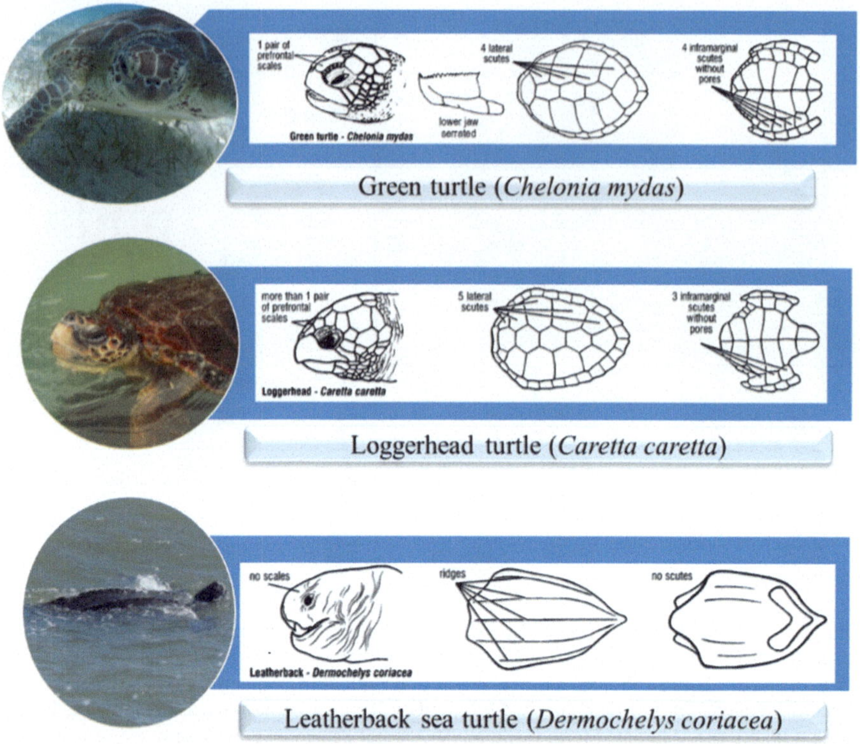

Fig. 12.2 Morphological characteristics of the three species of sea turtles present in the Argentine Sea. Modified and reprinted with permission from Wyneken, J. 2001. The Anatomy of Sea Turtles. NOAA Tech. Memo. NMFS-SEFSC-470

other species of the family, the hatchlings' shell color is mostly black or dark gray with a whitish plastron. As they grow, the shell changes to dark coffee or olive green. Colors in adults can vary from stained to striped brown, gray, black, or green tones (Fig. 12.2). The common name of the species came from the green color of the corporal fat (Pritchardand and Mortimer 1999) (Fig. 12.2).

Loggerhead turtle (*Caretta caretta*) has five pairs of costal shields and five central and two supracaudals. Another distinctive characteristic is that they show two pairs of prefrontal scales, three pairs of post-orbital scales, and a characteristic red-coffee dorsoventral coloration in subadults and adults. Also present are three or in rare occasions four inframarginal scales with no pores. The nape scale is in contact with the first two costal scales, different from the green and hawksbill turtles (Dodd 1988). The presence of asymmetric anomalies in the scale disposition is very frequent and can lead to mistakes in the specific determination (Rivilla et al. 2005). The head is proportionally bigger compared to the other species, hence the characteristic name. The frontal fins are relatively shorter in comparison with the other species of the family, showing two nails in each fin (Pritchard and Mortimer 1999) (Fig. 12.2).

Leatherback turtle (*Dermochelys coriacea*): Unlike the others species of sea turtles, this species has no keratinized shell. It was thought that the shell is covered by a thick skin, recovering a mosaic of dermic ossicles connected by cartilage (Wyneken 2001). The shell has a markedly tapered form, with seven prominent ridges that enhance swimming (Wyneken 2003). The head is triangular with two conspicuous maxillary cusps covered by skin (Pritchard and Mortimer, 1999). A distinctive sign of the specie is a rose stain in the back of the head that is used for photo identification (McDonald and Dutton 1996) (Fig. 12.2).

Each one of these species registered in Argentina is found in different stages of their life. The green turtles, for example, are in their early youth, recently raised from the oceanic environment to the neritic one. Their weight range is from 3 to 10 kg, with a shell length between 30 and 50 cm. Nevertheless the loggerhead turtles have a wider size range, between 45 and 90 cm, and a weight range between 15 and 40 kg, both young and adults. The leatherback turtles are subadults and adults with length up to 180 cm and more than 200 kg weight (González Carman et al. 2011).

The hawksbill turtles found correspond to young individuals (Prosdocimi et al. 2014a, b). The presence of sea turtles in our coast is seasonal, mainly during summer and fall. Most of the records are concentrated in Bahía Samborombón, Cabo San Antonio, and Bahía Blanca Estuary, though there are notices of fluvial raids in the La Plata River, the Paraná River, and the Uruguay River and also their presence in the colder waters of the north of Patagonia, as in San Matías, San José, and Golfo Nuevo (González Carman et al. 2011, 2012, 2016a, b).

The majority of the registers of sea turtles were made in the estuarial areas of Buenos Aires coast. Those areas are characterized for being highly productive and offer a great biomass formed by jellyfish, mollusks, and other organisms which are part of the sea turtles' food. This, added to the information obtained by the analysis of the stomach contents, allows to affirm that Argentine coastal waters are a feeding and development zones for these sea species. Nevertheless, thanks to the information obtained by the Marking Plan and the Satellite Following Program, we know that our waters are integrated to a great feeding zone, that is, the South Occidental Atlantic, also involving coastal and oceanic waters from Uruguay and Brazil. By the end of autumn, the sea turtles that were feeding all along Argentine coasts move to warmer waters in the south of Brazil, where they remain feeding. In some cases, they return to the same place they fed in our coast (López Mendilaharsu et al. 2009; Fossette et al. 2010; Gonzalez Carman et al. 2012, 2016a, b).

The origin of the sea turtles that arrived in Argentine waters and specifically in Bahía Blanca's zone has been recently unveiled through poblational genetic studies. Most of the green turtles come from nesting beaches in Ascension Island, a small volcanic island placed in front of the African coast, and, in a lesser percentage, from nesting beaches of Suriname and Trindade Island (Brazil) (Prosdocimi et al. 2012). In the case of leatherback turtles, the recovering of metal marks and genetics studies, let us identify its origin as nesting beaches in Occidental Africa (Gabon) (Billes et al. 2006; Prosdocimi et al. 2014b). Lastly, for the case of the loggerhead turtle, all the individuals came from Brazilian populations (Prosdocimi et al. 2015).

12.4 Identified Conservation Problems

The highly productive areas chosen by the sea turtles to feed and grow are also the ones chosen by independent and industrial fishing fleets from Argentina and Uruguay. The accidental bycatches of the three species (green turtle, loggerhead turtle, and leatherback turtle) have been registered mainly in gillnets and trawls, while in the shrimp nets only green turtles have been caught (Gonzalez Carman et al. 2011).

In a pilot study made at the port from San Clemente del Tuyú, Buenos Aires coast, it was estimated that an average of 100 green turtles are caught in gillnets, most of them were found dead in gillnets from suffocation (Albareda et al. 2007a). Although the fishing arts that interact with sea turtles in Argentina have been identified, currently there is no other information regarding the capture rates and the impact that this threat has on their populations.

On the other hand, in the stomach contents of green turtles analyzed in the last 5 years from the accidental capture in the Bahía de Samborombón, there are remains of anthropogenic residues, mostly plastic bags (Albareda et al. 2007b). These have also been found in live sea turtles that were in rehabilitation, who had defecated the residues for several days. Plastic bags and fecal matter are part of a kind of solid "framework" in the intestine, which can end up causing intestinal obstruction that could damage the mucosa and alter its normal functionality. A gas-filled intestine, unable to evacuate regularly, acts as a "life jacket," not allowing the sea turtles to dive, move, escape from predators, or feed. This causes a slow deterioration of their physical condition, finally leading to a long and agonizing death. In our latitude this chronical deterioration could affect their normal migration to warmer waters, failing to escape on time from the low winter temperatures of Buenos Aires coast and finally being dragged to the beach in a lethargic status, victims of a hypothermic deadly picture (Albareda et al. 2007b).

In addition to sea turtle populations, the accidental bycaught, and the ingestion of anthropic residues, we must add conservation problems shared with Uruguay and Brazil (Bugoni et al. 2001; Laporta et al. 2006) and the need to coordinate regionally the actions for conservation taken by each country.

12.5 Conservation Actions

Scientific research in the last 6 years has allowed us to understand the basic aspects of the biology of sea turtles in Argentina and, in that way, the main conservation problems, through the development of interdisciplinary and multidisciplinary activities, such as poblational genetics studies, satellite monitoring, marking plan, health assessment, feeding studies, and accidental bycatch assessment. The insertion of young investigators inside the national academic circuit (such as the Scientific and Technical Investigation National Council, Buenos Aires University, Marine and

Fishery Biology Institute Almirante Storni and the Fishery Investigation and Developing National Institute) to develop some investigation lines of PRICTMA guaranteed the ability to find and to consolidate management activities. On the other hand, thanks to the increasingly active participation of independent fishermen and joint effort with other research projects, such as the Franciscana Dolphin Conservation Program led by Aquamarina, different mitigation measures are currently being tested to reduce bycatch of these dolphins and sea turtles with gillnets.

The FRAAM foundation is a nonprofit entity (registration ID: PJ 40843) which works with veterinary professionals, biologists, and environmental educators. The main function of the foundation is the primary care mainly of sea turtles. In addition to that, the training of undergraduate and graduate students (volunteers), environmental education directed to general public, and students of all levels and related careers, as well as research support, are all goals of the foundation.

12.5.1 Sea Turtles in the Southwestern Atlantic: Red ASO and PRICTMA – Start of Research

A very valuable instrument that emerged in 2003 is the creation of the ASO, Southwestern Atlantic network. The ASO interacts with almost all fisheries in the Exclusive Economic Zones of Brazil, Uruguay, and part of Argentina and the adjacent international waters of the Atlantic Ocean between 5° N to 45° S and 20° to 65° W. This network has allowed the exchange of experiences between researchers and institutions in the region, as well as support to develop combined initiatives, thus strengthening actions for the conservation of sea turtles (Domingo et al. 2006). Regarding the conservation of biodiversity and in particular sea turtles, Argentina, Brazil, and Uruguay signed, approved, and ratified, through respective national laws, some of the following International Conventions:

- Convention on Wetlands of International Importance (RAMSAR CONVENTION 1971).
- Convention on International Trade in Endangered Species of Wild Fauna and Flora (CITES), Washington, DC, 1973.
- Convention on the Conservation of Migratory Species of Wild Animals (CMS), Bonn, 1979.
- United Nations Convention on the Law of the Sea (CONVEMAR), (New York, 1982).
- Convention on Biological Diversity (Rio de Janeiro, 1992).
- Inter-American Convention for the Protection and Conservation of Sea Turtles (IAC) (San José, 2001).

In 2010, Argentina approved the National Law 26.600, through which it adheres to the Inter-American Convention for the Protection and Conservation of Sea Turtles (CIT). In June 2011, the national government deposited its instrument of ratification. This meant among other measures to assume the commitment of protection and conservation methods for these species in our latitude, through the promotion of scientific research, environmental education, and the dissemination of information, also promoting participatory work and training of independent fishermen to minimize incidental catches, retention, damage, and killing of sea turtles during fishing activities. The National Ministry of Environment and Sustainable Development (SGA y DS) is the CIT enforcement authority. Going further in the conservation strategies, in 2014 the work began on the preparation of the National Action Plan for the Conservation of Sea Turtles in the Argentine Republic (PAN-TM). In November of the same year, a workshop was carried out with the participation of national and provincial government agencies, members of the scientific sector, and nongovernmental organizations. Since then they work on the issue together. Representatives of the National State, the provinces of Buenos Aires and Río Negro, and the Autonomous City of Buenos Aires participated in the preparation of the PAN-TM, since the sea turtles reach the marine coast of the Province of Río Negro from Río de la Plata. Research efforts were mainly invested in two National Action Programs: Reduce the Interaction of Sea Turtles with Marine Wastes (Approved by Resolution COFEMA 317/2015) and Reduce the Interaction of Sea Turtles with Fisheries (Approved by Resolution 14/2018 Federal Fisheries Council).

12.5.2 Our Work in the South of Buenos Aires Province: Wetlands of Bahía Blanca Estuary

In the estuary of Bahía Blanca, the multiple use reserve Bahía Blanca, Bahía Falsa, Bahía Verde (RNBB) is established, and adjacent to the protected area in the coastal town of Villa del Mar, the Foundation for Reception and Marine Animal Attendance and Environmental Education (FRAAM), both institutions are part of PRICTMA. Since 2003, action has been taken by different groups of people: outreach campaigns directed to the independent fishermen communities in the area, formal environmental education activities, articulating with educational establishments of all levels of education, and nonformal education directed to the general public.

Using different communication strategies and incorporating other knowledge, adding the valuable empirical knowledge of fishermen and the general community, the conservation work has focused on raising public awareness of the seasonal presence of sea turtles in the waters of the estuary of Bahía Blanca, recognizing and identifying problems of anthropic origin that different species face in our latitude, and from a holistic approach seeking solutions that minimize conflict with local actors. The varied range of activities has generated a gradual yet positive response from the different actors, allowing favorable results in the first scientific research tasks carried out in this coastal sector of Buenos Aires Province.

Actions:

- Initial survey of the problem
- Environmental education workshops for students and teachers
- Publicity talks
- Training for veterinarians and voluntary university students
- Interviews with fishermen
- General information talks to fishermen
- Training for fishermen

On our coasts, the direct observation of sea turtles is difficult mainly due to two factors: the turbidity of estuarial waters and also the animals' own behavior, which emerges briefly to breathe. To avoid this problem, an important project was launched through the Satellite Monitoring Program. The placement of satellite transmitters to several animals in four coastal locations, Magdalena, San Clemente del Tuyú, and Villa del Mar – estuary of Bahía Blanca – in Buenos Aires Province and San Antonio Oeste in Río Negro Province, has provided valuable information. The satellite monitoring of green turtles, loggerhead, and leatherback turtles allowed us to determine, for example, that Río de la Plata as the southern area of Buenos Aires Province and in the North of Rio Negro Province are areas of feeding, intensely used by these species. It was also known that sea turtles frequent the fishing areas of the coastal fleets of Buenos Aires Province and Uruguay. The common areas that juvenile sea turtles use and where fishing activities are carried out are also contaminated by solid garbage that comes from cities like Buenos Aires and Montevideo.

Throughout these years, constant monitoring efforts made it possible to determine that part of the Argentine continental shelf is the southernmost feeding and development zone of the Southwestern Atlantic for at least three species: green turtles, loggerhead turtles, and leatherback turtles (González Carman et al. 2011). Most of the bycatches have been of green turtles, small-sized individuals with an early youth development stage. The two hawksbill turtle individuals found correspond to juvenile animals (Prosdocimi et al. 2014a).

Regarding the distribution in Argentina, it is conditioned by the temperature of the sea.

Therefore, the waters overlying the northern sector of the Argentine continental shelf would be their main habitat. There the temperature of the sea ranges between 18° and 23 ° C in summer and averages 8 ° C in winter, unlike the waters overlying the southern sector of the platform that are colder. The presence of sea turtles, in turn, is registered from the end of spring until the beginning of autumn, showing a marked seasonal presence probably governed by the low water temperature of 8–10 °C in winter (González Carman et al. 2011, 2012).

A practical tool to gain quick insight into the behavior of marine turtles is satellite tracking. With these the migratory routes and foraging habitats can be estimated with accuracy (Godley et al. 2008).

From February 2008 to March 2009, nine green turtles (recovered from entanglement on the coast of Argentina) were instrumented with satellite tracking devices

Fig. 12.3 Juvenile green turtle equipped with satellite tracker being released in Bahía Blanca Estuary, carried out by the Bahía Blanca Nature Reserve staff, park rangers, and PRICTMA staff (OPDS 2009)

(Fig. 12.3), six of them in Bahía Blanca Estuary (Villa del Mar and El Rincón) (González Carman et al. 2012).

During summer and fall, tracked juveniles were located in the coastal waters of Argentina and Uruguay. In winter sea turtles migrated to warmer waters in southern Brazil. Migration started during the fall, as evidenced by an increase in mean movement rate and the northern position of the 20 ° C isotherm. In spring, most sea turtles remained in the waters of Brazil and Río de la Plata (Gonzalez Carman et al. 2012).

The use of this technique allowed to know the most intense foraging season in Bahía Blanca Estuary (El Rincón, Villa del Mar), in the coast of Buenos Aires, and in the La Plata River, an area of 11,282 km, during summer months and early autumn. While in winter and spring period, for the same purpose, sea turtles used the coasts of Uruguay and Brazil (Gonzalez Carman et al. 2012).

12.5.2.1 Summary of Actions for the Protection and Conservation of Marine Turtles in the Wetlands of Bahía Blanca Estuary

Bahía Blanca – OPDS Natural Reserve, Environmental Education Program and the Foundation for Reception and Marine Animal Attendance FRAAM, Villa del Mar. Our sustained work without interruption since 2003 to date has focused its efforts on training and divulgation. It has been designed and directed to different actors: independent fishermen's community (artisanal); educational community at all levels of education; connecting with academic institutions such as the South National University and government organizations such as Coronel Rosales Municipality and

Fig. 12.4 Training in primary assistance with sea turtles for Prefectura Naval Argentina (Coast Guard) staff and lifeguards of Villa del Mar and Arroyo Pareja (Coronel Rosales), carried out by staff of the FRAAM Foundation

its Honorable Deliberative Council, Bahía Blanca Municipality (Fig. 12.4); permanent support of the mass media for all the activities carried out:

- 2003–2015: We started training aimed at artisanal fishermen from the towns of Ing. White, Villa del Mar, and Coronel Rosales. We continue to interact today.
- 2004–2019: Uninterruptedly, from the Formal Environmental Education Program, we began to work on the topic Marine Turtles from the Initial Level to the Tertiary Level.
- From 2012 to date: World Sea Turtle Day Remembrance – traveling photo shows, plastic workshops, puppet theater, oral exhibitions by different specialists.
- 2012, 2013, and 2014: Beach campaigns in Monte Hermoso, Pehuén Co, and Villa del Mar during the summer season to inform tourists about the biology, ecology, and behavior of sea turtles on our coasts and particularly about the negative impacts they face as a consequence of inorganic waste left on the beach and in the sea.
- 2012, 2013, and 2014: Puppet Theater "S.O.S Who Can Help Us" – the story tells the problems faced by a green turtle and a mermaid in the sea and the beach because of anthropogenic waste. The work ends with songs and solutions shared in dialogue with the public.
- 2013: Inauguration of "Your Garbage Affects Us" Poster. Awareness and information about the environmental problems faced by the sea turtles; the poster was installed in Villa del Mar in a very busy sector, on a public access beach, during the development of the 5th edition of the Wetlands Festival. Proposed at the Education for Action Workshop, ASO Mar del Plata 2009.

- 2013: Approval of the first Municipal Ordinance No. 3408 "Emblematic Species of Coronel Rosales District." In the list of animals designated, the incorporation of the three species of sea turtles frequent on the southwest coasts of Buenos Aires was contemplated with special interest.
- 2013–2015: Video presentation – "Los Amigos de Las Tortugas" tells the story of the life of an independent fisherman and his connection with sea turtles. It was exhibited in the Cultural Historic Center of the South National University of Bahía Blanca, in Union y Fuerza hall in Punta Alta City, during the 5th, 6th, and 7th edition of the Wetlands Festival in Villa del Mar.
- 2013–2014: Second and third presentation at the Sustainable House of Culture Hall (UNS) with Sea Turtles, made of recycled material, the approval for the installation of the Clean Beaches Campaign Poster, located in the seaside of Villa del Mar Town, result of the 2009 ASO Workshop, being the second installed in the southwest of Buenos Aires Province.
- 2014: Participation to the first preparatory workshop of the National Sea Turtle Conservation Plan – the FRAAM Foundation was present during the 2-day workshop (26th and 27th of November of 2014), which took place at the headquarters of the National Ministry of Environment and Sustainable Development, the first preparatory workshop of the National Action Plan for the Conservation of Sea Turtles (PAN-TM) of the Argentine Republic. In the context of this workshop, we participate in the preparation of the National Action Program to reduce the interaction of sea turtles with marine debris in Argentina.
- 2014: Training of veterinarians and volunteer students in different endoparasite sampling and analyzing techniques.
- 2015: Signed Act Agreement for the non-use of plastic bags – in the general framework of the commitment assumed for the defense and protection of the environment, on February 2015 an agreement was signed between Bahía Blanca Municipality, Bahía Ambiental SAPEM, FRAAM Foundation, South Conservation Association TELLUS, and large supermarket companies (Cooperativa Obrera, Walmart Argentina, Carrefour, Burgos, and Super Vea). This agreement has the purpose of eliminating plastic bags from supermarkets.
- 2015: Necropsy and Research Techniques Workshop in Sea Turtles. National Action Program to reduce the interaction of sea turtles with marine debris in the Argentine Republic.
- 2015: Educational and Outreach Strategies Workshop for the reduction of marine litter in the area of the Río de la Plata and Buenos Aires Coast. National Action Program to reduce the interaction of sea turtles with marine debris in the Argentine Republic.
- 2017 Marine Garbage Workshop: Tools for a Better Impact on Public Policies and Cultural Change. National Action Program to reduce the interaction of sea turtles with marine debris in the Argentine Republic.
- 2019: First follow-up workshop on the National Action Plan for the Conservation of Sea Turtles (PAN Tortugas): ST Program – Fisheries and ST Program – Marine Residues.

12.5.2.2 Veterinary Actions

The veterinary staff has a protocol to follow in which different types of evaluations are carried out before the animal enters the Foundation if necessary. As previously mentioned, over the years, several training sessions have been carried out for different groups of people: university students, volunteers, and independent fishermen, among others. These fishermen, being the first ones to get in contact with sea turtles, will follow the instructions received, in addition to those that the corresponding personnel give them for the evaluation and management of the animal at the place. In some cases the sea turtle could be half drowned, they will make the first handlings by putting the animal in an oblique angle position, head down, in a soft and protected place, in order to prevent more injuries.

This coastal area is a migratory route for animals such as penguins, sea turtles, whales, fishes, and birds at different stages of their life cycle, from breeding areas to feeding places, and due to this it is possible to find a sea turtle (or other animals) stranded on the beach. In these cases, the action protocol is the same: the state of the animal must be evaluated as a first measure, and then the corresponding personnel should be contacted as soon as possible – FRAAM Foundation, Rangers, and Prefectura Naval Argentina (Coast Guard). Part of the protocol of action in these cases is to keep people at a good distance from the animal, avoid interacting with it in any way (feed, touch, etc.), to reduce stress. If the veterinary team is not in the area, the mentioned staff present at the scene will make an assessment of the animal's condition, if possible while on the telephone to the veterinarians, who will give instructions to verify if the animal presents some type of obvious injury or unusual behavior. If any type of anomaly is found at the anatomical or ethological level, it might be decided to transfer the animal to the Foundation where it will be given assistance with the necessary additional elements and corresponding equipment for the species in question.

The animal species that most frequently enter the Foundation are the green turtle (*Chelonia mydas*), the Magellanic penguin (*Spheniscus magellanicus*), and the two-haired sea lion (*Arctocephalus australis*). Other less frequent species are the loggerhead turtle (*Caretta caretta*) with two recorded assists and the one-haired sea lion (*Otaria flavescens*). There are other species, much less frequent, such as the Antartic fur seal (*Arctocephalus gazella*) and the Sub Antartic fur sea (*Arctocephalus tropicalis*).

Among the diagnoses made to the admitted animals, different factors are taken into account: if the injury requires surgery, if the animal must be placed in a dry area in case it has suffered drowning, or if it is very active, it is placed directly in a pool to observe its movements, reactions, and responses to the different stimuli administered. If the reaction is normal, we stimulate food intake, corresponding to the stage of the animal in question.

In this area, the sea turtles caught incidentally are at the juvenile stage of their cycle in which they have a non-selective diet, so they were fed with algae and arthropods.

When the sea turtle's responses are positive according to its behavior, it is released in the area where it was incidentally captured.

Before the release of these animals, the corresponding biometric data were taken: the length (straight and curved) and width (straight and curved) of the carapace and plastron and the length and width of the head, tail, and the four fins are measured (Workbook ASO 2004); the weight is measured upon entering the Foundation and upon release (Fig. 12.5). Blood samples are taken from the sinus occipital venous with a 25/8 needle and a 5 cc syringe. In addition to the above measurements, a micro-blood count can be made to see the proportion of blood cells by counting the percentage of proteins and blood staining to see abnormalities, as well as a small skin extract for gender determination.

All this biometric data is sent for scientific studies, for example, obtaining population data (to determine the origin). In addition to the above, the type of food administered, the one chosen by the animal itself, the quantity, the frequency, and whether this food was alive or dead are also measured and recorded. Once the turtle is ready to be released, it is marked with two metal rings, one in each fin (depending on the species), which consists a non-corrosive alloy printed with an alphanumeric identification, which will serve in the event that there is a recapture of the animal throughout the migratory route, to know where and when it was marked, and so compare the past biometric data with the current one. All this is developed following a national and international protocol. Some animals can be released in groups, as is the case with penguins. But sea turtles, which are independent and solitary animals, are released in the place where they were found or captured. Different species of birds have also been received, which after examination, if they cannot be released after recovery, will be sent to specialized centers or will remain in the

Fig. 12.5 Veterinary staff and FRAAM volunteers carrying out the biometrics on a juvenile green turtle. (Photo by V Massola)

Foundation facilities, to teach visitors about the problems of trying to domesticate wild animals. Any animal recovered and showing no remaining sequelae is released. Those who unfortunately do not survive (either because of the type of injuries they have had or because they did not arrive on time at the Foundation), their bodies are used for educational and scientific purposes.

Acknowledgments We would like to thank to the artisanal fishermen of Villa del Mar, especially to the memory of Mario Delgado, Coronel Rosales, and Ingeniero White; also the personnel of the Prefectura Naval Argentina (Coast Guard), Bahia Blanca Delegation, Coronel Rosales, and Monte Hermoso; SIPA Station (Rescue, Fire, and Environmental Protection); Oscar Liberman Alku, Carolina Parodi, the staff and park rangers of the Natural Reserve Bahía Blanca, Bahía Falsa, Bahía Verde and the FRAAM Foundation Volunteers.

References

Albareda DA, Bordino P, Prosdocimi L et al (2007a) Captura accidental de tortuga verde (*Chelonia mydas*) en la pesquería artesanal del sur de la Bahía Samborombón, Buenos Aires, Argentina. En: III Jornadas de Conservación e Investigación de Tortugas Marinas en el Atlántico Sur Occidental. Piriápolis, Uruguay. 37 p

Albareda D, Álvarez K, Iglesias M et al (2007b) ¡No te comas mi basura! Evidencia de ingestión de desechos antrópicos en tortugas verdes (*Chelonia mydas*) del estuario del Río de la Plata y la Bahía de Samboronbóm – Pcia. de Buenos Aires – Argentina. En: Libro de resúmenes de la III Jornadas de Investigación y Conservación de Tortugas Marinas del Atlántico Sur Occidental (ASO), Piriápolis, Uruguay, pp 39–40

Bailey H, Shillinger G, Palacios D et al (2008) Identifying and comparing phases of movement by leatherback turtles using state-space models. J Exp Mar Biol Ecol 356:128–135

Billes A, Fretey J, Verhage B et al (2006) First evidence of leatherback movement from Africa to South America. Marine Turtle Newsletter, IUCN/SSC MTSG News 111:13–14

Bolten AB (2003) Variation in sea turtle life history patterns: neritic versus oceanic developmental stages. In: Lutz PL, Musick JA, Wyneken J (eds) The biology of sea turtles, vol 2. CRC Press, Boca Raton, pp 243–457

Breed GA, Jonsen ID, Myers RA et al (2009) Sex-specific, seasonal foraging tactics of adult grey seals (*Halichoerus grypus*) revealed by statespace analysis. Ecology 90(11):3209–3221

Broderick AC, Godley BJ, Hays GC (2001) Trophic status drives inter- annual variability in nesting numbers of marine turtles. Proc R Soc Lond B 268:1481–1487

Broderick AC, Frauenstein R, Glen F, Hays GC et al (2006) Are green turtles globally endangered? Glob Ecol Biogeogr 15:21–26

Bugoni L, Krause L, Petry MV (2001) Marine debris and human impacts on sea turtles in southern Brazil. Mar Pollut Bull 41(12):1330–1334

Carpenter JW (2006) 3rd Edition. Formulary for Exotic Animals

Dodd CK Jr (1988) Synopsis of the biological data on the loggerhead sea turtle *Caretta caretta* (Linnaeus, 1758). US Fish Wild Serv Biol Rep 88(14):1–110

Domingo A, Bugoni L, Prosdocimi L, et al (2006) The impact generated by fisheries on Sea Turtles in the Southwestern Atlantic. WWF Progama Marino para Latinoamérica y el Caribe, San José, Costa Rica

Fossette S, Girard C, López-Mendilaharsu M, Miller P et al (2010) Atlantic leatherback migratory paths and temporary residence areas. PLoS One 5(11):e13908. https://doi.org/10.1371/Journal.pone.0013908.

Frazier J (1984) Las tortugas marinas en el Océano Atlántico Sur Occidental. La Plat: Asociación Herpetologíaca Argentina 22 p. ISSN 0326-5528

Frazier J (2003) Prehistoric and ancient historic interactions between humans and marine turtles. In: Lutz PL, Musick JA, Wyneken J (eds) The biology of sea turtles, vol II. CRC Press, Boca Raton, pp 1–38

Frazier J (ed) (2005) Marine turtles as flagships. MAST (Maritime Stud) 3(4):1–303

Freiberg MA (1938) Catálogo sistemático y descriptivo de las tortugas argentinas. Memorias del Museo de Entre Ríos, Zoología 9:1–25

Freiberg MA (1942) La tortuga laúd *Dermochelys coriacea* (L.) frente a las costas argentinas. Physis 19(52):263–265

Freiberg MA (1945) Observaciones sobre las tortugas de mar que se encuentran frente a las costas argentinas. Physis 20(55):50–53

Gallardo JM (1977) Reptiles de los alrededores de Buenos Aires. Buenos Aires Editoral EUDEBA:213

Gallo BMG, Macedo S, Giffoni BB et al (2006) Sea turtle conservation in Ubatuba, South Eastern Brazil, a feeding area with incidental capture in coastal fisheries. Chelonian Conserv Biol 5(1):93–101

Godley BJ, Blumenthal JM, Broderick AC et al (2008) Satellite tracking of sea turtles: where have we been and where do we go next? Endanger Species Res 4:3–22

González Carman V, Álvarez K, Prosdocimi L et al (2011) Temperate SW Atlantic: a feeding and developmental hábitat for endangered sea turtles. Mar Biol Res 7:500–508

González Carman V, Falabella V, Maxwell S et al (2012) Revisiting the ontogenetic shift paradigm: the case of juvenile green turtles in the SW Atlantic. J Exp Mar Biol Ecol 429:64–72

González Carman V, Mandiola A, Alemany A et al (2016a) Distribution of mega faunal species in the Southwestern Atlantic: key ecological areas and opportunities for marine conservation. ICES J Mar Sci. https://doi.org/10.1093/icesjms/fsw019

González Carman V, Bruno I, Maxwell S et al (2016b) Habitat use, site fidelity and conservation opportunities for juvenile loggerhead sea turtles in the Río de la Plata, Argentina. Mar Biol 163:20

Hart KM, Fujisaki I (2010) Satellite tracking reveals habitat use by juvenile green sea turtles *Chelonia mydas* in the Everglades, Florida, USA. Endanger Species Res 11:221–232

Hart KM, Lamont MM, Fujisaki I (2011) Common coastal foraging areas for loggerheads in the Gulf of Mexico: opportunities for marine conservation. Biol Conserv 145(1):185–194

IUCN (2014) IUCN Red List of Threatened Species. Version 2014.2. http://www.iucnredlist.org/. Accessed 13 Mar 2020

Jonsen ID, Myers RA, James MC (2007) Identifying leatherback turtle foraging behaviour from satellite telemetry using a switching state-space model. Mar Ecol Prog Ser 337:255–264

Laporta M, Miller P, Ríos M et al (2006) Conservación y manejo de tortugas marinas en la zona costera uruguaya. In: Menafra R, Rodríguez-Gallego L, Scarabino F et al (eds) Bases para la conservación y manejo de la costa uruguaya. Vida Silvestre Uruguay, Montevideo, pp 259–269

López-Mendilaharsu M, Rocha CFD, Miller P et al (2009) Insight son leatherback turtle movements and high use areas in the Southwest Atlantic Ocean. J Exp Mar Biol Ecol 378:31–39. https://doi.org/10.1016/j.jembe.2009.07.010

Luschi P, Hays GC, Del Seppia C et al (1998) The navigational feats of green sea turtles migrating from Ascension Island investigated by satellite telemetry. Proc R Soc Lond B 265:2279–2284

Makowski C, Seminoff JA, Salmon M (2006) Home range and habitat use of juvenile Atlantic green turtles (*Chelonia mydas* L.) on shallow reef habitats in Palm Beach, Florida, USA. Mar Biol 148:1167–1179

Maxwell SM, Breed GA, Nickel BA et al (2011) Using satellite tracking to optimize protection of long-lived marine species: Olive Ridley sea turtle conservation in Central Africa. PLoS One 6(5):e19905

Mc Clenachan L, Jackson JBC, Newman MJH (2006) Conservation implications of historic sea turtle nesting beach loss. Front Ecol Environ 4(6):290–296

McClellan CM, Read AJ (2009) Confronting the gauntlet: understanding incidental capture of green turtles through fine-scale movement studies. Endanger Species Res 10:165–179

McDonald DL, Dutton PH (1996) Use of PIT tags and photo identification to revise remigration estimates of leatherback turtles (*Dermochelys coriacea*) eating in St. Croix, U.S. Virgin Island, 1979-1995. Chel Cons Biol 2(2):148–152

Patterson TA, Thomas L, Wilcox C et al (2008) Statespace models of individual animal movement. Trends Ecol Evol 23(2):87–94

Piola AR, Matano RP (2001) Brazil and Falklands (Malvinas) currents. In: Steele JH, Thorpe SA, Turekian KK (eds) Encyclopedia of ocean sciences. Academic, London, pp 340–349

Polovina JJ, Kobayashi DM, Seki MP et al (2000) Turtles on the edge: movement of loggerhead turtles (*Caretta caretta*) along oceanic fronts, spanning longline fishing grounds in the central North Pacific, 1997–1998. Fish Oceanogr 9:71–82

Pritchard PCH, Mortimer JA (1999) Taxonomy, external morphology, and species identification. In: Eckert KL, Bjorndal KA, Abreu-Grobois FA et al (eds) Research and management techniques for the conservation of sea turtles. IUCN/SSC Marine Turtle Specialist Group. 235 p

Prosdocimi L, González Carman V, Albareda D et al (2012) Genetic composition of green turtle feeding grounds in coastal waters of Argentina based on mitochondrial DNA. J Exp Mar Biol Ecol 412:37–45

Prosdocimi L, Bruno I, Diaz L, González-Carman V, Albareda DA, Remis MI (2014a) Southernmost reports of the Hawksbill Sea turtle, *Eretmochelys imbricata* (Linnaeus, 1766), in Argentina. Herpetol Rev 45(1):1–5

Prosdocimi L, Dutton PH, Albareda DA et al (2014b) Origin and genetic diversity of leatherbacks (*Dermochelys coriacea*) at argentine foraging grounds. J Exp Mar Biol Ecol 458:13–19

Prosdocimi L, Bugoni L, Albareda DA et al (2015) Are stocks of immature loggerhead sea turtles always mixed? J Exp Mar Biol Ecol 466:85–91

Rivilla JC, Alís S, Alís L et al (2005) Ejemplar de tortuga boba (*Caretta caretta*) con anomalías morfológicas en los escudos del espaldar. Bol Asoc Herpetol Esp 15:98–100

Seaman DE, Powell RA (1996) An evaluation of the accuracy of kernel density estimators for home range analysis. Ecology 77(7):2075–2085

Seminoff JA (2004) Global status assessment green turtle (*Chelonia mydas*). Marine Turtle Specialist Group. The World Conservation Union IUCN, Gland

Seminoff JA, Resendiz A, Nichols WJ (2002) Home range of green turtles *Chelonia mydas* at a coastal foraging area in the Gulf of California. Mexico Mar Ecol Prog Ser 242:253–265

Work Booklet ASO (2004) Data notes standardization. Project, Tamar

Worton BJ (1989) Kernel methods for estimating the utilization distribution in homerange studies. Ecology 70(1):164–168

Wyneken J (2001) The anatomy of Sea Turtles. U. S. Department of Commerce NOAA Technical Memorandum NMFS-SEFSC-470. 172 p

Wyneken J (2003) In: Lutz P, Musick JA, Wyneken J (eds) The external, musculoskeletal and neuro-anatomy of sea turtles: the biology of sea turtles (II). CRC Press, Boca Raton, pp 39–77

Chapter 13
Shorebirds and Seabirds' Ecology and Conservation

Natalia S. Martínez-Curci, Germán O. García, Leandro Marbán, Pía Simonetti, and Sergio M. Zalba

13.1 Introduction

Shorebirds and seabirds belonging to the order Charadriiformes are probably some of the most conspicuous animals inhabiting Bahía Blanca Estuary. This extremely diverse order of birds is second only to passerines in number of families and species (Lovette 2016). Having a global distribution that ranges from the Arctic to the Antarctic, they can be found in almost any terrestrial habitat type, as well as in the sea. Consistently, they exhibit remarkably varied breeding strategies and migration

N. S. Martínez-Curci (✉)
Grupo Vertebrados, Instituto de Investigaciones Marinas y Costeras, Universidad Nacional de Mar del Plata, Consejo Nacional de Investigaciones Científicas y Técnicas (CONICET), Buenos Aires, Argentina

Coastal Solutions Fellows Program, Cornell Lab of Ornithology, Cornell University, Ithaca, NY, USA

Bird Ecology Lab, Instituto de Ciencias Marinas y Limnológicas, Universidad Austral de Chile, Valdivia, Chile
e-mail: nsm85@cornell.edu

G. O. García
Grupo Vertebrados, Instituto de Investigaciones Marinas y Costeras, Universidad Nacional de Mar del Plata, Consejo Nacional de Investigaciones Científicas y Técnicas (CONICET), Buenos Aires, Argentina

L. Marbán · S. M. Zalba
GEKKO, Grupo de Estudios en Conservación y Manejo, Departamento de Biología, Bioquímica y Farmacia, Universidad Nacional del Sur, Bahía Blanca, Argentina

Consejo Nacional de Investigaciones Científicas y Técnicas (CONICET), Buenos Aires, Argentina

P. Simonetti
Instituto Argentino de Oceanografía (IADO), Universidad Nacional del Sur, CONICET, Buenos Aires, Argentina

S. M. Fiori, P. D. Pratolongo (eds.), *The Bahía Blanca Estuary*,
https://doi.org/10.1007/978-3-030-66486-2_13

systems. Although the monophyly of Charadriiformes is well established, the relationship of the families within this order has been controversial. According to the most recent evidence (Paton et al. 2003; Thomas et al. 2004; Hackett et al. 2008), the order consists of three major groups: Charadrii (includes plovers, oystercatchers, avocets, stilts, thick-knees, and sheathbills); Scolopaci (includes sandpipers, seedsnipes, jacanas, and painted-snipes); and Lari (includes skuas, skimmers, gulls, and terns). The former two groups constitute what are commonly known as shorebirds, while the latter consists of seabirds. This chapter is organized into three sections in which we will describe some relevant ecological aspects of shorebirds (Sect. 13.2) and seabirds (Sect. 13.3) with emphasis on the species of Bahía Blanca Estuary. Finally, we will analyze their main conservation challenges both at global and local contexts (Sect. 13.4).

13.2 Shorebirds

Shorebirds, also called waders, are small- to medium-sized birds closely associated with open environments, especially wetlands that inhabit all continents. Most of them spend at least part of their lives on tidal flats at estuarine or marine shorelines, while others use inland habitats such as grasslands, rivers, lakes, and lagoons. As a consequence of their cosmopolitan distribution and habitat associations, these birds exhibit diverse morphological adaptations that include marked variations in body size and bill morphologies. For example, the white-rumped sandpiper (*Calidris fuscicollis*), one of the smallest shorebird species in Bahía Blanca Estuary, is 15–18 cm in length and 40–60 g in mass and has a short, fairly straight bill (Fig. 13.1a; Parmelee 2020). At the opposite end, the Hudsonian godwit (*Limosa haemastica*), one of the largest shorebird species, is 36–42 cm in length and 200–350 g in mass and has a long, upturned bill (Fig. 13.1b; Walker et al. 2020). These morphological variations allow shorebirds to exploit different trophic resources while coexisting in large numbers at key sites along their flyways (Martínez-Curci et al. 2015). Flyways are routes, used by several species, that encompass the full range of breeding, stopping, staging, and non-breeding areas occupied by a population during the annual cycle (Boere and Stroud 2006). As we will see, Bahía Blanca Estuary constitutes one of the most important sites identified within the Atlantic Americas Flyway.

There are about 215 recognized shorebird species in the world, distributed unevenly among 14 families (Colwell 2010). Records in Bahía Blanca account for 37 species in 9 families (Table 13.1). The largest number of species is hosted by Charadriidae and Scolopacidae. These two families show quite marked morphological and behavioral differences. Members of Charadriidae (i.e., plovers and allies) have relatively large heads with short bills and large eyes (Fig. 13.1c) that facilitate visual foraging for what they use a "run-stop-peck" strategy (Colwell and Haigh 2019) that is considered the ancestral condition in shorebirds (Barbosa and Moreno 1999). On the other hand, Scolopacidae members (i.e., sandpipers and allies) have long bills relative to their body size that are varied in shapes (e.g., upturned in

Fig. 13.1 Some of the most common shorebirds in Bahía Blanca Estuary: (**a**) white-rumped sandpiper (*Calidris fuscicollis*); (**b**) Hudsonian godwits (*Limosa haemastica*) with basic (left) and alternate plumage (right); (**c**) two-banded plover (*Charadrius falklandicus*); (**d**) red knots (*Calidris canutus*) with basic, intermediate, and alternate plumage. (Photos by Natalia S. Martínez-Curci (**a**, **b**, **d**), Lautaro Rodríguez Astorino (**c**))

godwits, straight in knots, decurved in curlews, wedge-shaped in turnstones; Figs. 13.1a,b,d). Most members of this family feed on buried prey using tactile strategies (Barbosa and Moreno 1999) that in many cases rely on mechanoreceptors located at the tip of the bill (Piersma et al. 1998); others use their brush-like tongues to feed on biofilm (Elner et al. 2005).

Another notable difference between Scolopacidae and Charadriidae is related to their breeding areas and migration strategies. Most sandpipers are northern-latitude breeders that in many cases undertake amazing migrations to the southern extreme of the globe. Most plovers, however, breed on temperate and tropical latitudes and are less prone to migrate long distances (Colwell and Haigh 2019). These patterns are reflected in the two migratory currents (norther vs. austral migrants), with opposite chronologies, that converge in Bahía Blanca Estuary. Northern migrants (i.e., birds that breed in North America during the boreal summer/austral winter and then migrate to the southern hemisphere) are the most abundant shorebirds in Bahía

Table 13.1 List of Charadriiformes recorded in Bahía Blanca Estuary (for the Charadrii and Scolopaci suborders, we followed Martínez-Curci and Petracci (2016, Petracci and Delhey 2005); for Lari we followed Petracci and Sotelo (2013); Petracci and Delhey (2005) and authors' experience). The migratory pattern indicates the following: summer visitor (Sv), winter visitor (Wv), northern migrant (NM), austral migrant (AM), or resident (R). The probability of observation (Prob Obs) refers to the frequency with which species can be observed within the estuary: abundant (A: frequent observations of large abundances), common (C: frequent observations), uncommon (U: infrequent but mostly annual observations), occasional (O: very infrequent observations, which generally do not occur every year). Conservation status (Cons Stat) shows the categorization of the species according to national (Nat) and international (Int) criteria. For the national categorization, we followed MAyDS and Aves Argentinas (2017) keeping the acronyms of the original document published in Spanish: critically endangered (EC), endangered (EN), threatened (AM), vulnerable (VU), and not threatened (NA). For the international categorization, we followed IUCN (2020): near threatened (NT) and low concern (LC). We also followed IUCN (2020) for population trends (Pop Trend): increasing (I), stable (S), decreasing (D), unknown (U)

		Migratory	Prob	Cons Stat		Pop
		pattern	Obs	Nat	Int	Trend
Charadrii						
Charadriidae						
American golden plover	*Pluvialis dominica*	Sv, NM	C	NA	LC	D
Grey plover	*Pluvialis squatarola*	Sv, NM	O	NA	LC	U
Tawny-throated dotterel	*Oreopholus ruficollis*	Wv, AM	C	NA	LC	D
Southern lapwing	*Vanellus chilensis*	R	C	NA	LC	I
Semipalmated plover	*Charadrius semipalmatus*	Sv, NM	C	NA	LC	S
Collared plover	*Charadrius collaris*	R	U	NA	LC	D
Two-banded plover	*Charadrius falklandicus*	Wv, AM	A	NA	LC	S
Rufous-chested plover	*Charadrius modestus*	Wv, AM	U	NA	LC	U
Haematopodidae						
American oystercatcher	*Haematopus palliatus*	R	A	NA	LC	S
Blackish oystercatcher	*Haematopus ater*	Wv, AM	O	NA	LC	U
Magellanic oystercatcher	*Haematopus leucopodus*	Wv, AM	V	NA	LC	S
Recurvirostridae						
Black-necked stilt	*Himantopus mexicanus*	R	A	NA	LC	U
Chionidae						
Snowy sheathbill	*Chionis albus*	Wv, AM	O	NA	LC	S
Pluvianellidae						
Magellanic plover	*Pluvianellus socialis*	Wv, AM	O	EN	NT	S

Scolopaci						
Scolopacidae						
Upland sandpiper	*Bartramia longicauda*	Sv, NM	O	VU	LC	I
Whimbrel	*Numenius phaeopus*	Sv, NM	U	NA	LC	D
Hudsonian godwit	*Limosa haemastica*	Sv, NM	A	NA	LC	D
Ruddy turnstone	*Arenaria interpres*	Sv, NM	U	NA	LC	D
Red knot	*Calidris canutus*	Sv, NM	C	EC	NT	D
Sanderling	*Calidris alba*	Sv, NM	U	NA	LC	U
White-rumped sandpiper	*Calidris fuscicollis*	Sv, NM	A	NA	LC	D
Baird's sandpiper	*Calidris bairdii*	Sv, NM	C	NA	LC	S
Pectoral sandpiper	*Calidris melanotos*	Sv, NM	U	NA	LC	S
Stilt sandpiper	*Calidris himantopus*	Sv, NM	O	NA	LC	I
Buff-breasted sandpiper	*Calidris subruficollis*	Sv, NM	U	AM	NT	D
Short-billed Dowitcher	*Limnodromus griseus*	Sv, NM	V	NA	LC	D
South American snipe	*Gallinago paraguaiae*	R	U	NA	LC	S
Wilson's phalarope	*Phalaropus tricolor*	Sv, NM	U	NA	LC	I
Spotted sandpiper	*Actitis macularius*	Sv, NM	O	NA	LC	D
Greater yellowlegs	*Tringa melanoleuca*	Sv, NM	C	NA	LC	S
Willet	*Tringa semipalmata*	Sv, NM	O	NA	LC	S
Lesser yellowlegs	*Tringa flavipes*	Sv, NM	C	NA	LC	D
Thinocoridae						
Least seedsnipe	*Thinocorus rumicivorus*	R	O	NA	LC	S

(continued)

Table 13.1 (continued)

		Migratory	Prob	Cons Stat		Pop
		pattern	Obs	Nat	Int	Trend
Charadrii						
Charadriidae						
Jacanidae						
Wattled jacana	*Jacana jacana*	R	O	NA	LC	S
Rostratulidae						
South American painted-snipe	*Nycticryphes semicollaris*	R	U	NA	LC	D
Lari						
Stercorariidae						
Parasitic jaeger	*Stercorarius parasiticus*	Sv, AM	O	NA	LC	S
Rynchopidae						
Black skimmer	*Rynchops niger*	Sv, NM	C	NA	LC	D
Laridae						
Brown-hooded gull	*Chroicocephalus maculipennis*	R	C	NA	LC	I
Gray-hooded gull	*Chroicocephalus cirrocephalus*	R	U	NA	LC	S
Olrog's gull	*Larus atlanticus*	R	A	VU	NT	S
Kelp gull	*Larus dominicanus*	R	A	NA	LC	I
Gull-billed tern	*Gelochelidon nilotica*	R	C	NA	LC	D
South American tern	*Sterna hirundinacea*	Wv, AM	U	NA	LC	D
Snowy-crowned tern	*Sterna trudeaui*	R	A	NA	LC	S
Sandwich tern	*Thalasseus sandvicensis*	R	U	NA	LC	S
Royal tern	*Thalasseus maximus*	R	U	NA	LC	S

Blanca Estuary. This group dominates the assemblage from austral spring to late summer/early fall. Some species, such as the white-rumped sandpiper (Fig. 13.1a), which is the most abundant shorebird in the area, spend the whole non-breeding season at Bahía Blanca. However, its abundance may be higher during the passage of individuals that spend the non-breeding season in southern latitudes (Petracci and Sotelo 2013, Belenguer et al. 1992). Other species, like the red knot (*Calidris canutus*; Fig. 13.1d), use the area for fueling during northward migration and can be found almost exclusively from March to May (Petracci and Sotelo 2013). With the arrival of the autumn, and during the winter, the assemblage becomes dominated by austral migrants (i.e., birds that breed in southern South America, mainly in Patagonia, during the austral spring and summer and then migrate north reaching southern Brazil). Among austral migrants, the two-banded plover (*Charadrius falklandicus;* Fig. 13.1c) is the most abundant species with some individuals remaining in the area throughout the year (Petracci and Sotelo 2013). Residents (i.e., birds that remain in the same area year-round), such as the American oystercatcher (*Haematopus palliatus*; Fig. 13.5a, in Box 13.2) or the black-necked stilt (*Himantopus mexicanus*), represent a smaller fraction of the assemblage (Martínez-Curci and Petracci 2016).

13.2.1 Migration

Perhaps one of the most fascinating aspects of shorebirds is their ability to migrate. Over 60% of shorebird species undertake short- to long-distance migrations (Colwell 2010). Several members make some of the most extreme migratory flights in the world exceeding 30,000 km per year. Migration annually takes shorebirds from breeding to non-breeding grounds where they may spend as much as 8 to 10 months of the year (Colwell 2010). From an evolutionary point of view, migration strategies may arise if the benefits of moving seasonally are greater than the benefits of staying in one region (Lack 1968). However, to take advantage of the world's seasonality, birds must perform energetically demanding flights.

Different species have developed different migration strategies, undertaking journeys of varying length and duration (Piersma 1987). Most species travel in many short periods of migratory flight interspersed with short periods of refueling (that strategy is called "hop"). Others may do some long-distance flights interspersed with resting and fueling periods ("skip"). Finally, in the most extreme cases, shorebirds can travel in one very long migratory flight ("jump"). Following the latter strategy, some of the species that inhabit Bahía Blanca Estuary make nonstop endurance flights of about 8000 km in 6 days in the case of red knots (Niles et al. 2010) and more than 10,000 km in 7 days in the case of Hudsonian godwits (Senner et al. 2014).

To accomplish these long-distance migratory flights, shorebirds rely on energy and metabolic water that is mainly stored subcutaneously in the form of lipids but also as proteins in flight muscles (Jenni and Jenni-Eiermann 1998; Jenni-Eiermann et al. 2002). Immediately before starting the trip, they deposit huge amounts of triglycerides at the same time they hypertrophy their flight muscles without any noticeable increase in exercise (Lindström and Piersma 1993). In this way, migratory shorebirds are capable of doubling their mass prior to migration (Piersma and Gill 1998) for which they spend considerable time foraging while increasing their consumption rate in a period of hyperphagia (Burger and Olla 1984).

13.2.2 Dependence on Key Sites

To achieve high intake rates and fat loads, shorebirds depend on a limited number of key sites that serve as staging areas (i.e., locations where birds fuel up in a pause during migration), stopover sites (i.e., locations where migrant birds take a short break to rest a feed), and non-breeding grounds (i.e., locations where migrant birds spend most of the non-breeding season). Most of these sites are wetlands that provide large amounts of high-quality food. Shorebirds' diet consist mainly of invertebrates; in Buenos Aires province, some of their major prey items are polychaetes, crabs, mollusks, and insects (Iribarne and Martínez 1999; Petracci 2002; Ieno et al. 2004; Ribeiro et al. 2004; Martínez-Curci et al. 2015; Martínez-Curci and Petracci 2016).

Several shorebird species exhibit philopatry to the mentioned key sites, as well as aggregative behaviors during the non-breeding season and migration. This tendency to return in large flocks to the same stopping, staging, and non-breeding areas confers several benefits, such as familiarity with resources and conditions or decreased predation risk (Shuter et al. 2011). However, since significant percentages of several populations are concentrated in a limited number of sites, it also makes shorebirds more susceptible to habitat loss or degradation (Myers 1983; Shuter et al. 2011). About 50% of shorebirds with known population trends are declining, and the conservation of this highly migratory species requires international cooperation. In pursuit of this, Bahía Blanca Estuary has joined the Western Hemisphere Shorebird Reserve Network (WHSRN), an international effort aimed to conserve shorebirds and their habitats across the Americas through action at a network of key sites. Holding more than 20,000 shorebirds per year and more than 1% of the biogeographical populations of American oystercatcher (Fig. 13.5 in Box 13.2), two-banded plover (Fig. 13.2c), Hudsonian godwit (Fig. 13.2b), red knot (Fig.13.1d), and white-rumped sandpiper (Fig. 13.2a), Bahía Blanca Estuary was designated as a site of regional importance within the WHSRN in 2016 (https://whsrn.org/whsrn_sites/estuario-de-la-bahia-blanca/).

Fig. 13.2 Representative seabirds of the order Charadriiformes in Bahia Blanca Estuary: (**a**) black skimmers (*Rynchops niger*); (**b**) flock of terns, South American tern (*Sterna hirundinacea*; left), sandwich terns (*Thalasseus sandvicensis*; back), and snowy-crowned terns (*Sterna trudeaui*; front); (**c**) Olrog's gulls (*Larus atlanticus*). (Photos by Hugo Gribman (**a**), Claudio Rodríguez (**b**), Pablo Fernández (**c**))

13.3 Seabirds

Seabirds are defined as those birds living in and making their living from the marine environment, which includes coastal areas, estuaries, wetlands, and oceanic islands (Schreiber and Burger 2001). Within Charadriiformes, seabirds are grouped in the suborder Lari that comprises about 127 species. In Bahía Blanca Estuary, this group is mostly represented by gulls and terns (see Table 13.1). From a morphological perspective, Lari constitutes an heterogeneous group. In Bahia Blanca these birds range from small size (e.g., South American tern *Sterna hirundinacea*, 170–220 g in weight and has an 84–86 cm wingspan; Gochfeld et al. 2020) to large species (e.g., kelp gull *Larus dominicanus*, 810–1335 g in weight and has a 128–142 cm wingspan, Burger et al. 2020). They are also heterogeneous from an ecological point of view. Some members feed on a few types of prey so they are considered specialists, while others are able to use a wide range of resources being considered generalists. Furthermore, there are resident species but also trans-equatorial migrants that travel large distances across several oceanographic systems.

Seabirds are equally at home on land, in the air, and in the water, changing from one to the other, often daily. Such flexibility requires unique morphological and physiological adaptations to an environment that has also exerted selective forces on their behavior, ecology, and demography. Bills, feet, and body shapes show innumerable adaptations to various lifestyles, allowing seabirds to swim and dive using webbed feet or even their wings to propel themselves in the water. Although all seabirds use their bills to capture and handle food, they exhibit adaptations for different types of feeding. An emblematic example is that of skimmers, which are represented in Bahía Blanca by the black skimmer (*Rynchops niger*; Fig 13.2a). They have a specialized bill in which the lower mandible is compressed laterally and is longer than the upper mandible. This provides a greater surface area that, along with their flight mechanics, allows skimmers to catch fish by efficiently skimming the water surface with their lower mandible as they search for prey with tactile cues (Zusi 1996). Adaptations for feeding in marine environments are not just restricted to bills. By feeding in the sea, these birds also must deal with high physiological loads of salt. To accomplish this, they rely on salt glands (i.e., organs for excreting excess salts) that are found in shallow depressions around the orbit, above the eye, which help to cope with a diet overloaded with salt (Schmidt-Nielsen 1960). In addition, just like shorebirds, seabirds have the capacity to cope with fattening periods as a physiological adaptation to migration.

13.3.1 Foraging

Seabirds are at the higher trophic levels of the marine food web. When breeding, they must return to their nest after every fishing trip to carry food for their chicks, a pattern referred by ecologist as central-place foraging (Orians and Pearson 1979). During the non-breeding season, social foraging is widespread, and it usually involves several species (Thiebault et al. 2014). Foraging activity can occur during the day or night, and presumably these different foraging habits have evolved in response to the behavior of their preys. Seabird diet consists mainly of fish, crustaceans, and/or mollusks. However, some species have incorporated anthropogenic items in their trophic spectrum. In Argentina, for example, coastal gulls such as the Olrog's gull (*Larus atlanticus*) and the kelp gull have been registered preying on resources facilitated by fishery activities (Berón et al. 2013; Marinao et al. 2019).

To obtain food, different foraging strategies, generally linked to morphological and/or physiological features, are used. Gulls fed by methods that include picking up items from the ground, surface dipping, jump-plunging, and other forms of dipping. In terns (Fig. 13.2b), plunge-diving, diving-to-surface, dipping, and hawking (to catch insects) are well-documented feeding techniques (Cabot and Nisbet 2013). Although most seabirds catch live prey, scavenging is a feeding method employed by a small proportion of seabirds, especially gulls and skuas (Furness et al. 2007). Many scavenging species have increased dramatically in numbers, and these increases have often been attributed to the feeding opportunities presented over many decades by fishery waste (Oro et al. 2013). Kleptoparasitism (i.e., parasitic interaction in which one animal steals food from another; Rothschild and Clay 1952) is another very special technique for acquiring prey. Skuas have often been considered to be highly evolved as specialized kleptoparasites, that is, they may use kleptoparasitism for all or most of their energy acquisition (Brockmann and Barnard 1979). Specialist kleptoparasites exhibit adaptations that are apparently absent in opportunist species (e.g., ability to detect and attack hosts carrying food concealed in the proventriculus, to sustain prolonged aerial chases, and to adapt their breeding cycle to match that of their host; Furness 1987). Within Bahía Blanca Estuary, there are some records where the Arctic skua (*Stercorarius parasiticus*) has been observed parasitizing on sandwich terns (*Thalasseus sandvicensis*; Fig. 13.2b) (Petracci and Sotelo 2013). Unlike skuas, gulls and terns may resort to food theft opportunistically, using a range of foraging tactics. In these cases, kleptoparasitism is context-dependent and often occurs during periods of low availability of primary food sources (García et al. 2010). Even in opportunistic kleptoparasites, some individuals may be more specialized as kleptoparasites than others, and this behavior may be related to reproductive performance (García et al. 2011, 2013). Recent studies on terns showed that in individuals specialized as kleptoparasites, the energetic reward of a kleptoparasitic event is sex-specifically related to the tactic used by the parasite, showing signs of within-individual improvement with age in both sexes (García et al. 2020).

13.3.2 Breeding

Most seabird species are social and nest in colonies that can include hundreds, thousands, tens of thousands, or even hundreds of thousands of pairs (Schreiber and Burger 2001). Almost all the species included in this chapter nest seasonally, being at breeding colonies for 3–5 months. After breeding, most of them migrate or disperse to non-breeding areas where they spend more than half the year. Migration patterns can include a variety of strategies, durations, and distances covered. The dominant mating system is monogamous with cooperation of both parents for a successful incubation and chick rearing. During breeding, there is some differentiation of sex roles. In terns, for example, males establish the nest territory and take the leading role in defending it and feed their mates during the period of egg formation and egg-laying. Some species show, as well, high mate fidelity. Courtship feeding can occur early in the season, as a way for females to assess how good food providers their partners might be. It is also important to allow proper egg formation during the pre-incubation stage, relieving females from the acquisition of food that consume time and energy. A nest of varying structure is usually built to protect the eggs. Some are placed on cliffs and islands, away from predators, while others can be placed in cavities to prevent attacks from aerial predators. Seabirds grouped as Charadriiformes lay more than one egg; mechanisms related to egg-size asymmetry and laying/hatching asynchrony may lead to reduced survival of the last chicks if the environmental variables are not favorable. Chick-rearing period is about 25–30 days.

Among the seabirds of Bahía Blanca Estuary (Table 13.1), only two species of gulls reproduce regularly in the area (Yorio et al. 1998). One of them is Olrog's gull (Fig. 13.2c), an endemic species to the Atlantic coast of southern South America, which is one of the few *Larus* species in the world with adverse conservation status (see Sect. 13.3). There are scant identified Olrog's gull breeding sites, and most of the breeding population is concentrated in Bahía Blanca Estuary. The largest colony reported to date is located at "Islote de la Gaviota Cangrejera" Natural Reserve (Delhey et al. 2001a; Yorio et al. 2005, Petracci et al. 2008). After the breeding season, this species disperses along the Atlantic coast reaching southern Brazil. Although the species remains throughout the year in the estuary and is considered a local resident (Table 13.1), recent evidence indicates that the population that breeds in Bahía Blanca Estuary may have a partial migration behavior as observed in the San Blas bay population (Copello et al. 2020). Olrog's gull is considered one of the few gulls specialized in feeding on crabs (Escalante 1966; Herrera et al. 2005; Petracci et al. 2004). However, they can occasionally feed on other items (Copello and Favero 2001; Petracci et al. 2007; Berón et al. 2013); recent studies conducted in non-breeding areas try to understand the ways in which they access to novel food sources (García et al. 2019; Castano et al. 2020). The other species that breeds in the area is the kelp gull. It has generalist habits, an extremely large population size, and a breeding distribution that spans the coasts and islands of much of the southern hemisphere (including South America, Africa, New Zealand, Australia, Subantarctic Islands, and Antarctic Peninsula; Burger and Gochfeld 1996a, b). Although there is no published information on the number of breeding pairs inhabiting Bahía Blanca Estuary, it is known that the Argentine population has increasing trends. It has been argued that, in some coastal areas, population growth would be related to the use of anthropogenic food subsidies (Lisnizer et al. 2011).

13.4 Conservation

13.4.1 Main Threats

The last centuries have been marked by a rapid loss of biodiversity (Pimm et al. 2014; Ceballos et al. 2015), being habitat loss and degradation one of the main identified causes (Sala et al. 2000; Hoekstra et al. 2005). This biodiversity crisis not only involves extinctions but also a decrease in the number of individuals in local populations, which can result in changes in the composition of communities and the function of ecosystems (Dirzo et al. 2014; Rosenberg et al. 2019). Accordingly, both shorebird and seabird species have been experiencing drastic population declines (Croxall et al. 2012; Colwell 2010; Rosenberg et al. 2019).

Decreasing trends are caused by a variety of threats including climate change among the most important on a global scale (Lascelles et al. 2016; van Gils et al. 2016; Kubelka et al. 2018). The effects of this threat are particularly relevant to species inhabiting coastal habitats such as estuaries and breeding in high latitudes, where climate change has already significantly modified the ecosystem. Some of the most important land-based threats include the presence of invasive alien species, problematic native species (e.g., those that have become super-abundant), human disturbance, changes in land use and land cover, habitat deterioration, commercial and residential development, and hunting (Sutherland et al. 2012a, b). Finally, at the sea, some of the main problems are associated with bycatch (in gillnet, trawl, and other fisheries; Paz et al. 2018), pollution (oil spills, chemical contaminants, plastic and marine debris; Quadri Adrogué et al. 2019), noise (busy shipping lanes, seismic surveys, and sonar; Pichegru et al. 2017), prey depletion caused by overfishing, energy production, and mining (Tasker et al. 2000; Croxall et al. 2012).

Bahía Blanca Estuary has been subject to novel threats and increasing levels of pollution during the last decades. Global data indicates that although estuarine and coastal systems provide vital habitats for many plants and animals, as well as a wide variety of goods and services for millions of people, they are some of the most heavily used and threatened natural systems in the world (Barbier et al. 2011). The proximity to Bahía Blanca Estuary to urban areas, industrial parks, and ports has led to contamination by coliforms, hydrocarbon derivatives, pesticides, and heavy metals as some of the major threats (Martínez-Curci and Petracci 2016). In recent years, problems associated with the expansion of invasive alien species have also become relevant. One of these cases is the potential threat given by the Pacific oyster (*Magallana gigas*). Although this alien mollusk does not seem to negatively affect the feeding habits of some shorebirds (Escapa et al. 2004), it might cause structural changes in the ecosystem (Luckenbach 1984; Kelly et al. 1996). On the other hand, the establishment and expansion of an alien plant species named opposite-leaved saltwort (*Salsola inermis*) is threatening the largest breeding colony of Olrog's gull at "Islote de la Gaviota Cangrejera" Natural Reserve, causing the displacement of nests to suboptimal sites.

13.4.2 Shorebirds and Seabirds' Conservation Status

Among the 46 shorebird and seabird species recorded at Bahía Blanca, 35% have decreasing population trends, 41% are stable, only 13% are increasing, and the remaining 11% have unknown trends (Table 13.1). According to the categorization of the International Union for Conservation of Nature, four of the Charadriiformes species inhabiting Bahía Blanca Estuary are near threatened (IUCN 2020). These are three shorebirds, the Magellanic plover (*Pluvianellus socialis*), the red knot, and the buff-breasted sandpiper (*Calidris subruficollis*), and one seabird, Olrog's gull. The latter three species are listed in the Appendix I of the Convention on Migratory

Species (CMS), while the Magellanic plover is listed in the Appendix II along with other families of Charadriiformes.

The Magellanic plover is an austral migrant that might be occasionally observed in brackish lagoons and salt ponds such as "La Vidriera" (Petracci and Sotelo 2013). Having a small but stable population estimated in 1500–7000 mature individuals (BirdLife International 2016), it is considered endangered in Argentina (MAyDS and Aves Argentinas 2017). The red knot is a northern migrant that is commonly seen in Bahía Blanca during northward migration (Petracci and Sotelo 2013). It has a Holarctic breeding distribution with six recognized subspecies of which only *C. canutus rufa* reaches the coast of South America up to Tierra del Fuego (Baker et al. 2020). Having numbered 67,500 individuals in the mid-1980s, dropped to around 13,000 by 2011 (Morrison and Ross 1989; Andres et al. 2012), and remained in the range 10,000–15,000 up to 2020 (R.I.G. Morrison, pers. comm.), the population that spends the non-breeding season at Tierra del Fuego is considered critically endangered in Argentina (MAyDS and Aves Argentinas 2017). The buff-breasted sandpiper, a northern migrant that mostly spends the non-breeding season in the Pampas region (McCarty et al. 2020), is considered an uncommon species in Bahía Blanca. It can be found in saltmarshes dominated by chickenclaws (*Sarcocornia perennis*) and short grass areas (Delhey et al. 2001b; Delhey and Petracci 2004; Petracci and Sotelo 2013). Its population is estimated in 15,300–56,000 mature individuals (BirdLife International 2017) being considered endangered in Argentina (MAyDS and Aves Argentinas 2017). Finally, Olrog's gull breeds in Bahía Blanca; its largest colony is located at "Islote de la Gaviota Cangrejera" with about 3500 nests (Yorio et al. 2013). Its population is estimated in 9800–15,600 mature individuals, and although it has a stable trend (BirdLife International 2018), it is subject to different impacts within the estuary such as coastal development, unregulated tourism, recreational fisheries, invasive alien species, and pollution (Yorio et al. 2013, Boxes 13.1 and 13.2).

13.4.3 Conservation Efforts

Despite the aforementioned threats and species with adverse conservation status, a positive aspect toward the conservation of Bahía Blanca Estuary and its birds is that most of the ecosystem is legally protected (Martínez-Curci and Petracci 2016). There are four protected areas: "Bahía Blanca" Natural Reserve under municipal administration; "Bahía Blanca, Bahía Falsa, and Bahía Verde" and "Islote de la gaviota cangrejera" Natural Reserves administered by the System of Natural Protected Areas of Buenos Aires Province; and "Defensa Baterías – Charles Darwin" Natural Reserve under the jurisdiction of national parks and the Argentine navy. As already mentioned, the area was included among the WHSRN key sites. In addition, "Bahía Blanca, Bahía Flasa, and Bahía Verde" Natural Reserve was recognized as

Box 13.1 Invasive Coastal Plants and the Conservation of Seabirds

Seabirds are one of the most threatened groups among birds, with almost a third of them categorized under some degree of threat (i.e., critically endangered, endangered, or vulnerable), and another 11% considered as near threatened (Croxall et al. 2012). Additionally, nearly half of seabird species are known or suspected to be declining (Croxall et al. 2012). Additionally, and given that most of seabirds breed on islands, the rise in sea level projected by scientists represents a novel threat associated with the loss and retraction of breeding habitat (Reynolds et al. 2015) and will also produce changes in the intertidal areas where they feed.

Colonial nesting is widespread among almost all species of seabirds, with monospecific or mixed reproductive groupings ranging from a few pairs up to millions of birds breeding together and commonly exhibiting a strong fidelity to natal sites. Although these high concentrations are temporary, bringing together high densities of individuals increases their exposure to potential threats.

Invasive exotic species (IES) are recognized as one of the principal threats to almost half the species of seabirds, mostly represented by land mammals that prey on their breeding colonies (Dias et al. 2019). Much less known is the impact of invasive plants that alter the structure of the breeding habitat. Natural processes like trampling by breeding birds, the mobilization of materials for the construction of the nest, and the eutrophication caused by the deposition of guano produce important physicochemical changes in the soil (Otero et al. 2018) that can open invasion windows for the establishment of opportunistic plant species. Their propagules can arrive dispersed by the tide, by the wind, or by the birds themselves. However, stress factors associated with coastal habitats, mainly salinity, represent a strong environmental barrier for the establishment of new species. That is probably the reason why there are very few records of plants impacting seabirds (Dias et al. 2019) and also why this relationship has been poorly explored in the literature. However, the increase in international maritime traffic, with the arrival of increasingly larger ships, in greater numbers and from a greater diversity of regions, significantly increases the chances of arrival of species with the ability to establish and invade this type of habitats.

Not Just Another Gull

Located on Bahía Blanca Estuary, the "Islote de la gaviota cangrejera" Nature Reserve (also named "Isla del Puerto," Fig. 2.2; Chap. 2) hosts the largest nesting site for Olrog's gull (*Larus atlanticus*). The name of the reserve, in fact, refers to this species that concentrates 70% of the total of its reproductive population, estimated in 4860–7790 pairs, in the south-east tip of the island (Yorio et al. 2013). This nesting site covers an area of less than 24 hectares, and it is just 3 linear kilometers from the Port of Ingeniero White and the associated petrochemical and industrial hub. Yorio et al.'s (2013) review indicates that industries are important source of persistent organochlorine pesticides, hydrocarbons, and heavy metals in Bahía Blanca Estuary, where the impact would be even greater due to its accumulation in crabs, the key prey of Olrog's gulls on their breeding grounds. Periodic operations of maintenance dredging of the channels increase this risk due to resuspension of polluted sediments on the water (Roberts 2012). The increase in the populations of the kelp gull (*Larus dominicanus*), a much more opportunistic species and capable of taking advantage of food resources associated with human activity, including domestic waste and fishing discards, seems to be another factor threatening Olrog's gull due to eventual competition for breeding areas. Lesions by sport fishing lines and deaths following the ingestion of discarded baited hooks have also been reported for the species in its wintering grounds (Berón and Favero 2009). Although larids are one of the largest and least threatened groups of seabirds (Croxall et al. 2012), Olrog's gull has a set of characteristics that make it a vulnerable species. Its low population size, restricted distribution range, migratory behavior, and the abovementioned threats led to its classification as globally near threatened (BirdLife International 2018), vulnerable in Argentina (MAyDS and Aves Argentinas 2017), and threatened in Uruguay (Azpiroz and Caballero-Sadi 2017).

A Saltmarsh Menace: A Case of Coastal Plant Invasion

Recently, a new menace was detected in the Bahía Blanca Estuary: the establishment and spread of a novel coastal plant species that colonizes the species breeding grounds. *Salsola inermis* (Amaranthaceae) is an annual halophyte plant native to the Mediterranean, reported in the region of Bahía Blanca eight decades ago, but that apparently managed to advance on the coastal sector just in the last few years. It grows into practically monospecific blocks that replace native shrubs of *Sarcocornia perennis* and *Heterostachys* sp., which are lower and typically grow more scattered, leaving bare ground spaces that disappear with the advance of *S. inermis*. Invasive plants preferentially colonize areas used by gulls for nesting, which are located in sectors slightly higher than the surrounding environment. As the plants grow, they cover the nesting area with densities that prevent their use by gulls. Dead and dry plants remain in the area until the following year, making this interference even more intense (Fig. 13.3). This change results in the movement of nesting sites toward the perisphere, increasing nest exposure to extraordinary tides (Fig. 13.4).

Invasion of Coastal Environments: A Special Management Challenge

Managing invasive species in coastal environments offers particular challenges, including difficulties in detecting population centers and accessing invaded areas. From this point of view, it is key to identify the vectors and dispersal routes, as well as the environments with the greatest susceptibility to being invaded (Zalba et al. 2000).

(continued)

Fig. 13.3 Traditional nesting areas of Olrog's gull (*Larus atlanticus*) colonized by *Salsola inermis*. Dry specimens from the previous year are observed in the foreground, while bright green plants below and among them are young specimens from the current season. Photo by Leandro Marbán

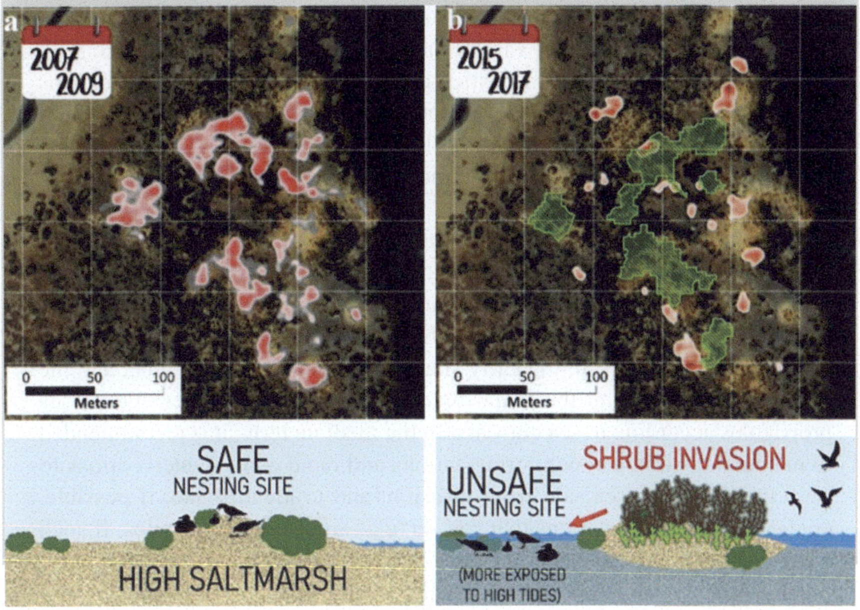

Fig. 13.4 Heat maps representing the spatial distribution of nests in the largest breeding colony of Olrog's gull (*Larus atlanticus*) **a** before and **b** after the *Salsola inermis* invasion in the "Isla del Puerto" nature reserve. Green polygons indicate the distribution of the invasive species. The drawings below the maps are a schematic representation of the invasion process and their impact. (Modified from Marbán et al. 2019)

Each plant of *S. inermis* can produce more than 16,000 seeds (Marbán and Zalba 2019), and, once it is mature and dry, its aerial part can detach from its root and roll with the wind, such as many tumbleweeds of the same genus, expanding its dispersal distance. The discontinuous distribution of this species along the coast, with great distances between populations, and the record of entire mature specimens floating adrift with the currents of water suggest the hydrochory as a principal way of long-distance dispersal. Additionally, seeds themselves are capable of staying floating up to a week and able to germinate even after a long exposure to seawater (Marbán and Zalba 2019). Hydrodynamic models combining information about tidal currents and the distribution of the species in the area further support this hypothesis (Marbán, unpublished data). Once they reach the coast, in order to germinate and successfully establish, seeds need to reach the supralittoral zone or areas slightly above the tidal line, where floods are sporadic or absent. In mean, just like the sites that Olrog's gulls use to nest.

Given the high vulnerability of the breeding environments of Olrog's gulls to be reached and colonized by *Salsola inermis*, and considering the impact that can be projected for this process on the population dynamics and survival of this endemic gull, in 2017 control actions of the invasive plants were initiated. The area covered by the species was about 3360 square meters, and 6 control campaigns were required with a group of 12 volunteers to remove the plants growing in the area. Manual removal of specimens took place during the months of February and March in order to minimize interference with Olrog's gulls' reproductive activities and to remove *S. inermis* specimens before they release their seeds. Monitoring of the effects of this intervention on the breeding activity of Olrog's gulls only started recently, and no results are available yet, but the changes in the structure of the breeding habitat are remarkable.

Beyond the aforementioned restrictions, a set of traits of the invasive species and the characteristics of the invasion process in the area allow us to propose its eventual eradication: the annual life cycle of Salsola and the short persistence of its seeds in the soil (Marbán and Zalba 2019), the ease of detection and manual removal, the early stage of the invasion in the area, and the possibility of modeling the main directions and dispersal intensities of its propagules according to sea currents. In the medium term, it is recommended to maintain continuous monitoring actions and rapid control interventions to keep the breeding area safe from invasion, and to identify and, if possible, eradicate nearby populations that can serve as sources of propagules. If such a strategy is not implemented, the persistence of Olrog's Gulls, and of a set of other shorebird species in this area, will possibly face their main obstacle.

Box 13.2 Contribution to the Study of Contamination in Shorebirds and Seabirds from the Bahía Blanca Estuary

Anthropogenic contaminants from industrial, domestic, and agricultural run-off are a major harmful issue for shorebirds and seabirds. Estuarine environments act as filters for the large amount of organic and inorganic compounds that are transported by rivers, streams, underground drainage, and effluents from terrestrial systems (Caçador et al 2012; Fu et al. 2013) toward the open sea. Among these substances, some elements are of particular interest due to their persistence in the environment, their biogeochemical recycling, and the risk they pose to living beings. Regarding this, industrial and domestic discharges are one of the most problematic issues since they carry several pollutants, such as heavy metals, microplastics, different persistent organic compounds (OCs, OPs, PAHs), and other emerging contaminants (Arias et al. 2010; Spetter et al. 2015; Fernández Severini et al. 2019; Villagrán et al. 2019).

As consumers, shorebirds and seabirds are found at most trophic levels of the food web playing a key role in marine and estuarine ecosystems. Because they are long-lived and are often at the top of their food chains, they are particularly vulnerable to a wide range of contaminants associated with the bioaccumulation of pollutants throughout their whole life (Burger 1993).

Estuarine sediments constitute the most important sink of pollutants in aquatic ecosystems. In this sense, shorebirds and seabirds may be particularly vulnerable to these contaminants as they forage primarily on benthic invertebrates (White et al. 1993; Braune and Noble 2009).

Trace Metals in Shorebirds and Seabirds from the Bahía Blanca Estuary

Trace metals, although they may have an anthropogenic origin, are also natural components in the environment and can be found in terrestrial and aquatic systems, as well as in the atmosphere. Shorebirds and seabirds are exposed to different nonessential (and thus potentially toxic beyond a threshold) elements such as silver (Ag), cadmium (Cd), lead (Pb), and mercury (Hg) and also essential elements that could be toxic at high levels such as arsenic (As), cobalt (Co), chromium (Cr), copper (Cu), iron (Fe), manganese (Mn), nickel (Ni), selenium (Se), and zinc (Zn) (Lucia et al. 2014).

Although the bioavailability of metals is associated with physicochemical factors such as pH, salinity, redox potential, organic matter content, and sediment grain size (Eggleton and Thomas 2004; Du Laing et al. 2009; Acosta et al. 2011; Zhao et al. 2013), bioaccumulation is also strongly influenced by the various accumulation strategies of aquatic organisms. For birds, there are several factors that influence uptake, accumulation, and biomagnification of metals, including exposure pathways, species of the metal, and bioavailability, as well as a number of host factors, such as trophic status, location, foraging behavior, nutrition, body condition, gender, size, genetic variability, and age (Burger et al. 2008).

Shorebirds and seabirds acquire metals both through water, which they drink, and through the diet, which in most cases includes marine invertebrates. Once the metals enter the bird's body, they can be accumulated in the tissues or eliminated by deposition through the feathers during periods of rapid growth (after molting), by the uropygeal gland and by the salt gland (Burger and Gochfeld 1996a, b; Burger et al. 1993, 2000). In turn, females can sequester some metals such as Mn, Hg, Cr, Pb, and Cd in egg content and eggshells, which could put embryonic development at risk.

There is a rich worldwide literature on heavy metals in shorebirds and seabirds. These works can be categorized according to whether they are laboratory studies, measurements of residues in sick or dead birds, surveys of metal levels in a given species, and finally, biomonitoring studies, which have been increasing over the years given that waterbirds are valuable for environmental monitoring because they are long-lived predators and are capable of integrating pollutant levels over a large area by bioaccumulation (Furness 1993; Burger and Gochfeld 2001; Zhang and Ma 2011; Lucia et al. 2014; Saalfeld et al. 2016; Dias et al. 2019).

Although there is an extensive database on the levels of heavy metals in the Bahía Blanca Estuary (see Chap. 4), there are no sufficient studies of heavy metals carried out specifically on the marine and coastal birds in this wetland.

The American oystercatcher (*Haematopus palliatus*; Fig. 13.5) is a resident shorebird that reproduces and feeds in the Bahía Blanca Estuary (Delhey and Petracci 2004; Simonetti 2012). The foraging behavior of this species has been described on a variety of intertidal benthic prey such as crabs, clams, oysters, limpets, mussels, and polychaetes (Goss-Custard and Durell 1988; Nol and Humphrey 1994; Bachmann and Martinez 1999; Daleo et al. 2005). One of its main prey items in the Bahía Blanca Estuary is the burrowing crab (*Neohelice granulata*), which is considered a key species within this ecosystem, given that it could play a major role in the transference of pollutants to higher trophic levels. Previous studies on this benthic invertebrate in the Bahía Blanca Estuary showed low to medium concentrations of several trace metals (Ferrer 2001; Beltrame et al. 2009, 2011; Simonetti et al. 2012, 2013, 2018).

Both eggs and eggshells of shorebirds and seabirds have been used in various studies on levels of heavy metals because they have certain advantages over other tissues. Sampling eggs takes less time, is easy to handle, can be collected with little interference, and its extraction puts less pressure on a population than that of adults or juveniles, especially if only one egg is removed from a given clutch (Dev and Bhattacharjee 2010).

As previously mentioned, females can sequester several metals in eggs and eggshells (Burger and Gotchfeld 2002; Lam et al. 2005; Barata et al. 2010; Dev and Bhattacharjee 2010; Hashmi et al. 2013), where metal concentrations represent recent exposure as well as mobilization from stored materials in females at the time of egg formation (Burger and Gochfeld 1995; Burger and Gochfeld 1996a, b).

(continued)

Fig. 13.5 (**a**) American oystercatcher (*Haematopus palliatus*), on the tidal plain of the Bahia Blanca Estuary, (**b**) clutch during the reproductive season (Photos by Natalia S. Martínez-Curci (**a**) Pia Simonetti (**b**))

One of the few studies made on trace metals in shorebirds from the Bahía Blanca Estuary was the work of Simonetti et al. (2015). In this work, concentration of Cu, Cd, Cr, Pb, Zn, and Ni were determined in eggshells of American oystercatchers. Only infertile eggs (i.e., not hatched) and eggs from abandoned nests were selected to avoid affecting the reproductive success. All metals showed detected levels, with exceptionally high Cd concentrations (Table 13.2).

Cu, Cr, Zn, and Ni are four essential metals which play an important role in the metabolism of birds. The values obtained in this study were within the values described in the literature for different species of birds like egrets, herons, albatrosses, and passerines (Dauwe et al. 2005; Ikemoto et al. 2005; Dev and Bhattacharjee 2010; Al-Obaidi et al. 2012; Hashmi et al. 2015; Dolci et al. 2017).

Regarding Cd and Pb, two nonessential metals, although the concentrations obtained for Pb were within the values described in the literature, the Cd

Table 13.2 Trace metal concentrations (μg/g dry weight) in eggshells of American oystercatcher (*Haematopus palliatus*), from the Bahía Blanca Estuary

	Cu	Cd	Cr	Ni	Zn	Pb
Range	0.52–2.63	1.70–14.87	0.03–0.84	0.89–7.59	1.02–3.86	2.13–9.79
Std	0.52	3.38	0.03	0.89	1.13	2.33
Median	1.97	11.62	0.78	6.06	2.08	8.14

values found in this study were surprisingly higher (Ayas 2007; Dev and Bhattacharjee 2010; Hashmi et al. 2013; Avazpour et al. 2014; Dolci et al. 2017; van Aswegen et al. 2019).

La Sala et al. (2011) studied the levels of Hg in feathers in adults and live and dead chicks of Olrog's gull (*Larus atlanticus*), from the Bahia Blanca Estuary. As for eggs, feathers are another pathway for elimination of metals that are accumulated during the inter-molt periods (Monteiro and Furness 1995). Moreover, they are useful indicators of heavy metal contamination because birds sequester heavy metals in their feathers, and the proportion of the body burden that is in feathers is relatively constant for any metal (Burger 1993). The results of this study showed detectable concentrations of Hg in both adults and chicks. For breeding adults Hg in feathers ranged between 2.20 and 16.97 μg/g dry weight. According to Burger and Gotchfeld (2002), these values largely fall within the range that is considered toxic, and there is a potential risk about long-term effects of Hg pollution on this population. The values recorded for feathers of dead and living chicks were lower than adults, ranging between and 1.12–3.00 μg/g dry weight for dead chicks and 0.77–3.02 μg/g dry weight for living chicks. These differences could be explained by (1) a fairly simultaneous growth of all feather types in pre-fledglings, (2) Hg concentrations in adult feathers reflecting metal input throughout a vast geographical area and a substantially longer accumulation period, and (3) a wider array of prey items during the non-breeding season which might be exposing the adults to higher Hg concentrations from other food resources such as fish discards (La Sala et al. 2011).

During the chick-rearing period, Olrog's gull adults prey mainly on the crabs *N. granulata* and *Cyrtograpsus angulatus*, which they feed to their young. Although there are no data on the concentration of Hg in any of these invertebrates, as previously mentioned, there is vast information about the bioaccumulation of heavy metals in *N. granulata*, which suggests the possibility that this metal is also bioaccumulated in these organisms, thus making Olrog's gull adults, and especially small chicks, susceptible to the effects of Hg (La Sala et al. 2011).

From these results, added to the state of knowledge of the presence of trace metals in one of the prey items for oystercatchers and gulls, metals that are bioavailable in the environment may also be transferred to other seabirds and shorebird species through food chains. Given the toxicity of these elements, it seems a potential risk in these species, since many of these metals can have effects on the development and nervous system of birds.

Microplastics and Other Contaminants

Plastic pollution in marine and estuarine environments is a worldwide concern and has been studied since the 1970s. The durability of the plastic implies that it is retained for years or centuries, while if it is not exposed to bacterial activity or UV radiation, degradation does not occur (Lambert et al. 2014). Large plastic debris, called "macroplastics," generate a significant environmental impact due to the injury and death of seabirds, mammals, fish, and reptiles as a result of plastic entanglement and ingestion (Derraik 2002; Gregory 2009; Lozano and Mouat 2009). Marine animals often mistake these fragments for food by ingesting them, which can cause damage to internal organs, digestive tract blockage, and ulcers among other affections (Blais et al. 2005; Gregory 2009).

In seabirds, the effect of plastic ingestion has been a cause of concern primarily for two reasons, firstly because of the frequency with which seabirds ingest plastic and secondly because of emerging evidence of impacts on body condition and transmission of toxic chemicals, which could lead to changes in mortality or reproduction (Wilcox et al. 2015). In recent years, "microplastics," referred as tiny plastic particles with a diameter smaller than 5 mm (Thompson et al. 2004), have become a subject of great concern at the environmental level. Because of their composition and relatively large surface area, microplastic fragments can be vectors for the diffusion of organic contaminants. Ingestion of microplastics may therefore be introducing toxins to the base of the food chain, from where there is potential for bioaccumulation (Teuten et al. 2009).

Again, although there are a considerable number of studies on plastic pollution in seabirds worldwide, to date there are no published studies for shorebirds and seabirds in the Bahía Blanca Estuary. The only registry available and whose investigation is currently under study is that of La Sala (pers. comm.), where plastic debris were detected in the stomachs of chicks of Olrog's gull.

Considering the limited information available as well as the possible adverse effects of these pollutants on shorebirds and seabirds, a comprehensive long-term research covering relevant disciplines is necessary to assess the effects of this and other pollutants on the health of populations of these species in the Bahía Blanca Estuary.

one of the Important Bird and Biodiversity Areas (IBA-BA15) in Argentina (Di Giacomo 2005), it was included within the Valuable Grassland Areas (VGAs; Bilenca and Miñarro 2004), and it is also one of the high priority areas for the conservation of Nearctic migratory birds in the southern cone grasslands of South America (Di Giacomo and Parera 2008).

Remarkably, migratory shorebirds and the red knot were declared emblematic species of Bahía Blanca and Coronel Rosales counties, respectively. The legal protection of the ecosystem, as well as the recognition of its importance by international and local communities, show signs of hope for Bahía Blanca Estuary to remain a healthy ecosystem in the long term and thus to continue supporting viable shorebirds and seabirds' populations.

References

Acosta JA, Jansen B, Kalbitz K et al (2011) Salinity increases mobility of heavy metals in soils. Chemosphere 85:1318–1324. https://doi.org/10.1016/j.chemosphere.2011.07.046

Al-Obaidi FA, Mehdi BA, Al-Shadeedi SM (2012) Identification of inorganic elements in egg shell of some wild birds in Baghdad. Adv Appl Sci Res 3(3):1454–1458

Andres BA, Smith PA, Morrison RIG et al (2012) Population estimates of North American shorebirds. Wader Study Group Bull 119:178–194

Arias AH, Fernández Severini MD, Delucchi F et al (2010) Persistent pollutants monitoring in a South Atlantic coastal environment. Case study: the Bahía Blanca Estuary. In: Ortiz AC, Griffin NB (eds) Pollution monitoring. Nova Science Publishers, New York, pp 1–30

Avazpour FG, Bakhtiari AR, Sajjadi MM (2014) Levels of vanadium in tern eggs and sediments of Shidvar and Bani Farour Islands, Persian gulf. Podoces 9(1)

Ayas Z (2007) Trace element residues in eggshells of grey heron (*Ardea cinerea*) and black-crowned night heron (*Nycticorax nycticorax*) from Nallihan bird paradise, Ankara-Turkey. Ecotoxicology 16:347–352

Azpiroz A, Caballero-Sadi D (2017) Gaviota cangrejera (Larus atlanticus). In: Azpiroz AB, Jiménez S, Alfaro, M (eds.) Libro Rojo de las Aves del Uruguay. Biología y conservación de las aves en peligro de extinción a nivel nacional. Categorías "Extinto a Nivel Regional", "En Peligro Crítico" y "En Peligro". DINAMA y DINARA, Montevideo.

Bachmann S, Martinez MM (1999) Feeding tactics of the American Oystercatcher (*Haematopus palliatus*) on Mar Chiquita Coastal Lagoon, Argentina. Ornitol Neotrop 10:81–84

Baker A, González P, Morrison RIG, Harrington BA (2020) Red Knot (*Calidris canutus*), version 1.0. In: Billerman SM (ed) Birds of the world. Cornell Lab of Ornithology, New York. Available via 10.2173/bow.redkno.01. Accessed 8 May 2020

Barata C, Fabregat MC, Cotín J et al (2010) Blood biomarkers and contaminant levels in feathers and eggs to assess environmental hazards in heron nestlings from impacted sites in Ebro basin (NE Spain). Environ Pollut 158(3):704–710

Barbier EB, Hacker SD, Kennedy C et al (2011) The value of estuarine and coastal ecosystem services. Ecol Monogr 81:169–193

Barbosa A, Moreno E (1999) Evolution of foraging strategies in shorebirds: an ecomorphological approach. Auk 116:712–725

Belenguer C, Delhey K, Di Martino S et al (1992) Observaciones de aves playeras migratorias de Bahia Blanca. Boletín Informativo Grupo Argentino de Limícolas Nro 10

Beltrame MO, De Marco SG, Marcovecchio JE (2009) Influences of sex, habitat, and seasonality on heavy-metal concentrations in the burrowing crab *(Neohelice Granulata)* from a coastal lagoon in Argentina. Arch Environ Contam Toxicol 58(3):746–756

Beltrame MO, De Marco SG, Marcovecchio JE (2011) The burrowing crab *Neohelice granulata* as potential bioindicator of heavy metals in estuarine systems of the Atlantic Coast of Argentina. Environ Monit Assess 172(1–4):379–389

Berón MP, Favero M (2009) Mortality and injuries of Olrog's Gull (Larus atlanticus) individuals associated with sport fishing activities in Mar Chiquita coastal lagoon, Buenos Aires Province. Hornero 24(2):99–102

Berón MP, Seco Pon JP, García GO et al (2013) The diet of Olrog's Gull (*Larus atlanticus*) reveals an association with fisheries during the non-breeding season. Emu 113:69–76

Bilenca D, Miñarro F (2004) Identificación de Áreas Valiosas de Pastizal (AVPs) en las Pampas y Campos de Argentina, Uruguay y sur de Brasil. Fundación Vida Silvestre Argentina, Buenos Aires

BirdLife International (2016) *Pluvianellus socialis*. In: The IUCN Red List of Threatened Species. Available via https://doi.org/10.2305/IUCN.UK.2016-3.RLTS.T22693570A93413261.en. Accessed 8 May 2020

BirdLife International (2017) *Calidris subruficollis*. In: The IUCN Red List of Threatened Species. Available via https://doi.org/10.2305/IUCN.UK.2017-1.RLTS.T22693447A111804064. Accessed 8 May 2020

BirdLife International (2018) *Larus atlanticus*. In: The IUCN Red List of Threatened Species. Available via https://doi.org/10.2305/IUCN.UK.2018-2.RLTS.T22694286A132538305.en. Accessed 8 May 2020

Blais JM, Kimpe LE, McMahon D et al (2005) Arctic seabirds transport marine-derived contaminants. Science 309(5733):445–445

Boere GC, Stroud DA (2006) The flyway concept: what it is and what it isn't. In: Boere GC, Galbraith CA, Stroud DA (eds) Waterbirds around the world. The stationery office, Edinburgh, pp 40–47

Braune BM, Noble DG (2009) Environmental contaminants in Canadian shorebirds. Environ Monit Assess 148(1–4):185–204

Brockmann HJ, Barnard CJ (1979) Kleptoparasitism in birds. Anim Behav 27:487–514

Burger J (1993) Metals in avian feathers: bioindicators of environmental pollution. Rev Environ Toxicol 5:203–311

Burger J, Gochfeld M (1995) Heavy metal and selenium concentrations in eggs of herring gulls (*Larus argentatus*): temporal differences from 1989 to 1994. Arch Environ Contam Toxicol 29(2):192–197

Burger J, Gochfeld M (1996a) Family Laridae (gulls). In: del Hoyo J, Elliott A, Sargatal J (eds) Handbook of the birds of the world, vol 3. Lynx Edicions, Barcelona, pp 572–623

Burger J, Gochfeld M (1996b) Heavy metal and selenium levels in Franklin's gull (*Larus pipixcan*) parents and their eggs. Arch Environ Contam Toxicol 30(4):487–491

Burger J, Gochfeld M (2001) On developing bioindicators for human and ecological health. Environ Monit Assess 66(1):23–46

Burger J, Gotchfeld M (2002) Effects of chemicals and pollution on seabirds. In: Schreiber EA, Burger J (eds) Biology of marine birds. CRC Press, Boca Raton, pp 485–525

Burger J, Olla BL (eds) (1984) Shorebirds: migration and foraging behavior. Plenum Press, New York, pp 73–123

Burger J, Rodgers JA, Gochfeld M (1993) Heavy metal and selenium levels in endangered wood storks *Mycteria americana* from nesting colonies in Florida and Costa Rica. Arch Environ Contam Toxicol 24(4):417–420

Burger J, Trivedi CD, Gochfeld M (2000) Metals in herring and great black-backed gulls from the New York bight: the role of salt gland in excretion. Environ Monit Assess 64(3):569–581

Burger J, Gochfeld M, Jeitner C et al (2008) Assessment of metals in down feathers of female common eiders and their eggs from the Aleutians: arsenic, cadmium, chromium, lead, manganese, mercury, and selenium. Environ Monit Assess 143(1–3):247

Burger J, Gochfeld M, Garcia EFJ et al (2020) Kelp Gull (*Larus dominicanus*), version 1.0. In: del Hoyo J, Elliott A, Sargatal J, Christie DA, de Juana E (eds) Birds of the world. Cornell Lab of Ornithology, New York. Available via 10.2173/bow.kelgul.01. Accessed 8 May 2020

Cabot D, Nisbet I (2013) Tern. Harper Collins, London

Caçador I, Costa JL, Duarte B et al (2012) Macroinvertebrates and fishes as biomonitors of heavy metal concentration in the Seixal Bay (*Tagus estuary*): which species perform better? Ecol Indic 19:184–190

Castano MV, Biondi LM, Zumpano F et al. (2020). Behavioral responses to a novel feeding problem in the Olrog's Gull *Larus altanticus*. Mar Ornithol (in press)

Ceballos G, Ehrlich PR, Barnosky AD et al (2015) Accelerated modern human–induced species losses: entering the sixth mass extinction. Sci Adv 1:e1400253. https://doi.org/10.1126/sciadv.1400253

Colwell MA (2010) Shorebird ecology, conservation, and management. University of California Press, California

Colwell MA, Haigh SM (2019) An overview of the world's plovers. In: Colwell MA, Haigh SM (eds) The population ecology and conservation of *Charadrius* plovers. CRC Press Taylor/Francis Group, London, pp 3–15

Copello S, Favero M (2001) Foraging ecology of Olrog's Gull *Larus atlanticus* in Mar Chiquita Lagoon (Buenos Aires, Argentina): are there age-related differences? Bird Conserv Int 11:175–188

Copello S, Suárez N, Yorio P et al (2020) Distribution of Olrog's Gull *Larus atlanticus* from Bahía San Blas during the non-breeding period: signals of partial migration. Bird Conserv Int:1–10. https://doi.org/10.1017/S0959270920000234

Croxall JP, Butchart SHM, Lascelles B et al (2012) Seabird conservation status, threats and priority actions: a global assessment. Bird Conserv Int 22:1–34

Daleo P, Escapa M, Isacch JP et al (2005) Trophic facilitation by the oystercatcher Haematopus palliatus Temminick on the scavenger snail *Buccinanops globulosum* Kiener in a Patagonian bay. J Exp Mar Biol Ecol 325(1):27–34

Dauwe T, Janssens E, Bervoets L et al (2005) Heavy metal concentrations in female laying great tits (*Parus major*) and their clutches. Arch Environ Contam Toxicol 49:249–256

Delhey, JKV, Petracci, PF (2004) Aves marinas y costeras. In: Ecosistema del estuario de Bahía Blanca. Piccolo MC, Hoffmeyer M (eds). Edi UNS, pp 203–220

Delhey JKV, Carrete M, Martínez M (2001a) Diet and feeding behaviour of Olrog's Gull *Larus atlanticus* in Bahía Blanca, Argentina. Ardea 89:319–329

Delhey K, Petracci PF, Pérez CHF (2001b) Observaciones de algunos Charadriiformes en el sur de la provincia de Buenos Aires, Argentina. Nuestras Aves 42:14–16

Derraik JGB (2002) The pollution of the marine environment by plastic debris: a review. Mar Pollut Bull 44:842–852

Dev B, Bhattacharjee PC (2010) Heavy metals in egg shells of six species of Ardeidae (Aves) from Barak Valley, Assam. ADBU J Eng Technol 5(1):48–52

Di Giacomo AS (2005) Áreas importantes para la conservación de las aves en Argentina. Sitios prioritarios para la conservación de la biodiversidad. Aves Argentinas/Asociación Ornitológica del Plata, Buenos Aires. Available via https://www.avesargentinas.org.ar/aica. Accessed 8 May 2020

Di Giacomo AS, Parera AF (2008) 20 Áreas prioritarias para la conservación de las aves migratorias neárticas en los pastizales del cono sur de Sudamérica. Aves Argentinas / Asociación Ornitológica del Plata, Buenos Aires

Dias MP, Martin R, Pearmain EJ et al (2019) Threats to seabirds: a global assessment. Biol Conserv 237:525–537

Dirzo R, Young HS, Galetti M et al (2014) Defaunation in the Anthropocene. Science 345:401–406. https://doi.org/10.1126/science.1251817

Dolci NN, Sá F, da Costa ME et al (2017) Trace elements in feathers and eggshells of brown booby *Sula leucogaster* in the Marine National Park of Currais Islands, Brazil. Environ Monit Assess 189(10):496

Du Laing G, Rinklebe J, Vandecasteele B et al (2009) Trace metal behaviour in estuarine and riverine floodplain soils and sediments: a review. Sci Tot Environ 407:3972–3985. https://doi.org/10.1016/j.scitotenv.2008.07.025

Eggleton J, Thomas KV (2004) A review of factors affecting the release and bioavailability of contaminants during sediment disturbance events. Environ Int 30:973–980. https://doi.org/10.1016/j.envint.2004.03.001

Elner RW, Beninger PG, Jackson DL et al (2005) Evidence of a new feeding mode in western sandpiper (*Calidris mauri*) and dunlin (*Calidris alpina*) based on bill and tongue morphology and ultrastructure. Mar Biol 146:1223–1234

Escalante R (1966) Notes on the Uruguayan population of *Larus belcheri*. Condor 68:507–510

Escapa M, Isacch JP, Daleo P et al (2004) The distribution and ecological effects of the introduced Pacific Oyster *Crassostrea gigas* (Thunberg, 1973) in Northern Patagonia. J Shellfish Res 23:765–772

Fernández Severini MD, Villagrán DM, Buzzi NS et al (2019) Microplastics in oysters (*Crassostrea gigas*) and water at the Bahía Blanca Estuary (Southwestern Atlantic): an emerging issue of global concern. Reg Stud Mar Sci 32:100829. https://doi.org/10.1016/j.rsma.2019.100829

Ferrer LD (2001) Estudio de los diversos metales pesados en sedimentos del estuario de Bahia Blanca y sus efectos toxicos sobre el cangrejo *Chasmagnathus granulata*. Doctoral thesis, Universidad Nacional del Sur, Argentina

Fu J, Tang XL, Zhang J et al (2013) Estuarine modification of dissolved and particulate trace metals in major rivers of East-Hainan, China. Cont Shelf Res 57:59–72

Furness RW (1987) Kleptoparasitism in seabirds. In: Croxal JP (ed) Seabirds, feeding biology and role in marine ecosystem. Cambridge University Press, Cambridge, pp 77–99

Furness RW (1993) Birds as monitors of pollutants. In: Birds as monitors of environmental change. Springer, Dordrecht, pp 86–143

Furness RW, Edwards AE, Oro D (2007) Influence of management practices and of scavenging seabirds on availability of fisheries discards to benthic scavengers. Mar Ecol Prog Ser 350:235–244. https://doi.org/10.3354/meps07191

García GO, Favero M, Vassallo AI (2010) Factors affecting kleptoparasitism by gulls in a multispecies seabird colony. Condor 112:521–529

García GO, Becker PH, Favero M (2011) Kleptoparasitism during courtship in *Sterna hirundo* and its relationship with female reproductive performance. J Ornithol 152:103–110

García GO, Favero M, Becker PH (2013) Intraspecific kleptoparasitism improves chick growth and reproductive output in common terns *Sterna hirundo*. Ibis 155:338–347

García GO, Nicoll A, Castano M et al (2019) Evaluation of neophobia in a threatened seabird: Olrog's Gull (*Larus atlanticus*) as a case study. Emu 119:166–175

García GO, Paterlini CA, Favero M et al (2020) Age-, sex- and tactic-specific kleptoparasitic performance in a long-lived seabird. J Ornithol 161:183–188

Gochfeld M, Burger J, de Juana E et al (2020) South American tern (*Sterna hirundinacea*), version 1.0. In: del Hoyo J, Elliott A, Sargatal J et al (eds) Birds of the world. Cornell Lab of Ornithology, New York. Available via 10.2173/bow.soater1.01. Accessed 8 May 2020

Goss-Custard JD, Durell SLVD (1988) The effect of dominance and feeding method on the intake rates of oystercatchers, *Haematopus ostralegus*, feeding on mussels. J Anim Ecol:827–844

Gregory MR (2009) Environmental implications of plastic debris in marine settings: entanglement, ingestion, smothering, hangers-on, hitch-hiking and alien invasions. Philos T R Soc B 364:2013–2025

Hackett SJ, Kimball RT, Reddy S et al (2008) A phylogenomic study of birds reveals their evolutionary history. Science 320:1763–1768

Hashmi MZ, Malik RN, Shahbaz M (2013) Heavy metals in eggshells of cattle egret (*Bubulcus ibis*) and little egret (*Egretta garzetta*) from the Punjab province, Pakistan. Ecotox Environ Safe 89:158–165

Hashmi MZ, Abbasi NA, Tang X et al (2015) Egg as a biomonitor of heavy metals in soil. In: Heavy metal contamination of soils. Springer, Cham, pp 127–143

Herrera G, Punta G, Yorio P (2005) Diet specialization of the threatened Olrog's gull *Larus atlanticus* during the breeding season at Golfo San Jorge, Argentina. Bird Conserv Int 15:89–97
Hoekstra JM, Boucher TM, Ricketts TH et al (2005) Confronting a biome crisis: global disparities of habitat loss and protection. Ecol Lett 8:23–29. https://doi.org/10.1111/j.1461-0248.2004.00686.x
Ieno E, Alemany D, Blanco DE et al (2004) Prey size selection by red knots feeding on mud snails at Punta rasa (Argentina) during migration. Waterbirds 27:493–498
Ikemoto T, Kunito T, Tanabe S et al (2005) Non-destructive monitoring of trace element levels in short-tailed albatrosses (*Phoebastria albatrus*) and black-footed albatrosses (*Phoebastria nigripes*) from Torishima Island, Japan, using eggs and blood. Mar Pollut Bull 51:889–895
Iribarne OO, Martínez MM (1999) Predation on the southwestern Atlantic fiddler crab (*Uca uruguayensis*) by migratory shorebirds (*Pluvialis dominica, P. squatarolla, Arenaria interpres* and *Numenius phaeopus*). Estuaries 22:47–54
IUCN (2020) The IUCN Red List of Threatened Species. Version 2020–1. Available via https://www.iucnredlist.org. Accessed 8 May 2020
Jenni L, Jenni-Eiermann S (1998) Fuel supply and metabolic constraints in migrating birds. J Avian Biol 29:521–528
Jenni-Eiermann S, Jenni L, Kvist A et al (2002) Fuel use and metabolic response to endurance exercise: a wind tunnel study of a long-distance migrant shorebird. J Exp Biol 205:2453–2460
Kelly JP, Evens JG, Stallcup RW et al (1996) The effects of aquaculture on habitat use by wintering shorebirds. California Fish and Game 82:160–174
Kubelka V, Šálek M, Tomkovich P et al (2018) Global pattern of nest predation is disrupted by climate change in shorebirds. Science 362:680–683. https://doi.org/10.1126/science.aat8695
La Sala LF, Petracci PF, Smits JE et al (2011) Mercury levels and health parameters in the threatened Olrog's Gull (*Larus atlanticus*) from Argentina. Environ Monit Asses 181(1–4):1–11
Lack D (1968) Bird migration and natural selection. Oikos 19:1–9
Lam JC, Tanabe S, Lam MH et al (2005) Risk to breeding success of waterbirds by contaminants in Hong Kong: evidence from trace elements in eggs. Environ Pollut 135(3):481–490
Lambert S, Sinclair C, Boxall A (2014) Occurrence, degradation, and effect of polymer based materials in the environment. Rev Environ Contam Toxicol 227(227):1–53
Lascelles BG, Taylor PR, Miller MGR et al (2016) Applying global criteria to tracking data to define important areas for marine conservation. Divers Distrib 22:422–431
Lindström A, Piersma T (1993) Mass changes in migrating birds: the evidence for fat and protein storage re-examined. Ibis 135:70–78
Lisnizer N, García Borboroglu P, Yorio P (2011) Spatial and temporal variation in population trends of kelp gulls in northern Patagonia, Argentina. Emu 111:259–267
Lovette IJ (2016) Avian diversity and classification. In: Lovette IJ, Fitzpatrick JW (eds) The Cornell lab of ornithology handbook of bird biology, 3rd edn. Wiley, West Sussex, pp 7–61
Lozano RL, Mouat J (2009) Marine litter in the North-East Atlantic region: Assessment and priorities for response KIMO international
Lucia M, Bocher P, Chambosse M et al (2014) Trace element accumulation in relation to trophic niches of shorebirds using intertidal mudflats. J Sea Res 92:134–143
Luckenbach MW (1984) Biogenic structure and foraging by five species of shorebirds (Charadrii). Estuar Coast Shelf S 196:691–696
Marbán LM, Petracci P, Sotelo M, Zalba S (2019) Plantas vs. gaviotas: colonizando una colonia. Un problema de invasión. In: Libro de resúmenes de la XVIII.
Marbán LM, Zalba SM (2019) When the seeds go floating in: A salt marsh invasion. Estuar Coast Shelf S 231:106442. https://doi.org/10.1016/j.ecss.2019.106442.
Marinao C, Suárez N, Yorio P (2019) Trophic interactions between the Kelp Gull (*Larus dominicanus*) and royal and Cayenne terns (*Thalasseus maximus maximus* and *Thalasseus sandvicensis eurygnathus*, respectively) in a human-modified environment. Can J Zool 97:904–913. https://doi.org/10.1139/cjz-2019-0047
Martínez-Curci NS, Petracci P (2016) Aves playeras del litoral costero de la provincia de Buenos Aires: ecología y conservación. In: Athor J, Celsi CE (eds) La costa atlántica de Buenos Aires:

naturaleza y patrimonio cultural. Fundación de Historia Natural Felix de Azara, Buenos Aires, pp 204–233
Martínez-Curci NS, Azpiroz AB, Isacch JP et al (2015) Dietary relationships among Nearctic and Neotropical migratory shorebirds in a key coastal wetland of South America. Emu 115:326–334
MAyDS (Ministerio de Ambiente y Desarrollo Sostenible), Aves Argentinas (2017) Categorización de las Aves de la Argentina según su estado de conservación. Available via https://avesargentinas.org.ar/sites/default/files/Categorizacion-de-aves-de-la-Argentina.pdf. Accessed 8 May 2020
MAyDS, AA (Ministerio de Ambiente y Desarrollo Sustentable and Aves Argentinas) (2017). Categorización de las Aves de la Argentina (2015). Informe del Ministerio de Ambiente y Desarrollo Sustentable de la Nación y de Aves Argentinas, edición electrónica. Buenos Aires, Argentina
McCarty JP, Wolfenbarger LL, Laredo CD, Pyle P, Lanctot RB (2020) Buff-breasted sandpiper (*Calidris subruficollis*), version 1.0. In: Rodewald PG (ed) Birds of the world. Cornell Lab of Ornithology, Ithaca, New York. Available via 10.2173/bow.bubsan.01. Accessed 8 May 2020
Monteiro LR, Furness RW (1995) Seabirds as monitors of mercury in the marine environment. Water Air Soil Pollut 80:851–870
Morrison RIG, Ross RK (1989) Atlas of Nearctic shorebirds on the coast of South America. Canadian Wildlife Service, Ottawa
Myers JP (1983) Conservation of migrating shorebirds: staging areas, geographic bottlenecks, and regional movements. Am Birds 37:23–25
Niles LJ, Burger J, Porter RR et al (2010) First results using light level geolocators to track red knots in the Western hemisphere show rapid and long intercontinental flights and new details of migration pathways. Wader Study Group Bull 117:123–130
Nol E, Humphrey RC (1994) American Oystercatcher (*Haematopus palliatus*) In: Poolea and Gill (eds) The birds of North America. Academy of Natural Sciences and American Ornithologists' Union, Filadelfia y Washington DC, pp 1–24
Orians GH, Pearson NE (1979) On the theory of central place foraging. In: Horn DJ, Mitchell RD, Stairs GR (eds) Analysis of ecological systems. Ohio State University Press, Columbus, pp 154–177
Oro D, Genovart M, Tavecchia G et al (2013) Ecological and evolutionary implications of food subsidies from humans. Ecol Lett 16:1501–1514. https://doi.org/10.1111/ele.12187
Otero XL, De la Peña-Lastra S, Pérez Alberti A et al (2018) Seabird colonies as important global drivers in the nitrogen and phosphorus cycles. Nat Commun 9:246. https://doi.org/10.1038/s41467-017-02446-8.
Parmelee DF (2020) White-rumped sandpiper (*Calidris fuscicollis*), version 1.0. In: Poole AF, Stettenheim PR, Gill FB (eds) Birds of the world. Cornell Lab of Ornithology, Ithaca, New York. Available via https://doi-org.proxy.library.cornell.edu/10.2173/bow.whrsan.01. Accessed 8 May 2020
Paton TA, Baker AJ, Growth JG et al (2003) RAG-1 sequences resolve phylogenetic relationships within Charadriiform birds. Mol Phylogenet Evol 29:268–278
Paz JA, Seco Pon JP, Favero M et al (2018). Seabird interactions and by-catch in the anchovy pelagic trawl fishery operating in northern Argentina. Aquat Conserv. https://doi.org/10.1002/aqc.2907
Petracci PF (2002) Diet of sanderling in Buenos Aires province, Argentina. Waterbirds 25:366–370
Petracci, PF, Delhey JKV (2005) Guía de las aves marinas y costeras de la ría de Bahía Blanca. In: Petracci PF, Delhey JKV (eds). Bahía Blanca
Petracci PF, Sotelo M (2013) Aves del Estuario de Bahía Blanca: Una herramienta para su conocimiento y conservacion. Grupo Editorial Muelle Sur, Bahía Blanca
Petracci PF, La Sala L, Aguerre G et al (2004) Dieta de la gaviota cocinera (*Larus dominicanus*) durante el período reproductivo en el Estuario de Bahía Blanca, Buenos Aires, Argentina. El Hornero 19(1):23–28

Petracci PF, Delhey KV, Sotelo MR (2007) Hábitos granívoros en la Gaviota Cangrejera (*Larus atlanticus*): implicancias sobre su estatus de especialista. Hornero 22:51–54

Petracci PF, Sotelo MR, Díaz L (2008) Nuevo registro de nidificación de la gaviota cangrejera (*Larus atlanticus*) en la Reserva Natural Bahía Blanca, Bahía Falsa y Bahía Verde, Buenos Aires, Argentina. Hornero 23(1):37–40

Pichegru L, Nyengera R, McInnes AM et al (2017) Avoidance of seismic survey activities by penguins. Sci Rep 7:16305. https://doi.org/10.1038/s41598-017-16569-x

Piersma T (1987) Hop, skip or jump? Constraints on migration of arctic waders by feeding, fattening, and flight speed. Limosa 60:185–194

Piersma T, Gill RE (1998) Guts don't fly: small digestive organs in obese Bar-tailed godwits. Auk 115:196–203

Piersma T, van Aelst R, Kurk K et al (1998) A new pressure sensory mechanism for prey detection in birds: the use of principles of seabed dynamics? Proc R Soc Lond B 265:1377–1383

Pimm SL, Jenkins CN, Abell R et al (2014) The biodiversity of species and their rates of extinction, distribution, and protection. Science 344:1246752. https://doi.org/10.1126/science.1246752

Quadri Adrogué A, Miglioranza KSB, Copello S et al (2019) Pelagic seabirds as biomonitors of persistent organic pollutants in the Southwestern Atlantic. Mar Pollut Bull 149:110516. https://doi.org/10.1016/j.marpolbul.2019.110516

Reunión Argentina de Ornitología, Universidad Nacional del Centro de la Provincia de Buenos Aires, Tandil, 4-6 September 2019. https://doi.org/10.13140/RG.2.2.24074.41923

Reynolds MH, Courtot KN, Berkowitz P et al (2015) Will the Effects of Sea-Level Rise Create Ecological Traps for Pacific Island Seabirds?. PLOS ONE 10(9):e0136773. https://doi.org/10.1371/journal.pone.0136773.

Ribeiro PD, Iribarne OO, Navarro D et al (2004) Environmental heterogeneity, spatial segregation of prey, and the utilization of Southwest Atlantic mudflats by migratory shorebirds. Ibis 146:672–682

Roberts DA (2012) Causes and ecological effects of resuspended contaminated sediments (RCS) in marine environments. Environ Int, 40:230–243. https://doi.org/10.1016/j.envint.2011.11.013.

Rosenberg KV, Dokter AM, Blancher PJ et al (2019) Decline of the North American avifauna. Science 366:120–124. https://doi.org/10.1126/science.aaw1313

Rothschild M, Clay T (1952) Fleas, flukes, and cuckoos: a study of bird ectoparasites. Collins Son and Co, London

Saalfeld DT, Matz AC, McCaffery BJ et al (2016) Inorganic and organic contaminants in Alaskan shorebird eggs. Environ Monit Assess 188(5):276

Sala OE, Chapin FS III, Armesto JJ et al (2000) Global biodiversity scenarios for the year 2100. Science 287:1770–1774. https://doi.org/10.1126/science.287.5459.1770

Schmidt-Nielsen K (1960) The salt-secreting gland of marine birds. Circulation 21:955–967

Schreiber EA, Burger J (2001) Seabirds in the marine environment. In: Schreiber EA, Burger J (eds) Biology of marine birds. CRC Press, Florida, pp 1–15

Senner NR, Hochachka WM, Fox JW et al (2014) An exception to the rule: carry-over effects do not accumulate in a long-distance migratory bird. PLoS One 9:e86588. https://doi.org/10.1371/journal.pone.0086588

Shuter JL, Broderick AC, Agenw DJ et al (2011) Conservation and management of migratory species. In: Milner-Gulland EJ, Fryxell JM, Sinclair ARE (eds) Animal migration: a synthesis. Oxford University Press Inc, New York, pp 172–206

Simonetti P (2012) El ostrero pardo, *Haematopus palliatus*, en el Estuario de Bahía Blanca: Estudio de la biología reproductiva, uso del hábitat y el potencial condicionamiento ambiental como consecuencia del impacto antrópico Doctoral Thesis Universidad Nacional del Sur, Bahía Blanca, p. 142

Simonetti P, Botté SE, Fiori SM et al (2012) Heavy-metal concentrations in soft tissues of the burrowing crab *Neohelice granulata* in Bahía Blanca Estuary, Argentina. Arch Environ Contam Toxicol 62:243–253

Simonetti P, Botté SE, Fiori SM et al (2013) Burrowing crab (*Neohelice granulata*) as a potential bioindicator of heavy metals in the Bahía Blanca Estuary, Argentina. Arch Environ Contam Toxicol 64:110–118

Simonetti P, Botté SE, Marcovecchio JE (2015) Exceptionally high Cd levels and other trace elements in eggshells of American oystercatcher (Haematopus palliatus) from the Bahía Blanca Estuary, Argentina. Mar Pollut Bull, 100(1):495–500

Simonetti P, Botté SE, Marcovecchio JE (2018) Heavy metal bioconcentration factors in the burrowing crab Neohelice granulata of a temperate ecosystem in South America: Bahía Blanca estuary, Argentina. Environmental Science and Pollution Research, 25(34):34652–34660

Spetter CV, Buzzi NS, Fernández EM et al (2015) Assessment of the physicochemical conditions sediments in a polluted tidal flat colonized by microbial mats in Bahía Blanca Estuary (Argentina). Mar Pollut Bull 9:491–505. https://doi.org/10.1016/j.marpolbul.2014.10.008

Sutherland WJ, Alves JA, Amano T et al (2012a) A horizon scanning assessment of current and potential future threats to migratory shorebirds. Ibis 154:663–679

Sutherland WJ, Mitchell R, Prior SV (2012b) The role of 'conservation evidence' in improving conservation management. Conservation Evidence 9(11):1–2

Tasker ML, Camphuysen CJ, Cooper J et al (2000) The impacts of fishing on marine birds. ICES J Mar Sci 57:531–547. https://doi.org/10.1006/jmsc.2000.0714

Teuten EL, Saquing JM, Knappe DRU et al (2009) Transport and release of chemicals from plastic to the environment and to wildlife. Philos T R Soc B 36:2027–2045

Thiebault A, Mullers RHE, Pistorius P et al (2014) Local enhancement in a seabird: reaction distances and foraging consequence of predator aggregations. Behav Ecol 25:1302–1310

Thomas GH, Wills MA, Székely T (2004) A supertree approach to shorebird phylogeny. BMC Evol Biol 4:28

Thompson RC, Olsen Y, Mitchell RP et al (2004) Lost at sea: where is all the plastic? Science 304:838

van Aswegen JD, Nel L, Strydom NA, Minnaar K et al (2019) Comparing the metallic elemental compositions of kelp Gull *Larus dominicanus* eggs and eggshells from the Swartkops Estuary, Port Elizabeth, South Africa. Chemosphere 221:533–542

van Gils JA, Lisovski S, Lok T, Meissner W et al (2016) Body shrinkage due to Arctic warming reduces red knot fitness in tropical wintering range. Science 352:819–821. https://doi.org/10.1126/science.aad6351

Villagrán DM, Fernández Severini MD, Biancalana F et al (2019) Bioaccumulation of heavy metals in mesozooplankton from a human-impacted south western Atlantic estuary (Argentina). J Mar Res 77:217–241. https://doi.org/10.1357/002224019826887362

Walker BM, Senner NR, Elphick CS et al (2020) Hudsonian godwit (*Limosa haemastica*), version 1.0. In: Poole AF, Stettenheim PR, Gill FB (eds) Birds of the world. Cornell Lab of Ornithology, New York. Available via https://doi-org.proxy.library.cornell.edu/10.2173/bow.whrsan.01. Accessed 8 May 2020

White DC, Smith GA, Guckert JB et al (1993) Nearshore benthic marine sediments. In: Antarctic microbiology. Wiley-Liss, New York, pp 235–240

Wilcox C, Van Sebille E, Hardesty BD (2015) Threat of plastic pollution to seabirds is global, pervasive, and increasing. Proc Natl Acad Sci 112(38):11899–11904

Yorio P, Rábano D, Rabuffetti F et al (1998) Distribución reproductiva y abundancia de las aves marinas de la Provincia de Buenos Aires: de Bahía Blanca a Punta Redonda. In: Yorio P, Frere E, Gandini P et al (eds) Atlas de la distribución reproductiva de aves marinas en el litoral patagónico argentino. Fundación Patagonia Natural and Wildlife Conservation Society, Buenos Aires, pp 19–28

Yorio P, Bertellotti M, García Borboroglu P (2005) Estado poblacional y de conservación de gaviotas que se reproducen en el litoral marítimo argentino. Hornero 20:53–74

Yorio P, Petracci P, García Borboroglu P (2013) Current status of the threatened Olrog's Gull *Larus atlanticus*: global population, breeding distribution and threats. Bird Conserv Int 23:477–486

Zalba SM, Sonaglioni MI, Compagnoni CA, Belenguer CJ (2000) Assessing the risk of invasion by an exotic plant in a coastal nature reserve. Biol Conserv 93(2):203–208. https://doi.org/10.1016/S0006-3207(99)00146-9

Zhang WW, Ma JZ (2011) Waterbirds as bioindicators of wetland heavy metal pollution. Procedia Environ Sci 10:2769–2774

Zhao S, Feng C, Wang D et al (2013) Salinity increases the mobility of Cd, Cu, Mn, and Pb in the sediments of Yangtze Estuary: relative role of sediments' properties and metal speciation. Chemosphere 91:977–984. https://doi.org/10.1016/j.chemosphere.2013.02.001

Zusi RL (1996) Family rynchopidae (Skimmers). In: del Hoyo J, Elliott A, Sargatal J (eds) Handbook of the birds of the world, vol 3. Lynx Edicions

Chapter 14
Marine Mammals: Is the Bahía Blanca Estuary and Its Area of Influence Important for Their Conservation?

Gisela Giardino, Estela M. Luengos Vidal, Victoria Massola, M. Agustina Mandiola, Joaquín C. M. Gana, Diego Rodríguez, and Ricardo Bastida

14.1 Role of Marine Mammals in Marine Ecosystems

Marine mammals include around 134 extant species that are primarily dwelling or dependent on the ocean for food. This group comprises three mammalian orders: Sirenia, 4 species (manatees and dugongs); Carnivora, 41 species (polar bears, sea otters, and pinnipeds: sea lions and fur seals, true seals, and walruses); and Cetacea, 89 species (baleen whales and toothed whales such as dolphins) (Berta et al. 2015). Marine mammals are not uniformly distributed among the oceanographic areas. Some of them inhabit deep water, marine shelf areas, and areas of open ocean, and few species live in freshwater. They live in several habitats and ecosystems from the Arctic to the Antarctic. Only in the Argentine Sea, 50 species of marine mammals have been cited so far, some of them are very frequent and well known, while others are little known and of circumstantial presence (Bastida and Rodríguez 2009).

Marine mammals are considered as top or apex predators whose food habits are in many cases specialized, feeding on krill, fish, or even birds and mammals.

G. Giardino (✉) · M. A. Mandiola · J. C. M. Gana · D. Rodríguez · R. Bastida
Instituto de Investigaciones Marinas y Costeras (IIMyC), FCEyN, UNMdP – CONICET, Mar del Plata, Argentina

E. M. Luengos Vidal
INBIOSUR (Instituto de Ciencias Biológicas y Biomédicas del Sur), CONICET-UNS, Bahía Blanca, Argentina

Grupo de Ecología Comportamental de Mamíferos, Laboratorio de Fisiología Animal, Departamento de Biología, Bioquímica y Farmacia, Universidad Nacional del Sur, Bahía Blanca, Argentina

V. Massola
Fundación para la Recepción y Asistencia de Animales Marinos (FRAAM). Sarmiento esquina Pichincha Villa del Mar S/N°. Reserva Natural de Usos Múltiples, Bahía Blanca, Bahía Falsa y Bahía Verde, Organismo Provincial para el Desarrollo Sostenible, Bahía Blanca, Argentina

S. M. Fiori, P. D. Pratolongo (eds.), *The Bahía Blanca Estuary*,
https://doi.org/10.1007/978-3-030-66486-2_14

They are foremost consumers at most trophic levels, from primary producers to predatory fish and even to other marine mammals. Because of their large body size and abundance, they are thought to have a great influence on the structure and function of some marine communities. In addition, as top predators, they can serve as key indicator species of marine ecosystem dynamics through changes in their abundance, behavior, and health. Consequently, marine mammals could have five main functions: as consumers, as preys, as vectors of nutrient and material flux, as a particular and new "habitat," and as environmental sentinels (Roman et al. 2014).

14.1.1 Marine Mammals as Consumers

Mammals may function as keystone species in some marine communities. As a result, a serious depletion in their numbers can cause major changes in species status, starting a chain of extinctions through the food web. Croll et al. (2006) estimated that 65% (range 53–86%) of the North Pacific Ocean's primary production was required to sustain the large whale populations prior to commercial whaling. The total metabolic rate of a whale is high, as an endotherm, but because of its immense size, it has a low mass-specific metabolic rate, relative to smaller animals. The amount of food required to sustain 1 blue whale (*Balaenoptera musculus*) could support 7 smaller minke whales (*Balaenoptera acutorostrata*) or 1500 penguins, but their collective biomass reach just 50% or 8% of the biomass of a blue whale, respectively, because of the relatively higher metabolic rates of these smaller animals (Roman et al. 2014).

If primary production is held constant, reducing whale populations lowers the potential for marine ecosystems to retain carbon, both in living biomass and in carcasses that sink to the ocean floor (Pershing et al. 2010). After whale biomass was removed from the Antarctic system (approximately 84%), it is estimated that 150 million tons of krill go uneaten each year. Crabeater seals (*Lobodon carcinophaga*), Antarctic fur seals (*Arctocephalus gazella*), leopard seals (*Hydrurga leptonyx*), and penguins, all krill-eating predators, began to increase, thus moving the Antarctic marine ecosystem to new equilibrium levels. These species directly benefited from the removal of whales and may now be hindering the recovery of whale stocks (Trites et al. 1997).

In addition, pinnipeds can affect their ecosystem, influencing the benthic fauna and community structure. Estes et al. (1998) explain the effect of the abundance of sea otters (*Enhydra lutris*) on the structure of kelp communities, and Anderson (1995) suggested that the overhunting of local sea otter populations could have led to the extinction of Steller's sea cows (*Hydrodamalis gigas*). How are related sea otters and Steller's sea cows? Sea otters feed on sea urchins, keeping their number low, which is essential for the maintenance of the dense kelp beds that, in turn, are the main food source for sea cows.

In British Columbia the annual diet of harbor seals (*Phoca vitulina*) contains about 4% of salmon and 43% of hake. Contrary to expectations, the harbor seals may be benefiting salmon because they affect the abundance of hake, which is one of the largest predators of young salmons (Trites 1997).

14.1.2 Marine Mammals as Prey

Cetaceans and pinnipeds are also eaten by several species. White sharks (*Carcharodon carcharias*) have been observed killing and feeding on small odontocetes and scavenging on carcasses of large cetaceans (Cockcroft et al. 1989). Shark predation attempts on large marine animals are rare, but recent studies suggest that they may occur more often than previously thought (Bornatowski et al. 2012). There are several reports of killer whales (*Orcinus orca*) attacking greater whales. In Península Valdés (Patagonia), the stranding behavior of orcas to capture both, sea lions (*Otaria flavescens*) and elephant seals (*Mirounga leonina*), is known worldwide. Also in Patagonia, killer whales were observed feeding on dusky dolphins (*Lagenorhynchus obscurus*) and chasing common dolphins (*Delphinus delphis*) (Coscarella et al. 2015), while in Brazil and in Península Valdés, they were observed attacking Southern right whale (*Eubalaena australis*) calves (Bastida and Rodríguez 2009; Ott et al. 2017).

In Antarctica, groups of killer whales cause a wave to make the seal pups fall from the floating ice and thus feed on them. In the Buenos Aires Province, killer whales have been observed attacking franciscanas (*Pontoporia blainvillei*), bottlenose dolphins (*Tursiops truncatus*), and fur seals (*Arctocephalus australis*) (Bastida and Rodríguez 2009). Other indirect evidence that whales are victims of predation are the high frequency of rake mark scars found on their flukes (Reeves et al. 2006).

Opportunistic terrestrial predators of pinnipeds (especially pups) include wolves, dogs, foxes, jackals, hyenas, and pumas, in addition to some birds that feed on pinniped pups (Weller 2018). At least five species of pinnipeds have been documented to feed on other pinniped species (Harcourt 1993). The strong tendency of pinnipeds to haul out on ice or islands limits the impact of terrestrial predators on these populations. The main prey of polar bears (*Ursus maritimus*) throughout their range is the ringed seal (*Pusa hispida*). Polar bears hunt ringed seals on both fjord and open sea ice. Other marine predators that can have serious effects on some pinniped populations include adult male sea lions and leopard seals, killer whales, and several species of large sharks. Stomach content analyses indicate that white sharks prefer pinnipeds or whales to other prey such as birds or sea otters. This selective preference for marine mammals with extensive lipid stores may be necessary to meet their elevated muscle temperatures and high growth rates in the cold waters where their attacks on pinnipeds are concentrated (Ainley et al. 1985).

14.1.3 Marine Mammals as Vectors of Nutrient and Material Flux

Whales facilitate nutrient transfer by releasing fecal plumes near the surface after feeding at depth, so nutrients are moved from highly productive to less productive areas. During urine and fecal plume production, nitrogen and iron is released, indirectly increasing prey productivity and abundance. These plumes enhance phytoplankton blooms and carbon sequestration (Lavery et al. 2010). Furthermore, during whale migration nutrients transfer through urea and placentas from areas of high to low productivity. Whale carcasses sequester carbon to the deep sea, where they provide a massive pulse of organic enrichment as well as habitat and food for many endemic species, including chemosynthetic bacteria and invertebrate hosts (Roman et al. 2014). Similarly, the excrement of pinnipeds contributes to the turnover and recycle nutrients, mainly in large colonies (Trites 1997). When foraging, whales and pinnipeds can locally influence the ocean physical environment. Through diving and surfacing, whales can enhance the upward transport of nutrient-rich deep waters, as they pass through density gradients during feeding sessions (Dewar et al. 2006). Humpback whales (*Megaptera novaeangliae*) intentionally disturb sandy bottoms and shell-hash beds (a mix of mud, sand, and broken shells) to flush sand lance, a prey, from their burrows (Hain et al. 1995), contributing mechanical energy to the ocean. In addition, under certain conditions, the bubble nets that humpback whales make to catch schools of small fish, along with bottom disturbance, play an important role in the flux of materials, since these activities also can break the thermocline, facilitating the transport of nutrients from colder to warmer layers in the ocean. Stranded whales can subsidize terrestrial food webs (Chap. 15 in this book).

14.1.4 Marine Mammals as a Particular and New "Habitat"

The ocean floor is rich on detritus due to the contribution made by the carcasses of the great whales that fall from the surface. Dead whales contribute with proteins and lipids, yielding massive pulses of organic enrichment to a realm that is often nutrient and energy impoverished. Only one gray whale (*Eschrichtius robustus*) of 40 tons, provides nearly 2 million g carbon, equivalent to more than 2000 years of the background carbon flux for an area equivalent to the carcass surface (Smith 2006). Whale falls influence deep-sea environments, by altering local food availability, providing habitat structure, and supporting diverse biotic assemblages (Lundsten et al. 2010). Dead bodies of marine mammals undergo ecological succession from a stage dominated by mobile scavengers that actively consume the soft tissues, passing through the enrichment-opportunist stage and getting to a sulphophilic stage where chemosynthetic bacteria dominate (Smith 2006; Lundsten et al. 2010; Amon et al. 2013). The persistent food-rich conditions and widespread occurrence of whale carcasses has led to ecological and evolutionary opportunities on the deep-sea floor, in a manner similar to that of hydrothermal vents and cold seeps (Smith 2006). In the North Pacific, more than 60 macrofaunal species have been associated with one single whale fall. Regarding whale carcasses stranded on coastlines, after death, it is a mechanism for transporting marine biomass to the sea–land interface. Although the number of stranded whales is small as compared with those that sink (Smith 2006), carcasses can attract and nourish many terrestrial consumers.

14.1.5 Marine Mammals as Sentinels

As marine mammals have long lives and move over great distances, they can register ecological variation across large spatial and long temporal scales. Thus, they can be considered sentinels of marine ecosystem change. Selecting the appropriate marine mammal species to use as sentinel of change depends on the ecological alteration of concern (Moore 2008, 2018). Apex predators as whales, seals, and sea lions can indicate environmental changes and degradation. For example, if we want to know broad scale shifts in ecosystems, migratory mysticete whales may be investigated, whereas polar cetaceans are more useful for assessing the effects of rapid changes in sea ice conditions and its impacts on food webs in these strongly seasonal ecosystems. On the other hand, coastal dolphins are good indicators of pollutant or disease vectors in nearshore habitat (Moore 2008).

Harwood (2001) says that critical habitat for marine mammals can be defined in terms of the ecological units that provide safe areas to breed and forage. The layer of blubber for insulation that marine mammals have is suitable for the accumulation of lipophilic pollutants. The information obtained from this layer is useful for indicating the prevalence and persistence of pollutants in marine ecosystems (Gil et al. 2006; Panebianco et al. 2012; Perez-Venegas et al. 2018; Muñoz-Arnanz et al. 2019). In addition, profiles of stable carbon isotopes and fatty acids in blubber can be used to infer the diet of marine mammals, thereby providing evidence of changes in food webs within marine ecosystem (Budge et al. 2008; de Castro et al. 2016).

Changes in individual body condition can demonstrate shifts in the prey base and food web structure as well as alterations in pathogen transmission. Indeed, to explore variability of ecosystem productivity and health, it seems essential to incorporate the biology and ecology of marine mammals and other apex predators in multidiscipline programs of research (Moore 2008). An understanding of the role of marine mammals in marine ecosystems is important because it provides a context within which to evaluate the potential impact of their predation on prey populations and community structure, and the impact of variation in prey populations, due to harvesting by humans, and environmental change on the dynamics of marine mammals. Mammals may function as keystone species in some marine communities such as in the famous example about the effect of the abundance of sea otters (*Enhydra lutris*) on the structure of kelp communities (Estes et al. 1998).

Indeed, marine mammals can guide human stewardship activities, reflect the ocean's role in climate interactions across regions, demonstrate ecosystem vulnerabilities and health, and thereby lead to ways to enhance human health (Moore 2008). Moreover, as charismatic megafauna, marine mammals capture the attention and concern of the public. This capability provides clear opportunities for education and outreach on oceanic and environmental themes.

14.2 Marine Mammal in the Bahía Blanca Estuary

Stranded marine mammals allow researchers to gather valuable information about many elusive species. They offer a unique opportunity to learn about a species' life history, population structure, occurrence, disease prevalence, and anthropogenic causes of mortality and to understand fossil assemblages. It is important to consider that stranded animals are not a random sample of the free-living population as most of them are unhealthy specimens. On the other hand, the stranding range detection is a dependent variable of the human effort and the possibilities of access to coastal sites. Beyond the health status of animals, stranding episodes may also reflect changes in environmental conditions and diverse human activities, shipping traffic collisions, oil exploration, etc. (Leeney et al. 2008). Cetaceans may strand alive for different reasons including behavioral tendencies of particular species to follow a leader, disorientation caused by geographical anomalies in the earth's magnetic field and acoustical "dead zones," coastal areas where echolocation signals are distorted, anthropogenic factors such as bycatch in fisheries or contaminants, and infectious diseases (Peltier et al. 2013). Pinnipeds seem to be particularly susceptible to entanglement in marine debris, perhaps because of their exploratory nature. Entanglement lesions could develop into infected chronic wounds and the entangled pinnipeds may live for months or years with a plastic line, or net, cutting into their skin and muscle tissues (Butterworth 2016). Interaction with fisheries is another source of animal death or injury. Impacts may be direct consequence of bycatch or shooting by fishermen, as well as indirect competition for fish resources and fisheries-induced changes to ecosystems that cause nutritional stress among pinnipeds (Kovacs et al. 2012).

The increase of global concern regarding the status of marine mammals promoted government programs for facilitating the public record of stranding episodes and fostered investigations into the causes of mortality. In some places, there is a long history of systematic records, for example, in the UK after the ASCOBANS (Agreement on the Conservation of Small Cetaceans of the Baltic and North Seas) signature in 1993, a stranding reporting scheme was established along with a research program to investigate death causes. One of the longest-term stranding projects is the one carried out by Leeney et al. (2008) that covers a record period of 96 years on the Cornwall coast (UK) and islands of Sicily (Italy). In Argentina there are some initiatives like Museo Acatushún in Tierra del Fuego, LAMAMA Lab in Puerto Madryn, Marine Mammal Lab at Mar del Plata University, Fundación Mundo Marino in San Clemente del Tuyú, Fundación Aquamarina in Pinamar and FRAAM (Marine Animal Rescue and Assistance Foundation) in Bahía Blanca

Estuary. All these research groups record systematic data on stranding marine mammals.

We made a list of marine mammals that could potentially be present in Bahía Blanca Estuary and surroundings, following Bastida and Rodríguez (2009) and the recently submitted Red List of Threatened Species for Argentina (SADSN-SAREM 2019). The information about some marine mammals is scarce and incomplete, but approximately 37 species have their distribution range within the area, 29 cetaceans and 8 pinnipeds (Table 14.1). According to the IUCN Red List of Threatened Species (IUCN, International Union for the Conservation of Nature) and Red List of Threatened Species for Argentina (SADSN-SAREM 2019), almost 39% of the species are "Data Deficient." Some Data Deficient species may be well studied and its biology be well known, but still lack appropriate data on its abundance and/or distribution. All the pinnipeds in the area are considered Least Concern. Among the cetaceans of possible occurrence in Bahía Blanca Estuary, 17% of the species belong to the threatened categories ("Endangered" or "Vulnerable"); any taxon in those categories faces a high or very high risk of extinction in the wild. Finally, the remaining marine mammal species belong to the "Data Deficient" (38%) or "Least Concern" categories (22%). This last category is implying that the population exist or spreads through its entire range. We also analyzed local information from two sources Fidalgo (2004) and FRAAM marine mammal database. Fidalgo (2004) reported 10 different marine mammal species from sightings, entanglements or stranding episodes, and other 6 species from records than needed confirmation (Table 14.1).

FRAAM marine mammals' database has records that come mostly from stranded animals both alive and dead, or that arrived to the coast in poor health conditions. These records span from 2004 to 2019 in the geographical area from Balneario Marisol to Bahía San Blas. FRAAM recorded 71 marine mammals, most of them pinnipeds (79%), particularly belonging to the Otariidae family (sea lions and fur seals) (75%) (Fig. 14.1). Four species of pinnipeds were registered in the area, but the South American sea lion (*Otaria flavescens*, see species data sheet a) and the South American fur seal (*Arctocephalus australis*, see species data sheet b) were the most frequent (Fig. 14.1). In addition, *O. flavescens* has an historic assessment in the area near Punta Lobos, on the Trinidad Island, in the Bahía Blanca Estuary (see box text in this chapter). Currently the colony is still active, although no studies have been reported recently. The presence of *A. australis* is expected due to the current population growth (Crespo et al. 2015; Mandiola 2015). SADSN-SAREM (2019) has listed both species as "Least Concern" and their main conservation problem is their interaction with anthropic activities (Romero et al. 2011; Mandiola et al. 2017). Most individuals were juvenile and subadult animals in good health condition, that leave the water to rest or because they were molting. Another pinniped recorded in Bahía Blanca Estuary was the Southern elephant seal (*Mirounga leonina*), which was registered three times, always as solitary subadult specimens. The presence of this species has also been increasing in other coastal areas of the Buenos Aires Province and in the harbor colonies of Mar del Plata and Necochea (Bastida and Rodríguez 2009; Group of Biology, Ecology and Conservation of Marine

Table 14.1 List of potential species of marine mammals surrounding the Bahía Blanca Estuary based on the information of Bastida and Rodríguez (2009) and SADSN-SAREM (2019), national conservation status and Fidalgo (2004)

Order	Group	Family	Common Name	Scientific Name	Cat. SAREM	FRAAM data	Fidalgo (2004)
Carnivora	Pinnipeds	Otariidae (sea lions and fur seals)	South American sea lion	*Otaria flavescens*	LC	P	P
			South American fur seal	*Arctocephalus australis*	LC	P	P
			Antarctic fur seal	*Arctocephalus gazella*	LC	P	
			Subantarctic fur seal	*Arctocephalus tropicalis*	LC		
		Phocidae (true seals)	Southern elephant seal	*Mirounga leonina*	LC	P	
			Weddell seal	*Leptonychotes weddellii*	LC		
			Crabeater seal	*Lobodon carcinophaga*	LC		
			Leopard seal	*Hydrurga leptonyx*	LC		
Cetacea	Mysticeti (baleen whales)	Balaenidae	Southern right whale	*Eubalaena australis*	LC	P	P
		Neobalaenidae	Pygmy right whale	*Caperea marginata*	DD		
		Balaenopteridae	Blue whale	*Balaenoptera musculus*	EN		
			Fin whale	*Balaenoptera physalus*	EN		
			Sei whale	*Balaenoptera borealis*	EN	P	
			Bryde's whale	*Balaenoptera edeni*	DD		
			Antarctic minke whale	*Balaenoptera bonaerensis*	DD		P
			Dwarf minke whale	*Balaenoptera acutorostrata*	DD		
			Humpback whale	*Megaptera novaeangliae*	LC	P	
	Odontoceti (toothed whales)	Physeteridae	Sperm whale	*Physeter macrocephalus*	VU		NC
		Kogiidae	Pygmy sperm whale	*Kogia breviceps*	DD		

Order	Group	Family	Common Name	Scientific Name	Cat. SAREM	FRAAM data	Fidalgo (2004)
			Dwarf sperm whale	*Kogia sima*	NA	P	
		Pontoporiidae	La Plata river dolphin	*Pontoporia blainvillei*	VU	P	P
		Delphinidae	Common dolphin	*Delphinus delphis*	LC		P
			Lahille's bottlenose dolphin	*Tursiops truncatus gephyreus*	VU	P	P
			Southern right whale dolphin	*Lissodelphis peronii*	DD		
			Dusky dolphin	*Lagenorhynchus obscurus*	LC		NC
			Risso's dolphin	*Grampus griseus*	LC		
			Killer whale	*Orcinus orca*	LC		P
			False killer whale	*Pseudorca crassidens*	DD		P
			Long-finned pilot whale	*Globicephala melas*	LC		P
		Phocoenidae	Burmeister's porpoise	*Phocoena spinipinnis*	DD		P
			Spectacled porpoise	*Phocoena dioptrica*	LC		NC
		Ziphiidae	Cuvier's beaked whale	*Ziphius cavirostris*	DD		P
			Southern bottlenose whale	*Hyperoodon planifrons*	DD		
			Arnoux's beaked whale	*Berardius arnuxii*	DD		
			Gray's beaked whale	*Mesoplodon grayi*	DD	P	NC
			Héctor's beaked whale	*Mesoplodon hectori*	DD		
			Strap-toothed whale	*Mesoplodon layardii*	DD		P

Ref:* discrepancies between Bastida and Rodríguez (2009) and SADSN-SAREM (2019). Categories of conservations status *DD* data deficient, *LC* Least Concern, *VU* vulnerable, *NA* without application, *P* presence, *NC* need confirmation

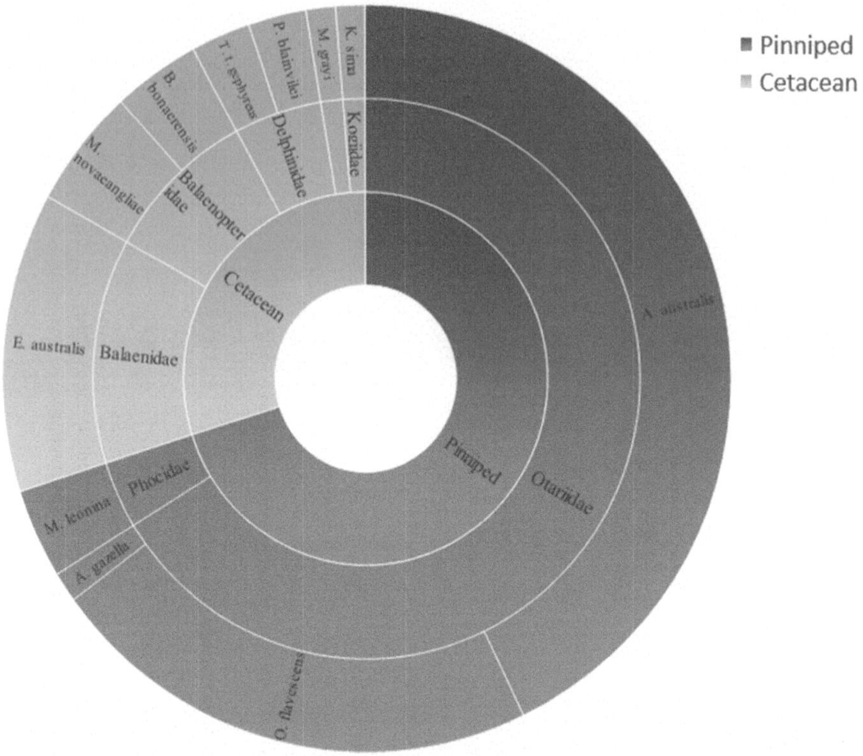

Fig. 14.1 Marine mammals recorded in Bahía Blanca Estuary and surrounding areas by the environmental non-governmental organization Marine Animals Assistance and Rescue Foundation (FRAAM)

Mammals, UNMdP, unpublished data). One young Antarctic fur seal (*Arctocephalus gazella*) has also been recorded as a wandering species with occasional records in Argentine coasts.

The cause of stranding was determined based on a combination of results from clinical examinations or gross necropsy. Animals, that were sick or injured and possibly admitted for rehabilitation, were provided supportive treatment and released afterwards if their condition improved (see Sect. 4 in this chapter). Dead individuals represented 27% of all pinnipeds, and from those alive, 66% were released after a short period of rehabilitation. Similar number of cases was registered in each year, decreasing slightly in autumn.

One of the toothed whales (Odontoceti) more frequently recorded was the small dolphin *Pontoporia blainvillei*, (see species data sheet c) mostly adult and subadult females, followed by males and less frequently newborns. Most of them were found with skin injuries related to entanglement. In the last 10 years, the information on the biology and ecology of this dolphin in Argentina has increased exponentially;

however, updated information about main population parameters, such as mortality, abundance, and population trends, is missing for a more accurate assessment (Denuncio et al. 2019). The annual incidental mortality of this species in Argentina reaches levels between 3.5% and 5.6%, far exceeding the 2% suggested as maximum sustainable level by the International Whaling Commission (Crespo et al. 2010). Nowadays this dolphin is considered as Vulnerable according to the last Red List of Threatened Species for Argentina (Denuncio et al. 2019).

Lahille's bottlenose dolphins (*Tursiops truncatus gephyreus*, see species data sheet d), is another frequent Odontoceti species registered by fishermen or people that use the estuary in a recreational way, but individuals do not show a stranding occurrence higher than other cetaceans. According to the last categorization, *T. t. gephyreus* is considered an endangered species (EN) (Vermeulen et al. 2019). Bottlenose dolphins coming from San Antonio Bay, Río Negro, are genetically isolated from those in Uruguay and southern Brazil and were identified as an evolutionarily significant unit within the Southwestern Atlantic (Fruet et al. 2014). Information on bottlenose dolphins in the country is dispersed in time and space (Vermeulen et al. 2018), but the Argentine population of bottlenose dolphins may have decreased at an estimated rate of 20% in two generations, and there may be less than 250 dolphins alive (Vermeulen et al. 2019).

In addition, killer whales (*Orcinus orca*) are frequently seen alive in the area, mainly during the southern summer. Other toothed whales occasionally found stranded and dead in the Bahía Blanca Estuary belong to the Ziphiidae and Kogiidae families, both of them of pelagic habits.

Baleen whales (Mysticeti) have been also recorded in the area. The first humpback whale (*Megaptera novaeangliae*) was recorded in 2011 (Angeletti et al. 2014), followed by two whales in 2018 and one in 2019. Humpback whales move annually from their main breeding grounds in Abrolhos (Brazil) to their temperate and polar summer feeding areas (Jefferson et al. 2015; Andriolo et al. 2010). Thus, Buenos Aires Province is in the middle of their journey. Estimations performed in the Abrolhos archipelago showed that population densities of humpback whales were increasing at a rate of approximately 7.4% per year and whales were re-occupying old distribution areas (Zerbini et al. 2004; Andriolo et al. 2010). Another baleen whale very common in the area is the Southern right whale (*Eubalaena australis*) observed from April to October. Their records in Buenos Aires Province have increased gradually since 1970 (Mandiola et al. 2020).

The Bahía Blanca Estuary is located between two main reproductive areas (Península Valdés Argentina and Santa Catarina-Brazil) of the Southern right whale, so this site, like the entire Buenos Aires coast, is considered as transit area. Whales recorded in autumn and winter season could be individuals travelling from their southern feeding areas to the breeding areas of southern Brazil (Mandiola et al. 2020). Nevertheless, recent satellite-tracking studies from Golfo San Matías and Península Valdés show variable individual movement patterns. Most tracked whales made coastal and offshore migrations to feeding grounds after the breeding season with no clear displacement pattern (Zerbini et al. 2016, 2018).

Most of the right whale records in the Bahía Blanca Estuary correspond to live whales' sightings. Only one Southern right whale was recorded stranded in the area with clear scars of vessel collision.

Antarctic minke whales (*Balaenoptera bonaerensis*) were also recorded three times; the first specimen was a calf which died after a short time, another was found alive and released (in 2006), and the last one in 2007 was found dead ashore.

Table 14.1 and Fig. 14.1 summarize all marine mammals recorded in the area. Considering the previous information, basic information about the four species most likely to be observed in Bahía Blanca Estuary is presented:

- South American sea lion – *Otaria flavescens* (Shaw, 1800)
- South American fur seal – *Arctocephalus australis* (Zimmerman, 1783)
- La Plata River dolphin – *Pontoporia blainvillei* (Gervais and d' Orbigny, 1844)
- Lahille's bottlenose dolphin – *Tursiops truncatus gephyreus* (Montagu, 1821)

14.2.1 South American Sea Lion **Otaria flavescens** *(Shaw, 1800)*

Common name: Lobo marino de un pelo Sudamericano, León marino (Spanish), South American sea lion (English)

14.2.1.1 Description

Medium to large size, adult males between 2.1–2.8 m in length and weight around 300–350 kg (Fig. 14.2); females between 1.5–2 m and 170 kg (Bastida and Rodríguez 2009). It is clearly different from other sea lions because adult males have an aspect more similar to a lion, and because it is the largest Otariidae species in the region. They exhibit strong sexual dimorphism, adult males have their neck and chest region covered with longer, thicker, coarser guard hairs, which can give the impression of a mane. Females lack mane, and their body is more stylized. The color is highly variable, ranging from reddish brown lighter to yellowish tones, especially in females (Bastida and Rodríguez 2009). At birth, pups weigh 10–15 kg and are 75–85 cm long. Pups are born black above and paler below. They undergo their first molt 1–2 months after birth, becoming dark brown. The pelage has a single hair layer (Jefferson et al. 2015). The ears are small, and their canine teeth are very large and strong.

Fig. 14.2 South American sea lion, *Otaria flavescens*, male assessment in Trinidad Island (Bahía Blanca Estuary). (Photo by Gisela Giardino)

14.2.1.2 Distribution and Habitat

The South American sea lion is the most abundant marine mammal occurring along the southern part of South America and it is distributed along the Atlantic and Pacific coast of South American. On the northern coast of Argentina, there are only four haulouts (about 2500 individuals) formed only by males, while the Patagonian region has both reproductive and non-reproductive colonies (about 120,700 individuals). An additional 7500 animals are found in the Malvinas-Falkland Islands (Dans et al. 2012). Although *Otaria flavescens* is considered as not migratory species and it remains somewhat concentrated on the coastal zone all around the year, males are able to travel hundreds of kilometers for breed (Giardino et al. 2016). Some sea lions habit of coming up the rivers, as in the case of Rio Negro basins (Bastida and Rodríguez 2009).

The coastal islands of Uruguay historically represented a major portion of its population in the Atlantic Ocean, but this site has shown a decline of about 2% per year during the last few decades, with actual numbers of about 10,000 animals (Páez 1996; Ponce de León 2000). South American sea lion haulout areas include sandy beaches, flat surfaced slab of rocks, flat bases of cliff, and places with big boulders (Vaz Ferreira 1981). Some individuals used to rest inside harbors, like in Rio Grande do Sul (Brazil) (Rosas et al. 1994; Pavanato et al. 2013), in Mar del Plata, and Puerto Quequén harbors in the northern coast of Argentina (Buenos Aires Province). Sea lions that settled in port areas may lead to negative human interaction if there are no policies of management (Giardino 2014).

14.2.1.3 Behavior

Terrestrial walking is performed by using all four flippers, while swimming is fundamentally powered by the forelimbs. Their breeding colonies are occupied from middle of December by small number of adult males that take positions and delimit territories through vocalizing, posturing, and fighting, prior to the arrival of females several days later. The Southern sea lion is a highly polygynous species. Only one dominant male can hold 4–10 adult females (harem) although some solitary couples are found disperse (Campagna 1985; Campagna and Le Boeuf 1988). Once the breeding season ends, sea lions change their distribution into haulouts of different age composition, location, and stability throughout the rest of the year (Lewis and Ximenez 1983). Mother–pup pairs maintain their social bonds ashore until weaning in late austral spring and remain close to the natal area (Grandi et al. 2008), whereas males do not provide parental care, have prolonged sexual maturation, and young males are excluded from breeding opportunities; therefore they tend to disperse farther away from the breeding grounds. Male sea lions from Buenos Aires Province travel to Patagonia and Uruguay during the austral summer in order to mate (Giardino et al. 2016) contributing to around 18% of the gene stock of this species in the breeding area (Giardino et al. 2017). At sea, South American sea lions frequently raft alone or in small to large groups. They have been also reported in association with feeding cetaceans and seabirds. On the Atlantic coast, most lactating females have been described as benthic divers and forage in shallow water within the continental shelf (Campagna et al. 2001; Rodríguez et al. 2013). As generalist feeders, South American sea lions take a wide variety of prey that varies by location. Their diet includes many species of benthic and pelagic fishes and invertebrates, some of them of commercial value (Koen Alonso et al. 2000; Bustos et al. 2012). Several authors recommended avoiding the physical contact with this species, as they are, like other wild animals, possible disease vectors (Bernardelli et al. 1996; Beron Vera et al. 2004; Kiers et al. 2008; Bastida et al. 2011; Arbiza et al. 2012; Bos et al. 2014; Timi et al. 2014; Dans et al. 2017).

14.2.1.4 Threats and Conservation Status

South American sea lion numbers are increasing in Argentina, and in Buenos Aires Province, the population grew in recent years (Giardino et al. 2017). The interaction with different anthropic activities is the main conservation problem such as competition with artisanal and industrial fisheries (Romero et al. 2011; Mandiola et al. 2017). Moreover, they interact negatively with aquaculture in open sea, due to the entanglement, use of space, chemical and acoustic contamination, industrial waste, and vessel traffic (Romero et al. 2019).

The Red List IUCN and SADSN-SAREM categorization of *Otaria flavescens* is Least Concern (Cardenas-Alayza et al. 2016; Romero et al. 2019).

14.2.2 South American Fur Seal Arctocephalus australis *(Zimmerman, 1783)*

Common name: Lobo marino de dos pelos Sudamericano, lobo fino austral (Spanish), South American fur seal (English)

14.2.2.1 Description

Medium size fur seal; male lengths between 1.80 and 2 m. and weight between 150 and 200 kg.; females between 1.20 and 1.40 m and weight between 50 and 60 kg. The snout has quite a pointed shape, thin and long ears pinnae, and very long and light-colored vibrissae (Fig. 14.3). Relative long forelimbs with poorly developed nails. Short and dense fur with a double hair layer, dark grayish brown in the back of body and a lighter-colored belly. This species shows sexually dimorphic; adult males are about 1.3 times the length and 3.3 times the weight of females. The canine teeth of adult males are larger and thicker than females. As their Spanish name says, they have two types of hair, one very dense inner layer with soft, fine, and short hair cover by another thicker and longer called guard hair (Bastida and Rodríguez 2009). This species could be confused, in our area, with Antarctic and Subantarctic fur seals. Adult male Antarctic fur seals are the same size and almost same color of

Fig. 14.3 South American fur seal, *Arctocephalus australis*, recorded in Bahía Blanca Estuary. (Photo by Victoria Massola)

Southern fur seal but have a shorter muzzle and proportionately longer fore- and hind flippers. Subantarctic fur seals, on the other hand, are unique with a pale blond face and neck (Jefferson et al. 2015).

14.2.2.2 Distribution and Habitat

South American fur seal has two recognized subspecies, the South American and Peruvian subspecies. The South American subspecies is distributed along all Patagonian Sea, from western South Atlantic (southern Brazil) to eastern South Pacific (southern Chile) coasts. For their distribution at sea, this species inhabit both the coastal zone and the entire Patagonian platform, reaching the edge of the slope (Mandiola et al. 2015; Baylis et al. 2018a, b). On the Atlantic side, haulouts can be found along the coasts of Rio Grande do Sul in Brazil (Muelbert and Oliveira 2006), although the breeding colonies goes from Islas del Castillo, Uruguay, continue to Tierra del Fuego-Isla de los Estados (Túnez et al. 2008; Crespo et al. 2015). They used to travel long distance (Mandiola et al. 2015; Baylis et al. 2018b).

14.2.2.3 Behavior

Males are polygynous and territorial (between 2 and 13 females for each male). Their fights can result in serious wounds and scars. Individual bulls can occupy territories for up to 60 days until most of the females are mated (Cappozzo 1995). After mating, female begins to make foraging trips punctuated by time attending the pup ashore (Pavés and Schlatter 2008). Only a few adult males achieve mating while a large proportion is excluded to peripheral or male exclusive areas. Pupping peaks take place in middle of November to middle of December, and mating occurs 1–6 days after the female gives birth (Franco-Trecu 2005; Pavés and Schlatter 2008). Colonies are generally found along rocky coasts, on edges above the shoreline or in boulder-strewn areas. The Southern fur seal food habits vary according to prey availability; it is an opportunist generalist predator, which feeds mainly on pelagic and demersal preys, both coastal and continental shelf species. They eat mainly on prawn, shrimp, squids, and several fish, as croakers, anchovy, and mackerel in Buenos Aires waters (Bastida and Rodríguez 2009; Pérez Salles 2015). Studies on stable isotopes from Patagonia showed that juvenile fur seal feed more pelagically than subadults and adults (Vales et al. 2015).

Dassis et al. (2012) observed that the most frequent behavior, recorded in Mar del Plata, was passive flotation, followed by prolonged coastal dives. This behavior is strongly influenced by the sea state, since when the sea is more "choppy," fur seals tend to concentrate while when the sea is calm, they tend to disperse. In addition, the wind direction and the surface current affect their behavior. This fact could be a strategy to maintain certain areas with food availability, and for forage energy optimization.

14.2.2.4 Threats and Conservation Status

Among its natural predators are the killer whale, *Orcinus orca* (Bastida et al. 2007), and the cat shark, *Notorynchus cepadianus* (Crespi-Abril et al. 2003). There is a large overlap between the areas used by fur seal and fisheries on the Patagonian platform (Mandiola et al. 2015; Baylis et al. 2018b), although operational interactions are sporadic (Crespo et al. 1997; Mandiola 2015). Actually, individuals have been observed feeding during trawling fishing maneuvers in Buenos Aires waters (Mandiola et al. 2017). The limited number of breeding areas could make this species particularly vulnerable to the effects of epidemics and several human activities that could have bad consequences on the population (Cardenas-Alayza et al. 2016). The intake of marine litter (mainly plastics derived from fishing activity and remains of bags) has been recorded in juvenile fur seal stranded on the Buenos Aires coast, although no lesions were observed in the digestive tract (Denuncio et al. 2017). In addition, the oil activity carried out on the Patagonian platform (transport route of oil ships, exploration and exploitation areas) is always a risk.

Arctocephalus australis is categorized as Least Concern by the IUCN Red List and SADSN-SAREM (Cardenas-Alayza et al. 2016; Vales et al. 2019). Listed in Appendix II of CITES (Convention on International Trade in Endangered Species of Wild Fauna and Flora).

14.2.3 La Plata River Dolphin or Franciscana **Pontoporia blainvillei** *(Gervais and d' Orbigny, 1844)*

Common name: Franciscana, Delfín del Plata (Spanish) Franciscana, La Plata River dolphin (English)

14.2.3.1 Description

Small-sized dolphin (128–175 cm) that weigh between 35 and 55 kg. Quickly distinguishes itself from the rest of the dolphins due to their extremely long and narrow beak (about 12–15% of total length in adults). The long beak is lined with 50–62 fine, pointed teeth per row, more than in nearly any other species of cetacean. The forehead is steep and rounded with flexible neck. The flippers are broad and spatulate, sometimes with an undulating trailing edge; meanwhile dorsal fin is relatively rounded and small. Newborns have proportionately larger flippers, dorsal fins, and flukes. The body color is uniform yellowish brown, in some cases lighter on the belly, which serves to camouflage itself in the murky water where it lives (Bastida and Rodríguez 2009; Jefferson et al. 2015). The species has a slightly inverted sexual dimorphism, with females being larger than males (Kasuya and Brownell 1979; Botta et al. 2010; Panebianco et al. 2012) (Fig. 14.4).

Fig. 14.4 La Plata River Dolphin (Franciscana dolphin) *Pontoporia blainvillei*. (Photo by Ricardo Bastida)

14.2.3.2 Distribution and Habitat

The Franciscana Dolphin is found only along the east coast of South America (Brazil, Uruguay, and Argentina), from Golfo San Matías, central Argentina (42°35′S), to Espirito Santo, southeastern Brazil (18°25′S) (Bastida and Rodríguez 2009; Jefferson et al. 2015). They live mainly in coastal marine waters with a maximum distribution to the isobath of 50 m, and it is possible to see them swimming behind the surfing zone (Bastida et al. 2000; Crespo et al. 2010). On the Argentine coast, abundance was estimated by aerial censuses in the marine area in almost 15,000 animals with a greater density in the northern coastal area of the Province of Buenos Aires, from Cabo San Antonio to Claromecó (Crespo et al. 2010). In this region, the highest concentration of dolphins was found between the coastline and the 30 m deep isobath. Secchi et al. (2003) proposed to divide the distribution area into four population management units (Franciscana Management Areas, FMAs). The FMA IV represents the coastal waters of Argentina, including the Buenos Aires Province, Río Negro, and Chubut. However, the arguments presented for this subdivision are not rigid and are subject to change, as more and better information will be recorded (Secchi et al. 2003). Subsequent studies, based on new genetic analyses, have suggested at least seven management units should be considered (Mendez et al. 2008; Cunha et al. 2014; Gariboldi et al. 2016). Studies performed in northern Buenos Aires Province revealed that in the Samborombón Bay, there would be a genetically isolated population and suggest that at least there are two populations only in Buenos Aires coast (Mendez et al. 2008, 2010). Recent studies suggest that there could be between three and five subpopulations in Argentina (Cunha et al. 2014; Gariboldi et al. 2016).

14.2.3.3 Behavior

Franciscanas feed mainly on bottom dwelling fish of the family Sciaenidae (Rodríguez et al. 2002; Denuncio et al. 2017), but crustacean and mollusks are also important. They feed mostly near the bottom and appear to be opportunistic, with at least 58 fish species, six cephalopod species, and six crustacean species known from the diet. They show cooperative feeding and foraging, and on average, they dedicated three quarters of their time searching for their preys. This dolphin lives in small groups up to 15 individuals with an average of two to five (Bastida and Rodríguez 2009). There is no evidence of large seasonal movements and little is known about daily displacement (Bordino et al. 1999; Bordino 2002). Bordino et al. (2008), using satellite telemetry, showed that this species has a relatively small home range (150 km^2–345 km^2) which barely exceeds 20 linear km to its maximum extent. In spite of its difficulty to be seen in open sea due to their color, behaviors, and small size, it is possible to see groups during the summer along the entire coast of Buenos Aires Province.

Franciscana's breeding season begin in late spring and summer, and newborn calves are recorded between the end of October and early April in Bahía Samborombón and Bahía Anegada with a gestation period between 10 and 11 months (Bordino et al. 1999). Killer whale (*Orcinus orca*) are known natural predators and some sharks have attacked franciscanas trapped in fishing nets. They are seldom observed close to motor boats, suggesting that they are scared and avoid them (Bastida and Rodríguez 2009). They produce high-frequency narrowband echolocation clicks with a maximum frequency recorded 139 kHz and a bandwidth of 19 kHz (Melcón et al. 2012) with lower frequencies in calves (Melcón et al. 2016).

14.2.3.4 Threats and Conservation Status

The main problem this species is facing is incidental mortality in gillnet fisheries. Every year between 2000 and 3000 dolphins die in coastal fisheries in Brazil, Uruguay, and Argentina (Bastida and Rodríguez 2009). Only in Buenos Aires coast, about 500–650 dolphins are entangled every year (Denuncio et al. 2019). Other threats include various forms of habitat degradation and pollution. Heavy metals and several organic chlorine compounds have been detected in individuals of different regions (Gerpe et al. 2002; Panebianco et al. 2011, 2012; Romero et al. 2018). Moreover, more than 30% of the franciscanas which were studied in the Rio de la Plata estuary and Cabo San Antonio contained plastic debris in their stomach (Denuncio et al. 2011, 2016).

The Red List of IUCN and SADSN-SAREM categorized this species as "Vulnerable" (Zerbini et al. 2017; Denuncio et al. 2019). It is the most endangered cetacean species in the Southwest Atlantic Ocean, listed in Appendix II of CITES.

14.2.4 *Lahille's Bottlenose Dolphin* Tursiops truncatus gephyreus *(Montagu, 1821)*

Common name: Delfin nariz de botella de Lahille, delfin Mular de Lahille (Spanish), Lahille's common bottlenose dolphin (English)

14.2.4.1 Description

Robust and large-sized dolphin (max, length 4 m). The adult weight of this South Western Atlantic subspecies ranges between 200 and 350 kg, but there are maximum records of 600 kg. At birth calves measure between 85 and 140 cm and weight between 14 and 30 kg. These general body values are much higher than those of the other subspecies, *Tursiops truncatus truncatus,* of the Caribbean region, where they behave more dynamic. Body color is uniform grey with lighter belly. Large, bulbous, and well-differentiated melon. Short and wide beak. Lower jaw slightly exceeding the upper one. Between 20 and 26 pairs of well-developed teeth in the upper jaw and between 18 and 24 pairs in the lower jaw. Developed subtriangular dorsal fin in *gephyreus* subspecies, and falcate dorsal fin in *truncatus* subspecies; in both placed at midback, well-developed typical pectoral fins with a convex anterior edge and concave posterior edge (Bastida and Rodríguez 2009) (Fig. 14.5).

14.2.4.2 Distribution and Habitat

Lahille's common bottlenose dolphin is a subspecies distributed along the coastal waters of Patagonia and Buenos Aires Province; it also inhabits coastal areas of Uruguay and southern Brazil. In Uruguay and Brazil, the subspecies *truncatus* is generally found in offshore waters of the continental shelf. Common bottlenose dolphins tend to be primarily coastal, but they can also be found in pelagic waters (Wells and Scott 1999). Individuals that primarily use inshore waters frequents estuaries, bays, lagoons, and other shallow coastal regions and occasionally can swim far up into rivers. Individuals of this ecotype tend to maintain definable, long-term multi-generational home ranges, but in some locations near the extremes of the species range, they show migratory behaviors. On the other hand, the offshore ecotype is apparently less restricted in range and movement. Some offshore dolphins are residents around oceanic islands.

14.2.4.3 Behavior

Bottlenose dolphins are commonly associated with many other cetaceans, including both large whales and other dolphin species (Wells and Scott 1999). This dolphin species consumes a wide variety of prey, mostly fish and squid (Barros and Odell

Fig. 14.5 Lahille's bottlenose dolphin, *Tursiops truncatus gephyreus*, in Bahía Blanca Estuary. (Photo by Agustina Mandiola)

1990; Barros and Wells 1998; Blanco et al. 2001; Santos et al. 2001), and sometimes squids, shrimps, and other crustacean. In Buenos Aires Province, Lahille's bottlenose dolphins' diet is mainly based in demersal fish as white mouth croaker, striped weakfish, and *Mugil* spp., among other bony fish species (Bastida and Lichtschein 1986; Mermoz 1977). Lahille's dolphins use a high variety of prey capture strategies and techniques, one of these, observed in small streams of Bahía Samborombón, is to pull mullet fish schools out of water and eat them with the dolphin body partially stranded (Bastida and Rodríguez 2009). Probably such behaviors also take place in the shallow areas of Bahía Blanca estuary. This coastal dolphin can eventually be caught by killer whales; in Argentina, cases of attack have already been reported in Bahía Samborombón and in Villa Gesell (Buenos Aires Province) (Bastida and Rodríguez 2009). Probably these attacks may also take place in the outer zone of the Bahía Blanca estuary. Bottlenose dolphins produce pulsed signals (echolocation) as well as tonal whistles. These sounds comprise a wide range of frequencies (40–130 kHz). Moreover, individual sounds known as signature whistles used in individual recognition and in maintaining group cohesion have been described. Maritime traffic seems to affect the acoustic behavior of the species (Buckstaff 2004). This vulnerable dolphin species lives in Argentina in coastal waters, from Buenos Aires Province to Northern Chubut Province. Although few records have been made as far south as the provinces of Santa Cruz and Tierra del Fuego (Goodall et al. 2011). Most behavioral studies conducted on bottlenose dolphins in Argentina (e.g., Würsig and Würsig 1979) were discontinued in the 1980s because of noted population decline and the subsequent lack of sightings (Bastida and Rodríguez 2009). One of the last remaining resident populations of the country is suggested to reside in Bahía San Antonio (Río Negro Province) (Vermeulen and Cammareri 2009). Research conducted in this area described this population as small, essentially closed, declining (Vermeulen and Bräger 2015), and highly resident to the study area, indicating this bay as the core region within the larger home range of this population (Vermeulen and Cammareri 2009).

14.2.4.4 Threats and Conservation Status

The abundance of this species is dramatically decreasing in Argentina, based on average numbers of sightings since the 1970s, even in areas with increasing or constant survey effort (Bastida and Rodríguez 2009; Vermeulen et al. 2017). For the Patagonia Sea, it is estimated that there are less than 1000 mature individuals, across at least two genetically distinct subpopulations, showing genetic evidence of population fragmentation (Fruet et al. 2014). Causes of declines in the world – and probably also in Argentina – are often related to either habitat degradation, prey depletion, or contamination (Vermeulen and Bräger 2015).

This species is categorized as "Least Concern" by the Red List IUCN (Wells et al. 2019) and "Vulnerable" by SADSN-SAREM (Vermeulen et al. 2019), listed in Appendix II of CITES. However, in a preliminary way, the Lahille's bottlenose dolphin can be considered the most threatened cetacean currently in Argentina and probably from Uruguay.

14.2.5 Importance of the Estuary for Marine Mammals

Estuaries have plenty of food and offer coastal protection and habitat for a wide variety of species, including seabirds, fish, and mammals. Moreover, estuaries are nursery areas for many fishes (Costanza et al. 1997; Martinho et al. 2007) which are common prey of marine mammals. In the Bahía Blanca Estuary, 30 fish species have been reported, where striped weakfishes (*Cynoscion guatucupa*), whitemouth croakers (*Micropogonias furnieri*), and narrownose smooth-hounds (*Mustelus schmitti*) are the most important fishing resources (Lopez Cazorla et al. 2004).

Franciscana dolphin, Lahille's bottlenose dolphin, South American sea lions, and South American fur seal, the four marine mammals most frequent in Bahía Blanca Estuary, mainly feed on whitemouth croakers and striped weakfishes of different sizes. As Failla et al. (2004) mentioned, the distribution and the accessibility of food may have been a determining factor for the establishment of franciscana dolphins. Paso-Viola et al. (2014) found that franciscana dolphins from the Bahía Blanca Estuary eat striped weakfishes 2.8–28.1 cm long (smaller than commercial size) and whitemouth croakers 4.1–7.8 cm (far smaller than commercial size). These mammals also eat horse mackerels (*Trachurus lathami*), Argentine anchovies (*Engraulis anchoita*), lantern midshipmans (*Porichthys porosissimus*), cusk-eels (*Raneya brasiliensis*), and Brazilian flatheads (*Percophis brasiliensis*). Besides fishes, they prey on invertebrates as the longfin inshore squids (*Loligo sanpaulensis*), Tehuelche octopuses (*Octopus tehuelchus*), marine shrimps (*Artemesia longinaris*), and Argentine red shrimps (*Pleoticus muelleri*).

Lahille's bottlenose dolphins from the Buenos Aires Province feed on marine and estuarine fishes such as the whitemouth croaker, striped weakfish, king weakfish (*Macrodon ancylodon*), drum (*Paralonchurus brasiliensis)*, Argentine menhaden (*Brevoortia pectinate*), Brazilian codling (*Urophycis brasiliensis*), and longfin inshore squid (*Loligo sanpaulensis*) (Bastida and Rodríguez 2009). Moreover, Vermeulen et al. (2015) had observed bottlenose dolphins feeding on silversides (Odontesthes sp.). All these species detected as prey for bottlenose dolphins are frequent fishes of the Bahía Blanca Estuary (see Chap. 11 in this book).

On the other hand, the otariids recorded in Bahía Blanca Estuary are considered generalist feeders. South American sea lions from Puerto Quequén mostly eat cusk-eels, striped weakfishes, horse mackerels, Argentine croakers (*Umbrina canosai*), flounders (*Paralichthys* sp.), and skates (Giardino 2014). South American fur seals from Buenos Aires Province eat whitemouth croakers, Argentine anchovies, king weakfishes, striped weakfishes, and squids (Pérez Salles 2015). Except for anchovies, otariids eat fishes smaller than those valuable for commercial interest, indicating that sea lions and fur seals feed mainly on juveniles (Giardino 2014; Pérez Salles 2015).

In addition, the Bahía Blanca Estuary offers a sheltered environment. It is known that relatively small communities of bottlenose dolphins, living in protected coastal environments with predictable availability of resources, often display a high degree of residency and long-term site fidelity, as Vermeulen et al. (2017) confirmed for Argentine groups. Franciscanas have low mobility, with prolonged associations (Wells et al. 2013) in small groups (Crespo et al. 1998; Failla et al. 2004). This dolphin needs turbid waters, depths ranging between 5 and 35 m, favorable conditions for feeding, and protection against natural predators (Bordino et al. 1999; Failla et al. 2004), thus the estuary is an ideal environment. Molecular analyses, carried out in franciscanas from Buenos Aires Province, allowed to recognize a reduced mobility of the species and a possibly high level of population isolation (Mendez et al. 2008, 2010). Sea lions, on the other hand, need a wintering area where they rest and feed, away from females and their puppies, thus avoiding competition for food. Thus, the Bahía Blanca Estuary provides a suitable wintering habitat for them.

There are also negative aspects that have to be considered, as the estuary accumulates large amounts of pollutants, becoming a potential threat to resident species. The Bahía Blanca Estuary has experienced a marked human population increase as well as industrial development during the past decades (Marcovecchio et al. 2008). This coastal area also supports an intensive anthropogenic activity, including five large harbors and one of the biggest industrial parks in South America comprising refineries, oil terminals, tanks for storing oil products, and multiple docks (Limbozzi and Leitao 2008; Oliva et al. 2017). Pollution, environmental contaminants, marine noise, plastic debris, offshore oil and gas activities, shipping, and commercial fisheries affect marine mammals. Individuals can die or be negatively impacted by entanglement or the ingestion of plastic litter (Reijnders et al. 2018). At local level plastic debris were detected in franciscanas (Denuncio et al. 2011) and in fur seals (Denuncio et al. 2017). In franciscanas high levels of metals, such as mercury, zinc, and cooper, were found (Gerpe et al. 2002; Panebianco et al. 2012). In addition, bottlenose dolphins, South American sea lions, and fur seals from Buenos Aires Province have mercury, zinc, cadmium, and copper in their tissues (Marcovecchio et al. 1990; Moreno et al. 1984; Marcovecchio et al. 1994). High concentration of metals and organic pollutants in the tissues of marine mammals has been associated with organ anomalies, impaired reproduction, and immune function and, as a consequence of the latter, with the occurrence of large die-offs among seals and cetacean species (Reijnders et al. 2018). However, a clear cause and effect relationship between residual levels of contaminants and observed effects has been demonstrated in only a few studies.

No less important is noise pollution. Underwater noise can interfere with key life functions of marine mammals (e.g., foraging, mating, nursing, resting, migrating) by impairing hearing sensitivity, masking acoustic signals, eliciting behavioral responses, or causing physiological stress (Erbe et al. 2018). Although in Argentina no studies regarding noise pollution in local species have been developed, worldwide disturbances of shipping traffic, military tests, and oil extraction, among others were found (Rolland et al. 2012; Tyack and Janik 2013; Fouda et al. 2018; Simonis et al. 2020). As the Bahía Blanca Estuary has a significant traffic vessel, this type of pollution also should be considered.

14.2.6 Local Activities for the Conservation of Marine Mammals

Stranding records provide valuable information about spatial distribution, seasonal movements, and factors related to mortality (Moore et al. 2018). Observations of stranded marine fauna on the beach are not new, but are more frequent nowadays. Many causes, natural or anthropogenic, or both, could lead to the stranding of live, dying, or dead animals. Natural causes include failure to thrive, predation, disease, parasites, injuries, and exposure to biotoxins. Other natural threats, such as prey distribution changes, can be driven by environmental fluctuations or overfishing (Moore et al. 2018). Directly and indirectly, humans are seriously affecting marine life. The main anthropic disturbances, among others, that cause pain, suffering, and often death include incidental entanglement, both in coastal and commercial fisheries; collision with boats, drift nets, and other fishing gear; oil spills; solid waste floating in the sea; and ocean noise pollution (Moore et al. 2018). In the last three decades, these events started to be considered by national and provincial government agencies, and different public policies were adopted. At the same time, private rehabilitation centers have emerged, for rescue, assistance, and research. These institutions have been coordinating efforts even inter-jurisdictionally, making progress in management and conservation along the extensive Atlantic coast of Argentina.

Government agencies are responsible for marine fauna. At the national level, the Ministry of Environment and Sustainable Development (SAyDS) has the competence to promote regulations and management actions for the conservation of aquatic resources. In Buenos Aires Province, the Provincial Direction of Natural Resources – Provincial Agency for Sustainable Development (OPDS) has a network that oversees the actions of collection centers, share scientific information with researchers, and promote a better knowledge of ecology, biology, and veterinary aspects necessary for the conservation of the species.

Locally, in the 1980s, Bahía Blanca Estuary was declared a Protected Provincial Natural Area; therefore, protection and conservation measures began to be taken through the legal figure as Bahía Blanca, Bahía Falsa, and Bahía Verde Multiple Use Nature Reserve. Following work began to be done at ecosystem level. A few years later, the environmental non-governmental organization Marine Animals Assistance and Rescue Foundation (FRAAM) was created. Biologists and park rangers started to work on the environmental awareness with coastal populations adjacent to the Natural Reserve area, the education continues nowadays. Through environmental education and interpretation strategies, FRAAM offers training to different social actors involved in the occasional discovery and assistance of marine fauna, such as the Argentine Naval Prefecture, park rangers, artisanal fishermen, recreational fishermen, lifeguards, and rescue center volunteers. Primary care assistance includes direct measures focused on animal health and welfare: systematic data collection, biometrics, epibionts, necropsies, parasites, and pathological, genetics, and contaminants studies.

In the last decade, work on data collection has been strengthened along the Atlantic coast of Buenos Aires Province, coordinated efforts with national, provincial, multidisciplinary academic groups of national universities, non-governmental rehabilitation centers, as well as artisanal and recreational fishermen. Altogether, they use the same protocols for each taxonomic group. All the information recorded is shared among all the organisms to the National Action Programs for the Conservation of each one of these species (e.g., PAN-Marine Mammal Conservation).

Accidental observers, in coincidence with the summer touristic season, reported most of the records. In this sense, FRAAM has an Environmental Education Program to teach school children and teachers of all levels. In addition, dissemination and information campaigns are carried out through the media, reinforced with billboards and brochures. Nowadays, social media contribute and allow direct communication with the informant. However, there is still much to do and direct intervention by the public without knowledge is discouraged. For example, it is not recommended to feed the animals, force them to return to sea, water, or try to catch them. Instead, keeping distance and giving immediate notice to the nearest beach authority helps. In addition, if possible people that discover a stranded animal should take photographs, keep dogs away, and await the arrival of the authority without putting themselves at any risk.

Not always animals on the beach are sick, injured, or in trouble. In the case of pinniped (seals and sea lions), for example, as they alternate periods of their life at sea and periods in land, they often go out to rest in places far from their settlements, simply to recover energy. Opposite, cetacean live all their life in the sea, so their stranding is not a normal condition. Cetaceans can single or mass strand and it is difficult to know the real reason behind it. However, through examinations, necropsy, and sampling of deceased animals, it is possible to understand direct and indirect threats to marine mammal populations (Moore et al. 2018). Each stranding requires logistics and participatory collaboration of different institutions and professionals. The faster an animal can be examined, the more accurate the diagnosis can be.

In the Bahía Blanca Estuary, when an animal is found alive, a first diagnosis is made on its body condition, following the corresponding sanitary protocol. Next, all possible samples and information are collected (biometrics, sex, epibionts, stage of development, geographic location, pictures, etc.). Meanwhile, authorities delimit an area to prevent disturbance from people and dogs. If the beach is full of people, authorities move the animal to a more remote place. If it is necessary due to the health condition and if the animal size allows it, it is transferred to the rehabilitation center for veterinary assistance. If the finding is a *post-mortem* animal, recommended proceeding is the same, followed by a necropsy and carcass recovery. Species found along the beach may be residents or transients. Transient species pass through out provincial limits through their lives, or even cross-national and international boundaries. For that reason, it is necessary to have coordinated work between governmental and non-governmental organizations with economic and human resources optimization.

Understanding how ecosystems function is essential; therefore, it is necessary to work in coordination, both from official organizations such as universities and non-governmental organizations, like FRAAM, for the species conservation and habitat preservation. The best way to do this is sharing methodologies, protocols, and scientific information.

Box 14.1 Bahía Blanca Estuary: A Point of Connection Between the South American Breeding Colonies of Sea Lions

South American sea lions (*Otaria flavescens*) live along the coast of South America, from Torres, southern Brazil (29° 20′ S; 49° 43′ W) in the Atlantic Ocean, to Punta Brava, Ecuador (02° 12′ S; 81° 00′ W) in the Pacific Ocean (Vaz Ferreira 1982; Bastida et al. 2007). Along the Argentine coast, there are 60 settlements, with an estimated total population of more than 200,000 individuals (Romero et al. 2019); number that is still far from historical populations (Bastida et al. 2007). The Buenos Aires coast has four rookeries, located at almost equal distances (ca. 1000 km) from two of the most important focal breeding areas of this species in the southwest Atlantic (Uruguay and Northern Patagonia). These four colonies are non-breeding rookeries, two of them located in Mar del Plata (38° 02′ S, 57° 31′ W) and in Quequén (38° 35′ S; 58° 42′ W) harbors (Fig. 1.2; Chap. 1) and the other two in Isla Trinidad (39° 13′ 59″ S; 61° 51′ 14″ W) and Banco Culebra (40° 24′ 30″ S; 61° 58′ 30″ W) (Fig. 2.1; Chap. 2).

Isla Trinidad is within the limits of the Bahía Blanca, Bahía Falsa, and Bahía Verde Multiple Use Natural Reserve. The South American sea lions and South American fur seals are commonly observed within the reserve (Petracci et al. 2010; Giardino 2014; Martín Sotelo and Victoria Massola pers. obs.). However, the area of Isla Trinidad has three large bays and an intricate network of channels and streams with extensive sandy beaches in the outermost islands. Because of the difficult access, surveys must be performed by boat or plane, and details on the fauna inhabiting Isla Trinidad is largely unknown.

The status of the haulout in Isla Trinidad is almost unknown. Its existence has been a subject of controversy, because of confusing quotes and erroneous locations in the literature published over the past few years. By 1935 one historical record mentions the killing of up to 100 sea lions per day in Isla Trinidad, by the Salvador Di Meglio and Minujin Society (Fidalgo 2004). Dr. Raúl Arámburu, during a census conducted on December 1973, quantified 900 sea lions in the area (Vaz Ferreira 1982). One of the most recent references on the status of the population remarks that this rookery shelters on average 76 South American sea lions, out of the breeding season with a maximum of 150 individuals (Petracci et al. 2010; Giardino 2014). Like Mar del Plata and Quequén rookeries, Isla Trinidad, have, predominantly, juvenile (3–5 years) and subadult (5–7 years) males (Petracci et al. 2010; Giardino 2014). Sea lions from Isla Trinidad are also connected with the other wintering grounds in the Buenos Aires coast. In April 2009, one of the sea lions bleached (Giardino et al. 2013) in Puerto Quequén, between July and December 2008, was resighted in Isla Trinidad confirming that connection (Giardino 2014).

Why estuaries are so important for sea lions? Estuaries play an important role in coastal regions. Their plant communities provide protection against the erosion of water and wind. In addition, bottom communities allow for sediment oxygenation and many species of fish and invertebrates carry out part of their life cycles in these environments. Several of these fish species are of commercial interest and are prey for apex predators such as many marine mammals and seabirds. Among the services provided, estuaries commonly act as nursery areas for fish (Bortolus 2008). Adults and juveniles of the most frequent preys of *Otaria flavescens* as the Brazilian menhaden (*Brevoortia aurea*), the whitemouth croaker (*Micropogonias furneri*), and the striped weakfish (*Cynoscion guatucupa*) (Lopez Cazorla et al. 2014; Giardino 2014) inhabit the Bahía Blanca Estuary.

Female sea lions remain close to the breeding site (Rodríguez et al. 2013; Grandi et al. 2008, 2018) nursing their calves, but males do not have any parental investment. Males tend to disperse away from their mating sites toward unisexual haulouts. Thus, male sea lions remain in remote places (as the Bahía Blanca Estuary) during most of the year in order to avoid competing with females on feeding resources (Giardino et al. 2016).

Like in the other male haulouts from Buenos Aires, sea lions from Isla Trinidad would be a functional part of the northern Patagonia and Uruguay breeding aggregations. Every year, during the austral summer, sexually and socially mature sea lions from Buenos Aires travel to the breeding grounds, returning once the breeding season is over (Giardino et al. 2016). With these movements, males maintain genetic flow between the two different genetic stocks of females (Szapkievich et al. 1999; Túnez et al. 2007, 2010; Feijoo et al. 2011; Giardino et al. 2016; de Oliveira et al. 2017). This connection is particularly important for the long-term persistence of this species (Frankham et al. 2002; Crooks and Sanjayan 2006). Furthermore, these wintering grounds are also relevant as passage, rest, feeding, training, and maturation areas.

Regarding the South American fur seal (*Arctocephalus australis*), this species is in a process of redistribution and recolonization, and it is observed with increasing frequency in the Bahía Blanca Estuary, settling on buoys close to Ingeniero White Port (Mandiola 2015). Even though there are no current studies on this fur seal in Bahía Blanca, these otariids may also be connected with Uruguayan colonies, as proven by one animal bleached in Mar del Plata and sighted afterward in Punta de Lobos, Uruguay.

Connectivity and dispersal are key factors for the long-term persistence of species, particularly in those with reduced populations or fragmented habitats (Frankham et al. 2002; Crooks and Sanjayan 2006). Based on the available information, Isla Trinidad, and in consequence the Bahía Blanca Estuary, would represent a key point of connection and should be seriously considered for regional conservation strategies.

Acknowledgments All this work would be impossible without the help of artisanal fishermen from Villa del Mar, Puerto Rosales, Ingeniero White, Mariano Natali, Prefectura Naval Argentina (from Monte Hermoso, Coronel Rosales, Bahía Blanca), park rangers as Martin Sotelo, Ezequiel Matías Chiatti Heinz, Ariel Tombo, Daniel Jofre and Ariadna Mamani, Dra. Suldrup, Club Nautico, Oscar Liberman, Ing. Adrían Daño, Verónica Lombad, Sr. Mansi, and all the local people and tourists who help to recover this valuable information.

References

Ainley DG, Henderson RP, Huber HR et al (1985) Dynamics of white shark/pinniped interactions in the Gulf of the Farallones. South Calif Acad Sci Mem 9:109–122

Amon DJ, Glover AG, Wiklund H et al (2013) The discovery of a natural whale fall in the Antarctic deep sea. Deep Sea Res Part II 92:87–96. https://doi.org/10.1016/j.dsr2.2013.01.028

Anderson PK (1995) Competition, predation, and the evolution and extinction of Steller's sea cow. Mar Mamm Sci 11:391–394. https://doi.org/10.1111/j.1748-7692.1995.tb00294.x

Andriolo A, Kinas PG, Engel MH et al (2010) Humpback whales within the Brazilian breeding ground: distribution and population size estimate. Endanger Species Res 11:233–243

Angeletti S, Cervellini PM, Massola V (2014) Nuevo registro de ballena jorobada (*Megaptera novaeangliae*) para el Mar Argentino y notas sobre sus epibiontes. Mastozool Neotrop 21(2):319–324

Arbiza J, Blanc A, Castro-Ramos M et al (2012) Uruguayan pinnipeds (*Arctocephalus australis* and *Otaria flavescens*): evidence of influenza virus and *Mycobacterium pinnipedii* infections. In: Romero A, Keith EO (eds) New approaches to the study of marine mammals. Tech Open, Rijeka, p 248

Barros NB, Odell DK (1990) Food habits of bottlenose dolphins in the southeastern United States. In: Leatherwood S, Reeves RR (eds) The bottlenose dolphin. Academic, New York, pp 309–328

Barros NB, Wells RS (1998) Prey and feeding patterns of resident bottlenose dolphins (*Tursiops truncatus*) in Sarasota Bay, Florida. J Mammal 79(3):1045–1059. https://doi.org/10.2307/1383114

Bastida R, Lichtschein V (1986) Capturas incidentales de pequeños cetáceos en el área de Mar del Plata (Provincia de Buenos Aires-Argentina). In: Actas, I Reunión de Trabajo de Expertos en Mamíferos Acuáticos de América del Sud, 24–29 June, Buenos Aires, pp 14–22

Bastida R, Rodríguez DH (2009) Mamíferos marinos de Patagonia y Antártida. Vazquez Mazzini, Buenos Aires

Bastida R, Rodriguez D, Moron S (2000) Avistajes costeros y tamaño grupal de Pontoporia blainvillei en el sudeste de la Provincia de Buenos Aires (Argentina). In: Crespo EA (ed) Report of the 3rd workshop for coordinated research and conservation of the Franciscana dolphin (*Pontoporia blainvillei*) in the Southwestern Atlantic. Bonn, UNEP/CMS Secretariat, pp 51–54

Bastida R, Rodríguez D, Secchi E et al (2007) Mamíferos Acuáticos de Sudamérica y Antártida. Vazquez Mazzini Editores, Buenos Aires

Bastida R, Quse V, Guichón R (2011) La tuberculosis en grupos de cazadores recolectores de Patagonia y Tierra del Fuego: nuevas alternativas de contagio a través de la fauna silvestre. R Arg Antrop Biol 13:83–95

Baylis AM, Tierney M, Orben RA et al (2018a) Geographic variation in the foraging behaviour of South American fur seals. Mar Ecol Prog Ser 596:233–245. https://doi.org/10.3354/meps12557

Baylis AM, Tierney M, Staniland IJ et al (2018b) Habitat use of adult male South American fur seals and a preliminary assessment of spatial overlap with trawl fisheries in the South Atlantic. Mamm Biol 93(1):76–81. https://doi.org/10.1016/j.mambio.2018.07.007

Bernardelli A, Bastida R, Loureiro J et al (1996) Tuberculosis in sea lions and fur seals from the south-western Atlantic coast. Rev Sci Tech 15:985–1005

Beron Vera B, Crespo EA, Raga A et al (2004) *Uncinaria hamiltoni* (Nematoda: Ancylostomatidae) in South American sea lions, *Otaria flavescens*, from northern Patagonia, Argentina. J Parasitol 90:860–863. https://doi.org/10.1645/GE-182R

Berta A, Sumich JL, Kovacs KM (2015) Chapter 1: Introduction. In: Berta A, Sumich JL, Kovacs KM (eds) Marine mammals, 3rd edn. Academic, San Diego, pp 1–14

Blanco C, Salomon O, Raga JA (2001) Diet of the bottlenose dolphin (*Tursiops truncatus*) in the western Mediterranean Sea. J Mar Biol Assoc UK 81:1053–1058. https://doi.org/10.1017/S0025315401005057

Bordino P (2002) Movement patterns of franciscana dolphins (*Pontoporia blainvillei*) in Bahia Anegada, Buenos Aires, Argentina. LAJAM 1(1):71–76. https://doi.org/10.5597/lajam00011

Bordino P, Thompson G, Iñiguez M (1999) Ecology and behaviour of the franciscana (*Pontoporia blainvillei*) in Bahía Anegada, Argentina. J Cetacean Res Manag 1:213–222

Bordino P, Wells RS, Stamper MA (2008) Satellite tracking of Franciscana dolphins Pontoporia blainvillei. In: Argentina: preliminary information on ranging, diving and social patterns. International whaling commission scientific committee meeting, SC60/SM14, Santiago de Chile, p 10

Bornatowski H, Wedekin LL, Heithaus MR et al (2012) Shark scavenging and predation on cetaceans at Abrolhos Bank, eastern Brazil. J Mar Biol Assoc UK 92(8):1767–1772. https://doi.org/10.1017/S0025315412001154

Bortolus A (2008) Influencia de los ambientes costeros patagónicos sobre los ecosistemas marino-oceánicos: las marismas como caso de estudio. In: Foro para la Conservación del Mar Patagónico y Áreas de Influencia (eds) Estado de conservación del Mar Patagónico y áreas de influencia, pp 80–104

Bos KI, Harkins KM, Herbig A et al (2014) Pre-Columbian mycobacterial genomes reveal seals as a source of New World human tuberculosis. Nature 514:494–497. https://doi.org/10.1038/nature13591

Botta S, Secchi ER, Muelbert MMC et al (2010) Age and growth of franciscana dolphins, *Pontoporia blainvillei* (Cetacea: Pontoporiidae) incidentally caught off southern Brazil and northern Argentina. J Mar Biol Assoc UK 8:1493–1500. https://doi.org/10.1017/S0025315410001141

Buckstaff KC (2004) Effects of watercraft noise on the acoustic behavior of bottlenose dolphins, *Tursiops truncatus*, in Sarasota Bay, Florida. Mar Mamm Sci 20(4):709–725

Budge SM, Springer AM, Iverson SJ et al (2008) Blubber fatty acid composition of bowhead whales, *Balaena mysticetus*: implications for diet assessment and ecosystem monitoring. J Exp Mar Biol Ecol 359:40–46. https://doi.org/10.1016/j.jembe.2008.02.014

Bustos RL, Daneri GA, Volpedo AV et al (2012) The diet of the South American sea lion (*Otaria flavescens*) at Río Negro, Patagonia, Argentina, during the winter-spring period. Iheringia Sér Zool 102:394–400

Butterworth A (2016) A review of the welfare impact on pinnipeds of plastic marine debris. Front Mar Sci 3:149. https://doi.org/10.3389/fmars.2016.00149

Campagna C (1985) The breeding cycle of the Southern sea lion, *Otaria byronia*. Mar Mamm Sci 1:210–218. https://doi.org/10.1111/j.1748-7692.1985.tb00010.x

Campagna C, Le Boeuf BJ (1988) Reproductive behaviour of southern sea lions. Behaviour 104:233–261. https://doi.org/10.1163/156853988X00539

Campagna C, Werner R, Karesh W et al (2001) Movements and location at sea of South American sea lions (*Otaria flavescens*). J Zool 257:205–220. https://doi.org/10.1017/S0952836901001285

Cappozzo HL (1995) Comportamiento Reproductivo en dos Especies de Otáridos de América del Sur. Tesis de Doctorado, Universidad de Buenos Aires, Buenos Aires, Argentina

Cardenas-Alayza S, Crespo EA, Oliveira L (2016) *Otaria byronia*, South American sea lion. The IUCN Red List of Threatened Species 2016:e.T41665A61948292. https://doi.org/10.2305/IUCN.UK.2016-1.RLTS.T41665A61948292.en. Accessed 3 Mar 2020

Cockcroft VG, Cliff G, Ross JB (1989) Shark predation on Indian Ocean bottlenose dolphins *Tursiops truncatus* off Natal, South Africa. S Afr J Zool:24305–24310. https://doi.org/10.1080/02541858.1989.11448168

Coscarella MA, Bellazzi G, Gaffet ML et al (2015) Technique used by killer whales (*Orcinus orca*) when hunting for dolphins in Patagonia, Argentina. Aquat Mamm 41:192–197. https://doi.org/10.1578/am.41.2.2015.192

Costanza R, d'Arge R, De Groot R et al (1997) The value of the world's ecosystem services and natural capital. Nature 387(6630):253–260. https://doi.org/10.1038/387253a0

Crespi-Abril AC, Garcia NA, Crespo EA et al (2003) Consumption of marine mammals by broadnose sevengill shark Notorynchus cepedianus in the northern and central Patagonian shelf. Lat Am J Aquat Mamm 2(2):101–107

Crespo EA, Pedraza SN, Dans SL et al (1997) Direct and indirect effects of the highseas fisheries on the marine mammal populations in the northern and central Patagonian coast. J Northwest Atl Fish Sci 22:189–207

Crespo EA, Harris G, González R (1998) Group size and distributional range of the franciscana, *Pontoporia blainvillei*. Mar Mamm Sci 14(4):845–849. https://doi.org/10.1111/j.1748-7692.1998.tb00768.x

Crespo EA, Pedrasa SN, Grandi MF et al (2010) Abundance and distribution of endangered Franciscana dolphins in Argentine waters and conservation implications. Mar Mamm Sci 26:17–35. https://doi.org/10.1111/j.1748-7692.2009.00313.x

Crespo EA, Schiavini AC, García NA et al (2015) Status, population trend and genetic structure of South American fur seals, Arctocephalus australis, in southwestern Atlantic waters. Mar Mamm Sci 31:866–890. https://doi.org/10.1111/mms.12199

Croll DA, Kudela R, Tershy BR (2006) Ecosystem impacts of the decline of large whales in the North Pacific. In: Estes JA, De Master DP, Doak DF et al (eds) Whales, whaling and ocean ecosystems, University of California Press, Berkeley Connectivity Conservation. Cambridge University Press, Cambridge, pp 202–2014

Crooks KR, Sanjayan M (2006) Connectivity conservation, vol 14. Cambridge University Press, Cambridge

Cunha HA, Medeiros BV, Barbosa LA et al (2014) Population structure of the endangered franciscana dolphin (*Pontoporia blainvillei*): reassessing management units. PLoS One 9(1). https://doi.org/10.1371/journal.pone.0085633

Dans SL, Sielfeld W, Aguayo A et al (2012) Estado y tendencia de las poblaciones. In: Crespo EA, Oliva D, Dans SL et al (eds) Estado desituación del lobo marino común *Otaria flavescens* en su área de distribución. Universidad de Valparaiso, Valparaiso, pp 18–35

Dans SL, Crespo EA, Coscarella MA (2017) Wildlife tourism: underwater behavioral responses of South American sea lions to swimmers. Appl Anim Behav Sci 188:91–96. https://doi.org/10.1016/j.applanim.2016.12.010

Dassis M, Farenga M, Bastida R (2012) At-sea behavior of South American fur seals: influence of coastal hydrographic conditions and physiological implication. Mamm Biol 77(1):7–52. https://doi.org/10.1016/j.mambio.2011.07.003

De Castro RL, Saporiti F, Vales DG et al (2016) What are you eating? A stable isotope insight into the trophic ecology of short-beaked common dolphins in the Southwestern Atlantic Ocean. Mamm Biol 81:571–578. https://doi.org/10.1016/j.mambio.2016.07.003

de Oliveira LR, MCM G, Fraga LD et al (2017) Ancient female philopatry, asymmetric male gene flow, and synchronous population expansion support the influence of climatic oscillations on the evolution of South American sea lion (*Otaria flavescens*). PLoS One 12:e0179442. https://doi.org/10.1371/journal.pone.0179442

Denuncio PE, Bastida R, Dassis M et al (2011) Plastic debris ingested by franciscana dolphins, *Pontoporia blainvillei*. Mar Pollut Bull 62:1836–1841. https://doi.org/10.1016/j.marpolbul.2011.05.003

Denuncio PE, Panebianco MV, Del Castillo V et al (2016) Beak deviations in the skull of Franciscana dolphins (*Pontoporia blainvillei*) from Argentina. Dis Aquat Org 120:1–7. https://doi.org/10.3354/dao03012

Denuncio PM, Mandiola MA, Pérez Salles SB et al (2017) Marine debris ingestion by the South American Fur Seal from the Southwest Atlantic Ocean. Mar Pollut Bull 122:420–425. https://doi.org/10.1016/j.marpolbul.2017.07.013

Denuncio PE, Paso Viola N, Cáceres-Saez I et al (2019) *Pontoporia blainvillei*. Categorización 2019 de los mamíferos de Argentina según su riesgo de extinción. Lista Roja de los mamíferos de Argentina. Versión digital: http://cma.sarem.org.ar

Dewar WK, Bingham RJ, Iverson RL et al (2006) Does the marine biosphere mix the ocean? J Mar Res 64:541–561. https://doi.org/10.1357/002224006778715720

Erbe C, Dunlop R, Dolman S (2018) Effects of noise on marine mammals. In: Slabbekoorn H et al (eds) Effects of anthropogenic noise on animals. Springer handbook of auditory research, vol 66. Springer, New York, pp 277–309

Estes JA, Tinker MT, Williams TM et al (1998) Killer whale predation on sea otters linking oceanic and nearshore ecosystems. Science 282:473–476. https://doi.org/10.1126/science.282.5388.473

Failla M, Iñiguez MA, Tossenberger V et al (2004) Bioecology of Franciscana (*Pontoporia blainvillei*) in Northern Patagonia, Argentina. In: 56th annual meeting of the Scientific Committee of the International Whaling Commission, Sorrento, p 4

Feijoo M, Lessa EP, Loizaga de Castro R et al (2011) Mitochondrial and microsatellite assessment of population structure of South American sea lion (*Otaria flavescens*) in the Southwestern Atlantic Ocean. Mar Biol 158:1857–1867. https://doi.org/10.1007/s00227-011-1697-4

Fidalgo GL (2004) Mamíferos Marinos. In: Piccolo MC, Hoffmeyer MS (eds) Ecosistema del Estuario de Bahía Blanca. Instituto Argentino de Oceanografía, Bahía Blanca, pp 221–227

Fouda L, Wingfield JE, Fandel AD et al (2018) Dolphins simplify their vocal calls in response to increased ambient noise. Biol Lett 14(10):20180484

Franco-Trecu V (2005) Comportamiento maternal y aspectos reproductivos de *Arctocephalus australis* en Isla de Lobos – Uruguay. Tesis de grado, Universidad de la República, Uruguay

Frankham R, Ballou JD, Briscoe DA (2002) Introduction to conservation genetics. Cambridge University Press, Cambridge

Fruet PF, Secchi ER, Daura-Jorge F et al (2014) Remarkably low genetic diversity and strong population structure in common bottlenose dolphins (*Tursiops truncatus*) from coastal waters of the Southwestern Atlantic Ocean. Conserv Genet 15(4):879–895. https://doi.org/10.1007/s10592-014-0586-z

Gariboldi MC, Túnez JI, Failla M et al (2016) Patterns of population structure at microsatellite and mitochondrial DNA markers in the franciscana dolphin (*Pontoporia blainvillei*). Ecol Evol 6(24):8764–8776. https://doi.org/10.1002/ece3.2596

Gerpe MS, Rodriguez D, Moreno VJ et al (2002) Accumulation of heavy metals in the Franciscana (*Pontoporia blainvillei*) from Buenos Aires province, Argentina. LAJAM 1:95–106. https://doi.org/10.5597/lajam00013

Giardino G (2014) Estructura y dinámica de las colonias de lobos marinos de un pelo de la Provincia de Buenos Aires, y su relación con pesquerías de la región. PhD thesis, Universidad Nacional de Mar del Plata

Giardino G, Mandiola A, Bastida J et al (2013) Técnica de marcado por decoloración de pelo en el lobo marino *Otaria flavescens*: descripción y evaluación del método. Mastozool Neotrop 20(2):393–398

Giardino GV, Mandiola MA, Bastida J et al (2016) Travel for sex: long range breeding dispersal and winter haulout fidelity in southern seal lion males. Mamm Biol Zeitschrift für Säugetierkunde 81:89–95. https://doi.org/10.1016/j.mambio.2014.12.003

Giardino G, Bastida J, Mandiola MA et al D (2017) Estimated population size of two South American sea lion male haulouts from the northern coast of Argentina. Mammalia 81(2):197–202 doi:https://doi.org/10.1515/mammalia-2015-0147

Gil MN, Torres A, Harvey M et al (2006) Metales pesados en organismos marinos de la zona costera de la Patagonia Argentina continental. R Biol Mar Ocean 41:167–176. https://doi.org/10.4067/S0718-19572006000200004

Goodall RNP, Marchesi MC, Pimper LE et al (2011) Southernmost records of bottlenose dolphins, *Tursiops truncatus*. Polar Biol 34(7):1085–1090. https://doi.org/10.1007/s00300-010-0954-1

Grandi MF, Dans SL, Crespo EA (2008) Social composition and spatial distribution of colonies in an expanding population of South American sea lions. J Mammal 89:1218–1228. https://doi.org/10.1644/08-MAMM-A-088.1

Grandi MF, Loizaga de Castro R, Terán E et al (2018) Is recolonization pattern related to female philopatry? An insight into a colonially breeding mammal. Mamm Biol 89:21–29. https://doi.org/10.1016/j.mambio.2017.12.002

Hain JHW, Ellis SL, Kenney RD et al (1995) Apparent bottom feeding by humpback whales on Stellwagen Bank. Mar Mamm Sci 11(4):464–479. https://doi.org/10.1111/j.1748-7692.1995.tb00670.x

Harcourt R (1993) Individual variation in predation on fur seals by southern sea lions (*Otaria byronia*) in Peru. Can J Zool 71(9):1908–1911. https://doi.org/10.1139/z93-273

Harwood J (2001) Marine mammals and their environment in the twenty-first century. J Mammal 82:630–640. https://doi.org/10.1644/1545-1542(2001)082<0630:mmatei>2.0.co;2

Jefferson TA, Webber MA, Pitman RL (2015) Marine mammals of the world – a comprehensive guide to their identification. Academic, London

Kasuya T, Brownell RL Jr (1979) Age determination, reproduction, and growth of the franciscana dolphin, *Pontoporia blainvillei*. Sci Rep Whales Res Inst 31:45–67

Kiers A, Klarenbeek A, Mendelts B et al (2008) Transmission of *Mycobacterium pinnipedii* to humans in a zoo with marine mammals. Int J Tuberc Lung Dis 12:1469–1473

Koen Alonso M, Crespo EA, Pedraza SN et al (2000) Food habits of the South American sea lion, *Otaria flavescens*, off Patagonia, Argentina. Fish Bull 98:250–263

Kovacs KM, Aguilar A, Aurioles D et al (2012) Global threats to pinnipeds. Mar Mamm Sci 28:414–436. https://doi.org/10.1111/j.1748-7692.2011.00479.x

Lavery TJ, Roudnew B, Gill P et al (2010) Iron defecation by sperm whales stimulates carbon export in the Southern Ocean. Proc R Soc B Biol Sci 277:3527–3531. https://doi.org/10.1098/rspb.2010.0863

Leeney RH, Amies R, Broderick AC et al (2008) Spatio-temporal analysis of cetacean strandings and bycatch in a UK fisheries hotspot. Biodivers Conserv 17:2323. https://doi.org/10.1007/s10531-008-9377-5

Lewis M, Ximenez I (1983) Dinámica de la población de *Otaria flavescens* en el área de Península Valdés zonas adyacentes (Segunda parte). Contribución No. 79, Centro Nacional Patagónico, p 21

Limbozzi F, Leitao TE (2008) Characterization of Bahía Blanca main existing pressures and their effects on the state indicators for surface and groundwater quality. In: Neves R, Baretta J, Mateus M (eds) Perspectives on integrated coastal zone management in South America. IST Press, Lisboa, pp 333–349

López Cazorla, A. (2004). Peces. In: Piccolo MC, Hoffmayer MS (eds) El Ecosistema del Estuario de Bahía Blanca. IADO, Bahía Blanca, Argentina, 191–201

Lopez Cazorla A, Molina JM, Ruarte C (2014) The artisanal fishery of *Cynoscion guatucupa* in Argentina: exploring the possible causes of the collapse in Bahía Blanca estuary. J Sea Res 88:29–35. https://doi.org/10.1016/j.seares.2013.12.016

Lundsten L, Schlining KL, Frasier K et al (2010) Time-series analysis of six whale-fall communities in Monterey Canyon, California, USA. Deep Sea Res I 57:1573–1584. https://doi.org/10.1016/j.dsr.2010.09.003

Mandiola MA (2015) Evaluación de la presencia de lobos marinos de dos pelos sudamericanos (*Arctocephalus australis*) en aguas de la Provincia de Buenos Aires. Mastozool Neotrop 22:419–420

Mandiola MA, Giardino GV, Bastida J et al (2015) Summer records of marine mammal on the Brazil-Malvinas confluence on Argentine sea shelf break during a seismic survey. Mastozool Neotrop 22:397–402

Mandiola MA, Blanco G, Rodríguez D (2017) Evaluación de interacciones con Mamíferos Marinos en la pesquería de anchoíta certificada bajo estándares del Marine Stewardship Council. Vinculación Tecnológica: de la Universidad Nacional de Mar del Plata al medio socio-productivo. vol IV. UNMdP, Mar del Plata, Argentina

Mandiola MA, Giardino G, Bastida J et al (2020) Half a century of sightings data of southern right whales in Mar del Plata (Buenos Aires, Argentina). J Mar Biol Assoc UK 100:165–171

Marcovecchio JE, Moreno VJ, Bastida RO et al (1990) Tissue distribution of heavy metals in small cetaceans from the Southwestern Atlantic Ocean. Mar Pollut Bull 21:299–304. https://doi.org/10.1016/0025-326X(90)90595-Y

Marcovecchio JE, Gerpe MS, Bastida RO et al (1994) Environmental contamination and marine mammals in coastal waters from Argentina: an overview. Sci Total Environ 154:141–151. https://doi.org/10.1016/0048-9697(94)90084-1

Marcovecchio J, Botté S, Delucchi F et al (2008) Pollution processes in Bahía Blanca estuarine environment. In: Neves R, Baretta J, Mateus M (eds) Perspectives on integrated coastal zone management in South America. IST Press, Lisboa, pp 301–314

Martinho F, Leitão R, Neto J et al (2007) The use of nursery areas by juvenile fish in a temperate estuary, Portugal. Hydrobiologia 587:281–290. https://doi.org/10.1016/0025-326X(90)90595-Y

Melcón ML, Failla M, Iñíguez MA (2012) Echolocation behavior of franciscana dolphins (*Pontoporia blainvillei*) in the wild. J Acoust Soc Am 131(6):448–453. https://doi.org/10.1121/1.4710837

Melcón ML, Failla M, Iñíguez MA (2016) Towards understanding the ontogeny of echolocation in franciscana dolphins (*Pontoporia blainvillei*). Mar Mamm Sci 32(4):1516–1521. https://doi.org/10.1111/mms.12336

Mendez M, Rosenbaum HC, Bordino P (2008) Conservation genetics of the franciscana dolphin in Northern Argentina: population structure, by-catch impacts, and management implications. Conserv Genet 9(2):419–435. https://doi.org/10.1007/s10592-007-9354-7

Mendez M, Rosenbaum HC, Subramaniam A et al (2010) Isolation by environmental distance in mobile marine species: molecular ecology of franciscana dolphins at their southern range. Mol Ecol 19(11):2212–2228. https://doi.org/10.1111/j.1365-294X.2010.04647.x

Mermoz JF (1977) Sobre el varamiento de un delfín nariz de botella, *Tursiops truncatus*, en la desembocadura del Río de la Plata (Buenos Aires, Argentina). Physis Sección C 93:227–235

Moore SE (2008) Marine mammals as ecosystem sentinels. J Mammal 89:534–540. https://doi.org/10.1644/07-mamm-s-312r1.1

Moore SE (2018) Climate change. In: Würsig B, Thewissen JGM, Kovacs KM (eds) Encyclopedia of marine mammals, 3rd edn. Academic, New York, pp 194–197

Moore KM, Simeone CA, Brownell RL (2018) Strandings. In: Würsig B, Thewissen JGM, Kovacs KM (eds) Encyclopedia of marine mammals, 3rd edn. Academic, New York, pp 945–951

Moreno V, Pérez A, Bastida R, Aizpún de Moreno JE et al (1984) Distribución del mercurio total en los tejidos de un delfín nariz de botella (*Tursiops gephyreus* Lahille, 1908) de la provincia de Buenos Aires (Argentina).[Total mercury distribution in tissues of Bottlenose dolphin (*Tursiops gephyreus* Lahille, 1908) from Buenos Aires Province (Argentine)]. Revista de Investigación y Desarrollo Pesquero 4:93–102

Muelbert MMC, Oliveira LD (2006) First records of stranded pregnant female South American fur seals, *Arctocephalus australis*, in the southern Brazilian coast. LAJAM 5(1):67–68. https://doi.org/10.5597/lajam00094

Muñoz-Arnanz J, Chirife AD, Vernazzani BG et al (2019) First assessment of persistent organic pollutant contamination in blubber of Chilean blue whales from Isla de Chiloé, Southern Chile. Sci Total Environ 650:1521–1528. https://doi.org/10.1016/j.scitotenv.2018.09.070

Oliva AL, Arias AH, Quintas PY et al (2017) Polycyclic aromatic hydrocarbons in mussels from a South American estuary. Arch Environ Contam Toxicol 72(4):540–551. https://doi.org/10.1007/s00244-017-0392-y

Ott PH, Sucunza F, Wickert J et al (2017) Evidences of attack of a killer whale on a calf southern right whale in Southern Brazil. Mastozool Neotrop 24:235–240

Páez E (1996) Simulaciones estocásticas en la población de *Otaria flavescens* en Uruguay. In: 7° Reunión de Trabajo de Especialistas en Mamíferos Acuáticos de América del Sur, Viña del Mar, p 116

Panebianco MV, Negri MF, Botte S et al (2011) Metales pesados en el riñón del delfín franciscana, *Pontoporia blainvillei* (Cetacea: Pontoporiidae) y su relación con parámetros biológicos. Lat Am J Aquat Res 39:526–533. https://doi.org/10.3856/vol39-issue3-fulltext-12

Panebianco H, Botteb S, Negria M et al (2012) Heavy metals in liver of the Franciscana dolphin, *Pontoporia blainvillei*, from the southern coast of Buenos Aires, Argentina. Ecotoxicol Environ Contam 7(1). https://doi.org/10.5132/jbse.2012.01.006

Paso-Viola MN, Denuncio P, Negri MF et al (2014) Diet composition of franciscana dolphin *Pontoporia blainvillei* from southern Buenos Aires, Argentina and its interaction with fisheries. Rev Biol Mar Oceanogr 49(2):393–400

Pavanato H, Silva KG, Estima SC et al (2013) Occupancy dynamics of South American sea-lions in Brazilian haul-outs. Braz J Biol 73(4):855–862. https://doi.org/10.1590/S1519-69842013000400023

Pavés HJ, Schlatter RP (2008) Temporada reproductiva del lobo fino austral, Arctocephalus australis (Zimmerman, 1783) en la Isla Guafo, Chiloé, Chile. Revi Chi His Nat 81:137–149. https://doi.org/10.4067/S0716-078X2008000100011

Peltier H, Baagøe HJ, Camphuysen KC et al (2013) The stranding anomaly as population indicator: the case of harbour porpoise *Phocoena phocoena* in North-Western Europe. PLoS One 8(4). https://doi.org/10.1371/journal.pone.0062180

Pérez Salles S (2015). Estudio preliminar de la dieta del lobo marino de dos pelos sudamericano, *Arctocephalus australis*, en el sector costero de la provincia de Buenos Aires, Argentina. Tesis de grado Universidad Nacional de Mar del Plata, Mar del Plata

Perez-Venegas DJ, Seguel M, Pavés H et al (2018) First detection of plastic microfibers in a wild population of South American fur seals (*Arctocephalus australis*) in the Chilean Northern Patagonia. Mar Pollut Bull 136:50–54. https://doi.org/10.1016/j.marpolbul.2018.08.065

Pershing AJ, Christensen LB, Record NR et al (2010) The impact of whaling on the ocean carbon cycle: why bigger was better. PLoS One 5:e12444. https://doi.org/10.1371/journal.pone.0012444

Petracci PF, Sotelo M, Massola V et al (2010) Actualización sobre el estado del apostadero de lobo marino de un pelo sudamericano (*Otaria flavescens*) en la isla Trinidad, estuario de Bahía Blanca, Argentina. Mastozool Neotrop 17(1):175–182

Ponce de León A (2000) Taxonomía, sistemática y sinopsis de la biología y ecología de los pinipedios de Uruguay. In: Rey M, Amestoy F (eds) Sinopsis de la biología y ecología de las poblaciones de lobos finos y leones marinos de Uruguay. Pautas para su manejo y Administración. Parte I. Biología de las especies. Montevideo. p 6–35

Reeves RR, Berger J, Clapham PJ (2006) Killer whales as predators of large baleen whales and sperm whales. Whales, whaling and ocean ecosystems: 174–187

Reijnders PJH, Borrell A, Van Franeker JA et al (2018) Pollution. In: Würsig B, Thewissen JGM, Kovacs KM (eds) Encyclopedia of marine mammals, 3rd edn. Academic, London, pp 746–753

Rodríguez D, Rivero L, Bastida R (2002) Feeding ecology of the franciscana (*Pontoporia blainvillei*) in marine and estuarine waters of Argentina. LAJAM 1(1):77–94. https://doi.org/10.5597/lajam00012

Rodríguez DH, Dassis M, Ponce de León A et al (2013) Foraging strategies of Southern sea lion females in the La Plata River Estuary (Argentina–Uruguay). Deep-Sea Res Part II 88:120–130. https://doi.org/10.1016/j.dsr2.2012.07.012

Rolland RM, Parks SE, Hunt KE et al (2012) Evidence that ship noise increases stress in right whales. Proc R Soc B Biol Sci 279(1737):2363–2368

Roman J, Estes JA, Morissette L et al (2014) Whales as marine ecosystem engineers. Front Ecol Environ 12:377–385. https://doi.org/10.1890/130220

Romero MA, Dans S, González R et al (2011) Solapamiento trófico entre el lobo marino de un pelo *Otaria flavescens* y la pesquería de arrastre demersal del golfo San Matías, Patagonia, Argentina. Lat Am J Aquat Res 39:344–358. https://doi.org/10.3856/vol39-issue2-fulltext-16

Romero MB, Polizzi P, Chiodi L et al (2018) Preliminary assessment of legacy and current–use pesticides in Franciscana dolphins from Argentina. Bull Environ Contam Toxicol 101:14–11. https://doi.org/10.1007/s00128-018-2352-2

Romero MA, Grandi MF, Túnez JI et al (2019) *Otaria flavescens*. Categorización 2019 de los mamíferos de Argentina según su riesgo de extinción. Lista Roja de los mamíferos de Argentina. Versión digital: http://cma.sarem.org.ar

Rosas FWC, Pinedo MC, Marmontel M et al (1994) Seasonal movements of the South American sea lion (*Otaria flavescens* Shaw) off the Rio Grande do Sul Coast, Brazil. Mammalia 58:51–59. https://doi.org/10.1515/mamm.1994.58.1.51

Santos MB, Pierce GJ, Reid RJ et al (2001) Stomach contents of bottlenose dolphins (*Tursiops truncatus*) in Scottish waters. J Mar Biol Assoc UK 81(5):873–878. https://doi.org/10.1017/S0025315401004714

Secchi ER, Ott PH, Danilewicz D (2003) Effects of fishing by-catch and conservation status of the franciscana dolphin, *Pontoporia blainvillei*. In: Gales N, Hindell M, Kirkwood R (eds) Marine mammals: fisheries, tourism and management issues. CSIRO Publishing, Collingwood, pp 174–191

SADSN-SAREM 2019 (Secretaría de Ambiente y Desarrollo Sustentable de la Nación y Sociedad Argentina para el estudio de los mamíferos) (eds) (2019) Categorización 2019 de los Mamíferos de Argentina según su riesgo de extinción. Lista Roja de los mamíferos de Argentina. www.cma.sarem.org.ar

Simonis AE, Brownell RL Jr, Thayre BJ et al (2020) Co-occurrence of beaked whale strandings and naval sonar in the Mariana Islands, Western Pacific. Proc R Soc B 287(1921):20200070. https://doi.org/10.1098/rspb.2020.0070

Smith CR (2006) Bigger is better: the role of whales as detritus in marine ecosystems. In: Estes JA, Demaster DP, Doak DF et al (eds) Whales, whaling and ocean ecosystems. University of California Press, Berkeley, pp 286–301

Szapkievich VA, Cappozzo HL, Crespo EA et al (1999) Genetics relatedness in two Southern sea lion (*Otaria flavescens*) rookeries in the southwestern Atlantic. Mamm Biol 64:1–5. https://doi.org/10.1111/j.1748-7692.1999.tb00814.x

Timi JT, Paoletti M, Cimmaruta R et al (2014) Molecular identification, morphological characterization and new insights into the ecology of larval *Pseudoterranova cattani* in fishes from the Argentine coast with its differentiation from the Antarctic species, *P. decipiens* sp. E (Nematoda: Anisakidae). Vet Parasitol 199:59–72. https://doi.org/10.1016/j.vetpar.2013.09.033

Trites AW (1997) The role of pinnipeds in the ecosystem. In: Stone G, Goebel J, Webster S (eds) Pinniped populations, eastern north Pacific: status trends and issues. A symposium of the 127th annual meeting of the American Fisheries Society. New England Aquarium, Conservation Department, Central Wharf, Boston, pp 31–39

Trites AW, Christensen V, Pauly D (1997) Competition between fisheries and marine mammals for prey and primary production in the Pacific Ocean. J Northwest Atl Fish Sci 22:173–187

Túnez JI, Centrón D, Cappozzo HL et al (2007) Geographic distribution and diversity of mitochondrial DNA haplotypes in South American sea lions (*Otaria flavescens*) and fur seals (*Arctocephalus australis*). Mamm Biol 72:193–203. https://doi.org/10.1016/j.mambio.2006.08.002

Túnez JI, Cappozzo HL, Cassini MH (2008) Natural and anthropogenic factors associated with the distribution of South American sea lion along the Atlantic coast. Hydrobiology 598:191–202. https://doi.org/10.1007/s10750-007-9150-x

Túnez J, Cappozzo H, Nardelli M et al (2010) Population genetic structure and historical population dynamics of the South American sea lion, *Otaria flavescens*, in north-central Patagonia. Genetica 138:831–841. https://doi.org/10.1007/s10709-010-9466-8

Tyack PL, Janik VM (2013) Effects of noise on acoustic signal production in marine mammals. In: Brumm H (ed) Animal communication and noise. Springer, Berlin, pp 251–271

Vales DG, Cardona L, García NA et al (2015) Ontogenetic dietary changes in male South American fur seals *Arctocephalus australis* in Patagonia. Mar Ecol Prog Ser 525:245–260. https://doi.org/10.3354/meps11214

Vales DG, Mandiola A, Romero MA et al (2019) *Arctocephalus australis*. Categorización 2019 de los mamíferos de Argentina según su riesgo de extinción. Lista Roja de los mamíferos de Argentina. Versión digital: http://cma.sarem.org.ar

Vaz Ferreira R (1981) South American sea lion, *Otaria flavescens*. In: Ridgway SH, Harrison RJ (eds) Handbook of marine mammals V1. Academic, London, pp 39–65

Vaz Ferreira R (1982) *Otaria flavescens* (Shaw), South American sea lion. In: FAO-UNEP (ed) Mammals in the seas IV. FAO, Rome, pp 477–496

Vermeulen E, Bräger S (2015) Demographics of the disappearing bottlenose dolphin in Argentina: a common species on its way out? PLoS One 10(3). https://doi.org/10.1371/journal.pone.0119182

Vermeulen E, Cammareri A (2009) Residency, abundance and social composition of bottlenose dolphins (*Tursiops truncatus*) in Bahía San Antonio, Patagonia, Argentina. Aquat Mamm 35:379–386. https://doi.org/10.1578/AM.35.3.2009.379

Vermeulen E, Holsbeek L, Das K (2015) Diurnal and seasonal variation in the behaviour of bottlenose dolphins (*Tursiops truncatus*) in Bahía San Antonio, Patagonia Argentina. Aquat Mamm 41(30):272–183. https://doi.org/10.1578/AM.41.3.2015.272

Vermeulen E, Balbiano A, Belenguer F et al (2017) Site-fidelity and movement patterns of bottlenose dolphins (*Tursiops truncatus*) in central Argentina: essential information for effective conservation. Aquat Conserv 27(1):282–292. https://doi.org/10.1002/aqc.2618

Vermeulen E, Bastida R, Berninsone LG et al (2018) A review on the distribution, abundance, residency, survival and population structure of coastal bottlenose dolphins in Argentina. LAJAM 12:2–16. https://doi.org/10.5597/00233

Vermeulen I, Failla M, Loizaga de Castro R et al (2019) *Tursiops truncatus*. Categorización 2019 de los mamíferos de Argentina según su riesgo de extinción. Lista Roja de los mamíferos de Argentina. Versión digital: http://cma.sarem.org.ar

Weller DW (2018) Predation on marine mammals. In: Würsig B, Thewissen JGM, Kovacs KM (eds) Encyclopedia of marine mammals, 3rd edn. Academic, New York, pp 772–780

Wells RS, Scott MD (1999) Bottlenose dolphin *Tursiops truncatus* (Montagu, 1821). Handbook of marine mammals: the second book of dolphins and porpoises 6:137–182
Wells RS, Fauquier DA, Gulland FM et al (2013) Evaluating postintervention survival of free-ranging odontocete cetaceans. Mar Mamm Sci 29(4):463–483. https://doi.org/10.1111/mms.12007
Wells RS, Natoli A, Braulik G (2019) Tursiops truncatus. The IUCN Red List of Threatened Species. https://doi.org/10.2305/IUCN.UK.2019-1.RLTS.T22563A50377908.en
Würsig B, Würsig M (1979) Behavior and ecology of the bottlenose dolphin, *Tursiops truncatus*, in the South Atlantic. Fish Bull 77(2):399–412
Zerbini AN, Andriolo A, Da Rocha JM et al (2004) Winter distribution and abundance of humpback whales (*Megaptera novaeangliae*) off Northeastern Brazil. J Cetacean Res Manag 6:101–107
Zerbini A, Rosenbaum H, Mendez M, Sucunza F et al (2016) Tracking southern right whales through the Southwest Atlantic: an update on movements, migratory routes and feeding grounds. Scientific Committee annual meeting IWC SC/66a/BRG22, pp 1–10
Zerbini AN, Secchi E, Crespo E et al (2017) *Pontoporia blainvillei* (errata version published in 2018). The IUCN Red List of Threatened Species. https://doi.org/10.2305/IUCN.UK.2017-3.RLTS.T17978A50371075
Zerbini A, Fernandez Ajos A, Andriolo A et al (2018) Satellite tracking of southern right whales (*Eubalaena australis*) from Golfo San Matias, Rio Negro Province, Argentina. Scientific Committee of the International Whaling Commission SC67b, Bled, Slovenia

Chapter 15
Use of Coastal Area Habitats by Land Mammals

Estela M. Luengos Vidal, Nicolás Caruso, Sabrina Martinez, Emma Casanave, and Lucherini Mauro

15.1 Introduction

The ocean-land interface forms a major ecosystem, the coastal ecotone, where marine and terrestrial abiotic and biotic factors intensively interact in a wide array of forms. At this ecotone, terrestrial mobile species (e.g., mammals, Carlton and Hodder 2003; reptiles, Lillywhite et al. 2008; ants, Garcia et al. 2011) and freshwater organisms have the ability to get into nearshore marine and estuarine waters or fringe intertidal and marginal habitats (i.e., supralittoral zone) to exploit marine-derived food resources. These species create connections between marine and terrestrial food chains that can have far-reaching implications, including changes in terrestrial plant communities (Croll et al. 2005) and effects on the circulation of pollutants between food webs (Dehn et al. 2006). The inputs from land to the sea have been extensively studied and proven to promote high primary and secondary productivity on the coastal zone (Polis and Hurd 1996a). It is less clear how the commonly large inputs of marine material to adjacent lands affects productivity in the surrounding area. Although this influence has been traditionally considered small, this topic has recently increase in importance, and a growing body of work has focused on the effects of marine inputs to the coastal land (Polis and Hurd 1996a; Spiller et al. 2010; Barreiro et al. 2013).

Besides the growing academic interest related with this topic, assessing the strength of marine transfers to the coastal land has implications not only for the ecological theory but also for wildlife management actions, and there is limited and

E. M. Luengos Vidal (✉) · N. Caruso · S. Martinez · E. Casanave · L. Mauro
INBIOSUR (Instituto de Ciencias Biológicas y Biomédicas del Sur), CONICET-UNS, Bahia Blanca, Argentina

Grupo de Ecología Comportamental de Mamíferos, Laboratorio de Fisiología Animal, Departamento de Biología, Bioquímica y Farmacia, Universidad Nacional del Sur, Bahia Blanca, Argentina
e-mail: luengos@criba.edu.ar

S. M. Fiori, P. D. Pratolongo (eds.), *The Bahía Blanca Estuary*,
https://doi.org/10.1007/978-3-030-66486-2_15

disperse information in our region. In the first section of this chapter, we set the theoretical framework for a greater understanding of the ecological influence and relevance of marine material inputs to the coastal lands, dismissed until recently. In the second section, we speculate on the base of the existing information, which terrestrial mammal species would use coastal food sources ("maritime mammals"; see definition later), and provided detailed information on the most relevant species in nine species data sheet (Sect. 15.4). In the third section, we analyze a study case focusing on the mammalian carnivores that use a sandy coastal area close to the Bahía Blanca Estuary. Finally, given the relevance of anthropogenic influences on this coastal region, we review the effects of human pressures on marine mammal populations in a text box.

15.2 Role of Coastal Ecosystems in the Ecology of Land Mammals

The effects of cross-habitat transport of allochthonous marine materials are most obvious in extreme habitats. In places where the difference between the primary productivity of the sea and land is ample, such phenomenon is more evident and has been the focus of a greater number of studies. This is the case, for example, of the interactions between the Benguela Current and the Namibian Desert in Africa, the Peru Current and the Atacama Desert in South America, or the effect of the water of the Gulf of California on the Baja California Peninsula in Mexico (Polis and Hurd 1996a). In the absence or poor production of endogenous energy, the terrestrial communities found in these regions are largely dependent on marine resources. However, this process is far more widespread; in fact, it has been found to be nearly ubiquitous (Polis and Hurd 1996a; Barreiro et al. 2013; Schlacher et al. 2017).

Seaweed, seagrasses, and carcasses of marine vertebrates are commonly washed ashore and utilized by terrestrial consumers. Strandings of carrion, which are the evident results of mortality events happening at sea, occur when positively buoyant carcasses disperse from the original location of death and drift toward the coastline. The causes of these events are diverse and can be related to biotoxins, bacteria, parasites, predators, human interactions, and oceanographic adverse conditions. Typically, these events are spatially unpredictable phenomena, in which marine mammal carrions strand in a delimited space and during a limited time and their occurrence is also dependent on oceanography and coastal topography (Leeney et al. 2008). The drifting algae depend largely on their residence time in a specific location, which is mostly conditioned by the flow velocities of currents at that site and the roughness of the substrate, but a number of physical, chemical, and biological factors influence their behavior (Arroyo and Bonsdorff 2016).

In general, marine input supports a great abundance and diversity of terrestrial or semi-terrestrial arthropods that feed on drift in the interface between land and sea. These remains are consumed by detritivores and scavengers, which in turn can be used by predatory arthropods (secondary consumers) of the supralittoral zone (Heatwole 1972; Ruiz-Delgado et al. 2016). The work of Colombini et al. (2003) provides an overview of the importance of the coastal accumulations of macrophytes and other organic beach-cast material on the ecology of sandy beach ecosystems. Carrions stranded on the shore represent a different system and constitute a suitable material for the study of arthropod invasion, utilization, and succession. The scavenger arthropod community develops primarily as a continuum of gradual change. Classically, the activity of dipterans and ants has been considered a major facilitation mechanism on behalf of the early invaders, but the activity of other taxa, such as vertebrate scavengers (i.e., vultures), can act as facilitation mechanisms. Shorebirds also prey opportunistically on larvae, flies, and other insects associated with carrion. Other secondary consumers connected to marine-generated sources of subsidies are represented by spiders, scorpions, lizards, rodents, and carnivores. In many cases, allochthonous resources are ephemeral and present seasonal and spatial fluctuations that probably affect all levels of the food web of a coastal beach. In the Baja California system of Mexico, Polis and Hurd (1996a) found a complex interaction of marine and terrestrial trophic webs. Terrestrial secondary consumers (spiders, scorpions, and lizards) primarily prey on arthropod species from the marine food web, which made up 95–99% of their diet. A high marine productivity makes the abundance of potential preys significantly higher in the supralittoral zone compared to adjacent terrestrial habitats. Consequently, the population of spiders found along the coast was six times more abundant than that of inland areas (Polis and Hurd 1995), and lizard populations were estimated to be four times more abundant in the supralittoral zone than in inland areas (Polis and Hurd 1996b).

For the same geographical area, Rose and Polis (1998) showed that the coyote (*Canis latrans*), an opportunist carnivore mammal that lives in coastal habitats, not only eats a variety of living marine species (arthropods such as crustaceans and larvae of insects that forage on algal drift, algae, and mollusks) but also consumes many terrestrial creatures (arthropods, lizards, land birds, and coastal rodents) exploiting marine resources. This diverse and conspicuous marine input, together with the in situ terrestrial food supply, has the effect of increasing both the dietary spectrum and the food intake of coastal coyotes compared with those of inland populations, favoring coastal population of those secondary consumers.

In the forementioned study, three additional carnivore species were considered: two canids, the gray fox (*Urocyon cinereoargenteus*) and the kit fox (*Vulpes macrotis*), and one member of the Procyonidae family, the ringtail (*Bassariscus astutus*), which were very rarily recorded (1% of all observations). There were no significant differences in the abundance of those species in coastal vs. inland areas. Both canids occurred at almost all sites, but ringtails were spotted only at one coastal site and one inland site. Nevertheless, Rose and Polis (1998) observed that all these species foraged on marine resources within the littoral and supralittoral zone (e.g., gray foxes were recorded eating sea cucumbers, order Holothuroidea).

Bristol Bay, Alaska, is one of the most productive marine ecosystems in the world, with extensive seabird colonies on land, but ungulate prey limited to low densities of moose (*Alces alces*) and migratory caribou (*Rangifer tarandus*). Watts et al. (2010) inferred the influence of marine items among wolf preys by combining direct observation, VHF, and GPS radio-tracking. The strong positive selection of marine habitats by wolves (*Canis lupus*) suggests that marine-derived subsidies may provide an important food source to the coastal populations of this canid (Watts et al. 2010). Wolves predominantly feed on ungulates, but they are opportunistic-generalist carnivores and exhibit considerable dietary plasticity both among and within populations. Previous studies have documented that wolves can make extensive use of a wide variety of food items, including intertidal organisms (e.g., marine invertebrates, Klein 1995) and carrion, which may constitute an important food source in some areas (Lewis and Lafferty 2014). Also, Lewis and Lafferty (2014) reported evidence of wolves feeding on large marine mammal carcasses (e.g., humpback whales) and hunting sea otters (*Enhydra lutris*). They observed a strong positive association with the coastline on the base of the intensity of use of this habitat, although it represented 1% of the total area available to pack members. These subsidies of food are likely to affect the fitness and survival of wolves in these areas and even increase wolf densities along coastlines, similar to what was observed for coyote populations by Rose and Polis (1998). This information would be of interest not only for the ecological theory but also for the planning of management actions of carnivore populations, especially those related to low-density populations of ungulates.

Thus, in many cases, marine energy may subsidize land consumers, promoting a numerical response and leading to population densities that would be impossible without such subsidy. In other cases, inputs enable some mammal populations to persist in low productivity areas such as islands.

Marine-based organic materials also enter land food web via colonies of sea birds nesting on the coast in the form of fish scrape, carcasses of dead chick, bird feathers, eggs, and guano, in addition to an array of arthropods. This input increases the abundance of many consumer species (e.g., arthropods) and forms the base of unique food webs, particularly in small island where land resources are limited and mammalian and reptilian predators are not present (Polis and Hurd 1996a). In islands where the native animal communities have not been modified by the anthropogenic introduction of exotic mammals, secondary consumers are typically predatory birds.

Islands may also provide ample opportunity for omnivorous mammalian mesopredators (i.e., medium-sized predators subordinate to larger predators) to exploit marine prey (Carlton and Hodder 2003). It has been shown that mesopredators may benefit substantially from such marine subsidies (Rose and Polis 1998). Consumption of marine preys has been suggested to exacerbate the impact of mesopredators on terrestrial communities. Although they are still scarce, some data-based studies demonstrate that mammalian mesopredators, whose diet is substantially subsidized by marine items, have a significant impact on terrestrial prey population (Polis and Strong 1996; Polis and Hurd 1996a; Rose and Polis 1998).

The limited evidence available suggests that such allochthonous input from the ocean has important effects on the distribution and abundance of omnivorous mammals in many adjacent terrestrial systems and that the use of marine resources by carnivorous mammals is widespread worldwide. The Arctic fox (*Vulpes lagopus*) is an opportunistic predator and scavenger distributed in tundra regions and Arctic islands of Eurasia and North America. Their main preys throughout most of its range are small mammals such as lemmings (*Dicrostonyx* spp. and *Lemmus* spp.; e.g., Kennedy 1980; Elmhagen et al. 2000). However, Arctic foxes living in the islands of the high Arctic Svalbard archipelago, a restricted geographical area, where no resident rodents are present mostly consume reindeer (*Rangifer tarandus*), rock ptarmigan (*Lagopus mutus*), geese, seabirds, and eggs, but inland and coastal foxes have different food habits (Prestrud 1992; Frafjord 1993). In coastal areas, seabirds and eggs constitute the majority of the diet of foxes (Prestrud 1992; Hersteinsson and Macdonald 1996). Coastal foxes which depend almost entirely on ocean cliff birds confined their activity to very small areas, illustrating their strong affinity to the large concentrated seabird colonies (Jepsen et al. 2002). However, once again, the effect of the allochthonous energy subsidy provided by the ocean is not limited to Artic fox populations but goes beyond the sea-land ecotone. On the Bylot Island of the Canadian Territory of Nunavut, the top-down control of lemming populations by Arctic foxes is likely strengthened by this allochthonous subsidy (Legagneux et al. 2012).

Studies considering the potential impacts of mammalian mesopredators on intertidal preys are even less numerous. Experimental studies on the importance of terrestrial mammals as intertidal predators are scarce, and thus the use of intertidal resources has been presumed to be insignificant both in terms of energy and flow pathway and as driving force of intertidal community structures. However, some studies suggest that this assumption may be incorrect. Suraci et al. (2014) concluded that terrestrial mammalian mesopredators on islands may directly impact both the intertidal and shallow subtidal marine communities. They showed that raccoon (*Procyon lotor*) populations in the Gulf Islands of British Columbia, Canada, with access to marine resources impacted the local abundance of their marine prey, where top predators are extirpated or are naturally absent. Under these circumstances, raccoons have been promoted to the top of both the terrestrial and nearshore marine food chains and thereby have a direct influence on terrestrial birds, intertidal vertebrates and invertebrates, and intertidal and shallow subtidal crab populations. This direct effect by raccoons may in turn initiate trophic cascades within both the terrestrial and intertidal communities. A similar effect has been reported by Kurle et al. (2008), who found that predation by introduced brown rats (*Rattus norvegicus*) on nesting seabirds on Alaska's Aleutian Islands indirectly affected intertidal communities through a trophic cascade.

This demonstrates the potential strength of the impact of small mammals on island ecosystems and, in general, of land mammals on coastal marine ecosystems. Given these examples and the paucity of information on this topic, it is not surprising that the effects of terrestrial mammals on the diversity, abundance, and distribution of marine preys have been identified as a conspicuous gap in the ecological literature (Carlton and Hodder 2003).

The review of Moore (2002) was a first attempt to join scattered literature about mammals that use seashores and maritime terrestrial environments. This review focused on mammalian contribution to ecological processes more than on ethological research. Moore (2002) found ecological information on the effect of mammals on marine environments in 11 orders (Marsupialia, Insectivora, Chiroptera, Lagomorpha, Rodentia, Cetacea, Carnivora, Sirenia, Perissodactyla, Artiodactyla, and Primates). Moreover, introduced mammals like cats (*Felis catus*), dogs (*Canis familiaris*), rats (*Rattus norvegicus* and *Rattus rattus*), and minks (*Mustela vison*) showed to be important predators of seabirds. Domestic grazers such as goats and sheep have been found to feed on seagrasses and seaweed. Salt and minerals could be a good motivation, but seaweeds and seagrasses are often cited as alternative protein sources from terrestrial agricultural resources by these domestic animals. But the grazing pressure in some opportunities causes coastal habitat degradation. Carlton and Hodder (2003) conducted a more specific review on terrestrial mammals intentionally entering the coastal zone at low tide to prey on living marine invertebrates, fish, algae, and seagrasses. Those authors introduced the term "maritime mammals," to indicate coastal mammalian predators or consumers that utilize living intertidal resources and transfer their energy to the land. They documented 135 records of consumption across 45 species of terrestrial mammals in 8 orders (Didelphimorphia, Dasyuromorphia, Insectivora, Primates, Lagomorpha, Rodentia, Carnivora, Artiodactyla). Maritime mammals were found to occur on all continental coastlines of the world except Antarctica. Most predation events were realized by carnivores (59%, mostly by raccoon, mink, black bear *Ursus americanus*, and Arctic fox), followed by rodents (20%) and artiodactyls (14%), but introduced populations of 17 species of mammals were also recorded as maritime predators.

Carlton and Hodder (2003) identified 228 different prey taxa, representing 12 phyla of marine organisms including 8 animal phyla (mainly bivalve and gastropod mollusks, crabs, and fish), 3 algae phyla, and 1 plant phylum. Because their review reports several observations across a wide variety of taxa, they suggest that predation by maritime mammals is a rarely studied phenomenon rather than a rare phenomenon. They conclude that there is a need for quantitative observations on predation and experimental studies on mammals as consumers of intertidal energy resources. Such studies should be facilitated by the use of new technologies, including advanced infrared night vision technology, radio telemetry of individuals, camera traps, and the use of stable isotopes for trophic web analysis.

To our knowledge, since the publication of reviews by Moore (2002) and Carlton and Hodder (2003), there have been no new global revisions on this subject, but numerous specific studies have been published that contribute to the ecological knowledge of these processes (e.g., Watts et al. 2010; Gaydos and Pearson, 2011; Tarroux et al. 2012; Lafferty et al. 2014; Lewis and Lafferty 2014; Broekhuis et al. 2014; Moss 2017; Lei et al. 2017).

15.3 The Maritime Mammals Surrounding Bahía Blanca Estuary

From this section on, we will extend the term "maritime mammals" proposed by Carlton and Hodder (2003) to include those species that can potentially predate on seabirds or other vertebrates that use the coastal zone and those that can scavenge carrion and detritus from marine origin.

To identify the species that may classify as maritime mammals for the region, we made a literature search on the potential mammals species that, according to their distribution ranges, could use the coastal and surrounding area of the Bahía Blanca Estuary, from mouth of the Quequén Salado River in the North-East to the mouth of Colorado River in the South and from the sea coast up to 1 km inland (Fig. 2.1; Chap. 2). The actual distribution ranges of many of these species are poorly known and are based on polygons that connect specific record points (CMA 2019). Additionally, the information published on their ecology in the study region is extremely limited. Thus, we collected and summarized in Table 15.1 any further evidence (including unpublished personal observations and personal communications from reliable sources such as researchers or park rangers) of the use of coastal habitats (Table 15.1).

The potential maritime mammal community of the study area comprises 43 species (Table 15.1), belonging to 5 out of the 12 orders of wild mammals occurring in Argentina (Montero and Autino 2018). Rodentia is the most abundant order, with approximately 20 species. Although there is no literature on the local use of the intertidal area by any of these native species, we do know that *Ctenomys australis* (see species data sheet 15.4.1; Fig. 15.1a) is an endemic species spatially restricted to the first ridge of dunes in the coastal zone of the Buenos Aires Province, with a southern limit coincident with the beginning of the estuarine area, in Punta Alta (Contreras and Reig 1965). The conservation status of this rodent is Endangered (Austrich et al. 2019), based on evidence of the fragmentation of its habitat due to the influence of anthropic activity, in addition to the ecological characteristics of the species (low dispersion rates and relatively low reproduction rates with a high degree of genetic structuring) (Mora and Mapelli 2010). *Ctenomys talarum* is another rodent strictly limited to the coastal zone specifically to the second and third ridges of dunes. Although this species is listed as Least Concern by the IUCN (International Union for Conservation of Nature) (Bidau 2016), it was considered as Vulnerable in Argentina by Fernandez et al. (2019). The capybara (see species data sheet 15.4.2; Fig. 15.1b) *Hydrochoerus hydrochaeris* is another rodent that could use the coastal zone close to Bahía Blanca. It is typical of wetlands and riparian environments, and its food habit is mainly herbivorous. In recent years, this rodent has been regularly registered in coastal areas of southern Buenos Aires where there are freshwater streams or lagoons (e.g., Laguna de Sauce Grande, near the city of Monte Hermoso) and there is a record of individuals entering the sea (https://www.

Table 15.1 List of the potential species that could use the coastal and surrounding area of the Bahía Blanca Estuary, from the town of Marisol in the North to the end of Colorado R River in the South and from the sea coast up to 1 km toward the continent, the scientific name, the local name in Spanish and the name in English, its conservation category according to CAM (2019) (LC, Least concerned; NT, near threatened; VU, vulnerable; DD, data deficient; EXO, exotic), and if it has been cited by reviews of 1 Carlton and Hodder (2003) and 2 Moore (2002)

	CC	Species	Spanish name	English name	1	2
	LC	*Lycalopex gymnocercus*	Zorro gris	Pampas fox	x	
	LC	*Leopardus geoffroyi*	Gato montés	Geoffroy's cat		
	VU	*Leopardus colocolo*	Gato de los pajonales	Pampas cat		
Carnivora	LC	*Herpailurus yagouaroundi*	Yaguarundi	Jaguarundi		
	LC	*Puma concolor*	Puma	Cougar		
	LC	*Galictis cuja*	Hurón menor	Lesser grison		x
	NT	*Lyncodon patagonicus*	Huroncito patagónico	Patagonian weasel		
	LC	*Conepatus chinga*	Zorrino	Molina's hog-nosed skunk		
	LC	*Chaetophractus villosus*	Peludo	Large hairy armadillo		
	NT	*Dasypus hybridus*	Mulita pampeana	Southern long-nosed armadillo		
Cingulata	LC	*Chaetophractus vellerosus*	Piche llorón	Screaming hairy armadillo		
	DD	*Chlamyphorus truncatus*	Pichiciego menor	Pink fairy armadillo		
	NT	*Zaedyus pichiy*	Piche	Pichi		
Lagomorpha	EXO	*Lepus europaeus*	Liebre europea	European hare		x
	LC	*Lama guanicoe*	Guanaco	Guanaco		
Cetartiodactyla	EXO	*Dama dama*	Ciervo dama	Fallow deer		x
	EXO	*Sus scrofa*	Jabalí	Wild boar	x	x
	LC	*Thylamys pallidior*	Marmosa pálida	White-bellied fat-tailed mouse opossum		
Didelphimorphia	LC	*Monodelphis dimidiata*	Colicorto pampeano	Yellow-sided opossum		
	LC	*Lutreolina crassicaudata*	Comadreja colorada	Red opossum		
	LC	*Didelphis albiventris*	Comadreja overa	White-eared opossum	x	
	LC	*Oxymycterus rufus*	Ratón hocicudo rojozo	Red hocicudo		
	LC	*Akodon azarae*	Ratón de pastizal pampeano	Azara's grassland mouse		x
	LC	*Akodon dolores*	Ratón cordobés	Córdoba akodont		x
	LC	*Akodon iniscatus*	Ratón Patagónico	Patagonian akodont		x

(continued)

Table 15.1 (continued)

	CC	Species	Spanish name	English name	1	2
	LC	*Calomys laucha*	Laucha chica	Little Laucha		
	LC	*Calomys musculinus*	Ratón maicero	Corn mouse		
	LC	*Eligmodontia typus*	Laucha colilarga baya	Lowland gerbil mouse		
	LC	*Graomys griseoflavus*	Pericote comun	Common pericote		
	LC	*Holochilus vulpinus*	Rata colorada	Crafty marsh rat		
	LC	*Oligoryzomys longicaudatus*	Ratón colilargo	Long-tailed colilargo	x	x
Rodentia	LC	*Reithrodon auritus*	Rata conejo	Bunny rat		
	LC	*Lagostomus maximus*	Vizcacha	Plains viscacha		
		Galea musteloides	Cuís común	Common yellow-toothed cavy		
	LC	*Microcavia australis*	Cuis chico	Southern mountain cavy		
	VU	*Dolichotis patagonum*	Mara	Patagonian hare		
	VU	*Ctenomys talarum*	Tuco-tuco de los talares	Tuco-tuco		
	NT	*Ctenomys australis*	Tuco-tuco de los medanos	Southern tuco-tuco		
	LC	*Myocastor coypus*	Coipo	Coipo		
	EXO	*Rattus norvegicus*	Rata noruega	Brown rat	x	x
	EXO	*Rattus rattus*	Rata negra	Black rat		
	LC	*Hydrochoerus hydrochaeris*	Carpincho	Capybara		
	LC	*Cavia aperea*	Cuis	Pampas cavy		

lacapitalmdp.com/sorpresa-en-mar-del-sud-aparecio-un-carpincho-nadando-en-la-playa). It is worth mentioning that this order includes also three exotic cosmopolitan species (*Mus musculus*, *R. norvegicus*, and *R. rattus*). In particular, the brown rat has been reported to make an intensive use of intertidal areas in certain sites of Central Chile (Navarrete and Castilla 1993).

The information about the four species of the order Didelphimorphia potentially behaving as maritime mammals is too scarce to provide any hint about the possible use of the coastal zone and its resources, except for the case of *Didelphis albiventris*. The white-eared opossum is listed as Least Concern by IUCN (Costa et al. 2015) and under the same category in Argentina (Chimesquy and Martin 2019) is omnivorous and extremely adaptable, which enables it to take great advantage of disturbance and changes of land use associated with human presence. Additionally, *D. virginiana*, a very similar congeneric species, has been reported to prey on crabs (*Uca pugilator*) in the Gulf of Mexico (Rathbun 1918).

Fig. 15.1 Photos of maritime mammals possible to find in areas surrounding the Bahía Blanca Estuary: **a** *Ctenomys australis*, southern tuco-tuco (English), tuco-tuco de las dunas (Spanish), and **b** *Hydrochoerus hydrochaeris* capybara (English), carpincho, capibara (Spanish). (Photos by Federico Becerra (a) and Irma Gamarra (b))

The only lagomorph (order Lagomorpha), the European hare (*Lepus europaeus*), is an exotic species that has dispersed across most of the Argentine territory and appears to be relatively abundant in the southern Buenos Aires Province, including the coastal zone. The European hare may graze at on salt marshes and, occasionally, on mudflats (Moore 2002).

The order Cingulata is represented by five species, but *Chaetophractus villosus* (large hairy armadillo; see species data sheet 15.4.3; Fig. 15.2a) is the one that most likely uses marine food resources due to its omnivorous diet and especially its scavenger habits (Arriagada et al. 2017). In accordance to this hypothesis, this armadillo was relatively frequently recorded in camera trap and sign surveys in a coastal protected area south of Marisol (unpubl. Data and following section of this chapter).

There are two species of the Cetartiodactyla order that occur in our area of interest. Although there is no site-specific ecological information about the guanacos (*Lama guanicoe*), a small population of this native camelid, which is currently very rare in Buenos Aires Province, persists in islands of the estuary of Bahía Blanca.

Fig. 15.2 Photos taken by camera trap of maritime mammals possible to find in areas surrounding the Bahía Blanca Estuary: **a** *Chaetophractus villosus*, large hairy armadillo (English), peludo (Spanish), and **b** *Sus scrofa*, wild boar (English), chancho jabali (Spanish). (Photos by Grupo de Ecologia Comportamental de Mamiferos)

Because the vegetation on the islands is scarce, it is possible that guanacos use cordgrass (Bortolus et al. 2015) as supplementary source of food. The wild boar (*Sus scrofa*) (see species data sheet 15.4.4; Fig. 15.2b) is an exotic species which is becoming increasingly common in the region. Because of their wide food niche and opportunistic habits, wild boars could predate on seabird nests and also consume other biological resources provided by the sea, as observed in the coastal area of Doñana National Park, Spain (Fernández-Llario et al. 1996).

Carnivores (order Carnivora) that use the ecotone between terrestrial and marine habitats may act as a strong and dynamic link connecting terrestrial and marine food webs. There are eight native species from the order Carnivora that reportedly live in the region of interest. Three of them are rare and/or difficult to observe, including two small felids, jaguarundi *Herpailurus yagouaroundi* (categorized as Near Threatened by IUCN, Caso et al. 2015, and in Argentina as Least Concern by

Bisceglia et al. 2019) and Pampas cat *Leopardus colocolo* (categorized as Vulnerable by IUCN, Lucherini et al. 2016, and the same category in Argentina by Lucherini et al. 2019), and a mustelid, the Patagonian weasel *Lyncodon patagonicus* (listed as Near Threatened by IUCN, Kelt et al. 2016, and the same category in Argentina by Schiaffini et al. 2019). Although the most recent distribution assessment includes the southern Buenos Aires Province into the range of the Patagonian ferret, the latest records of its occurrence are from 1881, and its present-day presence in the area is exclusively based on a distribution model that predicted a low probability of occurrence (Schiaffini et al. 2013a). Although the presence of this carnivore would need to be confirmed, the coastal habitats of southwestern Buenos Aires appear to represent a shelter for the populations of carnivores that are rarer elsewhere, most likely because the comparatively low fertility of the sandy soils along the coast has prevented the development of agriculture and limited livestock farming. This appears to be especially true for the Pampas cat and the puma (*Puma concolor*) (see species data sheet 15.4.5, Fig. 15.3a). Although little data is available (see the next section of this chapter), local inhabitants of the rural areas report that pumas, which had almost completely disappeared from the province, have greatly increased their numbers in the semi-natural habitats found along the coast in the last 10–15 years (Canevari and Balboa 2003). The puma is a generalist top predator that could potentially interact with marine and maritime mammals. There are reports of pumas feeding on penguin colonies (Frere et al. 2010; Martínez et al. 2012), and it cannot be excluded that they prey on other seabirds and that these events have gone undetected.

The Pampas fox (*Lycalopex gymnocercus*; see species data sheet 15.4.6; Fig. 15.3b) and Geoffroy's cat (*Leopardus geoffroyi*; see species data sheet 15.4.7; Fig. 15.3c) are two of the most common carnivores in the entire province. The Pampas fox is a generalist canid that can be found in all the habitats of our region, being capable of living in areas modified by extensive ranching and agriculture human activities (Luengos Vidal et al. 2012; Caruso et al. 2016). This carnivore has a very wide trophic niche (Lucherini and Luengos Vidal 2008). In spite of the fact that there is no study on Pampas foxes proving that they use the coastal zone or intertidal habitats, some evidence suggest that they effectively feed in coastal areas. García and Kittlein (2005) analyzed the summer diet of this fox in Bahıa San Blas and Isla Gama Provincial Reserve, located in southern Buenos Aires Province, and found that although fish carrion and crustaceans are not frequent, these items are present in its diet. They also observed a high opportunistic consumption of fruit probably obtained in sand dune habitats. Based on their findings, they conclude that foxes utilize sand dunes and scrublands as feeding sites, and grasslands for other activities unrelated to food search, such as refuge and breeding (García and Kittlein 2005; Bossi et al. 2019), found that the diet of the Pampas fox from the Brazilian Pampas included fish, crustaceans of the Decapoda family, and marine crabs (Brachyura), thus providing evidence that this canid uses marine preys. It is also worth mentioning that the Darwin fox (*Lycalopex fulvipes*), a very close relative of the Pampas fox that inhabits the Chiloe Island (Chile), was observed digging in the intertidal fringe of sandy beaches where the macrofauna consists of crabs, talidrid

Fig. 15.3 Photos of maritime mammals possible to find in areas surrounding the Bahía Blanca Estuary: **a** *Puma concolor*, cougar (English), puma (Spanish); **b** *Lycalopex gymnocercus*, Pampas fox (English), zorro pampeano (Spanish); and **c** *Leopardus geoffroyi* Geoffroy's cat (English), gato montes (Spanish). (Photos by Grupo de Ecologia Comportamental del Mamiferos (a, c) and Nicolás Mariano Chiaradía (b))

amphipods, and isopods (Elgueta et al. 2007). The above-reported findings suggest that the Pampas fox is a strong candidate to qualify as maritime mammal.

Geoffroy's cat, as most felids, has a specialized diet, especially if compared to that of sympatric canids (Kasper et al. 2016). Nevertheless, fish remains have been found in fecal samples of this cat from Campos del Tuyú (Manfredi et al. 2004), a coastal protected area adjacent of northern Buenos Aires Province, and aquatic birds have been found in Mar Chiquita Biosphere Reserve (Canepuccia et al. 2007), another protected area on the Buenos Aires coast.

Molina's hog-nosed skunk (*Conepatus chinga*; see species data sheet 15.4.9; Fig. 15.4a) is the only member of the Mephitidae family found in our region. It is typically an insect feeder (Castillo et al. 2014) that could potentially benefit from the trophic chain initiated by seasonal marine items (see Sect. 15.1 in this chapter) but is also clearly an opportunistic predator, and several other food items may be included in its diet, such as mammals, reptiles, and amphibians. Interestingly, in Southern Brazil, Kasper et al. (2012) recorded several events of Molina's hog-nosed skunks predating on turtle and bird nests to eat eggs. The lesser grison (*Galictis cuja*) (see species data sheet 15.4.8; Fig. 15.4a) has a diet similar to that of a small wildcat and is considered strictly carnivorous. The limited data available on this mustelid indicates that it preys mostly on rodents, but can probably use different foraging strategies (Kasper et al. 2016) and frequently consumes birds and their eggs (Borboroglu and Yorio 2004; Valenzuela et al. 2013; Galende and Raffaele 2016).

Fig. 15.4 Photos of maritime mammals possible to find in areas surrounding the Bahía Blanca Estuary: **a** *Galictis cuja*, lesser grison (English), hurón menor (Spanish), and **b** *Conepatus chinga* Molina's hog-nosed skunk (English), zorrino comun (Spanish). (Photos by Luciano Lapolla (a) and Grupo de Ecologia Comportamental del Mamiferos (b))

15.4 Species Data Sheet of Some Relevant Maritime Mammals

Ctenomys australis, Hydrochoerus hydrochaeris, Sus scrofa, Chaetophractus villosus, Puma concolor, Lycalopex gymnocercus, Leopardus geoffroyi, Conepatus chinga, Galictis cuja

15.4.1 Ctenomys australis *Rusconi, 1943 (Order: Rodentia)*

Common name: tuco-tuco de las dunas (Spanish), southern tuco-tuco (English).

Description: It has a chunky and plump body. Total adult length is between 210 and 370 mm and weight is between 248 and 500 g. The head is bulky, short, and flattened and the neck is short and thick; it has large ears, underdeveloped but not tiny eyes, and four large incisors left outside the mouth. The short, well-muscled legs finish in five fingers with nails that are used to dig. The color is brown yellow, clear than other species of the some genera (Gomez Villafañe et al. 2005) (Fig. 15.1a).

Distribution and habitat: *C. australis* is a territorial rodent that lives exclusively, in the narrow strip of coastal dunes, characterized by the presence of sandy, poorly compacted soils and little or no vegetation cover. Its distribution range extends approximately 300 km between the cities of Necochea (38°37'S, 58°50'O) and Punta Alta (39°30'S, 61°40'O). This rodent is a strict habitat specialist and shows low dispersion rates and relatively small population sizes with a high degree of genetic structuring (Zenuto and Busch 1998; Cutrera et al. 2006).

Behavior: *C. australis* is solitary, highly territorial, and most likely polygynous (Zenuto and Busch 1998). It is crepuscular and diurnal. It vocalizes like the other tuco-tucos, with drums like beats but only in the hottest hours of the day (Gomez Villafañe et al. 2005). Radio telemetry data suggest that home ranges are greater in males than females, averaging 1300 m^2 for males and 580 m^2 for females (Mora 2008). Individuals build large, exclusive burrow systems in coastal sand dunes. The habit of underground life of these animals contributes to the dynamics of plant communities by altering their composition and maintaining succession over pioneer states without altering total biomass (Monserrat et al. 2012).

Conservation status: This species was listed in the "Endangered" category under both the IUCN Red List (Bideau 2016) and the Argentine National Assessment (Austrich et al. 2019).

Threats: Its peculiar distribution makes this rodent extremely susceptible to the changes generated in its habitat as a result of tourism, urbanization, forestry development, and the progressive advance of artificial pastures. Only 2% of the area of the sandy dune ecosystem in which the species is found is protected by some type of legal conservation status (Monserrat et al. 2012).

15.4.2 Hydrochoerus hydrochaeris *(Linnaeus, 1766) (Order: Rodentia)*

Common name: capybara (English), carpincho, capibara (Spanish).

Description: The capybara is a large, semi-aquatic rodent that is found in regions of Central and South America. The body is large (approx. 1200 cm) and heavy (35–70 kg) with a descendent croup; tail is vestigial; limbs are short; feet are perissodactyle, with digits united by moderate webbing; forefeet possess four digits. The head is broad; ears are short and rounded; snout is large; nostrils are small and widely separated. Hair is coarse, dark brown to reddish, and light brown to light yellow (Mones and Ojasti 1986) (Fig. 15.1b).

Distribution and habitat: The species has a wide distribution in Argentina, where it is found in at least 10 provinces and 25 protected areas, although with varying degrees of protection. In protected sites, it is usually abundant, but outside of them, densities drop dramatically. Since 2005, there has been a range expansion in the Buenos Aires Province where it currently covers the entire eastern and southern fringe of the province to the proximity of Bahía Blanca, where capybaras were historically found (Doumecq Milieu et al. 2012).

Behavior: These rodents are strictly herbivores. They reproduce throughout the year, without a specific birth season. They can give birth to between one and eight offspring, with an average of four offspring, frequently in relation to the environment and nutritional conditions of the female (Cueto 1999). It is a species active during the day and night, usually in groups of between 2 and 30 individuals with a dominant male, and several females, young and male subalterns. The size of their home ranges between 11 and 27 ha, approximately (Corriale et al. 2013).

Conservation status: Capibara is included in the "Least Concern" category under both the IUCN Red List (Reid 2016) and the Argentine National Assessment (Bolkovic et al. 2019).

Threats: The loss of wetlands is one of the most important threatening factors for this species. The leather of this rodent has great commercial value. Thus, for decades, direct hunting was the main threat to the species in Argentina because the extraction was carried out with poor management plans. Although the leather trade is gradually decreasing, even today, many local communities within their range use capybaras to meet their protein needs. Population fragmentation is an additional threat and exotic species can affect capybaras. Finally, capybaras can be affected by vehicle traffic on roads: on Argentine National Route 174 (connecting the cities of Rosario and Victoria), 95% of reports of wildlife accidents involving vehicles corresponded to capybaras (Bolkovic et al. 2019).

15.4.3 Chaetophractus villosus *(Desmarest, 1804) (Order: Cingulata)*

Common name: peludo (Spanish), large hairy armadillo (English).

Description: It is a dark-colored armadillo, with a body length of 291 mm approximately without the tail and the tail of 145 mm (Redford and Eisenberg 1992). The weight ranges between 2 and 5 kg (Gallo et al. 2019). The male is generally larger than the female. Armadillos possess a characteristic protective armor that consists in a small shield between the ears on the back of the neck and a carapace that protects the shoulders, back, sides, and rump. This species has hairs projecting from between the scales of the body armor, and the limbs and belly are covered with whitish or light brown hairs. Some individuals in this species have 3–4 holes in the pelvic region of the armor that open to glandular pits (Redford and Eisenberg 1992) (Fig. 15.2a).

Distribution and habitat: This species has its origin in the Pampas region (Poljak et al. 2010). It is currently distributed throughout virtually the entire country, and it is documented that it invaded the Patagonia in historical times. It can be found in Argentina, Bolivia, Paraguay, and Chile (Abba et al. 2014).

Behavior: *C. villosus* is an omnivorous species that eats mainly invertebrates and digs under and into carcasses to eat carrion. It preys potentially on imperial cormorants (*Phalacrocorax atriceps*), rock shags (*Phalacrocorax magellanicus*), and kelp gulls (*Larus dominicanus*) in Patagonia, Argentina (Punta et al. 2002; Borboroglu and Yorio 2004). This armadillo is solitary except when rearing young or mating. It can build and use extensive burrow systems. Gestation lasts 60–75 days and litters are usually 1–3 offspring. Typically, females produce a litter per year. Birth occurs inside a burrow, in which the female usually builds a nest with plant material. Weaning occurs at 55–61 days of age (Gallo et al. 2019).

Conservation status: *C. villosus* is listed in the "Least Concern" category under both the IUCN Red List (Abba et al. 2014) and the Argentine National Assessment (Gallo et al. 2019).

Threats: It is hunted to be used as food by local people and to make musical instruments and handicrafts. It is persecuted as a pest in agricultural areas for damaging the soil with its burrows and breaking the grain storage bags. It is also accused of being a nuisance for poultry and for sheep farming during the lambing periods and is subjected to retaliatory/preventive hunting. Animals may also be accidentally killed on roads and by dogs.

15.4.4 Sus scrofa *Linnaeus, 1758 (Order: Cetartiodactyla)*

Common name: chancho jabali (Spanish), wild boar (English).

Description: It is a wild pig whose head and body lengths are between 1100 and 1650 mm, shoulder height is 550–1000 m, and weight is 10–190 kg. Females are generally smaller in size and weigh less than males. Body color varies from gray to brown or black or a mixture of colors. The tusks in males may reach 150 mm (Long 2003) (Fig. 15.2b).

Distribution and habitat: Wild boars are native to large parts of Europe, Asia, and North Africa, and they have become extinct over much of their former range. Introduced populations, descended from domestic pigs, wild boars, or a combination of the two, are present in many parts of the world. It is considered by the IUCN as one of the 100 invasive species most damaging to biodiversity in the world. *S. scrofa* was initially introduced in Buenos Aires in 1536 by colonizers as a food source. Only at the beginning of the twentieth century the wild boar was introduced from Europe for hunting purposes. Its presence has been confirmed in different ecoregions of Argentina with preponderance toward environments with high water availability and vegetation cover (Ballari et al. 2019).

Behavior: The capacity of wild boars to colonize different habitats and become an invasive species is related not only to their generalistic habitat requirements but also to its biological characteristics, such as the high reproduction rate, omnivorous diet, and behavioral plasticity. A research conducted in the southern part of the Buenos Aires Province Espinal in a non-protected area found that wild boars exhibited a crepuscular and nocturnal activity pattern that peaked around midnight and that their habitat use was significantly and negatively affected by the proportion of grass cover but temperature and water availability were two additional and important factors affecting distribution and abundance (Caruso et al. 2018). In Argentina, one or two annual reproductive events have been commonly recorded, although there are warm climate areas where it is possible to find pregnant females all year round (Ballari et al. 2019).

Conservation status: Wild board is included in the "Least Concern" category under the IUCN Red List (Oliver and Leus 2008). The Argentine National Assessment assigned the same threat category but also listed it as "Invasive species" (Ballari et al. 2019).

Threats/impacts: In Argentina, numerous negative impacts by wild boars on native species have been described, and the species is an important vector of parasites and diseases that can potentially affect the native mammals which it coexists with. Additionally, the wild boar causes damage in agricultural plantations and preys on young livestock, particularly sheep, while its hoofs destroy and deteriorate grassland and pasture.

15.4.5 Puma concolor *(Linnaeus, 1771) (Order: Carnivora)*

Common name: puma, leon americano (Spanish), puma, mountain lion, cougar (English).

Description: Body length is between 85 and 160 cm and weight between 22 and 100 kg. This cat presents a large variation in size across its distribution range. The coat is fairly uniform in color and unmarked, but color varies from a buff or sandy brown to reddish color, through to light silver and slate gray. The head is small with dark brown to black patches on the muzzle. The ears are short and rounded. The tail is long gradually darkening toward the tip (Sunquist and Sunquist 2002) (Fig. 15.3a).

Distribution and habitat: *Puma concolor* has a very extensive geographic distribution in the Americas and is characterized by a wide ecological plasticity. Across its distribution, the puma is found from high mountains to deserts (Nowell and Jackson 1995), including landscapes dominated by human activities (Caruso et al. 2016).

Behavior: It is a very adaptable species, with great leaping and climbing abilities. The activity pattern is primarily nocturnal and crepuscular. Puma food habits have been broadly studied across its distribution, showing a generalist foraging behavior with a diet composition reflecting the specific prey community found in each region. The preferred size of puma prey ranges from 70 to 165 kg (Carbone et al. 2014). Large kills are typically covered with scraped-over vegetation and dirt, and pumas return frequently to feed on the carcass. These felids are solitary carnivores, although recent studies have found a preciously unreported degree of sociality (Elbroch and Quigle 2017). Litters are between 1 and 3 kitties and can reach 6; reproductive age begins at more than 18 months, and the gestation time is approximately 91.5 days (Logan and Sweanor 2001).

Conservation status: Puma has been listed in the "Least Concern" category under both the IUCN Red List (Nielsen et al. 2015) and the Argentine National Assessment (De Angelo et al. 2019). It is listed in Appendix II of CITES.

Threaten: Puma populations are thought to be declining globally primarily due to anthropogenic pressure (De Angelo et al. 2011). This felid is considered harmful to livestock, and for that reason, palliative and/or preventive hunting is a common practice that may lead to population declines (Guerisoli et al. 2017). In some Argentine provinces, puma killing is legal, and bounty systems are in place, but this is not the case of Buenos Aires Province, where hunting of pumas is illegal. Despite the ability of the puma to adapt to anthropic changes, there are factors that cause the loss and degradation of its habitat and may lead to a decrease in the abundance of the species. Car-struck pumas on routes have been recorded in different regions, leading to considering this source of mortality as a potential conservation threat. Mascotism and transmission of diseases from domestic animals could also pose a threat to the species (Foley et al. 2013).

15.4.6 Lycalopex gymnocercus *Fisher, 1814 (Order: Carnivora)*

Common name: zorro pampeano (Spanish), Pampas fox (English).

Description: *L. gymnocercus* is a medium-sized fox. Body size varies geographically. Mean measurements of body mass are between 4.21 and 5.95 kg, and total body length for adults is between 505 and 800 mm (Crespo 1971; Luengos Vidal et al. 2009). Pelage on the top and sides of the head is reddish and on the dorsal rostrum is reddish to black. The ventral surface of the head is pale gray to white. Back, shoulders, and flanks are gray. A blackish line runs along the center of the back and tail. The tail is relatively long (50% of the length of head and body), bushy, and gray with a black tip (Lucherini and Luengos Vidal 2008) (Fig. 15.3b).

Distribution and habitat: The Pampas fox inhabits the Southern Cone of South America and is one of the most common and widespread carnivores within its geographic range. This canid prefers open habitats but also occurs in areas of Pampas grassland modified by extensive ranching and agricultural activities (Lucherini and Luengos Vida 2008; Caruso et al. 2016).

Behavior: Pampas foxes are considered either abundant or common in most areas. It is a generalist and adaptable predator. Diet varies geographically and includes both domestic and wild vertebrates, as well as fruit, insects, carrion, and garbage. Gestation lasts 55–60 days. It is a monoestric species with a well-defined period of reproduction (Crespo 1971). Observations of free-ranging foxes in Buenos Aires Province indicate that litter sizes vary between two and four individuals and that both parents participate in pup care (Sassola 2006). In the center of Argentina, home range sizes average 213.3 ± 136.8 ha, without significant variations between females and males or variations related to natural or modified environments (Luengos Vidal 2009).

Conservation status: The Pampas fox is included in the "Least Concern" category under both the IUCN Red List (Lucherini 2016) and the Argentine National Assessment (Luengos Vidal et al. 2019). It is listed in Appendix II of CITES.

Threats: This canid is apparently capable to adapt to strong alterations of natural habitats, and the main threat to its conservation is direct illegal hunting related to the high level of conflict with livestock, mainly sheep. Although this practice is illegal, a large part of the ranchers in the center of Argentina uses poison to control the population of this species, deploying poisoned baits before the sheep shedding stage, with reportedly very efficient results (Luengos Vidal et al. 2019). There are reports of predation and harassment of dogs on gray foxes.

15.4.7 Leopardus geoffroyi *(d'Orbigny and Gervais, 1844)*

Common name: gato montes (Spanish), Geoffroy's cat (English).

Description: This felid is a medium-sized (430 and 880 cm in length) spotted cat that weights between 2.2 and 7.8 kg. Males are larger than females (Lucherini et al. 2016). Thus its size is similar to a domestic cat. Its coat color and size vary. The fur is from gray to ochraceous with numerous black spots about 15 to 20 mm in diameter which tend to group in the central parts of the dorsum two by two so as to enclose lighter areas. The tail is shorter than that of other Neotropical small cat species (Sunquist and Sunquist 2002) (Fig. 15.3c).

Distribution and habitat: *L. geoffroyi* is widespread in the southern part of South America. It is also found in a variety of habitats, open and closed and wet and dry. It is tolerant to a certain degree of habitat alteration and degradation and is present in well-preserved areas as well as in sites highly modified by anthropic activity, either by agricultural activity or by urbanization (Caruso et al. 2016, but see Castillo et al. 2008).

Behavior: It feeds mainly on small rodents and sometimes birds and lagomorphs. It has been shown that the relative importance of waterbirds in the diet changed with seasonal fluctuations in their availability (Manfredi et al. 2004, Canepuccia et al. 2007). This wildcat is mainly, but not exclusively, nocturnal; human disturbance did not affect the patterns of activity, but cats appeared to be forced to spend more time active and move over greater distances in more disturbed areas (Manfredi et al. 2011). Geoffroy's cats show a very similar use of the space despite the largely diverse habitats that they inhabit (Tirelli et al. 2018). In Buenos Aires Province, home range size varied from 248 ha to 342 ha for a wetland area (Manfredi et al. 2006) and from 220 to 280 ha for the Espinal (Castillo et al. 2019), with male home ranges approximately 2.5–4.1 larger than those of females.

Conservation status: This cat is listed in the "Least Concern" category under the IUCN Red List (Pereira et al. 2015) as well as under the Argentine National Assessment (Pereira et al. 2019). It is also listed in Appendix II of CITES.

Threats: This cat is primarily threatened by habitat loss and fragmentation, but also by the hunting by farmers because of its attacks to domestic animals. Predation by domestic dogs and vehicle collisions are additional sources of mortality (Pereira et al. 2010; Caruso et al. 2017). Although international trade has dramatically decreased after it was declared illegal, the commercialization of Geoffroy's cat pelts can be locally important. Individuals of the species exposed to various pathogens and parasites shared with domestic dogs and cats have been reported, which constitutes a health risk for the affected populations (Vega et al. 2018).

15.4.8 **Galictis cuja** ***(Molina, 1782) (Order: Carnivora)***

Common name: hurón menor (Spanish), lesser grison (English).

Description: The lesser grison is a small mustelid that weighs between 800 and 2400 g, with a thin, elongate body with a long neck, narrow chest, short legs, and a short and bushy tail. Head is small and flat with short, broad, rounded ears. Body length is between 425 and 667 mm, and there is a clear pattern of sexual size dimorphism, with females smaller than males. The coat is gray-yellowish in the upper parts and black on the face, with a diagonal narrow strip that runs from the forehead to the shoulder (Redford and Eisenberg 1992; Yensen and Tarifa 2003) (Fig. 15.4a).

Distribution and habitat: Its distribution range includes Argentina, Peru, Bolivia, Paraguay, and Chile. Although no estimates of population density are available, it is considered common in grassland, savanna, and steppe habitats, while it appears to be naturally rare in some high altitude, arid, very dense vegetation habitats.

Behavior: The lesser grison is active mostly during daylight (Tellaeche et al. 2014; Luengos Vidal et al. 2016). It is not uncommon to observe two or three individuals feeding or moving together, probably from the same family groups. The litter size is 2–5 offspring. This mustelid is an active predator. Its diet resembles that of small felid and includes reptiles, small birds, lagomorphs, and rodents (Redford and Eisenberg 1992; Punta et al. 2002; Galende and Raffaele 2016).

Conservation status: The lesser grison is listed in the "Least Concern" category under both the IUCN Red List (Helgen and Schiaffini 2016) and the Argentine National Assessment (Aprile et al. 2019).

Threats: Habitat degradation due to agricultural activities, excessive overgrazing, and soil erosion by livestock are the most important threat they face. Harassment and predation by dogs (feral or domestic) were also documented. Likewise, the transmission of pathogens by dogs, such as distemper, could also affect this mustelid. Lesser grisons are occasionally hunted for considering it a good luck charm. They are kept as pets or trained to eliminate rodents in sheds and hatcheries (Aprile et al. 2019).

15.4.9 Conepatus chinga *(Molina, 1782) (Order: Carnivora)*

Common name: zorrino comun (Spanish), Molina's hog-nosed skunk (English).

Description: *C. chinga* is a small-sized mephitid, weighing approximately between 1 and 3 kg and with a body length from 410 to 685 mm from nose to tail. It has the characteristic skunk coloring with generally black fur and two white stripes running from the top of the head down the sides of the body to a mostly white tail, although in some cases the two stripes are absent (Redfor and Eisenberg 1992) (Fig. 15.4b).

Behavior: It is an omnivore species whose diet includes mainly insects but also reptile and bird and plats bulbs (Castillo et al. 2014; Kasper et al. 2016). It is mostly active at night and twilight. Although Molina's hog-nosed skunks are solitary animals, data from Buenos Aires Province indicate that males and females may be able to share dens, presumably in the breeding season (Castillo et al. 2013). The reproductive period of *C. chinga* seems related to climatic seasons (Kasper et al. 2009). The gestation period is close to 40 days. Litters that vary from two to three offspring stay between 4 and 5 months with the mother (Kasper et al. 2009). The home ranges vary significantly in size between males (243.7 ± 76.5 ha) and females (120.4 ± 77.6 ha) in Buenos Aires Province and averaged 163 ± 1.17 ha in Southern Brazil (Kasper et al. 2012). Overlap between home range and core area was extensive between and within sexes (Castillo et al. 2011, Kasper et al. 2004). It digs its own borrow but may also use borrows dug by other animals (Redford and Eisenberg 1992).

Distribution and habitat: It is distributed in Argentina, Bolivia, Brazil, Chile, Paraguay, Peru, and Uruguay. In Argentina, it is present in all provinces, except Tierra del Fuego. Currently, only one species is recognized throughout Argentina (Schiaffini et al. 2013b).

Conservation status: *C. chinga* is included in the "Least Concern" category under both the IUCN Red List (Emmons et al. 2016) and the Argentine National Assessment (Castillo and Schiaffini 2019). It is listed in Appendix II of CITES.

Threats: Although the abundance of its population is poorly known (Castillo et al. 2011; Kasper et al. 2012), this carnivore appears to be relatively common throughout its range. Molina's hog-nosed skunks were heavily hunted for their fur in Argentina during the 1970s and early 1980s. Additionally, extensive areas of skunk habitat have been severely degraded through overgrazing and soil erosion by livestock (primarily sheep) and feral, exotic species. It is often used as a pet and persecuted by rural inhabitants for the consumption of eggs and poultry (Castillo and Schiaffini 2019).

15.5 A Study Case: Maritime Carnivores of a Coastal Area of the Buenos Aires Province

Buenos Aires is the second largest and one of the most developed provinces in Argentina. It concentrates most livestock and cropland activities. This province once hosted the natural grasslands typical of the Pampas ecoregion, but since the arrival of the European colonizers, it has been going through a human-driven process of fragmentation, which caused the loss of more than 60% of its original habitats (Brown et al. 2006). This situation has had a great impact on the ecology, behavior, and conservation of several vertebrate species and particularly of mammalian carnivores, along with the role that they play in the ecosystems (see previous section).

Although the basic ecological information available for carnivores in the province has increased during the last decades, most of the projects studying these species focused on inland areas; almost no data is available for coastal areas. One of those projects was carried out on Campos del Tuyú Wildlife Reserve (36°20′S, 56°50′W) and mainly focused on Geoffroy's cat. Data on its food habits were reported (Manfredi et al. 2004), along with characterization of defecation sites (Soler et al. 2009), and spatial ecology (home range size and habitat use) (Manfredi et al. 2006), and the general activity pattern (Manfredi et al. 2011) of radio-tracked Geoffroy's cat individuals. Mar Chiquita Biosphere Reserve (37°46′S, 57°27′W) is another coastal area represented in the literature. Manfredi et al. (2004) and Canepuccia et al. (2007) studied Geoffroy's cat food habits in general and in relation to waterbird abundance, size, and distance before attack; and Farias and Kittlein (2008) reported data on Pampas fox food habits. Finally, Caruso et al. (2012) published the first record of occurrence of Geoffroy's cat in the Bahía Blanca, Bahía Falsa, and Bahía Verde Natural Reserve (38°56'S; 62°14'W).

Given this clear paucity of information, we present the ecological data on the carnivore community obtained in a coastal area that represents a relic of grasslands in the Buenos Aires Province. Because the natural grasslands have almost disappeared due to their transformation into croplands, these relicts have a high conservation value (Miñarro 2004).

Fieldwork was carried out in the "Arroyo Los Gauchos Nature Reserve," located in the southwestern Buenos Aires Province (Coronel Dorrego county, 38°56′1.93"S, 60°45′9.37"O; Fig. 2.1; Chap. 2 of this book). The "Arroyo Los Gauchos Nature Reserve" is a coastal protected area of 7.07 km^2 that belongs to the Pampas ecoregion and is characterized by the presence of dunes, both non-vegetated and covered by psammophyte vegetation, and a general low topography with interdune depressions occupied by relatively small wetlands and halophyte grasses. The climate is temperate, the annual mean temperature is 14.1 °C, and the annual mean precipitation is approximately 850 mm (Monserrat et al. 2012; Celsi and Giussani 2019).

Although the study area is formally protected by the provincial government since 2011, it lacks implementation, and its habitats are preserved primarily because of the difficulty of access. Until now, human disturbance has had a low impact in comparison to the other coastal areas of the province. Activities such as cattle raising, forestry, and tourism are still limited, and urban development is scarce (Monserrat and Celsi 2009; Monserrat and Codignotto 2013). The area was proposed as a valuable grassland area (Miñarro 2004) well before its establishment as a protected area due to its high biodiversity (more than 72 vascular plant species) and the presence of endemic rare or endangered vertebrate species, among which are *Ctenomys australis* and a lizard, *Liolaemus multimaculatus* (Monserrat and Celsi 2009).

To characterize the carnivore community of the study area, we conducted a survey using sign counts and camera trapping from January 2009 to March 2009.

To study the habitat use by carnivores, we conducted 1-km-long linear transects, looking for indirect evidences of the presence of mostly feces and tracks. Both the starting point and the bearing of transects were placed randomly over the study area. We sampled each transect walking at a constant speed along 1 km and recorded the GPS position of each evidence found. We characterized the habitat surrounding each evidence within a radius of 20 m by estimating visually the proportion of each habitat type. Seven habitat types were identified as the most characteristics in the study area: bare ground, mostly non-vegetated dunes or any other portion of land without vegetation cover; low grassland, grasses (Poaceae) less than 30 cm tall; medium-tall grassland, areas covered by grassy vegetation (Poaceae) taller than 30 cm; areas covered by *Juncus* spp.; areas covered by *Cortaderia* spp.; areas covered by *Hyalis argentea*; and shrubland, mostly dominated by *Lycium chilense*, *Baccharis divaricata*, and *Discaria americana*. Additionally, we used camera traps to increase the chances of recording rarer species that are more difficult to detect by sign counts and thus complete the information on the composition of the carnivore community.

In 14 line transects, we recorded 67 mammal evidences from 5 different taxa: Pampas fox, Geoffroy's cat, skunk, armadillo (probably large hairy armadillo), and rodents. Feces represented 62.8% of the evidences, followed by tracks (34.3%). The Pampas fox was the most frequent species, with 33 records, followed by Geoffroy's cat, with 15 records. Given the lack of reliable information for the rest of species on the base of their scarce evidence, we focused our analysis on the two most common carnivore species: the Pampas fox and Geoffroy's cat. We used logistic regressions to study if any of the variables registered affected the probability of finding evidences of these species. Evidences were treated as a dummy variable (i.e., evidence corresponding to *L. gymnocercus* or evidence corresponding to *L. geoffroyi*), and the habitat variables were used as predictors. Using the function "dredge" of the package MuMIn in R (R Core Team 2013), we created a list of all possible models given the list of predictors and assuming only additive effects. We ranked all models using Akaike's information criterion adjusted for small samples (AICc) and used the ΔAICc and the AICc weights to evaluate the relative importance of each model within the final set of models (Burnham and Anderson 2002). We used a multimodel inference approach (Burnham and Anderson 2002) and averaged the set of models with ΔAICc <2. The regression coefficients (β) were considered to have a significant effect when the corresponding 95% confidence intervals (CI) excluded zero (Zeller et al. 2011). With respect to the habitat surrounding each evidence, five models formed the set with ΔAICc <2 (Table 15.2). Proportion of *Cortaderia* spp. and distance to shoreline were the only variables included in the five models. The averaged model and the confidence intervals suggested that a greater proportion of *Cortaderia* spp. reduced the chance of detecting the presence of *L. geoffroyi* in relation to *L. gymnocercus*, which would indicate that Pampas foxes use more intensively the sites with more *Cortaderia* spp. than Geoffroy's cats. The distance to the shoreline increased the chance of presence of *L. gymnocercus* signs, thus indicating that Geoffroy's cat tended to use site closer to the shoreline.

Camera trapping produced records for five carnivore species: Pampas fox, Geoffroy's cat, Molina's hog-nosed skunk *Conepatus chinga*, puma *Puma concolor*, and lesser grison *Galictis cuja*. These are the most common carnivores in the Pampas agroecosystems (Luengos Vidal et al. 2005) and also in a sandy dune area in the Espinal ecoregion located not very far from our study area (Caruso et al. 2012). *Herpailurus yagouaroundi* and *Leopardus colocolo* were not detected during the study and are presumably absent from our study area because the local people

Table 15.2 Parameters of the generalized linear models (GLMs) contrasting the habitat use of *L. gymnocercus* and *L. geoffroyi*. Only parameters for the set of models with ΔAICc <2 are reported

1	Int.	S	BG	C	D	J	df	logLik	AICc	delta	weight
1	0.76			−0.68	−0.77		3	−27.85	62.23	0.00	0.29
2	0.47	−2.46		−0.70	−0.76		4	−26.80	62.51	0.28	0.25
3	0.48	−2.55	−0.56	−0.98	−1.02		5	−25.90	63.20	0.98	0.18
4	0.76		−0.41	−0.88	−0.97		4	−27.32	63.55	1.32	0.15
5	0.76			−0.78	−0.95	0.35	4	−27.35	63.62	1.39	0.14
Avr.	0.64	−2.50	−0.49	−0.78	−0.87	0.35					
CI (2.5%)	−63.59	−693.10	−1.35	−1.55	−1.71	−0.37					
CI (97.5%)	64.87	688.10	0.36	−0.01	−0.02	−1.07					

Int intercept, *S* proportion of shrubland, *BG* proportion of bare ground, *C* proportion of *Cortaderia* spp., *D* distance to shoreline, *J* proportion of *Juncus* spp., *d.f.* degrees of freedom, *logLik* = log likelihood function, *ΔAICc* = difference in value of Akaike's information criterion between the focal model and the top-ranked model, *Avr.* average model coefficients, *CI* confidence interval did not recognize those (Caruso et al. 2020). This is not surprising given the rarity of both species in more typical cropland areas far from the coast (Benzaquín et al. 2009; Luengos Vidal et al. 2017). The following non-carnivore species were also found in the study area: the red deer *Cervus elaphus*, European hare *Lepus europaeus*, wild boar *Sus scrofa*, greater rhea *Rhea americana*, and several rodents and armadillos. *L. gymnocercus* and *L. geoffroyi* were the most frequently recorded species, followed by *C. chinga*, *P. concolor*, and *G. cuja*.

With a relatively limited surveying effort, camera traps enabled us to detect five of the seven species of carnivores that may potentially occur in this coastal area. They also confirmed that the Pampas fox and Geoffroy's cat were the most common species in the carnivore community of southern Buenos Aires coastal habitats.

There are many factors that can affect the presence and use of habitat of a species, and it is very difficult to draw conclusions with a relatively brief survey in a heterogeneous habitat. For example, the stranding of marine vertebrate carcasses on the coast, which did not occur in the period during which the survey was carried out, could have substantially changed the results, given the importance of carcasses on the ecology of coastal scavengers such as canids or vultures (see the previous sections of this chapter). However, our results suggest a prevalence of foxes in environments with *Cortaderia*. This plant grows in coarse widespread patches, is associated with lightly humid soils, and is used as shelter and food resources by different species of rodents (Bonaventura et al. 2003) and birds (Pretelli et al. 2013). Thus, the use by a trophic generalist species like the Pampas fox (García and Kittlein 2005; Lucherini and Luengos Vidal 2008; Bossi et al. 2019) may be associated with the presence of these potential sources of food. Geoffroy's cat is more specialized in its food habits and may find the areas close to the shoreline more suitable for hunting birds, as reported in another coastal area (Canepuccia et al. 2007). However, because intraguild competition is a structuring factor in carnivores (e.g., May et al. 2008; Di Bitetti et al. 2010), based on our data, we cannot exclude the possibility that interspecific competition plays an important role in explaining carnivore habitat use in addition to the distribution of resources.

Although further studies would be needed to confirm and complete the information reported here, our findings suggest that the coastal dune areas to the east of Bahía Blanca are inhabited by a relatively diverse vertebrate community and shelter species that are considered rare in the human-dominated habitats that characterize today's Pampas landscapes, such as *P. concolor*, *L. geoffroyi*, and *R. americana*, thus supporting the conservation.

15.6 Final Remarks

Understanding the relevance of the role played by the contribution of the sea toward the terrestrial environment, as we have tried to show in this chapter, is of great utility not only in providing information that allows changing the ecological theoretical framework on the influence of the ocean on terrestrial trophic networks but also for the management and conservation of coastal biota. Our understanding of the interaction between marine and terrestrial ecosystems is very scarce for the area of the estuary of Bahía Blanca, and we have only been able to lay down some baseline information here to guide future investigations that are urgently needed to contribute to the conservation of the coasts of the southern part of Buenos Aires Province and all the biodiversity that they host.

Box 15.1 The Impact of Coastal Urbanization on Land Mammals

Urbanization (defined as the natural growth of urban centers associated with an increased in their population or with a high density of inhabitants and the need to cover their requirement) is particularly concentrated along the world's coasts and is growing rapidly, which makes of these areas represent very important hotspots of habitat modification (Martínez et al. 2007). Globally, one of the main factors affecting coastal biodiversity is urbanization, because its main consequences are habitat loss and fragmentation, but also because it causes overexploitation of fish stocks, pollution, and depletion of populations of coastal wildlife (McKinney 2002; McDonald et al. 2008).

The coastal infrastructure associated with recreational activities may be vast, including roads and paths to access beaches, waterfront housing development, car parking, picnic spots, swimming pools, service areas, beach facilities (toilets, lifeguard facilities), navigation canals, drainage, and stream canalization to reclaim coastal wetlands. These infrastructures lead to varied conflicts since they may alter the quality of the landscape and disturb ecosystem functioning, as they modify the coastline; cause erosion; interrupt sediment transport; alter dunes, reefs, and wetlands; increase traffic intensity, noise, and the contamination of air and water; and alter the patches of dune vegetation for the construction of houses which leads to the loss of native plant species and favors their replacement by exotic (Jędrzejczak 2004).

Although the information about the impacts of this infrastructure on the coastal mammals is still scarce, some effects have been demonstrated and others can be predicted.

Barriers to Movement of Animals and Wrack

Along with habitat loss, the alteration of the physical processes that affect the deposition and retention of sediments on modificated coasts may also affect the deposition and retention of buoyant material, including macrophyte wrack, driftwood, and other natural allochthonous debris, which can be important to biota as food or habitat. Predators, such as shorebirds or maritime mammals, which usually move between the dunes and the beach, mostly to feed on stranded material (see Sect. 15.1 in this chapter), can respond to a combination of habitat loss, decreased accessibility at particular sites, and reduced food availability. For instance, structures such as seawalls that are placed on the coastline negatively affected some animals such as small rodents and other land mammals (Carlton and Hodder 2003; Bird et al. 2004).

Changes in Species Distribution or Abundance

Negative effects of coastal urbanization such as the decrease in the abundance of species that are threatened of extinction due to loss of habitats have been reported in different vertebrate taxa. Schmidt et al. (2012) found evidence of the synergistic impacts from rising sea level and coastal development in Lower Florida Keys (USA) on population decline in an endangered mammal, the Lower Keys marsh rabbit (*Sylvilagus palustris hefneri*), and that several bird species modified the selection of nesting sites, due to anthropic disturbances on the beaches, such as vehicle access. Meager et al. (2012) observed that human activities occurred mostly in the mid to lower intertidal zone of the beach in Moreton and North Stradbroke Island, eastern Australia, overlapping closely with the preferred habitats of silver gulls (*Larus novaehollandiae*), pied oystercatchers (*Haematopus longirostris*), red-capped plovers (*Charadrius ruficapillus*), and endangered little terns (*Sternula albifrons*).

Wrack and Carrion Removal

Animal carcasses are fundamental resources in many food webs, supporting a diverse and abundant suite of scavengers and structuring ecosystems (e.g., Wilson and Wolkovich 2011; Moleón et al. 2014). In fact, carcasses of marine animals washing up on sandy beaches are a critical input for scavengers feeding at the sea-land interface (see Sect. 15.1 in this chapter). The consumption of beach-cast marine carrion on sandy shorelines is rapid and often complete, suggesting evolutionary pathways resulting in a pivotal role of carrion in food webs of these ecosystems. Moreover, carrion is utilized by a variety of species, both native and introduced, that are highly adapted to changes in resource supply and capable of reacting rapidly to these changes. A comparison between scavenging dynamics on sandy beaches found at two extreme levels of urbanization in Australia showed that invasive mammals replaced native raptors as scavengers on urban beaches, resulting in a significant decline in scavenging efficiency (Huijbers et al. 2013; Schlacher et al. 2015). The coastal areas of Buenos Aires Province are probably no exception to this kind of conflict. The removal of carrion on the coasts by dogs affects, through competition for resources, different local scavenger species such as the Pampas fox (obs. Pers.).

Introduction of Exotic Species

In general, the presence of dogs (*Canis familiaris*) and domestic cats (*Felis catus*) in urban areas represents one of the greatest negative threats to local biodiversity (Hughes and Macdonald 2013; Weston and Stankowich 2014; Loss et al. 2013). They can be predators or efficient competitors of native species and carriers of diseases that can affect wildlife. Hughes and Macdonald

(2013) found that mammals were the most studied taxa regarding the interaction between dogs and wildlife. Predation was one of the greatest impacts caused by dogs, followed by the transmission of diseases, the disturbance on wild species (which carries energetic costs for them), hybridization, and competition for food. The transmission of dog diseases to wildlife may be caused by direct interactions between species, but the fecal contamination of dogs in waterways has indirect negative effects also on marine mammal health (Holderness-Roddam 2011). Holderness-Roddam's (2011) review determined that feral and domestic cat predation on small vertebrates is significant, but both in poorly modified urban areas and on beaches and their surroundings, domestic dogs prey even more significantly on some species. Some damage may be caused by people walking through these areas with their pets, but the impact is even greater if dogs are allowed to go unleashed.

Rats are other highly invasive species whose impact over the coastal fauna has been unambiguously documented. Several authors have suggested that in Madagascar, *Rattus rattus* competes for resources with rodents of the endemic subfamily Nesomyinae and subsequently replaces them (Goodman 1995). Jones et al. (2008) examined 94 manuscripts that demonstrated effects of exotic rats on seabirds and found that they affect 75 species of seabirds and that the consequences of rat predation on seabirds are independent of time since rat introduction. Martin et al. (2000) studied islands in the Mediterranean and discovered that storm petrels (*Hydrobates pelagicus*) are limited to rat-free islands. Finally, rats may cause extinctions of endemic mammals on Australian islands through introduction of novel diseases and competition (Smith and Banks 2014).

Although there are essentially no studies on the impact of coastal urbanization on terrestrial mammals in the Buenos Aires Province, based on our review of the literature on this topic, we conclude that there is a great need to assess not only the changes in habitat but also the effects of urban development processes on native species and use this information to design plans for reducing this type of environmental impact.

As a recommendation, strategies that preserve the size and connectivity of patches of native vegetation and limit the introduction of non-native predators such as rats and domestic dogs and cats, in coastal urban areas where man and terrestrial mammals cohabit, should be considered priorities for the expected future advance of cities on the coasts.

References

Abba AM, Poljak S, Superina M (2014) *Chaetophractus villosus*. The IUCN Red List of Threatened Species: e.T4369A47438745. https://doi.org/10.2305/IUCN.UK.2014-1.RLTS.T4369A47438745.en. Accessed 13 Feb 2020

Aprile G, Cirignoli S, Varela D et al (2019) *Galictis cuja*. In: SAyDS–SAREM (eds) Categorización 2019 de los mamíferos de Argentina según su riesgo de extinción. Lista Roja de los mamíferos de Argentina. http://cma.sarem.org.ar. Accessed 13 Feb 2020

Arriagada A, Baessolo L, Saucedo C et al (2017) Hábitos alimenticios de poblaciones periféricas de *Zaedyus pichiy* y *Chaetophractus villosus* (Cingulata, Chlamyphoridae) en la Patagonia chilena. Iheringia Sér Zool 107:1–8. https://doi.org/10.1590/1678-4766e2017013

Arroyo NL, Bonsdorff E (2016) The role of drifting algae for marine biodiversity In: E Olafson (ed) Mar Macrophytes as Found Species CRC Press Boca Raton p 100

Austrich A, Mapelli FJ, Mora M et al (2019) *Ctenomys australis*. In: SAyDS–SAREM (eds) Categorización 2019 de los mamíferos de Argentina según su riesgo de extinción. Lista Roja de los mamíferos de Argentina. http://cma.sarem.org.ar. Accessed 13 Feb 2020

Ballari SA, Cirignoli S, Winter M et al (2019) *Sus scrofa*. In: SAyDS–SAREM (eds) Categorización 2019 de los mamíferos de Argentina según su riesgo de extinción. Lista Roja de los mamíferos de Argentina. http://cma.sarem.org.ar. Accessed 13 Feb 2020

Barreiro F, Gómez M, López J et al (2013) Coupling between macroalgal inputs and nutrients outcrop in exposed sandy beaches. Hydrobiologia 700:73–84. https://doi.org/10.1007/s10750-012-1220-z

Benzaquín M, Manfredi C, Luengos Vidal EM et al (2009) Estudio de la comunidad de felinos en una región del espinal del sudoeste de la Provincia de Buenos Aires.In: Cazzaniga NJ, Arelovich H (eds) Ambientes y recursos naturales del sudoeste bonaerense: Producción, contaminación y conservación. (Actas de las V Jornadas Interdisciplinarias del SO Bonaerense). EdiUNS, Bahía Blanca

Bidau C (2016) *Ctenomys talarum*. The IUCN Red List of Threatened Species: e.T5828A22195175. https://doi.org/10.2305/IUCN.UK.2016-2.RLTS.T5828A22195175.en. Accessed on 14 Feb 2020

Bird BL, Branch LC, Miller DL (2004) Effects of coastal lighting on foraging behavior of beach mice. Conserv Biol 18:1435–1439. https://doi.org/10.1111/j.1523-1739.2004.00349.x

Bisceglia S, Palacios R, Quiroga VA et al (2019) *Herpailurus yagouaroundi*. In: SAyDS–SAREM (eds) Categorización 2019 de los mamíferos de Argentina según su riesgo de extinción. Lista Roja de los mamíferos de Argentina. Versión digital: http://cma.sarem.org.ar. Accessed 13 Feb 2020

Bolkovic ML, Quintana R, Cirignoli S et al (2019) *Hydrochoerus hydrochaeris*. In: SAyDS–SAREM (eds) Categorización 2019 de los mamíferos de Argentina según su riesgo de extinción. Lista Roja de los mamíferos de Argentina. http://cma.sarem.org.ar

Bonaventura SM, Pancotto V, Madanes N et al (2003) Microhabitat use and density of sigmodontine rodents in *Spartina densiflora* freshwater marshes, Argentina. Mammalia 67(3):367–378. https://doi.org/10.1515/mamm.2003.67.3.367

Borboroglu PG, Yorio P (2004) Habitat requirements and selection by kelp gulls (*Larus dominicanus*) in central and northern Patagonia, Argentina. Auk 121:243–252. https://doi.org/10.1093/auk/121.1.243

Bortolus A, Carlton JT, Schwind E (2015) Reimagining South America coasts: unveiling the hidden invasion history of an iconic ecological engineer. Divers Distrib 21(11):1267–1283. https://doi.org/10.1111/ddi.12377

Bossi MAS, Migliorini RP, Santos TG et al (2019) Comparative trophic ecology of two sympatric canids in the Brazilian Pampa. J Zool 307:215–222. https://doi.org/10.1111/jzo.12636

Broekhuis F, Grünewälder S, McNutt JW (2014) Optimal hunting conditions drive circalunar behavior of a diurnal carnivore. Behav Ecol 25:1268–1275. https://doi.org/10.1093/beheco/aru122

Brown A, Martinez Ortiz U, Acerbo M et al (2006) La situación ambiental argentina 2005. Fundación Vida Silvestre, Buenos Aires

Burnham K, Anderson D (2002) Model selection and multi-model inference, 2nd edn. Springer, New York

Canepuccia AD, Martinez MM, Vassallo AI (2007) Selection of waterbirds by Geoffroy's cat: effects of prey abundance, size, and distance. Mamm Biol 72:163–173. https://doi.org/10.1016/j.mambio.2006.07.003

Canevari M, Balboa F (2003) 100 Mamíferos argentinos. Editorial Albatros, Buenos Aires

Carbone C, Codron D, Scofield C et al (2014) Geometric factors influencing the diet of vertebrate predators in marine and terrestrial environments. Ecol Lett 17:1553–1559. https://doi.org/10.1111/ele.12375

Carlton JT, Hodder J (2003) Maritime mammals: terrestrial mammals as consumers in marine intertidal communities. Mar Ecol Prog Ser 256:271–286. https://doi.org/10.3354/meps256271

Caruso NC, Sotelo M, Luengos Vidal EM (2012) Primer registro de gato montés, *Leopardus geoffroyi*, en la reserva Bahía Blanca, Bahía Falsa, Bahía Verde (RNBBBFBV), provincia de Buenos Aires. BioScriba 5:54–59

Caruso N, Lucherini M, Fortin D et al (2016) Species-specific responses of carnivores to human-induced landscape changes in Central Argentina. PLoS One 11(3). https://doi.org/10.1371/journal.pone.0150488

Caruso N, Luengos Vidal EM, Lucherini M et al (2017) Carnivores in the southwest of the province of Buenos Aires: ecology and conflicts with farmers. RIA, Rev Inves Agropec 43:165–174

Caruso N, Valenzuela AEJ, Burdett CL et al (2018) Summer habitat use and activity patterns of wild boar Sus scrofa in rangelands of Central Argentina. PLoS One 13:e0206513. https://doi.org/10.1371/journal.pone.0206513

Caso A, de Oliveira T, Carvajal SV (2015) *Herpailurus yagouaroundi*. The IUCN Red List of Threatened Species: e.T9948A50653167. https://doi.org/10.2305/IUCN.UK.2015-2.RLTS.T9948A50653167.en. Accessed on 15 Feb 2020

Castillo DF, Schiaffini M (2019) *Conepatus chinga*. In: SAyDS–SAREM (eds) Categorización 2019 de los mamíferos de Argentina según su riesgo de extinción. Lista Roja de los mamíferos de Argentina. http://cma.sarem.org.ar. Accessed 13 Feb 2020

Castillo D, Luengos Vidal EM, Lucherini M et al (2008) First report on the geoffroyi cat in a highly modificated area of the argentine areas. Cat News 49:27–28

Castillo DF, Lucherini M, Luengos Vidal EM et al (2011) Spatial organization of Molina's hog-nosed skunk (*Conepatus chinga*) in two landscapes of the pampas grassland of Argentina. Can J Zool 89:229–238. https://doi.org/10.1139/Z10-110

Castillo DF, Luengos Vidal EM, Caruso N et al (2013) Denning ecology of *Conepatus chinga* (Carnivora: Mephitidae) in a grassland relict of Central Argentina. Mastozool Neotrop 20:373–379

Castillo DF, Luengos Vidal EM, Casanave EB (2014) Feeding habits of Molina's hog-nosed skunk in the pampas grassland of Argentina. Mammalia 78:473–447. https://doi.org/10.1515/mammalia-2013-0020

Castillo DF, Luengos Vidal EM, Caruso NC et al (2019) Spatial organization and habitat selection of Geoffroy's cat in the Espinal of Central Argentina. Mamm Biol 94(1):30–37. https://doi.org/10.1016/j.mambio.2018.12.003

Caruso NC, Sotelo M, Luengos Vidal EM (2012) Primer registro de gato montés, Leopardus geoffroyi, en la reserva Bahía Blanca, Bahía Falsa, Bahía Verde (RNBBBFBV), provincia de Buenos Aires. BioScriba 5:54–59

Caruso NC, Luengos Vidal EM, Manfredi MC, Araujo MS, Lucherini M, Casanave EB (2020). Spatio-temporal interactions of carnivore species in a coastal ecosystem in Argentina. Ocean Coast Manag 198:105311. https://doi.org/10.1016/j.ocecoaman.2020.105311

Celsi CE, Giussani LM (2019) Geographical distribution and habitat characterization of *Poa schizantha* (Poaceae), a narrow endemic of the coastal sand dunes of the southern pampas, Argentina. Bot J Linn Soc 192:296–313. https://doi.org/10.1093/botlinnean/boz069

Chemisquy MA, Martin GM (2019) *Didelphis albiventris*. In: SAyDS–SAREM (eds) Categorización 2019 de los mamíferos de Argentina según su riesgo de extinción. Lista Roja de los mamíferos de Argentina. Versión digital: http://cma.sarem.org.ar. Accessed 13 Feb 2020

CMA-Categorización 2019 de los Mamíferos de Argentina según su riesgo de extinción. Lista Roja de los mamíferos de Argentina (2019) SAREM http://cma.sarem.org.ar

Colombini I, Chelazzi L, Gibson RN et al (2003) Influence of marine allochthonous input on sandy beach communities. Oceanogr Mar Biol An Annu Rev 41:115–159

Contreras JR, Reig OA (1965) Datos sobre la distribución del género Ctenomys (Rodentia, Octodontidae) en la zona costera de la Provincia de Buenos Aires comprendida entre Necochea y Bahía Blanca. Physis 25:169–186

Corriale MJ, Muschetto E, Herrera EA (2013) Influence of group sizes and food resources in home-range sizes of capybaras from Argentina. J Mammal 94:19–28. https://doi.org/10.1644/12-MAMM-A-030.1

Costa LP, Astua de Moraes D, Brito D et al (2015) *Didelphis albiventris*. The IUCN Red List of Threatened Species: e.T40489A22176404. https://doi.org/10.2305/IUCN.UK.2015-4.RLTS.T40489A22176404.en. Accessed on 15 Feb 2020

Crespo JA (1971) Ecología del zorro gris *Dusicyon gymnocercus antiquus* (Ameghino) en la provincia de La Pampa. Rev Mus Arg Cs Nat 1:147–205

Croll DA, Maron JL, Estes JA et al (2005) Introduced predators transform subarctic islands from grassland to tundra. Sci 307:1959–1961. https://doi.org/10.1126/science.1108485

Cueto GR (1999) Biología reproductiva y crecimiento del carpincho (*Hydrochoerus hydrochaeris*) en cautiverio: Una interpretación de las estrategias poblacionales. PhD Thesis. Universidad de Buenos Aires, Buenos Aires

Cutrera AP, Antinuchi CD, Mora MS et al (2006) Home-range and activity patterns of the south American subterranean rodent *Ctenomys talarum*. J Mammal 87:1183–1191. https://doi.org/10.1644/05-MAMM-A-386R1.1

De Angelo C, Paviolo A, Rode D et al (2011) Participatory networks for large-scale monitoring of large carnivores: pumas and jaguars of the upper Paraná Atlantic Forest. Oryx 45:534–545. https://doi.org/10.1017/S0030605310000840

De Angelo C, Llanos R, Guerisoli MM et al. (2019) *Puma concolor*. In: SAyDS–SAREM (eds) Categorización 2019 de los mamíferos de Argentina según su riesgo de extinción. Lista Roja de los mamíferos de Argentina. http://cma.sarem.org.ar

Dehn LA, Follmann EH, Thomas DL et al (2006) Trophic relationships in an Arctic food web and implications for trace metal transfer. Sci Total Environ 362:103–123. https://doi.org/10.1016/j.scitotenv.2005.11.012

Di Bitetti MS, De Angelo CD, Di Blanco YE et al (2010) Niche partitioning and species coexistence in a Neotropical felid assemblage. Acta Oecol 36(4):403–412. https://doi.org/10.1016/j.actao.2010.04.001

Doumecq Milieu RE, Morici A, Nigro NA (2012) Ampliación de la distribución austral del carpincho (*Hydrochoerus hydrochaeris*) en la provincia de Buenos Aires, Argentina. Nótulas Faunísticas 92:1–10

Elbroch LM, Quigle H (2017) Social interactions in a solitary carnivore. Current zoology 63(4):357–362. https://doi.org/10.1093/cz/zow080

Elgueta EI, Valenzuela J, Rau JR (2007) New insights into the prey spectrum of Darwin′ s fox (*Pseudalopex fulvipes* Martin, 1837) on Chiloé Island, Chile. Mamm Biol 72:179–185. https://doi.org/10.1016/j.mambio.2006.07.004

Elmhagen B, Tannerfeldt M, Verucci P et al (2000) The arctic fox (*Alopex lagopus*): an opportunistic specialist. J Zool 251:139–149. https://doi.org/10.1111/j.1469-7998.2000.tb00599.x

Emmons L, Schiaffini M, Schipper J (2016) *Conepatus chinga*. The IUCN Red List of Threatened Species: e.T41630A45210528. https://doi.org/10.2305/IUCN.UK.2016-1.RLTS.T41630A45210528.en. Accessed on 13 Feb 2020

Farias AA, Kittlein MJ (2008) Small-scale spatial variability in the diet of pampas foxes (Pseudalopex gymnocercus) and human-induced changes in prey base. Ecol Res 23:543–550. https://doi.org/10.1007/s11284-007-0407-7

Fernández GP, Carnovale CS, Mora MS (2019) *Ctenomys talarum*. En: SAyDS–SAREM (eds) Categorización 2019 de los mamíferos de Argentina según su riesgo de extinción. Lista Roja de los mamíferos de Argentina. http://cma.sarem.org.ar

Fernández-Llario P, Carranza J, De Trucios SJH (1996) Social organization of the wild boar (Sus scrofa) in Doñana National Park. Misc Zoo 19:9–18

Foley JE, Swift P, Fleer KA et al (2013) Risk factors for exposure to feline pathogens in California mountain lions (*Puma concolor*). J Wildl Dis 49:279–293. https://doi.org/10.7589/2012-08-206

Frafjord K (1993) Food habits of arctic foxes (*Alopex lagopus*) on the western coast of Svalbard. Arctic 49–54

Frere E, Millones A, Morgenthaler A et al (2010) High predation rates by pumas on Magellanic Penguin adults: new conflicts in coastal protected areas in Argentina. 1st World Seabird Conference Poster Session

Galende GI, Raffaele E (2016) Predator feeding ecology on Patagonian rocky outcrops: implications for colonies of mountain vizcacha (*Lagidium viscacia*). Stud Neotrop Fauna Environ 51:104–111. https://doi.org/10.1080/01650521.2016.1185270

Gallo JA, Poljak S, Abba, AM et al (2019) *Chaetophractus villosus*. In: SAyDS–SAREM (eds) Categorización 2019 de los mamíferos de Argentina según su riesgo de extinción. Lista Roja de los mamíferos de Argentina. http://cma.sarem.org.ar

García VB, Kittlein MJ (2005) Diet, habitat use, and relative abundance of pampas fox (Pseudalopex gymnocercus) in northern Patagonia, Argentina. Mamm Biol 70:218–226. https://doi.org/10.1016/j.mambio.2004.11.019

Garcia EA, Bertness MD, Alberti J et al (2011) Crab regulation of cross-ecosystem resource transfer by marine foraging fire ants. Oecologia 166:1111–1119. https://doi.org/10.1007/s00442-011-1952-x

Gaydos JK, Pearson SF (2011) Birds and mammals that depend on the Salish Sea: a compilation. Northwest Nat 92:79–94. https://doi.org/10.1898/10-04.1

Gomez Villafañe IE, Miño M, Cavia R et al (2005) Guía de Roedores de la provincia de Buenos Aires. Editor LOLA (Literature Lat Am Argentina) Buenos Aires

Goodman SM (1995) Rattus on Madagascar and the dilemma of protecting the endemic rodent fauna. Conserv Biol 9:450–453. https://doi.org/10.1046/j.1523-1739.1995.9020450.x

Guerisoli M d l M, Luengos Vidal EM, Franchini M et al (2017) Characterization of puma–livestock conflicts in rangelands of Central Argentina. R Soc Open Sci 4:170852. https://doi.org/10.1098/rsos.170852

Heatwole H (1972) Marine-dependent terrestrial biotic communities on some cays in the Coral Sea. Ecology 52:363–366. https://doi.org/10.2307/1934598

Helgen K, Schiaffini M (2016) *Galictis cuja*. The IUCN Red List of Threatened Species: e.T41639A45211832. https://doi.org/10.2305/IUCN.UK.2016-1.RLTS.T41639A45211832.en. Accessed on 13 Feb 2020

Hersteinsson P, Macdonald DW (1996) Diet of arctic foxes (*Alopex lagopus*) in Iceland. J Zool 240:457–474. https://doi.org/10.1111/j.1469-7998.1996.tb05298.x

Holderness-Roddam B (2011) The effects of domestic dogs (*Canis familiaris*) as a disturbance agent on the natural environment. PhD Thesis. University of Tasmania

Hughes J, Macdonald DW (2013) A review of the interactions between free-roaming domestic dogs and wildlife. Biol Conserv 157:341–351. https://doi.org/10.1016/j.biocon.2012.07.005

Huijbers CM, Schlacher TA, Schoeman DS et al (2013) Urbanisation alters processing of marine carrion on sandy beaches. Landsc Urban Plan 119:1–8. https://doi.org/10.1016/j.landurbplan.2013.06.004

Jędrzejczak MF (2004) The modern tourist's perception of the beach: is the sandy beach a place of conflict between tourism and biodiversity. J Coast Res 2:109–119

Jepsen JU, Eide NE, Prestrud P et al (2002) The importance of prey distribution in habitat use by arctic foxes (*Alopex lagopus*). Can J Zool 80:418–429. https://doi.org/10.1139/z02-023

Jones HP, Tershy BR, Zavaleta ES et al (2008) Severity of the effects of invasive rats on seabirds: a global review. Conserv Biol 22:16–26. https://doi.org/10.1111/j.1523-1739.2007.00859.x

Kasper CB, Fontoura-Rodrigues ML, Cavalcanti GN et al (2004) Recent advances in the knowledge of Molina's hog-nosed skunk Conepatus chinga and striped hog-nosed skunk C. semistriatus in South America. Small Carniv Conserv 41:25–28

Kasper CB, Soares JBG, Freitas TRO (2012) Differential patterns of home-range, net displacement and resting sites use of *Conepatus chinga* in southern Brazil. Mamm Biol 77:358–362. https://doi.org/10.1016/j.mambio.2012.03.006

Kasper CB, Peters FB, Christoff AU et al (2016) Trophic relationships of sympatric small carnivores in fragmented landscapes of southern Brazil: niche overlap and potential for competition. Mammalia 80:143–152. https://doi.org/10.1515/mammalia-2014-0126

Kasper CB, Fontoura-Rodrigues ML, Cavalcanti GN, Freitas TRO, Rodrigues, FHG, Oliveira TG, Eizirik E (2009). Recent advances in the knowledge of Molina's hog-nosed skunk Conepatus chinga and striped hog-nosed skunk C. semistriatus in South America. Small Carnivore Conservation 41:25–28

Kelt D, Pardiñas U, Schiaffini M et al (2016) *Lyncodon patagonicus*. The IUCN Red List of Threatened Species: e.T41647A45212747. https://doi.org/10.2305/IUCN.UK.2016-1.RLTS.T41647A45212747.en. Accessed on 15 Feb 2020

Kennedy AJ (1980) Site variation in summer foods of arctic fox, Prince of Wales Island. Northwest Territories Arctic:366–368

Klein DR (1995) The introduction, increase, and demise of wolves on Coronation Island, Alaska. Ecol Conserv wolves a Chang world 275:280

Kurle CM, Croll DA, Tershy BR (2008) Introduced rats indirectly change marine rocky intertidal communities from algae-to invertebrate-dominated. Proc Natl Acad Sci 05:3800–3804. https://doi.org/10.1073/pnas.0800570105

Lafferty DJR, Belant JL, White KS et al (2014) Linking wolf diet to changes in marine and terrestrial prey abundance. Arctic:143–148

Leeney RH, Amies R, Broderick AC et al (2008) Spatio-temporal analysis of cetacean strandings and bycatch in a UK fisheries hotspot. Biodivers Conserv 17:2323

Legagneux P, Gauthier G, Berteaux D et al (2012) Disentangling trophic relationships in a high Arctic tundra ecosystem through food web modeling. Ecology 93:1707–1716. https://doi.org/10.1890/11-1973.1

Lei J, Booth DT, Dwyer RG (2017) Spatial ecology of yellow-spotted goannas adjacent to a sea turtle nesting beach. Aust J Zool 65:77–86. https://doi.org/10.1071/ZO17006

Lewis TM, Lafferty DJR (2014) Brown bears and wolves scavenge humpback whale carcass in Alaska. Ursus 25:8–13. https://doi.org/10.2192/URSUS-D-14-00004.1

Lillywhite HB, Sheehy CM III, Zaidan III (2008) Pitviper scavenging at the intertidal zone: an evolutionary scenario for invasion of the sea. Bioscience 58:947–955. https://doi.org/10.1641/B581008

Logan KA, Sweanor L (2001) Desert puma: evolutionary ecology and conservation of an enduring carnivore. Island press, Washington

Long JL (2003) Introduced mammals of the world: their history, distribution and influence. CABI Publishing, xxi, Wallingford

Loss SR, Will T, Marra PP (2013) The impact of free-ranging domestic cats on wildlife of the United States. Nat Commun 4(1):1–8. https://doi.org/10.1038/ncomms2380

Lucherini M (2016) *Lycalopex gymnocercus*. The IUCN Red List of Threatened Species: e.T6928A85371194. https://doi.org/10.2305/IUCN.UK.2016-1.RLTS.T6928A85371194.en. Accessed on 19 Feb 2020

Lucherini M, Luengos Vidal EM (2008) *Lycalopex gymnocercus* (Carnivora: Canidae). Mamm Species 2008:1–9. https://doi.org/10.1644/820.1

Lucherini M, Eizirik E, de Oliveira T et al (2016) *Leopardus colocolo*. The IUCN Red List of Threatened Species: e.T15309A97204446. https://doi.org/10.2305/IUCN.UK.2016-1.RLTS.T15309A97204446.en. Accessed on 15 Feb 2020

Lucherini M, Cuyckens GAE, Morales MM et al (2019) *Leopardus colocolo*. In: SAyDS–SAREM (eds.) Categorización 2019 de los mamíferos de Argentina según su riesgo de extinción. Lista Roja de los mamíferos de Argentina. Versión digital: http://cma.sarem.org.ar. Accessed 13 Feb 2020

Luengos Vidal EM (2009) Ecología espacial de *Pseudalopex gymnocercus* en el pastizal pampeano. PhD Thesis. Universidad Nacional del Sur, Bahia Blanca

Luengos Vidal EM, Manfredi C, Castillo D et al (2005) Variaciones en la composición del gremio de carnívoros en la región pampeana. In: Vaquero M, Cernadas de Bulnes M (eds) Producción, recursos y medioambiente en el SO Bonaerense. Actas III Jornadas Interdisciplinarias del Sudoeste Bonaerense. EdiUNS, Bahía Blanca, pp 97–106

Luengos Vidal EM, Lucherini M, Casanave E et al (2009) Morphometrics of pampas foxes (*Pseudalopex gymnocercus*) in the argentine pampas. Mammalia 73(1):63–67. https://doi.org/10.1515/MAMM.2009.013

Luengos Vidal EM, Sillero-Zubiri C, Marino J (2012) Spatial organization of the pampas fox in a grassland relict of central argentina: a flexible system. J Zool 287:133–141. https://doi.org/10.1111/j.1469-7998.2011.00896.x

Luengos Vidal EM, Castillo DF, Caruso NC et al (2016) Field capture, chemical immobilization, and morphometrics of a little-studied south American carnivore, the lesser grison. Wildl Soc Bull 40:400–405. https://doi.org/10.1002/wsb.654

Luengos Vidal EM, Guerisoli M d l M, Caruso N et al (2017) Updating the distribution and population status of jaguarundi *Puma yagouaroundi* (É. Geoffroy saint-Hilaire, 1803) Mammalia, Carnivora, Felidae, in the southernmost part of its distribution range. Check List: J Biodivers 13(4):75–79

Luengos Vidal EM, Farías A, Valenzuela AEJ et al (2019) *Lycalopex gymnocercus*. En: SAyDS–SAREM (eds) Categorización 2019 de los mamíferos de Argentina según su riesgo de extinción. Lista Roja de los mamíferos de Argentina. http://cma.sarem.org.ar

Manfredi C, Lucherini M, Canepuccia AD et al (2004) Geographical variation in the diet of Geoffroy's cat (*Oncifelis geoffroyi*) in pampas grassland of Argentina. J Mammal 85:1111–1115. https://doi.org/10.1644/BWG-133.1

Manfredi C, Soler L, Lucherini M et al (2006) Home range and habitat use by Geoffroy's cat (*Oncifelis geoffroyi*) in a wet grassland in Argentina. J Zool 268:381–387. https://doi.org/10.1111/j.1469-7998.2005.00033.x

Manfredi C, Lucherini M, Soler L (2011) Activity and movement patterns of Geoffroy's cat in the grasslands of Argentina. Mamm Biol 76:313–319. https://doi.org/10.1016/j.mambio.2011.01.009

Martin J, Thibault J, Bretagnolle V (2000) Black rats, island characteristics, and colonial nesting birds in the Mediterranean: consequences of an ancient introduction. Conserv Biol 14:1452–1466. https://doi.org/10.1046/j.1523-1739.2000.99190.x

Martínez Zanon JI, Travaini A, Zapata S et al (2012) The ecological role of native and introduced species in the diet of the puma Puma concolor in southern Patagonia. Oryx 46:106–111. https://doi.org/10.1017/S0030605310001821

Martínez ML, Intralawan A, Vázquez G et al (2007) The coasts of our world: ecological, economic and social importance. Ecol Econ 63:254–272. https://doi.org/10.1016/j.ecolecon.2006.10.022

May R, Van Dijk J, Wabakken P et al (2008) Habitat differentiation within the large-carnivore community of Norway's multiple-use landscapes. J Appl Ecol 45(5):1382–1391. https://doi.org/10.1111/1365-2664.12581

McDonald RI, Kareiva P, Forman RT et al (2008) The implications of current and future urbanization for global protected areas and biodiversity conservation. Biol Conserv 141:1695–1703. https://doi.org/10.1016/j.biocon.2008.04.025

McKinney ML (2002) Urbanization, biodiversity, and conservation. Bio Sci 52:883–890. https://doi.org/10.1641/0006-3568(2002)052[0883:UBAC]2.0.CO;2

Meager JJ, Schlacher TA, Nielsen T (2012) Humans alter habitat selection of birds on ocean-exposed sandy beaches. Divers Distrib 18:294–306. https://doi.org/10.1111/j.1472-4642.2011.00873.x

Miñarro F (2004) Identificación de áreas valiosas de pastizal en las pampas y campos de Argentina, Uruguay y sur de Brasil. Fundación Vida Silvestre Argentina, Buenos Aires

Moleón M, Sànchez-Zapata JA, Selva N et al (2014) Inter-specific interactions linking predation and scavenging in terrestrial vertebrate assemblages. Biol Rev 89:1042–1054. https://doi.org/10.1111/brv.12097

Mones A, Ojasti J (1986) Hydrochoerus hydrochaeris. Mamm Species 1986:1–7. https://doi.org/10.2307/3503784

Monserrat AL, Celsi CE (2009) Análisis regional de la costa pampeana austral en el marco del sistema de áreas protegidas y caracterización de un área clave como reserva, en el partido de coronel dorrego. Bioscriba 2:1–23

Monserrat AL, Codignotto J (2013) Geodiversidad pampeana: Geomorfología y conservación de los paisajes de dunas costeras. Comunicacoes Geologicas 100:21–32

Montero R, Autino A (2018) Sistemática y filogenia de los Vertebrados. Con énfasis en la fauna argentina. Editorial Independiente. San Miguel de Tucuman.

Monserrat AL, Celsi CE, Fontana SL (2012) Coastal dune vegetation of the southern pampas (Buenos Aires, Argentina) and its value for conservation. J Coast Res 279:23–35. https://doi.org/10.2112/JCOASTRES-D-10-00061.1

Moore PG (2002) Mammals in intertidal and maritime ecosystems: interactions, impacts and implications. Oceanogr Mar Biol Annu Rev 40:491–608

Mora MS (2008) Metapopulation biology of the sand dune tuco–tuco (*Ctenomys australis*): effects of habitat spatial structure on the ecology and population genetics. PhD Thesis. Universidad Nacional de Mar del Plata. Argentina

Mora MS, Mapelli FJ (2010) Conservación en médanos: Fragmentación del hábitat y dinámica poblacional del tuco–tuco de las dunas. In: Isla F, Lasta C (eds) Manejo barreras medanosas para la Prov Buenos Aires, Editor la Univ Nac Mar del Plata, Mar del Plata, p 161

Moss B (2017) Marine reptiles, birds and mammals and nutrient transfers among the seas and the land: an appraisal of current knowledge. J Exp Mar Bio Ecol 492:63–80. https://doi.org/10.1016/j.jembe.2017.01.018

Navarrete SA, Castilla JC (1993) Predation by Norway rats in the intertidal zone of Central Chile. Mar Ecol Ser 92:187

Nielsen C, Thompson D, Kelly M et al (2015) *Puma concolor* (errata version published in 2016). The IUCN Red List of Threatened Species: e.T18868A97216466. https://doi.org/10.2305/IUCN.UK.2015-4.RLTS.T18868A50663436.en. Accessed on 05 Feb 2020

Nowell K, Jackson P (1995) Wild cats: status survey and conservation action plan. IUCN, Gland

Oliver W, Leus K (2008) Sus scrofa. The IUCN Red List of Threatened Species: e.T41775A10559847. https://doi.org/10.2305/IUCN.UK.2008.RLTS.T41775A10559847.en. Accessed on 12 Feb 2020

Pereira JA, Fracassi NG, Rago V et al (2010) Causes of mortality in a Geoffroy's cat population a long-term survey using diverse recording methods. Eur J Wildl Res 56:939–942. https://doi.org/10.1007/s10344-010-0423-8

Pereira J, Lucherini M, Trigo T (2015) *Leopardus geoffroyi*. The IUCN Red List of Threatened Species: e.T15310A50657011. https://doi.org/10.2305/IUCN.UK.2015-2.RLTS.T15310A50657011.en. Accessed on 05 Feb 2020

Pereira JA, Lucherini M, Cuyckens GAE et al (2019) *Leopardus geoffroyi*. Categorización 2019 de los mamíferos de Argentina según su riesgo de extinción. Lista Roja de los mamíferos de Argentina. http://cma.sarem.org.ar

Polis GA, Hurd SD (1995) Extraordinarily high spider densities on islands: flow of energy from the marine to terrestrial food webs and the absence of predation. Proc Natl Acad Sci 92:4382–4386. https://doi.org/10.1073/pnas.92.10.4382

Polis GA, Hurd SD (1996a) Allochthonous input across habitats, subsidized consumers, and apparent trophic cascades: examples from the ocean-land interface. In: Polis GA, Winemiller KO (eds) Food webs. Springer, Boston, p 275

Polis GA, Hurd SD (1996b) Linking marine and terrestrial food webs: allochthonous input from the ocean supports high secondary productivity on small islands and coastal land communities. Am Nat 147:396–423

Polis GA, Strong DR (1996) Food web complexity and community dynamics. Am Nat 147:813–846

Poljak S, Confalonieri V, Fasanella M et al (2010) Phylogeography of the armadillo *Chaetophractus villosus* (Dasypodidae Xenarthra): post-glacial range expansion from pampas to Patagonia (Argentina). Mol Phylogenet Evol 55:38–46. https://doi.org/10.1016/j.ympev.2009.12.021

Prestrud P (1992) Food habits and observations of the hunting behaviour of Arctic foxes, *Alopex lagopus*, in Svalbard. Can Field-Naturalist Ottawa 106:225–236

Pretelli MG, Isacch JP, Cardoni DA (2013) Year-round abundance, richness and nesting of the bird assemblage of tall grasslands in the south-east pampas region, Argentina. Ardeola 60(2):327–343. https://doi.org/10.13157/arla.60.2.2013.327

Punta G, Yorio P, Saravia J, Borboroglu PG (2002) Breeding habitat requirements of the Imperial cormorant and rock shag in Central Patagonia, Argentina. Waterbirds 26:176–183. https://doi.org/10.1675/1524-4695(2003)026[0176:BHROTI]2.0.CO;2

R Development Core Team (2013) R: A Language and Environment for Statistical Computing. Vienna, Austria: R Foundation for Statistical Computing. www.R-project.org

Rathbun MJ (1918) The grapsoid crabs of America. US Nat Mus Bull 97:461

Redford KH, Eisenberg JF (1992) Mammals of the Neotropics. The southern cone. Chile, Argentina, Uruguay, Paraguay. University of Chicago Press, Chicago

Reid F (2016) *Hydrochoerus hydrochaeris*. The IUCN Red List of Threatened Species: e.T10300A22190005.https://doi.org/10.2305/IUCN.UK.2016-2.RLTS.T10300A22190005.en. Accessed on 12 Feb 2020

Rose MD, Polis GA (1998) The distribution and abundance of coyotes: the effects of allochthonous food subsidies from the sea. Ecology 79:998–1007. https://doi.org/10.1890/0012-9658(1998)079[0998:TDAAOC]2.0.CO;2

Ruiz-Delgado MC, Vieira JV, Reyes-Martínez MJ et al (2016) Colonisation patterns of supralittoral arthropods in naturally stranded wrack debris on Atlantic sandy beaches of Brazil and Spain. Mar Freshw Res 67:1634–1643. https://doi.org/10.1071/MF14407

Sassola N (2006) Caracterización del comportamiento del zorro pampeano (*Pseudalopex gymnocercus*) en estado silvestre en el período de crías postemergentes, en la región de la pampa Argentina. Licenciatura Thesis Univesidad Nacional del Sur, Bahía Blanca

Schiaffini MI, Gabrielli M, Prevosti FJ et al (2013a) Taxonomic status of southern south American Conepatus (Carnivora: Mephitidae). Zool J Linnean Soc 167:327–344. https://doi.org/10.1111/zoj.12006

Schiaffini MI, Martin GM, Giménez AL (2013b) Distribution of *Lyncodon patagonicus* (Carnivora, Mustelidae): changes from the last glacial maximum to the present. J Mammal 94:339–350. https://doi.org/10.1644/12-MAMM-A-155.1

Schiaffini MI, Ercoli MD, Díaz, G (2019) *Lyncodon patagonicus*. In: SAyDS–SAREM (eds) Categorización 2019 de los mamíferos de Argentina según su riesgo de extinción. Lista Roja de los mamíferos de Argentina. Versión digital: http://cma.sarem.org.ar. Accessed 13 Feb 2020

Schlacher TA, Weston MA, Lynn D et al (2015) Conservation gone to the dogs: when canids rule the beach in small coastal reserves. Biodivers Conserv 24(3):493–509. https://doi.org/10.1007/s10531-014-0830-3

Schlacher TA, Hutton BM, Gilby BL et al (2017) Algal subsidies enhance invertebrate prey for threatened shorebirds: a novel conservation tool on ocean beaches? Estuar Coast Shelf Sci 191:28–38. https://doi.org/10.1016/j.ecss.2017.04.004

Schmidt JA, McCleery R, Seavey JR et al (2012) Impacts of a half century of sea-level rise and development on an endangered mammal. Glob Chang Biol 18:3536–3542. https://doi.org/10.1111/gcb.12024

Smith HM, Banks PB (2014) Disease and competition, not just predation, as drivers of impacts of the black rat (*Rattus rattus*) on island mammals. Glob Ecol Biogeogr 23:1485–1488. https://doi.org/10.1111/geb.12220

Soler L, Lucherini M, Manfredi C et al (2009) Characteristics of defecation sites of the Geoffroy's cat *Leopardus geoffroyi*. Mastozoología Neotropical 16:485–489. Sociedad Argentina para el Estudio de los Mamíferos

Spiller DA, Piovia-Scott J, Wright AN et al (2010) Marine subsidies have multiple effects on coastal food webs. Ecology 91:1424–1434. https://doi.org/10.1890/09-0715.1

Sunquist M, Sunquist F (2002) Wild cats of the world. University of Chicago Press, Chicago

Suraci JP, Clinchy M, Zanette LY et al (2014) Mammalian mesopredators on islands directly impact both terrestrial and marine communities. Oecologia 176:1087–1100. https://doi.org/10.1007/s00442-014-3085-5

Tarroux A, Bety J, Gauthier G et al (2012) The marine side of a terrestrial carnivore: intra-population variation in use of allochthonous resources by arctic foxes. PLoS One 7(8)

Tellaeche CG, Reppucci JI, Luengos Vidal EM et al (2014) New data on the distribution and natural history of the lesser grison (*Galictis cuja*), hog-nosed skunk (*Conepatus chinga*), and culpeo (*Lycalopex culpaeus*) in northwestern Argentina. Mammalia 78(2):261–266. https://doi.org/10.1515/mammalia-2013-0005

Tirelli FP, Trigo TC, Trinca CS et al (2018) Spatial organization and social dynamics of Geoffroy's cat in the Brazilian pampas. J Mammal 99:859–873. https://doi.org/10.1093/jmammal/gyy064
Valenzuela AEJ, Rey AR, Fasola L et al (2013) Trophic ecology of a top predator colonizing the southern extreme of South America: feeding habits of invasive American mink (*Neovison vison*) in Tierra del Fuego. Mamm Biol 78:104–110. https://doi.org/10.1016/j.mambio.2012.11.007
Vega RM, Prous CG, Krivokapich S et al (2018) Toxocariasis in Carnivora from Argentinean Patagonia: species molecular identification, hosts, and geographical distribution. Int J Parasitol Parasites Wildl 7:106–110. https://doi.org/10.1016/j.ijppaw.2018.02.004
Watts DE, Butler LG, Dale BW et al (2010) The Ilnik wolf *Canis lupus* pack: use of marine mammals and offshore sea ice. Wildlife Biol 16:144–149
Weston MA, Stankowich T (2014) Dogs as agents of disturbance. In: Gompper ME (ed) Free-ranging dogs and wildlife conservation. Oxford University Press, Oxford, p 94. https://doi.org/10.2981/09-040
Wilson EE, Wolkovich EM (2011) Scavenging: how carnivores and carrion structure communities. Trends Ecol Evol 26:129–135. https://doi.org/10.1016/j.tree.2010.12.011
Yensen E, Tarifa T (2003) Galictis cuja. Mamm Species 2003:1–8. https://doi.org/10.1644/728
Zeller KA, Nijhawan S, Salom-Pérez R et al (2011) Integrating occupancy modeling and interview data for corridor identification: a case study for jaguars in Nicaragua. Biol Conserv 144(2):892–901. https://doi.org/10.1016/j.biocon.2010.12.003
Zenuto R, Busch C (1998) Population biology of the subterranean rodent *Ctenomys australis* (tuco-tuco) in a coastal dunefield in Argentina. Zeitschrift fur Saugetierkd 63:357–367

Chapter 16
Coastal Wetlands of the Bahía Blanca Estuary: Landscape Structure and Plant Associations

Paula Daniela Pratolongo, Flavia Funk, María Julia Piovan, Carla Celleri, and Vanesa L. Negrín

16.1 Introduction

Coastal wetlands include a wide variety of environments located within coastal watersheds. As coastal watersheds can extend many kilometers inland from the coast, the associated wetlands can be tidal or non-tidal and freshwater or saltwater. These types of ecosystems support several natural functions providing essential services to human societies. For instance, they can protect upland areas from flooding due to sea level rise and storms. Vegetation can also prevent coastline erosion due to their ability to dissipate the energy of waves and ocean currents, which would otherwise damage human infrastructure. Coastal wetlands provide habitat to waterfowl and shorebirds, especially many globally threatened and endangered species. Many commercial fish and shellfish species also rely on coastal wetlands. From a biogeochemical point of view, coastal wetlands filter pollutants and excess of nutrients out

P. D. Pratolongo (✉) · F. Funk
Centro de Recursos Naturales Renovables de la Zona Semiárida, CERZOS (Universidad Nacional del Sur-CONICET), Bahía Blanca, Argentina

Departamento de Biología, Bioquímica y Farmacia, Universidad Nacional del Sur, Bahía Blanca, Argentina

M. J. Piovan
Instituto Argentino de Oceanografía, IADO (Universidad Nacional del Sur-CONICET), Bahía Blanca, Argentina

C. Celleri
Centro de Recursos Naturales Renovables de la Zona Semiárida, CERZOS (Universidad Nacional del Sur-CONICET), Bahía Blanca, Argentina

V. L. Negrin
Departamento de Biología, Bioquímica y Farmacia, Universidad Nacional del Sur, Bahía Blanca, Argentina

Instituto Argentino de Oceanografía, IADO (Universidad Nacional del Sur-CONICET), Bahía Blanca, Argentina

S. M. Fiori, P. D. Pratolongo (eds.), *The Bahía Blanca Estuary*,
https://doi.org/10.1007/978-3-030-66486-2_16

of the water before it reaches the ocean. They also sequester and store large amounts of carbon due to their rapid growth, high sedimentation rates, and slow decomposition (Barbier 2019).

In temperate regions, tidal salt marshes are the dominant vegetation type in the intertidal zone of protected coasts. Instead, extensive mangroves characterize the tidal fringe in tropical latitudes. Salt marsh plants are mostly herbaceous and salt-tolerant macrophytes. They are well adapted to face regular inundation by marine water, especially those species occupying the lowest portion of a marsh, flooded twice a day by tides. At higher elevations, as tidal influence becomes less frequent, the terrestrial hydrology dominates, and a different type of vegetation may appear in response to varying salt stress (Cronk and Fennessy 2001). In arid and semiarid coastal locations, high evaporation rates may produce muddy salt flats instead of a typical high marsh. At the other end of the scale, large volumes of freshwater discharging into the upper salt marsh may support less salt-tolerant or even freshwater wetland plants (Pratolongo et al. 2019).

The coastal wetland continuum encompasses a broad spectrum of ecosystems, from salt marshes and tidal flats to non-tidal wetlands landward. The unifying attribute along all these contrasting ecosystems is that sea level acts as a significant driver of their hydrology. While sea level distinctly sets the limits of the tidal fringe, its influence is less obvious landward. According to Hageman (1969), the *perimarine* zone is the area where non-tidal wetlands persist under control of the relative sea level. In humid climates, freshwater seepage and high groundwater levels keep the saturated conditions necessary for the persistence and growth of perimarine freshwater swamps, non-tidal marshes, and fens (Waller et al. 1999; Gardner et al. 2000). On the contrary, a combination of high evaporation and low freshwater inputs characterizes the perimarine zone of arid climates. In these areas, soils develop extremely high salinities, and salt flats commonly replace freshwater wetlands (Pratolongo et al. 2009).

16.1.1 Habitat Zonation and Response to Sea Level Changes

Within the intertidal zone, salt marsh plants commonly display distinct patterns of zonation, driven by the individual species tolerance to physical stress and biological interactions acting across abiotic gradients (Pennings and Bertness 2001). The elevation gradient within the marsh establishes flood frequency, depth, and duration (the *hydroperiod*). Thus, abiotic factors linked to hydroperiod, like soil moisture content, redox state, nutrient limitation, and salt concentration, commonly correlate with elevation (Silvestri et al. 2005; Moffett et al. 2012). Significant biological interactions, like competition and facilitation, also perform in close association with abiotic stress (Bertness 1991; Pennings et al. 2005). The interplay between stress gradients and biological responses results in the typical shore-parallel zonation of plants, which may be more complex and spatially variable depending on the microtopography of the marsh surface.

At the seaward margin of the elevation gradient, saltwater regularly inundates the low marsh. A few resistant plant genera such as *Sarcocornia*, *Suaeda*, *Aster*, and *Spartina* can tolerate these stressful abiotic conditions imposed by tidal influence (Doody 1992). In low marsh areas, frequent tidal flooding prevents the accumulation of salts, and soil salinity is comparatively low. Instead, substrate stability, oxygenation, and sulfide toxicity may control plant establishment and survival at the seaward margin (Adam 1990; Mendelssohn and Morris 2000). Landward, soil salinity levels vary across marsh elevations, and the shape of the salt gradient depends on climate. In warm, humid regions, freshwater input from rain and upland sources moderates soil salinity in the high marsh. Thus, salts concentrate by evaporation at intermediate marsh elevations, where tidal flushing is inefficient. Mid-marshes characterized by higher salinities support distinctive salt-tolerant vegetation, and salt accumulations may also lead to the development of bare areas known as salt pans (Davy and Costa 1992). In colder climates, on the other hand, salinity decreases with increasing elevation (Pennings and Bertness 1999), and the opposite pattern prevails in arid climates, that is, increasing salt concentration along with elevation (Zedler 1982; Callaway et al. 1990).

Besides the salt marsh zonation of vascular plants, another biota also occupies distinct zones across the elevation gradient. For sessile or low mobility benthic species, their position within the tidal range reflects the relative ecological tolerance to a combination of abiotic and biotic stressors (Zapperi et al. 2016). Flooding duration, which depends on elevation, also drives the distribution of microorganisms on salt marsh sediments (Gehrels 2000), allowing for paleoecological studies and reconstructions of past relative sea levels (e.g., Horton and Edwards 2006). The elevation gradient across the marsh also creates different stress gradients to mobile aquatic and terrestrial animal species. Different species must either tolerate or avoid submergence and emergence, constraining their distributions (Pennings and Bertness 2001).

The transition from the coastal wetland continuum to the upland is also dependent on the regional patterns of rainfall or freshwater discharge. In perimarine environments with enough freshwater inputs, waterlogging is caused by freshwater seepage, rainfall, and poor drainage due to the low elevation respect to the mean sea level. Here, there is a progressive transition from the upper salt marsh into freshwater wetland communities. Depending on the nutrient status, an array of floristically different wetland communities may develop, including freshwater marshes and forested swamps. In arid climates with low freshwater inputs, evaporation increases with elevation, and soils develop extremely high salinities in the perimarine zones, precluding vegetation development. In arid tropical and subtropical regions, where evaporation by far exceeds precipitation, extensive, barren, salt-encrusted flats characterize coastal lowlands (Barth and Böer 2002). Barren salt flats, covered by evaporite accumulations, also appear in arid coastal settings within temperate climates (e.g., the Bahía Blanca Estuary and the Shark Bay, in Western Australia). Coastal salt flats in the perimarine zone of arid environments are above the limits of the tidal inundation, but sea level keeps shallow groundwater near the surface. From a hydrogeomorphic point of view, these salt flats are comparable to the perimarine freshwater fens and swamps typical of humid regions (Pratolongo et al. 2016).

A central process in coastal wetland development is ecosystem response to changes in the relative sea level. A changing sea level produces a modification in the hydroperiod and, thus, in the ecological state of wetlands. In response to the different hydrologic conditions, the different plant zones within the coastal wetland continuum are expected to migrate. In coastal settings along the mid-Atlantic and Pacific coasts of North America, for instance, the relative sea level was continually rising from the last glacial maximum to the present due to the combination of isostatic and eustatic effects. In these regions, Holocene estuaries drowned, and new wetlands formed landward. The process of wetlands migration upslope, in response to a rising relative sea level during the Holocene, has been widely studied along the eastern coast of North America. Coastal barrier ecosystems along the seaward margin of the Delmarva Peninsula, on the Atlantic coast of North America, underwent a sustained sea level rise during the Holocene (Oertel et al. 1989). Changes in sea level have been driving a succession of state changes in wetlands along the mainland edge of the Peninsula in which high marshes replace former upland forests. Following this sequence, the low marsh replaces a high marsh, and bare mudflats replace the former low marshes (Brinson et al. 1995; Christian et al. 2002).

An opposite pattern was described in high latitude coastal regions (e.g., Fennoscandia, Finland, Labrador), where the relative sea level was continually falling from the last glacial maximum to the present. In these regions, the different plant zones within the coastal wetland continuum have been typically migrating seaward. Along with the falling sea level, new land emerged, and pioneer low marsh plants colonized mudflats. The coastal area of western Finland rose during the Holocene as a result of rapid land uplift that followed the last glacial maximum. In the northern Gulf of Bothnia, the relative sea level fell and the shoreline displaced seaward. Shoreline retreat still occurs on this relatively flat coast, at a rate exceeding 8 mm a year, along with downward migration of plant zones (Vartiainen 1988; Ecke and Rydin 2000) and seaward expansion of pioneer plant communities (Zobel and Kont 1992). A more complex dynamics characterizes wetland environments in the temperate Atlantic coast of South America, where the relative sea level reached a transgressive maximum during the Holocene. In these systems, the late Holocene marine regression resulted in significant low-lying coastal landforms inherited from the former estuarine dynamics, which are presently occupied by extensive perimarine wetlands.

16.2 The Bahía Blanca Estuary

16.2.1 Coastal Landscape Evolution

The Holocene marine transgression has deeply modeled the coastal landscape around the Bahía Blanca Estuary. On the northern coasts of Argentina, the relative sea level reached around 6 m above present, about 6000 years ago (Isla 1989; Violante and Parker 2000; Cavallotto et al. 2004). The falling sea level after the highstand (i.e., the regressive phase) resulted in large low-lying areas of former estuarine environments. Typical regressive forms, like extensive plains composed of beach ridge and lagoonal deposits, characterize many coastal settings in Argentina, from Río de la Plata Coastal Plain to San Sebastián Bay, in Tierra del Fuego.

Raised Holocene deposits in the Bahía Blanca Estuary (Fig. 16.1) formed after the mid-Holocene highstand (ca. 6000 BP) during the regressive phase. According to González and Weiler (1983), these deposits correspond to relatively short periods of high energy identified as *Transgressive Stages* (e.g., pulses of rising sea level within the general falling trend). On the northern shore of the Principal Channel, there is a succession of Holocene beach ridges and tidal flat deposits, corresponding to high- and low-energy periods (González 1989). The oldest and highest deposits, located in the inner section of the Principal Channel, form a spit composed of several sand-shell ridges up to 10 m above the present sea level (González et al. 1983; Aliotta and Farinati 1990). Based on ^{14}C dates, González (1989) described at least five beach ridges representing significant episodes of high wave energy within the regressive phase, after the highstand. These episodes were named *Transgressive Stages I* to *V* and dated between 5990 ± 115 and 3560 ± 100 years BP. The inland, oldest beach ridge in this sequence represents the maximum transgressive episode.

Holocene deposits also appear further East, close to Punta Alta City. These deposits form a relatively continuous sand ridge parallel to the coast located about 6 to 7 m above the mean sea level (^{14}C age of 4615 ± 110 years BP). This ridge indicates the end of the maximum transgressive event before a regressive pulse (Aliotta and Farinati 1990). In this area, there is a second (seaward) ridge, at a lower elevation, assigned to a younger transgressive pulse (Aliotta and Farinati 1990). Extensive coastal flats prograded seaward from the second ridge during the late Holocene, under a falling sea level. Their radiocarbon ages range between 3300 and 3900 ^{14}C years BP, determined from fossils in life position (Farinati et al. 1992).

The raised Holocene deposits in the northern margin of the Principal Channel define the base of a scarp that divides two different domains. *La Vidriera Salt Flat* is a narrow, depressed landform that extends inland, aligned with the Principal Channel (Fig. 16.1). The axis along Principal Channel-La Vidriera Salt Flat separates the elevated Positive of Ventania to the North and the Colorado River Basin to the South (González-Uriarte 1984). South of the alignment Principal Channel-La Vidriera Salt Flat, the northwestern margin of the Bahía Blanca Estuary belongs to

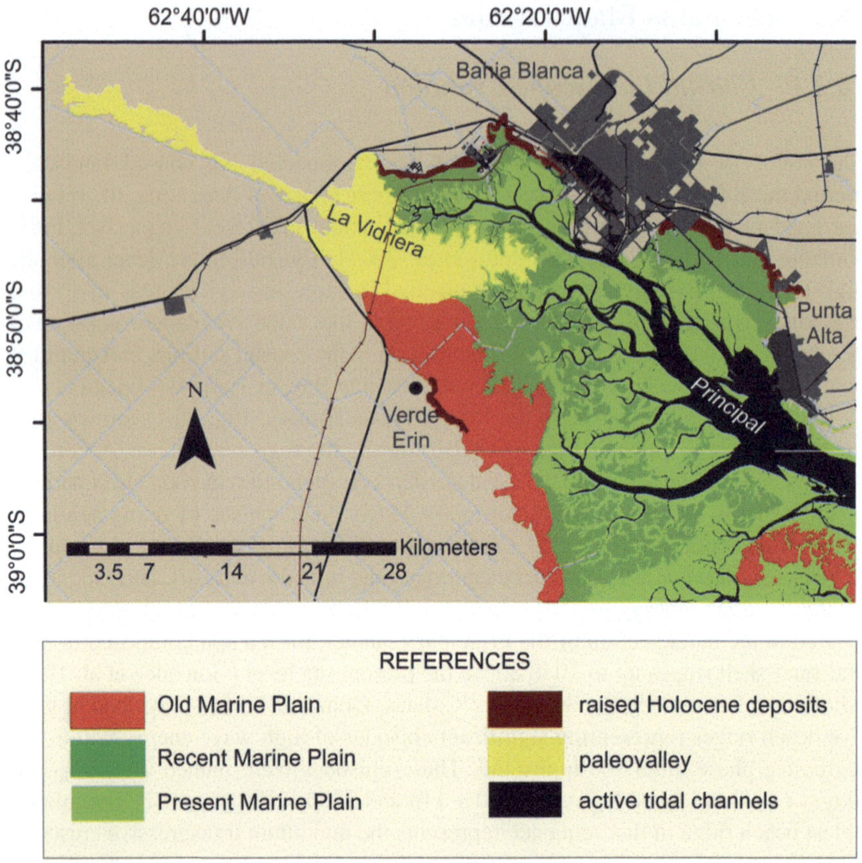

Fig. 16.1 Major landscape units in the coastal zone of Bahía Blanca. (Modified from Pratolongo et al. 2016, 2017)

the Colorado River Basin and has a gentler slope. In *Verde Erin*, a cattle ranch located 20 km southwest from Bahía Blanca (Fig. 16.1), Farinati (1983) described a shelly ridge at the base of a paleo-cliff, which would correspond to storm deposits indicating the inland limits of the marine transgression, at a ^{14}C age of 5406 ± 227 years BP. In the coastal plain extending from the ridge toward Principal Channel, two successive terrace levels formed at different ages during the regressive phase (González-Uriarte 1984). The *Old Marine Plain* (OMP), at an average elevation of 5 m above the present sea level, is a nearly continuous flat surface covered by different types of halophytic shrub communities. The *Recent Marine Plain* (RMP) is 2–3 m above the mean sea level. The unit is a mosaic of topographic highs and elongated depressions, corresponding to former tidal channels. In the RMP, the gentle slope creates a gradual transition to the *Present Marine Plain* (PMP), composed of active tidal channels, mudflats, and salt marshes, currently affected by the estuarine dynamics.

In a more recent paleoenvironmental reconstruction, Pratolongo et al. (2017) found well-preserved mollusks at the base of a core obtained in the OMP (530 cm deep), with ages between 5660 ± 30 and 5470 ± 30 years BP. It suggests that, by the time of the Holocene transgressive maximum, the presently OMP was a sandy bottom at a subtidal elevation. Thinner laminated sediments lay on top of the sand deposits, containing abundant shells of a brackish gastropod species. It suggests that, after the transgressive peak, the estuarine system progressively infilled during a low-energy period. A thick deposit of massive, grayish, organic-rich muds overlies the laminated sediments (Pratolongo et al. 2017). This sequence would correspond to rapid sedimentation during a high sea level stillstand or slow regression. A series of creeks, west of the Bahía Blanca Estuary, showed a similar evolution, with infill estuarine sequences composed of grayish muds forming flat plains in estuaries rapidly filled during the Holocene transgression (Isla et al. 1996; Marquez et al. 2016). In the Bahía Blanca Estuary, González-Uriarte (1984) described a paleodrainage that occupied the aligned depressions along La Vidriera Salt Flat-Principal Channel (Fig. 16.1). It indicates the former presence of a large river, which brought sediments to the region within a deltaic environment. Enhanced sedimentation under low-energy conditions may have filled the estuarine area and deposited the top massive mud layer. This model agrees with fluvial paleochannel structures described along the Principal Channel (Spagnuolo 2005; Giagante et al. 2008, 2011) and particularly with the fluvial-deltaic environment described by Aliotta et al. (2014).

In contrast with the uninterrupted, elevated OMP, there is fragmented RMP at a lower elevation, with old tidal channels dissecting topographic highs. The RMP formed at a later stage during the regressive phase, after sea level dropped to about 3 m above the present mean tidal level (the approximate highest elevation of the RMP). Then, the flat and continuous surface corresponding to the OMP emerged, and a more recent coastal dynamics shaped the fragmented landscape pattern of the RMP. The old tidal channels formed during this later stage, after tidal working dissected the older surface of mud deposits. Conversely, there is no well-defined boundary between the RMP and the PMP, but a gradual transition. The differentiation between the RMP and PMP would be then a response to the increased tidal influence on the RMP, because of the recent rising trends in the relative sea level (Pratolongo et al. 2013).

16.2.2 Coastal Wetlands and Landscape Patterns

Plant associations in the study area are reliable indicators of the dominant hydrogeomorphic conditions imposed by landscape position (Pratolongo et al. 2013, 2016). From a hydrogeomorphic perspective, coastal wetlands in Bahía Blanca classify into intertidal and inland (i.e., perimarine) wetlands. Within the intertidal zone (i.e., the area below the elevation of mean high tides), extensive barren mudflats are the dominant land cover type. Pure stands of *Spartina alterniflora* are commonly restricted to low marshes in the middle reach of the estuary but are extremely rare in the inner zone. *Spartina densiflora* marshes occasionally appear as pure stands that form a transition zone between *S. alterniflora* and *Sarcocornia ambigua* in places influenced by freshwater discharges.

Through the shallow inner section of the Principal Channel, vegetated marshes constrain to elevations close to the mean high tide level, with *S. ambigua* as the dominant species. Seasonally hypersaline conditions, because of the higher evaporation rates, would impose physiological limitations to vegetation development, precluding the establishment of less tolerant species (Pratolongo et al. 2010). A supralittoral zone develops above the elevation of the mean high tide and below the limits of the highest tides (spring high tides and storm surges). Seawater inundates this elevational fringe irregularly, which enhances evaporation and soil salinization. Vegetation in the supralittoral zone is sparse, forming a mosaic of salt flats, halophytic steppes, and shrubs. Plants and bare soil commonly assemble in mound-intermound complexes (Piovan et al. 2014).

Soils in these intertidal and supralittoral wetlands classify as *Aquisalids* (formerly *Salorthids*). They usually have an A horizon, enriched in organic matter compared to the underlying strata. The organic-rich layer may extend up to 25–50 cm depth in highly productive intertidal marshes. In supratidal halophytic steppes, A horizons rarely exceed 8–10 cm depth. In barren areas, the A horizon may not be present at all, and the soil surface exposes parental material (Kruger 1989). More elevated landscape positions in the supralittoral zone are less affected by tidal inundation, but soil properties are influenced by a highly saline groundwater table. In situ determinations of groundwater salinity ranged from 41 to 45 (seawater salinity is around 35). During dry years, groundwater level fluctuates from 3 to 0.3 m below the soil surface, but it rises to the soil surface during wet years, leaving soils ponded for several months. Soluble salt crystals form surface salt crusts when the soil is dry. Salt crusts disappear after tidal flooding and rain events but reappear after a few days of high evaporation (Piovan 2016).

Intertidal and supralittoral wetlands form a gently sloped continuum that extends from PMP through the RMP. The narrow depressions that typically dissect the landscape pattern in the RMP correspond to former tidal channels that presently drain rainfall excess. Within channels, soils are permanently flooded or saturated, even in dry years. The water table fluctuates around the soil surface (from below 0.4 to above 0.5 m), rapidly responding to rainfall events in the catchment area. Continuous records of groundwater levels in wells showed increases of more than 50 cm within hours after a single precipitation event (Piovan et al. 2014; Piovan 2016). Groundwater salinity in these channels is significantly lower (23–26) than in more elevated environments within the unit. The lower salinity allows for the development of highly productive *S. densiflora* marshes and soils with a deeper layer enriched with organic matter.

The OMP is not affected by tidal flooding. Beyond tidal influence, an irregularly humid zone develops close to the inland limit of the OMP. In this landscape position, a saline shallow water table fluctuates close to the soil surface, fed by seepage from the surrounding uplands. The permanent field indicator of a dryland saline seepage in the area is salt accumulation in a fringe at the base of the scarp in the inland limits of the marine transgression. During humid periods with anomalously high rainfall, a saturated zone develops in this area due to the water table rising to the soil surface. Under dry conditions, especially in hot summer months, evaporation intensifies the capillary rise of saline groundwater, leading to salt accumulations in the soil surface (Celleri et al. 2019). Soils in the seepage zone classify as *Petrogypsids* (formerly *Petrogypsic Gypsiorthids*) because of the presence of a subsurface soil horizon cemented by gypsum (Kruger 1989). Gypsum is an evaporite mineral that frequently results from concentration and crystallization by evaporation of a fluctuating water table enriched in calcium and sulfate (Gómez-Miguel et al. 1984). Soils in this saline seepage area develop a loose cover of salt crystals that form occasionally when the capillary fringe rises to the soil surface. In the peripheral zone downslope from the discharge area, soils classify as *Aquic Ustifluvents* and show typical characteristics indicating saturation.

Wetlands in the OMP are dry-end wetlands, that is, wetlands that occur in landscape positions where flooding is irregular and soil saturation to the surface occurs occasionally (Whigham 1999). These wetlands are subjected to episodic disappearance and may get dry for several years, relying on the occurrence of rainy years during humid periods to resurge. Plant communities in these environments are very reactive to small changes in climate and quickly respond to exceptionally humid or dry periods. Because of their highly dynamic nature, and the climatic settings of the area, these wetlands are sensitive indicators of climate change and variations in large-scale circulation patterns. Out of the seepage area, soils in the OMP are well-drained and have lower contents of soluble salts.

16.2.3 Wetlands Indicators in Arid Environments

In arid and semiarid environments, coastal wetlands beyond the limits of the tidal inundation, like those in the OMP of the Bahía Blanca Estuary, may have a pulsed ephemeral nature that makes them difficult to identify (Cintron-Molero and Schaeffer Novelli 2002). Thus, soils are useful instruments for wetland identification and delineation (Tiner 1999). Extended flooding has a significant effect on soil-forming processes, which produce a set of unique and noticeable soil properties. When porewater replaces the air in the soil pore spaces, the supply of oxygen depletes, and chemical processes change (Henderson and Patrick 1982). Dissolved oxygen is virtually absent in wetland soils that are continuously saturated with water. These soils chemically reduce and develop an *aquic* moisture regime (Soil-Survey-Staff 1975). Aquic soils (Vepraskas 1996) have morphological indicators of saturation and reduced conditions (i.e., redoximorphic features). Common redoximorphic indicators of wetland soil include gray colors, iron, and manganese concretions near the surface (Tiner 1999).

Wetland soils in arid climates concentrate carbonates, gypsum, and even more soluble salts on their surface. The combination of high salinities and low organic matter content hampers typical processes expected in humid climates, such as microbial activity and chemical reduction of iron. If redoximorphic features and hydric soil indicators do not form, seasonally dry or intermittent wetlands would be challenging to identify. Boettinger (1997) suggests that salts more soluble than gypsum in the upper 30 cm indicate a hydric soil. Thus, the presence of a salt crust should be enough as a field indicator of wetland soil, subject to periodic saturation from a saline water table.

16.2.4 Plant Communities

From the physiognomic-floristic point of view, the spontaneous vegetation of the Bahía Blanca region lies within three Phytogeographic Provinces: *Monte*, *Pampeana*, and *Espinal* (Canepuccia et al. 2013; Oyarzabal et al. 2018). This ecotone is characterized by the dominance of shrubs and grasses, mostly perennial. Due to climatic constraints, this type of vegetation is typically composed by a high number of C_4 grasses and xerophytic species (León et al. 1998; Paruelo et al. 1998). Plant leaves are frequently small, succulent, reduced to thorns, or absent. Photosynthetic stems are another common adaptation to arid conditions. Vegetation in the area has a patchy structure, with vegetation islands that contain different species and even different growth forms, but with low species richness.

Several authors described plant associations in the coastal zone of Bahía Blanca (e.g., Verettoni 1974; Kruger and Peinemann 1996; Nebbia and Zalba 2007). Based on this previous information and extensive field work carried out in the area, Piovan et al. (2014) defined nine major land cover classes, which were used to derive a land cover map at the landscape level. The methodological approach involved the processing of a 30-year-long time series of Landsat satellite images to obtain surface reflectance data, and the implementation of numerical classification algorithms, masking, and reclassification, to produce the final land cover map presented in Fig. 16.2. Examples of the more representative land cover classes are presented in Fig. 16.3.

Southwestern Atlantic salt marshes are characterized by the presence of the halophytic genera *Spartina* and *Sarcocornia* (Isacch et al. 2006). In Argentina, *Sarcocornia* appears all along the coast, from Buenos Aires to Tierra del Fuego, commonly forming marshes below the elevation of the highest astronomical tides (González et al. 2019). Within the intertidal fringe of the Bahía Blanca Estuary, bare mudflats are the dominant land cover type (836 km^2). Salt marshes of *Spartina alterniflora* (196 km^2) appear as monospecific associations with variable plant heights and densities, from sparse plants, up to 20 cm tall (González Trilla et al. 2013), to dense stands taller than 1 m in places subject to high sedimentation rates (Pratolongo et al. 2010). Intertidal marshes of *Sarcocornia ambigua* are less represented (72 km^2), and the dominant species commonly forms circular mounds, sometimes in association with *Heterostachys ritteriana* or *S. densiflora* at higher elevations (Pratolongo et al. 2013). Vegetation cover of this type of marshes is

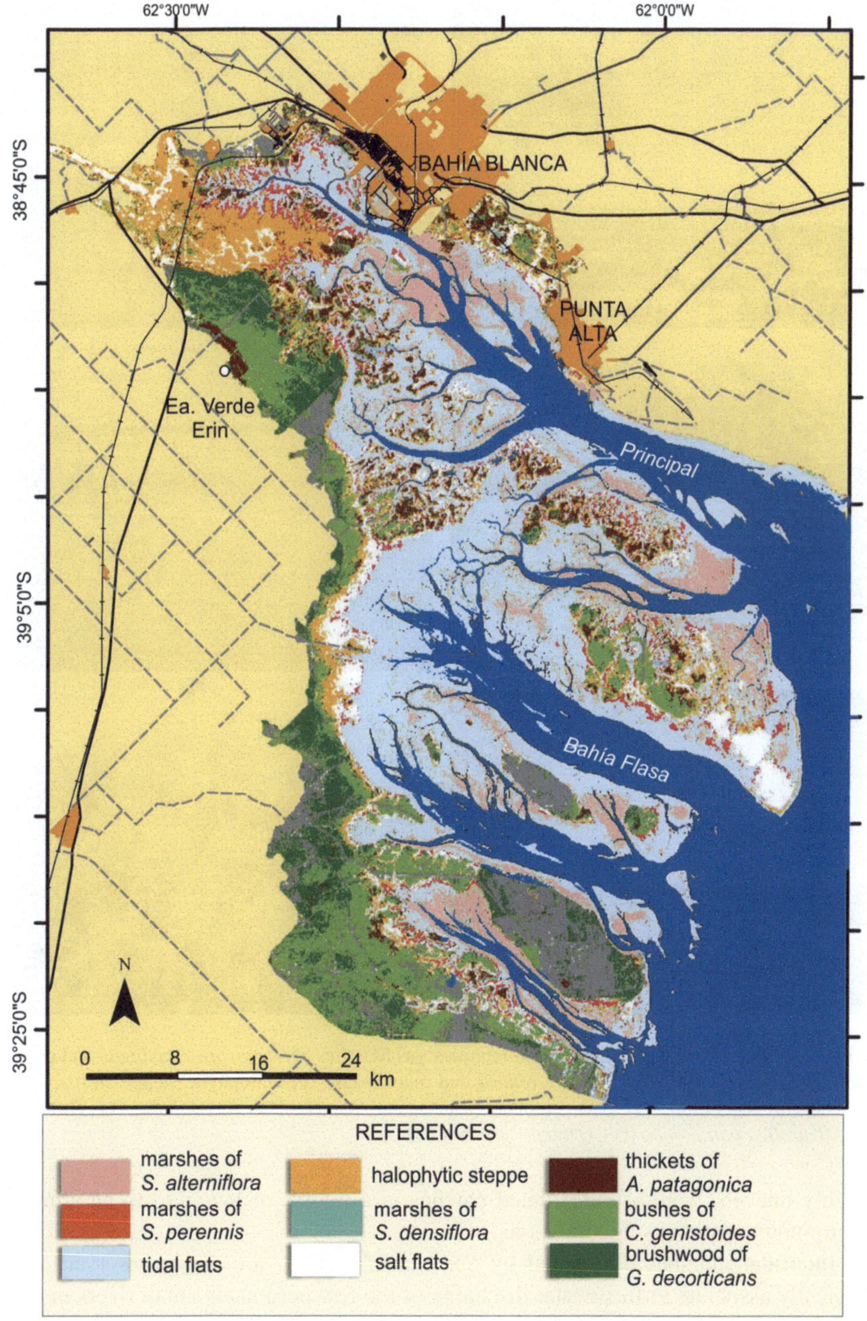

Fig. 16.2 Land cover map showing major plant associations in the coastal zone of Bahía Blanca. Land cover classes were previously defined, based on field surveys, and the map was further obtained through digital classification of satellite images from the Landsat series. (Modified from Pratolongo et al. 2016, 2017)

Fig. 16.3 Most representative plant communities. (**a**) Marshes of *Sarcocornia ambigua* and plant detail. (**b**) Marshes of *Spartina alterniflora* and plant detail. (**c**) Halophytic steppe with sparse patches of vegetation. (**d**) Bare mudflats with sparse plants of *S. alterniflora*. (**e**) Thickets of *Allenrolfea patagonica* (**f**) *Cyclolepis genistoides*

highly uneven, from a few isolated patches in a matrix of bare soil to an almost continuous carpet at some locations.

Intertidal marshes dominated by *Spartina densiflora* are rarely observed and typically associate to freshwater discharge of the few permanent small rivers in the estuary (e.g., Napostá, Sauce Chico, and Maldonado Channel). In the supralittoral zone, however, *S. densiflora* marshes occupy the old tidal channels of the RMP, covering 1.8 km^2. *S. densiflora* is the clear dominant species in these later marshes, forming dense stands about 50 cm tall, and *S. ambigua* may be also present with relatively high percent cover (Piovan 2016).

Tidal inundation becomes less frequent upslope from the mean high tide level, and there is a spatial transition from *S. ambigua* marshes to halophytic steppes. Vegetation in these steppes keeps a similar structure, with plant patches forming vegetation islands. Barren areas increase their cover close to the elevation of the highest tides, and soils develop more permanent salt crusts. At some locations, vegetation patches are sparse enough to allow for the distinction of salt flats as a pure land cover type. *S. ambigua* is also a dominant species in halophytic steppes, but, as tidal inundation decreases, the number of species within vegetation patches increases. Typical accompanying species are *H. ritteriana* and *S. densiflora*, with higher abundances than those observed in marshes downslope. *H. ritteriana* is a salt-tolerant species whose presence is considered as an indicator of soils with high concentrations of chlorides and sodium (Cantero et al. 1998). At higher elevations, vegetation is more diverse, and species like *Atriplex undulata*, *Limonium brasiliensis*, and *Frankenia juniperoides* may also appear in vegetation islands.

Three different types of woody associations can be distinguished in the coastal zone of Bahía Blanca. Dense thickets dominated by *Allenrolfea patagonica* form a continuous fringe in the inland limits of the OMP. This association covers 45 km^2 within the OMP. Vegetation height is about 1 m, and *A. patagonica* has an average 60% cover, sometimes in association with *Cyclolepis genistoides*. The plant association of *A. patagonica* and *C. genistoides* commonly denotes soils with high concentrations of sulfates and calcium (Cantero et al. 1998). Most species in the undergrowth appear in response to episodic rainfall, but *S. ambigua* and *Lycium chilense* are always present, even in dry periods. *Grahamia bracteata* and cactus from the genus *Trichocereus* are also commonly observed.

Downslope from thickets of *A. patagonica*, vegetation is slightly taller, and the amount of bare soil increases. This landscape position is typically covered by bushes dominated by *C. genistoides* (261 km^2). In this plant association, *A. undulata* appears as a common species in the underbrush, replacing *A. patagonica*. Other common species in this lower stratum are *S. ambigua*, *F. juniperoides*, *Cressa truxillensis*, and *L. brasiliensis*. Bushes dominated by *C. genistoides* also occur in inland saline environments of central Argentina (Cantero et al. 1998; Karlin et al. 2012; Cabido et al. 2018). In the RMP, these two plant associations (thickets of *A. patagonica* and bushes of *C. genistoides*) appear as a discontinuous mosaic jointly mapped as halophytic shrubs in Fig. 16.2. In this landscape unit, *A. patagonica*, *C. genistoides*, *A. undulata*, and *S. ambigua* are all common species, whose relative abundance is determined by specific site conditions.

The last woody association described for the Bahía Blanca Estuary is the brushwood dominated by *Geoffroea decorticans* (150 km^2), which is indicative of well-drained soils and lower salinities. The dominant species, *G. decorticans*, is a tall shrub, exceeding 2 m height, which forms dense thorny bushes. This plant community is characterized by the predominance of medium to tall shrub associations with several strata and relatively high vegetation cover, typical of the Monte Phytogeographic Province (Gaitan et al. 2019). The most frequent shrub species are

Prosopidastrum globosum, *Monttea aphylla*, and *Condalia microphylla*. The diverse underbrush is composed of *L. chilense*, *A. undulata*, *Atriplex heterosperma*, *Grindelia brachystephana*, *Ephedra triandra*, and several grasses (*Poa lanuginosa*, *Trichloris crinita*, *Jarava* spp., and *Nassella* spp., among others).

Suaeda divaricata is an emblematic shrub species in the area, which usually appears within woody stands in highly saline soils. Locally called *Vidriera*, the extensive salt flat La Vidriera was named after this species. Endemic species of the Cactaceae family also occur in vegetated patches within the salt flat, like *Echinopsis leucantha*, *Opuntia sulphurea*, *Opuntia penicilligera*, and *Trichocereus candicans*. Other frequent species are *Sesuvium portulacastrum*, *A. patagonica*, *Heterostachys olivascens*, *H. ritteriana*, *Nitrophila australis*, *Suaeda argentinensis*, and *G. bracteata* (Pérez Cuadra 2008).

Along the salinity and flooding gradient, it is possible to identify changes in vegetation structure from isolated and cushion plants to steppes of greater height and higher species diversity. Salinity and flooding act as ecological filters that reduce plant diversity in the salt marsh (Rogel et al. 2000). Therefore, the environmental gradient and its unique physicochemical characteristics set an ecotone in the transition between intertidal marshes and upland plant communities. Much of the variation in vegetation structure between intertidal marshes, perimarine wetlands, and upland communities can be related to a gradient in elevation, salinity, soil water content, and severity of hypoxia (Traut 2005).

16.2.5 Plant Adaptations to Life in Salt Marshes

Salt marshes are exposed to extreme environmental conditions as high salinity, regular tidal flooding, and the mechanical impact of waves and storms (González et al. 2019). Salt marsh environments are stressful, and plant species must survive intervals of complete inundation, as well as changes in water and soil salinity. The spatial distribution of vegetation within salt marshes is not random or spatially uncorrelated, and the role of adaptations to soil salinity and tidal inundation in determining this distribution has been widely studied (Traut 2005; Bao-Shan et al. 2011). Several adaptations allow salt marsh plants to thrive in high salinity and low oxygen environments. For some species, morphological adaptations or anatomical structures provide a strategy to tolerate environmental stress. Morphological adaptations include smaller leaves, fewer stomata, increased succulence, thick cuticle, and deposition of wax. Anatomical changes include salt secretory trichomes and glands, located on leaves, and well-developed water storing tissues (Wahid 2003).

Oxygen availability is essential for plant growth, and it is governed by soil type, topographic position, and the frequency and the duration of salt marsh flooding (Silvestri et al. 2005). Various adaptations allow salt marsh plants to survive under low oxygen levels imposed by tidal flooding. The most common adaptation is the formation of aerenchyma in leaves, stems, and roots. This tissue enables the vertical gas transport within plants, allowing oxygen to reach the flooded roots (Cronk and

Box 16.1 Most Emblematic Native Plant Species in the Bahía Blanca Estuary
Family Amaranthaceae

At present, the Amaranthaceae and Chenopodiaceae form a monophyletic group (Kadereit et al. 2003) and are considered as a single family Amaranthaceae (Stevens 2001). This family comprises about 180 genera and around 2500 species widely distributed. Preferred habitats are warm temperate and tropical grasslands, savannas, sand dunes, salt marshes, semideserts, and deserts (Kadereit et al. 2003). Most species in this family have xeromorphic characteristics. In Argentina, 14 genera are represented with nearly 121 species (Múlgura and Marticorena 2008).

Sarcocornia ambigua

Subfamily	Salicornioideae
Genus	*Sarcocornia*
Species	*ambigua*
Sigla sp.	(Michx.) M.A. Alonso & M.B. Crespo
Common name	Jume

Plants of the *Sarcocornia* genus grow in saline areas, usually near the coast where they form salt marshes. *Sarcocornia* genus comprises about 28 succulent perennial species (Steffen et al. 2010). Five South American species of *Sarcocornia* have been identified, and four different morphological types are found in South America and the Mediterranean countries. All *Sarcocornia* species from South America were first described as *Salicornia*. Recent studies indicate that *Sarcocornia* and *Salicornia* are paraphyletic groups that show strong morphological and ecological similarity (Alonso and Crespo 2008). *Sarcocornia* and *Salicornia* can be distinguished only by inflorescence characters and life form (perennial and annual, respectively) (Steffen et al. 2015). The taxonomy and nomenclature of South American species have recently been updated, and they are now referred as *Sarcocornia* spp. (Alonso and Crespo 2008). *Sarcocornia ambigua* is the species currently accepted for the area of Bahía Blanca. However, most of the existing bibliography regarding marshes in Bahía Blanca refers to this species as *Sarcocornia perennis*.

Sarcocornia ambigua is a perennial erect shrub. The stems are succulent and articulated, leaves opposite, and flowers hidden in cavities of the inflorescence axis. This is a widespread species and grows in areas with highly variable salinity, which makes it a potential candidate for the development of a novel crop (Freitas and Costa 2014). The aerial parts of *Salicornia* and *Sarcocornia* species have been introduced into the European market for human consumption, due to their high nutritional value in terms of minerals and antioxidant vitamins (Ventura et al. 2011; Bertin et al. 2016).

Allenrolfea patagonica

Subfamily	Salicornioideae
Genus	*Allenrolfea*
Species	*patagonica*
Sigla sp.	(Moq.) Kuntze
Common name	Black jume

The genus *Allenrolfea* has three species: *Allenrolfea occidentalis* is found in North America, while *Allenrolfea patagonica* and *Allenrolfea vaginata* are exclusive from Argentina. *A. patagonica* grows in a wide range of salinities, in soils not affected by tidal flooding, from Salta to southern Buenos Aires (Kruger and Peinemann 1996). *A. patagonica* is a 1-m-tall perennial shrub, with pyramidal leaves and flowers grouped in terminal inflorescences. This species is used for the preparation of bleach (del Vitto et al. 1997).

Atriplex undulata

Subfamily	Chenopodioideae
Genus	*Atriplex*
Species	*undulata*
Sigla sp.	(Moq.) D. Dietr.
Common name	Wavy-leaf saltbush

The genus *Atriplex* is represented in Argentina by 34 species. This is a widespread genus that colonizes many arid and semiarid regions (Giusti 1997). *Atriplex* species can be used to increase productivity in arid or semi-arid lands because of their salt tolerance and high productivity. Several species have been planted as a foraging shrub in marginal agricultural lands in many countries (Salem et al. 2010). *Atriplex undulata* is a perennial shrub, native to arid and semiarid rangelands of central Argentina (Piovan et al. 2014). *A. undulata* is a 1-m-high shrub with ramified erect stems. Leaves are alternate, obovate-oblong, grayish-white on both sides, and with undulate edges. Male and female flowers are on separate plants (dioecious). Each round and soft fruit contains one single seed that matures in autumn.

Suaeda divaricata

Subfamily	Chenopodioideae
Genus	*Suaeda*
Species	*divaricata*
Sigla sp.	Moq.
Common name	Vidriera

The genus *Suaeda* is globally distributed in saline or alkaline habitats. In Argentina, there are three native species: *Suaeda argentinensis*, *Suaeda divaricata*, and *Suaeda neuquenensis* (Alonso et al. 2004). *S. divaricata* is a perennial shrub 1–3 m tall that inhabits saline soils from Jujuy to Chubut. Leaves are succulent, semi-cylindrical, and positioned perpendicular to stems. They are dioecious plants with sessile, axillary flowers that bloom in spring and fructify in summer (Gates et al. 2018). *S. divaricata* has vacuoles containing a high salinity solution, which facilitates survival in arid environments. These salts are partially eliminated by crystal formation in stomata, what makes their leaves reflect sunlight, and the name *Vidriera* (crystal window) alludes to this optical property. Its ashes are rich in sodium carbonate and can be employed for traditional saponification.

Heterostachys ritteriana

Subfamily	Salicornioideae
Genus	*Heterostachys*
Species	*ritteriana*
Sigla sp.	(Moq.)Ung.-Sternb.
Common name	Jumecillo

The genus *Heterostachys* comprises two species, *Heterostachys olivascens*, found from southern Buenos Aires to Río Negro, and *Heterostachys ritteriana*, from Central America to South America (Soriano 1947). *H. ritteriana* is a halophyte shrub 0.3–0.5 m tall that grows in arid environments. This is the most frequent species in inland saline steppes of Central and South America and tolerates saline soils and periodic flooding events (Karlin et al. 2012). The small leaves are succulent, alternate, and caduceus. Young stems are succulent and turn woody afterward (Pérez Cuadra and Hermann 2014). This species has solitary flowers located in the axil of succulent bracts.

Family Asteraceae

The largest family of flowering plants, Asteraceae, has around 1535 genera and 23,000 species (Bremer 1994). This family includes many economically important species (sunflower, lettuce), as well as many ornamentals. This group has a cosmopolitan distribution and is highly diversified in its habitat preference and life forms. It includes aquatics, herbs, and shrubby trees in temperate, tropical, and arid environments, as well as trees in tropical rainforests (Jansen and Palmer 1987).

Cyclolepis genistoides

Subfamily	Gochnatioideae
Genus	*Cyclolepis*
Species	*genistoides*
Sigla sp.	Gillies ex D. Don
Common name	Palo Azul

Cyclolepis genistoides is a native shrub distributed from Chaco to Patagonia regions, in arid environments and saline soils (Cabrera 1978). *C. genistoides* is 1–2.5 m tall, with grayish-green striated stems and branches perpendicular to the stem. Leaves are alternate, entire, linear-lanceolate, and pubescent on both sides. The capitulum is yellowish-white, with 10–15 flowers. It blooms in September–October. It is widely used in folk medicine as a diuretic, anti-rheumatic, and antispasmodic agent (Sosa et al. 2011).

Family Fabaceae

The Fabaceae is a large and economically important family of angiosperms, with 730 genera and over 19,400 species (Schrire et al. 2005). It includes trees, shrubs, and herbaceous plants, perennials or annuals, which are easily recognized by their fruits (legume). The group is widely distributed and becomes the most common family in tropical rainforests and dry forests of the Americas and Africa (Lavin et al. 2005).

Geoffroea decorticans

Subfamily	Faboideae
Genus	*Geoffroea*
Species	*decorticans*
Sigla sp.	(Gillies ex Hook. &Arn.) Burkart
Common name	Chañar

Geoffroea decorticans is a native South American tree, distributed from Jujuy to northern Patagonia (Silva et al. 2004). It occurs in dry forests to regularly flooded areas, as well as saline soils. *G. decorticans* trees can rise up to 12 m tall and form extensive uniform colonies due to their reproduction by gemmiferous roots (Costamagna et al. 2013). The bark detaches itself in strips uncovering lighter layers of the same hue. Their leaves are compound, imparipinnate, and alternate or arranged in fascicles. The species blooms in September–October, with shiny yellow flowers arranged in dense 3–8 flower racemes. Fruits mature in January (Eynard and Galetto 2002). In folk medicine, flowers, fruit, bark, and leaves are used for their emollient, balsamic, antitussive, and expectorant properties (Silva et al. 2004).

Fennessy 2001). Several salt marsh plants can have as much as 60% of air space in their root, shoot, and leaf cortex, and species such as *S. alterniflora* have a continuous air space that extends from leaves to roots (Maricle and Lee 2002; Callaway et al. 2007). The ability of some salt marsh plants to allow gas transport might have implications not only for the root itself but also to the surrounding sediments. Oxygen can be released out of the roots in a process called “radial oxygen loss” which oxidizes the rhizosphere (Armstrong 1979; Colmer 2003) and changes the redox conditions and chemical forms of several elements (Sundy et al. 2003). Another adaptation of many salt marsh species is the development of extensive root systems to increase contact with oxygen-rich water o with the air. For example, root and rhizome growth would contribute 50–90% of *S. alterniflora* productivity (Redelstein et al. 2018), and *S. ambigua* has stems with adventitious roots at some of their nodes (Alonso and Crespo 2008).

Excess in soil salinity has an immediate effect on plant growth and development (Munns and Tester 2008). For instance, a decline in biomass and plant height with increasing salt stress has been observed in *S. alterniflora* (Wang et al. 2006; Canepuccia et al. 2013). Halophytes are plants that have specific adaptations to survive and grow under high soil salinities. However, different adaptations lead to large differences in salt tolerance among species (Kruger and Peinemann 1996). Halophytes can avoid or tolerate salts through exclusion, secretion, shedding, and succulence (Cronk and Fennessy 2001). Salt exclusion is the most important mechanism to cope with high salt concentrations. *S. alterniflora* excludes at the root level 91–97% of the ions present in saltwater (Bradley and Morris 1991). However, for most species, salt exclusion at the roots is not enough, and other mechanisms arise. For example, *F. juniperoides* and *L. brasiliensis* have salt glands on their leaves for salt secretion, and *A. undulata* secretes salt through salt secretory trichomes located on leaves and stems (Pérez Cuadra and Cambi 2016). These secreted or excreted salts are released back to the environment by leaf shedding or by wind and rain (Wahid 2003).

A great challenge for plants in saline soils is water acquisition. In addition, high concentrations of salts within the plant can be toxic. Some plants can concentrate salts in their cell solutions, which results in a high osmotic pressure. Through accumulation of Na^+ and Cl^- inside the cell, these plants maintain an osmotic gradient that allows water uptake from saline sediments (Bradley and Morris 1991), but this increased internal solute concentration can damage the plant. Thus, to avoid toxic concentrations within the cytoplasm, there is a compartmentalization of the cell, and Na^+ and Cl^- get sequestered in vacuoles (Munns and Tester 2008). Succulence is an increase in cell size induced by salts, due to the large vacuole volume. Well-developed water storing tissues in the cortex and pith characterize succulent stems (Dickison 2000). Succulence is widespread in halophytes and occurs in *Heterostachys*, *Allenrolfea*, and *Sarcocornia* species in the Bahía Blanca Estuary. Succulence provides these species with the advantage of diluting their internal media and decreasing the negative effects of salts (Box 16.1).

16.3 Landscape Dynamics and Changes in Ecosystem Functions

Despite the limited extent of the area covered by marine vascular plants (<2% of the ocean surface), salt marshes, mangroves, and seagrass beds play a significant role in sequestering carbon dioxide. The carbon sequestered in these vegetated coastal ecosystems has been termed *blue carbon*. Tidal saline wetlands account for an important fraction of the carbon storage in the ocean (Chmura et al. 2003; Mcleod et al. 2011). Although their global area is one to two orders of magnitude smaller than that of terrestrial forests, the contribution of tidal saline wetlands per unit area to long-term carbon sequestration is much greater. It is in part due to their generally high primary productivity and their efficiency in trapping suspended matter and associated organic carbon from outside their ecosystem boundaries during tidal inundation. Despite their importance, these blue carbon sinks are being lost at critical rates in response to multiple stressors (Theuerkauf et al. 2015). Global estimates of carbon released by salt marsh land-use change are large, ranging from 0.02 to 0.24 Pg CO_2 year^{-1} (Pendleton et al. 2012), and shoreline erosion is a significant mechanism for current global salt marsh loss, enhanced by relative sea level rise and human activities (Mariotti et al. 2010).

Based on historical aerial photographs and high-resolution satellite images, Pratolongo et al. (2013) quantified changes in land cover in four sites along Principal Channel, for the years 1967, 1996, and 2005. This work was a first attempt to analyze changes in size and position of the plant communities within the coastal zone of Bahía Blanca, as well as primary human modifications like dredge spoil deposition and landfilling. Results showed that the influence of anthropic transformations decreases to the west of the port area, that is, through the head of the Principal Channel. The loss of *S. ambigua* marshes was a recurring pattern through the four sites, and, to a lesser extent, the area covered by halophytic steppes also reduced. In the case of *S. ambigua* marshes, most of the area changed to mudflats. This type of replacement involves lowering surface elevation and losing vegetation cover. The loss of halophytic steppes was mainly due to the replacement of the original land cover by human land uses (e.g., landfilling and dredge spoil deposits). Through the four sites, most of the original halophytic steppes area was replaced by landfills and dredge spoil deposits, and a smaller fraction of these natural environments eroded to mudflats and channels. Besides these general patterns, the erosion of *S. ambigua* marshes dominated landscape dynamics through the head of Principal Channel and direct human transformations in the port area. While *S. ambigua* marshes and halophytic steppes reduced, *S. alterniflora* marshes expanded their cover. The expansion of *S. alterniflora* marshes was widespread west of the port area, in a process entailing sediment accretion, surface elevation, and colonization by plants.

More recent work based on these previously identified trends, aimed at quantifying the impacts of land cover replacements on ecosystem function. Maps based on aerial photographs (year 1967) and high-resolution satellite images (years 2005 and

2017) served to quantify changes in area between years, and the updated land cover maps, over a larger area, confirmed most of the observed tendencies (Fig. 16.4). In the North section of the Bahía Blanca Estuary (i.e., the area represented in Fig. 16.4), erosion rates of *S. ambigua* marshes (% loss per year) accelerated from 0.8% year $^{-1}$ (from 1967 to 2005) to 2.0% year $^{-1}$ (from 2005 to 2017). For this later period, *S. ambigua* marshes lost to mudflats at an average rate of 93 ha year $^{-1}$. Halophytic shrubs and steppes also reduced their cover, but erosion was less pronounced and steady through time. For the entire period (1964–2017), 1750 ha of halophytic steppes, equivalent to 14% of the original area, eroded at an average rate of 40 ha year $^{-1}$. In turn, halophytic shrubs eroded at an average rate of 11 ha year $^{-1}$ (545 ha for the entire period).

Elevation profiles obtained in the field were used to estimate how deep the marsh soil was excavated during shoreline transgression. To assess the volume of material eroded from *S. ambigua* marshes, the total area of salt marshes replaced by mudflats for a given pair of dates was considered, along with the depth of erosion at the shoreline. Considering the average erosion depths and soil organic carbon densities, the northern section of the Bahía Blanca Estuary exported soil organic carbon at a rate of 2893 t C year $^{-1}$, because of marsh erosion between 2005 and 2017. The soil organic material that is being eroded presents a very low C:N ratio (8–10), and its bulk isotopic composition ($\delta^{13}C$ −24.7 to −16.8‰; $\delta^{15}N$ + 8.7 to +11.3‰) reflects a dominance of organic matter of marine origin (unpublished data). Along with the loss of salt marsh soils, erosion involves a large amount of above- and belowground biomass of *S. ambigua*, which may represent an additional export of between ~450 and 1400 t C year^{-1} (based on biomass data and carbon contents in Negrin et al. 2016).

According to the landscape evolution described for the area, the presently intertidal platform occupied by *S. ambigua* marshes is composed of marine deposits that formed under a higher relative sea level, after the transgressive maximum in the Holocene. The present relative sea level trend in the Bahía Blanca Estuary has not been evaluated, but the closest estimation is 0.85 mm year^{-1}, with a 95% confidence interval of ±0.31 mm year^{-1}, based on tide gauge station measurements obtained from 1918 to 1982 in Puerto Quequén (290 km north-east from Bahía Blanca). In Puerto Madryn (about 450 km south-west from Bahía Blanca), the mean relative sea level trend is 1.5 mm year^{-1} with a 95% confidence interval of ±0.79 mm year^{-1} based on data from 1944 to 2000 (Permanent Service for Mean Sea Level, available through NOAA's *Sea Levels Online* website at http://tidesandcurrents.noaa.gov/sltrends/sltrends.html). Under the current rates of relative sea level rise, there is accelerated erosion, and salt marsh soils act as a significant net source of organic carbon to estuarine waters.

These estimations do not discern the proportion of labile or refractory carbon in the eroded material. Moreover, the fate of eroded carbon is complex and can follow multiple pathways that are highly dependent on the individual characteristics of the estuary. Some eroded material would be transported in suspension and deposited elsewhere in the estuary. Relatively young and bioavailable carbon would be respired within the estuary and metabolized by microbes (Canuel et al. 2012). On

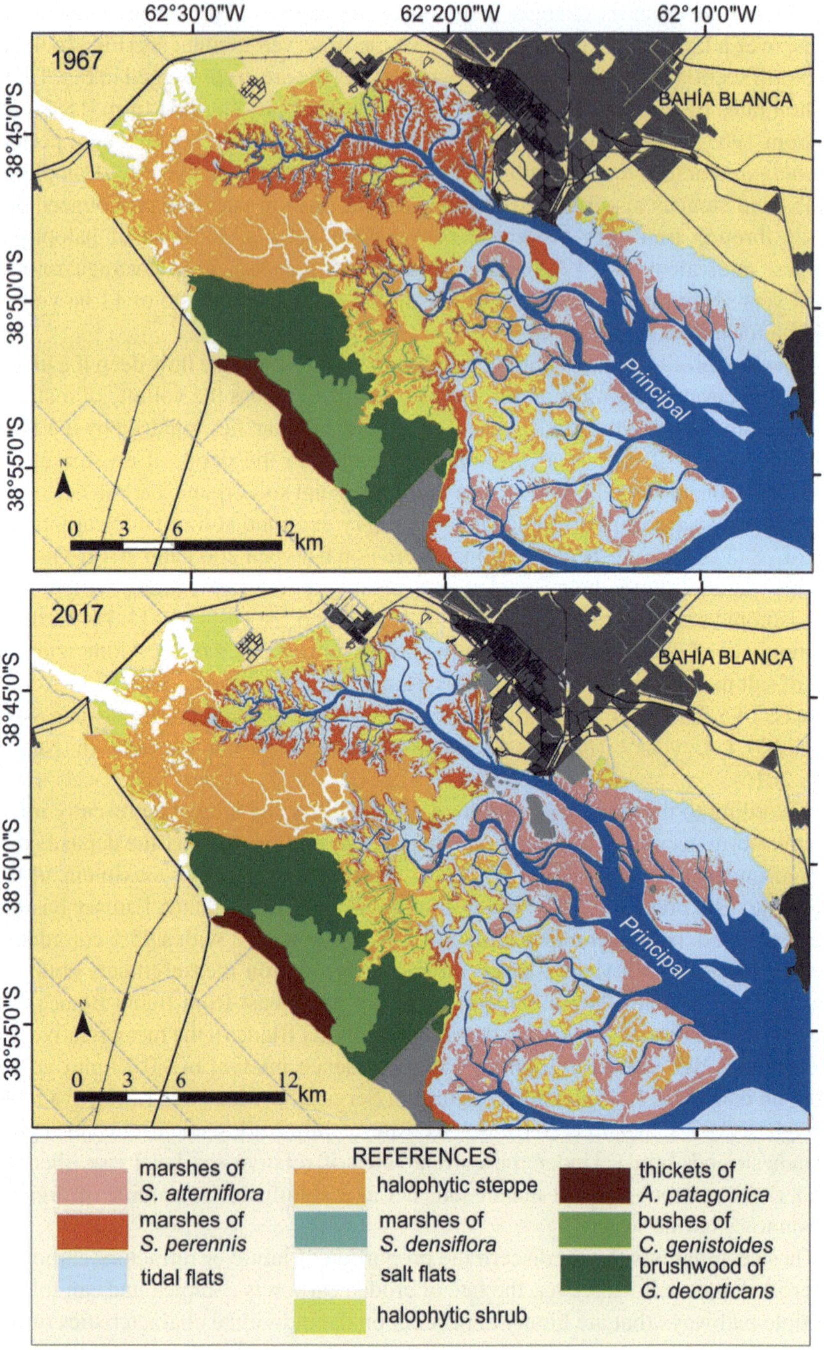

Fig. 16.4 Land cover changes between years 1967 and 2017

the contrary, older and more refractory soil carbon that is not processed within the estuary would be exported to open waters, where it may enter detrital ocean webs or the microbial loop or it may sink to the ocean floor (Cai et al. 2003). Thus, the relative importance of carbon export from marshes for the global carbon budget depends on the fate and biochemical characteristics of the eroded carbon.

While *S. ambigua* marshes are eroding, *S. alterniflora* marshes are expanding their distribution, and part of the eroded material may be redeposited in these expanding marshes. In 1967, for the same region in the northern section of the estuary, *S. alterniflora* marshes covered 4454 ha. The area increased at accelerated rate from 1967 to 2005 (62 ha year^{-1}), but the expansion slowed down more recently. *S. alterniflora* marshes covered 7090 ha by 2017, and their expansion rate for this later period was 22.5 ha year^{-1}. *Islote del Puerto* is an island in the harbor area that hosts a large breeding colony of the endangered species Olrog's gull (*Larus atlanticus*). In this island, *S. alterniflora* marshes covered 178 ha by 1967. Vegetation cover also encompassed *S. ambigua* marshes and halophytic shrubs (208.2 and 70.8 ha, respectively), but these areas were completely covered with a massive dredge spoil deposit in 1989. In 2005, *S. alterniflora* covered 566 ha occupying most of the intertidal fringe, and marshes did not expand from 2005 to 2017.

Spartina alterniflora is presently increasing in abundance and expanding its regional extent in many South American countries, including several coastal locations in Argentina. This species is globally regarded as a powerful ecosystem engineer, based on the ability to cause significant modifications to the abiotic environment. Worldwide, the establishment and expansion of *S. alterniflora* marshes fostered large-scale alterations in ecosystem processes (see Box 16.2), including the extinction of native species, loss of functional native diversity, changes in nutrient cycling and organic matter storage, and loss of habitat (Ayres et al. 2004). The consequences of *S. alterniflora* expansion in South America are largely unexplored.

Besides carbon sequestration, there are essential ecosystem services provided by salt marshes that relate to elemental cycling (Barbier et al. 2011). For instance, for biomass production and growth, plants take up nutrients from the sediments. Both essential and toxic metals can also be absorbed during plant growth, providing tools for mitigation of eutrophication and metal pollution in coastal waters (Tangahu et al. 2011). The uptake of nutrients and metals depends mainly on the elemental concentrations in the porewater and sediment characteristics like grain size, organic matter content, pH, and redox potential (Wang et al. 2013). Although mineral uptake occurs at the root level, elements can be further translocated to leaves and stems, setting a distinctive distribution within aerial and belowground tissues and linking their cycling to the fate of detritus. After senescence, plant tissues release elements back to the environment through decomposition, whose rate depends on environmental conditions like climate, but also on detritus quality. Thus, decomposition rates usually differ between species and between organs within a species (Simões et al., 2011; Tong et al. 2011).

Negrin et al. (2016) reviewed the available information on ecological and biogeochemical processes in salt marshes of the Bahía Blanca Estuary. Evaluations of the belowground and aboveground biomass dynamics, decomposition, and nutrient

Box 16.2 Spartina alterniflora: An Invasive Exotic or a Native Increasing Its Range?

Invasive species are those introduced to an ecosystem that they did not occupy previously and establish a population that spreads autonomously (Simberloff 2010). Once established in the non-native environment, invasive species can affect the ecosystem in many ways, including modifications that affect most of the originally resident species. Occasionally, native plant species spread into formerly unoccupied habitats and become invasive (Valéry et al. 2009). An example is the large expansion of *Phragmites australis* in North American coastal marshes, which spread rapidly into new habitats after the mid-nineteenth century. The introduction of Old-World genotypes in North America has triggered the invasive behavior of this native species (Saltonstall 2002). Native species may also become invasive after human modifications of the environment. For instance, *Elymus athericus* was commonly present in low abundances in high marshes of western Europe. Since the mid-1980, this native grass has been aggressively invading the middle and low marshes, where it often forms dense monospecific stands replacing natural *Atriplex portulacoides* low marshes. This seaward expansion is facilitated by increased anthropogenic nitrogen in aerial depositions and runoff (Valéry et al. 2016).

The grass group *Spartina* encompasses several successfully invasive species of intertidal mudflats and salt marshes worldwide (Ainouche and Gray 2016). The critical ecological role of several *Spartina* species as "ecosystem engineers" on coastal salt marshes and their remarkable history punctuated by natural or human-mediated introductions outside their native range, rapid expansion, and propensity to interspecific hybridization and polyploid speciation have long captured the attention of researchers and institutions involved in land management. In 2014, a DNA-based phylogenetic study confirmed the paraphyly of the grass subtribe Sporobolinae, and the creation of a large monophyletic genus *Sporobolus* was proposed, which includes species previously included in the genus *Spartina* (Peterson et al. 2014). The name *Spartina* designates a morphologically well-circumscribed group of grasses, representing a monophyletic clade. *Spartina* species have a tremendous impact in many scientific and non-scientific fields, provided their global ecological importance (Bortolus et al. 2019). Therefore, for the sake of simplicity, the newly designated subgenus *Spartina* will be treated here as the genus *Spartina*.

Spartina anglica appeared by the end of the 1880s in Southampton Water, UK, by chromosomal doubling from *Spartina* x *townsendii*, a hybrid between the introduced *Spartina alterniflora* and the native *Spartina maritima* (Ayres and Strong, 2001). *S. anglica* invaded large areas during the first 30 years (Raybould, 1997), and the species continues to expand along the north-east and north-west coasts of England, despite the aggressive control methods employed (Lacambra et al. 2004). *S. anglica* is also increasing in abundance and spreading in marshes of the Wadden Sea, possibly due to warmer spring temperatures (Nehring and Hesse 2008). *Spartina anglica* was later introduced to Australia in the 1920s. The species was planted across many states, but populations established in Port Gawler, South Australia, Bass River and Western Port Bay, and Victoria, as well as northern and eastern Tasmania. After management for control and eradication of the species, the largest infestations persist in the Tamar River and the Rubicon Estuary, in Tasmania, as well as the Anderson Inlet, in Victoria (Beasy and Ellison 2013).

In Willapa Bay, in the Pacific coast of North America, *S. alterniflora* invaded nearly one third of the original mudflat area. This species was introduced during the late 1800s but was not reported until the 1940s. During the first 50 years, salt marshes gradually expanded throughout the bay, resulting in radical changes to the ecosystem functions (Simenstad and Thom 1995). During the past decades, after an aggressive eradication campaign that included widespread application of herbicides, *S. alterniflora* has virtually disappeared from Willapa Bay (Strong and Ayres 2016).

Spartina alterniflora was first transplanted in coastal China in 1979, to stabilize tidal flats. Since the mid-1990s, this species began to spread in the Yangtze Estuary (Ouyang et al. 2013) gradually invading native marshes (Cheng et al. 2006). Before the introduction of *S. alterniflora*, coastal wetlands in the Jiangsu Province typically contained *P. australis* and *Suaeda salsa* high marshes, with extensive bare mudflats at lower elevations. At present, *S. alterniflora* characterizes plant zonation in the Jiangsu Province. It is the dominant species forming marshes in the upper intertidal zone, while marshes of *S. salsa* and *P. australis* persist landward. In estuarine areas of the Yangtze Delta, *S. alterniflora* occupies an elevation range that overlaps with the native salt marsh species, often leading to its disappearance (Gao et al. 2014).

In Atlantic South America, *Spartina alterniflora* increased its abundance and expanded its regional extent over the twentieth century. Early records and indirect observations indicate that *S. alterniflora* may have expanded from Rio de Janeiro, where it was first collected in 1817. To the North, the species was not registered on the coasts of Suriname and Guyana until the 1830–1840s. To the South, *S. alterniflora* was first detected in 1880, in Uruguay. In Argentina, this species was first collected in the Bahía Blanca Estuary in 1902, near Punta Alta. Noteworthy, *S. alterniflora* was never reported in the many significant botanical accounts made across the Argentine coast during the 1800s (Bortolus et al. 2015). More recently, this species has increased in abundance in many South American countries. In Argentina, salt marshes of *S. alterniflora* are aggressively expanding along mudflats of the Bahía Blanca Estuary. In San Antonio Bay (Río Negro province), extensive marshes appeared after 1914 (Willis 1914), and marsh expansions were also documented in Península Valdés (42°24’ S), at the southernmost limit of its distribution (Bortolus et al. 2015).

Although *S. alterniflora* has been historically described as a species native to the Atlantic coasts of South America, recent work proposed that this species would be either an exotic invader in this region or an infrequent native plant, whose modern abundance and geographic expansion are due to habitat changes (Bortolus et al. 2015; Schwindt et al. 2018). In Argentina, *S. alterniflora* was traditionally regarded as a native species (e.g., Zuloaga et al. 2019). Accordingly, *S. alterniflora* does not appear in the national database of exotic species (http://www.inbiar.uns.edu.ar/). Yet, the native status of many species may have been wrongly assigned, leading to substantial misinterpretations of ecosystem change (Carlton 2009). Based on several historical, environmental, biological, ecological, geographic, vector, physiological, and morphological criteria, Bortolus et al. (2015) proposed that *S. alterniflora* would be a non-native species, which was accidentally introduced from either North America or Europe, sometime prior to 1817. In case that additional evidence corroborates that *S. alterniflora* is introduced to South America, this would be one of the largest biological invasions involving this species, with these alien marshes entirely reshaping coastal systems (Bortolus et al. 2015).

contents allowed for the identification of major pathways in the biogeochemical cycling of carbon, nitrogen, and phosphorus. Comparing result obtained in *S. ambigua* and *S. alterniflora* marshes, located close to the elevation of the mean high tide, both types of marshes would be similar in terms of their net aerial primary production. Thus, they would play similar roles in the sequestration of carbon and nitrogen. For both types of marshes, published values on sequestration rates are in the range 181–247 g C m^{-2} $year^{-1}$ and 4.9–6.7 g N m^{-2} $year^{-1}$. Regarding phosphorus sequestration, however, marshes of *S. alterniflora* would be more efficient because of the higher phosphorus concentration in aboveground tissues. While the average phosphorus sequestration in *S. ambigua* is 0.33 g P m^{-2} $year^{-1}$, values reported for *S. alterniflora* marshes range 0.42–0.57 g P m^{-2} $year^{-1}$.

Both types of marshes also differ in their decomposition rates. Although aboveground tissues decompose at relatively similar rates, belowground tissues of *S. ambigua* decompose faster than *S. alterniflora* biomass (Negrin et al. 2016). The average difference of 70 versus 29% during the first year for belowground material suggests that both types of marshes have a differential impact on elemental cycling. Accordingly, the loss of elements bound to fine particulate and dissolved organic and inorganic compounds would be higher in *S. ambigua* marshes, but marshes of *S. alterniflora* would be more efficient for the accumulation in situ of elements bound to undecomposed detritus. The nature of the exported material (particulate or dissolved, organic or inorganic) defines the ultimate fate of elements and their ability to re-enter grazing and detrital food webs, but *S. ambigua* marshes would be more efficient in the recycling of elements, while *S. alterniflora* marshes would play a more substantial role in the long-term storage of elements, and the building up of elevation, through the accumulation of undecomposed plant material.

Considering the different biogeochemical roles of *S. alterniflora* and *S. ambigua* marshes, the observed land cover changes may have a large impact on the overall elemental cycling within the Bahía Blanca Estuary, and the actual rates of change may accelerate in response to global warming and sea level rise (Kirwan and Murray 2008). Direct human modifications derived from dredging of navigation channels, urban expansion, and industrial development in the coastal zone might also modify salt marsh cover. Moreover, net primary productivity and decomposition rates are also sensitive to climate change (Kirwan and Blum 2011). Under this changing scenario, further research is needed to completely understand the complex dynamics of biogeochemical cycling in these salt marshes.

16.4 Conclusions

This chapter offers an insight on the factors shaping landscape structure and wetland dynamics in the Bahía Blanca Estuary. The Holocene marine transgression and subsequent regressive stages had a remarkable influence on the coastal landscape structure and define hydrogeomorphic wetland presence and characteristics. Vegetation is dominated by halophytic species, like *S. alterniflora*, *S. ambigua*, *A. patagonica*, *C. genistoides*, and *A. undulata*. These species associate to form salt marshes, steppes, and shrubs. Barren surfaces are also extensive, with tidal mudflats covering most of the intertidal fringe. Salt flats and bare soil in sparsely vegetated steppes thoroughly spread beyond the limits of tidal influence. *S. alterniflora* has been increasing its abundance over the last decades, and new marshes appeared in the area. As a counterpart, *S. ambigua* marshes eroded at accelerated rate, with a large exportation of particulate organic carbon from the marsh soil pool to estuarine waters. *S. alterniflora* and *S. ambigua* marshes play different roles in sequestration and remineralization of C, N, and P. Thus, the observed land cover changes may have a large impact on the overall elemental cycling within the Bahía Blanca Estuary.

References

Adam P (1990) Saltmarsh ecology. Cambridge University Press, New York

Ainouche M, Gray A (2016) Invasive *Spartina*: lessons and challenges. Biol Inv18:2119–2122

Aliotta S, Farinati EA (1990) Stratigraphy of Holocene sand-shell ridges in the Bahía Blanca Estuary, Argentina. Mar Geol 34:353–360

Aliotta S, Ginsberg SS, Giagante D et al (2014) Seismic stratigraphy of Pleistocene deltaic deposits in Bahía Blanca Estuary, Argentina. An Acad Brase Cienc 86:649–662

Alonso MÁ, Crespo MB (2008) Taxonomic and nomenclatural notes on South American taxa of *Sarcocornia* (Chenopodiaceae). Ann Bot Fenn 5:241–255

Alonso MA, Conticello L, Cerazo MB (2004) *Suaeda neuquenensis* (Chenopodiaceae), a new species from Argentina. Novon 14:1–5

Armstrong W (1979) Aeration in higher plants. Advances in Botanical Research 7:225–332

Ayres DR, Strong DR (2001) Origin and genetic diversity of *Spartina anglica* (Poaceae) using nuclear DNA markers. Am J Bot 88:1863–1867

Ayres DR, Zaremba K, Strong DR (2004) Extinction of a Common Native Species by Hybridization with an Invasive Congener. Weed Technol 18:1288–1292

Bao-Shan CUI, Qiang HE, Yuan AN (2011) Community structure and abiotic determinants of salt marsh plant zonation vary across topographic gradients. Estuar Coast 34:459–469

Barbier EB (2019) The Value of Coastal Wetland Ecosystem Services. In: Perillo GME, Wolanski E, Cahoon DR, Hopkinson CS (eds) Coastal Wetlands, 2nd edn. Elsevier, pp 947–964

Barbier EB, Hacker SD, Kennedy C et al (2011) The value of estuarine and coastal ecosystem services. Ecol Monogr 81:169–193

Barth HJ, Böer B (2002) Introduction. In: Barth HJ, Böer B (eds) Sabkha Ecosyst. Vol. I Arab. Penins. Adjac. Ctries. Kluwer Academic Publishers, Dordrecht, Netherlands, pp 1–5

Beasy KM, Ellison JC (2013) Comparison of three methods for the quantification of sediment organic carbon in salt marshes of the Rubicon Estuary, Tasmania. Australia. Int J Biol 5:1–13

Bertin RL, Maltez HF, de Gois JS et al (2016) Mineral composition and bioaccessibility in Sarcocornia ambigua using ICP-MS. J Food Compos Anal 47:45–51

Bertness MD (1991) Zonation of *Spartina patens* and *Spartina alterniflora* in New England salt marsh. Ecology 72:138–148

Boettinger JL (1997) Aquisalids (Salorthids) and other wet saline and alkaline soils: Problems identifying aquic conditions and hydric soils. In: Vepraskas MJ, Sprecher SW (eds) Aquic Cond. hydric soils Probl. soils. SSSA Spec. Publ. No. 50. Soil Science Society of America, Madison, WI, pp p79–p97

Bortolus A, Carlton JT, Schwindt E (2015) Reimagining South American coasts: unveiling the hidden invasion history of an iconic ecological engineer. Divers Distrib 21:1267–1283

Bortolus A, Adam P, Adams JB et al (2019) Supporting *Spartina*: Interdisciplinary perspective shows *Spartina* as a distinct solid genus. Ecology 100:e02863

Bradley PM, Morris JT (1991) Relative importance of ion exclusion, secretion and accumulation in Spartina alterniflora Loisel. J Exp Bot 42:1525–1532

Bremer K (1994) Asteraceae: cladistics and classification. Timber Press, Portland, Ore

Brinson MM, Christian RR, Blum LK (1995) Multiple states in the sea-level induced transition from terrestrial forest to estuary. Estuaries 18:648–659

Cabido M, Zeballos SR, Zak M et al (2018) Native woody vegetation in central Argentina: classification of Chaco and Espinal forests. Appl Veg Sci 21:298–311

Cabrera AL (1978) Flora de la provincia de Jujuy, República Argentina: Compositae. INTA, Buenos Aires

Cai WJ, Wang Y, Krest J et al (2003) The geochemistry of dissolved inorganic carbon in a surficial groundwater aquifer in North Inlet, South Carolina, and the carbon fluxes to the coastal ocean. Geochim Cosmochim Ac 67:631–639

Callaway RM, Jones S, Ferren WR et al (1990) Ecology of a mediterranean-climate estuarine wetland at Carpinteria, California: plant distributions and soil salinity in the upper marsh. Can J Botany 68:1139–1146

Callaway JC, Parker VT, Vasey MC et al (2007) Emerging issues for the restoration of tidal marsh ecosystems in the context of predicted climate change. Madrono 54:234–249

Canepuccia AD, Pérez CF, Farina JL et al (2013) Dissimilarity in plant species diversity between salt marsh and neighboring environments decreases as environmental harshness increases. Mar Ecol Prog Ser 494:135–148

Cantero JJ, Cisneros JM, Zobel M et al (1998) Environmental relationships of vegetation patterns in saltmarshes of central Argentina. Folia Geobot 33:133–145

Canuel EA, Cammer SS, McIntosh HA et al (2012) Climate change impacts on the organic carbon cycle at the land-ocean interface. Ann Rev Earth Pl Sc 40:685–711

Carlton JT (2009) Deep Invasion Ecology and the Assembly of Communities in Historical Time. In: Rilov G, Crooks JA (eds) Biological Invasions in Marine Ecosystems. Springer, Berlinp, pp 13–56

Cavallotto JL, Violante RA, Parker G (2004) Sea-level fluctuations during the last 8600 years in the de la Plata river (Argentina). Quat Int 114:155–165

Celleri C, Zapperi G, González Trilla G et al (2019) Assessing the capability of broadband indices derived from Landsat 8 Operational Land Imager to monitor above ground biomass and salinity in semiarid saline environments of the Bahía Blanca Estuary, Argentina. Int J Remote Sens 40:4817–4838

Cheng X, Luo Y, Chen J et al (2006) Short-term C4 plant *Spartina alterniflora* invasions change the soil carbon in C_3 plant-dominated tidal wetlands on a growing estuarine Island. Soil Biol Biogeochemestry 38:3380–3386

Chmura GL, Anisfeld SC, Cahoon DR et al (2003) Global carbon sequestration in tidal, saline wetland soils. Global Biogeochem Cy 17(4). https://doi.org/10.1029/2002GB001917

Christian RR, Stasavich LE, Thomas CR et al (2002) References is a Moving Target in Sea-Level Controlled Wetlands. In: Weinstein MP, Kreeger DA (eds) Concepts and Controversies in Tidal Marsh Ecology. Springer, Dordrecht, pp 805–825

Cintron-Molero G, Schaeffer Novelli Y (2002) A guide to wetlands on arid and semiarid zones. USFWS Divisi6n of International Conservation, USAColmer TD (2003) Long-distance transport of gases in plants: a perspective on internal aeration and radial oxygen loss from roots. Plant. Cell and Environ 26:17–36

Colmer TD (2003) Long-distance transport of gases in plants: a perspective on internal aeration and radial oxygen loss from roots. Plant Cell Environ 26:17–36

Costamagna MS, Ordóñez RM, Zampini IC et al (2013) Nutritional and antioxidant properties of Geoffroea decorticans, an Argentinean fruit, and derived products (flour, arrope, decoction and hydroalcoholic beverage). Food Res Int 54:160–168

Cronk JK, Fennessy MS (2001) Wetland plants: biology and ecology. Lewis Publishers, CRC Press, Boca Raton, USA

Davy AJ, Costa CSB (1992) Development and Organization of Saltmarsh Communities. In: Seeliger U (ed) Coastal plant communities of Latin America. Academic Press, pp 157–178

del Vitto LA, Petenatti EM, Petenatti ME (1997) Recursos herbolarios de San Luis (República Argentina) primera parte: plantas nativas. Multequina 6:49–66

Dickison WC (2000) Integrative Plant Anatomy. Harcourt Academic Press, San Diego

Doody JP (1992) The conservation of British saltmarshes. In: Allen JR, Pye K (eds) Saltmarshes: Morphodynamics, conservation and engineering significance. Cambridge University Press, New York, pp 80–114

Ecke F, Rydin H (2000) Succession on a land uplift coast in relation to plant strategy theory. Ann Bot Fennici 37:163–171

Eynard C, Galetto L (2002) Pollination ecology of *Geoffroea decorticans* (Fabaceae) in central Argentine dry forest. J Arid Environ 51:79–88

Farinati EA (1983) Paleontología, Paleoecología y Paleogeografía de los sedimentos marinos de los alrededores de Bahía Blanca. PhD Thesis. Universidad Nacional del Sur, Bahía Blanca, Argentina

Farinati EA, Aliotta S, Ginsberg SS (1992) Mass mortality of a Holocene *Tagelus plebeius* (Mollusca, Bivalvia) population in the Bahía Blanca Estuary, Argentina. Mar Geol 106:301–308

Freitas RF, Costa CS (2014) Germination responses to salt stress of two intertidal populations of the perennial glasswort *Sarcocornia ambigua*. Aquat bot 117:12–17

Gaitan JJ, Bran DE, Oliva GE et al (2019) Patagonian desert. In: Reference module in Earth systems and environmental sciences. Elsevier, Amsterdam. https://doi.org/10.1016/B978-0-12-409548-9.11929-3

Gao S, Du Y, Xie W et al (2014) Environment-ecosystem dynamic processes of *Spartina alterniflora* salt-marshes along the eastern China coastlines. Sci China Earth Sci 57:2567–2586

Gardner LR, Reeves HW, Thibodeau PM (2000) Groundwater dynamics along forest-marsh transects in a southeastern salt marsh, USA: description, interpretation, and challenges for numerical modeling. Wetl Ecol Manag 10:145–159

Gates MW, Torrens J, Fidalgo P et al (2018) The gall associates of Asphondylia poss. swaedicola Kieffer & Jörgensen (Diptera: Cecidomyiidae) on *Suaeda divaricata* Moq.(Amaranthaceae) in the semiarid Argentina and summary of parasitic Hymenoptera associated with Suaeda worldwide. Neotrop entomol 47:598–609

Gehrels WR (2000) Using foraminiferal transfer functions to produce high-resolution sea-level records from salt-marsh deposits, Maine, USA. The Holocene 10:367–376

Giagante DA, Aliotta S, Ginsberg SS (2008) Análisis sismoestratigráfico de paleocanales en el subsuelo marino del estuario de Bahía Blanca. Rev de la Asociación Geológica Argentina 63:65–75

Giagante DA, Aliotta S, Ginsberg SS et al (2011) Evolution of a coastal alluvial deposit in response to the last Quaternary marine transgression, Bahía Blanca estuary, Argentina. Quat Res 75:614–623

Giusti L (1997) Chenopodiaceae. In: Hunziker AT (ed) Flora Fanerogámica Argentina. Proflora Conicet, Buenos Aires, pp 1–53

Gómez-Miguel V, Peres Arias J, Guerrero F, et al (1984) The soils and water table properties of the Polder area "Castillo de Dona Blanca", Puerto de Santa Maria, Cadiz Spain. In: Polders of the world. International symposium, Netherlands, pp 374–383

González MA (1989) Holocene levels in the Bahía Blanca Estuary, Argentina Republic. J Coast Res 5:65–77

González Trilla G, Pratolongo P, Beget ME et al (2013) Relating Biophysical Parameters of Coastal Marshes to Hyperspectral Reflectance Data in the Bahía Blanca Estuary, Argentina. J Coast Res 286:231–238

González MA, Weiler NE (1983) Ciclicidad de niveles marinos holocénicos en Bahía Blanca y en el Delta del Río Colorado (Provincia de Buenos Aires), en base a edades de Carbono – 14. Oscil. del Niv. del mar durante el último hemiciclo deglacial en la Argentina. Mar del Plata, Bahía Blanca, pp 69–90

González MA, Panarello HO, Marino H et al (1983) Niveles marinos del Holoceno en el estuario de Bahía Blanca (Argentina). Isótopos estables y microfósiles calcáreos como indicadores paleoambientales. Oscil. del Niv. del mar durante el último hemiciclo deglacial en la Argentina. Mar del Plata, Bahía Blanca, pp 48–68

González E, Gonzalez Trilla G, San Martin L et al (2019) Vegetation patterns in a South American coastal wetland using high-resolution imagery. J Maps 15:642–650

González-Uriarte M (1984) Características geomorfológicas de la porción continental que rodea la Bahía Blanca, Provincia de Buenos Aires. IX Congr. Argentino Geol. Bariloche, Argentina, pp 556–576

Hageman BP (1969) Development of the western part of the Netherlands during the Holocene. Geol en Mijnb 48:373–386

Henderson RE, Patrick JR (1982) Soil aeration and plant productivity. In: Rechcigl M Jr (ed) Handb. Agric. Product. CRC Press, Boca Raton, USA, pp 51–69

Horton BP, Edwards RJ (2006) Quantifying Holocene sea level change using intertidal foraminifera: lessons from the British Isles. Departmental Papers http://repository.upenn.edu/ees_papers/50. Accessed 21 Dec 2019

Isacch JP, Costa CSB, Rodríguez-Gallego L et al (2006) Distribution of saltmarsh plant communities associated with environmental factors along a latitudinal gradient on the south-west Atlantic coast. J Biogeogr 33:888–900

Isla FI (1989) Holocene sea-level fluctuation in the southern hemisphere. Quat Sci Rev 8:359–368

Isla FI, Cortizo LC, Schnack EJ (1996) Pleistocene and Holocene beaches and estuaries along the Southern Barrier of Buenos Aires, Argentina. Quat Sci Rev 15:833–841

Jansen RK, Palmer JD (1987) A chloroplast DNA inversion marks an ancient evolutionary split in the sunflower family (Asteraceae). P Natl Acad Sci USA 84:5818–5822

Kadereit G, Borsch T, Weising K et al (2003) Phylogeny of Amaranthaceae and Chenopodiaceae and the evolution of C_4 photosynthesis. Int J Plant Sci 164:959–986

Karlin MS, Buffa EV, Karlin UO et al (2012) Relations among soil properties, plant communities and cattle receptivity in saline environments (Salinas Grandes, Catamarca, Argentina). Rev Latinoamericana de Recursos Naturales 8:30–45

Kirwan ML, Blum LK (2011) Enhanced decomposition offsets enhanced productivity and soil carbon accumulation in coastal wetlands responding to climate change. Biogeosciences 8:987–993

Kirwan ML, Murray AB (2008) Ecological and morphological response of brackish tidal marshland to the next century of sea level rise: Westham Island, British Columbia. Global Planet Change 60:471–486

Kruger HR (1989) Suelos halomórficos de la Bahía Blanca. Dinámica de sales y relación con la vegetación. PhD Thesis. Universidad Nacional del Sur, Bahía Blanca, Argentina

Kruger HR, Peinemann N (1996) Coastal plain halophytes and their relation to soil ionic composition. Vegetatio 122:143–150

Lacambra C, Cutts N, Allen J, et al. (2004) *Spartina anglica*: a review of its status, dynamics and management. English Nature Research Reports No. 527, English Nature, Peterborough, UK

Lavin M, Herendeen PS, Wojciechowski MF (2005) Evolutionary rates analysis of Leguminosae implicates a rapid diversification of lineages during the tertiary. Syst Biol 54:575–594

León RJC, Bran D, Collantes M et al (1998) Grandes unidades de vegetación de la Patagonia extra andina. Ecología Austral 8:123–141

Maricle BR, Lee RW (2002) Aerenchyma development and oxygen transport in the estuarine cordgrasses Spartina alterniflora and S-anglica. Aquat Bot 74:109–120

Mariotti G, Fagherazzi S, Wiberg PL et al (2010) Influence of storm surges and sea level on shallow tidal basin erosive processes. J Geophys Res. 115(C11). https://doi.org/10.1029/2009JC005892

Marquez M, Ferrero L, Cusminsky GC (2016) Holocene palaeoenvironmental evolution of the Pampa coastal plain (Argentina) based on calcareous microfossils. Rev Bras Paleontol 19:25–40

Mcleod E, Chmura GL, Bouillon S et al (2011) A blueprint for blue carbon: toward an improved understanding of the role of vegetated coastal habitats in sequestering CO_2. Front Ecol Environ 9:552–560

Mendelssohn IA, Morris JT (2000) Eco-physiological controls on the primary productivity of *Spartina alterniflora*. Concepts and Controversies in Tidal Marsh. Ecology:59–80

Moffett KB, Gorelick SM, McLaren RG et al (2012) Salt marsh ecohydrological zonation due to heterogeneous vegetation–groundwater–surface water interactions. Water Resour Res 48(2):W02516. https://doi.org/10.1029/2011WR010874

Múlgura ME, Marticorena A (2008) Chenopodiaceae. In: Zuloaga FO, Morrone O, Belgrano MJ (eds) Catálogo de las Plantas Vasculares del Cono Sur, vol II. Missouri Botanical Garden Press, Missouri, pp 1909–1929

Munns R, Tester M (2008) Mechanisms of salinity tolerance. Annu Rev Plant Biol 59:651–681

Nebbia AJ, Zalba S (2007) Comunidades Halófilas de la costa de la Bahía Blanca (Argentina): Caracterización, mapeo y cambios durante los últimos cincuenta años. Boletín la Soc Argentina Botánica 42:261–271

Negrin VL, Botté SE, Pratolongo PD et al (2016) Ecological processes and biogeochemical cycling in salt marshes: synthesis of studies in the Bahía Blanca estuary (Argentina). Hydrobiologia 774:217–235

Nehring S, Hesse KJ (2008) Invasive alien plants in marine protected areas: the *Spartina anglica* affair in the European Wadden Sea. Biol. Invasions 10:937–950

Oertel GF, Kearney MS, Leatherman SP et al (1989) Anatomy of a barrier platform: outer barrier lagoon, southern Delmarva Peninsula, Virginia. Mar Geol 88:303–318

Ouyang ZT, Gao Y, Xie X et al (2013) Spectral discrimination of the invasive plant *Spartina alterniflora* at multiple phenological stages in a saltmarsh wetland. PLoS One 8:e67315

Oyarzabal M, Clavijo JR, Oakley LJ et al (2018) Unidades de vegetación de la Argentina. Ecología Austral 28:40–63

Paruelo JM, Beltrán B, Jobbagy EG et al (1998) The climate of Patagonia: general patterns and controls on biotic processes. Ecología Austral 8:85–101

Pendleton L, Donato DC, Murray BC et al (2012) Estimating global "blue carbon" emissions from conversion and degradation of vegetated coastal ecosystems. PloS One 7(9). https://doi.org/10.1371/journal.pone.0043542

Pennings SC, Bertness MD (1999) Using latitudinal variation to examine effects of climate on coastal salt marsh pattern and process. Current Topics in Wetland Biogeochemistry 3:100–111

Pennings SC, Bertness MD (2001) Salt marsh communities. In: Bertness MD, Gaines SD, Hay ME (eds) Marine community ecology. Sinauer Associates, Sunderland, pp 289–316

Pennings SC, Grant MB, Bertness MD (2005) Plant zonation in low-latitude salt marshes: disentangling the roles of flooding, salinity and competition. J Ecol 93:159–167

Pérez Cuadra V (2008) Salitral de la Vidriera, un refugio para especies vegetales con características particulares. Boletín de la Sociedad Latinoamericana y del Caribe de Cactáceas y otras Suculentas 5:6–8

Pérez Cuadra V, Cambi VN (2016) Caracteres epidérmicos de 30 especies xero-halófilas:¿ es el ambiente el principal factor determinante? Lilloa 53:282–298

Pérez Cuadra V, Hermann R (2014) Anatomía foliar y caulinar de tres Salicornieae (Chenopodiaceae) halófilas argentinas. Phyton-Int J Exp Bot 83:369–377

Peterson PM, Romaschenko K, Herrera Arrieta Y (2014) A molecular phylogeny and classification of the Cteniinae, Farragininae, Gouiniinae, Gymnopogoninae, Perotidinae, and Trichoneurinae (Poaceae: Chloridoideae: Cynodonteae). Taxon 63:275–286

Piovan MJ (2016) Controles geomorfológicos sobre la presencia y estructura de humedales costeros en el estuario de Bahía Blanca. PhD Thesis. Universidad Nacional del Sur, Bahía Blanca, Argentina

Piovan MJ, Zapperi GM, Pratolongo PD (2014) Seed germination of *Atriplex undulata* under saline and alkaline conditions. Seed Sci Technol 42:286–292

Pratolongo P, Kirby JR, Plater A et al (2009) Temperate coastal wetlands: morphology, sediment processes, and plant communities. In: Perillo GME, Wolanski E, Cahoon DR, Brinson MM (eds) Coastal Wetlands 1 st edn. Elsevier, pp 89–118

Pratolongo P, Perillo GME, Píccolo MC (2010) Combined effects of waves and plants on a mud deposition event at a mudflat-saltmarsh edge in the Bahía Blanca estuary. Estuar Coast Shelf Sci 87:207–212

Pratolongo P, Mazzon C, Zapperi G et al (2013) Land cover changes in tidal salt marshes of the Bahía Blanca estuary (Argentina) during the past 40 years. Estuar Coast Shelf Sci 133:23–31

Pratolongo PD, Piovan MJ, Cuadrado DG (2016) Coastal Environments in the Bahía Blanca Estuary, Argentina. In: Khan M, Boër B, Őzturk M et al (eds) Sabkha Ecosystems. Tasks for Vegetation Science, vol 48. Springer, Cham, pp 205–224

Pratolongo P, Piovan MJ, Cuadrado DG et al (2017) Coastal landscape evolution on the western margin of the Bahía Blanca Estuary (Argentina) mirrors a non-uniform sea-level fall after the mid-Holocene highstand. Geo-Mar Lett 37:373–384

Pratolongo P, Leonardi N, Kirby JR et al (2019) Temperate coastal wetlands: morphology, sediment processes, and plant communities. In: Perillo GME, Wolanski E, Cahoon DR, Hopkinson CS (eds) Coastal Wetlands, 2nd edn. Elsevier, pp 105–152

Raybould AF (1997) The history and ecology of *Spartina anglica* in Poole Harbour. Proc. Dorset Nat. Hist. Archaeol. Soc. 119:147–158

Redelstein R, Dinter T, Hertel D et al (2018) Effects of inundation, nutrient availability and plant species diversity on fine root mass and morphology across a saltmarsh flooding gradient. Front Plant Sci 9. https://doi.org/10.3389/fpls.2018.00098

Rogel JA, Ariza FA, Silla RO (2000) Soil salinity and moisture gradients and plant zonation in Mediterranean salt marshes of Southeast Spain. Wetlands 20:357–372

Salem HB, Norman HC, Nefzaoui A et al (2010) Potential use of oldman saltbush (*Atriplex nummularia* Lindl.) in sheep and goat feeding. Small Ruminant Res 91:13–28

Saltonstall K (2002) Cryptic invasion by a non-native genotype of the common reed, *Phragmites australis*, into North America. Proc. Natl. Acad. Sci. 99:2445–2449

Schrire BD, Lavin M, Lewis GP (2005) Global distribution patterns of the Leguminosae: insights from recent phylogenies. Biol Skrif 55:375–422

Schwindt E, Battini N, Giachetti C et al (2018) Especies exóticas marino-costeras (Argentina). Vázquez-Mazzini, Buenos Aires

Silva RA, López de Ruiz RE, Ruiz SO (2004) Estudio fitoquímico de flores de *Geoffroea decorticans* (Gill. ex Hook. et Arm.) Burk, Leguminoseae (Fabaceae). Acta Farmacéutica Bonaerense 23:524–526

Silvestri S, Defina A, Marani M (2005) Tidal regime, salinity and salt marsh plant zonation. Estuar Coast Shelf S 62:119–130

Simberloff D (2010) Invasions of Plant Communities. More of the Same, Something Very Different, or Both? The Am Mid Nat 163:220–233

Simenstad C, Thom R (1995) *Spartina alterniflora* (smooth cordgrass) as an invasive halophyte in Pacific northwest estuaries. Hortus Northwest 6:9–13

Simões MP, Calado ML, Madeira M et al (2011) Decomposition and nutrient release in halophytes of a Mediterranean salt marsh. Aquat. Bot. 94:119–126

Soil-Survey-Staff (1975) Soil taxonomy: a basic system of soil classification for making and interpreting soil surveys, Agricultural handbook no. 436. Soil Conservation Service, U.S. Department of Agriculture, U.S. Government Printing Office, Washington, DC

Soriano A (1947) Las Quenopodiáceas de la Tribu "Salicornieae" en la República Argentina. Rev Argentina de Agronomía 14:148–172

Sosa A, Fusco MR, Rossomando P et al (2011) Anti-inflammatory properties from isolated compounds of Cyclolepis genistoides. Pharm biol 49:675–678

Spagnuolo JO (2005) Evolución geológica de la región costera-marina de Punta Alta, provincia de Buenos Aires. PhD Thesis. Universidad Nacional del Sur, Bahía Blanca, Argentina

Steffen S, Mucina L, Kadereit G (2010) A revision of Sarcocornia in South Africa, Namibia and Mozambique. Syst Bot 35:390–408

Steffen S, Ball P, Mucina L et al (2015) Phylogeny, biogeography and ecological diversification of *Sarcocornia* (Salicornioideae, Amaranthaceae). Ann Bot-London 15:353–368

Stevens PF (2001 onwards). Angiosperm Phylogeny Website. Version 14, July 2017. http://www.mobot.org/MOBOT/research/APweb/. Acceded Diciembre 2019

Strong DR, Ayres DA (2016) Control and consequences of *Spartina spp.* invasions with focus upon San Francisco Bay. Biol Invasions 18:2237–2246

Tangahu BV, Abdullah S, Rozaimah S et al (2011) A review on heavy metals (As, Pb, and Hg) uptake by plants through phytoremediation. International Journal of Chemical Engineering. https://doi.org/10.1155/2011/939161

Theuerkauf EJ, Stephens JD, Ridge JT et al (2015) Carbon export from fringing saltmarsh shoreline erosion overwhelms carbon storage across a critical width threshold. Estuar Coast Shelf S 164:367–378

Tiner R (1999) Wetland Indicators; A Guide to Wetland Identification, Delineation, Classification, and Mapping. Lewis Publishers/CRC Press, Boca Raton, USA

Tong C, Zhang L, Wang W et al (2011) Contrasting nutrient stocks and litter decomposition in stands of native and invasive species in a sub-tropical estuarine marsh. Environ. Res. 111:909–916

Traut BH (2005) The role of coastal ecotones: a case study of the salt marsh/upland transition zone in California. J Ecol 93:279–290

Valéry L, Fritz H, Lefeuvre JC, Simberloff D (2009) Invasive species can also be native.... Trends Ecol Evol 24:585

Valéry L, Radureau A, Lefeuvre JC (2016) Spread of the native grass *Elymus athericus* in salt marshes of Mont-Saint-Michel bay as an unusual case of coastal eutrophication. J Coast Conserv 21:421–433

Vartiainen T (1988) Vegetation development on the outer islands of the Bothnian Bay. Vegetatio 77:149–158

Ventura Y, Wuddineh WA, Myrzabayeva M et al (2011) Effect of seawater concentration on the productivity and nutritional value of annual *Salicornia* and perennial *Sarcocornia* halophytes as leafy vegetable crops. Sci Hortic 128:189–196

Vepraskas MJ (1996) Redoximorphic features for identifying aquic conditions. North Carolina Agr. Res. Serv. Tech. Bull. 301. North Carolina State University, Raleigh, NC

Verettoni HN (1974) Las comunidades vegetales de la región de Bahía Blanca. UNS, Bahía Blanca

Violante RA, Parker G (2000) El Holoceno en las regiones marinas y costeras del nordeste de Buenos Aires. Rev la Asoc Geológica Argentina 55:337–351

Wahid A (2003) Physiological significance of morpho-anatomical features of halophytes with particular reference to Cholistan flora. Int J Agric Biol 5:207–212

Waller MP, Long AJ, Long D et al (1999) Patterns and processes in the development of coastal more vegetation: multi-site investigations from Walland Marsh, Southeast England. Quat Sci Rev 18:1419–1444

Wang Q, Wang CH, Zhao B et al (2006) Effects of growing conditions on the growth of and interactions between salt marsh plants: implications for invisibility of habitats. Biol Invasions 8:1547–1560

Wang Y, Zhou L, Zheng X et al (2013) Influence of *Spartina alterniflora* on the mobility of heavy metals in salt marsh sediments of the Yangtze River Estuary, China. Environ Sci Pollut R 20:1675–1685

Whigham DF (1999) Ecological issues related to wetland preservation, restoration, creation and assessment. J Total Environ 240:31–40

Willis B (1914) Northern Patagonia, character and resources. Scribner Press, New York, USA

Zapperi G, Pratolongo P, Piovan MJ et al (2016) Benthic-Pelagic Coupling in an Intertidal Mudflat in the Bahía Blanca Estuary (SW Atlantic). J Coast Res 32:629–637

Zedler JB (1982) The ecology of southern California coastal salt marshes: a community profile. Fish and Wildlife Service, Washington

Zobel M, Kont A (1992) Formation and succession of alvar communities in the Baltic land uplift area. Nord J Bot 12:249–256

Zuloaga FO, Belgrano MJ, Zanotti CA (2019) Actualización del Catálogo de las Plantas Vasculares del Cono Sur. Darwiniana, nueva serie 7:208–278

Chapter 17
Environmental Diagnosis of the Protected Coastal Areas of the Bahía Blanca Estuary

M. Elizabeth Carbone, María Ángeles Speake, and Walter Daniel Melo

17.1 Introduction

The interrelation between natural environments and the various human activities in coastal areas generates conflicts between conservation and development. For this reason, the compatibility of both concepts is very important in areas with high territorial occupation pressure and presence of vulnerable natural resources. Recreational and productive activities are currently being developed in coastal areas, which directly or indirectly affect natural resources. Human pressure on natural environments is imposed by space demand for new settlements, installation of infrastructure and equipment, sand extraction, and commercial fishing among others. These human activities are the main stressors on coastal ecosystems (Charlier and Bologa 2003). Sea level variations, erosion, and sedimentation are major natural processes also involved in the present evolution of coastal areas. All these transformations can damage coastal environments when they are not accompanied by an adequate system of orderly and sustainable management guidelines. The formulation and implementation of appropriate coastal planning and management measures arise from a concrete current diagnosis that can project future changes and modifications in this kind of environments (Maelfait et al. 2006). Coastal protected areas contribute to the conservation of the natural and cultural heritage of a region and reduce pressures caused by human activities on these environments. The proposal and implementation of programs and policies for a sustainable coastal management of these areas are strengthened with an adequate monitoring of the environmental

M. E. Carbone (✉) · M. Á. Speake · W. D. Melo
Instituto Argentino de Oceanografía IADO (Universidad Nacional del Sur-CONICET), Bahía Blanca, Argentina

Departamento de Geografía y Turismo, Universidad Nacional del Sur, Bahía Blanca, Argentina
e-mail: ecarbone@criba.edu.ar

S. M. Fiori, P. D. Pratolongo (eds.), *The Bahía Blanca Estuary*,
https://doi.org/10.1007/978-3-030-66486-2_17

quality of the site and the integrated knowledge of the current processes that occur in these environments.

The coastal areas of Latin America encompass a large variety of geo-environmental units that include sandy beaches, estuaries, mangroves, coral reefs, marshes, coastal lagoons, islands, and fjords. Likewise, the sea that bathes its coasts is very important in terms of biodiversity and productivity of the regional ecosystems. Coastal waters also hold a high productivity, which supports one of the five major commercial fisheries worldwide (Food and Agricultural Organization 2012). The coastal and marine areas of Latin America showed for more than 20 years the effects caused by human settlements. Urban expansion implies a significant change in land uses, especially on coastal ecosystems (Barragán Muñoz 2014). The 60% of the population is concentrated less than 100 km from the coast, and this behavior induces overexploitation of marine resources (Food and Agricultural Organization 2014).

Recreational activities and tourism also exacerbate impacts on coastal areas, along with uncontrolled waste discharge into the oceans, and the expansion of aquaculture. Non-compliance with the objectives of creating coastal protected areas is another negative impact on these coastal areas (Aldana Mazorra and Hernández Zanuy 2018). The most degraded coastal ecosystems in Latin America are coastal wetlands and coral reefs, which implies the loss of valuable ecosystem services essential to many economies in the region. These coastal habitats also play an important role of coastal protection and shoreline stabilization, in the face of extreme meteorological phenomena, occurring at increasing frequency and with greater intensity (Food and Agricultural Organization 2012). The decreasing abundance of fishing resources is noticeable at the local scale, and also regional extinctions of some commercial fish species were reported, along with profound changes in the structure and functioning of ecosystems. The current policies of coastal territorial planning through the creation of protected areas, as an instrument of control and regulation of the use of the coastline, are no longer sufficient to address the complexity and magnitude of the problems that arise in the coastal ecosystems of different countries of Latin America (Day et al. 2012; Barragán Muñoz 2014).

Marine protected areas in Latin American countries are the product of a relatively recent conservation management. Although there are experiences since the 1970s of the twentieth century. Until 2011, more than 700 marine and coastal protected areas have been established in Latin America covering 250,000 km^2 or 1.5% of its coastal waters. These numbers reflect the fact that the region has a significant delay in meeting the goal of the working program on protected areas of the Convention on Biological Diversity that recommends conserving at least 10% of the ocean surface (Aldana Mazorra and Hernández Zanuy 2018). Latin America has made considerable progress in aspects related to the protection of its marine and coastal biodiversity. It is possible to verify that all the countries of the region have protected coastal areas, although not all of them are structured in the same way. The problems that are noticed in these geographical spaces are the low level of implementation of protected areas in the marine environment in relation to the continental sector, less knowledge of the adjacent marine environment, high costs for the management of coastal areas, insufficient and not specialized staff, and budget restrictions (Day et al. 2012; Barragán Muñoz 2014; Aldana Mazorra and Hernández Zanuy 2018).

Latin American countries such as Argentina and Uruguay host some of the most biodiverse areas in the world and intensify the protection of their coastal seas and the surrounding oceans. It is done by creating coastal protected areas and by extending the conservation objectives to adjacent waters. The initiative aligned with the overall objective of safeguarding at least 10% of the marine and coastal areas worldwide by 2020. Marine coastal protected areas (MPAs), extensions of water managed for conservation, recently expanded and now cover 8.4% of the oceanic territory of Latin America (Food and Agricultural Organization 2014).

The jurisdiction of the Bahía Blanca Estuary is complex (Fig. 2.1; Chap. 2), covering areas of national, provincial, and municipal level of public and private domain. Six sites are analyzed in this chapter: three of these geographical areas have management categories: (1) Bahía Blanca, Bahía Falsa, and Bahía Verde Nature Reserve; (2) Islote del Puerto Nature Reserve (both natural reserves are provincials); and (3) Bahía Blanca Coastal Reserve (Fig. 17.1). The following sites are of preservation and conservation interest due to their predominant ecological characteristics: (4) Arroyo Parejas-Isla Cantarelli Natural Area, (5) Villa del Mar Wetlands, and (6) the Puerto Cuatreros Wetland. Due to the ornithological importance of these coastal areas in 2008, the Bahía Blanca, Bahía Falsa, and Bahía Verde Multiple Use Nature Reserve was designated as a Bird Conservation Area (IBA) by BirdLife International, and in 2016 the six sub-sites were designated Site of Regional Importance by the Western Hemisphere Shorebird Reserve Network (Fig. 17.2).

17.2 Geo-environmental Characteristics of Coastal Protected Areas

The Bahía Blanca Estuary is an extensive and complex system of natural marshes in the Buenos Aires Province, Argentina (Fig. 2.1; Chap. 1). It is a coastal wetland system with unique geographical and biological characteristics. It is a diverse and very dynamic geographical space that has an important natural capital. The intertidal ecotone is the spine of coastal zones and represents the area of greatest interest in coastal area management (Barragán Muñoz 2014). The legalization of these geographical spaces through the creation of coastal protected areas is one of the most versatile tools for the preservation and conservation of the natural physical system of a coastal region.

The intertidal zone that characterizes the Bahía Blanca Estuary is the main object of interest in terms of integrated management of natural coastal areas, besides the islands and tidal channels. The jurisdiction of the ecosystems of the Bahía Blanca Estuary is complicated by the different levels of administration that involve both the local, provincial, and national level. It should be noted that in the coastal study area,

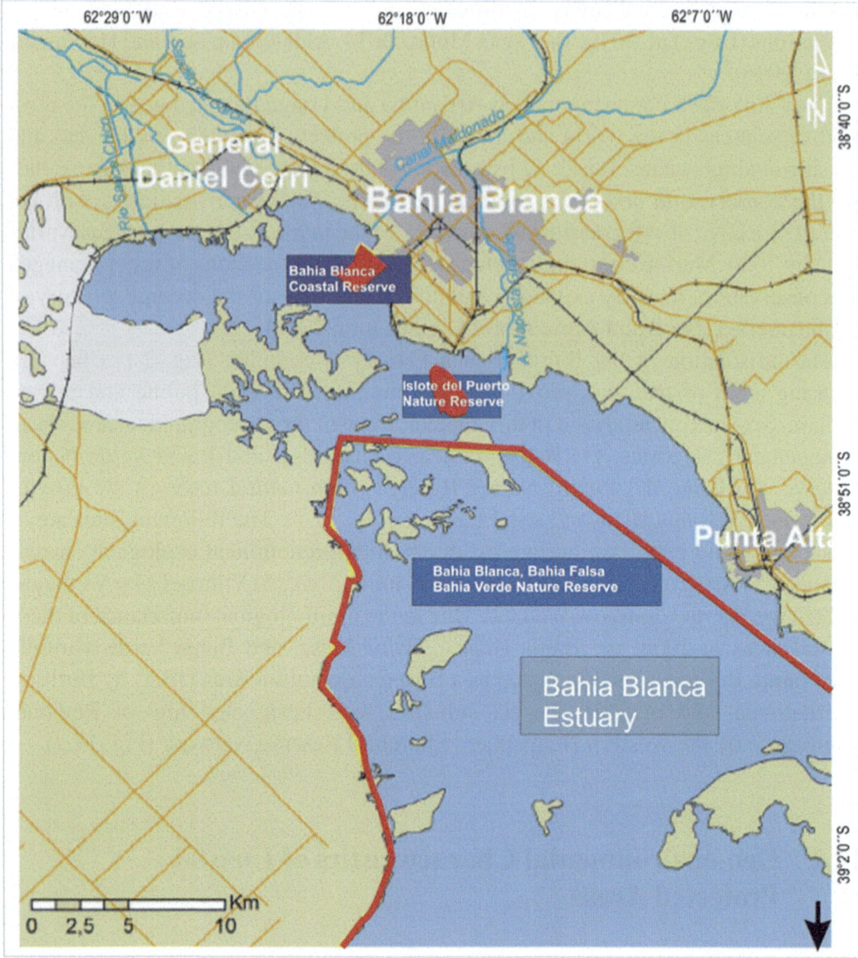

Fig. 17.1 Location of the natural reserves of the Bahía Blanca Estuary. In detail, the limits of the Bahía Blanca Coastal Reserve and the Islote de la Gaviota Cangrejera Nature Reserve are observed. The total area of the Bahía Blanca, Bahía Falsa, and Bahía Verde Multiple Use Nature Reserve provincial is shown in Fig. 2.1; Chap. 2. (Modified from Speake et al. 2018)

some sites have been declared of international interest due to their importance for migratory birds.

The geomorphological units that present the Bahía Blanca, Falsa, and Verde Multiple Use Nature Reserve are made up of tidal channels, intertidal zones, and islands. Elongated tidal channels have different dimensions, the smaller ones being those at the west end of the Bahía Blanca Estuary head, and the predominant orientation of them is northwest-southeast and represents 39% of the study area. The drainage design is dendritic and subparallel and drains into the Principal Channel (CP) which is 97 km long in total. In the southern sector of this area, the design is

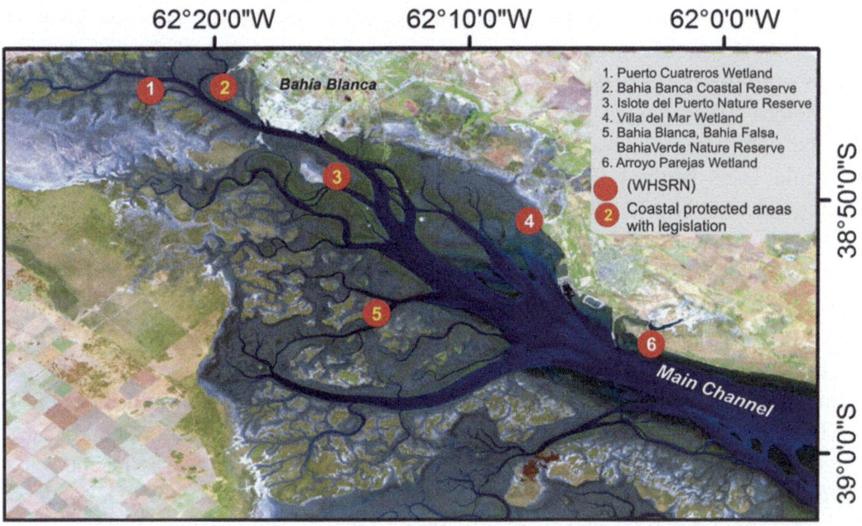

Fig. 17.2 Areas of ecological interest in the Bahía Blanca Estuary, location of the Western Hemisphere Shorebird Reserve Network sub-sites and nature reserves

anastomosed presenting various meanders, which separate plains and islands (Carbone et al. 2016).

The Bahía Blanca, Bahía Falsa, and Bahía Verde Nature Reserve has islands, tidal channels, and intertidal zones represented by silt-clay tidal flats and vegetated marshes (Fig. 2.1; Chap. 2). The intertidal environments occupy 67.3% of the area, with tidal plains being the predominant ones with 38.1%, and the marshes of *Spartina alterniflora* and *Sarcocornia perennis* occupy 10.2% and 19%, respectively, while the islands and tidal channels of various dimensions occupy 32.7% of the reserve.

The islands are made up of two different highly fragmented morphological groups, and these occupy the outermost sector of the Bahía Blanca Estuary. In the north, the Zuraita, Bermejo, and Trinidad islands are the largest, while in the south, the Monte and Ariadna islands are smaller, characterized by being more compact than those located to the north of the reserve (Melo 2007). These geo-environmental units have very complex dynamics. Inside these there are abandoned tidal channels, while in the southeast sector of Trinidad Island, a tidal plain develops with more than 10 km of development with a predominance of silty clay silt.

The southern islands have tidal flats with a high percentage of fine sand (Melo 2007). The current coastal processes generated the growth of geoforms, which give protection to these environments with intense dynamics.

The exposed surfaces are covered with scrubland typical of the shrub steppe and halophilic scrubs (Angeles 2001). The phytogeographic provinces the Espinal and the Monte predominate. The vegetation is the xerophilous and halophilic shrub steppe and grasslands. The vegetation that can be found on the islands is dominated by argentine *Spartina densiflora*, *Atriplex undulata*, and perennial glasswort

Sarcocornia perennis, which do not usually exceed the meter. This environment represents the breeding and feeding habitat of the crab gull and the breeding of *Larus atlanticus*, *Larus dominicanus*, *Egretta ardesiaca*, and *Ardea alba* (Petracci and Sotelo 2013).

The nature reserve was created on March 21, 1991, and then regulated as a Multiple Uses Natural Reserve by Law N° 11.074, ratified latter in 1998 by Provincial Law N°. 12.101. The reserve covers an area of 30.000 ha of land and 180.000 ha of water. The reserve has a basic implementation of regularization and control of permitted activities compatible with the conservation of the area and its integration into the management plan in force since 2007. As part of the implementation, the reserve has a park ranger and infrastructure service, with a service office, checkpoints and surveillance in island sectors, a motor vehicle, semi-rigid boat for nautical control, and radio communication means (Massola and Cinti 2012).

The Islote de la Gaviota Cangrejera Nature Reserve (115 ha) protects a very large breeding colony of *Larus atlanticus*, with 3800 nests censored in 2005 and with similar population values recorded for 2006 (Petracci et al. 2008). This species is in risk of extinction, and it is only found on the Southwestern Atlantic coast (BirdLife International 2019). For this reason, the purpose of this reserve is the conservation and protection of this colony. It was declared of provincial interest by Decree 469/11. This species in Argentina is considered "Threatened" by Resolution N° 348/10 of the Environment and Sustainable Development Ministry of the Nation (Petracci and Sotelo 2013). In Bahía Blanca, it was also declared as "Emblematic Species" by Ordinance N°. 12671/04. The greatest impacts on this species are caused by the reduction of the nesting and breeding area, the dumping of hydrocarbons in the Principal Channel, illegal subtraction of eggs and pigeons, and alteration of the surrounding island environment through urban and sewage wastes thrown into the sea (Sotelo and Mazzola 2008; Petracci and Sotelo 2013; Carbone et al. 2016). The species is also currently protected by the International Convention on Migratory Species (CMS) through Law N° 23918/91.

It should be noted that due to the dredging activities to deepen the Principal Channel of the Bahía Blanca Estuary, these gulls stopped nesting in 1989 and returned in 1999, in the sectors formed by the dredged material. Due to its importance for the conservation of the species, the Government of the Buenos Aires Province declared of "Provincial Interest" the conservation and protection activities carried out in the Islote del Puerto Nature Reserve y Resolution N°. 04/05 (Mazzola and Cinti 2012; Petracci and Sotelo 2013).

The Bahía Blanca Coastal Reserve (Ordinance N° 13.892/2006) has an area of approximately 320 ha. The main objective is the conservation and protection of the coastal environment, and environmental research and education are allowed. The typical geomorphological units of this reserve belong to a marine-continental transition environment, with plains and tidal channels, some of the plains covered by halophilic vegetation (Angeles 2001; Nebbia and Zalba 2007). The fauna stands out for the presence of crabs in the tidal plains and the prairies of *Spartina densiflora*. Petracci and Sotelo (2013) in the nearby marine sector recorded the presence of adult and juvenile individuals with *Thalassarche melanophris*. In rural areas peripheral to the reserve, they observed flocks of *Sturnella defilippii*. Also in the vicinity of the area, the *Porzana spiloptera* has been registered.

17.3 Human Occupation of the Coastal Areas of the Bahía Blanca Estuary Before Being Reserves

The islands that make up the Bahía Blanca, Bahía Falsa, and Bahía Verde Multiple Use Natural Reserve were used by man since 1890 through productive and extractive activities. The occupation of the islands was generated by taking advantage of the natural resources that they generated, such as the exploitation of *Geoffroea decorticans* that were on the islands and constituted a source of cheap and abundant energy for those boats that roamed the area, activity regulated by the State, imposing the payment of 10% of the profits obtained (Amarfil 2006).

From the beginning of the twentieth century on the islands, facilities were built, which changed their original ecosystem of herbaceous vegetation with the introduction of sheep, goats, cows, and horses. Isla Verde was olso exploited was also exploited with sheep farming reaching a production of up to 5000 sheep, by the year 1852 also in Bermejo Island and Trinidad (Amarfil 2006), and there are currently remains of these past activities. From 1944, the Trinidad Island was in private domain, and the same families administered the Bermejo Island, dedicating mainly to the fishing of *Galeorhinus galeus*. Later they installed gouges, mills, and pens, to raise sheep and grow barley. In 1961, the concession to the first settlers ceased; on that occasion, the new tenants were engaged in the breeding of goats and sheep (Cinti 2017). The administration of these islands depended on the agency of Fiscal Lands of La Plata (Buenos Aires Province) and then passed to the Ministry of Agrarian Affairs, which declared the protected area that included these islands, in 1998. Livestock activities continued in those environments through the lease of land to those who inhabited the islands. During 2007, the Provincial Organization for Sustainable Development is created, which currently manages this coastal nature reserve.

The Ariadna, Monte, and Verde islands have sandy beaches, low tidal plains, and cliffs of reddish-colored siltstones with herbaceous steppe and halophilic scrublands, subsequently replaced by agricultural crops (Melo 2007). Ariadna Island has an inn where overnight is possible. Some of the islets near this area have been colonized by *Larus atlanticus* and *Larus dominicanus*, registering up to 300 nests of the former (Petracci and Sotelo 2013). The permitted uses of the reserve include scientific research activities, inter-island walks, guided tours of islands, and interpretive trails where sightings of marine and terrestrial fauna are made. In addition to activities related to tourism and recreation, artisanal and sport fishing is carried out, which must be controlled.

The Bahía Blanca Coastal Reserve is located on municipal land, which from 1950 to 2006 was used as a solid waste deposit in the open. The usual practice consisted of the massive deposit of waste, both organic and inorganic, and their burning. Despite this situation and facing the need for more recreational spaces on the city's waterfront, in 1978, the location of a new urban park in this sector called Almirante Brown Maritime Park was decided (Speake and Carbone 2017).

17.4 Environmental Quality Index (EQI)

The results presented in this section are part of an investigation work where the protected areas of the southwest of Buenos Aires Province and the north of the Patagonian region are evaluated to compare the naturalness and state of the different geo-environments of the different reserves present throughout these coastal zones. The methodological procedure used to determine the EQI for the natural reserves of the Bahía Blanca Estuary was carried out through 36 indicators for the year 2018, obtained from:

(a) Topographic charts scale 1:100,000, satellite image charts 1:250,000, aerial photographs scale 1:20,000, and Landsat satellite images 5 and 7 processed
(b) Institutional environmental monitoring data
(c) Specific bibliographic background
(d) Statistical information from the 2010 Census of Home and Housing, Agricultural Census 2004, Meteorological Statistics 2010–2015, Information from the Municipality of Bahía Blanca, and the Provincial Organization for Sustainable Development

The DPSIR model (Organiztion for Economic Cooperation and Development 1993; EEA 1998) is applied, which provides a framework for organizing information to structure the indicators. It implies elaborating in a general way a causal progression of human actions that cause pressure on the environment and natural resources, which lead to a change in the state of the environment and to which society responds with measures or actions to reduce or prevent the impact. The degree of naturalness (NI) of a region, the absence of modifications introduced by man, the role of the physical environment can be conceived as a source of various resources (RI), as a sink for different types of waste (WI) generated for human activities (SSI) and as support for these activities (Organization for Economic Cooperation and Development 1993). During the second stage, the data of the selected environmental indicators were normalized, and a nominal scale of 0 to 1 was assigned, where 0 corresponds to the worst possible situation with respect to the indicator in question and 1 to the best possible situation with respect to the indicator analyzed according to the following expression:

$$\mathrm{V} = (Im - \mathrm{Imin}) / (\mathrm{Imax} - \mathrm{Imin}) \tag{17.1}$$

where V is the normalized value; Im, value of the indicator; Imax, maximum value in the study area; and Imin, minimum value in the zone. The third stage consists in the calculation of the aggregate indices of pressure, state, and response (in turn, indicators of motive forces and impacts are included within them) for each of the four functions mentioned according to:

$$NIp = \Sigma Vi \cdot Wi \tag{17.2}$$

where NIp is the pressure index on the natural condition of the environment, Vi is the standardized value of the indicator, and Wi is the weight of the indicator; i = (1 – n), where n is the total number of pressure indicators. Similarly, the calculation of status and response indices is carried out, both for naturalness and for the source, sink, and activities-services functions. The pressure, status, and response indices are integrated into indices of each of the functions:

$$NI = \left(NIp + NIs + NIr\right)/3 \tag{17.3}$$

where NI is the index of the naturalness function, NIp the index of pressure on naturalness, NIs the index of state of naturalness, and NIr the index of response on naturalness.

Proceeding also for each of the functions addressed, RI is the index of the resource source function; WI, index of the waste sink function; and SSI, index of the activities-services function. Finally, the environmental quality index (EQI) obtained from the following expression is calculated:

$$EQI = \left(NI + RI + WI + SSI\right)/4 \tag{17.4}$$

where IN is the naturalness index, RI the resource source index, WI the waste sink index, and SSI the index of activities/services. The interpretation of the final value of the environmental quality index is expressed on a scale of five classes whose maximum value is one and the minimum zero, the highest values corresponding to the most optimal environmental situation.

The different indices obtained from the indicators in the complete list present significant differences between the different coastal protected areas. It should be noted that each of them has an administrative legal jurisdiction of different levels, namely, the Bahía Blanca, Bahía Falsa, and Bahía Verde Multiple Use Natural Reserve and the Islote de la Gaviota Cangrejera Nature Reserve belong to the provincial scope, while the Bahía Blanca Coastal Nature Reserve is under the municipal domain. This characteristic is stated as the response indicators correspond to government actions with different criteria for the coastal protected areas (CPA) mentioned.

The values correspond to the interaction of human activities with the physical environment in each of the geographical spaces analyzed in the Bahía Blanca Estuary area and that interrelation in each of the geo-environmental units of each of the reserves also refers to the function and execution of the management guidelines through the management plans that regulate each of the reserves. IN values refer to the natural and pristine state of the ecosystems involved, being 0.27 for the Bahía Blanca Coastal Reserve, while the value of the Islote del Puerto Nature Reserve is

0.20, and the minimum corresponds to the first reserve with a value of 0.18. The presence of the large intertidal extension in the first mentioned reserve and the difficulty in accessing and connecting to it are reflected in the maximum obtained with respect to the NI. The values of the NIs and NIr for the same reached values of 0.33 and 0.39 representing the maximums for this index in the analyzed areas. The pressure and response dimensions obtained their maximum and minimum values in the Bahía Blanca Coastal Reserve and Islote de la Gaviota Cangrejera Nature Reserve, respectively.

The minimums with respect to the response obtained for the coastal reserve can be analyzed from the complex situation that has characterized this sector that appropriate sanitation measures have not yet been taken since it is still used as a solid *waste deposit* (Speake and Carbone 2017). The natural environments of this area are vulnerable to this type of anthropic intervention, which is reflected in the decrease in the ecological condition of these environments.

In the Islote de la Gaviota Cangrejera Nature Reserve, the naturalness index has the intermediate value with 0.19 composed of the NIp and NIr values of 0.17 and 0.23, respectively. It is worth mentioning that this sector was used as a deposit of dredging material after the deepening of the Principal Channel between 1989 and 1992. On that occasion, 50 million cubic meters of material were extracted that were deposited on the island that today in this reserve.

The minimum values of the IN for the Bahía Blanca Coastal Reserve ranged between 0.13 and 0.15 for the status and response indicators, respectively. In the three coastal areas, the highest value of the index corresponds to the state of the ecosystems involved in two environments, being 0.39 for the Bahía Blanca, Bahía Falsa, and Bahía Verde Multiple Use Nature Reserve, 0.25 for the Islote del Puerto Nature Reserve, and 0.15 for the Bahía Blanca Coastal Reserve. The lower values of the naturalness index in terms of the state dimension reflect the loss of the natural condition of these environments, mainly associated with land uses in areas adjacent to the reserves. The productive and industrial economic activities are located in the north sector of the Principal Channel. The response index values are closely related to the control and control measures of productive and industrial activities near natural reserves. These measures are particularly difficult in the Bahía Blanca, Bahía Falsa, and Bahía Verde Multiple Use Nature Reserve given the extension and variety of geo-environmental units that integrate it as detailed in the previous section.

The NI values indicate the intermediate nature condition considering that the natural coastal protected areas in this sector do not have a marked demographic pressure and extractive activities are prohibited, as are the old productive activities. It should be noted that in the places where these practices were currently allowed, the cattle that live on the Bahía Blanca, Bahía Falsa, and Bahía Verde Multiple Use Nature Reserve islands are beginning to be extracted.

The index that involves the environment, as a source of IR resources, presents the highest value for the Bahía Blanca Coastal Reserve, where the sector with the greatest presence of pre-existing human activities with a value of 0.36 for the pressure variable is concentrated. The lowest data corresponds to the state indicators with 0.12 referring to the diversity of land uses of this reserve and adjacent area (agricultural activities, fishing). One of the most important objectives in the management plan is to make the concept of extracting the mentioned resources compatible with conservation and sustainability. The highest values of the RI are evidenced in the Bahía Blanca, Bahía Falsa, and Bahía Verde Nature Reserve in the response category, the fishing activities are regulated and controlled, and the species of commercial interest are the *Micropogonias furnieri* and *Artemesia longinaris*.

Currently, the fishing activity has suffered a significant decrease; the main causes are associated with the increase in port activity, maintenance and dredging of the ports, sewage discharges, and the overexploitation of the resource (Conde et al. 2009; López Cazorla et al. 2014). The status and response indicators ranged from 0.38 to 0.09 due to the increase in the importance of conservation management in this area. The highest values were observed in the Bahía Blanca, Bahía Falsa, and Bahía Verde Multiple Use Nature Reserve, while the minimums are observed in the Bahía Blanca Coastal Reserve; IR values range between 0.35 and 0.19, respectively. The dimensions that reached intermediate values correspond to the pressure and response indicators with values ranging between 0.46 and 0.36.

The analysis of this index for the sectors adjacent to the urbanizations also revealed lower numbers with respect to the state of the same, throwing values of 0.16 and 0.22; it should be noted that the reserves of the Islote de la Gaviota Cangrejera Nature Reserve and the Bahía Blanca Coastal Reserve were located near the industrial zone and urban waste discharges, respectively. The indices with the highest value correspond to those of pressure with a maximum of up to 0.38 the reserve of multiple uses. It should be noted that the islands that make up the Bahía Blanca, Bahía Falsa, and Bahía Verde Multiple Use Nature Reserve before being declared protected areas were provided for agricultural practices by provincial organizations; today, there are vestiges of these activities as part of the infrastructure and equipment (Angeles 2001; Massola and Cinti 2012).

The waste index that considers the environment as a waste disposal IW, the maximum value corresponds to the Bahía Blanca Coastal Reserve; it should be noted that in this environment it has not yet been completely remedied and household waste continues to be deposited in the study area. The dimensions that formed this index yielded higher values corresponding to pressure and response reaching values of for the second variable mentioned.

The variables that stand out mainly in this index correspond to the pressure and response, with values ranging from for the first reserve and from 0.35 to 0.20 for Islote de la Gaviota Cangrejera Nature Reserve and Bahía Blanca Coastal Reserve. The state variable ranged between 0.22 and 0.18, respectively (Fig. 17.3), maintaining a lower range of values which represents approaching an optimal situation according to the scale considered. With respect to places adjacent to the analyzed reserves, although there are no human settlements inside the perimeter of the reserves, it is worth mentioning that the proximity to the city of Bahía Blanca with more than 300,000 inhabitants (Indec 2010), the industrial port area, the urban and industrial effluent discharge sites, and waste treatment plants act as direct driving forces and impact their natural environment (Spetter et al. 2019).

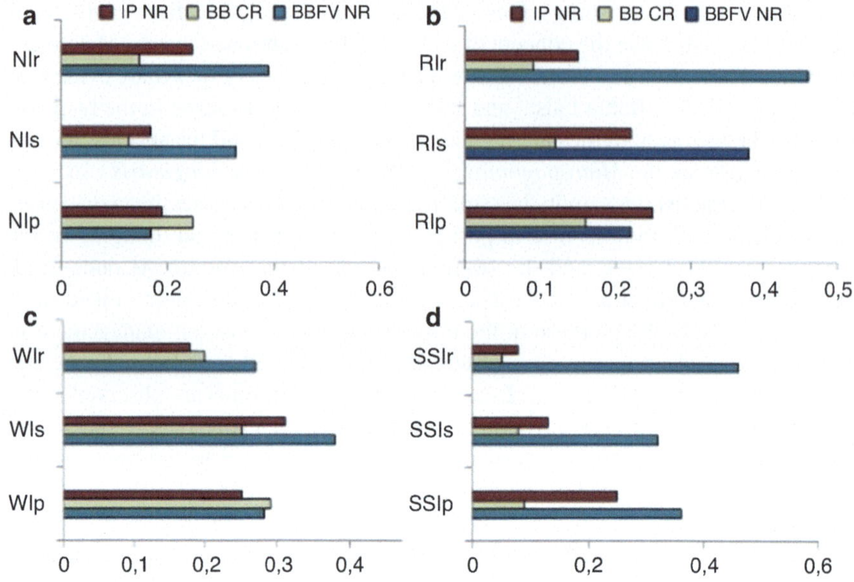

Fig. 17.3 (**a**) Naturalness indices, (**b**) response, (**c**) waste sink and activities, and (**d**) services for the Bahía Blanca Estuary of Islote de la Gaviota Cangrejera Nature Reserve (IP NR), Bahía Blanca Coastal Reserve (BB CR), and Bahía Blanca, Bahía Falsa, and Bahía Verde Multiple Use Nature Reserve (BBFV NR)

The capacity of the medium as a support for activities and services through the ISS presented a significant variability of values mainly among coastal reserves. In this case, activities and services derived from tourism were considered considering the restrictions on its compatibility with the legislation of the areas and management plans. The main tourist activities that are carried out are excursions embarked to Bermejo Island, to Puerto Cuatreros and tour along the seafront under the ecotourism modality, with specialized guides.

In order to analyze this index, the relation of the tourist activities with the state of the environments that make up the areas of the coastal reserves is also considered, and the degree of deterioration of the environments is an indicator of their status. The Bahía Blanca Coastal Reserve currently presents a significant deterioration of the ecological system whose main causes are clandestine final disposal of urban solid waste generated by the inhabitants of the city of Bahía Blanca, open burning of waste, land and water pollution by leaching of the same, and presence of rodents and predators of conservation species (Speake and Carbone 2017). Another problem is the expansion of invasive alien species, such as *Magallana gigas*, which represents a threat to the conservation of the biodiversity of nature reserves exacerbated by the absence of a management plan.

For the state variables, the better indices for the Bahía Blanca, Bahía Falsa, and Bahía Verde Multiple Use Nature Reserve are observed with values ranging between 0.34 and 0.46, while for the municipal natural reserve, the index has much lower values ranging between 0.12 and 0.09. It should be noted that this reserve response rate is 0.05 and it is related to what has been detailed above about the current environmental deterioration processes and lack of effective planning. The maximum value of the SSI is presented in the multipurpose reserve with a value of 0.38, and the minimum value of 0.07 is of the municipal natural reserve, while the Islote de la Gaviota Cangrejera Nature Reserve has a value of 0.15 (Fig. 17.3).

The aggregate indices for these areas were also obtained from simplified lists of indicators. The results of the application of 12 previously selected indicators presented higher values with respect to those previously analyzed. The naturalness index for the Bahía Blanca, Bahía Falsa, and Bahía Verde Multiple Use Nature Reserve was 0.52. State and response are the predominant dimentions. For the Islote de la Gaviota Cangrejera Nature reserve, it was 0.25 and 0.14 for the municipal reserve (Fig. 17.3). Indices that involve the environment as a source of resources, as waste sinks, and as support for different activities and services varied among the three coastal reserves.

The values of the dimensions that constitute these indices in the three reserves varied significantly, with pressure and status prevailing in the majority of the response involved. This condition arises because the weight given to the response indicators as to the other dimensions is 1 and considering that these sites depend on different administrative units, where public investments are different for the three reserves.

Obtaining disaggregated indices allowed zoning the study area according to the EQI values in three classes. Each of these belongs and represents Class 1 to Class 3 (low, moderately low, and medium) of the study area. The highest value corresponds to the Bahía Blanca, Bahía Falsa, and Bahía Verde Multiple Use Nature Reserve with 0.39, followed by the Islote del Puerto reserve with a value of 0.21 and the municipal reserve with a value of 0.09 which has the Class 1 that represents a low environmental quality. These results arise from the interrelation of all the indices analyzed above using the complete list of indicators where all the dimensions of the DPSIR model were calculated.

The values obtained allowed the area to be zoned in the following way: the reserve with the largest areal dimension presents values corresponding to Class 3 that represents an average environmental quality, which reflects a higher NI related to the state of the geo-environmental units that comprise it. The pressure exerted on the physical environment from the extraction of fishing resources, the discharge of sewage effluents without treatment in the surrounding area, and the location of areas of urban household waste deposits in the municipal coastal area alters and deteriorates the natural condition of the environments. The use of the response indicators showed lower quality values that highlight the current concern to begin to counteract the pressures on the environment.

The different jurisdictional levels of administration of these reserves make it difficult to apply appropriate remediation and mitigation measures in this geographical area. This is reflected in the state indices that marked a better condition of naturalness among the most external estuary reserves.

The coastal reserve has lower values of environmental quality, and they fit it into the lower category. The indices obtained in this ecosystem show an environmental problem accentuated by poor planning, the lack of a diagnosis of its current situation, and delay in the implementation of appropriate management guidelines. The response indicators do not show the formulation and application of public policies to mitigate impacts in the coastal zone.

The various geomorphological units that make up the reserves (tidal plains, marshes, secondary tidal channels, salt, islands, etc.) have difficult accessibility which restricts access to them. This condition also hinders the tasks of surveillance and control of the protected areas analyzed; another problem with the reserves is the limited staff and means available for patrolling them.

The continental sector adjacent to the multi-use reserve is characterized by the concentration of livestock farming activities. Likewise, in each of the islands that make it up, it is possible to differentiate conflict points where direct actions should be implemented to restore, mitigate, conserve, and rehabilitate sectors vulnerable to the pressures exerted regarding the presence of cattle and rabbits, mainly on the Trinidad Island and Isla de los Conejos and Ariadna.

The integrated management guidelines that should be applied in this coastal zone are related to ecosystem conservation, rehabilitation of areas used, and uses and appropriation of renewable natural resources with low-impact techniques with efficient and sustainable management of them.

The indices obtained through the indicators made possible the organization, systematization, quantification, and communication of the relative information of different aspects of this coastal zone, which is essential for making decisions about environmental policies that will be applied in the short term. The variability of the site, the infrastructure, the equipment, and the natural and anthropic processes that act in each of the units considered allowed zoning the area according to the values of the environmental index obtained.

The application of the environmental quality index determined that the Bahía Blanca, Bahía Falsa, and Bahía Verde Multiple Use Nature Reserve has an average environmental quality according to the results obtained from the aggregated indices (NI, RI, WI, SSI). The Islote del Puerto reserve has the intermediate value (0.21), while the municipal reserve has the lowest values, between 0.12 and 0.09. The most optimal environmental quality was presented by the two reservations mentioned above. Likewise, the three natural reserves acquire maximum value from the point of view of ecological and geomorphological importance when evaluating the geo-environmental units that characterize them.

The interrelation of the DPSIR indicators and the various functions of the medium allowed the zoning obtained from the EQI where it was determined which are the sectors in this area that need the rapid implementation of strategic management guidelines. The values indicate a high antthropic pressure on the natural quality of these ecosystems, generating an overexploitation of the physical environment, pollution of the same and in turn the low application of environmental and urban policies in the near past could be inferred. According to the response rate, some indicators mark a change in the formulation and application of sustainable measures to be able to modify the present and past situations through different levels of administration.

Continuous monitoring of the environmental condition of coastal areas is essential to guide the actions and policies to be implemented. In the Bahía Blanca Estuary reserves, the application of the EQI allowed us to observe a vulnerable and environmental risk area, but it also arises from this study that there is an incipient change in conservation policy and adequate management of resources when the conservation of these is included in the government agenda spaces. These types of studies must be carried out continuously due to the dynamics of this coastal environment. The application of appropriate environmental policies will perhaps be reflected in the following assessment of the environmental quality of this coastal zone.

17.5 Areas of Ecological Interest in the Northern Sector of the Bahía Blanca Estuary

The estuary was formally included in Western Hemisphere Shorebird Reserve Network (WHSRN) during 2016 due to its importance for the conservation of migratory shorebirds and its unique wetland character due to the ecosystem services they provide to human beings (WHSRN 2016). The areas described below do not have an official management category but are sites in the WHSRN and represent areas of ecological interest and are being requested to be included in the protected area system.

Puerto Cuatreros is located in both coast of the main estuary channel between Puerto Galvan and the La Vidriera Salt Flat, has areas of tidal, island, and shrubland plains (Fig. 2.1, Chap. 2). The objectives set are intended to preserve the structure, composition, and functioning of the estuarial ecosystem and promote the implementation of sustainable productive activities that allow a socially fair and environmentally sustainable local and regional development (Langhoff 2015).

The sector was called a site of interest by the Western Hemisphere Shorebird Reserve Network (WHSRN). The presence of certain species indicates the good state of health of ecosystems. Birds migrate from Canada and the Arctic area and can fly up to 11,000 km without descent. The birds feed and rest in the Bahia Blanca wetlands, in Tierra del Fuego, and in different parts of Brazil (Petracci et al. 2008).

Different social actors of the Municipality of Bahía Blanca and Coronel Rosales, of the Management Consortium of Puerto Rosales, of the OPDS, of the Yacht and Fishing Club of Gral. Daniel Cerri, and of the Management Consortium of the Port of Bahía Blanca participate in the management of the integral ecosystem to promote not only the conservation of these birds but the ecosystem as a whole.

Villa del Mar the continental sector is dominated by the salt fats, and in the tidal plains, the channels have a parallel design to the Principal Channel. In the internal sectors of the district in general, the modelers of the coastal processes are the reflux currents, while in the ambit the active processes are waves and coastal drift. The presence of old bars and spikes was observed that reveal the previous marine activity in the sector.

Wetlands the predominant environments of this area are the tidal plains and the marshes. Both cover approximately 47% in this sector. The tidal plains are areas that have little slope and lack of vegetation. In turn, they are crossed by numerous tidal channels that interconnect with each other and lead to larger channels. At low tide, they have a greater air exposure than the plains, allowing the formation of plant populations (mainly by the *Spartina* sp. and *Sarcocornia* sp.) tolerant to partial immersion, anoxia, and edaphic hypersalinization (Cabrera 1971). Below the lower limit of the vegetation, there are extensive tidal plains covering approximately 4.5 km^2, and in some sectors, they are interrupted by rocky outcrops, as in the eastern sector of the wetland. Another geomorphological feature to highlight is the underwater dunes in the bottoms in front of Villa del Mar, geoforms of mesomareal estuaries (Melo 2007; Spagnuolo 2005).

In the Arroyo Parejas Wetland located south of the town of Punta Alta with an area of 753 ha, the internal tidal plains belong to an old marine inlet, while the external plains correspond to current sedimentation processes. The dominant water supply is the sewer discharge channel of the mentioned locality, which in its final route is incorporated into the old tidal channel (Melo 2007). The Arroyo Parejas Wetland-Cantarelli Island has a marine coastal environment populated by *Sarcocornia perennis* and *Spartina alterniflora* forming large marshes that are exposed at low tide, where crab caves are found (Petracci and Sotelo 2013); in addition to these species, this site is characterized by the presence of great biodiversity, with birds being noted for their great abundance and diversity (Petracci et al. 2008). Every year there are birds that arrive from the north of the American continent to rest and feed; some of these species are threatened, and such is the case of the *Calidris canutus* that flies from the Canadian Tundra to Tierra del Fuego, across the entire American continent (Petracci et al. 2008; Petracci and Sotelo 2013). *Hirundo rustica* is another bird commonly observed using tidal flats and the semi-industrial zone located in Cantarelli Island to feed and nest (Larracoechea et al. 2012). A project has been presented to request to incorporate this natural environment into the provincial protected areas.

17.6 Evaluation of the Management Effectiveness of the Protected Coastal Areas of the Bahía Blanca Estuary

The declaration of the three coastal protected areas in the Bahía Blanca Estuary has been of utmost importance to conserve and protect the environments involved. In order to make a comprehensive diagnosis of the management of the mentioned coastal areas, the evaluation of their effectiveness was carried out, and in this section, the partial results of this work are presented. The evaluation was carried out using the framework proposed by the International Union for Conservation of Nature (IUCN) and the World Commission on Protected Areas (WCPA), adapted to the particular conditions of the study area. From it, the five fundamental elements of the management cycle were considered: context, planning, inputs, processes, and results.

The Rapid Assessment Prioritization Protected Area Management evaluation questionnaire was applied for each of the reservations. The objective is to evaluate the management effectiveness of the coastal protected areas of the Bahía Blanca Estuary region through variables and indicators grouped into five areas: human resources infrastructure and equipment, financial resources: planning and political-Institutional referred to the legislation in the coastals protected areas.

The criteria and The indicators evaluated in these natural reserves were biological importance, socioeconomic importance and vulnerability for the context variable, for the planning variable the objectives, legal security and site design are considered, for the inputs variable the resources are considered human and financial assigned to each reserve, for the process variables are involved the management plans, the operational plans and the monitoring of the results are evaluated through the restoration, planning and management of the natural resources of the reserves.

With respect to the context indicators, the highest value is held by the multipurpose reserve with 42% (Fig. 17.4), and the one that acquires the lowest value is the municipal reserve (24%), where the vulnerability of its ecosystem acquires greater prominence (becomes more important since out door landfill has not been reused (Speake et al. 2018).

All reserves have a high biological importance due to the presence of emblematic species that represent part of the ecological criteria used for the foundation of their creation. In the natural reserves of the estuary, sites of high value for the conservation of certain species have been identified; hence, there is an importance of the indicators that make up the context variable.

The extension of the protected areas of the estuary is not adequate to maintain natural processes at the landscape level; this variable is mainly intensified in the

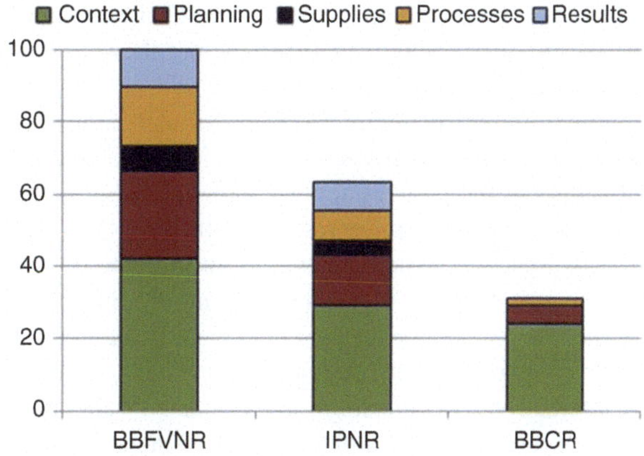

Fig. 17.4 Management effectiveness of Bahía Blanca Estuary natural reserves [Islote de la Gaviota Cangrejera Nature Reserve (IP NR), Bahía Blanca Coastal Reserve (BB CR), and Bahía Blanca, Bahía Falsa, and Bahía Verde Multiple Use Nature Reserve (BBFV NR)] (Modified from Speake et al. 2018)

Bahía Blanca, Bahía Falsa, and Bahía Verde Multiple Use Nature Reserve. The area bordering the north of the Principal Channel must be preserved, and controls must be carried out in port and industrial uses incompatible with the conservation objectives of these areas. The main threats to these areas are urban effluent drains, industrial discharges, oil spills, and untreated waste (Spetter et al. 2015; Carbone et al. 2016).

It should be noted that the Bahía Blanca, Bahía Falsa, and Bahía Verde Multiple Use Nature Reserve created in 1991 has a management plan in force since 2007. It was developed from inter- and multidisciplinary work with the participation of public officials, representatives of private institutions, civil associations, and NGO responding to a strategic planning process with public participation mechanisms, has not been updated, and has partial approval. The smaller reserves, the Islote de la Gaviota Cangrejera Nature Reserve and the Bahía Blanca Coastal Reserve, do not have a management plan. In 2008, the Bahía Blanca, Bahía Falsa, and Bahía Verde Multiple Use Nature Reserve was designated as an Important Area for Bird Conservation (AICA) by BirdLife International, and in 2016, most of the Bahía Blanca Estuary was listed as a "Regional Importance Site" by the Western Hemisphere Shorebird Reserve Network (WHSRN), becoming the eighth site in Argentina to obtain such recognition as mentioned above (Speake et al. 2018). In the same vein, at the local level, the Municipality of Bahía Blanca declared a list of "emblematic" species through ordinance 12.671/2004, including the *Larus atlanticus*, *Pontoporia blainvillei*, *Carcharhinus brachyurus*, and Charadriidae, while the Municipality of Coronel Rosales declared a list of endemic species of the region by Decree 3408/2013, becoming one of the first municipalities in the country to protect sea turtles (Petracci et al. 2008; Matamala 2013).

The indicators where the variable inputs of the protected areas referred to the human and economic resources assigned to each of the Bahía Blanca Estuary natural reserves were evaluated varied between 7.3 and 4.1% for the multiple use nature reserve and Islote de la Gaviota Cangrejera Nature Reserve. The number of assigned park rangers is scarce for the two areas mentioned and null in the case of the municipal reserve. Administrative and dissemination tasks are carried out in collaboration with non-governmental organizations. The areal dimension of the reserves makes it difficult to monitor and control these environments by assigned personnel. The media and infrastructure are insufficient. Likewise, the financial resources allocated are also insufficient.

Regarding the planning variables, these include objectives and design of the area and cover the legal aspects, and the values obtained varied between 24 and 7%. The indicators of the policies applied show little follow-up of the objectives and operational plans in the estuary reserves, although they are present in the Bahía Blanca Coastal Reserve, given that it has more staff and resources allocated. The municipal reserve currently has no resources or monitoring of operational plans since it does not have a management plan either. Nor is a complete inventory of the biological biodiversity of the entire region. There is only one list of species that were surveyed by international organizations that evaluated Bahía Blanca Estuary to be included as a site of regional interest in the WHSRN and a list of interest from the municipality.

The surveys of flora and fauna species were carried out when the management plan of the Bahía Blanca, Bahía Falsa, and Bahía Verde Multiple Use Nature Reserve, currently outdated, was prepared. The management plan of these area proposes operational objectives to carry out an environmental biological monitoring; regulate the activities of artisanal, sport fishing, and recreational tourism; monitor exotic species such as the *Magallana gigas*, in addition to implementing an education and dissemination program; and also intensify marine turtle conservation studies as a species of accidental capture by fishermen, in addition to other objectives related to emblematic species of the aforementioned natural reserves. This management plan should be updated continuously due to the complexity and dynamism of the ecosystems that are conserved and protected in these coastal areas.

There is a slight direct and continuous communication between government agencies that manage natural reserves in the estuary. Likewise, there is no efficient application of the laws and ordinances related to the protected area, given that in the islands that are part of the Bahía Blanca, Bahía Falsa, and Bahía Verde Multiple Use Nature Reserve, there are irregularities with respect to the livestock use observed in them. With respect to the Islote de la Gaviota Cangrejera Nature Reserve, its limits are not well defined, and the legislation in force has a lack of representation. The effective application of current legislation would allow the detection of illegal activities to identify and penalize those responsible.

The environmental problems that directly affect the integrity of the estuarine protected area system also arise from the insufficient application of the laws that regulate the different activities allowed and not allowed in natural reserves and adjacent areas. This condition is observed in the indicators of the process variables that refer to management practices and its planning. The values varied between 14 and 2%, with the lowest value for the municipal reserve.

The laws related to protect area must be complementary to their objectives and promote management effectiveness. Laws related to protected areas must include land use planning and laws to regulate agriculture and regulations and policies at national, regional, and local levels. With respect to legal security, the reservations are analyzed by having different jurisdictional levels present delays in the application of the legislation. Environmental protection goals should be incorporated into all aspects of the legislation of protected coastal areas. Environmental impact assessments must be carried out for the development of pre-existing infrastructure; to the expansion of the economic activities incompatible with the conservation measures of the reserves, controlled by land use planning guidelines that in these environments make implementation difficult, although the assigned human resources are highly trained, the allocated resources are not sufficient.

The comparative analysis allowed to identify systemic strengths and weaknesses of the natural protected area system of the estuary. The main problems are linked to inputs (insufficient infrastructure and financing) and processes (lack of adequate monitoring programs). The Bahía Blanca Coastal Reserve operates in the absence of equipment and infrastructure, while the Bahía Blanca, Bahía Falsa, and Bahía Verde Multiple Use Nature Reserve and Islote de la Gaviota Cangrejera Nature Reserve present significant deficiencies.

On the other hand, although the staff is highly trained to carry out critical management activities, the number is insufficient. As for planning, all reservations have legal protection. However, only one of them has a partially approved management plan. Finally, the indicators that detected the highest score are those related to the context (biological and socioeconomic importance).

The study area stands out for its great biodiversity and the provision of important ecosystem services. The reserves protect the habitat of numerous animal and plant species of high social, cultural, and economic importance, as well as geoforms and natural processes of interest, holding great tourist, educational, and scientific value. The use of this methodology is presented as a tool of great relevance for the determination of aspects to strengthen and prioritization of the actions, in order to improve the effective management of protected areas (Speake et al. 2018; Speake and Carbone 2017).

The results highlight an index with values of 10.2 and 7.9% for provincial reserves, for the education and management of visitors, which could indicate that within the management of Bahía Blanca Estuary protected areas, greater emphasis is given to tourist use and the management of the resources to be protected. This situation is also evident in the environmental education programs that are developed for the dissemination and knowledge of the estuary reserves.

The multipurpose nature reserve achieved the best results, with the highest overall driving effectiveness index. The inputs are more scarce than in the other reserved considered. However, in terms of defining objectives and management planning, they outperform the other areas. This area has the limits and management objectives better defined than the areas declared later.

The strengths identified in the estuary coastal protected areas refer to the natural physical system where biological and socioeconomic importance are highlighted. Most of these units of analysis cover a size that is not sufficient to ensure effective long-term conservation. The system of the analyzed reserves also has conservation objectives, both at the level of provincial policy and at the level of each conservation area, although at the municipal level there is a lack of setting specific application objectives. The majority of the units that make up the system have adequate legal support and the necessary basic infrastructure, in accordance with the established objectives, although it is the municipal reserve that does not have adequate legal support. Regarding the processes, the lack of complete and updated inventory of natural resources and the collaboration between protected area personnel, communities, and other organizations stand out.

The weaknesses observed for planning are inadequate limits, including inappropriate size and land uses in adjacent areas of protected areas, which are the greatest threats to Bahía Blanca Estuary reserves. For effective management, inputs constitute the aspect that needs the greatest improvement, highlighting the lack of funds and the low number of personnel assigned to conservation units.

The scarcity of sustainable financial resources for the management of protected areas and the administration of the coastal protected area system complicates their effectiveness. The results also show the limited availability of direct personnel assigned to protected areas, inadequacies in threat control, and the results of recent investigations carried out by various government agencies such as the Instituto Argentino de Oceanografía and the Universidad Nacional del Sur that should be incorporated into the plans of handling these reserves. The pressures that most affect the coastal protected areas of the Bahía Blanca Estuary are garbage, liquid waste, invasive species, and the development of port and industrial activities.

The values of the indicators of the different areas analyzed allowed the classification of reserves into two categories. They represent the lowest ranges with values below 25% (Islote de la Gaviota Cangrejera Nature Reserve 13% and Bahía Blanca Coastal Reserve 7%). These areas lack the minimum resources necessary for basic management, and only one of the reserves reaches the value of 29% (Bahía Blanca, Bahía Falsa, and Bahía Verde Multiple Use Nature Reserve) reaching the second category that represents an unsatisfactory effectiveness; the area has resources and means that are indispensable for its management, but it lacks elements to reach an acceptable minimum level.

17.7 Importance of the Wetlands of the Bahía Blanca Estuary in Compliance with the Sustainable Development Goals (SDG)

Given the advance of threats to wetlands, the international approach is to implement the SDGs proposed in the 2030 Agenda to achieve harmony between the natural physical subsystem, the socioeconomic subsystem, and the administrative political subsystem. To address the study of coastal areas in this case of Bahía Blanca Estuary wetlands in an integrated manner, the subsystems mentioned must be taken into account.

The objective of the Convention on Wetlands is "the conservation and wise use of all wetlands through local, regional and national action and international cooperation as a means of achieving sustainable development throughout the world". The term "rational use" refers to the sustainable use of wetlands to keep their ecological functions as pristine as possible for future generations.

The advance of populations over coastal areas with 600 million people lives in coastal areas that are less than 10 m above sea level shows the importance of conservation and protection of coastal environments (Barragán Muñoz 2014). Economic and extractive activities should be managed in such a way that spawning areas that are often found in coastal marine areas are protected to ensure that populations are restored (Barragán Muñoz 2014). Considering the effects that climate change is likely to have on coastal ecosystems in the medium and short term, the governing bodies of the different administrative legal levels should generate integrated management legislation to mitigate those effects.

"Wetlands play an important role in disaster risk reduction if they are effectively managed and restored if necessary." Coastal ecosystems and, in particular, saltwater marshes called "blue carbon ecosystems" play a role important in mitigating climate change. This carbon is stored in the long term in wetland soils (Barragán Muñoz 2014).

The International Alliance on Blue Carbon – announced during the Climate Change Conference held in Paris in 2015 – aims to bring together different partners, from government organizations to non-governmental and research organizations, to conserve coastal ecosystems. Particularly the estuary wetlands are of great biological importance and acquire a fundamental function as regulators of natural and ecological processes.

The planning of coastal protected areas should include the sustainable development objectives that consider not only the environmental quality of the natural capital of coastal ecosystems but human well-being; not only the objective of the quality of life of marine ecosystems should be considered but also the objectives for the increase of the ecosystem services of the wetlands.

17.8 Conclusion

The creation of three marine coastal protected areas in the estuary sets a precedent for the importance attached to the value of this environment and the need to preserve it. The creation of coastal protected areas (CPA) constitutes a very versatile strategy for the conservation of natural and cultural heritage. Although there are problems that put the long term viability of the ecological systems they preserve at risk. The diagnosis updated in adaptive management processes and continuous monitoring can achieve the objectives of its creation. However, there are problems of execution where the legal creation rule is the only measure promoted by the administrator in favor of the area. In order to achieve effective management, it is imperative that the respective management plans be developed and updated, given that it is not only a territorial planning instrument that clearly indicates the permitted and prohibited activities (FAO 2014). Likewise, it should be noted that the distinctions granted by international organizations to these wetland environments (WHSRN and AICA) do not have a legislated management category, which makes it difficult to lodge protection against land uses incompatible with the protection of these areas. The lack of adequate financing and equipment is unleashed (Speake and Carbone 2017) which represents a direct threat against biodiversity conservation objectives. The same situation occurs with assets of ecological interest that do not have laws or management strategies for their value and conservation.

The main pressures received by the natural reservoirs of the estuary and that favor the degradation of the estuary are the concentration of infrastructure on the northern coastal edge (port and industrial), environmental pollution, dredging with the subsequent artificial filling of wetlands, and the presence of invasive species.

The direct consequences on human well-being are derived from the decrease in food from fishing, the loss of first-class tourist resources, degradation of the landscape, loss of biodiversity and health problems due to air quality, and conflicts between human activities and the physical environment.

References

Aldana Mazorra O, Hernández Zanuy A (2018) La Planificación Espacial Marina: marco operativo para conservar la diversidad biológica marina y promover el uso sostenible del potencial económico de los recursos marinos en el Caribe. En: Hernández Zanuy A. C. (Ed.). Adaptación basada en Ecosistemas: alternativa para la gestión sostenible de los recursos marinos y costeros del Caribe. Red CYTED 410RT0396. (E. Book). Editorial Instituto de Oceanología, La Habana. p171

Amarfil R (2006) Actividades económicas puntaltenses en la Ría de la Bahía Blanca, en revista *El Archivo*. Número 16:6–9

Angeles G (2001) Estudio Integrado del Estuario de Bahía Blanca. Tesis Doctoral. Departamento de Geografía. Universidad Nacional del Sur, p 166

Barragán Muñoz JM (2014) Política, gestión y litoral. Una nueva visión de la Gestión Integrada de Áreas Litorales. Madrid: Ed. Tébar Flores p 456

BirdLife International (2019) Important Bird Areas (IBA) factsheet: Reserva de Uso Múltiple de Bahía Blanca, Bahía Falsa y Bahía Verde. http://www.birdlife.org Consultado el 28/11/2019

Cabrera AL (1971) Fitogeografía de la República Argentina. Sociedad Argentina Botánica 14(1):1–42

Carbone ME, Spetter CV, Marcovecchio JE (2016) Seasonal and spatial variability of macronutrients and Chlorophyll a based on GIS in the South American estuary Bahía Blanca, Argentina. Env Earth Sci 75:1–15

Charlier RH, Bologa AS (2003) Coastal zone under siege – is there realistic relief available? J Coast Res 19(4):884–889

Cinti S (2017) Las islas de la Bahía Blanca y Los forjadores de su historia, Ed Vacasagrada p 302

Conde AA, Piccolo MC, Pizarro N (2009) Análisis histórico de las capturas de la flota costera en el puerto de Bahía Blanca. Período 1983–2007. Actas del 12° Encuentro de Geógrafos de América Latina, p 123

Day J, Dudley N, Hockings M, Holmes G, Laffoley D, Stolton S, Wells S (2012) Guidelines for applying the IUCN protected area management categories to marine protected areas. IUCN, Gland, p 36

European Environmental Agency EEA (1998). Europe's environment: the second assessment. Copenhagen 132 p

Food and Agricultural Organization (2012) Estado de las áreas marinas y costeras protegidas en América Latina. Elaborado por Aylem Hernández Avila. REDPARQUES Cuba. Santiago de Chile, p 620

Food and Agricultural Organization (2014) Manejo de las Áreas Marino Costeras Protegidas para garantizar medios de vida sustentables y seguridad alimentaria. Elaborado por A. Rocío Motrán Ferrándiz y Vanessa Dávila. REDPARQUES Santiago de Chile

INDEC Instituto Nacional de Estadísticas y Censos (2010) Censo Nacional de Población, Hogares y Viviendas 2010. Buenos Aires p 345

Langhoff ML (2015) La reacción de la comunidad local ante el proyecto de dragado en el área de Puerto Cuatreros. ¿Es posible recuperar la relación comunidad naturaleza en Bahía Blanca? En: V Jornadas de investigación en Humanidades, 76–89

Larracoechea G, Durán H, D'Acunto C (2012) Nidificación de la Golondrina Tijerita (*Hirundo Rustica*) en el Balneario Arroyo Pareja, Buenos Aires, Argentina. Nuestras Aves 57:18–19

López Cazorla A, Molina JM, Ruarte C (2014) The artisanal fishery of Cynoscion guatucupa in Argentina: exploring the possible causes of the collapse in Bahía Blanca estuary. J Sea Res 88:29–35

Maelfait H, Belpaeme K, Lescrauwaet AK, Mees J (2006) Indicators as reliable guides for Integrated Coastal Zone Management. In: Forkiewicz, M. 2006. Integrated Coastal Zone Management: theory and practice. EuroCoast – Littoral 180–186

Massola MV, Cinti S (2012) Valoración del humedal costero del sudoeste bonaerense desde el abordaje holístico IX Jornadas Nacionales de Geografía Física Bahía Blanca 20–26

Matamala R (2013). Ecoturismo accesible como alternativa turístico-recreativa para personas con discapacidad visual en la localidad de Villa del Mar, partido de Coronel Rosales. Departamento de Geografía y Turismo, Universidad Nacional del Sur p 145

Melo, W. D. (2007). Orígenes morfológicos. En: M. C. Piccolo, M. C. y M. Hoffmeyer (Eds.). Ecosistema del estuario de Bahía Blanca 21–27

Nebbia AJ, Zalba SM (2007) Comunidades halófitas de la costa de la Bahía Blanca (Argentina): caracterización, mapeo y cambios durante los últimos cincuenta años. Boletín de la Sociedad Argentina de Botánica 42:161–171

Organization for Economic Cooperation and Development (1993) Core set of indicators for environmental performance reviews. A synthesis report by the Group on the State of the Environment. Environment Monographs, p 83

Petracci P, Sotelo M (2013) Aves del estuario de Bahía Blanca: Una herramienta Para su conocimiento y conservación. Grupo Editorial Muelle Sur, Bahía Blanca, p 192

Petracci PF, Sotelo MR, Díaz LI (2008) Nuevo registro de nidificación de la Gaviota Cangrejera (Larus atlanticus) en la Reserva Natural Bahía Blanca, Bahía Falsa y Bahía Verde, Buenos Aires, Argentina. Hornero 023:037–040

Sotelo M, Massola V (Eds.) (2008) Propuesta Plan de Manejo Reserva Natural Provincial de Uso Múltiple Bahía Blanca, Bahía Falsa, Bahía Verde. Bahía Blanca, p 167

Spagnuolo JO (2005) Evolución geológica de la región costera-marina de Punta Alta, provincia de Buenos Aires. Bahía Blanca Tesis de Doctorado Departamento de Geología UNS p 269

Speake MA, Carbone ME (2017) Evaluación y priorización rápida del manejo de las áreas protegidas costeras en el estuario de la Bahía Blanca. En: Cenizo M, Celsi C. (Eds.). Libro de resúmenes de Segundas Jornadas Bonaerenses sobre Conservación de Ambientes y Patrimonio Costero. Villa Gesell, Buenos Aires, p 453

Speake MA, Carbone ME, Spetter CV (2018) Ocurrencia de eventos de emergencia ambiental en el área costera de Bahía Blanca, provincia de Buenos Aires. En: Libro de resúmenes de las XII Jornadas Nacionales de Geografía Física 91–95

Spetter CV, Buzzi NS, Fernández EM, Cuadrado DG, Marcovecchio JE (2015) Assessment of the physicochemical conditions sediments in a polluted tidal flat colonized by microbial mats in Bahía Blanca Estuary (Argentina). Mar Poll Bull 9:491–505

Spetter CV, Fernández EM, Carbone ME, Negrin V et al (2019) Estudio de línea de base sobre la dinámica de nutrientes en una planicie de marea del estuario de Bahía Blanca previo a la instalación de una planta de tratamiento de desechos cloacales. V Reunión Argentina de Geoquímica de la Superficie La Plata, Asociación Argentina de Sedimentología, pp 330–333

Chapter 18
Small-Scale Artisanal Fishers and Socio-environmental Conflicts in Estuarine and Coastal Wetlands

Daniela M. Truchet and M. Belén Noceti

18.1 Introduction

From the paradigmatic constitution of what we call modernity, the relationship nature-culture has been introduced as a dichotomic form in the scientific area, probably as a strategy of legitimation of the extractive practices on nature (Svampa 2012; Giarraca and Teubal 2013). In academic environments, such dichotomy has been overcome with the contributions to the social sciences of authors like Descola and Pálsson (2003), Ingold (2000, 2003), Ellen (2003), and Latour (2004), among others. For these authors, the nature/environment is not only a simple scenario where the social actors and actresses unfold their survival activities, but it is more like a sociocultural resource. In this sense, nature has a *plus value* in the policy design and execution oriented to the habitat conformation and organization (Carman 2011). These different forms of inhabiting nature are defined as continuous disputes and may fall under the category of *socio-environmental conflicts* (Wagner 2016): conflicts where different actors and actresses struggle for the access and control of territories and natural resources, which are referred as divergent interests and values in scenarios with strong politic asymmetries. In these conflicts, there are also established meanings and feelings that acquire the notions of progress and development

D. M. Truchet (✉)
Área de Oceanografía Química. Instituto Argentino de Oceanografía (IADO), Universidad Nacional del Sur (UNS), CCT-CONICET, Buenos Aires, Argentina

Departamento de Biología, Bioquímica y Farmacia, Universidad Nacional del Sur (UNS), Buenos Aires, Argentina
e-mail: dmtruchet@iado-conicet.gob.ar

M. B. Noceti
Instituto de Investigaciones Económicas y Sociales del Sur (IIESS), Universidad Nacional del Sur (UNS) – CONICET Bahía Blanca, Buenos Aires, Argentina

Departamento de Economía, Universidad Nacional del Sur (UNS), Buenos Aires, Argentina

S. M. Fiori, P. D. Pratolongo (eds.), *The Bahía Blanca Estuary*,
https://doi.org/10.1007/978-3-030-66486-2_18

and, in a certain way, of democracy. Occasionally, the ways in which the state decides about the organization and exploitation of spaces are imperative, that is, imposed without the population consensus, assigning *sacrificial territories* or *socially empty* spaces (Sack 1986; Quist 2019), a situation that enhances the process of social fragmentation. This fragmentation produces, in one hand, the restriction and gradual exclusion of certain social sectors from the access to and the availability of natural resources (Latour 2004), and on the other hand, it produces inequality in the distribution of the benefits and externalities of certain productive activities (Girado 2012; Merlinsky 2013).

This context of social, ecological, and cultural crisis finds no answers on the mechanist paradigms. Complexity thus becomes the new emergent boarding perspective because environmental risks do not know about social classes. However, risk distribution is unequal and follows the logics of a class inequality, typical of the neoliberalism, known as "environmental racism." By denying identities and cultures, the environmental racism deepens the fragmentation and widens the gap between different social sectors (Machado Aráoz 2010). The so-called "neo-extractivism" in South American countries raised in dependence on the global economies. Commonly, underdeveloped South American countries that produce primary goods (*commodities*) have a peripheral position in the world economy and economic dependence on the rich developed countries. In this scenario, Latin America becomes a territory of widespread loss of natural resources, along with the displacement of several communities and cultures. Social resistance to displacement and inequality tends to create a spiral of violence, criminalization, and repression (Svampa 2012), with large power asymmetry: large empowered companies and political lobbies against indigenous communities and unpowered populations.

In South America, ecological damages raise social and territorial demands, and the ecological legislation often tends to consider these conflicts as "creative transformations" based on the tensions and antagonisms of the different actors and actresses (Merlinsky 2013). In other words, the struggles that arise within the kaleidoscope of actors and actresses create opportunities for making the popular demands more visible.

In several regions of this continent, small-scale artisanal fishers are actors often related to socio-environmental conflicts, as they are affected by the consequences of climate change, industrial development and large fisheries, and, sometimes, tourism and large aquaculture projects (Charles 1992; Coulthard et al. 2011; DuBois and Zografos 2012; Noceti 2017; McGregor 2018). For example, augmented emissions of CO_2 and climate change are expected to diminish fishing stocks in emerging countries, coinciding with the present socio-economic vulnerable areas (Cheung et al. 2010). In these changing scenarios, artisanal fisheries are at serious risk because entire families could be driven into poverty and social destitution (Allison et al. 2009). Thus, poverty and vulnerability are, and will continue to be, endemic conditions in small-scale fishing communities.

Thus, artisanal fishing keeps entire families away from marginal conditions and stands for a set of values over the profit-oriented and capitalist enterprises typical of large and industrial fisheries (Johnson 2018). These values are context-specific, committed to a community and a place and their socio-economic and historical relations. They are also according and respectful to what nowadays is called the "rights of nature" (Truchet 2018): small-scale artisanal fishers have an intrinsic relation with nature and are invaluable actors in contributing to alternative forms of development in harmony with human and nonhuman beings.[1] Although, in most cases, they lack formal education and specialized training, fishers exhibit a great empirical knowledge about biology and ecology, which has proven to be as valuable as scientific information and provides significant tools for ecological academic research and animal conservation (García-Quijano and Valdés-Pizzini 2015). Furthermore, the arts of artisanal fishing are based on the interaction with what we call nonhuman beings, and their function is to protect and respect the different species that develop their life cycle in the seas and coastal wetlands. This form of knowledge is not only based on biological academic knowledge but also in learnings acquired through ancestral practices and traditional ontologies.

In addition, several studies have demonstrated that fishers' knowledge about ecology (hereafter, FEK) is priceless for ecological studies, since fishers have been navigating and observing aquatic ecosystems for decades, acquiring a unique comprehension of ecological interactions (Johannes et al. 2000). Although the empirical knowledge of fisher communities should be well appreciated, Johannes et al. (2000) found that marine ecological researches often neglect local knowledge and dismiss fishers' culture. In our experience, the dismissal of FEK and their cosmology is the main reason for which biologists and ecologists often arrive at uncertain or wrong conclusions about some marine species traits and the status of coastal ecosystems. In this sense, Quist (2019) also argues that scientific indetermination about the causes of depleting fish populations and the weakness of environmental legislation result from the exclusion of FEK from politics. Industrial knowledge, such as that coming from the petroleum industry, is often included in policy and lawmaking and is, therefore, considered valid. Thus, considering FEK perspectives into research, conservation, and management strategies represents a new approach that includes local knowledge.

From what we exposed so far, the main objectives of this chapter are a) to understand which socio-ecological conflicts take place in the Bahía Blanca Estuary; b) to comprehend the different conservation paradigms coexisting in the Bahía Blanca Estuary through the voice of different actors; c) to characterize small-scale artisanal fisheries in the Bahía Blanca Estuary from their past, present, and future status; and d) to assess fishers' ecological knowledge (FEK) in the Bahía Blanca Estuary and how it can be useful for future monitoring programs in estuaries that are presently seen as sacrificial territories.

[1] Descola (2005) studied indigenous societies (jíbaros) in the Amazonas describing their relationships with nature and was able to identify different forms of inhabitants in the world: human and nonhuman beings. In a reductionist and vague explanation, nonhuman beings are rocks, plants, animals, and spirits, and humans are also able to establish a relationship or a way to socialize with them. Thus, the relations are not only conceived as human-human like the occidental world perceives but also as humans-nonhumans. This was also observed in other local communities, such as fishers and hunters by Pálsson (2003).

18.2 The Bahía Blanca Estuary

18.2.1 Sea-Land Dispute and Maritime Possession: An Ontological and Territorial Perspective

The Bahía Blanca Estuary (southwest of the Buenos Aires Province) is an area in dispute regarding the use of its maritime-coastal resources since it was occupied for the government of the United Provinces of the Río de la Plata in the nineteenth century. This area was known as the "Liverpool of the South" or the "City of the Seven Ports" because Bahía Blanca was the only city in Argentina with seven docks by the first half of the twentieth century (Lugones 1883). Until the 1980s, the inhabitants of this region had free access to the coasts of the estuary, building their identities attached to the maritime-coastal territory. Following Peron and Rieucau (1996), this singular type of identity is called "maritimity," and it constitutes lifestyles in reference to the link human being-sea. The upcoming installation of the petrochemical complex of Bahía Blanca in the 1980s triggered a gradual privatization of the access to the estuary. The state policies were afterward oriented to alienate the estuary from the public sphere, which ended up with the sociability contexts that were carried for almost a century. This constitution of the territory in an economically productive and socially empty scenario and where the monetary income justifies the negative environmental externalities will be called from now on "maritimization" (Peron and Rieucau 1996; Noceti 2017). Therefore, in this section, we will use an anthropological and ethno-ecological perspective to explain the dispute for inhabiting the space of the Bahía Blanca Estuary, a conflict named as "fishers' conflict" (see Text Box 18.1: Fishers' conflict) in juridical and social media.

In the present point, we will try to give an account of the identity process built in the estuarine space, from an ontological perspective that denies the existence of a unique world and a single representation of itself (cosmovisions). Thus, we consider the existence of multiple forms of nature, and the alterity is in function of the existence of these compositions (Tola 2011). The ontological turn is built on the multinaturalism (Viveiros de Castro 1996) and might be helpful when it comes to explain how and why the socio-environmental conflicts persist in the Bahía Blanca Estuary (Noceti 2018). We will assume that the conflict is materialized in socio-environmental coordinates, but it hides a political-ontological discussion. Therefore, it is possible to explain why these conflicts persist in time and why it was not possible to find so far any solutions that allow the coexistence between actors. We hereby assume that it does not exist in one single and universal form of nature over which different actors develop their representations. Instead, the space is perceived as a territory to be inhabited (Yory 2013, 2015), and it emerges as an ontological composition raised toward different coordinates: ethics, cognitive, perceptual, social, cultural, politics, and affective, including topophilia developments (Tuan 1974). The debate is not about domain and control of a common good such as the sea and its coasts. More likely, it happens because of the identification of the existence of multiple territories (De la Cadena and Blaser 2018).

In the Bahía Blanca Estuary, we identified three territories disputing their existence in the same space and at the same time. Each one refers to different collectives who are in a conflictive interrelation: the fishers of the Bahía Blanca Estuary; seniors and state officials, along with businessmen of private companies of the Bahía Blanca Port Management Consortium (Consorcio General de Puertos de Bahía Blanca, CGPBB); and the conservationists (most of them environmental professionals of nongovernmental organizations – NGOs). These territories are, respectively, *the estuary-ria*, *the estuary-port*, and the *estuary-protected areas* (Fig. 18.1) and coexist in permanent conflict, trying to impose themselves in a scenario of strong political asymmetry in favor of the CGPBB and in detriment of the popular sectors, like the artisanal fishers. It reinforces the existence of a fragmentation process, a socio-spatial segmentation, and the alienation of the productive spaces, with a strong meaning for fishers. Through the ethnographic exploration, we were able to identify the following characteristics of each territorial moment that could help us to explain

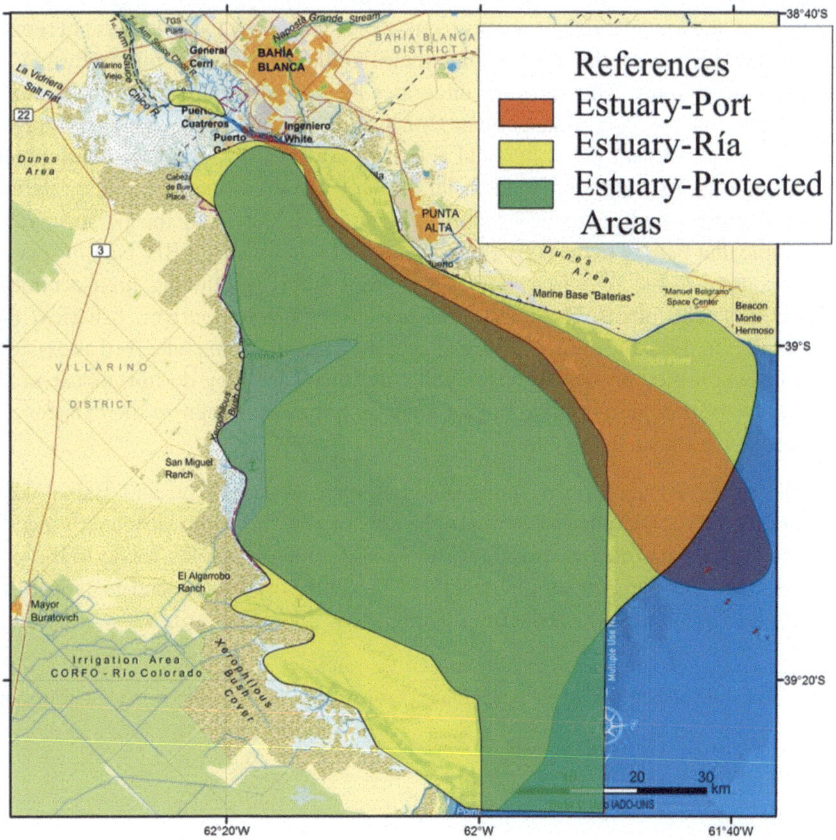

Fig. 18.1 The Bahía Blanca Estuary and the three different forms of territory that coexist in permanent conflict: the estuary-ria, the estuary-port, and the estuary-protected areas

the areas where the socio-environmental conflicts in the Bahía Blanca Estuary arose and why:

- The estuary-ria: this territorial form has its foundation in artisanal fishers, mostly from Italian immigrant families,[2] which named the estuary as a "ría," though, in terms of geomorphology, it is not correct. This territory denotes the forms in which the alterity is constructed between humans and between human and non-human beings (i.e., physical, spirits, animals, vegetation, other forms) (Descola 2005). Artisanal fishers have built their identities differentiating from "people that only fish" due to the forms in which they configure their lives in space and time. The space and time are in function of their fishing activity, which is seasonal. In terms of space, it is observed that the artisanal fishers have their own maps besides the official naval chart. These maps are built in function of the places that are identified as "fishes' hot spots" and are classified and named based on the experiences lived by their forefathers. As an example, the "pajarera," "palito," and "hundido" are places that are invisible for the eyes of people that do not inhabit the estuary nor belong to the history of fisheries in the region. Instead, they received their names from stories that every fisher knows but remain unknown for those that do not share their livelihood as an indentitary form.
- The estuary-port: This territory integrates the petrochemical complex of Bahía Blanca and the Port Management Consortium (CGPBB), because both have a similar speech and petrochemical activities are linked to the port development through the exportation of industrial products. In the estuary-port, the relationship human-nonhuman is virtually null or inexistent from the perspective of people that belong to the petrochemical complex and the CGPBB. Nature appears as a mere "object," where human activities (those producing high economic income) take place. The rest of the activities (sports, tourism, fisheries) that develop in the same area are ignored, and they are not even conceived as existent, even when they are easily observable, and therefore they do not appear in the estuary-port cognitive map. In fact, they do not even recognize the existence of beaches in the area.

The protected areas (especially Bahía Blanca, Bahía Falsa, and Bahía Verde Natural Reserve) are considered out of the territory and contact restricted to specific situations. Companies/enterprises of the petrochemical complex relate to the protected areas through money disbursement by their corporate social responsibility programs, pursuing tax reduction. Companies like Profertil, Dow Chemical,

[2] Small-scale artisanal fishers of the Bahía Blanca Estuary are fully described in the following point 18.2.2.

Petrobras, and YPF maintain economic cooperation agreements with city governments of the area and even with the local universities (Heredia Chaz 2014). Moreover, for these actors, the existence of the "other" is a stumbling block to be overcome, where the relationship mode, called *rapacity* by Descola (2003), is useful to describe the relationship not only between nonhuman and humans but also between humans belonging to the port system and fishers. Their discourse based on development makes evident the strong political asymmetry and deepens the differences between fishers and the port system.

– The estuary-protected areas: this scenario is represented by a series of actors that are dissimilar between each other but with a strong speech based on the defense of the nonhuman environment, above the existence of human productive activities, such as fisheries and the industrial context. In the view of conservationists, nature is an entity that must be protected from human actions that cause damage. Nature exists if it can be protected by these actors, and therefore we can verify a passive role of the nonhuman beings that differs from the human nature, making clear its non-belonging to nature.

This cognitive formation presumes in the social imaginary a great diversity of nonhumans that depend on humans for their well-being. This dependence is utilitarian: the protection of nonhumans ensures the benefits of humans that found their actions on the speeches of defending and protecting the environment. Under this protectionist and paternalist scenario, there is a place for the workers of the protected areas and the conservationists from the local NGOs. The ways in which these actors catalogue the rest of humans and nonhumans are in function of what they consider that must be controlled and what it is expected to be protected. This group of actors is not homogeneous when it comes to claim for the protection of nonhuman life and holding other actors responsible for the damage to the environment. While the workers of the protected areas develop their programs of reeducation oriented to modify the arts of fishing – considering this collective as the main responsible of the decrease of fishing stocks in the estuary – the conservationists of the environmental NGOs have accompanied artisanal fishers in their strikes and legal claims against the CGPBB and even against the Natural Reserve, because they adduce that the Argentine government is the main responsible for the loss of fishing stocks, since they do not control the human activities that affect the ecosystem.

Box 18.1 Fishers' conflict

The emergent marine pollution that was registered in the Bahía Blanca Estuary drove a constant and fierce socio-ecological conflict known as the "fishers' conflict." The main actors were represented by the artisanal fishers' families, which initiated a lawsuit against the industrial complex and a sewage water treatment plant, which, according to them, were the major causes of the decline in fishing. Noceti (2017) and, recently, explained that, during the 2009–2012 period, the fishers' families suffered from a drastic reduction in their incomes due to the collapse of the fisheries. The community claimed this was due to overfishing in El Rincón area and the pollution of the estuary, which resulted in a rapid impoverishment of the families. Fishers' communities started a series of multiple claims to the municipal government, demanding economic help to support basic survival conditions to sustain their livelihoods. Given this situation, both the municipality and the provincial government designed the fishery "reconversion program," which involved the purchase of fishing permits by the provincial government from those fishers who chose to stop fishing or the purchase of new, more powerful boats with outboard motors for fishing offshore (mostly in the external part of the estuary), where there was a greater biomass of commercial species. Most of the families chose to sell their permits to obtain money that they spent quickly and failed to reinvest in another type of productive activity given the recession periods that hit Argentina over the last few years (especially during 2015–2019).

These families had to start fishing again in situations of greater vulnerability, because most of them did (and still do it) under conditions of illegality, since they sold their permissions and use alternative places as "ports" to avoid being legally enrolled and to pay extensive taxes. On the other hand, fishers who chose larger boats failed to sustain the activity due to the amount of fuel needed to operate these boats and the continual increase in fuel prices, which led them to sell some of these boats and return to fishing in the inner part of the estuary. However, artisanal fishing in the Bahía Blanca Estuary is now an activity that is at risk of being lost, along with the fishers' culture, due to the decrease in the number of fleets. Nevertheless, the "illegality" in fishers' activities is a constant discussion. For example, Nahuelhual et al. (2018, 2019) explained that small-scale artisanal fishers that do their activities in an illegal form cause less damage to the ecosystems that "legal" activities such as the extractives large fisheries companies, that are sometimes – if not always – extent of high taxes.

In summary, fishers' conflict exists since 2009, and legal proceedings are still underway. At present, fishers' lawsuit against companies, the port, and the sewage plant was not fully resolved, and contrasting reports about pollutant levels in the estuary raise questions and concern in the fishers' community. Therefore, scientific commitment and responsibility are required to provide this vulnerable collective with a sound evaluation on the health of the environment that they are part of and where they develop their livelihoods.

18.2.2 Small-Scale Artisanal Fisheries in the Estuary: The Past, Present, and Future Status

For the Argentine law, small-scale artisanal fishers are defined as people who extract fish for their commercialization, using artisanal beach nets, such as trammel nets, gillnets, fixed nets, traps, hook and line, or any other fishing gear that is not forbidden, by manual extraction, diving, from the coast or boats, and/or manual collection from the intertidal zone (Noceti 2017). In this context, the artisanal fishers from the Bahía Blanca Estuary are linked to Italian immigrants, especially from the Island of Ponza, who arrived to the area around 1890–1920 and whose forefathers handed down sustainable fishing methods from "parent to child" (London et al. 2017; Noceti 2018). Some of the Italian traditions are still observed like their religious belief to San Silverio, who is considered in the Italian tradition "the keeper and guardian of the fishers." Also, some of the foods, fishing arts, vocabulary, and words are Italian and are present in their livelihoods. The fishers' circuits find their extreme situations in the child labor of male children and teenagers in families with extreme socioeconomic vulnerability and low levels of schooling, where kids work in order to earn money to sustain their families (Truchet et al. 2020a, b): "When we came from Ponza (Nápoles, Italy), I was ten years… At the school in Italy, I was in fifth grade, but when I came to Ingeniero White, they put me in third grade! The verbs, the language… they were very difficult for me. When I got to the fifth grade here, I missed my exams, and I had to take the year again. And, honestly, I didn't want to do it, and my mom sent me to fish. I was the only support to my family; my father earned too bad; he was a hairdresser. At that time, (Ingeniero) White was a small village and fifty percent Italians, we knew each other, so my mom immediately looked for a shipment for me and at the age of thirteen, I was already fishing. I remember we were seven boys at 'La Envidio', Antonio was the owner, and at that time it was different… Not like now, where the kids can earn their own money when they work. I worked all day and the money was earned by my mom and we built our home in that way. Then at the age of sixteen I started fishing with my uncle…" (SG, ex-fisher).

Despite traditions, artisanal fishing has also been an economic refuge for families during the economic crises that have hit Argentina during the last decades, especially in 2001 and from 2014 to the current days. In this way, unemployed and/or less experienced men learned the job from other fishers, in order to provide economical support for their families. As some of our interviewees claimed: "I started fishing in 2001 (Argentine economic crisis) at twenty-one… why? Because I lost my job at a supermarket. And mi brother in law brought me here, to the fishing. He had a boat and he used to fish. And he said to me 'let us try, you might like it'. And honestly, I did. I did like it. And slowly I started to buy the materials; it was a great effort… But I like it if you don't like this, then it's useless. And slowly I learned to fish by watching and paying attention to the environment…" (PB, fisher).

The most commercialized species in the estuary have always been the "lenguado" (*Paralichthys orbignyanus*), the silverside (*Odontesthes argentinensis*), the striped weakfish (*Cynoscion guatucupa*), the whitemouth croaker (*Micropogonias furnieri*), and the narrownose smooth-hound (*Mustelus schmitti*), which constitute the varied fishery, and also shellfishes, like the Argentine shrimp and prawn (*Pleoticus muelleri* and *Artemesia longinaris*). The arts of fishing have changed from those described by Lopez Cazorla (2004) and have been simplified. Nowadays fishers only use gillnets and trammel nets for the different species and a larger net for shellfishes. Also, few fishers use a modify version of the shellfishes' net for *C. guatucupa* in order to avoid the possible damage caused to the nets by the South American sea lion (*Otaria flavescens*). In the Bahía Blanca Estuary, fishing activities are done in groups of four in good seasons for shellfishes or in groups of two, when the economic conditions are not suitable to afford more sailors. In the past, the job was done by a single person, the owner of the boat. Overall, there are 12 working hours: 6 h during flood tide and 6 h during ebb, especially on shellfish seasons. In the past, fishers could afford to work less hours, but nowadays, they spent almost 2 days in the sea because of the high prices of the fuel.

Presently, there are 72 families of artisanal fishers of the Bahía Blanca Estuary distributed in the coastal towns and villages of the estuary, especially in Villa del Mar (alternative port), Punta Alta (Puerto Rosales), Ingeniero White (Port of Ingeniero White or Puerto Piojo, as it is named by the fishers' collective), and General Daniel Cerri (Puerto Cuatreros) (Noceti 2018) (Fig. 18.2). One hundred fifty families were recorded in 2013, which are less than a half of the 420 that were recorded in the 40s of the twentieth century (Noceti 2013). After the fishers' conflict (Box 18.1) between 2011 and 2012, the number has decreased, and some of them are fishing illegally; therefore, the information for the main artisanal fishers' ports (Port of Ingeniero White, Puerto Rosales, Villa del Mar, and Puerto Cuatreros) is quite outdated or obsolete. To our knowledge, almost 50% of the fishers retired after the reconversion plan. The unknown sociodemographic profile has been reconstructed from the several interviews that we carried out and public information on the social media (Truchet et al. under review).

In accordance to our data, the gross profit per fishing boat is from 500 to 700 dollars per week, under good climatic conditions in the varied fishing. From 1000 to 1700 dollars is the estimated profit for a good week of shellfishes. The owner of the boat gets the bigger share (almost 40%), and the rest is equally distributed between sailors. However, most of the profit has to be used to cover the expensive costs of fuel, repairing nets, and maintaining the equipment. Under bad weather or fishing conditions, weekly profit may be as low as 14 dollars, and it does not cover the fuel costs. At this point, it is important to point out that good weather conditions in the Bahía Blanca Estuary are not common, due to cold temperatures in winter and strong wind storms during summers. Therefore, fishers have a quite low average income, considering the amount of days with no activity.

Active fishers' ages go from 25 to 50 years, while more than 50-year-old fishers are mostly retired because they sell their permissions. It is a risky job, and fishers retire younger than the usual age of 65 years for the Argentine men (Fig. 18.3). Gendered roles and activities are described in Box 18.2, but in general, women are secluded at home, raising their children and doing unpaid domestic labors. Some of

Fig. 18.2 Small-scale artisanal fishers' ports: (**a**) Port of Ingeniero White-Puerto Piojo, (**b**) Villa del Mar alternative port, and (**c**) Puerto Cuatreros (General Daniel Cerri). (Photos by Lic. Rocío M. Truchet)

them are involved only in postharvest activities, and a few of them sail but only in bad economic periods, where men cannot afford for external help. The formal education of active and retired fishers includes only primary to high school, and none of them have achieved superior education. However, some of their children, especially the women, did: "For me, sailing wasn't dangerous… I sailed with storms and winds. I had four daughters and I had to give feed them; if I didn't go fishing, it was a lost day. I fished to eat and to feed my family. With that, I could educate and raise my daughters… the older one studied psychology, another one is a teacher… and people don't like us or blame us for overfishing, when we use nets that do no harm for nature" (AB, ex-fisher).

The high taxes, fuel costs, the overfishing of large ships at El Rincón area, the pollution of the estuary, and the policies that have been taken in the last years, favoring big companies and the port consortium, make small-scale artisanal fisheries a collective in risk of disappearing. Our recent research (Truchet et al. under review) showed that artisanal fishers reached a catch of 500 tons in 2008, while in 2016 (after the fishers' conflict) reported catches were less than 100 tons. Actual data on the present catches are not available, but we suggest that it could reach less than 50

Fig. 18.3 (**a**) and (**b**) encounters with artisanal fishers of the Bahía Blanca Estuary, (**c**) and (**d**) artisanal fishers arriving to the Port of Ingeniero White (Puerto Piojo) after a working day where they fished *Micropogonias furnieri*, and (**e**) recreational fishers in Puerto Cuatreros. (Photos by Lic. Rocío M. Truchet)

tons. On the other hand, large fishing ships continue operating in the outer area of the estuary with much higher landings per year. In this scenario, small-scale artisanal fishers are in risk to be lost in a nearby future, along with their ancestral culture and knowledge.

Box 18.2 Gender roles

Globally, women and men in fisheries have different but complementary roles and activities, but at the same time, they are influenced by the socio-economic and cultural contexts of their region and country. According to FAO (2017 a, b), in most regions worldwide, fishing is carried out by men, especially off-shore the coastal zones, while women, in a smaller proportion, are secluded to coastal areas in small precarious boats or canoes with rows. The international literature places women in artisanal fisheries as key actresses in the local economies and in the food security and sovereignty. According to Harper et al. (2013), the roles of women are variable according to the regions: in most coastal countries of the Central and North Pacific Ocean, women contribute with 56% of the total captures in small-scale artisanal fisheries, with a positive impact in the small economies, and help to reverse poverty situations. However, in other continents like Africa and Latin America, only 47% of the jobs promoted by artisanal fisheries are occupied by women, being the post-harvest jobs (cleaning, cutting, packing, selling), the jobs that are allowed for women, while sailing is usually forbidden, because they are symbol of bad luck in the maritime tradition (Rodríguez 2009; Luna López 2011; Vasconcellos et al. 2011; Oliveira and Silva 2012; Álvarez et al. 2017; FAO 2017b; Rojas Herrera 2018).

Concerning gendered fishing activities in the Bahía Blanca Estuary and other nearby coastal areas (i.e., Pehuen-có and Monte Hermoso), London et al. (2017) described fishing activities as being related to men, whereas the women are limited to domestic work and to repairing nets. A recent study by Truchet et al. (2020a, b) disagrees, pointing out that some women assist men in the postharvest and help with fishing when families cannot afford to pay for external help of another male sailor. However, women are not considered "fishers" in the Bahía Blanca Estuary (nor according to themselves or males), only fishers' "assistants" or "women of fishermen"; they are still restricted to their homes for processing the final product and not paid equally for their work: "You won't find any women that knows how to knit her own nets, it's weird, that's a men thing. Except for my niece, I taught her so she could help me sometimes, and she knows something. Pablo, her husband, is a fisher and also knows how to knit nets, but no so much because this's a family business, and that's my job. I sold Pablo my permission and the boat he has used to be mine. Now I'm helping to repair some nets. Anyway, regarding women, I've never seen a women fishing, except for a woman I've met one day, she was a widow and used to fish here… Otherwise, it's uncommon" (MF, ex-fisher).

Moreover, in the maritime tradition of the southwestern Atlantic, a strong patriarchal society, women are also a symbol of bad luck for the number of catches. As it was explained by one of our informants: "It's uncommon to see women fishing… we have that expression 'don't board a woman nor kill a seagull!', it's bad luck for the job, you know…'" (AB, ex-fisher). On the other

(continued)

Box 18.2 (continued)

hand, fishermen claim that the job is only for those that "have more balls" and "face the sea," being men the only ones considered as "fishers": "The only ones that fishes are the one that has balls, the ones that has guts; it's not for everyone" (MF, ex-fisher).

In this context, we corroborated that the gender condition of women is a sub-alterity to the point in which they are not considered fishers, even when they are fully involved in the activity. Fishermen assume that the sea is his territory because they develop their ontologies, focused in values such as the prestige and the virility.

18.2.3 A Clash of Cosmologies: The Conservation Paradigms

In order to study the different forms in which conservation was considered in many communities, often with different cosmovisions, Pálsson (2003) established three anthropological categories based on the contributions of human ecology and social sciences: *orientalism*, *paternalism*, and *communalism* (Fig. 18.4). According to this author, each paradigm represents a particular stance with respect to human-environment or a particular way in which humans relate with nonhuman beings. The communalism differs from others because it rejects the modern imposed radical separation of nature from society, arguing that a constant dialogue has always flowed between them (Pálsson 2003).

The orientalism and paternalism take for granted the division between nature and society and emphasize the contrast between domination, exploitation, and protection to the environment. This is to say, that one tends to see nature as "something" that is an "object" of human domination and exploitation, while the other one believes that nature should always be protected by humans. The *orientalism* suggests a negative relation, reciprocity, and break in human-nature relation, with a misbalance between them and a clear colonial regime. However, this paradigm not only belongs to the view of industrial and extractive companies and policy makers but also belongs to a large group of scientists that are unaffected by ethical considerations and are only present as mere analysts of the material world, without considering the human compound (Pálsson 2003; Carman and Gónzalez Carman 2016).

The *paternalism* assumes that the conservation of nature is a human responsibility, and it is characterized by protectionism instead of exploitation (Pálsson 2003). However, though they propose radical thoughts, in both paradigms, humans are the "masters of nature" just for believing they can exploit it or by thinking that they should always protect it. Thus, there are clear evidences of modern traces in the speeches of both paradigms.

Finally, the *communalism* suggests a reciprocity in human-nonhuman/environmental relations, with notions of participation and dialogue and where humans are not considered the masters of nature. This paradigm rejects any division between humans and the environment in ecological studies, an assumption that nowadays is widely recognized but still poorly studied and therefore poorly understood. Pálsson

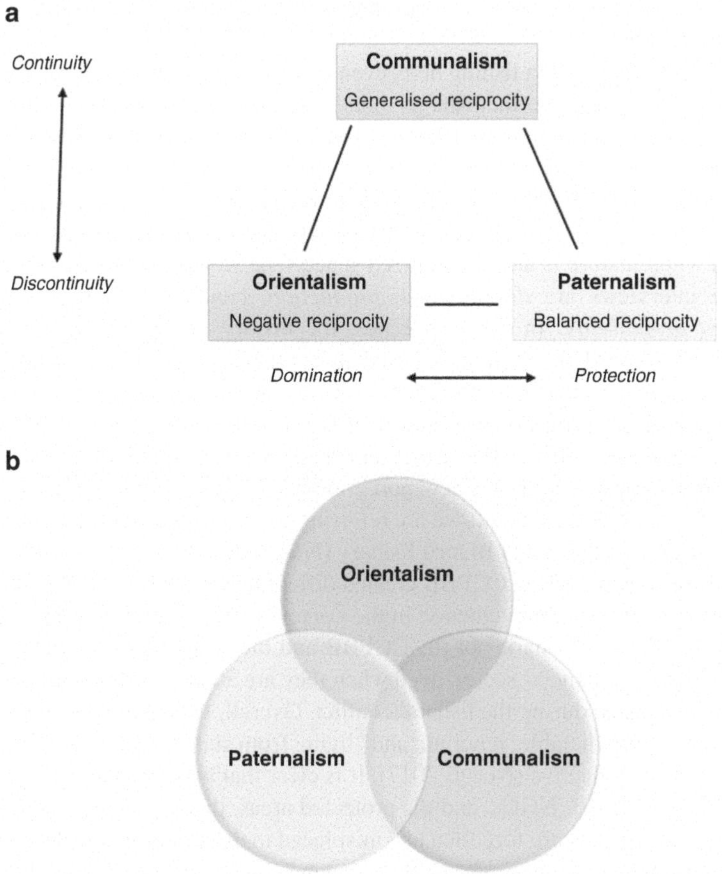

Fig. 18.4 (**a**) Types of human-environment relations (conservation paradigms) adapted from Pálsson (2003): orientalism, paternalism, and communalism. (**b**) A co-existence of paradigms in the speeches of small-scale artisanal fishers. Adapted from González de Carman and Carman (2018) for the Bahía Blanca Estuary

(2003) attributes this paradigm to indigenous and some traditional fishers and hunters' societies that have no fundamental distinction between nature and society.

At this point, our main question is if these paradigms are present in the arenas where the socio-environmental conflicts take place in the Bahía Blanca Estuary. For instance, we interviewed different actors like fishers, employees of the CGPBB, the petrochemical complex, and the protected areas of the Bahía Blanca Estuary. We found several dissimilarities between the speeches of the CGPBB, the petrochemical complex people, and some dialogues we carried out with workers of the nongovernmental organizations (NGOs) and protected areas (Noceti 2017; Truchet et al. in review): "The development of this region (the southwest of Buenos Aires province) depends on the port and that the society understands the importance of the foundation of the city-port of Bahía Blanca, which we used to call the 'third foundation of

Bahía Blanca'… fisheries here are marginal; it is expected that they disappear" (VC, CGPBB worker).

"They (fishers) insist in fishing here, even when the consortium bought their sailing permissions and give them larger boats with better motors, so they can fish outside of the estuary… they are a bunch of thieves" (EE, worker of a petrochemical company).

In words of Pálsson (2003), it is possible to verify an *orientalist* speech regarding the relations with the environment, where society and nature are extremely dissociated and where nature is also a colonized space. The words that overflow speeches of these interviews are *development*, *production*, *conquer of new spaces*, and *increased profitability*. In these speeches, the history of the port-petrochemical complex is founded in ontologies based in capitalists' values and ethics where nature is a simple scenario, where some actors are erased to achieve, first, the objectives and ideals of progressivism and then the development proposed by an elite congregated in the CGPBB. The strikes and protests are criminalized, and an enemy is built in the construction of a new port complex. This enemy is local, and it is a threat to the local development: we are referring to the collective of the small-scale artisanal fishers of the Bahía Blanca Estuary (Noceti 2018).

The government and the CGPBB created different devices for social control and fragmentation of social movements. In the context of these interviewees, it can be inferred that the functionaries of the CGPBB and the petrochemical complex tend to homogenize the fishers' sector, even when they are already fragmented by previous policies created during the fishers' conflict. Overall, fishers are considered abusive, violent, unpredictable, ignorant, and "living from state social plans," instead of a hard-working collective (Noceti 2017). It is clear that between the petrochemical complex, the CGPBB, NGOs, and the protected areas, there is a close relationship, where artisanal fishers are forgotten and misplaced in the arenas where decisions are taken regarding conservation strategies, as it can be inferred by their speeches: "Our company delivers a thousand dollars in cultural projects and also in conservation projects…several environmental NGOs asked us for financing for some promotional activity, and we're ok with it; we offer money to them, to FRAAM (a NGO) to be more specific. These people are really wise about the ecology of this area" (MC, worker of a petrochemical company).

"Fisheries here are a joke, there's no fishing here, fisheries are the ones from Mar del Plata, for example. But, here? They've never existed… We can't compare the magnitude of the money incomes by artisanal fisheries with what the port and the petrochemical activities delivers" (EE, worker of a petrochemical company).

On the other hand, in the speeches of the NGOs and the workers of the protected areas of the Bahía Blanca Estuary, there is a clear *paternalist* paradigm: humans are not part of the nature, and the role of humans is to protect a form of nature where there is no place for other human beings and cultures: "We (FRAAM, the NGO) constantly do actions for nature, which is constantly attacked by men and their different actions, like pollution due to untreated sewage waters and the actions of the dredge. During all these years, we've done several campaigns to raise awareness of the habitat and nesting of the Olrog's gull (*Larus atlanticus*), the American oystercatcher (*Haematopus palliatus*). Our last campaign was with the help of the Tamar Project for the sea turtles" (VM, NGO FRAAM).

Besides having a paternalist-environmental view, the NGOs and the workers of the protected areas of the Bahía Blanca Estuary want to reconvert the arts of fishing, and they do not recognize the intrinsic relation that fishers have with the environment, denying their culture and education, which might not be formal but come from several generations and more than 100 years of history. There is also a misinformation regarding the arts of fishing, because fishers elaborate manually their nets to avoid catching juveniles. It has also been proven that recreational and industrial fishing has a larger impact on the bycatch of juveniles of certain species like the narrownose smooth-hound (*Mustelus schmitti*) (Colautti et al. 2010; Llompart et al. 2017): "Our main goal is to transform some of the arts of fishing… I mean, fishers don't get how dangerous they're… they tend to catch juveniles with consequences of the reproduction of the species. Sometimes they understand it; sometimes they don't. It's complicated to work with them. On the other side, it's easier to work with the recreate and sport fishers because they have another education…" (MS, worker of a protected area).

Several conversations and dialogues carried out with fishers lead us to infer a coexistence of paradigms in the human-environmental relations. The *occidentalism* appears in the case of the profits they can obtain from shellfishes (*Pleoticus muelleri* and *Artemesia longinaris*) considering that they are "fool animals" and "easy to fish." There is a clear separation between these species and the artisanal fishers, who have an extractive behavior toward these animals: "The shrimp and prawn are silly animals, easy to catch, there isn't much to say about them. They came and you took them, and if you catch them, they're good profit. I remember that in the Easter of 2007; I could buy my last boat…you see? That's the profit we can obtain" (DG, ex-fisher).

Regarding the South American sea lion (*Otaria flavescens*), there is a coexistence of two paradigms: the *orientalism* and *communalism*. The *orientalism* because some fishers see these animals as a threat that they have the right to kill, in order to avoid losing their fishing and profits: "The worst enemy of the fishers is the South American sea lion (*O. flavescens*); they destroy our nets. That's a day of lost work. They ruin the fishing and you lose the fish…If we can kill them, we do it. But the thing is that we can't; I mean, they're really smart" (DG, fisher).

But also, the *communalism* is present, since they believe they share the same environment with sea lions and have to respect them: "Those who shot the South American sea lion (*Otaria flavescens*) or kill them with guns, they aren't real fisher; let's say it's a person that only fishes" (MD, ex-fisher).

"Being a fisher implies to respect the animals. If I'm already in the environment of the animal (the South American sea lion), why would I kill it? I'm already doing damage by taking its food" (MF, ex-fisher).

The *communalism* is also preset in speeches where fishers claim that the best fishing is the sole, because it is risky and they have to prove themselves as males. Thus fishers get into the mud, though profits are lower than those obtained with shellfishes. The virility of fishers of the Bahía Blanca Estuary is present along all of

Fig. 18.5 An artisanal fisher from Villa del Mar repairing a shellfish net that was manually knitted. (Photos by Lic. Rocío M. Truchet)

their speeches about loneliness. Virility is proved by getting into the sea in order to face any kind of risks including the situation of no fishing. Fishing means uncertainties, abilities, wisdom, courage, and boldness to cope with the inclemency of weather. It also means "freedom and no owner," living by themselves and for themselves. In such a way of life, women just wait on the coast, and they are not included in the fishing mud.

The *paternalism* is present along with the *communalism* in the dialogues about the net building (Fig. 18.5). Fishers build their net based on their wide knowledge about the geomorphology of the estuary, the species, and their life cycles: "I have nets for every species… we try to take care of the animals. We know their reproductive cycles, where we can find them, which sediment, which channel, we have references and take notes. We know where to fish. I do admit that we, the fishers, sometimes overexploit our resources. But nothing in comparison with those big boats out there (in reference to El Rincón area). We try to avoid overfishing by using these nets that are made and designed only for this estuary and the species we fish" (DG, fisher).

Finally, based on their speeches, it was possible to identify that they feel excluded by the conservationist, after the actions are taken in order protect the estuary, and on the other side, they consider that the petrochemical complex is the main cause of the ecosystem decline, in complicity with the CGPBB. Fishers do not see these actors as part of the estuary, and a great distance can be inferred between these actors in one side and, on the other side, fishers and the estuary: "Everything it's ok with the workers of the protected areas, but I want them far... When an endangered animal appears, we always call the people from FRAAM to take care of them, but they come and give us an endless speech of conservation. Anyway, they live a different reality in comparison to us. Unlike us, they have the food guaranteed every day" (DG, fisher).

"Well... they're good people, but they know everything, you know? Regarding the sea, turtles and the shellfishes they're kind of annoying, but we have to let them speak, they like to talk a lot, like a burrowing parakeet (laughs, in reference to the noise emitted by *Cyanoliseus patagonus*). Anyway, they're good people, but too idealistic... I don't know, they live in their own world, they're from a different social class and they understand too little about fishing" (LF, fisher).

"They're a bunch of vultures, where they can get benefits, they do it. They don't care about anything in order to fill their pockets with money. They don't care if the ria dies, if we sail in rotten mud. Nothing... If you don't speak Chinese or English, these guys won't sit and dialogue with you" (MF, ex-fisher in reference to the CGPBB).

18.3 Use of Fishers' Ecological Knowledge (FEK) in the Conservation of Coastal Systems: The Bahía Blanca Estuary Case

Over the past decades, many ecological studies were built using fishers' knowledge, but only at the end of the 1990s that fishers' knowledge was taking seriously in scientific studies. A study by Johannes et al. (2000) reveals several cases in which fishers' knowledge reflected doubt and was discarded by the scientific community, what led them to arrive to some uncertain and wrong conclusions about fishes and marine mammals' biology, ethology, and ecology. Therefore, the use of fishers' ecological knowledge (hereafter, FEK) in science is recent and still discussed among the most rational and modern sectors of science.

Leaving those sectors aside, FEK is now widely used in ecological studies in the North Hemisphere (Johannes 1993; Neis 1998; Neis et al. 1999; Zukowski et al. 2011; Ambrose et al. 2014; Anbleyth-Evans 2018; Anbleyth-Evans and Lacy 2019, among others) and has proven to be a priceless management tool in the South, especially in Brazil (Leite and Gasalla 2013; Silvano and Begossi 2016; Messias et al. 2019; Thykjaer et al. 2019, others).

In the last decades, marine environments started to show "symptoms" of the anthropic activities (i.e., dredging, oil spills, industrial development, tourism, bigger urban settlements with untreated sewage discharges, agricultural and cattle activities, among others) (Marcovecchio et al. 2008); eutrophication process; the inputs of toxics substances like PCB, PAHs, OC trace metals, and other emerging concern pollutants (Arias et al. 2010; Spetter et al. 2015; Buzzi and Marcovecchio 2018; Fernández Severini et al. 2019); and the acidification of the water and the increasing CO_2 emissions. Degradation affected trophic food webs and worldwide fisheries (Allison et al. 2009; López Abbate et al. 2017), and the complexity of these environmental problems stresses more the need to develop new ways of management and conservation of the marine ecosystems that combine scientific and local knowledge, specifically FEK.

Unlike the Brazilian coastal areas, studies exploring the possible application of FEK in conservation are few in Argentina, and they are still in development, especially in the Samborombón Bay (Carman and Gónzalez Carman 2016; González Carman and Carman 2018) and more recently in the Bahía Blanca Estuary (Truchet et al. in review). In the Samborombón Bay, FEK was used along with conservationists and NOGs to help in reducing the bycatch of endangered animals (*Caretta caretta*, *Chelonia mydas*, *Dermochelys coriacea*, *Pontoporia blainvillei*). Reconversion involved the implementation of a fishing device using hooks. It was not fully successful because fish biomass was lower with hooks than with nets, and hence, the profits were lower. The importance of these studies was that fishers demonstrated to have a wide ecological knowledge about the species and how to help endangered animals in case of possible bycatch.

On the other hand, as we stated before, in the Bahía Blanca Estuary, the studies are incipient. The characteristics of fisheries have been barely described (Noceti 2017; Truchet et al. 2020a, b), though fishers' ecological knowledge has been widely used in biological samplings, scientific studies, and doctoral thesis (Lopez Cazorla 1987; Lopez Cazorla 2004). Recently, demonstrated that fishers' perceptions and knowledge toward the Bahía Blanca Estuary as a human-impacted sacrificial zone are in accordance to several scientific studies (Marcovecchio et al. 2008; Arias et al. 2010), but there are some gaps in knowledge that could be fulfilled with fishers' experiences. There are some examples that reenforce what we stated before, like the case of the flat fish, the "lenguado" (*Paralichthys orbygnatus*), the mussel (*Brachidontes rodriguezii*), and the Pacific oyster (*Magallana gigas*). In the case of *P. orbygnatus*, the ethnobiological and anthropological research with fishers led us to think that they present malformations due to pollution: in this case, fishers recognize the healthy anatomical and morphometric body conditions of fishes: "I've observed soles with spots and protuberances as if they were tumors… I don't know if it's because of petroleum as many fishers believe. I think it's due to all the pollutants and the sewage water without treatment from Rosales that are thrown into the estuary" (DG, fisher).

Flat fishes' malformations have been ignored in the scientific literature until a study by Castellini et al. (2018); however these authors indicate that the abnormalities are due to natural constrains during the ontogeny and do not mention the possible effects of pollution on a highly impacted estuary. Nevertheless, fishers declare that they have not seen them in the past, before the industrial settlements and the untreated sewage plants. The same thing happened with the decrease in the populations of *B. rodriguezii*; though there are no studies regarding this problem, fishers strongly believe that constant dredging of the principal channel is a possible explanation to the decline of this species: "There was a mortality mass of the scorched mussel. I'm sure that was due to the dredging because the mud buried them and they couldn't breathe through their siphons" (DG, fisher). Recent studies on this species by Buzzi and Marcovecchio (2018) recorded medium levels of some toxic trace metals and PAHs that could be harmful for human consumption, but these studies do not give a perspective on why this species tended to decrease its populations in the estuary as fishers stated. In all these cases, where the scientific and fishers' observations might not agree, and when observations diverge, they should both be re-examined so that we can understand our ecosystem better (Johannes et al. 2000).

FEK and local knowledge have proven to be also a valuable tool to understand long-term modifications in coastal areas and the distribution and changes of the species pools (Sievanen 2014; Gonçalves et al. 2019; Thykjaer et al. 2019), because fishers have a sense of observation of the entire ecosystem since they navigate more than 12 h daily over most of their lives. This information is also useful to detect alien species and their possible ecological damage: "Decades ago, that grass (in reference to *Spartina alterniflora*) wasn't observed here in Villa del Mar, it's not native, people from FRAAM (the NGO) say it is, but I think it's not. I believe it came from the great cargo ships and the sediments removed from dredging the principal channel that was thrown here. Now you can see it everywhere, and we don't know what to do with it" (MD, ex fisher). A parallel was detected here because a study by Bortolus et al. (2015) demonstrated that *S. alterniflora* is a species original from North America, while the older fishers always knew it was an alien species that they did not observe in the estuary. It is also remarkable that through observation, fishers know that *S. alterniflora* is an alien species, while conservationists still deny it – probably through misinformation – and want to protect it, instead of trying to find different managing strategies to mitigate the possible effects on the ecosystem of invasive species.

Our recent unpublished data (Truchet et al. in review) also enhances the value of FEK in long-term studies for the Bahía Blanca Estuary, a scenario that is under the global warming change and industrial development (Arias et al. 2013; Marcovecchio et al. 2013; López Abbate et al. 2017). We found that the fishers have a wide knowledge about the diet, migration, and reproduction of fishes as well as the geomorphology and long-term climatic condition of the estuary. Nevertheless, their knowledge is dismissed in management programs, especially in the case of the industrial and port companies that see them as "annoying" and do not embrace their culture as they do with NGOs and scientists who represent an expert and imperialist rationalist knowledge. FEK has demonstrated to be a valuable tool that could change the modern epistemological and technocratic ways of management in human-impacted estuarine systems of developing countries, such as the Bahía Blanca Estuary.

18.4 Future Perspectives and Recommendations

Nowadays, thousands of communities live in coastal areas around the world. Daily, most of them cope with the consequences of degradation, pollution, overexploitation, and overcapitalization of natural resources. These socio-environmental conflicts are an invitation to re-think the ways in which conservation and management of estuarine and coastal areas have been developed. These problems are complex and have not been solved nor achieved through reductionist, disciplinary, and neutral knowledge models of science, where emotion and common sense are dissociated. This disassociation has deprived scientists of the valuable local knowledge acquired through peoples' emotions and experiences based on an intimate relationship with the sea.

The new ways of management and resolution of environmental conflicts in estuarine and coastal systems need to be solved with more tangible and participative strategies conceived from and for the communities that are attached to the environment they are part of. Thus, it is fundamental to generate critical research for management programs that integrate the human compound that is part of the nature and promote the collective creation of knowledge, offering not only a sustainability economy but also ecological and human justice for the most vulnerable sectors.

This is where small-scale artisanal fishers of the Bahía Blanca Estuary play a role, because their livelihoods are in risk to be lost due to neoliberal policies that favor big foreign companies, the port consortium, and protected areas that are the last link for denying the identities connected with the sea. Also, as scientist, we must recognize that rationalist ecological management did not consider their voices for the conservation of this ecosystem. Therefore, these subjects and their cultures live in a vulnerable scenario, where they do not take place in the arenas where management decisions are taken. Now, our main goal lies in not taken for granted what the different cultures inhabiting the estuary think; moreover, we need to work side by side in order to create dialogists and participative science as well as new decolonial paradigms in the management and conservation of this sacrificial estuarine territory.

Acknowledgments Authors would like to thank to all the fishers of the Bahía Blanca Estuary and their families and to Lic. Rocío M. Truchet (IHuCSo-Litoral, CONICET-UNL) for the photos of the fishers and ports of the Bahía Blanca Estuary.

References

Allison EH, Perry AL, Badjeck M-C, Neil Adger W, Brown K, Conway KD, Halls AS, Pilling GM, Reynolds JD, Andrew NL, Dulvy NK (2009) Vulnerability of national economies to the impacts of climate change on fisheries. Fish Fish 10(2):173–196. https://doi.org/10.1111/j.1467-2979.2008.00310

Álvarez MC, Ruiz G, Navia D, Cortes C (2017) La visualización femenina en la pesca artesanal: transformaciones culturales en el sur de Chile. Polis, Revista Latinoamericana 16(46):175–191

Ambrose WG, Clough LM, Johnson JC, Greenacre M, Griffith DC, Carroll ML, Whiting A (2014) Interpreting environmental change in coastal Alaska using traditional and scientific ecological knowledge. Front Mar Sci 1:680. https://doi.org/10.3389/fmars.2014.00040
Anbleyth-Evans J (2018) Aggregate dredging impacts in South East England: improving ecological health by integrating fisher ecological knowledge with scientific research. Mar Pollut Bull 55(9):34–64. https://doi.org/10.1016/j.marpolbul.2018.06.051
Anbleyth-Evans J, Lacy SN (2019) Feedback between fisher local ecological knowledge and scientific epistemologies in England: building bridges for biodiversity conservation. Mar Stud 18:189–203. https://doi.org/10.1007/s40152-019-00136-3
Arias AH, Fernández Severini MD, Delucchi F, Freije RH, Marcovecchio JE (2010) Persistent pollutants monitoring in a South Atlantic coastal environment. Case study: the Bahía Blanca Estuary. In: Ortiz AC, Griffin NB (eds) Pollution monitoring. Nova Science Publishers, New York, pp 1–30
Arias AH, Piccolo MC, Spetter CV, Freije H, Marcovecchio JE (2013) Lessons from multi decadal oceanographic monitoring at an estuarine ecosystem in Argentina. I J Environ Res 6(1):219–234. https://doi.org/10.22059/ijer.2011.488
Bortolus A, Carlton JT, Schwindt E (2015) Reimagining South American coasts: unveiling the hidden invasion history of an iconic ecological engineer. Div Distrib 21:1267–1283. https://doi.org/10.1111/ddi.12377
Buzzi NS, Marcovecchio JE (2018) Heavy metal concentrations in sediments and in mussels from Argentinean coastal environments, South America. Environ Earth Sci 77:321. https://doi.org/10.1007/s12665-018-7496-1
Carman M (2011) Las trampas de la Naturaleza. Medio ambiente y segregación. Fondo de Cultura Económica, Buenos Aires, p 285
Carman M, Gónzalez Carman V (2016) La fragilidad de las especies: tensiones entre biólogos y pescadores artesanales en torno a la conservación marina. Entnográfica 20:411–438. Available at: https://journals.openedition.org/etnografica/4333
Castellini DL, González-Castro M, Lopez Cazorla A, Díaz De Astarloa JM (2018) Abnormalities in flatfishes of the South-West Atlantic Ocean. J Fish Biol 92:1225–1234. https://doi.org/10.1111/jfb.13578
Charles AT (1992) Fishery conflicts. Mar Policy 16(5):379–393. https://doi.org/10.1016/0308-597x(92)90006-b
Cheung WWL, Lam VWY, Sarmiento JL, Kearney K, Watson R, Zeller D, Pauly D (2010) Large-scale redistribution of maximum fisheries catch potential in the global ocean under climate change. Glob Chang Biol 16(1):24–35. https://doi.org/10.1111/j.1365-2486.2009.01995.x
Colautti D, Baigun C, Lopez Cazorla A, Llompart F, Molina JM, Suquele P, Calvo S (2010) Population biology and fishery characteristics of the smooth-hound *Mustelus schmitti* in Anegada Bay, Argentina. Fish Res 106(3):351–357. https://doi.org/10.1016/j.fishres.2010.09.004
Coulthard S, Johnson D, McGregor JA (2011) Poverty, sustainability and human wellbeing: a social wellbeing approach to the global fisheries crisis. Glob Environ Chang 21(2):453–463. https://doi.org/10.1016/j.gloenvcha.2011.01.003
De la Cadena M, Blaser M (2018) A world of many worlds. Duke University Press, Durham, p 224
Descola P (2003) Constructing natures: symbolic ecology and social practice. In: Descola P, Pálsson G (eds) Nature and society. Anthropological perspectives. Taylor and Francis, New York, pp 82–102
Descola P (2005) Par-delà la nature et culture. Bibliothèque des sciences humaines, Gallimard, Paris, p 624
Descola P, Pálsson G (2003) Introduction. In: Descola P, Pálsson G (eds) Nature and society. Anthropological perspectives. Taylor and Francis, New York, pp 1–22
DuBois C, Zografos C (2012) Conflicts at sea between artisanal and industrial fishers: inter-sectoral interactions and dispute resolution in Senegal. Mar Policy 36(6):1211–1220. https://doi.org/10.1016/j.marpol.2012.03.007

Ellen RF (2003) The cognitive geometry of nature: a contextual approach. In: Descola P, Pálsson G (eds) Nature and society. Anthropological perspectives. Taylor and Francis, New York, pp 103–124
FAO (2017a) Promover la igualdad de género y el empoderamiento de las mujeres en la pesca y la acuicultura, p. 12. Available at: http://www.fao.org/3/a-i6623s.pdf
FAO (2017b) Towards gender-equitable small-scale fisheries governance and development – a handbook. In: Biswas N (Ed) Support of the implementation of the voluntary guidelines for securing sustainable small-scale fisheries in the context of food security and poverty eradication, Rome, Italy, p. 154. Available at: http://www.fao.org/3/a-i7419e.pdf
Fernández Severini MD, Villagrán DM, Buzzi NS, Chatelain Sartor G (2019) Microplastics in oysters (*Crassostrea gigas*) and water at the Bahía Blanca Estuary (Southwestern Atlantic): an emerging issue of global concern. Reg Stud Mar Sci 32:100829. https://doi.org/10.1016/j.rsma.2019.100829
García-Quijano CG, Valdés-Pizzini M (2015) Ecosystem-based knowledge and reasoning in tropical, multispecies, small-scale fishers' LEK: what can fishers LEK contribute to coastal ecological science and management? In: Fischer J, Jorgensen J, Josupeit H, Kalikoski D, Lucas C (eds) Fishers' knowledge and the ecosystem approach to fisheries: applications, experiences and lessons in Latin America. Publisher: FAO Fisheries and Aquaculture Technical Paper, Rome, p 591
Giarraca N, Teubal M (2013) Las actividades extractivas en la Argentina. In: Giarraca N, Teubal M (eds) Actividades extractivas en expansión. Reprimarización de la economía en la Argentina? Antropofagia, Buenos Aires, pp 19–44
Girado A (2012) Resistencias y conflictos socioambientales en Tandil. Un estudio de caso Revista Sociedad y Equidad 4. http://www.sye.uchile.cl/index.php/RSE/article/view/20930/22123
Gonçalves PH, Rego AE, Medeiros P (2019) "There was a virgin forest here; it was all woods": local perceptions of landscape changes in Northeastern Brazil. Ethnobio Conserv 8:4. https://doi.org/10.15451/ec2019-01-8.04-1-17
González Carman V, Carman M (2018) A coexistence of paradigms: understanding human–environmental relations of fishers involved in the bycatch of threatened marine species. Conserv Soc 16(2):205–216. https://doi.org/10.4103/cs.cs_17_45
Harper S, Dirk Z, Hauzer M, Pauly D, Sumaila U (2013) Women and fisheries: contribution to food security and local economies. Mar Policy 39:56–63. https://doi.org/10.1016/j.marpol.2012.10.018
Heredia Chaz E (2014) De la Responsabilidad a la Contaminación Social Empresaria: la ingeniería social del Polo Petroquímico de Bahía Blanca. Degree in History Thesis, Universidad Nacional del Sur, Available at: http://repositoriodigital.uns.edu.ar/bitstream/123456789/528/1/Heredia%20Chaz,%20Emilce.%20Tesina.pdf
Ingold T (2000) The perception of the environment. Essays on livelihood, dwelling and skill. Routledge, New York, p 465
Ingold T (2003) The optimal forager and economic man. In: Descola P, Pálsson G (eds) Nature and society. Anthropological perspectives. Taylor and Francis, New York, pp 25–44
Johannes RE (1993) Integrating traditional ecological knowledge and management with environmental impact assessment. In: Inglish JT (ed) Traditional ecological knowledge: concepts and cases. International Development Research Centre, Ottawa, pp 33–40
Johannes RE, Freeman MMR, Hamilton RJ (2000) Ignore fishers' knowledge and miss the boat. Fish Fish 1:257–271
Johnson DS (2018) The values of small-scale fisheries. In: Johnson DS, Acott TG, Stacey N, Urquhart J (eds) Social wellbeing and the values of small-scale fisheries. MARE Publication Series, Springer, Cham, pp 1–22
Latour B (2004) Politics of nature. How to bring the sciences into democracy? Harvard University Press, Cambridge, MA/London, p 320
Leite MCF, Gasalla MA (2013) A method for assessing fishers' ecological knowledge as a practical tool for ecosystem-based fisheries management: seeking consensus in Southeastern Brazil. Fish Res 145:43–53. https://doi.org/10.1016/j.fishres.2013.02.013

Llompart FM, Colautti DC, Baigún CRM (2017) Conciliating artisanal and recreational fisheries in Anegada Bay, Argentina. Fish Res 190:140–149. https://doi.org/10.1016/j.fishres.2017.01.011

London S, Rojas ML, Ibáñez MM, Scordo F, Huamantinco Cisneros MA, Bustos ML, Perillo GME, Piccolo MC (2017) Characterization of an artisanal fishery in Argentina using the social-ecological systems frameworks. I J Commons 11:11–69. https://doi.org/10.18352/ijc.534

López Abbate MC, Molinero JC, Guinder VA, Perillo GME, Sommer U, Spetter CV, Marcovecchio JE (2017) Time-varying environmental control of phytoplankton in a changing estuarine system. Sci Total Environ 609:1390–1400. https://doi.org/10.1016/j.scitotenv.2017.08.002

Lopez Cazorla A (1987) Contribución al conocimiento de la ictofauna marina del área de Bahía Blanca. Doctoral Thesis, Universidad Nacional de la Plata, pp. 247

Lopez Cazorla A (2004) Peces del estuario de Bahía Blanca. In: Píccolo MC, Hoffmeyer MS (eds) El ecosistema del estuario de Bahía Blanca. Bahía Blanca, Instituto Argentino de Oceanografía (IADO, CONICET-UNS), Bahía Blanca, pp 191–220

Lugones B (1883) Una excursión al sur, el puerto de Bahía Blanca. La Nación, Buenos Aires 14:3753

Luna López L (2011) Estudio de caso sobre el rol de las mujeres en comunidades pesqueras del Golfo de Fonseca, Marcovia, Choluteca. Degree Thesis, Escuela Agrícola Panamericana, Zamorano, Honduras, p. 33. Available at: https://bdigital.zamorano.edu/bitstream/11036/408/1/IAD-2011-T016.pdf

Machado Aráoz H (2010) Imperialismo ecológico y racismo ambiental. Una lectura EcoBiopolítica sobre las Industrias extractivas en el Sur. Aportes Científicos desde Humanidades 8:1897–1911

Marcovecchio JE, Botté SE, Delucchi F, Arias AH, Fernández Severini MD, de Marco SG, Tombesi N, Andrade S, Ferrer L, Freije HR (2008) Pollution processes in Bahía Blanca estuarine environment. In: Neves R, Barreta J, Mateus M (eds) Perspectives on integrated coastal zone management in South America. IST Press, Lisbon, pp 303–316

Marcovecchio JE, Buzzi NS, Tartara MN, Spetter CV, Simonetti P (2013) Potential effects of climate changes on the marine ecosystem stability: assessment of the water quality. In: Arias AH, Menéndez C (eds) Marine ecology in a changing world. CRC Press, London, pp 1–43

McGregor A (2018) Foreword: social wellbeing and the values of small-scale fisheries. In: Johnson DS, Acott TG, Stacey N, Urquhart J (eds) Social wellbeing and the values of small-scale fisheries. MARE Publication Series, Springer, Cham

Merlinsky G (2013) La cuestión ambiental en la agenda pública. In: Merlinsky G (ed) Cartografías del conflicto ambiental en Argentina. Buenos Aires, CLACSO-CICCUS, pp 19–60

Messias MA, Alves TIP, Melo CM, Lima M, Rivera-Rebella C, Rodrigues DF, Madi RR (2019) Ethnoecology of Lutjanidae (snappers) in communities of artisanal fisheries in Northeast Brazil. Ocean Coast Manag 181:104866. https://doi.org/10.1016/j.ocecoaman.2019.104866

Nahuelhual L, Saavedra G, Blanco G, Wesselink E, Campos G, Vergara XV (2018) On super fishers and black capture: images of illegal fishing in artisanal fisheries of southern Chile. Mar Policy 95:36–45. https://doi.org/10.1016/j.marpol.2018.06.020

Nahuelhual L, Saavedra G, Mellado MA, Vergara XV, Vallejos T (2019) A social-ecological trap perspective to explain the emergence and persistence of illegal fishing in small-scale fisheries. Mari Stud. https://doi.org/10.1007/s40152-019-00154-1

Neis B (1998) Fishers' ecological knowledge and the identification and management of localized populations of Atlantic Cod (*Gadus morhua*) in Northeast Newfound-land. In: von Herbin H, Kornfield II, Tupper M, Wilson J (eds) The implications of localized fisheries stocks. NRAES, Ithaca, pp 217–238

Neis B, Schneider DC, Felt L, Haedrich RL, Fischer J, Hutchings JA (1999) Fisheries assessment: what can be learned from interviewing resource users? Can J Fish Aquat Sci 756:1949–1963

Noceti MB (2013) Tejedores de redes, pescadores y anécdotas que se desvanecen. Miradas antropológicas en torno a saberes y modo de vida de pescadores artesanales del sudoeste bonaerense Conference presente at: VII Jornadas Santiago Wallace de Investigación en Antropología Social. Available at: http://www.11caas.org/conf-cientifica/comunicacionesDocGetfile.php

Noceti MB (2017) ¿Reserva, puerto o ría? Conflicto socioambiental en el estuario de Bahía Blanca, Argentina. Etnografías Contemporáneas 3(4):64–91

Noceti MB (2018) Maritimidad vs Maritimización. Ontologías y Territorialidades en disputa en el sudoeste bonaerense, Argentina. Conference Presented at the Universidad Autónoma de Nuevo León. Instituto de Investigación en Ciencias Sociales, Monterrey, Mexico, p 46

Oliveira OM, Silva VL (2012) ¿Pescadoras o "mujeres de pescador"? La regulación jurídica de la pesca artesanal en Brasil y el problema del no reconocimiento del trabajo profesional de las pescadoras en Santa Catarina-Brasil. Investigación y género, inseparables en el presente y en el futuro. Conference presented at: IV Congreso Universitario Nacional "Investigación y Género", Sevilla, pp. 1327–1352

Pálsson G (2003) Human-environmental relations: orientalism, paternalism and communalism. In: Descola P, Pálsson G (eds) Nature and society. Anthropological perspectives. Taylor and Francis, New York, pp 63–81

Peron F, Rieucau J (1996) La maritimité aujourd'hui. Ed. L'Harmattan, Paris, p. 102

Quist LM (2019) Fishers' knowledge and scientific indeterminacy: contested oil impacts in Mexico's sacrifice zone. Mar Stud 18:65–76. https://doi.org/10.1007/s40152-018-0123-7

Rodríguez S (2009) "A veces las mujeres también entramos al mar". La pesca de camarón en Machalilla. In: Paulson S, Poats S, Argüello M (eds) Huellas de género en el mar, el parque y el páramo. EcoCiencia, Corporación Grupo Randi Randi y Abya Yala, Quito, Ecuador, pp 12–33

Rojas Herrera S (2018) Significaciones identitarias asignadas al mar desde las mujeres que habitan territorios marinos costeros en Costa Rica y Nicaragua. Doctoral Thesis, Universidad Nacional de Costa Rica, Costa Rica, p 216. Available at: https://www.repositorio.una.ac.cr/bitstream/handle/11056/14797/Rojas%20Herrera%2C%20Silvia%20Elena.pdf?sequence=1&isAllowed=y

Sack R (1986) Human territoriality: its theory and history. Cambridge University Press, Cambridge, p 272

Sievanen L (2014) How do small-scale fishers adapt to environmental variability? Lessons from Baja California, Sur, Mexico. Mar Stud 13:9. https://doi.org/10.1186/s40152-014-0009-2

Silvano RAM, Begossi A (2016) From ethnobiology to ecotoxicology: Fisher's knowledge on trophic levels as indicators of bioaccumulation in tropical marine and freshwater fishes. Ecosystems 19:1310–1324. https://doi.org/10.1007/s10021-016-0002-2

Spetter CV, Buzzi NS, Fernández EM, Cuadrado DG, Marcovecchio JE (2015) Assessment of the physicochemical conditions sediments in a polluted tidal flat colonized by microbial mats in Bahía Blanca Estuary (Argentina). Mar Pollut Bull 9(2):491–505. https://doi.org/10.1016/j.marpolbul.2014.10.008

Svampa M (2012) Pensar el desarrollo desde América Latina. Available at: http://maristellas-vampa.net/archivos/ensayo56.pdf

Thykjaer V, Rodrigues L, Haimovici M, Cardoso LG (2019) Long-term changes in fishery resources of an estuary in southwestern Atlantic according to local ecological knowledge. Fisheries Manag Ecol. https://doi.org/10.1111/fme.12398

Tola F (2011) Reflexiones dislocadas. Facultad de Filosofía y Letras – UBA, Asociación Civil Rumbo Sur, Buenos Aires, Argentina, p 208

Truchet DM (2018) De espaldas al mar: conflictividad socioecológica en el Estuario de Bahía Blanca (Buenos Aires, Argentina). La pesquería artesanal frente a las políticas neoextractivistas. Papeles del Centro de Investigaciones 8(19):9–26

Truchet DM, Noceti MB, Villagrán DM, Truchet RM (2020a) Vindicating fishers' ecological knowledge in a scenario of climate change and industrial development: the Bahía Blanca Estuary, Southwestern Atlantic Ocean. Mar Stud (in revision)

Truchet DM, Truchet RM, Noceti MB (2020b) Roles y relaciones de género en contextos de pesca artesanal: una reconstrucción a partir de las narrativas orales de varones pescadores del Estuario de Bahía Blanca. Revista de Estudios Sociales y Marítimos 13(16):64–86

Tuan Y-F (1974) Topophilia: a study of environmental perception, attitudes, and values. Prentice-Hall, Englewood Cliffs, p 260

Vasconcellos M, Diegues A, Kalikosli D (2011) Coastal fisheries of Brazil. In: Salas S, Chuepadgee R, Charles A, Seijoo JC (eds) Coastal fisheries of Latin America and the Cari Bahía Blanca Estuaryan, FAO Fisheries and Aquaculture Technical Paper. No. 544. FAO, Rome, pp 73–116

Viveiros de Castro E (1996) Os pronomes cosmológicos e o perspectivismo amerindio. Mana 2(2):115–144

Wagner L (2016) Problemas ambientales y conflicto social en Argentina. Movimientos socioambientales en Mendoza. La defensa del agua y el rechazo a la megaminería en los inicios del Siglo XXI. Doctoral Thesis, Universidad Nacional de Quilmes, Argentina. Available at: http://ridaa.unq.edu.ar/handle/20.500.11807/192

Yory C (2013) El desarrollo territorial integrado: una estrategia sustentable de construcción social del territorio, en el contexto de la globalización, a partir del concepto de Topofilia. Universidad Piloto de Colombia, Bogotá, p 388

Yory C (2015) La construcción social del hábitat como estrategia de integración social. Universidad Piloto de Colombia, Bogotá, p 434

Zukowski S, Curtis A, Watts RJ (2011) Using fisher local ecological knowledge to improve management: the Murray crayfish in Australia. Fish Res 110(1):120–127. https://doi.org/10.1016/j.fishres.2011.03.020

Chapter 19
Estuarine Environmental Monitoring Programs: Long-Term Studies

Jorge E. Marcovecchio, Sandra E. Botté, Silvia G. De Marco, Andrea Lopez Cazorla, Andrés H. Arias, Mónica Baldini, María Amelia Cubitto, Sandra M. Fiori, Ana L. Oliva, Noelia La Colla, Gabriela Blasina, Juan Manuel Molina, Pia Simonetti, Analía V. Serra, Vanesa L. Negrín, Ana C. Ronda, and Marcelo T. Pereyra

19.1 What Is an Environmental Monitoring Program?

A continuous observation and control system of measures and evaluations for a defined purpose is called "monitoring." This is an important tool within the impact assessment process and in any vigilance and control program (Pali and Swaans 2013; Valle Junior et al. 2015). There is currently a strong consensus that environmental monitoring is not an end-point in itself but an essential step in environmental management processes (Stelzenmüller et al. 2013). Taking into account the previously mentioned concepts, the importance of monitoring within different processes

J. E. Marcovecchio (✉)
Instituto Argentino de Oceanografía (IADO – CONICET/UNS). CCT CONICET Bahía Blanca, Bahía Blanca, Argentina

Universidad Tecnológica Nacional, Facultad Regional Bahía Blanca (UTN-BHI), Bahía Blanca, Argentina

Facultad de Ingeniería, Universidad FASTA (FI-UFASTA), Mar del Plata, Argentina

Academia Nacional de Ciencias Exactas, Físicas y Naturales (ANCEFN), Ciudad Autónoma de Buenos Aires, Argentina
e-mail: jorgemar@iado-conicet.gob.ar

S. E. Botté · A. L. Cazorla · S. M. Fiori · G. Blasina · J. M. Molina · V. L. Negrín
A. C. Ronda
Instituto Argentino de Oceanografía (IADO – CONICET/UNS). CCT CONICET Bahía Blanca, Bahía Blanca, Argentina

Departamento de Biología, Bioquímica y Farmacia, Universidad Nacional del Sur (DBByF-UNS), Bahía Blanca, Argentina

S. G. De Marco
Facultad de Ingeniería, Universidad FASTA (FI-UFASTA), Mar del Plata, Argentina

Departamento de Biología, Facultad de Cs.Exactas y Naturales, Universidad Nacional de Mar del Plata (UNMdP), Mar del Plata, Argentina

S. M. Fiori, P. D. Pratolongo (eds.), *The Bahía Blanca Estuary*,
https://doi.org/10.1007/978-3-030-66486-2_19

of human activity can be observed. In addition, and as it is rightly mentioned, it is a fundamental tool within all that development or procedure that is desired to be carried out in a controlled and safe way (Collins et al. 2012).

The monitoring programs consist of carrying out a permanent surveillance of a natural system, controlling the state and evolution of its variables, quantifying the variations that occur, and identifying the reasons that generated them. There are different work strategies, both spatial and temporal, and can be developed either exclusively based on field data or based on experimental data and even combining both (Lindenmayer and Likens 2010). When these programs are carried out for a long period of time (e.g., decades), the set of information they provide supports a solid basis on which it is possible to accurately characterize the structure and operation of the system under study (Biber 2013). As long as the continuity of monitoring is maintained, it can be argued that as the program has more seniority (therefore more information), it is easier to identify a significant anomaly or variation in any of the parameters studied, taking into account that the distribution of "normal" values will be very dense, and therefore those that move away from that model will be easily distinguishable (Gray and Shimshack 2011).

Another type of alternative work is the application of environmental specimen banks (ESBs) that can be very useful, since they allow the establishment of environmental monitoring networks (Snyder et al. 2013; Hsu and Zomer 2014) with the objective of early detection of changes in the parameters evaluated (i.e., increasing concentrations of several contaminants). This process constitutes a development of real-time monitoring (Viana et al. 2010; Fang et al. 2014), which results from enormous application in management and control processes. In this case, and after having developed a monitoring program for a period long enough to have the system's

A. H. Arias
Instituto Argentino de Oceanografía (IADO – CONICET/UNS). CCT CONICET Bahía Blanca, Bahía Blanca, Argentina

Departamento de Química, Universidad Nacional del Sur (DQ-UNS), Bahía Blanca, Argentina

M. Baldini
Departamento de Agronomía, Universidad Nacional del Sur (DA-UNS), Bahía Blanca, Argentina

M. A. Cubitto
Departamento de Biología, Bioquímica y Farmacia, Universidad Nacional del Sur (DBByF-UNS), Bahía Blanca, Argentina

A. L. Oliva · N. La Colla · P. Simonetti · A. V. Serra
Instituto Argentino de Oceanografía (IADO – CONICET/UNS). CCT CONICET Bahía Blanca, Bahía Blanca, Argentina

M. T. Pereyra
Departamento de Química, Universidad Nacional del Sur (DQ-UNS), Bahía Blanca, Argentina

operation characterized, copies of one or several key species of the studied environment are stored and used as a permanent reference to compare values of the evaluated parameter obtained at different times (Küster et al. 2015).

A particular study situation is the use of environmental indicators (e.g., bioindicators, geoindicators, biomarkers, etc.) that allow monitoring of the system through variations recorded in a species, type of sediment, molecules, etc. (Martínez-Haro et al. 2015). It is a very practical and dynamic method, but it requires a very deep and consolidated knowledge of the evaluated environment. The incorrect, ill-founded, or inconsistent application of these techniques can lead to false conclusions and subsequently to unnecessary or inappropriate management measures (Brewin et al. 2015).

In the particular case of water quality control methods, it is fair to say that this type of monitoring has often been seen as a practical exercise by those who initiate a program. The underlying philosophy of why monitoring should be done tends to be neglected. Wondering why, however, leads to an essential step in formulating the objectives for which the information is obtained. The choice of how to monitor is then limited, considering that only some data sampling and processing actions allow the objectives of the information to be achieved (Behmel et al. 2016). In fact, both the choice of data analysis methods and statistical methods and procedures on how to make observations should be made before sampling. In addition, the concept of water quality must be defined a priori in terms of indicators that can be observed and quantified (Hounslow 2018).

19.1.1 Long-Term Monitoring Programs

In the specialized bibliography, it is possible to find numerous definitions and conceptual frameworks about the long-term monitoring program (LMP) process, and in this sense and for the purposes of this review, we will consider the LMP as "repeated empirical measurements based on field work, collected in continuously, and subsequently analyzed for at least 10 years" (Goldsmith 2012).

In some viewpoints, the scientific value of monitoring tends to be underestimated, without knowing that the structuring of an LMP necessarily requires excellent scientific research work (Luthardt 2010; Wolf et al. 2013). In principle it should be noted that good science and therefore good LMPs begin with good questions. Some of the considerations that we must keep in mind to ask "good questions" are the following:

- Think critically.
- Build strong conceptual models that represent the functioning of the ecosystems studied (i.e., Grace et al. 2010).
- Evaluate real questions (de Oliveira Roque et al. 2018) on policies relevant to management (Voget-Kleschin 2013).

- Promote open dialogue between scientists and management leaders (Lawton 2007; Likens et al. 2009).
- Critically evaluate the design and opportunism of the proposed study.

In this regard, it is important to clarify that new visits (or revisits) to a site after a prolonged absence are not considered as long-term monitoring ones (i.e., Smith et al. 2007), as well as to eventual measurements discontinued over time.

Depending on its structure and design, we can differentiate three main types of monitoring programs:

1. *Passive or curiosity-driven monitoring*. This is the monitoring devoid of specific questions or underlying design of the study and with little or no purpose other than curiosity. Its usefulness is very limited to address environmental problems or to discover how the world works, because it is not driven by hypotheses and because it lacks management interventions or different experimental treatments that facilitate the scientific understanding of these things in response to ecosystems to natural or human disturbances.
2. *Mandatory or directed monitoring*. In this type of monitoring, environmental data is collected as stipulated by some requirement of government or political legislation (i.e., a directive on monitoring the climate or river flow, etc.). The usual thing is that they have enormous quality requirements and that they must follow strict protocols to ensure the validity of the information collected. The design of this type of monitoring does not attempt to identify or understand the mechanism that produces a change in an ecosystem or an entity but only to detect its existence.
3. *Question-driven monitoring*. This is monitoring guided by a conceptual model and a rigorous study design. The use of a conceptual model will typically result in a priori predictions that can be tested as part of program development. Often, this type of learning is characterized by strong contrasting management interventions (Johnson 2012). In statistical language, these types of studies could be called "studies with longitudinal interventions" (i.e., Lindenmayer et al. 2008). Therefore, these approaches can lead to generating a strong predictive capacity and allow the research team to raise new questions. Thus, a monitoring program can evolve as the key questions change. Predictive ability can be of great value to researchers, resource managers, and decision makers, in contrast to the simple trend lines obtained from many targeted surveillance programs (Possingham et al. 2012).

19.2 Structure of a Long-Term Monitoring Program

It is important to be clear that an LMP is not simply limited to a large collection of data and its corresponding storage, but that it must fulfill a sequential series of steps that constitute a monitoring cycle. The greater the number of monitoring cycles that the LMP has, the greater its efficiency, the easier the interpretation of the generated

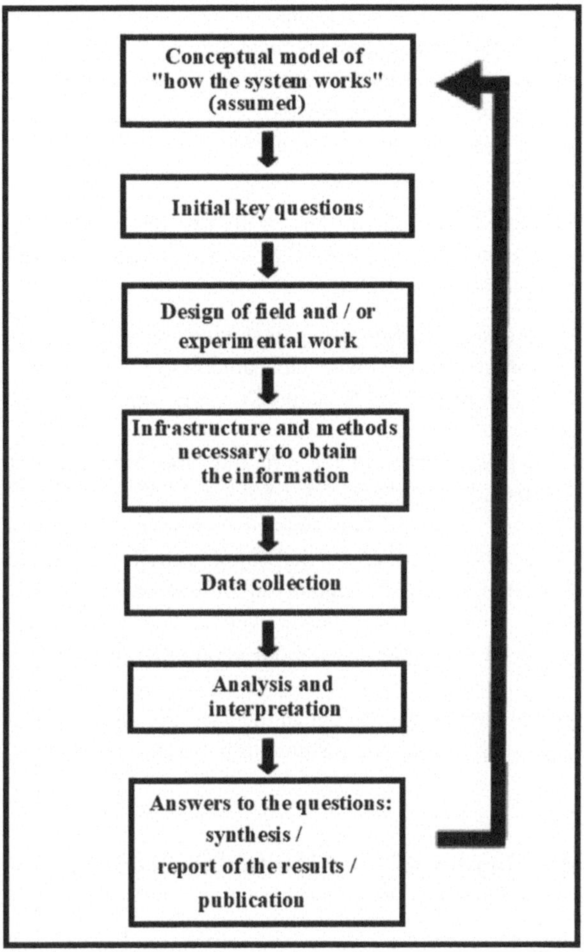

Fig. 19.1 General scheme of a long-term monitoring program

information will be, the anomalies will be identified with greater speed and certainty, the evaluation tasks will be simplified, and the obtained conclusions to be those that arrive will allow better management measures to be taken by the corresponding authorities (Smith et al. 2012; Converse and Royle 2012).

A tentative scheme of an LMP would be as follows: In Fig. 19.1, there are some properties of LMPs that distinguish them particularly and that frame many of their advantages. The main ones of these distinctive characteristics that deserve to be highlighted are:

- Based on good questions, which evolve
- Based on conceptual models
- Specified by active interlocutors

- Developed by well-led solid work teams
- Sustained by adequate and continuous financing
- Generators of frequent use of information
- Reservoirs of intense scientific productivity
- Qualified through appropriate validation processes of data quality and calibration of field and laboratory methods

When the long-term monitoring programs are executed according to the scheme mentioned above, the data generated and the reached conclusions are fundamentally valuable for many purposes, including:

- Document and provide benchmarks against which changes in conditions or endpoints can be evaluated (i.e., Mayne and Zapico-Goñi 2017).
- Evaluate ecological responses to natural or experimental disturbances (Seidl et al. 2016).
- Detect and quantify changes in the structure and function of the ecosystem (Wrona et al. 2016).
- Identify ecological "surprises" (unexpected events) (i.e., Zhan et al. 2006).
- Establish evidence-based principles that support environmental legislation (i.e., laws that control the levels of pollutants in air and water) (Davies and Mazurek 2014).
- Generate significant new questions about ecological dynamics (Persson et al. 2009).
- Provide empirical data to test ecological theory (McClure et al. 2016) and development models, such as numerical simulation models (Neves et al. 2008).

19.3 A Little History of the Monitoring Programs

Although for a long time scientific works were carried out basically in the field, with structures similar to monitoring programs, it can be considered as a pioneer case that carried out by Jack Pearce in 1967, when he completed a benthic research program in the marine basin from New York, with a view to determining the consequences of dredging materials and sewage sludge in that marine area for more than 2 years. The research objects for that initial phase were about 100 species of benthic and demersal organisms, a half-dozen toxic metal footprint, an indicator microorganism (*Escherichia coli*), and the levels of organic matter in the sediments affected by solid waste. In the early 1970s, reports were published with the data obtained, including distributions of trace metals in sediments, levels of bacteria in and around landfills, data on sick fish, and the main missing (or severely affected) benthic species, like some amphipods, in and around the discharge operations' centers (Pearce 1998).

At approximately the same time, oil spills produced in different parts of the planet (i.e., the English Channel, San Juan de Puerto Rico harbor, or in front of Santa Barbara, CA, among others) drove the development of monitoring programs to evaluate the effects and consequences on the affected natural systems. A few years later, scientists were reporting the arrival of agricultural pesticides to the Monterey Canyon Bay, based on activities developed in the great valleys of California. Others pointed to some ports as heavily polluted areas (i.e., London, in the UK or Boston, Seattle, and Washington in the USA). These studies, and many other reports developed in that period, significantly claimed long-term surveillance. Soon national and international workshops and study groups were convened to discuss the strategies for monitoring and evaluation of pollutants and outline models for the development of monitoring programs in both aquatic and terrestrial systems (US National Academy of Sciences 1975, 1977). By 1978, scientific concern focused on the preparation of long-term programs and the creation of specimen banks, as well as the standardization of procedures for the collection and handling of samples.

In the following decade, several national research and monitoring programs were carried out, and global evaluations began; among others, the United Nations Environment Program (UNEP) began to transfer the data and facilitated studies and reports on the state of the marine environment (GESAMP 1990).

These programs were evolving and being improved, and many efforts were mutating and focused on other previously ignored aspects, such as the measurement of ecological, genetic, pathological, and/or physiological values that allowed to identify the changes produced when coastal and marine life began to be impacted by pollutants and physical pressures. Many of the measurements used as tools were outlined in McIntyre and Pearce (1980), considering that they turned out to be more realistic indicators of habitat degradation than conventional measurement and the evaluation of the presence of toxins in biota, sediments, or water.

19.4 A Historical Turning Point: The Mussel Watch Program

The *Mussel Watch Program* was initially proposed as an alternative strategy to understand the quality of marine waters (Goldberg 1975). Bivalves had been recognized for their ability to concentrate four groups of pollutants from the waters within the system they inhabit: (i) transuranic elements; (ii) halogenated hydrocarbons (i.e., PCBs, DDTs); (iii) petroleum hydrocarbons; and (iv) heavy metals (Bryan 1973; Boyden 1974). These substances, mobilized by human activities, are introduced through air and/or water to the marine environment and, subsequently, can generate deleterious effects on human populations or marine life. The initial design of this program involved a little more than 100 sampling stations in coastal waters of the USA, analyzed quarterly and originally funded during the first 3 years (1976–1978) by the US Federal Environmental Protection Agency (US EPA) (Goldberg et al. 1978).

This concept of sentinel organism or *Mussel Surveillance* (Goldberg et al. 1978; Phillips 1980) is a method for assessing the current state of chemical contamination of coastal waters and identifying geographic areas of interest and new or renewed topics for scientific research.

The theoretical framework that allows the use of the common mussel (*Mytilus* sp.), several oyster species (i.e., *Crassostrea* sp., *Ostrea* sp.), and other bivalves as sentinel organisms includes the following premises (Goldberg et al. 1978; Phillips 1980; Farrington et al. 1983):

- Bivalves are cosmopolitan (widely distributed geographically). This feature minimizes the problems inherent in comparing the data between very different species, with different life histories and types of relationships within their habitat.
- They are sedentary and are therefore more useful than mobile species as integrators of chemical state of pollution in a given area.
- They can concentrate many chemicals with factors of 10^2–10^5 with respect to the contents of seawater in their habitat. This significantly facilitates the measurement of trace elements.
- When measuring the chemicals in bivalves, an evaluation of the biological availability of the compounds studied is being achieved.
- Unlike fish and crustaceans, bivalves have a very low or undetectable activity of enzyme systems that metabolize many xenobiotics (i.e., polycyclic aromatic hydrocarbons (PAHs) or polychlorinated biphenyls (PCBs), among others). This allows more precise assessments of the magnitude of xenobiotic contamination in the habitat of the bivalves studied.
- There are many local populations of bivalves that are relatively stable and large enough to be repeatedly sampled, providing data on short-term and long-term temporary changes in the concentrations of pollutants.
- They survive in pollution conditions that could often severely reduce or eliminate other species.
- They can be transplanted from anchor sites and successfully maintained in different coastal intertidal waters, even with very different conditions. This allows the identification of different toxic distribution processes between different areas, detoxification mechanisms, recovery times, etc.
- They are marine species that are used for human food in almost all cultures and ethnicities, which assign them high commercial value throughout the world. Therefore, the measurement of chemical contamination in these species is of interest to public health.

After that, Goldberg himself (1986) commented that another additional advantage of this work strategy is that the biological half-lives of contaminants in bivalves can last up to periods of months. There is, therefore, a spatial and temporal integration of pollutant levels in body tissues. Therefore, the use of bivalves offers advantages over seawater and sediments for the determination of the contaminant, where the integration period can vary from short (1 day or less) to long (usually 1 year or plus). This situation is very advantageous for the design of surveillance programs.

This excellent evaluation tool began to spread rapidly throughout the world, i.e., since 1979 a Mussel Watch Program has been carried out in bivalves of the French coastal zone: the National Network for the Observation of the Quality of the Marine

Environment (*Réseau National d'Observation*, RNO). Concentration levels of metals (Hg, Cd, Pb) and organic pollutants (PAHs, PCBs, DDT, DDD, and DDE) in mussel soft tissues and oysters sampled quarterly have been determined. The French Ministry of Environment has funded this monitoring program to determine levels of pollution in French coastal waters, using the mentioned organisms as quantitative indicator species. The planning and implementation have been coordinated by the French Research Institute for the Exploration of the Sea (IFREMER) (Claisse 1989). This program allowed to summarize the results of 10-year monitoring of mercury, cadmium, lead, polycyclic aromatic hydrocarbons (PAHs), polychlorinated biphenyls (PCBs), and dichloro-diphenyl-chloroethanes (p, p 'DDT, p, p′ DDD, p, p 'DDE) in the soft tissues of the mollusks studied: *Mytilus edulis*, *Mytilus galloprovincialis*, and *Crassostrea gigas*. The network consisted of 110 sampling stations along the entire length of the French coast, and samples were taken in February, May, August, and November of each year.

Despite the wide acceptance achieved by the concept of *Mussel Surveillance* (Goldberg et al. 1978), the use of sentinel organisms (mainly bivalve such as *Mytilus* sp.) for monitoring the concentration of pollutants in coastal areas, and as an indicator of its bioavailability, demonstrated through the establishment of such programs both nationally and internationally, an important problem was observed. Unfortunately, the most useful species, mussels and oysters, do not occur frequently in tropical waters, where pollution monitoring strategies were also required, and this was inconvenient for the implementation of global surveillance programs. To overcome this problem, a research program was carried out in the Todos os Santos Bay region (Bahia, Brazil) (Porte et al. 1990), which basically highlighted two central aspects: (i) the evaluation of the geographical distribution of hydrocarbons in edible bivalves of the bay's intertidal and (ii) the selection of organisms that indicate organic pollution in the tropical regions of the coast of Brazil. Particular emphasis was placed on the use of chemical markers for the recognition of the origin of hydrocarbons, considering that marine organisms can concentrate not only a variety of anthropogenic hydrocarbons in their tissues but also biogenic ones obtained through their diet (Albaigés et al. 1987). This was the first study of this type conducted in the South Atlantic. From the obtained results, it could be concluded that the concept of *surveillance through the Mussel* can also be applied in tropical areas using alternative bivalve species such as sentinel organisms. In that sense, preliminary data indicate that *Anomalocardia brasiliana* gives a satisfactory response to exposure to local changes in the concentration of pollutants. The bivalves selected for this study were *Anomalocardia brasiliana*, *Protothaca pectorina*, *Lucina pectinata*, and *Macoma constricta* (Porte et al. 1990).

The rapid increase in trade in chemical substances within developing Asian countries implied increased production and use of toxic products such as organochlorines (OCs) and organic tin compounds (BTs), with a potential increase in the exposure of beings, humans and wildlife, to those substances. Consequently, environmental problems associated with contamination by compounds such as BTs and OCs became of great concern (Tanabe et al. 2000). Considering that bivalves, particularly mussels, have been used as bioindicators for the monitoring of toxic substances in coastal waters due to their own characteristics previously mentioned (Phillips 1980; Goldberg 1986), it was decided to apply them to assess this conflicting environmental situation. For this case, the green mussel *Perna viridis* has been chosen, considering that it has a wide geographical distribution within the Asia-Pacific region and is also recognized as a seafood with high commercial value and intensively consumed by the populations of the region (Tanabe et al. 2000). The monitoring studies developed include a part of the Asia-Pacific Mussel Watch Program (APMWP) region, which aims to control marine pollution in this region using mussel as a bioindicator (Tanabe 1994). APMWP was under the umbrella of the International Mussel Watch Program, which mainly involved a coastal monitoring program with bivalves as sentries (mussels and oysters as bioindicators), in order to determine the quality of marine coastal waters. The initial phase of this program was carried out in South and Central America during the period from 1991 to 1993 and revealed serious contamination by organochlorine insecticides in third world countries within this region (Sericano et al. 1995). The second phase intended to cover the three zones of the mentioned region: Asia-Pacific, Northwest Pacific, and South Pacific. On the other hand, in addition to the traditional evaluation of persistent compounds such as PCBs, DDT and their metabolites, and HCH and HCB, other compounds of environmental interest such as organically bound metals (Cu and Sn), and highly toxic compounds (dioxins, furans, and coplanar PCBs) were also included to be monitored within this second phase. The APMWP project was a collaborative work of scientists from the Philippines, Thailand, India, and Japan, coordinated by Prof. Shinsuke Tanabe of Ehime University (Matsuyama, Japan) (Tanabe et al. 2000).

This environmental strategy to assess the quality of aquatic ecosystems has demonstrated to be extremely successful and has been fully applied all over the world. Briefly, several of the developed studies on the last decades are mentioned, i.e., Minier et al. (2006) in the Seine estuary, France; Frontalini and Coccioni (2008) in the Adriatic Sea coast, Italy; Ogata et al. (2009) within the five continents; Fang et al. (2009) in coastal areas of Hong Kong; Looi et al. (2013) in the Strait of Malacca, Malaysia; and Bat and Özkam (2019) in the Turkish Black Sea coast, among many others, including the extense and very detailed review by Beyer et al. (2017) on this topic.

19.5 Monitoring Programs in Argentina

A history of monitoring programs in different aquatic environments from Argentina exists, although those of long-term monitoring are very scarce. In general, existing programs are linked either to large urban centers or to environments conditioned by some particularity that potentially hinders their functionality.

One of the environments that historically has received successive programs for monitoring the quality of its waters, sediments, and/or organisms is the Río de la Plata, although these programs have had different coordination, management styles, and objectives over time. So, the works presented by AGOSBA/SHIN (1997) in which the environmental quality (physicochemical characteristics, bacteriology, nutrients, heavy metals, pesticides) in water and surface sediments of the south coastal strip of the Río de la Plata can be mentioned. This report synthesized the information of 3 years of work. In the same way, it is possible to mention numerous works of research groups that studied different environmental phenomena and reached very valuable conclusions. So, among others, the work of the Administrative Commission of the Río de la Plata (CARP 1989), which includes water quality data and some contaminant data, can be mentioned; or by Janiot et al. (1991) determining organochlorine pesticides in sediments of the Argentine coast and frontal zone of the Río de la Plata. Subsequently, works by other groups, such as those by Colombo et al. (2006) measuring concentrations and flows of aromatic hydrocarbons in sediments of the Río de la Plata; by Ronco et al. (2008) who presented information on a "screening" of contaminants in sediments of tributaries of the southwestern area of the Río de la Plata; by Lombardi et al. (2010) determining trace metal levels in a large fish species (*Prochilodus lineatus*) from Río de la Plata waters; by López et al. (2013) studying the microbiological pollution in surface waters of Reconquista River; by Sathicq et al. (2015) researching on the influence of El Niño phenomenon on environmental condition of Río de la Plata estuary; or by Castañé et al. (2015) assessing the water quality of Luján River near the metropolitan area of Buenos Aires city, among others. This environment is a clear example of systems that have a lot of information generated at different times over an extended period but cannot be considered a long-term monitoring program, because the data series has different origins, discontinuities in its execution, and different work methodologies in several stages of its concretion.

Another environment currently under monitoring is the Uruguay River. Given the problem of the existence of algal blooms occurring within this river and the possible appearance of cyanobacteria that can affect the localities using this water resource for the production and consumption of drinking water, as well as recreational and tourist use, organisms of the involved provinces signed two agreements, one in 2008 and the last in 2010, with a duration of 2 years, through which permanent controls of the water quality of the mentioned river were carried out. Together, the Institute of Water and Environment from Corrientes Province (ICAA) as well as the regulatory agency Administration of Current Sanitary Works (AOSC) carried out permanent monitoring of the Uruguay River since December 2008, establishing as sampling stations four locations in the province: Garruchos, Paso de los Libres, São Tomé, and Monte Caseros, determining parameters in situ and in the physicochemical, bacteriological, and protistological laboratory. The study allowed to determine early alerts for the implementation of contingency plans in the water treatment plants, before the presence of cyanobacteria, as well as to characterize effluents and to have truthful information regarding the loading capacity and self-purification within the Uruguay River (ICAA, 2019).

Many other environments from Argentina are eventually under surveillance through environmental monitoring programs which are carried out following different design and application models, i.e., Río Hondo Reservoir, Santiago del Estero Province (Secretaría de Recursos Hídricos de la Nación 2007); monitoring on San Roque Reservoir, Córdoba Province (Crema et al. 2014, Instituto Nacional del Agua); in the scope of the National Parks of Argentina (APN, Administración de Parques Nacionales, 2016); or Integrated Monitoring Program of the Matanza-Riachuelo Basin/authority of the Matanza-Riachuelo Basin, ACUMAR, Buenos Aires Province (ACUMAR 2019), among others.

Moreover it is important to highlight that the abovementioned monitoring programs are not included within the long-term monitoring ones, considering they are not continuously developed a long time and their designs are not the required as to be considered so.

19.6 A Nice Study Case: Long-Term Monitoring Within the Bahía Blanca Estuary, in Argentina

The existing information on the estuary of Bahía Blanca is very wide, and its history goes back to the second half of the nineteenth century, including reports by Darwin (1845) on the distribution and characterization of salts present in the Salitral de la Vidriera. In any case, it is important to point out that in its first stage, the information is fragmented and dispersed and has not been obtained following a systematic protocol. In spite of the existence of several works that included measurements of physicochemical parameters of the estuary water, only in 1974 a systematic biweekly sampling started at both Ingeniero White and Puerto Cuatreros ports (in the inner area of the estuary) and continues up to the present. The results obtained within this work program have allowed to define the interior area of the estuary as "vertically homogeneous" and "hypersaline" on the occasion of hot and dry summers (Freije and Gayoso 1988; Marcovecchio and Freije 2004; Freije et al. 2008). In addition, the influence of the precipitation produced not only on the tributary basin but also on the bay itself on salinity values was also identified (Freije and Marcovecchio 2004).

The identification of the behavior and natural distribution of the structural (i.e., temperature, salinity, pH, alkalinity), ecophysiological (i.e., inorganic nutrients, dissolved oxygen, organic matter, photosynthetic pigments), and physicochemical parameters of the system, determined during a sufficiently long period (as is the case in this study), allow us to fully characterize its baseline operating conditions, as well as to quickly detect deviations from normal behavior attributable to human activities carried out within the region.

19.6.1 Is 40 Years a Sufficiently Long Monitoring Period?

There is a history of long-term global research programs on the distribution of chemical substances, including pollutants, in large geographic areas on various continents; among them, the *International Mussel Watch Program* (Goldberg et al. 1978; Farrington and Tripp 1995) or the *Worldwide Persistent Organochlorine Compounds' Monitoring Program* (Tanabe et al. 1987, 1994; Tanabe and Tatsukawa 1991) have been mentioned, which they undertook to assess global trends in the distribution of some compounds in coastal marine systems around the world.

Unlike these, the study program developed in the inner zone of the Bahía Blanca Estuary includes the generation of a long time series of data obtained with high frequency; in fact, and since December 1974, biweekly samplings were carried out at two points in the aforementioned region, and they were even weekly throughout much of the indicated period (Marcovecchio et al. 2010). The systematic development of this study in the Bahía Blanca Estuary, carried out over the last four decades, has allowed the generation of a database, which allows to base some concepts that explain the operation and the physical-chemical condition of the estuary. On the other hand, the long data series makes it possible to differentiate deviations due to the natural variability of the samples from those produced by the incidence of human activities which separates the results from their normal distribution. This study program has allowed to significantly characterize both the environmental condition and the functioning of the Bahía Blanca Estuary.

In addition, and since the middle 1990s, a complementary monitoring program required by Bahía Blanca municipal government has started within Bahía Blanca Estuary, directed to fully characterize the environmental quality within the inner area of the estuary. This study includes not only the chemical aspect previously mentioned but also the surveillance of biological features (i.e., microbiology, biology of fishes as well as benthic communities within Bahía Blanca Estuary) which allow optimizing the obtained results.

Both monitoring programs continue to operate simultaneously and have proven to be an excellent tool to support decision-making on environmental issues within this system. The following is a brief summary of the main results obtained through these long-term monitoring programs:

19.6.2 Chemical Aspects

The distribution of the physical-chemical parameters of seawater, observed both in the stage prior to the strong industrialization of this region (1970s) and in the later period, shows that the system has not undergone significant changes that implied variations in its operation. In this sense, it is possible to comment on the relative stability shown by the structural parameters (i.e., temperature and salinity) as well as the trends of the ecophysiological parameters (i.e., nutrients, dissolved oxygen, photosynthetic pigments) sustained over time and observed over more than 40-year evaluation. Furthermore, the fact that the estuary fully conserves its synchronicity, by repeating with natural frequencies the natural cycles of the elements (i.e., N, P, Si) which sustain biological production (as well as its power and magnitude), must be highlighted. The horizontal distribution of these parameters did not present significant spatial variations, and the vertical one was totally homogeneous, which allows to characterize the water body as non-stratified (Marcovecchio and Freije 2004; Freije et al. 2008). All this indicates that the human activities that take place in the internal zone of the Bahía Blanca Estuary have not generated significant changes within the basal physical-chemical condition of the system, which maintains its properties' and parameters' distribution trends similar to the corresponding historical records (Marcovecchio and Freije 2013).

On the other hand, when the distribution of potentially pollutant substances was evaluated within this environment, a different situation could be recorded. Most of the studied pollutants were determined at clearly detectable levels in the estuary, including heavy metals, hydrocarbons, organochlorine pesticides, and microplastic particles. These contaminants were basically found in the sediments of the system, although some (i.e., heavy metals, PAHs) were also determined in the tissues of organisms, suspended in the particulate matter, and dissolved in water. Temporal distribution trends of these compounds were scattered, i.e., while trace metals have showed a decreasing trend since 1980s and up to now (IADO 1997, 1999, 2002, 2006, 2008, 2010, 2012, 2014, 2016, 2018), microplastic particles have been just recently determined and appeared to be increasing within the system (Arias et al. 2019; Ronda et al. 2019).

So the joint application of both monitoring programs has allowed to characterize the BB estuary as an environment that functions properly and without problems from a biogeochemical point of view, with a marked synchronicity in the production of nutrients as well as a corresponding high level of primary production. However, the presence and distribution of different types of potentially toxic substances in its waters, sediments, and organisms have also been verified, although without reaching critical levels for human health and ecosystem's one.

19.6.3 Microbiological Aspects

Marine and coastal resources represent strategic assets of great importance to diversify and improve regional economies. The settlement of large cities and important industries is also common in coastal areas, with the potential that they entail to generate pollution processes.

The increasing levels of pollution of estuaries and bays constitute an increasing risk to public health. They affect marine productivity and diversity and, at the same time, raise costs for tourism and aquaculture.

In the Bahía Blanca Estuary, coastal pollution comes through diffuse sources such as continental rain runoff and point sources such as sewage and industrial discharges. The last of them have increased in recent decades, due to intense human activity related to urban and industrial development and, above all, due to the constant and increasing dumping of raw or insufficiently treated sewage. For decades, the city's sewage effluents have been dumped, insufficiently treated through the Primera Cuenca collector (2500 $m^3.h^{-1}$). Toward the end of September 2008, the discharge of the Tercera Cuenca sewage liquid treatment was added with a tipping flow of 200 $m^3.h^{-1}$ (Streitenberger and Baldini 2010). This was designed to carry out secondary treatment of effluents and is located in the innermost area of the estuary.

Bacterial Indicators The number and type of bacteria present in a natural ecosystem are generally in balance (homeostasis). When environmental conditions are altered, changes occur in the bacterial community (Cabezalí et al. 1995). Therefore the bacterial populations present in marine and coastal ecosystems constitute an excellent early warning system, since they respond to changes in the environment, through qualitative or quantitative changes, more quickly than the rest of the biota. Hence, any environmental alteration can be detected in its initial stages, if carefully and appropriately chosen bacterial indicators are used.

Bacteria are key participants in the flow of heterotrophic energy and mineralization of organic matter. They actively contribute to the self-cleaning capacity of the environment, that is, they are directly involved in the removal and recycling processes.

Both in the water column and in the sediment, heterotrophic bacteria efficiently transform dissolved organic matter into particulate one, leaving this critical resource available for the rest of the food chain. The intensity and nature of the exchanges carried out by the microorganisms strongly depend on the qualitative and quantitative distribution of their various communities and their dynamics in the different ecological niches in which they are located. In this way, differences in the abundance of bacterial populations in different areas can be attributed to existing concentrations of easily assimilable organic matter.

The advantage of using biological indicators is that they reflect the cumulative impact of all stressors over a period. As such, they are a unique measure of the response of the environment and provide information that cannot be collected in any other way.

Coastal waters are the final receptacle for most of the waste caused by human activity on land. Although man's organic residues must be integrated into the biogeochemical cycles of ecosystems, it is not reasonable to believe that increasing amounts of organic matter and bacteria can be safely introduced. This input of material influences macro- and microscopic trophic chains, rapidly stimulating the microbial growth of indigenous groups and contributing new populations, associated with the type of substances that reach the environment.

It is a known fact that enteric infections in the population are favored by multiple deficiencies in basic sanitation and in the excreta final deposition systems. This contributes to maintain a level of human fecal contamination, particularly in aquatic ecosystems (Cabezalí et al. 1995). Epidemiological studies carried out since the 1950s demonstrate the relationship between fecal contamination of recreational waters and the adverse effect on human health including the appearance of gastrointestinal symptoms and eye, ear, nasal, respiratory, and skin infections (Wade et al. 2006; Soller et al. 2010).

Escherichia coli is the most routinely used microbial compliance parameter for confirming fecal (though not necessarily pathogen) contamination of the environment, and their use in water quality legislation around the world demonstrates their widespread utility (Oliver et al. 2016). It is assumed that their behavior is similar to that of other bacteria of the same origin when they are released into the environment (Anderson et al. 2005). Its detection alerts about the possible presence of pathogenic intestinal microorganisms, with the consequent hygiene-health risk for the population.

In the internal area of the estuary, since years, are recorded higher population densities than those internationally recommended for primary contact waters ($\leq$126 CFU.100 mL^{-1}) (USEPA 2003). The most compromised areas coincide with the sewage discharges (Primera and Tercera Cuenca) and with Ingeniero White Port. The fluctuations in the bacteria indicator counting show that the Tercera Cuenca sewage treatment plant fails to stabilize its operation; therefore there is a continuous contribution of effluents to the Bahía Blanca Estuary with parameters outside the acceptable range. The sustained bacteriological counts, the constant presence of fecal indicators, and the tendency to accumulate in sediments in the surrounding areas are evidence of the impact that the system has been suffering in recent years (Streitenberger and Baldini 2010; Pierini et al. 2012).

In Bahía Blanca Estuary waters, terrestrial heterotrophic bacteria (THB) densities fluctuate between 10^2 and 10^3 CFU.mL^{-1}, depending on the continental influences that the different zones receive. For example, the proximity to the coast determines the presence of THB coming from the continent and arriving at the estuary transported by the wind, the runoff produced by the rains, or the water courses that flow into the estuary. Marine heterotrophic bacteria always exceed THB by an order of magnitude (10^4–10^5 CFU.mL^{-1}) and with little spatial and temporal fluctuation. These are native bacteria that are better adapted to environmental conditions and that compete more successfully for nutrients. They are metabolically versatile, being able to use a wide range of N and P sources for their growth (Kirchman 2000).

Hydrocarbon-degrading microorganisms are ubiquitous in the world's oceans, and biodegradation mediated by indigenous microbial communities is the ultimate fate of the majority of oil hydrocarbon that enters the marine environment (Leahy and Colwell 1990; Prince 2010; Atlas and Hazen 2011). Hydrocarbon-degrading bacteria represent the first line of defense against oil pollution in the marine environment. In response to the complexity of hydrocarbon compounds found in petroleum deposits, diverse marine microorganisms have evolved with an equal complexity of metabolic pathways to take advantage of hydrocarbons as a rich carbon and energy source (Kostka et al. 2011, 2014). So it becomes significant to know whether microbial degraders of oil are present in water and sediments of the area to be studied. Enumeration of petroleum-degrading microorganisms is important to assess the magnitude of oil pollution that has occurred. So they are powerful indicators of the impacts that the environment has received or is receiving. As it was indicated above, the Bahía Blanca Estuary has several sources of pollutions such as continental rain runoff and point sources such as sewage and industrial discharges. During several years quantification of hydrocarbon-degrading bacteria in superficial sediments of Principal Channel of the estuary was carried out. These counts have been able to indicate the sites most affected by the impact of hydrocarbons, and counts greater about 10^4 ($MPN.g^{-1}$) were detected in major part of the sampled sites. These values indicated the impact of port activity and municipal sewages in the entire studied area. However, in recent years, the highest values have been coincident with the site where the insufficiently treated municipal sewages are dumped.

So the application of the monitoring program within Bahía Blanca Estuary allowed to identify quite significant points. Throughout the years, bacteriological studies demonstrated the significant anthropic impact received by the Principal Channel (Fig. 2.1, Chap. 2), as well as the urgent need for all the sewage effluents using the estuary as a receiving body to be properly treated and controlled according to current legislation. This was evidenced by the increased amount of bacteria indicating fecal contamination in the innermost stations after the establishment of a sewage treatment plant (2008), which worked poorly for a long time. It is of crucial importance to ensure the proper functioning of the Tercera Cuenca sewage treatment plant, as it is located in a particularly vulnerable area of the estuary with a notable tendency to accumulate contaminants. The waters and sediments of the areas near the sewage outlet showed the least variability in them, indicating the continuity and impact produced by municipal effluents without treatment.

The hydrocarbon-degrading bacteria counts indicated the existence of hydrocarbons' impact and their accumulations in the sediments in studied area, especially those affected by municipal sewage.

The results show the important impact of the dumping of insufficiently treated sewage effluents on the studied area of the Bahía Blanca Estuary. This situation generates an accumulation of biological and chemical contaminants in the sediments, which can be transformed into reservoirs of potentially dangerous microorganisms. In turn, there is a probability that they will be resuspended in the water column, due to weather conditions or dredging activities. If this situation persists, not only the ecosystem is put at risk but also the health of the population that uses the area for different purposes.

International experience has shown that the most effective results in maintaining adequate quality for recreational purposes are achieved with rigorous control of discharges and not with mere speculation about the self-cleaning power of receiving courses. The dumping of pollutants into the aquatic environment and the degree of impact they produce on the receiving body are technically controllable.

19.6.4 Biology of Fishes

Increased anthropic activities in estuaries negatively impact water quality and aquatic fauna, producing changes that can be observed in feeding ecology, loss of spawning and recruitment grounds, and reduction of biodiversity, which in turn affect the ecosystem as a whole (López Rojas and Bonilla Rivero 2000; Whitfield and Elliott 2002; Eddy 2005).

Feeding is one of the most important factors that control fish populations, directly affecting abundance, growth, mortality, and migrations (Sánchez and Prenski 1996). Traditionally, feeding ecology studies imply the analysis of fish gut content. The information produced by these studies has been key to our understanding of trophic relationships between species, as well as on the impact of fish feeding on other components of the aquatic communities (Hynes 1979, in Escalante 1987). It is therefore desirable to generate knowledge on estuarine trophic links to be able to better understand how anthropic disturbances can impact fish and other estuarine species (Elliot et al. 2007).

Fish are useful bioindicators and can provide relevant information on habitat quality evaluations (Hynes 1979, in Escalante 1987); however, not all fish species are suitable for this purpose, as many show migratory behaviors, meaning that they will migrate outside of estuaries and be exposed to other stressors, alien to the estuary itself. It is for that reason that in the present study, the fish species employed to assess water quality were juveniles of *Cynoscion guatucupa* (Lopez Cazorla 2000; Sardiña and Lopez Cazorla 2005a), *Micropogonias furnieri* (Sardiña and Lopez Cazorla 2005b), and *Mustelus schmitti* (Lopez Cazorla 1987, 2004), which remain in the estuary during their first year. Additionally, *Ramnogaster arcuata*, a fish species that completes its entire life cycle inside the waters of the estuary, was also used (Lopez Cazorla and Sidorkewicj 2009).

Starting in 2015, a biological monitoring of the abovementioned fish species has been seasonally conducted in two sites within Bahía Blanca Estuary. The general objective was to help understand how the fish species utilize the impacted areas of the estuary, in order to use the obtained knowledge to inform management plans and conservation actions. The specific objectives laid out to achieve the aforementioned general objectives were: a) generate size frequency and age composition distributions, b) estimate length-weight relationships and growth type, c) evaluate feeding intensity, and d) describe the feeding ecology and selectivity by species, season, and size class.

The biological monitoring allows us to follow the evolution of the changing biological parameters of the populations of the fish species under study throughout time and space and establish links and relationships with the results obtained by chemical analysis of muscle tissue of these species.

19.6.5 Benthic Communities

Biodiversity inventories and monitoring programs have increased markedly worldwide over the last decade in response to concerns about extinctions and the sustainability of natural ecosystems. In general, inventories are conducted to determine the distribution and composition of wildlife and wildlife habitats in areas where such information is lacking, and monitoring is typically used to understand the rates of change or the effects of management practices on wildlife populations and habitats

(Morrison et al. 2008). Making inventories and monitoring biodiversity are crucial to identifying the natural processes and human activities that affect ecosystems. Benthic communities are considered to be good indicators of ecosystem health because of their sedentary nature and longevity, providing long-term exposure to toxic substances, and also due to the presence of diverse taxa that can respond to multiple types of man-made perturbations (Pearson and Rosenberg 1978). In addition, benthic communities play an important role in the regulation of ecosystem processes; they are central to the maintenance of the links between benthic and pelagic systems. Moreover, carbon, oxygen, and nutrient cycling and the decomposition of dead matter or waste materials are important processes driven by the benthic species(Snelgrove et al. 1997).

The evaluation of benthic communities was included in the general environmental monitoring program of the Bahía Blanca Estuary during the last decade. Since then the inventory of macrofaunal invertebrates in the Principal Channel and other minor channels, affected by the discharge of urban and industrial effluents, has been carried out as well as monitoring of benthic communities associated with the port area. It took many years to complete the inventory, due to the large size of the area surveyed and the limited taxonomic knowledge of the species identified (see Chap. 9 for the description of the invertebrate communities). Sampling for the biodiversity inventory was carried out along the Principal Channel at the same sites where the annual measurements of the physical and chemical variables of the water column and sediments are taken periodically. In addition, environmental samples were taken in the minor channels together with the biological samples (IADO 2014, 2016, 2018). The comprehensive analysis of the biological and environmental data allowed the elaboration of a distribution map of the main biological assemblages, their characteristic habitats, and the main risks that affect the local biodiversity. The biological communities associated with minor channels have low diversity and biomass, and only a few polychaete species are dominant in the areas most affected by the discharge of sewage effluents (IADO 2018).

The introduction of exotic species is considered as one of the main threats to biodiversity at different scales, and the extent is causing serious and persistent changes in the marine and coastal environments (Bax et al. 2003; Souza et al. 2009). A generalized consensus points to prevention through the management of vectors and pathways as being the most effective line of defense against invasive marine species (Molnar et al. 2008). The monitoring of benthic communities associated with the port areas enables early detection of new exotic species since the most common pathway for invasive marine species includes shipping, ballast, and/or fouling. Monitoring of the port areas in the Bahía Blanca Estuary indicates that the artificial structures of this sector are dominated by few exotic species, mainly barnacles, most of which have not been detected elsewhere (IADO 2014, 2016). However, the Pacific oyster, which was detected thanks to port monitoring, has accelerated its expansion and abundance in recent years, generating serious environmental problems.

Consequently, and keeping in mind the previously commented results, several concluding points should be pointed out:

- Periodic monitoring of port areas is an effective and low-cost tool to alert about the entry and establishment of non-native species.
- The development of a management plan is necessary for recently established and expanding exotic species, such as the Pacific oyster.
- It is recommended to update the inventory of the benthic communities along the main channel every 5 years.
- It is recommended to start monitoring the communities associated with the minor channels in order to evaluate the environmental changes associated with the industrial and urban discharge areas.

19.7 Concluding Comments

The application of monitoring programs, particularly the long-term ones, has demonstrated to be an excellent and quite efficient tool not only to evaluate environmental quality within aquatic systems but also to assess their evolution along times and changes within their biological, physical, and chemical conditions, as well as to consider the possible and sustainable exploitation of their natural resources.

Sustaining this type of programs in the long term allows to observe consolidated tendencies of the environmental condition of the systems and helps to suggest better management decisions to the corresponding authorities. In this sense, the example of the BBE case study fully indicates the mentioned concept.

References

ACUMAR (Autoridad de Cuenca Matanza-Riachuelo) (2019) Plan Integral de Saneamiento Ambiental (PISA). http://www.acumar.gob.ar/monitoreo-ambiental/ (retrieved Nov.07/2019)

AGOSBA/SHIN (Administración General de Obras Sanitarias de la Pcia. de Buenos Aires/Servicio de Hidrografía Naval) (1997) Calidad de las aguas de la franja costera sur del Río de la Plata (San Fernando – Magdalena). Informe Consejo Permanente para el Monitoreo de la Calidad de las Aguas de la Franja Costera Sur del Rio de la Plata, Buenos Aires (Argentina), 173 p

Albaigés J, Farran A, Soler M et al (1987) Accumulation and distribution of biogenic and pollutant hydrocarbons, PCBs and DDT in tissues of Western Mediterranean fishes. Mar Environ Res 22:1–18. https://doi.org/10.1016/0141-1136(87)90078-X

Anderson KL, Whitlock JE, Harwood VJ (2005) Persistence and differential survival of fecal indicator bacteria in subtropical waters and sediments. Appl Environ Microbiol 71:3041–3048. https://doi.org/10.1128/AEM.71.6.3041-3048.2005

APN – Administración de Parques Nacionales (2016) Reglamento para la Evaluación de Impacto Ambiental en la Administración de Parques Nacionales. Ministerio de Ambiente y Desarrollo Sustentable, Administración de Parques Nacionales, Ley N°22351

Arias AH, Ronda A, Oliva AL et al (2019) Evidence of microplastic ingestion by fish from the Bahía Blanca estuary in Argentina, South America. B Environ Contam Tox 102(6):750–756. https://doi.org/10.1007/s00128-019-02604-2
Atlas RM, Hazen TC (2011) Oil biodegradation and bioremediation: a tale of the two worst spills in U.S. Environ Sci Technol 45:6709–6715. https://doi.org/10.1021/es2013227
Bat L, Özkam EY (2019) Heavy metal levels in sediments of the Turkish Black Sea Coast. Oceanography and Coastal Informatics: Breakthroughs in Research and Practice: 86–107, IGI Global. https://doi.org/10.4018/978-1-5225-7308-1.ch004
Bax N, Williamson A, Agüero M et al (2003) Marine invasive alien species: a threat to global biodiversity. Mar Policy 27:313–323. https://doi.org/10.1016/S0308-597X(03)00041-1
Behmel S, Damour M, Ludwig R et al (2016) Water quality monitoring strategies – a review and future perspectives. Sci Total Environ 571:1312–1329. https://doi.org/10.1016/j.scitotenv.2016.06.235
Beyer J, Green NW, Brooks S et al (2017) Blue mussels (*Mytilus edulis* spp.) as sentinel organisms in coastal pollution monitoring: a review. Mar Environ Res 130:338–365. https://doi.org/10.1016/j.marenvres.2017.07.024
Biber E (2013) The challenge of collecting and using environmental monitoring data. Ecol Soc 18(4):68–81. https://doi.org/10.5751/ES-06117-180468
Boyden CR (1974) Trace element content and body size in molluscs. Nature 251:311–314. https://doi.org/10.1038/251311a0
Brewin RJW, de Mora L, Jackson T et al (2015) On the potential of surfers to monitor environmental indicators in the coastal zone. PLoS One 10(7):e0127706. https://doi.org/10.1371/journal.pone.0127706
Bryan GW (1973) The occurrence and seasonal variation of trace metals in the scallops *Pecten maximus* (L.) and *Chlamys opercularis* (L.). J Mar Biol Assoc UK 53:145–166. https://doi.org/10.1017/S0025315400056691
Cabezali CB, Baldini MD, Cubitto MA et al (1995) Estudio bacteriológico de aguas marinas para uso recreacional. Rev Argent Microbiol 27:115–122
CARP (Comisión Administradora del Río de la Plata) (1989) Estudios para la evaluación de la contaminación del Río de la Plata. CARP/Servicio de Hidrografía Naval (Argentina)/Servicio de Oceanografía, Hidrografía y Meteorología de la Armada (Uruguay). Informe de Avance. Buenos Aires-Montevideo. 42 p
Castañé PM, Sánchez-Caro A, Salibián A (2015) Water quality of the Luján river, a lowland watercourse near the metropolitan area of Buenos Aires (Argentina). Environ Monit Assess 187:645–659. https://doi.org/10.1007/s10661-015-4882-y
Claisse D (1989) Chemical contamination of French coasts. The results of a ten years Mussel Watch. Mar Pollut Bull 20:523–528. https://doi.org/10.1016/0025-326X(89)90141-0
Collins A, Ohandja D-G, Hoare D et al (2012) Implementing the Water Framework Directive: a transition from established monitoring networks in England and Wales. Environ Sci Policy 17:49–61. https://doi.org/10.1016/j.envsci.2011.11.003
Colombo JC, Cappelletti N, Lasci J et al (2006) Sources, vertical fluxes, and equivalent toxicity of aromatic hydrocarbons in coastal sediments of the Río de la Plata Estuary, Argentina. Environ Sci Tecnol 40:734–740. https://doi.org/10.1021/es051672y
Converse SJ, Royle JA (2012) Dealing with incomplete and variable detectability in multi-year, multi-site monitoring of ecological populations. In: Gitzen RA, Millspaugh JJ, Cooper AB, Licht DS (eds) Design and analysis of long-term ecological monitoring studies. Cambridge Univ. Press, Cambridge, UK. Ch.19: 426–442 (586 pp). ISBN: 978-0-521-19154-8
Crema N, Fernández A, Pacharoni F et al (2014) Monitoreo de la calidad del agua del embalse San Roque. Efecto de contaminantes por descargas de efluentes. https://www.ina.gob.ar›ifrh-2014›Eje2›2.08.pdf Accessed on 14 Nov 2019
Darwin CR (1845) Journal of researches into the geology and natural history of the various countries visited by HMS Beagle under the command of Captain Fitzroy, from 1832 to 1836. Murray Ed, London, 738 p

Davies JC, Mazurek J (2014) Pollution control in the United States: evaluating the system. Resources for the Future Publ, New York. 59 pp. ISBN 0-915707-8-8
de Oliveira RF, Uehara-Prado M, Valente-Neto F et al (2018) A network of monitoring networks for evaluating biodiversity conservation effectiveness in Brazilian protected areas. Perspect Ecol Conser 16:177–185. https://doi.org/10.1016/j.pecon.2018.10.003
Eddy FB (2005) Ammonia in estuaries and effects on fish. J Fish Biol 67:1495–1513. https://doi.org/10.1111/j.1095-8649.2005.00930.x
Elliott, M., Whitfield, A. K., Potter, I. C., Blaber, S. J., Cyrus, D. P., Nordlie, F. G., & Harrison, T. D. (2007). The guild approach to categorizing estuarine fish assemblages: a global review. Fish and fisheries, 8(3), 241–268.
Escalante A (1987) Alimentación de *Bryconamericus iheringi* y *Jenynsia lineata lineata* (Osteichthyes) en Sierra de la Ventana (Argentina). Anales del Museo de Historia Natural de Valparaíso 18:101–108
Fang JKH, Au DWT, Wu RSS et al (2009) The use of physiological indices in rabbitfish *Siganus oramin* for monitoring of coastal pollution. Mar Pollut Bull 58:1229–1244
Fang S, Xu LD, Zhu Y et al (2014) An integrated system for regional environmental monitoring and management based on Internet of Things. IEEE T Ind Electron 10(2):1596–1605. https://doi.org/10.1109/TII.2014.2302638
Farrington JW, Tripp BW (1995) International Mussel Watch Project. Initial implementation phase: final report. NOAA Technical Memorandum NOS-ORCA 95, 136 p
Farrington JW, Goldberg ED, Risebrough RW et al (1983) U.S. Mussel Watch 1976–1978: An overview of the trace-metal, DDE, PCB, hydrocarbon and artificial radionuclide data. Environ Sci Technol 17:490–496. DOI: https://doi.org/10.1021/es00114a010
Freije RH, Gayoso AM (1988) Producción primaria en el estuario de Bahía Blanca. Informes sobre Ciencias del Mar – UNESCO 47:112–114
Freije RH, Marcovecchio JE (2004) Oceanografía química del estuario de Bahía Blanca. En: Piccolo MC, Hoffmeyer MS (eds) El ecosistema del estuario de Bahía Blanca, Instituto Argentino de Oceanografía (IADO–CONICET/UNS), Bahía Blanca (Argentina), Cap. 8:69–78. ISBN 987-9281-96-9
Freije RH, Spetter CV, Marcovecchio JE et al (2008) Water chemistry and nutrients of the Bahía Blanca Estuary. In: Neves R, Baretta J, Mateus M (eds) Perspectives on integrated coastal zone management in South America. Part B: from shallow water to the deep fjord: the study sites, IST Scientific Publishers, Lisbon. Chapter 23:243–256. ISBN: 978-972-8469-74-0
Frontalini F, Coccioni R (2008) Benthic foraminifera for heavy metal pollution monitoring: a case study from the Central Adriatic Sea coast of Italy. Estuar Coast Shelf Sci 76:404–417. https://doi.org/10.1016/j.ecss.2007.07.024
GESAMP (1990) GESAMP: the state of the marine environment. UNEP Regional Seas Reports and Studies N°115. Group of Experts on Scientific Aspects of Marine Pollution, Nairobi, 111 p
Goldberg ED (1975) The Mussel Watch – a first step in global marine monitoring. Mar Pollut Bull 6:111–114
Goldberg ED (1986) The Mussel Watch Concept. Environ Monit Assess 7:91–103. https://doi.org/10.1007/BF00398031
Goldberg ED, Bowen VT, Farrington JW et al (1978) The Mussel Watch. Environ Conserv 5:101–125. https://doi.org/10.1017/S0376892900005555
Goldsmith B (ed) (2012) Monitoring for conservation and ecology, vol 3. Springer Science & Business Media, Heidelberg. 396 p. ISBN: 978-0-412-35600-1
Grace JB, Anderson TM, Olff H et al (2010) On the specification of structural equation models for ecological systems. Ecol Monogr 80(1):67–87. https://doi.org/10.1890/09.0464.1
Gray WB, Shimshack JP (2011) The effectiveness of environmental monitoring and enforcement: a review of the empirical evidence. Rev Environ Econ Policy 5(1):3–24. https://doi.org/10.1093/reep/req017
Hounslow A (ed) (2018) Water quality data: analysis and interpretation, 1st edn. CRC Press–Taylor & Francis Publ. Co, Boca Raton. 416 pp. ISBN: 9780203734117

Hsu A, Zomer A (2014) Environmental performance index. Wiley StatsRef: Statistics Reference Online, 1–5. https://doi.org/10.1002/9781118445112.stat03789.pub2

IADO (1999) Programa de monitoreo de la calidad ambiental de la zona interior del estuario de Bahía Blanca. Informe Final, Instituto Argentino de Oceanografía: 25 + 68 pp. Available in: http://www.bahiablanca.gov.ar/cte/index.html

IADO (2002) Programa de monitoreo de la calidad ambiental de la zona interior del estuario de Bahía Blanca. Informe Final, Instituto Argentino de Oceanografía: 76 pp. Available in: http://www.bahiablanca.gov.ar/cte/index.html

IADO (2006) Programa de monitoreo de la calidad ambiental de la zona interior del estuario de Bahía Blanca. Informe Final, Instituto Argentino de Oceanografía: 82 pp. Available in: http://www.bahiablanca.gov.ar/cte/index.html

IADO (2008) Programa de monitoreo de la calidad ambiental de la zona interior del estuario de Bahía Blanca. Informe Final, Instituto Argentino de Oceanografía: 107 pp. Available in: http://www.bahiablanca.gov.ar/cte/index.html

IADO (2010) Programa de monitoreo de la calidad ambiental de la zona interior del estuario de Bahía Blanca. Addenda al Informe Final 2009, Instituto Argentino de Oceanografía: 92 pp. Available in: http://www.bahiablanca.gov.ar/cte/index.html

IADO (2012) Programa de monitoreo de la calidad ambiental de la zona interior del estuario de Bahía Blanca. Informe Final, Instituto Argentino de Oceanografía: 122 pp. Available in: http://www.bahiablanca.gov.ar/cte/index.html

IADO (2014) Programa de monitoreo de la calidad ambiental de la zona interior del estuario de Bahía Blanca. Informe Final, Instituto Argentino de Oceanografía: 242 pp. Available in: http://www.bahiablanca.gov.ar/cte/index.html

IADO (2016) Programa de monitoreo de la calidad ambiental de la zona interior del estuario de Bahía Blanca. Informe Final, Instituto Argentino de Oceanografía: 228 pp. Available in: http://www.bahiablanca.gov.ar/cte/index.html

IADO (2018) Programa de monitoreo de la calidad ambiental de la zona interior del estuario de Bahía Blanca. Informe Final, Instituto Argentino de Oceanografía: 364 pp. Available in: http://www.bahiablanca.gov.ar/cte/index.html

IADO (Instituto Argentino de Oceanografía, CONICET/UNS) (1997) Programa de monitoreo de la calidad ambiental de la zona interior del estuario de Bahía Blanca. Informe Final, Instituto Argentino de Oceanografía: 65 pp. Available in: http://www.bahiablanca.gov.ar/cte/index.html

ICAA (Instituto Correntino del Agua y del Ambiente) (2019) Monitoreo permanente del río Uruguay. http://icaa.gov.ar/ Accessed on 11 Nov 2019

Janiot LJ, Orlando AM, Roses OE (1991) Niveles de plaguicidas clorados en el Río de la Plata. Acta Farmacológica Bonaerense 10:15–23

Johnson DH (2012) Monitoring that matters. In: Gitzen RA, Millspaugh JJ, Cooper AB et al (eds) Design and analysis of long-term ecological monitoring studies. Cambridge University Press, Cambridge. Chapter 3:54–73, 586 p. ISBN: 978-0-521-19154-8

Kirchman DL (2000) Uptake and regeneration of inorganic nutrients by marine heterotrophic bacteria. In: Kirchman DL (ed) Microbial ecology of the oceans. Wiley & Sons, Birmingam. 528 p. ISBN: 978-0-470-04344-8

Kostka JE, Prakash O, Overholt WA et al (2011) Hydrocarbon-degrading bacteria and the bacterial community response in Gulf of Mexico Beach sands impacted by the Deepwater Horizon oil spill. Appl Environ Microbiol 77(22):7962–7974. https://doi.org/10.1128/AEM.05402-11

Kostka JE, Teske AP, Joye SB et al (2014) The metabolic pathways and environmental controls of hydrocarbon biodegradation in marine ecosystems. Front Microbiol 5:471. https://doi.org/10.3389/fmicb.2014.00471

Küster A, Becker PR, Kucklick JR et al (2015) The international environmental specimen banks—let's get visible. Environ Sci Pollut R 22:1559–1561. https://doi.org/10.1007/s11356-013-2482-3

Lawton JH (2007) Ecology, politics and policy. J Appl Ecol 44:465–474. https://doi.org/10.1111/j.1365-2664.2007.01315.x

Leahy JG, Colwell RR (1990) Microbial degradation of hydrocarbons in the environment. Microbiol Rev 54(3):305–315

Likens GE, Walker K, Davies P et al (2009) Ecosystem science: toward a new paradigm for managing Australia's inland aquatic ecosystems. Mar Freshw Res 60:271–279. https://doi.org/10.1071/MF08188

Lindenmayer DB, Likens GE (2010) The science and application of ecological monitoring. Biol Conserv 143:1317–1328. https://doi.org/10.1016/j.biocon.2010.02.013

Lindenmayer DB, Cunningham RB, McGregor C et al (2008) The changing nature of bird populations in woodland remnants as a pine plantation emerges: results from a large-scale "natural experiment" of landscape context effects. Ecol Monogr 78:567–590. https://doi.org/10.1890/07-0945.1

Lombardi PE, Peri SI, Verrengia Guerrero NR (2010) Trace metal levels in *Prochilodus lineatus* collected from the La Plata River, Argentina. Environ Monit Assess 160:47–59. https://doi.org/10.1007/s10661-008-0656-0

Looi LJ, Aris AZ, Johari WLW et al (2013) Baseline metals pollution profile of tropical estuaries and coastal waters of the Straits of Malacca. Mar Pollut Bull 74(1):471–476. https://doi.org/10.1016/j.marpolbul.2013.06.008

Lopez Cazorla A (1987) Contribución al conocimiento de la ictiofauna marina del área de Bahía Blanca. Tesis Doctoral Universidad Nacional de La Plata, Argentina

Lopez Cazorla A (2000) Age structure of the population of weakfish *Cynoscion guatucupa* (Cuvier) (Pisces: Sciaenidae) in the Bahía Blanca waters, Argentina. Fish Res 46:279–286. https://doi.org/10.1016/S0165-7836(00)00152-1

Lopez Cazorla A (2004) Peces del estuario de Bahía Blanca. In: Piccolo MC, Hoffmeyer MS (eds) El ecosistema del estuario de Bahía Blanca. Instituto Argentino de Oceanografía (CONICET-UNS), Bahía Blanca, pp 191–201. (233 p). ISBN 987-9281-96

Lopez Cazorla A, Sidorkewicj N (2009) Some biological parameters of Jenyns's sprat *Ramnogaster arcuata* (Pisces: Clupeidae) in South-Western Atlantic waters. Mar Biodivers Rec 2(e127):1–8. https://doi.org/10.1017/S1755267209001286

López Rojas H, Bonilla Rivero AL (2000) Anthropogenically induced fish diversity reduction in Lake Valencia Basin, Venezuela. Biodivers Conserv 6:757–765. https://doi.org/10.1023/A:1008945813101

López OCF, Duverne LB, Mazieres JO et al (2013) Microbiological pollution of surface water in the upper-middle basin of the Reconquista river (Argentina): 2010–2011 monitoring. Int J Environ Health Res 6(3):276–289. https://doi.org/10.1504/IJENVH.2013.054087

Luthardt V (2010) Monitoring of ecosystems: two different approaches – long-term observation vs success control. In: Müller F, Baessler C, Schubert H, Klotz S (eds) Long-term ecological research – between theory and application. Springer Sci Publ, Dordrecht. Chapter 22:317–325, (475 p). ISBN: 978-90-481-8781-2

Marcovecchio JE, Freije RH (2004) Efectos de la intervención antrópica sobre sistemas marinos costeros: el estuario de Bahía Blanca. Anales de la Academia Nacional de Ciencias Exactas, Físicas y Naturales, Buenos Aires (Argentina) 56:115–132

Marcovecchio JE, Freije RH (eds) (2013) Procesos Químicos en Estuarios, Editorial de la Univ Tecnológica Nacional (EdUTecNe), Buenos Aires (Argentina). E-book. ISBN 978-987-1896-16-5

Marcovecchio JE, Freije RH, Popovich CA et al (2010) Long-term observational system for oceanographic studies within Bahía Blanca estuary (Argentina): state of the art and perspectives. AQUASHIFT: life in warming waters, Christian Albrecht's University, Kiel

Martinez-Haro M, Beiras R, Bellas J et al (2015) A review on the ecological quality status assessment in aquatic systems using community based indicators and ecotoxicological tools: what might be the added value of their combination? Ecol Indic 48:8–16. https://doi.org/10.1016/j.ecolind.2014.07.024

Mayne J, Zapico-Goñi E (eds) (2017) Monitoring performance in the Public Sector future – Directions from international experience. Routledge – Taylor & Francis Group, London, p 52. ISBN 13: 978-1-4128-0632-9

McClure ML, Hansen AJ, Inman RM (2016) Connecting models to movements: testing connectivity model predictions against empirical migration and dispersal data. Landsc Ecol 31:1419–1432. https://doi.org/10.1007/s10980-016-0347-0

McIntyre AD, Pearce JB (1980) Biological effects of marine pollution and the problems of monitoring. ICES Rapports et Procès-Verbaux des Réunions 179, Copenhagen (Denmark), 346 p

Minier C, Abarnou A, Jaouen-Madoulet A et al (2006) A pollution-monitoring pilot study involving contaminant and biomarker measurements in the Seine estuary, France, using zebra mussels (*Dreissena polymorpha*). Environ Toxicol Chem 25(1):112–119. https://doi.org/10.1897/05-161r.1

Molnar JL, Gamboa RL, Revenga C et al (2008) Assessing the global threat of invasive species to marine biodiversity. Front Ecol Environ 6(9):485–492. https://doi.org/10.1890/070064

Morrison ML, Block WM, Strickland MD et al (2008) Inventory and monitoring studies. In: Morrison ML, Block WM, Strickland MD et al (eds) Wildlife study design, 2nd edn. Series on Environmental Management, Springer, New York, 238 p. https://doi.org/10.1007/978-0-387-75528-1_7

Neves R, Baretta J, Mateus M (eds) (2008) Perspectives on integrated coastal zone management in South America. IST Scientific Publ, Lisbon. 679 p. ISBN: 978-972-8469-74-0

Ogata Y, Takada H, Mizukawa K et al (2009) International Pellet Watch: Global monitoring of persistent organic pollutants (POPs) in coastal waters. 1. Initial phase data on PCBs, DDTs, and HCHs. Mar Pollut Bull 58:1437–1446. https://doi.org/10.1016/j.marpolbul.2009.06.014

Oliver DM, Porter KDH, Pachepsky YA et al (2016) Predicting microbial water quality with models: over-arching questions for managing risk in agricultural catchments. Sci Total Environ 544:39–47. https://doi.org/10.1016/j.scitotenv.2015.11.086

Pali P, Swaans K (2013) Guidelines for innovation platforms: facilitation, monitoring and evaluation. ILRI Manual 8, Nairobi. 42 p. ISBN 92-9146-311-6

Pearce JB (1998) A short history of marine environmental monitoring. Mar Pollut Bull 37(1–2):1–2

Pearson TH, Rosenberg R (1978) Macrobenthic succession in relation to organic enrichment and pollution of the marine environment. Oceanogr Mar Biol: Ann Rev 16:229–311

Persson I-L, Nilsson MB, Pastor J et al (2009) Depression of belowground respiration rates at simulated high moose population densities in boreal forests. Ecology 90:2724–2733. https://doi.org/10.1890/08-1662.1

Phillips DJH (ed) (1980) Quantitative Biological Indicators. Their use to monitor trace metal and organochlorine pollution. Applied Science Publ, Barking. 356 p. ISBN: 0853348847

Pierini JO, Streitenberger ME, Baldini MD (2012) Evaluación de la contaminación fecal en el estuario de Bahía Blanca (Argentina) aplicando un modelo numérico. Rev Biol Mar Oceanogr [online] 47:193–202. https://doi.org/10.4067/S0718-19572012000200003

Porte C, Barceló D, Tavares TM et al (1990) The use of the Mussel Watch and molecular marker concepts in studies of hydrocarbons in a tropical bay (Todos Los Santos, Bahia, Brazil). Arch Environ Contam Toxicol 19:263–274. https://doi.org/10.1007/BF01056096

Possingham HP, Fuller RA, Joseph LN (2012) Choosing among long-term ecological monitoring programs and knowing when to stop. In: Gitzen RA, Millspaugh JJ, Cooper AB et al (eds) Design and analysis of long-term ecological monitoring studies. Cambridge Univ. Press, Cambridge. Chapter 23:498–508 (586 p). ISBN: 978-0-521-19154-8

Prince RC (2010) Bioremediation of marine oil spills. In: Timmis KN (ed) Handbook of hydrocarbon and lipid microbiology. Springer-Verlag, Berlin. 2618–2626 p. ISBN 978-3-540-77587-4

Ronco A, Peluso L, Jurado M et al (2008) Screening of sediment pollution in tributaries from the southwestern coast of the Río de la Plata Estuary. Lat Am J Sedimentol Basin Anal 15:67–75. ISSN 1669 7316

Ronda AC, Arias AH, Oliva AL et al (2019) Synthetic microfibers in marine sediments and surface seawater from the Argentinean continental shelf and a Marine Protected Area. Mar Pollut Bull 149:110618. https://doi.org/10.1016/j.marpolbul.2019.110618

Sánchez F, Prenski LB (1996) Ecología trófica de peces demersales en el Golfo San Jorge. Revista de Investigación y Desarrollo Pesquero 10:57–71. ISSN: 0325-6375

Sardiña P, Lopez Cazorla A (2005a) Trophic ecology changes of the whitemouth croaker, *Micropogonias furnieri* (Pisces: Sciaenidae), in South-Western Atlantic waters. J Mar Biol Assoc UK 85:405–413. https://doi.org/10.1017/S0025315405011331h

Sardiña P, Lopez Cazorla A (2005b) Feeding habits of the juvenile striped weakfish, *Cynoscion guatucupa* Cuvier 1830, in Bahía Blanca estuary (Argentina): seasonal and ontogenetic changes. Hydrobiologia 532(1):23–38. https://doi.org/10.1007/s10750-004-8769-0

Sathicq MB, Bauer DE, Gómez N (2015) Influence of El Niño Southern Oscillation phenomenon on coastal phytoplankton in a mixohaline ecosystem on the southeastern of South America: Río de la Plata estuary. Mar Pollut Bull 98(1–2):26–33. https://doi.org/10.1016/j.marpolbul.2015.07.017

Secretaría de Recursos Hídricos de la Nación (2007) Programa de monitoreo Embalse Río Hondo (Diciembre de 2007). http://www.hidricosargentina.gov.ar/InformeMonitoreo_cap5.pdf

Seidl R, Spies TA, Peterson DL et al (2016) Searching for resilience: addressing the impacts of changing disturbance regimes on forest ecosystem services. J Appl Ecol 53:120–129. https://doi.org/10.1111/1365-2664.12511

Sericano JL, Wade TL, Jackson T et al (1995) Trace organic contamination in the Americas: an overview of the US National Status & Trends and the International "Mussel Watch" Programmes. Mar Pollut Bull 31:214–225. https://doi.org/10.1016/0025-326X(95)00197-U

Smith TB, Purcell J, Barino JF (2007) The rocky intertidal biota of the Florida Keys: fifty-two years of change after Stephenson and Stephenson (1950). Bull Mar Sci 80:1–19

Smith DR, Lei Y, Walter CA et al (2012) Incorporating predicted species distribution in adaptive and conventional sampling designs. In: Gitzen RA, Millspaugh JJ, Cooper AB et al (eds) Design and analysis of long-term ecological monitoring studies. Cambridge Univ. Press, Cambridge., Chapter17:381–396 (586 p). ISBN: 978-0-521-19154-8

Snelgrove PVR, Blackburn TH, Hutcjing P et al (1997) The importance of marine biodiversity in ecosystem process. Ambio 26:578–583

Snyder EG, Watkins TH, Solomon PA et al (2013) The changing paradigm of air pollution monitoring. Environ Sci Technol 47:11369–11377. https://doi.org/10.1021/es4022602

Soller JA, Schoen M, Bartrand T et al (2010) Estimated human health risks from exposure to recreational waters impacted by human and non-human sources of faecal contamination. Water Res 44:4674–4691

Sousa R, Gutiérrez JL, Aldridge DC (2009) Non-indigenous invasive bivalves as ecosystem engineers. Biol Invasions 11:2367–2385. https://doi.org/10.1007/s10530-009-9422-7

Stelzenmüller V, Breen P, Stamford T et al (2013) Monitoring and evaluation of spatially managed areas: a generic framework for implementation of ecosystem based marine management and its application. Mar Policy 37:149–164. https://doi.org/10.1016/j.marpol.2012.04.012

Streitenberger ME, Baldini MD (2010) Deterioro de un área recreacional por efectos del volcado de líquidos cloacales. Rev Argent Microbiol 42:307–310

Tanabe S (1994) International mussel watch in Asia-Pacific. Mar Pollut Bull 28(9):518–526. https://doi.org/10.1016/0025-326X(94)90057-4

Tanabe S, Tatsukawa R (1991) Persistent organochlorines in marine mammals. In: Jones KC (ed) Organic contaminants in the environment – environmental pathways and effects. Elsevier Science Publ, New York. Chapter 8: 275–289 (348 p). ISBN-I3: 978-94-010-8424-6

Tanabe S, Kannan N, Subramanian AN et al (1987) Highly toxic coplanar PCBs: occurrence, source, persistency and toxic implications to wildlife and humans. Environ Pollut 47:147–163. https://doi.org/10.1016/0269-7491(87)90044-3

Tanabe S, Iwata H, Tatsukawa R (1994) Global contamination by persistent organochlorines and their ecotoxicological impact on marine mammals. Sci Total Environ 154:163–177. https://doi.org/10.1016/0048-9697(94)90086-8

Tanabe S, Prudente MS, Kan-Atireklap S et al (2000) Mussel watch: marine pollution monitoring of butyltins and organochlorines in coastal waters of Thailand, Philippines and India. Ocean Coast Manag 43:819–839. https://doi.org/10.1016/S0964-5691(00)00060-0

U.S. National Academy of Sciences (1975) Assessing potential ocean pollutants: a report of the study panel on assessing potential ocean pollutants. US National Academy of Sciences, Washington DC, 438 p

U.S. National Academy of Sciences (1977) Environmental monitoring, a report to the U.S.EPA from the Study Group on Environmental Monitoring. US National Academy of Sciences, Washington DC, 181 p

USEPA (U.S. Environmental Protection Agency) (2003) Bacterial water quality standards for recreational waters (freshwater and marine waters) Status Report. US Environmental Protection Agency publication N° EPA- 823-R-03-008. Office of Water, Washington, DC, 76 p

Valle Junior RF, Varandas SGP, Sanches Fernandes LF et al (2015) Multi criteria analysis for the monitoring of aquifer vulnerability: a scientific tool in environmental policy. Environ Sci Policy 48:250–264. https://doi.org/10.1016/j.envsci.2015.01.010

Viana IG, Aboal JR, Fernández JA et al (2010) Use of macroalgae stored in an environmental specimen bank for application of some European framework directives. Water Res 44:1723–1744. https://doi.org/10.1016/j.watres.2009.11.036

Voget-Kleschin L (2013) Large-scale land acquisition: evaluating its environmental aspects against the background of strong sustainability. J Agric Environ Ethics 26(6):1105–1126. https://doi.org/10.1007/s10806-013-9448-9

Wade TJ, Calderón RL, Sams E (2006) Rapidly measured indicators of recreational water quality are predictive of swimming-associated gastrointestinal illness. Environ Health Perspect 114:24–28

Whitfield AK, Elliott M (2002) Fishes as indicators of environmental and ecological changes within estuaries: a review of progress and some suggestions for the future. J Fish Biol 61:229–250. https://doi.org/10.1111/j.1095-8649.2002.tb01773.x

Wolf B, Lindenthal T, Szerencsits M et al (2013) Evaluating research beyond scientific impact – how to include criteria for productive interactions and impact on practice and society. GAIA 22(2):104–114. https://doi.org/10.14512/gaia.22.2.9

Wrona FJ, Johansson M, Culp JM et al (2016) Transitions in Arctic ecosystems: ecological implications of a changing hydrological regime. Eur J Vasc Endovasc Surg 121:650–674. https://doi.org/10.1002/2015JG003133

Zhan X, Li MY, Zhang Z, Goossens B et al (2006) Molecular censusing doubles giant panda population estimate in a key nature reserve. Curr Biol 16:451–452. https://doi.org/10.1016/j.cub.2006.05.042

Chapter 20
Environmental Education: Mud and Salt Classrooms

Cristina Sanhueza and Paola Germain

20.1 Introduction

By the end of the 1960s and the beginning of the 1970s, the impact of human actions on the environment became the focus of attention for government agencies and entities worldwide (Zabala and García 2008). Issues such as air and water pollution, the loss of environments, and the extinction of species began to concern society, and educational actions were taken into account as possible ways, among others, to reduce the planet deterioration. Thus, in 1972 the United Nations convened the Stockholm Conference, considered the first world environment forum, where the environmental problems generated by the prevailing style of development, disproportionate industrialization, and accelerated population growth were considered (Zabala and García 2008). This conference has among its achievements the creation of the United Nations Environment Programme (UNEP 1972) and the Stockholm Declaration (1972) that establishes in its 19th principle: "Education in environmental matters, for the younger generation as well as adults, giving due consideration to the underpriviliged, is essential in order to broaden the basis for an enlightened opinion and responsible conduct by individuals, enterprises and communities in protecting and improving the environment in its full human dimension…." This principle is considered the formal beginning of environmental education, although, previously, there were initiatives from nongovernmental organizations (NGO) and educators.

There is a lot of definitions of environmental education, which is also characterized by its heterogeneity of practices, but all promote some kind of change and have the action in common, that is, it is education for action (Melillo et al. 2015), and it is oriented toward the training of active and committed citizens in building a more

C. Sanhueza (✉) · P. Germain
GEKKO – Grupo de Estudios en Conservación y Manejo, Departamento de Biología, Bioquímica y Farmacia, Universidad Nacional del Sur (UNS), Bahía Blanca, Argentina

S. M. Fiori, P. D. Pratolongo (eds.), *The Bahía Blanca Estuary*,
https://doi.org/10.1007/978-3-030-66486-2_20

just, democratic, and caring society. It is a pedagogical-political practice that tends toward a critical analysis of the socio-environmental reality in which its transformation in favor of responsible human development prevails (González-Gaudiano and Puente-Quintanilla 2010). This is why environmental education must be incorporated into educational systems worldwide.

20.2 Environmental Education in Latin America

As in the rest of the world, the first steps in environmental education in Latin America took place in the 1960s. Groups linked to popular education incorporated an environmental vision in the conception of development, although at the beginning they considered human beings as mere predators of nature. On the other hand, ecological education incorporated a systemic understanding that interrelated social and natural processes (Tréllez Solís 2006). Social problems such as poverty and inequality, widespread in Latin America, and the need to make contributions and involve communities in improving their living conditions crossed environmental education. Thus, Latin America laid the foundations for participatory approaches that were subsequently collected at the international level. Besides the initial conservationist proposals, it adds fundamental elements to involve the communities in the processes toward the improvement of the environmental situation, including natural, social, and economic components at the local level (Tréllez Solís 2006).

In the 1980s, a lot of NGOs related to conservation and environmental management emerged. Educational material was generated, with theoretical and reflective support for participatory action and with regional environmental thought. Different topics are discussed: the model of development, its impact on ecosystems and its link with the population impoverishment, the complex causality of the environmental problems, and the necessary understanding of the articulation of social and natural processes. All of this, in order to find concrete solutions to environmental problems, related to a new organizational capacity of society as a whole, based on the cultural values of communities, on popular creativity, and in its innovative potential (Tréllez Solís 2006).

In the 1990s, Latin American governments began to include environmental education in the curriculum of the different educational levels of formal education and to develop educational materials. However, teacher training is scarce, and in practice, most of the countries tend to equate traditional natural science instruction with the teaching of environmental education, and teachers use traditional methods and approaches (Arias-La Forgia 1994). On the other hand, on many occasions, environmental education does not represent an institutional interest but is the result of the interest of local actors, who promote innovative initiatives on their own (González-Gaudiano & Puente-Quintanilla 2010).

Besides possessing great biological diversity, Latin American countries are multicultural societies, where indigenous or mestizo people cultures as well as peoples of African and European ancestry coexist. Environmental education must integrate the different knowledge, generating new pedagogical possibilities, especially because the knowledge of indigenous peoples and people of African ancestry transcends the anthropocentric perspective of colonial European cultural heritage (García-Campos 2019).

Community interest in environmental issues has been increasing, which represents a great opportunity for environmental education. It must be a priority for political authorities in charge of the formal education system, who should be more actively involved. Considering humanity as part of nature, exchanging knowledge, generating proposals for community participation, incorporating artistic and awareness-raising activities, and highlighting the importance of ethics and reflection are some of the aspects to be addressed, to involve the community in actions of nature conservation and improvement of the quality of life.

20.3 Environmental Education in Argentina

In Argentina, in 2002, the General Law of the Environment (N° 25,675) was passed. This law establishes the minimum budgets for the achievement of a sustainable and adequate management of the environment, the preservation and protection of biological diversity, and the implementation of sustainable development in Argentina. It also establishes a general framework for information and participation in environmental matters, responsibility for environmental damage, and environmental education. In several subsections, it mentions environmental education as a tool and a means to achieve certain goals. For instance, Article 14 says: *Environmental education is the basic instrument to generate in citizens values, behaviors and attitudes that are consistent with a balanced environment, aim at the preservation of natural resources and their sustainable use, and improve the quality of life of the population.* In article 15, it refers to environmental education as *a continuous and permanent process, subjected to constant updating that, as a result of the orientation and articulation of different educational disciplines and experiences, should facilitate the comprehensive perception of the environment and the development of an environmental conscience. The competent authorities must coordinate with the Federal Councils for the Environment (COFEMA), Culture (CFC), and Education (CFE), the implementation of plans and programs in the education systems, formal and non-formal. The jurisdictions, depending on the determined basic contents, will implement the respective programs or curricula through the pertinent norms.*

Despite having a legal framework, environmental education in the different educational spaces is insufficient, low, or nonexistent. When it exists, it mostly refers to global environmental problems, and little is known at a regional or local level. This may be due to different factors, like the lack of information or the ignorance about regional or local natural, cultural, and social environment. Many teachers have not had such training and information, which is transferred to their students. This unknowledge of the local environment leads teachers to address global topics such as pollution or garbage under a general approach, without knowing the true values of the environment, the interactions that take place in it, the relation with human beings, and the factors that threaten the health of that environment, among others.

The current way of life, the technology, the rush, the comfort, and the consumerism have created a wall between nature and people. The societies forget or ignore the essential, the relationship of our ancestors with nature, our relationship with the environment, and that we are part of it. The rush leads to immediacy, loss of observation, contemplation, and search of answers through our means.

In a global context of loss of environments and biodiversity, and distance from the natural environments of which we are a part, particularly by people living in cities, environmental education is an essential tool for nature conservation, promoting a change of attitudes toward the environment and a life in harmony with it. As Leonardo Boff (1995) says "We do not live on Earth. We are Earth, part of Earth. Between living and inert beings, between the atmosphere, the oceans, the mountains, the earth's surface, the biosphere and the anthroposphere, interrelationships prevail. There is no addition of all these parts, but organicity between them."

20.4 Environmental Education in the Bahía Blanca Estuary

Environmental education can be approached from three perspectives: community, systemic, and interdisciplinary. This allows to achieve a comprehensive environmental education (Álvarez 2004).

From a *community perspective*, it aims to form environmental attitudes and values to reduce the crisis and transform the predatory attitude of the human being (Álvarez 2004). The environmental problems and their causes must be studied and analyzed from a local to a global perspective. At first, start from the local problem and from daily realities of the community, facing individuals with local environmental realities, and then move along regional or global situations. A clear example of this approach in the Bahía Blanca Estuary is the problem of plastic garbage in the sea. From this problem in the coastal area, we can see how this problem affects coastal activities and artisanal fishing or how it affects marine species. It is not necessary to work with that issue on a global level, in distant and foreign seas, since it is a local problem where the community could play an active role that can be known and acted on the local environment. Many NGOs conduct coastal garbage censuses and coastal cleanings, activities that involve the community and can generate awareness and responsible attitudes toward the environment.

Environmental education must develop in all individuals of all ages in a community the capacity for critical observation, understanding, and responsibility toward the environment. These skills must be acquired by all social actors. From a community perspective, everyone should know the situation of the environment where they live.

Following the community approach, to achieve a comprehensive environmental education, the focus must be not only on environmental problems but also on the value of the environment "without problems," the environment in balance, fulfilling all its ecological, social, and cultural functions with its values positively impacting the daily life of the local community.

All environmental problems necessarily have a *systemic constitution*, understanding the system as "a set of elements related to each other and to the environment, which constitute a specific integral formation" (Álvarez 2004). Therefore, understanding the environment as a system in which its elements are interrelated is a fundamental characteristic of the environmental dimension, where the components of the system are integrated into the physical, biotic, economic, and sociocultural environment, added to the group of beliefs, values, techniques, and worldview, shared by the members of the community (Álvarez 2004). This approach allows to take into account the different dimensions that cross a certain environmental topic, recognizing the interactions that exist between them and valuing each element that makes up these dimensions. An example from this perspective in the Bahía Blanca Estuary is the environmental conflict described later in this text, in Sect. 20.5. In that conflict, a great diversity of dimensions such as ecological, biological, social, cultural, economic, and social representations about the estuary of the community were revealed. All these dimensions and the relationships between them were interpellated by the activity, the dredging, that would be carried out in the environment. It was a case in which it would be decided whether or not to protect aquatic birds, marine fauna, or the ecosystem dynamic against job opportunities, but also it required a comprehensive analysis of all dimensions, its elements, and interrelations. In many arguments and opinions about this conflict, the estuary was approached from a systemic perspective. However, more simplistic discourses were also exists, in which the environment was summarized in only one dimension or one element, leaving aside the systemic vision.

Interdisciplinarity represents a set of interrelated disciplines and with defined relationships, so their activities do not occur in isolation, dispersed nor divided. The articulation of the different disciplines allows a global understanding of a process and then performing the analysis and solution of a particular problem (Álvarez 2004).

The incorporation of this interdisciplinary approach both in educational practice and in the treatment of environmental conflicts presupposes the realization of a planning that achieves an adequate organization of the analysis, which allows understanding the complex structure of the environment. The interaction of its physical, biological, social, and cultural aspects is demonstrated, as well as providing a clear awareness of the political, economic, and ecological interdependence of the environment.

In addition to the three approaches to environmental education discussed above, there is the symbolic vision of the environment. That refers to the different meanings that social groups have on the environment, which vary depending on cultural contexts (Pedroza and Argüello 2002). The coastline of the Bahía Blanca Estuary has undergone multiple transformation processes that have led to a loss of biodiversity and parts of cultural heritage. At the same time, port and industrial development practically hinder people's access to the coast, creating a kind of "collective amnesia" regarding the city's marine and coastal character. This local natural environment is generally unknown by the new generations. This has led to its transformation into a city facing away from the sea for many years. Frequently, the inhabitants of the area ignore the environmental values that the estuary contains, its scenic beauty, and its importance for regional development. The idea that it is a "lost" place from the point of view of contamination is also frequent. This symbolic vision of loss and deterioration does not promote attitudes of caring for the environment.

20.5 Social Conflicts That Triggered the Emergence of "Guardians of the Estuary"

Bahía Blanca is no stranger to the problem of globalization of the environment, of ignorance of the local natural heritage. Many of its citizens do not know its coasts. This is undoubtedly the most serious aspect that threatens the conservation of the environment's biodiversity, history, and culture.

The estuary coast has undergone multiple transformation processes associated both with the growth of the city and with the construction and development of industries and port terminals, leaving a few coastal remnants in a natural state. These processes lead to an impoverishment of biodiversity and the loss of parts of the cultural heritage resulting from the historical interaction between humans and their natural environment.

How is the wall that has been built between the citizens and the coastal environment, its nature, history, and culture demolished? The first step is undoubtedly the public recognition of the importance of the environment, both from a practical point of view, as a provider of resources and ecological services, and from the admiration, aesthetic enjoyment, and responsibility of the entire community as custodian of the regional nature.

The first step to begin to demolish that wall was in May 2011. That year rumors of a dredging project for the extension of the Principal Channel in the Bahía Blanca Estuary began to circulate. That extension would reach the area called Puerto Cuatreros and included the construction of a dock, on the north bank, to receive regasification ships and an interconnection gas pipeline. Puerto Cuatreros (Figs. 2.1 and 2.2; Chap. 2) is located in the innermost sector of the estuary, where small boats circulate, and is not subject to deep dredging. This information generated concern in neighbors and nongovernmental organizations, which began to make meeting and learn about this initiative. Finally, in July 2011, the responsible for the project formalized the presentation of the proposal, and in September 2011, they communicate it to the local media. Something that was not explicitly mentioned in the project, but that would surely occur in the future, was that the deepening of the dredging to Puerto Cuatreros would allow the development of new industries, nonexistent in the area until this moment. For a year and a half, local people organized into an assembly, which in January 2012 became the "Buenos Aires South Environmental Assembly" (AABAS). A great variety of activities were launched to inform the population, spread the value and importance of the ecosystem conservation, and express the opposition to the project: natural interpretation walks in the terrestrial area of the estuary, informative talks, debates, collection of signatures, bike riding, murals, photo exhibitions, film cycles, interventions in different public events, marches, recitals, celebrations for "World Wetlands Day," bird watching, kayaking, creation of the "From the mud" murga, teachers' training, etc. At first, a few people participated in the activities, but later they became massive, achieving a citizen participation never seen before in an environmental cause. At the same time, professionals specialized in biological, ecological, environmental, port, and urban planning issues from the National University of the South (UNS), the Argentine Institute of Oceanography (IADO), and the National Technological University (UTN), all academic institutions, expressed their disagreement with the project with numerous reasons related to the environmental impact of this undertaking. This commitment of academic institutions with an environmental-social problem was also unprecedented. Finally, for economic reasons, the project was rejected, but the spreading of the value of the Bahía Blanca Estuary continues, to be able to face future threats that endanger its conservation. With no doubt, the way forward was always environmental education, to transmit the value of the Bahía Blanca Estuary, its biodiversity, its stories, and the interrelationships between nature and the city.

20.6 Guardians of the Estuary

To meet this challenge, the GEKKO Group (Group of Studies in Conservation and Management), belonging to the Department of Biology, Biochemistry, and Pharmacy of the National University of the South, organized in 2013 the course workshop for the training of guides-interpreters of the Bahía Blanca Estuary, called "Guardians of the estuary." Through this project, the aim was to combine the work of university teachers and students with environmental conservation NGOs and residents of the towns of General Cerri, Ingeniero White, and Bahía Blanca in order to develop educational materials and activities that promote the approach of community to natural environments, promoting the feeling of belonging to the environment, without losing sight of the environmental and social reality of the community. In addition, those interested in developing educational and tourism ventures in contact with nature, a genuine alternative for sustainable local development, were trained. After this project, in 2014 a group of people interested in working in the conservation, valuation, and dissemination of the biological, social, and cultural wealth of the estuary is formed. This is how the NGO – Guardians of the estuary – in Bahía Blanca was born.

Guardians' different activities are carried out throughout the year, such as talks at the different educational levels, outings of natural interpretation along the estuary coast, awareness activities such as kayaking, walks to learn about the estuary environment, trail development self-guided to discover firsthand the value of the environment, as well as activities and dissemination material such as posters, brochures, murals, and photographic exhibitions.

Box 20.1 Estuarial Classrooms

Coastal marine areas generally undergo major anthropic changes that affect the functioning of the ecosystem and its biodiversity and are also particularly affected by climate change. On the other hand, they provide numerous ecosystem services that depend on their good state of conservation, such as the control of coastal erosion, the production of food, and the possibility of carrying out sports, tourism, and recreational activities. Thus, coastal marine environments represent a great pedagogical opportunity for environmental education.

In the next section, we will take a walk through the different classrooms that we can find in the estuary: classrooms that are not only to visit with an educational institution, in a formal education, but classrooms for the entire community, inclusive classrooms, and action learning classrooms. Here knowledge is within the reach of whoever wants to take it and whoever wants to discover or build it; you just have to enter and enjoy those classrooms. They are the appropriate space to achieve an integral, community, systemic, and interdisciplinary environmental education.

Classrooms as Pharmacies: History of the Native Peoples

The territory that we walk today in the estuary has its history from the time of the native peoples and from before them. Many of the plants and animals that we see today in the estuary were the source of food, remedies, and survival of the native peoples that inhabited this environment. Inquire how was the relationship between human beings and their natural environment in those times. It offers us a great opportunity to rebuild the identity of the community with its local environment and with its natural and cultural heritage. Many of the plants that grow today in the Bahía Blanca Estuary are recognized by popular wisdom for their medicinal uses, examples of which are *Cyclolepis genistoides*, *Schinus longifolius*, and *Jodina rhombifolia*, among others. There are also species like *Sarcocornia perennis* that was used and is still used today as food (Sanhueza et al. 2014). The skin of some animals, such as guanaco, was used for awnings and clothing. Some report that the clay from the estuary coast was used to make vessels and pots (personal communication from a renowned ceramist from the city).

Supply Classrooms

The sea as a provider of food and work continues to be of importance for many families of artisanal fishers in the Bahía Blanca Estuary; currently 72 families live on this resource. In the past, many more families practiced this art, as explained in Chap. 18 of this book; artisanal fishers have been protagonists of a strong socio-environmental conflict.

Knowing this rich source of food gives us the idea of an environment as a supplier, as a supermarket, both for humans and for many other animals. The importance of recognizing it as a finite resource, which must be cared for and not exploited, provides us with the right space to learn, develop educational programs, know the diversity that lives in the sea, know the dynamics of natural populations, know the interactions that take place there, recognize what are the possible threats that put this resource at risk, and thus make decisions to avoid its collapse.

This classroom also offers us the possibility of changing the paradigm and thinking about the estuary from ecocentrism and not from anthropocentrism, to think that the estuary and its coasts are not only a source of food for human beings but also for thousands of species that inhabit or pass through it, such as migratory birds. For several species of migratory birds, the Bahía Blanca Estuary represents a key feeding site on their journey. If this environment disappeared, they would be left without food to continue their journey.

Classrooms That Provide Shelter

Undoubtedly, all animals need a shelter, either to protect themselves from the weather, to survive predators, to reproduce, or simply to rest. The estuary seems to be a very hostile environment, a lot of wind, salt, and sun and little water; it seems to be an unfriendly environment to take refuge, but it still turns out to be a key place for many species. We can find caves, burrows, nests that belong to crabs, polychaetes, guinea pigs, birds, ants, spiders, and snake; all find the right conditions to make the estuary an ideal refuge.

Art Classrooms

Art and environmental education are two doors to wonder, to wonder at the natural and social environment that surrounds the individual, and this is the door to the question and therefore to meaningful learning. Art allows us to perceive and get to know the environment and its socio-environmental problems in a new way. The artistic gaze allows us to discover new elements and integrate other points of view. The German painter Paul Klee said that "Art does not reproduce the visible. It makes it visible" (López Abril et al. 2017). The NGO Guardians of the estuary used street art to make visible the great biodiversity that exists in the estuary by creating collective murals in different sectors of the city, making visible to the community the presence of flamingos, dolphins, sea turtles, sharks, and wildlife that the community is unaware of their existence (Figs. 20.2 and 20.3). Appealing to people's astonishment in order to bring them closer to the knowledge of the natural environment is a powerful tool.

Creativity and art allow us to perceive the elements of the landscape, as well as the landscape as a whole, in a different way, activating different perspectives that integrate intellectual and emotional learning. The Bahía Blanca Estuary has been a source of inspiration for literary works, drawings, paintings, music, and ceramics, among other areas. Many of these artistic expressions have been both individual and collective and have thus become multipliers of knowledge about the estuary.

Classrooms of Different Spatial and Temporal Scales

The estuary is a natural space in which a large number of species live together, occupying different environments (microenvironments and macro-environments) generating innumerable interactions between individuals of the same species and between different species, as well as interactions with the environment that surrounds them. Observing this space at different scales will give us opportunities to discover different points of view.

Used points of view (Arengo et al. 2002) in environmental education offer us a very powerful tool to perceive the impact that disturbances will have at different spatial and temporal scales. If a bush falls, it probably will not affect the flight of the *Cathartes*, but it will be a catastrophe for the firewood-gatherer who had his nest in it or the common yellow-toothed cavy with their burrows underneath; the soil conditions will be modified generating windows of opportunities for different plants species. The footsteps on crabs' land trail will be imperceptible to Olrog's gull but a terrible impact on crabs. We can also witness disturbances on a larger scale that affect many species, including humans, such as the dredging that occurs in the estuary. Actions of daily life can generate disturbances without people perceiving it, such as the waste that is generated in homes or an external drain from the laundry or kitchen that ends up in the estuary. Recognizing the different scales and their interactions offers endless opportunities for learning and reflection on the action of the human being on the environment.

Disturbance can be natural, such as a fire, a hurricane, and a flood, and after a disturbance natural or anthropic, we can ask ourselves: What happens next? Who does it affect? How affects? What were the impacts? How long will it take for the environment to recover? We can work many of these questions to develop formal or non-formal environmental education activities.

The tide, seasonal changes, day, and night offer changing scenarios and patterns that can be used as teaching and learning tools. We can inquire about the consequences of disturbances, of temporal changes at different spatial scales, or of different intensities.

Fig. 20.1 Educational walks of natural interpretation carried out by the NGO "Guardians of the estuary" in the Bahía Blanca Estuary. (Photo by Guardians of the estuary)

Fig. 20.3 (**a**) Collective green turtle mural made in a neighborhood of the city of Bahía Blanca. (**b**) Collective mural of the estuary biodiversity made in the multiple use salon of the fishing club in Puerto Cuatreros, General Daniel Cerri. (Photo by Guardians of the estuary)

The main objective of the group is to create a space where you can experiment with all the senses, reinforcing the identity with the local environment. As a final goal, it seeks to modify attitudes, acquiring new habits and achieving a change in values both from the exercise of sensitivity and rationality, being part of current historical social processes as active subjects and responsible for caring for the environment. They work under the motto "Together we can become pillars of a paradigm shift, turning our gaze towards nature and towards sustainable ways of life."

Between the months of September to November 2018 and March to May and September to November 2019, the NGO conducted guided tours of natural

Fig. 20.2 Collective murals made in the downtown area of the city of Bahía Blanca. Photos by Guardians of the estuary

interpretation with different educational institutions of private and public management of the initial, primary, and secondary levels, including the modalities of additional support needs and adults. In addition, the tertiary and university levels and groups of PAMI (Integral Medical Care Program, health insurance for retirees and pensioners) and open trips to all public were added. The total of visitors who enjoyed this environment, learning and being part of it, were 2500 people, making one or two weekly outings only (Fig. 20.1). If you take into account that the work carried out by this NGO is completely voluntary, without a contribution from the state, neither from the public nor the private sector, your work is highly valued and invites us to question the role of the state in environmental education that it preaches in its legislation.

The NGOs that carry out environmental education activities in the estuary do not meet the great demand of educational institutions and diverse groups that request a guided visit. This realizes the need that exists in the community to know its local environment in order to value it and thus preserve it (Box 20.1).

To learn more about the Guardians of the estuary and find out about the activities they carry out, you can visit:

Facebook – Guardianes del Estuario-Bahía Blanca
Instagram – guardianesdelestuariobahia

20.7 Final Comment

The challenge is to achieve an environmental education that teaches how to produce and consume responsibly, to live together to share life with other living beings, to get out of the noise in which we live submerged, and to listen to nature, which recognizes, conserves, and thanks services that natural ecosystems provide us, to teach us to be more sensible beings.

References

Álvarez CO (2004) Educación ambiental a partir de tres enfoques: comunitario, sistémico e interdisciplinario. Revista Iberoamericana de Educación, ISSN-e 1681-5653, ISSN 1022-6508, Vol. 35, N°. Extra 1

Arengo N, Elfi Chaves M, Feinsinger P (2002) Guía metodológica para la enseñanza de ecología en el patio de la escuela. Audobon Programa para America Latina y el Caribe

Arias-La Forgia A (1994) Environmental education in the school systems of Latin America and the Caribbean. Working Papers 4. Agency for International Development. Washington, USA

Boff L (1995) Nueva era. La civilización planetaria: desafíos a la sociedad y al cristianismo. Ed. Verbo Divino. España

García-Campos HM (2019) Environmental education from an intercultural approach: a glimpse into Latin America. Southern Afr J Environ Educ 35. https://doi.org/10.4314/sajee.v35i1.12

González-Gaudiano EJ, Puente-Quintanilla JC (2010) El perfil de la educación ambiental en América Latina y el Caribe: Un corte transversal en el marco del Decenio de la Educación para el Desarrollo Sustentable. Pesquisa em Educação Ambiental 5:27–45

López Abril M, Vega M, Loren L (2017) El Arte como herramienta para la Educación Ambiental. Boletín del Centro Nacional de Educación Ambiental

Melillo F, Belmes A, Priotto G et al (2015) Educación ambiental: Ideas y propuestas para docentes. Nivel secundario. Secretaría de Ambiente y Desarrollo Sustentable de la Nación y Ministerio de Educación de la Nación. Argentina

Pedroza R, Argüello F (2002) Interdisciplinariedad y transdisciplinariedad en los modelos de enseñanza de la cuestión ambiental. Cinta de Moebio: Revista Electrónica de Epistemología de Ciencias Sociales, ISSN-e 0717-554X, N°15

PNUMA. Programa de Naciones Unidas para el Medio Ambiente (1972) Resolución 2997/24

Sanhueza C, Germain P, Zapperi G et al (2014) Plantas nativas de Bahía Blanca y alrededores, descubriendo su historia, belleza y magia. TELLUS

Stockholm Declaration (1972) United nations conference on the human environment. Stockholm, Sweden

Tréllez Solís E (2006) Algunos elementos del proceso de construcción de la educación ambiental en América Latina. Revista Iberoamericana de Educación 41:69–81

Zabala I, García M (2008) Historia de la Educación Ambiental desde su discusión y análisis en los congresos internacionales. Rev Invest 63:201–218

Index

S. M. Fiori, P. D. Pratolongo (eds.), *The Bahía Blanca Estuary*,
https://doi.org/10.1007/978-3-030-66486-2

C

D

G

H

The manufacturer's authorised representative in the EU is Springer Nature Customer Service Centre GmbH, Europaplatz 3, 69115 Heidelberg, Germany. If you have any concerns regarding our products, please contact ProductSafety@springernature.com

Printed and bound by CPI Group (UK) Ltd, Croydon, CR0 4YY
30/09/2026
02225691-0012